FIFTEENTH EDITION

W9-AOX-044

Environmental
SCIENCE

A Global Concern

William P. Cunningham
University of Minnesota

Mary Ann Cunningham
Vassar College

ENVIRONMENTAL SCIENCE: A GLOBAL CONCERN, FIFTEENTH EDITION

Published by McGraw-Hill Education, 2 Penn Plaza, New York, NY 10121. Copyright © 2021 by McGraw-Hill Education. All rights reserved. Printed in the United States of America. Previous editions © 2018, 2015, and 2012. No part of this publication may be reproduced or distributed in any form or by any means, or stored in a database or retrieval system, without the prior written consent of McGraw-Hill Education, including, but not limited to, in any network or other electronic storage or transmission, or broadcast for distance learning.

Some ancillaries, including electronic and print components, may not be available to customers outside the United States.

This book is printed on acid-free paper.

2 3 4 5 6 7 8 9 LWI 21 20

ISBN 978-1-260-36382-1 (bound edition)
MHID 1-260-36382-1 (bound edition)
ISBN 978-1-260-48624-7 (loose-leaf edition)
MHID 1-260-48624-9 (loose-leaf edition)

Chief Product Officer, SVP Products & Markets: *G. Scott Virkler*
Portfolio Manager: *Michael Ivanov, PhD*
Product Developers: *Jodi Rhomberg*
Marketing Manager: *Britney Ross*
Content Project Managers: *Jessica Portz, Rachael Hillebrand, Sandra Schnee*
Buyer: *Laura Fuller*
Design: *Egzon Shaqiri*
Content Licensing Specialist: *Lorraine Buczek*
Cover Image: *Paul Souders/Getty Images*
Compositor: *SPi Global*

All credits appearing on page or at the end of the book are considered to be an extension of the copyright page.

Library of Congress Cataloging-in-Publication Data

Names: Cunningham, William P., author. | Cunningham, Mary Ann, author.
Title: Environmental science: a global concern/William P. Cunningham,
  University of Minnesota, Mary Ann Cunningham, Vassar College.
Description: Fifteenth Edition. | New York: McGraw-Hill Education, [2020] |
  Audience: Ages: 18+
Identifiers: LCCN 2019018227 | ISBN 9781260363821 (bound edition: acid-free paper) |
  ISBN 1260363821 (bound edition: acid-free paper) | ISBN 9781260486247
  (loose-leaf edition: acid-free paper) | ISBN 1260486249 (loose-leaf edition: acid-free paper)
Subjects: LCSH: Environmental sciences—Textbooks.
Classification: LCC GE105 .C86 2020 | DDC 304.2—dc23
LC record available at https://lccn.loc.gov/2019018227

The Internet addresses listed in the text were accurate at the time of publication. The inclusion of a website does not indicate an endorsement by the authors or McGraw-Hill Education, and McGraw-Hill Education does not guarantee the accuracy of the information presented at these sites.

logo applies to the text stock only

About the Authors

Source: Glow Images/SuperStock

© Tom Finkle

William P. Cunningham

William P. Cunningham is an emeritus professor at the University of Minnesota. In his 38-year career at the university, he taught a variety of biology courses, including Environmental Science, Conservation Biology, Environmental Health, Environmental Ethics, Plant Physiology, and Cell Biology. He is a member of the Academy of Distinguished Teachers, the highest teaching award granted at the University of Minnesota. He was a member of a number of interdisciplinary programs for international students, teachers, and nontraditional students. He also carried out research or taught in Sweden, Norway, Brazil, New Zealand, China, and Indonesia.

Professor Cunningham has participated in a number of governmental and nongovernmental organizations over the past 40 years. He was chair of the Minnesota chapter of the Sierra Club, a member of the Sierra Club national committee on energy policy, vice president of the Friends of the Boundary Waters Canoe Area, chair of the Minnesota governor's task force on energy policy, and a citizen member of the Minnesota Legislative Commission on Energy.

In addition to environmental science textbooks, he edited three editions of the *Environmental Encyclopedia,* published by Thompson-Gale Press. He has also authored or coauthored about 50 scientific articles, mostly in the fields of cell biology and conservation biology, as well as several invited chapters or reports in the areas of energy policy and environmental health. His Ph.D. from the University of Texas was in botany.

Professor Cunningham's hobbies include photography, birding, hiking, gardening, and traveling. He lives in St. Paul, Minnesota, with his wife, Mary. He has three children (one of whom is coauthor of this book) and seven grandchildren.

Both authors have a long-standing interest in the topics in this book. Nearly half the photos in the book were taken on trips to the places they discuss.

Mary Ann Cunningham

© Tom Finkle

Mary Ann Cunningham is a professor of geography at Vassar College. A biogeographer with interests in landscape ecology, geographic information systems (GIS), and climate impacts on biodiversity and food production, she teaches environmental science, natural resource conservation, land-use planning, and GIS. Field research methods, statistical methods, and data analysis and visualization are regular components of her teaching. Every aspect of this book is woven into, and informed by, her courses and her students' work. As a scientist and an educator, Mary Ann enjoys teaching and conducting research with both science students and non-science liberal arts students. As a geographer, she likes to engage students with the ways their physical surroundings and social context shape their world experience. In addition to teaching at a liberal arts college, she has taught at community colleges and research universities.

Professor Cunningham has been writing in environmental science for nearly two decades, and she has been coauthor of this book since its seventh edition. She is also coauthor of Principles of Environmental Science and an editor of the Environmental Encyclopedia (third edition, Thompson-Gale Press). She has published work on pedagogy in cartography, as well as instructional and testing materials in environmental science, and a GIS lab manual that introduces students to spatial and environmental analysis. She has also been a leader in sustainability programs and climate action planning at Vassar.

In addition to environmental science, Professor Cunningham's primary research activities focus on land-cover change, habitat fragmentation, and distributions of bird populations. This work allows her to conduct field studies in the grasslands of the Great Plains, as well as in the woodlands of the Hudson Valley. In her spare time she loves to travel, hike, and watch birds. Professor Cunningham holds a bachelor's degree from Carleton College, a master's degree from the University of Oregon, and a Ph.D. from the University of Minnesota.

Source: imageBROKER/Alamy Stock Photo

Brief Contents

Contents

Source: imageBROKER/Alamy Stock Photo

4 Evolution, Biological Communities, and Species Interactions 71

5 Biomes: Global Patterns of Life 97

6 Population Biology 116

23 Ecological Economics 518

24 Environmental Policy, Law, and Planning 544

25 What Then Shall We Do? 565

List of Case Studies

About the Cover

King penguins are one of the astonishing and exquisitely adapted species that inhabit our world. They are also among the many species urgently threatened by human activities. Climate change is expected to shift critical feeding areas far from breeding colonies, and rising seas will flood nesting areas; commercial fisheries capture a rising share of the marine food web to support fish farms far away; plastic pollution is growing in the world's marine environments. At the same time, hope for these and other species can be found in global policies and growing cooperation to protect marine reserves, to monitor fisheries, and to curb greenhouse gas emissions. Understanding interconnections in environmental systems is critical to protecting the extraordinary diversity of life around us, and to protecting the ecosystem services on which we also depend. Environmental science helps you explore these interconnections and make sense of this amazing complexity, and the ways survival of these living systems is tied to the well-being of our own communities. Enjoy the journey.

Preface

© Claudiad/Vetta/Getty Images

Environmental Science: A Search for Solutions

Environmental science focuses on understanding challenges that affect our lives, and on finding solutions to those challenges. Your decision to study environmental science is an important step. This field can help you find answers to some of the most important problems facing us today. Environmental science is an integrative field. It draws on diverse knowledge bases and skills to address issues: For example, preserving healthy ecosystems depends on strategies such as reducing greenhouse gas emissions, developing renewable energy systems, reducing pollution, improving social and environmental justice, improving sustainable farming systems, and reducing resource consumption.

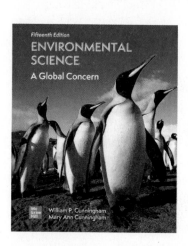

In many ways, environmental science is also an optimistic field. Although you will examine serious environmental challenges in this book, you will also explore many ways you can contribute to finding solutions. Understanding the nature of environmental problems is a first step to fixing them. Learning about new approaches gives you the power to help make a difference, no matter where you come from or what your interests are. For many of us, discovering ways to contribute makes this an exciting and engaging field.

As you will find in the "What Can You Do?" boxes in every chapter, there are countless practical opportunities to protect and sustain natural resources. It doesn't take a huge project to do important work for your local environment. Individuals and small groups have many opportunities to make positive change. As you read this book, look for ways to connect the issues and ideas to your other interests. Whether you are a biologist, a geologist, a chemist, an economist, a political scientist, a writer, or an artist or poet who can capture our imagination, you can find fruitful and interesting ways to connect with the topics in this book.

Sustainable development is a central theme

Several main themes run through this book. As you will read in chapter 1, these include **sustainable development** (including population growth, food production, environmental quality, energy, and resources), **climate change** and its impacts, and fundamentals of how **scientific methods** help us ask and answer questions about the world around us.

These and other themes show both continuing challenges and evidence of progress. **Human population growth** continues, for example, but it is slowing almost everywhere as women's education and economic opportunity allow for small, well-cared-for families. We remain addicted to fossil fuels, but **new energy technologies** now provide reliable alternatives in many countries. Solar, wind, biomass, geothermal energy, and conservation could supply all the energy we need, if we chose to invest in them. **Water quality** and **air pollution** remain dire problems in many areas, but we have shown that we can dramatically improve water quality, air quality, and environmental health, when we put our minds to it.

Governments around the world are acknowledging the costs of environmental degradation and are taking steps to reduce their environmental impacts. From China to Europe to North America and developing countries, policymakers have plans to restore forests, conserve water, reduce air and water pollution, and develop sustainable energy supplies. Public support for environmental protection has been overwhelmingly enthusiastic.

Businesses everywhere increasingly recognize the opportunities in conservation, recycling, producing non-toxic products, and reducing their ecological footprints. New jobs are being created in environmental fields. Public opinion supports environmental protection because voters see the importance of environmental health for the economy, society, and quality of life.

What Sets This Book Apart?

As practicing scientists and educators, we bring to this book decades of experience in the classroom, in the practice of science, and in civic engagement. This experience helps give students a clear sense of what environmental science is and why it matters. Throughout the book, we also provide recent data that underly and inform emerging ideas in the field.

Engaged and active learning

We've given particular attention to learning styles and active learning features in this edition, both in the text and in online **Connect** study materials and supplements. Throughout, the text promotes active, engaged learning practices. In each section heading, **key concepts** identify ideas for students to focus on as they read. **Section reviews** encourage students to check their learning at the end of each main section. These practices of active reading have been shown to improve retention of class topics, as well as higher-order thinking about concepts. **Key terms** at the end of each chapter encourage students to test their understanding. **Critical thinking and discussion questions** and **Data Analysis** exercises push students to explore further the concepts in the text.

A rich collection of online study resources is available on the **Connect** website. **LearnSmart** study resources, practice quizzes, animations, videos, and other resources improve understanding and retention of course material.

The book also engages course material with students' own lives: **What Can You Do?** sections help students identify ways to apply what they are learning to their own lives and communities. **What Do You Think?** readings ask students to critically evaluate their own assessments of a complex problem. We devote a special introduction (**Learning to Learn**) to the ways students can build study habits, take ownership of this course, and practice critical, analytical, and reflective thinking.

Many of these resources are designed as starting points for lectures, discussions in class, essays, lab activities, or projects. Some data analysis exercises involve simple polls of classes, which can be used for graphing and interpretation. Data analysis exercises vary in the kinds of learning and skills involved, and all aim to give students an opportunity to explore data or ideas discussed in the text.

Quantitative reasoning and methods of science

Quantitative reasoning is increasingly recognized as essential in many aspects of education, and this book has greater coverage of this topic, and provides more up-to-date data and graphs, than other books on the market. **Quantitative reasoning** questions in the text push students to evaluate data and graphs they have read about. Attention to statistics, graphing, graph interpretation, and abundant up-to-date data are some of the resources available to help students practice their skills with data interpretation.

Exploring Science readings show how science is done, to demystify the process of answering questions with scientific and quantitative methods. Throughout the text, we emphasize principles and methods of science through discussions of scientific methods, uncertainty and probability, and detailed examination of how scientists observe the world, gather data, and use data to answer relevant questions.

A positive focus on opportunities

Our intent is to empower students to make a difference in their communities by becoming informed, critical thinkers with an awareness of environmental issues and the scientific basis of these issues. Many environmental problems remain severe, but there have been many improvements in recent decades, including cleaner water and cleaner air for most Americans, declining rates of hunger and fertility, and increasing access to education. An entire chapter (chapter 13) focuses on ecological restoration, one of the most important aspects of ecology today. Case studies show examples of real progress, and What Can You Do? sections give students ideas for contributing to solutions. Throughout this text we balance evidence of serious environmental challenges with ideas about what we can do to overcome them.

A balanced presentation for critical thinking

Among the most important practices a student can learn are to think analytically about evidence, to consider uncertainty, and to skeptically evaluate the sources of information. This book offers abundant opportunities to practice the essential skills of critically analyzing evidence, of evaluating contradictory interpretation, and identifying conflicting interests. We ask students to practice critical and reflective thinking in What Do You Think? readings, in end-of-chapter discussion questions, and throughout the text. We present balanced evidence, and we provide the tools for students to discuss and form their own opinions.

An integrated, global perspective

Globalization spotlights the interconnectedness of environmental resources and services, as well as our common interest in how to safeguard them. To remain competitive in a global economy, it is critical that we understand conditions in other countries and cultures. This book provides case studies and topics from regions around the world, with maps and data illustrating global issues. These examples show the integration between environmental conditions at home and abroad.

Google Earth™ placemarks

Our global perspective is supported by placemarks and questions you can explore in Google Earth. This free, online program lets students view detailed satellite images of the earth that aid in understanding the geographical context of topics in the book. Through Connect, students can access placemarks, descriptions, and questions about those places. These stimulate a thoughtful exploration of each site and its surroundings. This interactive geographical exploration is a wonderful tool to give an international perspective on environmental issues.

What's New in This Edition?

This edition has thoroughly updated data, figures, and tables, as well as 16 new opening case studies that reflect new developments in the field, and over a dozen new "Exploring Science" or "What Do You Think?" boxed readings. We have enhanced our focus on climate action and environmental action, something students in our classes find especially valuable. Brief "benchmark data" tables provide reference values or comparisons that reflect key ideas in the chapter. Systematic discussions review topics such as uncertainty, graphing, statistics, experimental design, models, and systems. At the end of each chapter, we conclude with a new section, "Connecting the Dots," that draws together major themes of the chapter.

Specific chapter changes

The Introduction (Learning to learn) explains how each of us can engage with this field. Knowing what you care about is a good way to start connecting your interests to the study of our environment and how it works. We examine the nature of **critical thinking,** and we emphasize that learning to learn helps students not only in studying but in everyday life.

Chapter 1 presents **climate change** as an overarching concern. We introduce **sustainable development** as a topic that runs throughout the book as both a goal and a measure of progress. We discuss new environmental leaders, as well as the idea of **planetary boundaries,** which define limits of environmental services from major sectors of our environment.

Chapter 2 introduces a new case study on camera traps and **citizen science** to monitor migratory wildlife in Tanzania's Serengeti National Park. This example illustrates **study design** as well as ways each of us can contribute to original research. Continuing our discussion of the principles and applications of science, we discuss significance and confidence in data.

Chapter 3 opens with a new case study on the growing hypoxic "dead zone" in the Gulf of Mexico. This case illustrates interconnections in a vast ecological system and shows how chemical elements and energy transfers underlie pollution, wastewater treatment, eutrophication, and other processes. An "Exploring Science" reading reviews the **CRISPR** gene editing system, including ethics of human embryo editing, in this fast-moving field.

Chapter 4 introduces a new contributor to this book. Dr. Kimberly Byrd, a conservation biologist who has revised this crucial chapter. She has written a new case study on the ecological importance of seagrass meadows, including ideas of **ecosystem complexity** and "blue carbon." She has added a discussion of **complex adaptive systems** and system resilience. We hope readers will find her voice refreshing, interesting, and informative.

Chapter 5 has a new case study on **climate-driven shifts** in species ranges and biomes. These ecosystem changes directly affect lives and livelihoods. Recognizing the adaptations that allow species to adapt helps us understand survival factors for both humans and other species. A new section on human **disturbance** to biomes and ecosystems addresses the ways we are transforming the world.

Chapter 6 opens with a new case study on invasive Asian carp in the Mississippi watershed. Millions of dollars in sport fishing, recreation, and ecosystem services are at risk, as well as native species. We discuss growth patterns, life history strategies, and intrinsic and extrinsic factors that regulate growth. A new "Exploring Science" box describes methods for estimating population sizes for species, such as carp, that are difficult to count.

Chapter 7 uses a new case study on the rapid aging of China's population to discuss **population momentum** and factors that influence **birth rates**. China now has the largest number of senior citizens in the world, and has one of the largest percentages of old people of any country. This phenomenon is becoming global, as world population growth has fallen from about 2.1 percent in 1960 to 0.1 percent today. Half us now live in countries where the birth rate just replaces the death rate. We have long called for this shift, but its implications for societies are not entirely clear.

Chapter 8 has an updated case study on perflourocarbons, including an $850 million settlement in 2018 between the state of Minnesota and the 3M corporation for uncontrolled dumping of these persistent chemicals. Developments in **contagious diseases** among humans and wildlife have necessitated major chapter updates. A new section reviews growing transfer of **antibiotic resistance** from livestock that threaten human health. Building on the opening case study, we highlight four widely distributed persistent organic pollutants that threaten the health of millions of people. The "What Do You Think?" box on acceptable risk has also been revised.

Chapter 9 opens with a new case study on low-cost **food security** initiatives in Burkina Faso, one of the world's poorest countries. Farmers there are fighting land degradation and hunger using simple, traditional water conservation and farming techniques to improve food production. We also consider dietery diversity. We have new discussions of climate impacts on food production and on *Diet for a Small Planet*, and eating low on the food chain.

Chapter 10 has an updated opening case study on farming in Brazil's Cerrado. This case became even more urgent with the 2019 election of Jair Bolsonaro, who aims to expand soy production and reduce protections for Amazonian rainforest. Destruction of the world's largest tropical forest has dire implications for our climate and for survival of indigenous people. A new section discusses **carbon farming,** which could be part of the solution to controlling climate change. We also have updated the "What do you Think?" box on the environmental benefits of shade-grown coffee and cocoa.

Chapter 11 leads with a new case study on how the reintroduction of wolves, a top predator, has enhanced **biodiversity** in Yellowstone National Park, with cascading effects through both the food chain and the physical environment. We have emphasized the "climate" component of HIPPO factors in threats to species survival. We have enhanced discussion of the "sixth extinction" and added a boxed reading on the startling crisis of **disappearing insects**. Studies show losses of 80 percent of the flying insect fauna in some areas, with probably profound impacts on biodiversity more broadly.

Chapter 12 has a new case study on ecosystems in transition. Longer fire seasons and more extreme outbreaks of bark beetles threaten to alter western forests, as climate warming has produced the largest, most intense, and most damaging forest fires in U.S. history. Continuing our survey of landscapes in transition, we have added a new "Exploring Science" box on the effects of palm oil plantations on endangered orangutan populations on Borneo. A new "What Do You Think?" box examines new threats to U.S. national monuments from mining and other extractive industries.

Chapter 13 introduces **restoration ecology** with a new case study on the science and practice of restoring coral reefs. At least one-third of all coral reefs have been damaged beyond recovery by pollution, overharvesting, ocean acidification, and climate change. Some experts warn there may be no coral reefs anywhere in the world by the end of this century. But restoration ecologists are exploring innovative strategies for protecting and restoring these amazing systems. A new box on the "monarch highway" project describes both the threats to these charismatic insects and efforts to restore their populations.

Chapter 14 begins an **environmental geology** discussion with a new case study on the proposed Pebble Mine in headwater salmon streams of Alaska's Bristol Bay. This controversial project pits the fate of pristine wilderness and the world's largest sockeye salmon run against the estimated profits and likely environmental damage from a mammoth copper-nickel mine. On one side are about 850 high-paying mining jobs over the expected 20-year life of the mine compared to 12,000 permanent jobs for native people and Alaskan citizens in the salmon fishing industry. This struggle reflects issues in many controversies about earth resources.

Chapter 15 demonstrates leadership in **climate action** with a new case study on groundbreaking climate policy in California. Challenges are daunting, but solutions are diverse, creative, and exciting. We have enhanced the discussion of jet streams and **polar vortex** effects on local weather, as well as the latest IPCC report as well as current information about major greenhouse gases as well as the latest news about polar ice melting and warming seas. A new box illustrates the effects of **black carbon** emissions on climate systems. We also examine options for **carbon capture** and other efforts to combat climate change.

Chapter 16 provides updated data on air pollution as well as updated discussion of the Montreal Protocol on ozone-destroying substances—including the **Kigali Amendment,** which accelerates the phase out of refrigerants that are also critical greenhouse gases. This step alone could prevent 0.5 degrees of global warming by 2100. We increase emphasis of the dangers of air pollution particulates smaller than 2.5 um, and we discuss the problems of air pollution in developing countries.

Chapter 17 updates the opening case study, "When Will Lake Mead Go Dry?" and the demands for Colorado River water that exceed the river's flow. We provide recent data on looming **water shortages,** especially in regions dependent on glacial rivers, as in South Asia. Water is likely to be the most contentious natural resource in the future, but smarter **water conservation** policies, including pricing, irrigation and farming practices, and low-flow household appliances could reduce these risks. We also discuss China's expanding dam-building projects, especially on the Mekong River.

Chapter 18 continues the water resource discussion with the example of the Ganges River, on which nearly a billion people in South Asia depend. We know how to prevent **water pollution,** and we know how to capture and remove pollutants. But finding ways to implement policies and pay for treatment is difficult even in wealthy countries. These challenges are even steeper in developing regions as they struggle to improve health and quality of life.

Chapter 19 presents a new case study on the demise of one of the U.S. coal companies. We emphasize that while fossil fuels still provide most energy, the future of energy is not the past. We update data on production and consumption and discuss the shifting landscape of conventional energy, including growth in China. A new "Exploring Science" box discusses the growing importance of indigenous resistance to new pipelines across their land. We also highlight new debates about nuclear power, which is both expensive and low carbon.

Chapter 20 explores the fast-changing landscape of **renewable energy** with an updated case study on Germany's *Energiewende,* or **energy transition** from fossil fuels to renewable energy. This chapter is heavily revised to reflect new developments in technology and energy production. Explanations of new systems include a discussion of *efficiency* and *power capacity,* as well as battery storage. We examine analysis showing how sustainable energy systems could meet all our needs, often saving money as well as reducing pollution.

Chapter 21 includes an updated case study on the phenomenal amounts of **plastic pollution** in the world's oceans. A new section reviews the options for waste disposal and updates both the amounts and types of materials in our waste stream. We examine China's decision to reject U.S. recycling and what this means for waste management.

Chapter 22 opens with a new case study showing how cities are leading efforts to become environmentally, socially, and economically sustainable. We update data on **urban growth,** especially in African states, where some cities may have 100 million residents by the end of this century. How will these cities manage pollution, traffic, energy, food, and water supplies? We also examine the plight of sinking coastal cities amid rising seas. A final section discusses ways cities can be livable and sustainable.

Chapter 23 has an updated case study about British Columbia's **carbon tax** and notes that when Washington State tried to pass a similar tax, the fossil fuel industry spent $30 million to block the plan. Will other states be able to overcome this spending power? A new "Exploring Science" box notes that estimates of the value of global **ecosystem services** have increased from $33 trillion a few decades ago to $173 trillion today. Another boxed essay compares rapid job growth in sustainable energy compared to the fossil fuel industry. With interest growing in a Green New Deal, we have added a new "What Do You Think?" box to review this proposal.

Chapter 24 opens with a new case study about the recovery of North American green sea turtles with the help of the Endangered Species Act. In 1978 fewer than 300 sea turtles nested in Florida. By 2017, more than 39,000 turtles came ashore to nest, a major success in species protection. We review the provisions and successes of this and other major environmental policies. A new section discusses problems of **regulatory capture** in government agencies, as well as debates about how much regulation we want.

Chapter 25 presents a new case study on the history of Earth Day. It is critical that students understand how we got to where we are, and how public involvement with environmental issues has emerged. A new box discusses fossil fuel divestment debates at U.S. colleges and universities. We end the chapter with a review of sustainability as an overarching goal for environmental science.

Acknowledgments

We owe a great debt to the hardworking, professional team that has made this the best environmental science text possible. We express special thanks for editorial support to Michael Ivanov, PhD, and Jodi Rhomberg. We are grateful to Jessica Portz, Lora Neyens, Sherry Kane, Carrie Burger, Lorraine Buczek, and Tara McDermott, for their work in putting the book together, and marketing leadership by Noah Evans. We thank Tricia Lawrence for copyediting and Janet Robbins for excellent work on photographs.

The following individuals helped write and review learning goal–oriented content for LearnSmart for Environmental Science:

Central Washington University, Susan Kaspari

College of DuPage, Shamili Ajgaonkar Sandiford

Columbus State Community College, Morteza Javadi

Community College of Philadelphia, Christopher Murphy

Florida Atlantic University, Jessica Miles

Florida Gulf Coast University, Chad Evers

Georgia Southern University, J. Michelle Cawthorn

Indian Hills Community College, Chad V. Gatlin

John Tyler Community College, Joressia A. Beyer

Missouri State University, Kip R. Thompson

Moraine Valley Community College, Jennifer Kaye Sheppard

Northern Arizona University, Sylvester Allred

Ozarks Technical Community College, Michael S. Martin

Reading Area Community College, Heather A. Hinkle

Roane State Community College, Arthur C. Lee

Rock Valley Community College, Joseph E. Haverly

Rock Valley Community College, Megan M. Pease

Southeast Community College, Daniel D. Fogell

St. Cloud State University, Matthew Julius

State University of New York at Cortland, Noelle J. Relles

Temple College, Nathan Gardiner

The Community College of Baltimore County, Arthur C. Cage III

University of North Carolina Wilmington, Jeffery Hill

University of North Carolina at Chapel Hill, Trent McDowell

University of Wisconsin, Milwaukee, Tristan J. Kloss

University of Wisconsin, Milwaukee, Gina Seegers Szablewski

Waubonsee Community College, Dani Fischer

Winona State University, Jennifer L. Cochran

Input from instructors teaching this course is invaluable to the development of each new edition. Our thanks and gratitude go out to the following individuals who either completed detailed chapter reviews of *Environmental Science, A Global Concern*, fifteenth edition, or provided market feedback for this course.

American University, Priti P. Brahma

Antelope Valley College, Zia Nisani

Arizona Western College, Alyssa Haygood

Aurora University, Carrie Milne-Zelman

Baker College, Sandi B. Gardner

Baylor College, Heidi Marcum

Boston University, Kari L. Lavalli

Bowling Green State University, Daniel M. Pavuk

Bradley University, Sherri J. Morris

Broward College, Elena Cainas

Broward College, Nilo Marin

California Energy Commission, James W. Reede

California State University–East Bay, Gary Li

California State University, Natalie Zayas

Campbellsville University, Ogochukwu Onyiri

Central Carolina Community College, Scott Byington

Central State University, Omokere E. Odje

Clark College, Kathleen Perillo

Clemson University, Scott Brame

College of DuPage, Shamili Ajgaonkar Sandiford

College of Lake County, Kelly S. Cartwright

College of Southern Nevada, Barry Perlmutter

College of the Desert, Tracy Albrecht

College of the Desert, Candice Weber

College of the Desert, Kurt Leuschner

Columbia College, Jill Bessetti

Columbia College, Daniel Pettus

Community College of Baltimore County, Katherine M. Van de Wal

Connecticut College, Jane I. Dawson

Connecticut College, Chad Jones

Connors State College, Stuart H. Woods

Cuesta College, Nancy Jean Mann

Dalton State College, David DesRochers

Dalton State College, Gina M. Kertulis-Tartar

Deanza College, Dennis Gorsuch

East Tennessee State University, Alan Redmond

Eastern Oklahoma State College, Patricia C. Bolin Ratliff

Edison State College, Cheryl Black

Elgin Community College, Mary O'Sullivan

Erie Community College, Gary Poon

Estrella Mountain Community College, Rachel Smith

Farmingdale State College, Paul R. Kramer

Fashion Institute of Technology, Arthur H. Kopelman

Flagler College, Barbara Blonder

Florida State College at Jacksonville, Catherine Hurlbut

Franklin Pierce University, Susan Rolke

Galveston College, James J. Salazar

Gannon University, Amy L. Buechel

Gardner-Webb University, Emma Sandol Johnson

Gateway Community College, Ramon Esponda

Geneva College, Marjory Tobias

Georgia Perimeter College, M. Carmen Hall

Georgia Perimeter College, Michael L. Denniston

Gila Community College, Joseph Shannon

Golden West College, Tom Hersh

Gulf Coast State College, Kelley Hodges

Gulf Coast State College, Linda Mueller Fitzhugh

Holy Family University, Robert E. Cordero

Houston Community College, Yiyan Bai

Hudson Valley Community College, Daniel Capuano

Hudson Valley Community College, Janet Wolkenstein

Illinois Mathematics and Science Academy, C. Robyn Fischer

Illinois State University, Christy N. Bazan

Indiana University of Pennsylvania, Holly J. Travis

Indiana Wesleyan University, Stephen D. Conrad

James Madison University, Mary Handley

James Madison University, Wayne S. Teel

John A. Logan College, Julia Schroeder

Kentucky Community & Technical College System-Big Sandy District, John G. Shiber

Lake Land College, Jeff White

Lane College, Satish Mahajan

Lansing Community College, Lu Anne Clark

Lewis University, Jerry H. Kavouras

Lindenwood University, David M. Knotts

Longwood University, Kelsey N. Scheitlin

Louisiana State University, Jill C. Trepanier

Lynchburg College, David Perault

Marshall University, Terry R. Shank

Menlo College, Neil Marshall

Millersville University of Pennsylvania, Angela Cuthbert

Minneapolis Community and Technical College, Robert R. Ruliffson

Minnesota State College–Southeast Technical, Roger Skugrud

Minnesota West Community and Technical College, Ann M. Mills

Mt. San Jacinto College, Shauni Calhoun

Mt. San Jacinto College, Jason Hlebakos

New Jersey City University, Deborah Freile

New Jersey Institute of Technology, Michael P. Bonchonsky

Niagara University, William J. Edwards

North Carolina State University, Robert I. Bruck

North Georgia College & State University, Kelly West

North Greenville University, Jeffrey O. French

Northeast Lakeview College, Diane B. Beechinor

Northeastern University, Jennifer Rivers Cole

Northern Virginia Community College, Jill Caporale

Northwestern College, Dale Gentry

Northwestern Connecticut Community College, Tara Jo Holmberg

Northwood University Midland, Stelian Grigoras

Notre Dame College, Judy Santmire

Oakton Community College, David Arieti

Parkland College, Heidi K. Leuszler

Penn State Beaver, Matthew Grunstra

Philadelphia University, Anne Bower

Pierce College, Thomas Broxson

Purdue University Calumet, Diane Trgovcich-Zacok

Queens University of Charlotte, Greg D. Pillar

Raritan Valley Community College, Jay F. Kelly

Reading Area Community College, Kathy McCann Evans

Rutgers University, Craig Phelps

Santa Monica College, Dorna S. Sakurai

Shasta College, Morgan Akin

Shasta College, Allison Lee Breedveld

Southeast Kentucky Community and Technical College, Sheila Miracle

Southern Connecticut State University, Scott M. Graves

Southern New Hampshire University, Sue Cooke

Southern New Hampshire University, Michele L. Goldsmith

Southwest Minnesota State University, Emily Deaver

Spartanburg Community College, Jeffrey N. Crisp

Spelman College, Victor Ibeanusi

St. Johns River State College, Christopher J. Farrell

Stonehill College, Susan M. Mooney

Tabor College, Andrew T. Sensenig

Temple College, John McClain

Terra State Community College, Andrew J. Shella

Texas A&M University–Corpus Christi, Alberto M. Mestas-Nuñez

Tusculum College, Kimberly Carter

University of Akron, Nicholas D. Frankovits

University of Denver, Shamim Ahsan

University of Kansas, Kathleen R. Nuckolls

University of Miami, Kathleen Sullivan Sealey

University of Missouri at Columbia, Douglas C. Gayou

University of Missouri–Kansas City, James B. Murowchick

University of Nebraska, James R. Brandle

University of North Carolina Wilmington, Jack C. Hall

University of North Texas, Samuel Atkinson

University of Tampa, Yasoma Hulathduwa

University of Tennessee, Michael McKinney

University of Utah, Lindsey Christensen Nesbitt

University of Wisconsin–Stevens Point, Holly A. Petrillo

University of Wisconsin–Stout, Charles R. Bomar

Valencia College, Patricia Smith

Vance Granville Community College, Joshua Eckenrode

Villanova University, Lisa J. Rodrigues

Virginia Tech, Matthew Eick

Viterbo University, Christopher Iremonger

Waubonsee Community College, Dani DuCharme

Wayne County Community College District, Nina Abubakari

West Chester University of Pennsylvania, Robin C. Leonard

Westminster College, Christine Stracey

Worcester Polytechnic Institute, Theodore C. Crusberg

Wright State University, Sarah Harris

You're in the driver's seat.

Want to build your own course? No problem. Prefer to use our turnkey, prebuilt course? Easy. Want to make changes throughout the semester? Sure. And you'll save time with Connect's auto-grading too.

65%
Less Time Grading

Laptop: McGraw-Hill; Woman/dog: George Doyle/Getty Images

They'll thank you for it.

Adaptive study resources like SmartBook® 2.0 help your students be better prepared in less time. You can transform your class time from dull definitions to dynamic debates. Find out more about the powerful personalized learning experience available in SmartBook 2.0 at **www.mheducation.com/highered/connect/smartbook**

Make it simple, make it affordable.

Connect makes it easy with seamless integration using any of the major Learning Management Systems—Blackboard®, Canvas, and D2L, among others—to let you organize your course in one convenient location. Give your students access to digital materials at a discount with our inclusive access program. Ask your McGraw-Hill representative for more information.

Padlock: Jobalou/Getty Images

Solutions for your challenges.

A product isn't a solution. Real solutions are affordable, reliable, and come with training and ongoing support when you need it and how you want it. Our Customer Experience Group can also help you troubleshoot tech problems—although Connect's 99% uptime means you might not need to call them. See for yourself at **status.mheducation.com**

Checkmark: Jobalou/Getty Images

SUPPORT AT
every step

Effective, efficient studying.

Connect helps you be more productive with your study time and get better grades using tools like SmartBook 2.0, which highlights key concepts and creates a personalized study plan. Connect sets you up for success, so you walk into class with confidence and walk out with better grades.

Study anytime, anywhere.

Download the free ReadAnywhere app and access your online eBook or SmartBook 2.0 assignments when it's convenient, even if you're offline. And since the app automatically syncs with your eBook and SmartBook 2.0 assignments in Connect, all of your work is available every time you open it. Find out more at **www.mheducation.com/readanywhere**

> *"I really liked this app—it made it easy to study when you don't have your textbook in front of you."*
>
> - Jordan Cunningham, Eastern Washington University

No surprises.

The Connect Calendar and Reports tools keep you on track with the work you need to get done and your assignment scores. Life gets busy; Connect tools help you keep learning through it all.

Calendar: owattaphotos/Getty Images

Learning for everyone.

McGraw-Hill works directly with Accessibility Services Departments and faculty to meet the learning needs of all students. Please contact your Accessibility Services office and ask them to email accessibility@mheducation.com, or visit **www.mheducation.com/about/accessibility** for more information.

Top: Jenner Images/Getty Images, Left: Hero Images/Getty Images, Right: Hero Images/Getty Images

Key Elements

A global perspective is vital to learning about environmental science.

Case Studies

All chapters open with a real-world case study to help students appreciate and understand how environmental science impacts lives and how scientists study complex issues.

Exploring Science

Current environmental issues exemplify the principles of scientific observation and data-gathering techniques to promote scientific literacy.

What Do You Think?

Students are presented with challenging environmental studies that offer an opportunity to consider contradictory data, special interest topics, and conflicting interpretations within a real scenario.

Google Earth

Google Earth interactive satellite imagery gives students a geographical context for global places and topics discussed in the text. Google Earth icons indicate a corresponding exercise in Connect. In these exercises, students will find links to locations mentioned in the text, as well as corresponding assessments that will help them understand environmental topics.

Data Analysis

At the end of every chapter, these exercises give students further opportunities to apply critical-thinking skills and analyze data. These are assigned through Connect in an interactive online environment. Students are asked to analyze data in the form of documents, videos, and animations.

Learning Outcomes

Found at the beginning of each chapter, and organized by major headings, these outcomes give students an overview of the key concepts they will need to understand.

Learning Outcomes

After studying this chapter, you should be able to:

9.1 Describe patterns of world hunger and nutritional requirements.

9.2 Identify key food sources, including protein-rich foods.

9.3 Explain new crops and genetic engineering.

9.4 Discuss how policy can affect food resources.

Section Reviews

Section reviews are a series of content-specific questions that appear at the end of each section in the chapter. These questions encourage students to periodically review what they have read and offers an opportunity to check their understanding of key concepts.

Section Review

1. How many people in the world are chronically undernourished? What does chronically undernourished mean?

2. List at least five African countries with high rates of hunger (fig. 9.3; use a world map to help identify countries).

3. What are some of the health risks of overeating? What percentage of adults are overweight in the United States?

Connecting the Dots

This section summarizes the chapter by highlighting key ideas and relating them to one another.

Connecting the Dots

The potential location of biological communities is largely determined by temperature and moisture availability. Consequently, ecologists anticipate that changes in climate patterns will produce changes in biome distributions. For people accustomed to familiar ecosystems, such as the maple-rich northern hardwood forests of New England, these changes are likely to disrupt livelihoods and even cultural references.

Understanding the global distribution of biomes, and knowing the differences in what lives where and why, are essential to ____ nce. Plants and animals ____ hem to live in particular ____ alpine tundra, or chapar- ____ you understand limiting ____ man occupation and use ____ nt on the biomes found

in particular locations. Humans tend to prefer mild climates and the highly productive biological communities found in temperate zones. These biomes also suffer the highest rates of degradation and overuse.

While many of us pay most attention to terrestrial systems, oceans cover over 70 percent of the earth's surface. Marine biomes, such as coral reefs or mangroves, can be as biologically diverse and productive as any terrestrial biome. Freshwater ecosystems, too, are critically important, even though their extent is small overall. People have always depended on rich, complex ecosystems. In recent times, the rapid growth of human populations, coupled with more powerful ways to harvest resources, has led to extensive destruction of these environments. Awareness of emerging threats like climate change may help inspire more action to protect these living systems.

Critical Thinking and Discussion Questions

1. Do people around you worry about hunger? Do you think they should? Why or why not? What factors influence the degree to which people worry about hunger in the world?

2. Global issues such as hunger and food production often seem far too large to think about solving, but it may be that many strategies can help us address chronic hunger. Consider your own skills and interests. Think of at least one skill that could be applied (if you had the time and resources) to helping reduce hunger in your community or elsewhere.

3. Suppose you are a farmer who wants to start a confined animal feeding operation. What conditions make this a good strategy for you, and what factors would you consider in weighing its costs and benefits? What would you say to neighbors who wish to impose restrictions on how you run the operation?

4. Debate the claim that famines are caused more by human actions (or inactions) than by environmental forces. What kinds of evidence would be needed to resolve this debate?

5. Outline arguments you would make to your family and friends for why they should eat a mostly vegetarian diet, along the lines of *Diet for A Small Planet*. What reasons would be most compelling? What are some reasons why it is, or is not, fair to influence someone else's food practices?

6. Given what you know about GMO crops, identify some of the costs and benefits associated with them. Which of the costs and benefits do you find most important? Why?

7. Corn is by far the dominant crop in the United States. In what ways is this a good thing for Americans? How is it a problem? Who are the main beneficiaries of this system?

Critical Thinking and Discussion Questions

Brief scenarios of everyday occurrences or ideas challenge students to apply what they have learned to their lives.

What Can You Do?

This feature gives students realistic steps for applying their knowledge to make a positive difference in our environment.

What Can You Do?

Controlling Pests

Based on the principles of integrated pest management, the U.S. EPA releases helpful guides to pest control. Among their recommendations:

1. *Identify pests, and decide how much pest control is necessary.* Does your lawn really need to be totally weed-free? Could you tolerate some blemished fruits and vegetables? Could you replace sensitive plants with ones less sensitive to pests?

2. *Eliminate pest sources.* Remove from your house or yard any food, water, and habitat that encourages pest growth. Eliminate hiding places or other habitats. Rotate crops in your garden.

3. *Develop a weed-resistant yard.* Pay attention to your soil's pH, nutrients, texture, and organic content. Grow grass or cover varieties suited to your climate. Set realistic goals for weed control.

4. *Use biological controls.* Encourage beneficial insect predators such as birds, bats that eat insects, ladybugs, spiders, centipedes, dragonflies, wasps, and ants.

5. *Use simple manual methods.* Cultivate your garden and handpick weeds and pests from your garden. Set traps to control rats, mice, and some insects. Mulch to reduce weed growth.

6. *Use chemical pesticides carefully.* If you decide that the best solution is chemical, choose the right pesticide product, read safety warnings and handling instructions, buy the amount you need, store the product safely, and dispose of any excess properly.

Source: *Citizen's Guide to Pest Control and Pesticide Safety*: EPA 730-K-95-001

Quantitative Reasoning

Compare the estimates of known and threatened species in table 11.1. Are some groups overrepresented? Are we simply more interested in some organisms, or are we really a greater threat to some species?

Quantitative Reasoning

Quantitative reasoning questions in the text push students to evaluate data and graphs they have read about. Attention to statistics, graphing, graph interpretation, and abundant up-to-date data are some of the resources available to help students

Relevant Photos and Instructional Art Support Learning

High-quality photos and realistic illustrations display detailed diagrams,
graphs, and real-life situations.

Digital Vision/Getty Images
Stockbyte
simon22/123RF

Source: a -b. USDA Natural Resources Conservation
Service

Introduction

Learning to Learn

▲ Learning to learn is a lifelong skill.
William P. Cunningham

Learning Outcomes

After studying this introduction, you should be able to:

L.1 Form a plan to organize your efforts and become a more effective and efficient student.

L.2 Apply critical and reflective thinking in environmental science.

L.3 Identify logical errors, persuasive tricks, and biases used in popular media.

L.4 Describe issues that motivate you and consider ways they connect to environmental science.

"What kind of world do you want to live in? Demand that your teachers teach you what you need to know to build it."

– Peter Kropotkin

How can I do well in environmental science?

Case studies in environmental science examine a particular place or theme that draws together many of the themes in a chapter. For this chapter on learning to learn, a good case study to start with is you. You come to this course with particular backgrounds and ideas. You have expertise and skills. As you start reading this book, consider these two questions: How do you want to draw on your abilities and background and connect them to themes in this book? And how do you want to develop your knowledge and skills to answer questions that are important to you?

Responses to these question will differ for everyone. But the questions are relevant for everyone because environmental science is a field that involves a diversity of topics, with connections to basic ecology, natural resources, and policy questions that influence those systems. Topics in this course primarily involve our natural environment, but we also examine our human environment, including the built world of technology and cities, as well as human social or cultural institutions. All of these interrelated aspects of our life affect us, and, in turn, are affected by what we do.

Another way this chapter relates to you is that it gives suggestions for how you can organize your learning process as you study. This means being aware and intentional about your study habits. Take time as you read this chapter to consider what you do well as you study, and what you need to do better to be effective with study time. This is another skill set that will serve you well in other contexts.

FIGURE L.1 Knowing what you care about and why is a good start to connecting your interests to the study of our environment and how it works.
Hero Images/Image Source

Part of doing well in this course is to develop your habits of critical thinking, that is, assessing how and why we think about things as we do. Critical thinking is one of the most useful skills you can learn in any of your classes, and so it is a focus of this chapter. Many central topics in environmental science are highly contested: What kinds of energy are most important? Where should they come from? What is a resource? How should we manage and conserve water resources? Who should pay the cost of controlling air pollution? Answering these questions requires analysis of evidence. But evidence can depend on when and by whom it was gathered and evaluated. For every opinion there is an equal and opposite opinion. How can you make sense out of this welter of ever-changing information?

As you consider these sometimes contradictory views, pay attention to developing your capacity to think independently, systematically, and skillfully to form your own opinions (fig. L.1).

These qualities and abilities can help you in many aspects of life. Throughout this book you will find "What Do You Think?" boxes that invite you to practice your critical and reflective thinking skills.

Thinking about how we think is a practice that applies in ordinary conversation, as well as in media you encounter, and even in textbooks. Finding these patterns in arguments can be fun; it's also important. Paying attention to these sorts of argument strategies is also a good practice in any class you take. These are a few of the logical errors you can watch for:

- *Red herring:* Introducing extraneous information to divert attention from the important point.

- *Ad hominem attacks:* Criticizing the opponent rather than the logic of the argument.

- *Hasty generalization:* Drawing conclusions about all members of a group based on evidence that pertains only to a selected sample.

- *False cause:* Drawing a link between premises and conclusions that depends on some imagined causal connection that does not, in fact, exist.

- *Appeal to ignorance:* Because some facts are in doubt, a conclusion is impossible.

- *Appeal to authority:* It's true because someone says so.

- *Equivocation:* Using words with double meanings to mislead the listener.

- *Slippery slope:* A claim that some event or action will cause some subsequent action.

- *False dichotomy:* Giving either/or alternatives as if they are the only choices.

These skills are important to doing well in this class, and they are part of becoming a responsible and productive environmental citizen. Each of us needs a basis for learning and evaluating scientific principles, as well as some insights into the social, political, and economic systems that impact our global environment. We hope this book and the class you're taking will give you the information you need to reach those goals. As the noted Senegalese conservationist and educator Baba Dioum once said, "In the end, we will conserve only what we love, we will love only what we understand, and we will understand only what we are taught." The more you can connect ideas in this course to topics you care about, the better you can make use of them—and the more likely you will be to do well in the class.

L.1 How Can I Get an A in This Class?

- *Making a frank and honest assessment of your strengths and weaknesses will help you do well in this class.*
- *Reading in a purposeful, deliberate manner is an important part of productive learning.*

What do you need to know to succeed in a class on environmental science? This chapter provides an overview of some skills to keep in mind as you begin. As Henry Ford once said, "If you think you can do a thing, or think you can't do a thing, you're right."

One of the first things that will help you do well in this class—and enjoy it—is to understand that science is useful and accessible, if you just take your time with it. To do well in this class, start by identifying the ways that science connects with your interests and passions. Most environmental scientists are motivated by a love for something: a fishery biologist might love fishing; a plant pathologist might love gardening; an environmental chemist might be motivated by wanting to improve children's health in the city in which she lives. All these people use the tools of science to help them understand something they get excited about. Finding that angle can help you do better in this class, and it can help you be a better and happier member of your community (fig. L.2).

Another key to success is understanding what "science" is. Basically, science is about trying to figure out how things work. This means examining a question carefully and methodically. It means questioning your own assumptions, as well as the statements you hear from others. Understanding some basic ideas in science can be very empowering: Learning to look for evidence and to question your assumptions is a life skill, and building comfort with thinking about numbers can help you budget your groceries, prioritize your schedule, or plan your vacation. Ideas in this book can help you understand the food you eat, the weather you encounter, the policies you hear about in the news—from energy policy to urban development to economics.

FIGURE L.2 Finding the connections between your studies and the community, places, and ideas you care about can make this class more rewarding and fun.
Source: Gwen Bausmith, U.S. EPA

What are good study habits?

What are your current study skills and habits? Making a frank and honest assessment of your strengths and weaknesses will help you set goals and make plans for achieving them during this class. A good way to start is to examine your study habits. Rate yourself on each of the following study skills and habits on a scale of 1 (excellent) to 5 (needs improvement). If you rate yourself below 3 on any item, think about an action plan to improve that competence or behavior.

- How well do you manage your time (do you tend to run late, or do you complete assignments on time)?
- Do you have a regular study environment where you can focus?
- How effective are you at reading and note-taking (do you remember what you've read; do you take notes regularly)?
- Do you attend class regularly, listen for instructions, and participate actively in class discussions? Do you bring questions to class about the material?
- Do you generally read assigned chapters in the textbook before attending class, or do you wait until the night before the exam?
- How do you handle test anxiety (do you usually feel prepared for exams and quizzes or are you terrified of them? Do you have techniques to reduce anxiety or turn it into positive energy)?
- Do you actively evaluate how you are doing in a course based on feedback from your instructor and then make corrections to improve your effectiveness?
- Do you seek out advice and assistance outside of class from your instructors or teaching assistants?

Procrastination is something almost everyone does, but a few small steps can help you build better habits. If you routinely leave your studying until the last minute, then consider making a study schedule, and keep a written record of how much time you spend studying. Schedule time for sleep, meals, exercise, and recreation so that you will be rested and efficient when you do study. Divide your work into reasonable sized segments that you can accomplish on a daily basis. Carry a calendar to keep track of assignments. And find a regular study space in which you can be effective and productive.

How you behave in class and interact with your instructor also can have a big impact on how much you learn and what grade you get. Make an effort to get to know your instructor. Sit near the front of the room where you can see and be seen. Learn to ask questions: This can keep you awake and engaged in class. Practice the skills of good note-taking (table L.1). Attend every class and arrive on time. Don't fold up your papers and prepare to leave until after the class period is over. Arriving late and leaving early says to your instructor that you don't care much about either the class or your grade.

Practice active, purposeful learning. It isn't enough to passively absorb knowledge provided by your instructor and this textbook. You need to actively engage the material in order to really understand it. The more you invest yourself in the material, the easier it will be to comprehend and remember. It is very helpful to have a study buddy with whom you can compare notes and try out ideas (fig. L.3).

Table L.1   Learning Skills—Taking Notes

1. Identify the important points in a lecture and organize your notes in an outline form to show main topics and secondary or supporting points. This will help you follow the sense of the lecture.

2. Write down all you can. If you miss something, having part of the notes will help your instructor identify what you've missed.

3. Leave a wide margin in your notes in which you can generate questions to which your notes are the answers. If you can't write a question about the material, you probably don't understand it.

4. Study for your test under test conditions by answering your own questions without looking at your notes. Cover your notes with a sheet of paper on which you write your answers, then slide it to the side to check your accuracy.

5. Go all the way through your notes once in this test mode, then go back to review those questions you missed.

6. Compare your notes and the questions you generated with those of a study buddy. Did you get the same main points from the lecture? Can you answer the questions someone else has written?

7. Review your notes again just before test time, paying special attention to major topics and questions you missed during study time.

Source: Dr. Melvin Northrup, Grand Valley State University.

It's well known that the best way to learn something is to teach it to someone else. Take turns with your study buddy explaining the material you're studying. You may think you've mastered a topic by quickly skimming the text, but you're likely to find that you have to struggle to give a clear description in your own words. Anticipating possible exam questions and taking turns quizzing each other can be a very good way to prepare for tests.

How can you use this textbook effectively?

An important part of productive learning is to read assigned material in a purposeful, deliberate manner. Ask yourself questions as you read. What is the main point being made here? How does the evidence presented support the assertions being made? What personal experience have you had or what prior knowledge can you bring to bear on this question? Can you suggest alternative explanations for the phenomena being discussed? A study technique developed by Frances Robinson and called the **SQ3R** method can improve your reading comprehension. It's also helpful to have a study group (fig. L.4). After class and before exams, you can compare notes, identify priorities, and sort out points that are unclear. Try these steps as you read the first few chapters of this book, and see if they improve your recall of the material:

1. *Survey* the entire chapter or section you are about to read, so you can see how it fits together. What are the major headings and subdivisions?

2. Q*uestion* what the main points are likely to be in each of the sections. Which parts look most important or interesting? Where should you invest the most time and effort?

3. *Read* the material, taking brief notes as you go. Read in small segments and stop frequently for reflection and to make notes.

4. *Recite*: Stop periodically to recite to yourself what you have just read. Check your comprehension at the end of each major section. Ask yourself: Did I understand what I just read? What are the main points being made here? Summarize the information in your own words to be sure that you really understand and are not just depending on rote memory.

5. *Review:* Once you have completed a section, review the main points to make sure you remember them clearly. Did you miss any important points? Do you understand things differently

FIGURE L.3 Cooperative learning, in which you take turns explaining ideas and approaches with a friend, can be one of the best ways to comprehend material.
Prostock-studio/Shutterstock

FIGURE L.4 Talking through ideas with your peers is an excellent way to test your knowledge. If you can explain it, then you probably understand the material.
Tara Moore/Getty Images

the second time through? This is a chance to think critically about the material. Do you agree with the conclusions suggested by the authors?

Will this be on the test?

You should develop different study strategies depending on whether you are expected to remember and choose between a multitude of facts and details, or whether you will be asked to write a paragraph summarizing some broad topic. Organize the ideas you're reading and hearing in lecture. This course will probably include a great deal of information, so try to organize for yourself what ideas are most important. What's the big picture? As you read and review, ask yourself what might be some possible test questions in each section. Memorize some benchmark figures: Just a few will help a lot. Pay special attention to ideas, relationships, facts, and figures about which your instructor seemed especially interested. Usually those points are emphasized in class because your teacher thinks they are most important to remember. There is a good chance you'll see those topics again on a test.

Pay special attention to tables, graphs, and diagrams. They were chosen because they illustrate important points, and they are often easy to put on a test. Also pay attention to units. You probably won't be expected to remember all the specific numbers in this book, but you probably should know orders of magnitude. The world population is about 7.3 *billion* people (not thousands, millions, or trillions). It often helps to remember facts and figures if you can relate them to some other familiar example. The United States, for instance, has about 330 million residents. The populations of the European Union is slightly larger; India and China are each more than four times as large. Those general relationships are usually easier to remember and compare than detailed figures.

Section Review

1. What is your strongest learning style?
2. What are the five techniques of the SQ3R method for studying?

L.2 THINKING ABOUT THINKING

- Critical thinking is a valuable tool in learning and in life.
- Certain attitudes, skills, and approaches are essential for well-reasoned analysis.

Perhaps the most valuable skill you can learn in any of your classes is the ability to think clearly, creatively, and purposefully. Developing the ability to learn new skills, examine new facts, evaluate new theories, and formulate your own interpretations is essential to keep up in a changing world. In other words, you need to learn how to learn on your own.

Thinking about thinking means pausing to examine you are forming ideas, or how you interpret what you hear and read. A number of approaches can help us evaluate information and make decisions. **Analytical thinking** asks, "How can I break this problem down into its constituent parts?" **Creative thinking**

asks, "How might I approach this problem in new and inventive ways?" **Logical thinking** asks, "How can orderly, deductive reasoning help me think clearly?" **Critical thinking** asks, "What am I trying to accomplish here and how will I know when I've succeeded?" **Reflective thinking** asks, "What does it all mean?" As figure L.5 suggests, critical thinking is central in the constellation of thinking skills. Thinking critically can help us discover hidden ideas and means, develop strategies for evaluating reasons and conclusions in arguments, recognize the differences between facts and values, and avoid jumping to conclusions.

How do you tell the news from the noise?

With the explosion of cable channels, blogs, social networks, and e-mail access, most of us are interconnected constantly to a degree unique in history. There are well over 150 million blogs on the Web, and new ones are added every day. Most of us, even in low-income countries and regions, are linked in social networks. Every day several billion e-mails, tweets, text messages, online videos, and social media postings connect us to one another. As you participate in these networks, you probably already think about the sources of information you are exposed to on a daily basis.

One of the issues that has emerged with this proliferation of media is partisan journalism—reports that serve one viewpoint, rather than trying to weigh diverse evidence and perspectives. Partisan journalism has become much more prevalent since the deregulation of public media in 1988. From the birth of the broadcasting industry, the airwaves were regulated as a public trust. Broadcasters, as a condition of their licenses, were required to operate in the "public interest" by covering important policy issues and providing equal time to both sides of contested issues. In 1988, however, the

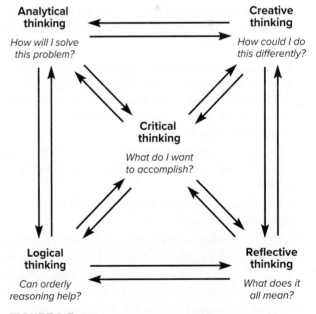

FIGURE L.5 Different approaches to thinking are used to solve different kinds of problems or to study alternate aspects of a single issue.

Federal Communications Commission ruled that the proliferation of mass media gives the public adequate access to diverse sources of information. Media outlets are no longer obliged to provide fair and balanced coverage of issues. Presenting a single perspective or even a deceptive version of events is no longer regarded as a betrayal of public trust.

An important aspect of partisan reporting is attack journalism. Commentators often ridicule and demean their opponents rather than weighing ideas or reporting objective facts and sources, because shouting matches are entertaining and sell advertising. Most newspapers have laid off almost all their investigative reporters and most television stations have abandoned the traditional written and edited news story. According to the Center for Journalistic Excellence, more than two-thirds of all TV news segments now consist of on-site "stand-up" reports or live interviews in which a single viewpoint is presented as news without any background or perspective.

Part of the reason for the growth of sensationalist media is that real news—topics that affect your community and your environment—often don't make exciting visuals. So they don't make it into TV coverage. Instead, crime, accidents, disasters, lifestyle stories, sports, and weather make up more than 90 percent of the coverage on a typical television news program. An entire day of cable TV news would show, on average, only 1 minute each about the environment and health care, 2 minutes each on science and education, and 4 minutes on art and culture. More than 70 percent of the segments are less than 1 minute long, which allows them to convey lots of emotion but little substance. People who get their news primarily from TV are significantly more fearful and pessimistic than those who get news from print media. And it becomes hard to separate rumor from truth. Evidence and corroboration take a backseat to dogma and passion.

How can you detect bias in blogs, social media, or news reporting? Ask the questions below as you look at media. Also ask these questions as you examine your own work, to avoid falling into these traps.

1. Are speakers discussing facts and rational ideas, or are they resorting to innuendo, name-calling, character assassination, and *ad hominem* (personal) attacks? When people start calling each other Nazi or communist (or both), civil discourse has probably come to an end.

2. What special interests might be involved? Who stands to gain presenting a particular viewpoint? Who is paying for the message?

3. What sources are used as evidence in this communication? How credible are they?

4. Are facts or statistics cited in the presentation? Are they credible? Are citations provided so you can check the sources?

5. If the presentation claims to be fair and balanced, are both sides represented by credible spokespersons, or is one simply a foil set up to make the other side look good?

6. Are the arguments presented based on evidence, or are they purely emotional appeals?

Applying critical thinking

In logic, an argument is made up of one or more introductory statements (called **premises**), and a **conclusion** that supposedly follows logically from the premises. Often in ordinary conversation, different kinds of statements are mixed together, so it is difficult to distinguish between them or to decipher hidden or implied meanings.

We all use critical or reflective thinking at times. Suppose a television commercial tells you that a new breakfast cereal is tasty and good for you. You may be suspicious and ask yourself a few questions. What do they mean by good? Good for whom or what? Does "tasty" simply mean more sugar and salt? Might the sources of this information have other motives in mind besides your health and happiness? Although you may not have been aware of it, you already have been using some of the techniques of critical analysis. Working to expand these skills helps you recognize the ways information and analysis can be distorted, misleading, prejudiced, superficial, unfair, or otherwise defective. Here are some steps in critical thinking:

Identify and evaluate premises and conclusions in an argument. What is the basis for the claims made here? What evidence is presented to support these claims and what conclusions are drawn from this evidence? If the premises and evidence are correct, does it follow that the conclusions are necessarily true?

Acknowledge and clarify uncertainties, vagueness, equivocation, and contradictions. Do the terms used have more than one meaning? If so, are all participants in the argument using the same meanings? Are ambiguity or equivocation deliberate? Can all the claims be true simultaneously?

Distinguish between facts and values. Are claims made that can be tested? (If so, these are statements of fact and should be able to be verified by gathering evidence.) Are claims made about the worth or lack of worth of something? (If so, these are value statements or opinions and probably cannot be verified objectively.) For example, claims of what we *ought* to do to be moral or righteous or to respect nature are generally value statements.

Recognize and assess assumptions. Given the backgrounds and views of the protagonists in this argument, what underlying reasons might there be for the premises, evidence, or conclusions presented? Does anyone have an "axe to grind" or a personal agenda in this issue? What do they think you know, need, want, or believe? Is there a subtext based on race, gender, ethnicity, economics, or some belief system that distorts this discussion? (fig. L.6).

Distinguish the reliability or unreliability of a source. What makes the experts qualified in this issue? What special knowledge or information do they have? What evidence do they present? How can we determine whether the information offered is accurate, true, or even plausible?

Recognize and understand conceptual frameworks. What are the basic beliefs, attitudes, and values that this person, group, or society holds? What dominating philosophy or ethics control their outlook and actions? How do these beliefs and values affect the way people view themselves and the world around them? If there are conflicting or contradictory beliefs and values, how can these differences be resolved?

FIGURE L.6 Often the conditions that lead to environmental problems like hazardous waste, and the explanations that surround them, are based on unspoken assumptions. Identifying underlying assumptions is a key step to finding solutions.
Source: Eric Vanceonse, U.S. EPA

As you read this book, you will have many opportunities to practice critical thinking. Every chapter includes facts, figures, opinions, and theories. Are all of them true? Probably not. They were the best information available when this text was written, but scientific knowledge is always growing. Data change constantly as does our interpretation of them. Environmental conditions change, evidence improves, and different perspectives and explanations evolve over time.

As you read this book or any book, try to distinguish between statements of fact and opinion. Ask yourself if the premises support the conclusions drawn from them. Although we have tried to present the best available scientific data and to represent the main consensus among environmental scientists, it is always important for you, as a reader, to think for yourself and utilize your critical and reflective thinking skills to find the truth.

Section Review

1. Describe seven attitudes needed for critical thinking.
2. List six steps in critical thinking.

Connecting the Dots

In each chapter, we try to help connect issues in the topic back to the case study. Sometimes the connections will be obvious, sometimes less so. You can try to make those connections for yourself, too, as you read and study.

There are many ways to do well in a course like this. Finding the ways topics are meaningful and useful for you will help make the work worthwhile. Doing well also involves paying attention to things like good study habits, setting realistic goals for yourself,

taking the initiative to look for interesting topics, finding an appropriate study space, and working with a study partner. We all have our own learning styles. You may understand and remember things best if you see them in writing, hear them spoken by someone else, reason them out for yourself, or learn by doing. By determining your preferred style, you can study in the way that is most comfortable and effective for you.

1 Understanding Our Environment

▲ Many of the most important challenges in environmental quality and sustainable development occur in informal settlements like Kibera, in Nairobi, Kenya.
Tatsiana Hendzel/Shutterstock

Learning Outcomes

After studying this chapter, you should be able to:

1.1 Explain what environmental science is, and how it draws on different kinds of knowledge.

1.2 Identify some early thinkers on environment and resources, and contrast some of their ideas.

1.3 Describe sustainable development and its goals.

1.4 Explain core concepts in sustainable development.

1.5 Identify ways in which ethics and faith might promote sustainability and conservation.

"Working together, we have proven that sustainable development is possible; that reforestation of degraded land is possible; and that exemplary governance is possible when ordinary citizens are informed, sensitized, mobilized and involved in direct action for their environment."

– Wangari Maathai (1940–2011)
Winner of 2004 Nobel Peace Prize

Sustainable Development Goals for Kibera

The idea of sustainable development is that we can improve well-being for poor populations, including reducing severe poverty, while maintaining or improving the environment on which we depend. These goals might seem contradictory, but increasing evidence shows that they can go together. In fact, as our resource consumption and population grow, it is increasingly necessary that they go together. Starting in 2016, the United Nations launched a new program to promote 17 Sustainable Development Goals, including access to education, health care, a safe natural environment, clean water, and other priorities, as well as conserving biodiversity and slowing climate change (fig. 1.1). Are all these goals possible?

Perhaps the greatest test case of this question is in fast-growing urban settlements of the developing world. One of the largest of these is a slum known as Kibera in Nairobi, Kenya. Every week, some 2,500 people arrive in Nairobi, drawn by hopes for better jobs and education. The city cannot build housing fast enough for this influx. Nor can it provide sanitary sewage, safe water systems, electric power, or other services. New arrivals build informal neighborhoods on the margins, using whatever materials are available to construct simple shelters of mud, brick, and tin roofing. Kibera is the largest of about 200 such settlements in Nairobi. These are home to over 2.5 million people, around 60 percent of the city's population (although reliable numbers are hard to come by).

Kibera grew on lowlands along the Nairobi River, in an area prone to flooding that periodically inundates houses and muddy informal streets. Because there is no system for managing sewage or garbage, both end up in the river, often entering homes with flood waters. Much of the time, a fetid odor of decomposing waste fills the air, and plastic shopping bags and other debris fill the corners of roadways and buildings. Occupying degraded outskirts of large cities, neighborhoods like Kibera suffer from the pollution produced by wealthy neighborhoods, and also create their own pollution and health hazards.

The city government has a complicated relationship with Kibera. The settlement provides much-needed housing, and residents contribute labor and consumer markets for growing businesses. But substandard housing is an embarrassment for city governments. Impoverished and unemployed populations turn to crime, even while they are the main victims of criminal activity. The city regularly tries to remove this and other slums, replacing them with modern housing, but the new flats are usually too expensive, and insufficient in supply, for the displaced residents.

Similar settlements exist in many of the world's fast-growing urban areas—Rio de Janeiro, Manila, Lagos, Cairo, Mumbai, Delhi, and many others—because global processes drive the growth of these vast slums. Rural population growth reduces access to farmland; forest destruction and soil degradation make traditional lifestyles difficult to maintain. Large landholders expand, displacing rural communities. Climate change threatens crop production. Declining water resources make farming difficult, and farmers are driven to the city.

In striving to enter the middle class, residents of Kibera also increase their environmental impacts. As they succeed, they consume more material goods, more energy, more cars and fuel, and electronics. All of these expand the environmental footprint of residents. On the other hand, the per capita energy and resource consumption of most Kibera residents is vanishingly small compared to consumption of their wealthy neighbors, who may have multiple cars and large houses, many appliances, and rich diets.

The global challenge of sustainable development is to find ways to improve the lives and the environment of people everywhere, including those in Kibera and other informal settlements. Slum residents have energy and ideas and are eager to improve the lives of their children, like people everywhere.

Environmental science is a discipline that seeks to understand both the natural systems we depend on and the ways we exploit or steward those resources. Sustainable development is central to environmental science, as we seek to protect resources and also support human well-being. As you read this book, you'll consider many issues of environmental systems, stewardship, and resource use. Ideally, a better understanding of these issues can help us find ways to address them, both locally and globally.

FIGURE 1.1 Sustainable development goals include access to education and electricity to study by at night.
Mark Boulton/Alamy Stock Photo

1.1 What Is Environmental Science?

- *This subject draws on many disciplines, skills, and interests.*
- *We face persistent challenges, as well as progress, in themes such as population growth, climate change, pollution, and biodiversity losses.*
- *Ecological footprints are a way to estimate our impacts.*

Humans have always inhabited two worlds. One is the natural world of plants, animals, soils, air, and water that preceded us by billions of years and of which we are a part. The other is the world of social institutions and artifacts that we create for ourselves using science, technology, and political organization. Both worlds are essential to our lives, but their intersections often cause enduring tensions: More than ever before, we have power to extract and consume resources, produce waste, and modify our world in ways that threaten both our continued existence and that of many organisms with which we share the planet. We also have better access than ever before to new ideas, efficient technologies, and opportunities to cooperate in finding sustainable strategies. To ensure a sustainable future for ourselves and future generations, we need to understand more about how our world works, what we are doing to it, and what we can do to protect and improve it.

Environment (from the French *environner:* to encircle or surround) can be defined as (1) the circumstances or conditions that surround an organism or group of organisms, or (2) the complex of social or cultural conditions that affect an individual or community. Because humans inhabit the natural world, as well as the "built" or technological, social, and cultural world, all constitute important parts of our environment (fig. 1.2).

Environmental science is the systematic study of our environment and the ways in which we both depend on it and influence it. Environmental science is interdisciplinary, integrating natural sciences, social sciences, and humanities in a broad, holistic study of the world around us. Much of environmental science focuses on understanding and resolving problems in our natural environment, such as pollution or lost biodiversity. But solutions have to do with how we consume resources and dispose of waste. This is why environmental science also includes discussion of policy, population, economics, and urbanization.

For an increasing number of environmental issues, the difficulty is not to identify remedies. Remedies are now well

FIGURE 1.2 Many kinds of knowledge contribute to solutions in environmental science. For a goal such as achieving clean and sustainable energy, strategies involve input from many disciplines.

understood. The problem is to make them socially, economically, and politically acceptable. Foresters know how to plant trees, but not how to establish conditions under which we can agree to let forests grow to maturity. Engineers know how to control pollution, but not how to persuade factories to install the necessary equipment. We even know how to address climate change and sustainable development, although we have not agreed to do so. These are complex problems, then, that require input from multiple perspectives.

As you study environmental science, you should aim to do the following:

- understand how natural systems function;
- understand ecological concepts that explain biological diversity;
- understand current environmental challenges, such as pollution and climate change; and
- use critical thinking to envision solutions to these challenges.

Environmental science is about understanding where we live

In this course, you will examine processes in our physical environment, including factors affecting biological diversity, biological productivity, sources of earth resources and energy, and circulation of climate and of water resources. You will also consider the ways resource use, policy, and practices influence those environmental systems. But as you read, also remember that the systems we discuss are amazing and beautiful. Imagine you are an astronaut

Benchmark Data	
Among the ideas and values in this chapter, the following are a few worth remembering.	
7.7 billion	Global human population, 2019
2.1	Replacement fertility rate (children/woman)
0.3 tons	CO_2 per person in least developed countries
11 tons	In most developed countries (table 1.1)
1.7	Number of planet Earths to satisfy our global footprint

FIGURE 1.3 Perhaps the most amazing feature of our planet is its rich diversity of life.
Fuse/Getty Images

returning to Earth after a trip to the moon or Mars. What a relief it would be to come back to this beautiful, bountiful planet after experiencing the hostile, desolate environment of outer space. We live in a remarkably prolific and hospitable world. Compared to the conditions on other planets in our solar system, temperatures on the earth are mild and relatively constant. Plentiful supplies of clean air, fresh water, and fertile soil are regenerated endlessly and spontaneously by geological and biological cycles (discussed in chapters 3 and 4).

Perhaps the most amazing feature of our planet is the rich diversity of life that exists here. Millions of beautiful and intriguing species populate the earth and help sustain a habitable environment (fig. 1.3). This vast multitude of life creates complex, interrelated communities where towering trees and huge animals live together with, and depend upon, tiny life-forms such as viruses, bacteria, and fungi. Together, all these organisms make up delightfully diverse, self-sustaining communities, including dense, moist forests, vast sunny savannas, and richly colorful coral reefs. From time to time, we should pause to remember that, in spite of the challenges and complications of life on earth, we are incredibly lucky to be here. We should ask ourselves: What is our proper place in nature? What *ought* we do and what *can* we do to protect the irreplaceable habitat that produced and supports us?

To really understand our environment, we also need to get outdoors and experience nature, in our backyard, a local park, or somewhere new. As author Ed Abbey said, "It is not enough to fight for the land; it is even more important to enjoy it. While you can. While it is still there. So get out there and mess around with your friends, ramble out yonder and explore the forests, encounter the grizz, climb the mountains. Run the rivers, breathe deep of that yet sweet and lucid air, sit quietly for a while and contemplate the precious stillness, that lovely, mysterious and awesome space. Enjoy yourselves, keep your brain in your head and your head firmly attached to your body, the body active and alive."

Major themes in environmental science

Environmental science covers several major themes. One of these is sustainable development, a question that runs through many chapters of this book. How do we foster healthy, rewarding livelihoods under conditions of population growth, shifting food systems, precarious water resources and water quality, air pollution, and growing demands for energy? Why is it important to protect resources and biodiversity, and how can they be protected, as human needs expand?

A second major theme that runs through most chapters is climate change and climate action. Climate change is the single most urgent issue of our time. It will affect human health, economies, poverty, and conflict, as well as dramatically changing environmental systems, including water resources, weather, food production, and biodiversity, on which we depend. Over the past 200 years, atmospheric CO_2 concentrations have increased about 35 percent. Climatologists warn that if current trends continue, by 2100 mean global temperatures will probably increase by 2° to 6°C (3.6° to 12.8°F) compared to temperatures in 1900 (fig. 1.5a). This warming explains the increasing severity and frequency of droughts, storms, and wildfires observed in recent years.

Fortunately, we know many ways to reduce climate change, from protecting forests to developing renewable energy and transforming food and farming systems. Global agreement on the importance of acting is nearly universal. At the 2015 Paris Climate Conference, and at multiple conferences since then, the world's nations have gradually strengthened efforts to take action. We are not acting fast enough, but we know many of the strategies that we need to pursue. As you read this book, one of your tasks is to understand solutions to this and other major challenges. You should also try to understand some of the many co-benefits to climate action and other environmental progress, including reductions in poverty, pollution, and illness.

In most topics we discuss, there are major challenges and also key areas of progress. It is important to recognize that both dire threats and exciting opportunities exist simultaneously. We tend to forget that in many cases, conditions were much worse in the past—air and water pollution in most U.S. cities, for example, were far more unhealthy in past decades, and global population growth has slowed dramatically from previous decades. Recognizing where conditions have improved over time also reminds us that the hard work of generations before us has been fruitful.

Here are a few examples of major topics in environmental science, with both challenges and encouraging progress.

Population and resource consumption

One of the most widely debated challenges is population growth. With some 7.7 billion humans on earth, we're adding about 80 million more every year. Family sizes have declined almost everywhere, from about five children per family 60 years ago to about two today, but still demographers project a population between 8 and 10 billion by 2050 (fig. 1.4a). The impacts of that many people on our natural resources and ecological systems is a serious concern. All high–birth rate countries are low-income, often

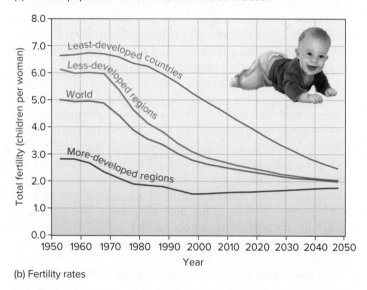

(a) Possible population trends: Where will we be in 2050?

(b) Fertility rates

FIGURE 1.4 Bad news and good news: Globally, populations continue to rise, but our rate of growth has plummeted. Nearly half of countries are below the replacement rate of about two children per woman.

war-affected areas. Of the 40 countries with the highest birth rates, all are in Africa except Afghanistan.

On the other hand, population growth has stabilized in nearly all industrialized countries and even in most poor countries where social security and democracy have been established. Over the last 20 years, the average number of children born per woman worldwide has decreased from 6.1 to 2.5 (fig. 1.4*b*). The UN Population Division predicts that by 2050 all developed countries and 75 percent of the developing world will experience a below-replacement fertility rate of 2.1 children per woman. This prediction suggests that the world population could stabilize sooner and lower than previously estimated.

Rising resource consumption per person is also an urgent concern. Poor populations consume very little energy, food, and other resources, compared to wealthy populations, which consume energy and goods from around the globe. As wealth rises around the world, people emerging from poverty desire the same high levels of consumption. Thus, both population and consumption rates are persistent questions in environmental science.

Hunger

In spite of population growth that added nearly a billion people to the world during the 1990s, the number facing food insecurity and chronic hunger during this period declined by about 40 million. Global food production has more than kept pace with human population growth, but hunger persists in many areas (fig. 1.5*b*). In a world of food surpluses, the United Nations estimates that some 800 million people are chronically undernourished, often because of drought, floods, displacement from land, or war. On the other hand, the *percentage* of people suffering from chronic hunger has declined, even in developing regions. Soil scientists report that about two-thirds of all agricultural lands show signs of degradation. Biotechnology and intensive farming techniques, responsible for much of our recent production gains, are too expensive for poor farmers, and they contaminate waterways and deplete soils. Small-scale farms still produce 80 percent of food consumed worldwide, according to the United Nations Development Programme. Can we ensure the sustainability of these farms without further environmental degradation?

Biodiversity loss and conservation efforts

Biologists report that habitat destruction, overexploitation, pollution, and the introduction of exotic organisms are eliminating species at a rate comparable to the great extinction that marked the end of the age of dinosaurs. The UN Environment Programme reports that, over the past century, more than 800 species have disappeared and at least 10,000 species are now considered threatened (fig. 1.5*c*). This includes about half of all primates and freshwater fish, together with around 10 percent of all plant species. Top predators, including nearly all the big cats in the world, are particularly rare and endangered. At least half of the forests existing before the introduction of agriculture have been cleared, and much of the diverse "old growth" on which many species depend for habitat is rapidly being cut and replaced by ecologically impoverished forest plantations.

Despite ongoing losses, we are also finding ways to conserve resources and use them more sustainably. Restoration ecology (chapter 13) has contributed to species monitoring and recovery. Improved monitoring of fisheries and networks of marine protected areas promote species conservation, as well as human development (fig. 1.5*d*). Brazil, which has the largest area of tropical rainforests in the world, reduced forest destruction by nearly two-thirds during just half a decade of conservation-minded policy. (Politics are fickle, though, and subsequent administrations have reversed this progress.) In addition to protecting endangered species, forest conservation is necessary to stabilize our global climate. Nature preserves and protected areas have increased sharply, from about 7 million km^2 in 1990 to nearly 25 million km^2 in

2 2.8 3.6 5 7 9 11 13 15 20°F
Projected winter temperature increase

(a) Climate change

(b) Hunger

(c) Biodiversity

(d) Resource management

FIGURE 1.5 Major environmental themes: (a) Climate change is projected to raise temperatures, especially in northern winter months. (b) Chronic hunger remains a persistent problem, especially in regions of political conflict. (c) Poaching and habitat loss threaten many species, including rhinos, but (d) sustainable resource use can safeguard vital resources, such as local fisheries.
a: Source: NOAA Geophysical Fluid Dynamics Laboratory; b: Jonas Gratzer/LightRocket/Getty Images; c: Tom Finkle; d: William P. Cunningham

2018, or around 12 percent of global land area. Marine preserves add another 33 million km² of designated reserves. Many of these areas are weakly protected, however, and rapidly expanding agriculture, forestry, mining, and urbanization make improved protection more critical than ever (see chapter 12).

Energy

How we obtain and use energy will determine our environmental future. Fossil fuels (oil, coal, and natural gas) presently provide around 80 percent of the energy used in industrialized countries. These resources have transformed our society and economies, giving us unprecedented access to new opportunities and goods. But our dependence on these resources has grown unsustainably high. Acquiring and using these fuels causes air and water pollution, mining damage, shipping accidents, and political conflict. Fossil fuels are also nonrenewable resources, with finite supplies, at least on a human time scale. Cleaner energy resources, including solar power, wind, geothermal, and biomass, are renewable (naturally replenished). These, together with conservation, could give us cleaner, less destructive options if we invest appropriately. Cities and regions everywhere are investing in renewable energy sources in order to protect energy security, employment, and the climate (fig. 1.6a).

Rapidly developing countries have the capacity to make real progress. China leads the world in solar energy, wind turbines, and biogas generation (from agricultural waste), and developing countries are increasingly investing in renewable energy sources. Progress in photovoltaic production has helped prices for solar panels in the United States drop by from $20 per watt in the 1980s to less than 50 cents today. The price of solar and wind is now competitive with fossil fuels. The European Union has pledged to get 20 percent of its energy from renewable sources by 2020. Improved permitting, financing, and installation strategies have been almost as important as improved technology. The United Kingdom aims to cut carbon dioxide emissions by 60 percent through energy conservation and a switch to renewables. Denmark and Sweden aim to eliminate most fossil fuel uses by 2050.

Pollution and environmental health

In developing areas, especially China and India, air quality has worsened dramatically in recent years. Over southern Asia, for example, satellite images recently revealed that a 3-km (2-mile)-thick toxic haze of ash, acids, aerosols, dust, and photochemical products regularly covers the entire Indian subcontinent for much of the year. At least 3 million people die each year from diseases triggered by air pollution. The United Nations estimates that, worldwide, more than 2 billion metric tons of air pollutants (not including carbon dioxide or windblown soil) are emitted each year, and many of these pollutants travel worldwide. Mercury, pesticides, perfluorocarbons, and other long-lasting pollutants accumulate in arctic ecosystems and native people after being transported by air currents from industrial regions thousands of kilometers to the south. And on some days, 75 percent of the smog and particulate pollution recorded in California can be traced to Asia.

The good news is that we know how to control air pollution. Metals, dust, even greenhouse gases can be captured before they leave the smoke stack. Most cities in Europe and North America are cleaner and healthier now than they were a half century ago. Clean technology benefits the economy and saves lives. The question is how to ensure that pollution controls are used where they are needed.

Water resources

Water may well be the most critical resource in the twenty-first century. Climate change is reducing irrigation supplies in many farming regions. Over 600 million people (9 percent of us) lack safe drinking water, and 2.4 billion (32 percent) don't have safe sanitation (fig. 1.6b). These figures are considerably better than 25 years ago, but polluted water and inadequate sanitation are

estimated to contribute to illness in more than a billion people annually, and to the death of over 5 million children per year. About 40 percent of the world population lives in countries where water demands now exceed supplies, and the UN projects that by 2025 as many as three-fourths of us could live under similar conditions. Water shortages and drought are frequently blamed for displacement of "climate refugees," who lack water for farming or basic subsistence.

Health and quality of life have improved in many regions, however, as safer water supplies have lowered incidence of waterborne illnesses. These and other infectious diseases have declined, while life expectancy has nearly doubled, on average (fig. 1.6c). Since 1990, more than 800 million people have gained access to improved water supplies and modern sanitation.

Information and education

Education, especially for girls, is now recognized to be the most powerful strategy for slowing population growth and reducing child mortality. In this and many other cases, increasing access to education and information are transforming lives around the

(a) Renewable energy

(b) Water resources

(c) Health care

(d) Education

FIGURE 1.6 Renewable energy (a) is a central theme. Water quality (b) continues to cause illness around the world, but there has been substantial progress in (c) health care, and (d) education.

a: Source: Dennis Schroeder/NREL/U.S. Dept. of Energy; b: Roger A. Clark/Science Source; c: Dimas Ardian/Getty Images News/Getty Images; d: Anjo Kan/Shutterstock world. Rates of illiteracy are falling in many areas, including very poor regions (fig. 1.6*d*). Because so many environmental issues can be fixed by new ideas, technologies, and strategies, expanding access to knowledge is essential to progress. The increased speed at which information now moves around the world offers unprecedented opportunities for sharing ideas. Developing countries may be able to avoid the mistakes made by industrialized countries and grow with new, efficient, and environmentally sustainable technologies.

Quantitative Reasoning

In the Ecological Footprint discussion, examine figure 1.7. Which factor shown has the largest effect? The second largest? Which has the smallest? Can you explain the idea of a "global hectare"? Finally, which factor has increased the most since 1963?

Think of a parent or grandparent who was an adult in 1963. In what ways was his or her energy use different from yours? Why?

What Do You Think?

Calculating Your Ecological Footprint

Can the earth sustain our current lifestyles? Will there be adequate natural resources for future generations? These questions are among the most important in environmental science today. We depend on our environment for food, water, energy, oxygen, waste disposal, and other life-support systems. For resource use to be sustainable, we cannot consume them faster than nature can replenish them. Degradation of ecological systems ultimately threatens everyone's well-being. Although we may be able to overspend nature's budget temporarily, future generations will have to pay the debts we leave behind.

To calculate your debts, you need a good accounting system. Organizations such as Redefining Progress provide tools to calculate an **ecological footprint,** a measure used to quantify the demands placed on nature by individuals or by nations. Online footprint calculators, such as the WWF Footprint Calculator, or the Redefining Progress calculator, let you assess your own footprint by answering a simple questionnaire about consumption patterns, such as electricity use, shopping, and driving habits.

Footprints are often calculated in terms of global hectares ("gha") of productive capacity, or the global area that would be needed to support one person. Part of the power of this metaphor is that we can visualize a specific area of land—one hectare is an area 100 m × 100 m—and we can use the numbers to compare overall consumption patterns among countries. The term "global hectares" also reminds us that we are always consuming resources from around the world.

According to Redefining Progress, the average world citizen has an ecological footprint equivalent to 2.7 gha, while the biologically productive land available is only 1.8 gha per person. How can this be? We're using nonrenewable resources (such as fossil fuels) to support a lifestyle beyond the productive capacity of our environment. It's like living by borrowing on your credit cards. You can do it for a while, but eventually you have to pay off the deficit. The imbalance is far more pronounced in wealthier countries. The average resident of the United States, for example, lives at a consumption level that requires 7.2 gha of biologically productive land. If everyone in the world were to adopt a North American lifestyle, we'd need about four more planets to support us all.

Like any model, an ecological footprint gives a usefully simplified description of a system. Also like any model, it is built on a number of simplifying assumptions: (1) Various measures of resource consumption and waste flows can be converted into the biologically productive area required to maintain them; (2) different kinds of resource use and dissimilar types of productive land can be standardized into roughly equivalent areas; (3) because these areas stand for mutually exclusive uses, they can be added up to a total—a total representing humanity's demand—that can be compared to the total world area of productive land.

Technological change sometimes can reduce our footprint: For example, world food production has increased about fourfold since 1950, mainly through advances in irrigation, fertilizer use, and higher-yielding crop varieties, rather than through increased croplands. How to sustain this level of production is another question, but this progress shows that land area isn't always an absolute limit. Similarly, switching to renewable energy sources such as wind and solar power can greatly reduce our ecological footprint. Note that in figure 1.7, carbon emissions (from energy consumption) make up the largest share of the calculated footprint globally. In Germany, which has invested heavily in wind, solar, small-scale hydropower, and public transportation, the ecological footprint is only 4.6 gha per person.

What are the most important steps your community could take to reduce its footprint? Are there things you could do to reduce your personal footprint? Is technological progress most important, or are there policy measures that could be just as important in helping developing areas grow without increasing their ecological footprint?

FIGURE 1.7 Humanity's ecological footprint has nearly tripled since 1961, when we began to collect global environmental data.
Source: WWF, Global Footprint Network, 2018.

Section Review

1. Why is population an important question in environmental science? In what ways is population less of a problem than in earlier years?

2. In what ways is pollution still a problem? Has it improved? Why?

3. What is an "ecological footprint"?

1.2 WHERE DO OUR IDEAS ABOUT OUR ENVIRONMENT COME FROM?

- *Utilitarian conservation seeks to protect useful resources.*

- *Wilderness preservation aims to preserve wilderness for aesthetic, intellectual, or philosophical value.*

- *Modern environmental movements have formed to fight pollution, injustice, and poverty.*

Debates about human misuse of nature have a long history. Plato complained in the fourth century B.C.E. that Greece once had been blessed with fertile soil and clothed with abundant forests of fine trees. After the trees were cut to build houses and ships, however, heavy rains washed the soil into the sea, leaving only a rocky "skeleton of a body wasted by disease." Springs and rivers dried up and farming became all but impossible (fig. 1.8). Many classical authors regarded Earth as a living being, vulnerable to aging, illness, and even mortality following the devastation of forest clearing, soil degradation, and other activities.

Some of the earliest *scientific* studies of environmental degradation were carried out in the eighteenth century by French and British colonial administrators. These administrators, some of whom were trained scientists, observed rapid soil loss and drying wells that resulted from intensive colonial production of sugar and other commodities. These early conservationists observed and understood the connection between deforestation, soil erosion, and

FIGURE 1.8 Nearly 2,500 years ago, Plato lamented land degradation that denuded the hills of Greece. Have we learned from history's lessons?
Ken Walsh/Alamy Stock Photo

local climate change, so they recognized that environmental stewardship was an economic necessity. The pioneering British plant physiologist Stephen Hales, for instance, suggested that conserving green plants preserved rainfall. His ideas were put into practice in 1764 on the Caribbean island of Tobago, where about 20 percent of the land was designated as "reserved in wood for rains."

Similarly, Pierre Poivre, an early French governor of Mauritius, an island in the Indian Ocean, was appalled at the environmental and social devastation caused by destruction of wildlife (such as the flightless dodo) and the felling of ebony forests on the island by early European settlers. In 1769, Poivre declared that one-quarter of the island was to be preserved in forests, particularly on steep mountain slopes and along waterways. Mauritius remains a model for balancing nature and human needs. Its forest reserves shelter a larger percentage of its original flora and fauna than most other human-occupied islands.

Current ideas have followed industrialization

Many of our current ideas about our environment and its resources were articulated by writers and thinkers in the past 150 years. Although many earlier societies had negative impacts on their environments, recent technological innovations have greatly accelerated our impacts. As a consequence of these changes, different approaches have developed for understanding and protecting our environment.

We can divide conservation history and environmental activism into at least four distinct stages: (1) pragmatic resource conservation, (2) moral and aesthetic nature preservation, (3) a growing concern about health and ecological damage caused by pollution, and (4) global environmental citizenship. Each era focused on different problems and each suggested a distinctive set of solutions. These stages are not necessarily mutually exclusive, however. Ideas from all these stages persist today, shaping our ideas and priorities about environmental resources and conservation. But it is useful to associate these ideas with particular stages in history that inspired their widespread adoption.

Stage 1. Resource waste inspired pragmatic, utilitarian conservation

Many historians consider the publication of *Man and Nature* in 1864 by geographer George Perkins Marsh as the wellspring of environmental protection in North America. Marsh, who also was a lawyer, politician, and diplomat, traveled widely around the Mediterranean as part of his diplomatic duties in Turkey and Italy. He read widely in the classics (including Plato) and personally observed the damage caused by the excessive grazing by goats and sheep and by the deforestation of steep hillsides. Alarmed by the wanton destruction and profligate waste of resources still occurring on the American frontier in his lifetime, he warned of its ecological consequences. Largely as a result of his book, national forest reserves were established in the United States in 1873 to protect dwindling timber supplies and endangered watersheds.

Among those influenced by Marsh's warnings were President Theodore Roosevelt (fig. 1.9a) and his chief conservation advisor,

(a) President Teddy Roosevelt

(b) Gifford Pinchot

(c) John Muir

(d) Aldo Leopold

FIGURE 1.9 Some early pioneers of the American conservation movement. President Teddy Roosevelt (a) and his main advisor Gifford Pinchot (b) emphasized pragmatic resource conservation, while John Muir (c) and Aldo Leopold (d) focused on ethical and aesthetic relationships.

a: Source: Library of Congress Prints & Photographs Division [LC-USZ62-77199];
b: Source: Library of Congress Prints & Photographs Division [LC-USZ62-103915];
c: Bettmann/Getty Images; d: AP Images

Gifford Pinchot (fig. 1.9*b*). In 1905, Roosevelt, who was the leader of the populist, progressive movement, moved the Forest Service out of the corruption-filled Interior Department into the Department of Agriculture. Pinchot, who was the first native-born professional forester in North America, became the founding head of this new agency. He put resource management on an honest, rational, and scientific basis for the first time in our history. Together with naturalists and activists such as John Muir, William Brewster, and George Bird Grinnell, Roosevelt and Pinchot established the framework of our national forest, park, and wildlife refuge systems, passed game protection laws, and tried to stop some of the most flagrant abuses of the public domain.

The basis of Roosevelt's and Pinchot's policies was pragmatic **utilitarian conservation.** They argued that the forests should be saved "not because they are beautiful or because they shelter wild creatures of the wilderness, but only to provide homes and jobs for people." Resources should be used "for the greatest good, for the greatest number for the longest time." "There has been a fundamental misconception," Pinchot said, "that conservation means nothing but husbanding of resources for future generations. Nothing could be further from the truth. The first principle of conservation is development and use of the natural resources now existing on this continent for the benefit of the people who live here now.

There may be just as much waste in neglecting the development and use of certain natural resources as there is in their destruction." This pragmatic approach still can be seen today in the multiple use policies of the Forest Service.

Stage 2. Ethical and aesthetic concerns inspired the preservation movement

John Muir (fig. 1.9*c*) was a geologist, author, and first president of the Sierra Club. He strenuously opposed Pinchot's utilitarian approach. Muir argued that nature deserves to exist for its own sake, regardless of its usefulness to us. Aesthetic and spiritual values formed the core of his philosophy of nature protection. This outlook has been called **biocentric preservation** because it emphasizes the fundamental right of other organisms to exist and to pursue their own interests. Muir wrote: "The world, we are told, was made for man. A presumption that is totally unsupported by the facts. . . . Nature's object in making animals and plants might possibly be first of all the happiness of each one of them. . . . Why ought man to value himself as more than an infinitely small unit of the one great unit of creation?"

Muir, who was an early explorer and interpreter of the Sierra Nevada Mountains in California, fought long and hard for the establishment of Yosemite and Kings Canyon National Parks. The National Park Service, established in 1916, was first headed by Muir's disciple Stephen Mather and has always been oriented toward preservation of nature in its purest state. It has often been at odds with Pinchot's utilitarian Forest Service.

In 1935, pioneering wildlife ecologist Aldo Leopold (fig. 1.9*d*) bought a small, worn-out farm in central Wisconsin. A dilapidated chicken shack, the only remaining building, was remodeled into a rustic cabin (fig. 1.10). Working together with his children, Leopold planted thousands of trees in a practical experiment in restoring the health and beauty of the land. Leopold argued for stewardship of the land. He wrote of "the land ethic," by which we should care for the land because it's the right thing to do—as well as the smart thing. "Conservation," he wrote, "is the positive exercise of skill and insight, not merely a negative exercise of abstinence or caution." The shack became a writing refuge and became the main focus of *A Sand County Almanac,* a much beloved collection of essays about our relation with nature. In it, Leopold wrote, "We abuse land because we regard it as a commodity belonging to us. When we see land as a community to which we belong, we may begin to use it with love and respect."

Quantitative Reasoning

Most successful pollution control and conservation efforts have succeeded only when it became possible to gather reliable data, such as the number of illnesses and economic losses resulting from air pollution. Discuss with other students: Are you more convinced of a problem when you see data? Why or why not? Do you, or those around you, feel that environmental data is easy to understand when you see it?

FIGURE 1.10 Aldo Leopold's Wisconsin shack, the main location for his *Sand County Almanac,* in which he wrote, "A thing is right when it tends to preserve the integrity, stability, and beauty of the biotic community. It is wrong when it tends otherwise." How might you apply this to your life?
William P. Cunningham

(a) Rachel Carson

(b) Bill McKibben

(c) Van Jones

(d) Wangari Maathai

FIGURE 1.11 Among many distinguished environmental leaders in modern times, (a) Rachel Carson, (b) Bill McKibben, (c) Van Jones, and (d) Wangari Maathai stand out for their dedication, innovation, and bravery.
a: Alfred Eisenstaedt/The LIFE Picture Collection/Getty Images; b: Cindy Ord/Getty Images; c: Ryan Rodrick Beiler/Shutterstock; d: s_bukley/Shutterstock

Stage 3. Rising pollution levels led to the modern environmental movement

The undesirable effects of pollution probably have been recognized at least as long as those of forest destruction. In 1273, King Edward I of England threatened to hang anyone burning coal in London, because of the acrid smoke it produced. In 1661, the English diarist John Evelyn complained about the noxious air pollution caused by coal fires and factories and suggested that sweet-smelling trees be planted to purify city air. Increasingly dangerous smog attacks in Britain led in 1880 to the formation of a national Fog and Smoke Committee to combat this problem.

The tremendous industrial expansion during and after the Second World War added a new set of concerns to the environmental agenda. *Silent Spring,* written by Rachel Carson (fig. 1.11*a*) and published in 1962, awakened the public to the threats of pollution and toxic chemicals to humans as well as other species. The movement she engendered has been called **modern environmentalism,** with concerns that include both environmental resources and pollution. Like many environmentalists, Carson was concerned with the ways pollution simultaneously endangered human health and the survival of other species.

As environmental concerns have expanded to climate action, one of the new leaders has been Bill McKibben (fig. 1.11*b*), an author, educator, and environmentalist who has written extensively about climate change and has led campaigns to demand political action on this existential threat. As scholar in residence at Middlebury College, he worked with a group of undergraduate students to create 350.org, an organization that has sponsored thousands of demonstrations around the world to raise public awareness about climate change. The group, which sparked actions for fossil fuel divestment in many campuses and communities, has been widely praised for its creative use of social media and public organization. McKibben and 350.org led the opposition to the Keystone XL pipeline project, which was designed to transport crude oil from Alberta's tar sands to export terminals in Texas. Many pipeline protests since then have helped push for alternatives to oil and gas.

Stage 4. Environmental quality is tied to social progress

Some have claimed that the roots of the environmental movement are elitist—promoting the interests of a wealthy minority who can afford to vacation in wilderness. In fact, most environmental leaders have seen social justice and environmental equity as closely linked. Gifford Pinchot, Teddy Roosevelt, and John Muir all strove to keep land and resources accessible to everyone, at a time when public lands, forests, and waterways were increasingly controlled by industrial interests, especially railroad, mining, and logging companies. The idea of national parks, one of our principal strategies for nature conservation, is to provide public access to natural beauty and outdoor recreation. Aldo Leopold, a founder of the Wilderness Society, promoted ideas of land stewardship among farmers, fishers, and hunters. Robert Marshall, also a founder of the Wilderness Society, campaigned all his life for social and economic justice for low-income groups. Many environmental leaders grew up in working-class families, so their sympathy with social causes is not surprising.

Issues of social justice are increasingly central to discussions of sustainablility and environmental policy. Anthony Kapel "Van" Jones (fig. 1.11c) is one who has spoken powerfully for social and environmental progress, helping to bring visibility to the power that people of color can have in environmental action. As both a social justice and environmental activist, Jones has fought poverty and racial injustice by creating hundreds of thousands of "green-collar" jobs installing solar systems and upgrading the energy efficiency of millions of American homes. He served as President Barack Obama's Special Advisor for Green Jobs and has worked to build a "green economy for everyone." He has also brought artists, athletes, and local leaders into national dialogues and engagement around social and environmental issues.

Some of today's leading environmental thinkers come from developing nations, where poverty and environmental degradation plague hundreds of millions of people. Dr. Wangari Maathai of Kenya (1940–2011) was a notable example. In 1977, Dr. Maathai (see fig. 1.11d) founded the Green Belt Movement in her native Kenya as a way of both organizing poor rural women and restoring their environment. Beginning at a small, local scale, this organization has grown to more than 600 grassroots networks across Kenya. They have planted more than 30 million trees, while mobilizing communities for self-determination, justice, equity, poverty reduction, and environmental conservation. Dr. Maathai was elected to the Kenyan Parliament and served as assistant minister for environment and natural resources. Her leadership helped bring democracy and good government to her country. In 2004, she received the Nobel Peace Prize for her work, the first time a Nobel has been awarded for environmental action. In her acceptance speech, she said, "Working together, we have proven that sustainable development is possible; that reforestation of degraded land is possible; and that exemplary governance is possible when ordinary citizens are informed, sensitized, mobilized and involved in direct action for their environment."

Photographs of the earth from space (fig. 1.12) provide a powerful icon for the fourth wave of ecological concern, which might be called **global environmentalism.** These photos remind us how small, fragile, beautiful, and rare our home planet is. We all share a common environment at this global scale. As our attention shifts from questions of preserving particular landscapes or preventing pollution of a specific watershed or airshed, we begin to worry about the life-support systems of the whole planet.

A growing number of Chinese activists are part of this global environmental movement. In 2006, Yu Xiaogang was awarded the Goldman Prize, the world's top honor for environmental protection. Yu was recognized for his work on Yunan's Lashi Lake, where he brought together residents, government officials, and entrepreneurs to protect wetlands, restore fisheries, and improve water quality. He also worked on sustainable development programs, such as women's schools and microcredit loans. His leadership was instrumental in stopping plans for 13 dams on the Nu River (known as the Salween in Thailand and Burma). Another Goldman Prize winner is Dai Qing, who was jailed for her book that revealed the social and environmental costs of the Three Gorges Dam on the Yangtze River.

FIGURE 1.12 The life-sustaining ecosystems on which we all depend are unique in the universe, as far as we know.
StockTrek/age fotostock

Section Review

1. Differentiate "conservation" and "preservation." Identify one person associated with each.
2. What was Rachel Carson's *Silent Spring* about? Why?
3. In what ways is environmental quality tied to social progress?

1.3 SUSTAINABLE DEVELOPMENT

- *Poverty affects quality of life in many ways.*
- *Poverty has declined dramatically in recent years, but so has environmental quality.*
- *Sustainable development aims to reduce poverty without damaging environmental resources.*

Policymakers are becoming aware that eliminating poverty and protecting our common environment are inextricably interlinked, because the world's poorest people are both victims and agents of environmental degradation. The poorest people are often forced to meet short-term survival needs at the cost of long-term sustainability. Desperate for croplands to feed their families, and for fuel, many clear forests or cultivate steep hillsides, where soil is rapidly eroded. Others migrate to the crowded shantytowns that surround most major cities in the developing world.

As we have noted, one of the core concepts of modern environmental thought is **sustainable development,** often defined as "meeting the needs of the present without compromising the ability of future generations to meet their own needs." In other words, we should be able to improve conditions for the world's poorest populations without devastating the environment. This definition was given in a 1987 report of the World Commission on Environment and Development, *Our Common Future*. This report is often called the Brundtland Report, after the chair of the commission, Norwegian prime minister Gro Harlem Brundtland.

The Brundtland Report led to the pivotal 1992 Earth Summit, a United Nations meeting held in Rio de Janeiro, Brazil. The Rio meeting was a pivotal event because it brought together many diverse groups. Environmentalists and politicians from wealthy countries, indigenous people and workers struggling for rights and land, and government representatives from developing countries all came together and became more aware of their common needs. The Rio meeting is largely credited with spreading the idea of sustainable development.

The World Bank estimates that 700 million people (nearly 10 percent of us) live below an international poverty line of (U.S.) $1.90 per day. This is less than a third as many severely impoverished people as there were in 1990, so it represents tremendous progress. Still, the human suffering engendered by this poverty is tragic. The very poor often lack access to an adequate diet, decent housing, basic sanitation, clean water, education, medical care, and other essentials for a humane existence (fig. 1.13). Seventy percent of those people are women and children.

Poverty affects many **quality-of-life indicators** (table 1.1). Infant mortality in the least-developed countries is about 25 times as high as in the most-developed countries. Modern sanitation, essential for controlling disease, is available to only 23 percent of

Table 1.1 Quality-of-Life Indicators

	Least-Developed Countries	Most-Developed Countries
GDP/Person[1]	$2,122	$41,395
Life Expectancy	63 years	81 years
Adult Literacy	58%	99%
Child Labor[2]	21.7	~0
Female Secondary Education	17%	95%
Total Fertility[3]	4.1	1.8
Infant Mortality[4]	55	5
Percent Urban	29.8%	81.9%
Electricity Access	34.2%	99.9%
CO_2/Capita[5]	0.3 tons	11 tons

[1]Annual gross domestic product per person, U.S. dollar equivalent
[2]Percent ages 5–14
[3]Average births/woman
[4]Per 1,000 live births
[5]Metric tons/yr/person
Source: UNDP Human Development Index, 2018.

residents in poorer countries. Meanwhile, carbon dioxide emissions (a measure of both energy use and contributions to global warming) are over 30 times greater in rich countries.

Global regions of extreme poverty and wealth are distinctly separate (fig. 1.14). The Human Development Index (HDI) is an index that aggregates factors such as health, education, and poverty, some of which appear in table 1.1. Having an index like this allows us to compare countries, and to see where need is greatest.

Affluence is a goal and a liability

About one-sixth of the world's population live in the richest countries, where the average per capita income is above (U.S.) $41,000 per year. Most of these countries are in North America or Western Europe, but Japan, Singapore, and Australia also fall into this group. Another one-sixth live in the least-developed countries, with per capita income averaging less than $2,500. The remaining two-thirds of the world's population live in middle- or low-income countries, where incomes average over $10,000, and where most people live above the extreme poverty line, by global standards.

Those of us in the richer nations enjoy unprecedented affluence and comfort. To do so, we consume an inordinate share of the world's resources and produce unsustainable amounts of pollution—although much of that pollution occurs far from where we live, in industrial cities and degraded mining regions and farmlands of distant countries. The United States, for example, with about 4.6 percent of the world's population, consumes about 20 percent of all oil, and produces about 15 percent of all carbon dioxide and half of all toxic wastes in the world (fig. 1.15).

FIGURE 1.13 People living in extreme poverty generally lack adequate food, housing, medical care, clean water, and safety.
William P. Cunningham

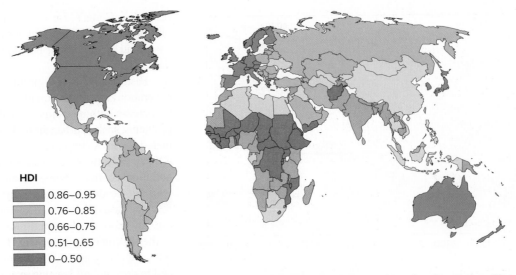

HDI
■	0.86–0.95
	0.76–0.85
	0.66–0.75
	0.51–0.65
■	0–0.50

FIGURE 1.14 Human Development Index. Values near 1 represent strong health, education, and quality of life indicators.
Source: UNEP 2016.

FIGURE 1.15 "And may we continue to be worthy of consuming a disproportionate share of this planet's resources."
©Lee Lorenz, The New Yorker Collection. ©The Cartoon Bank. Used with permission. All rights reserved

FIGURE 1.16 A rapidly growing economy has brought increasing affluence to China that has improved standards of living for many Chinese people, but it also brings environmental and social problems associated with Western lifestyles.
06photo/Shutterstock

Low-income countries hope to emulate this prosperity. Take the example of China, the world's most dramatic case of development. In the early 1960s, some 300 million Chinese suffered from chronic hunger, and perhaps 50 million starved to death in the worst famine in world history. Since then, China has experienced amazing economic growth and modernization. The national GDP has been increasing at about 5–10 percent per year. Hundreds of millions of people have been lifted out of extreme poverty. Chronic hunger has become uncommon. Average life expectancy has increased from 42 to 76 years, and infant mortality dropped from 150 per 1,000 live births in 1960 to 11 today. Annual per capita GDP has grown from less than (U.S.) $200 per year to more than $12,000 (fig. 1.16).

As a consequence of this growth, pollution has become more severe each year, as demand has exploded for resources, consumer goods, cars, and other luxuries. Most Chinese still consume far less than Americans or Europeans, though. In terms of ecological footprints (What Do You Think?, p. 15), it takes about 9.7 global hectares (gha, or hectares worth of resources) to support the

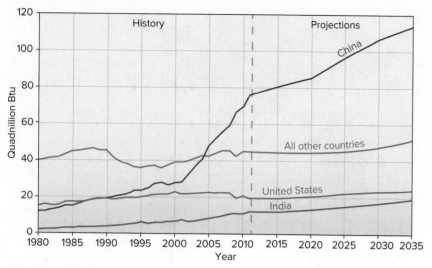

FIGURE 1.17 Coal consumption, most of it used for electricity generation, has fueled much of China's recent growth. Because coal is our primary source of air pollutants and greenhouse gases, projected increases would be disastrous.
Source: US Energy Information Agency 2013.

resources. China's use of coal-powered electricity, one of the most important sources of greenhouse gases and air pollution, has doubled in less than a decade (fig. 1.17).

Historically, some problems such as waste generation and carbon dioxide emissions have tended to increase with economic growth (fig. 1.18). Other problems, such as urban air pollution, tend to rise and then decline. A common explanation for the decline is that increasing wealth leads to better regulation and pollution controls. Some environmental costs can fall steadily with increasing development, as in the case of household sanitation: Communities can pay more to clean and protect water resources as they become wealthier (fig. 1.18). Some of these rules are changing, too. Economic growth has been tied to coal and oil burning ever since the Industrial Revolution. But in recent years, economic growth in many areas has proceeded despite regionally falling greenhouse gas emissions (see chapter 15).

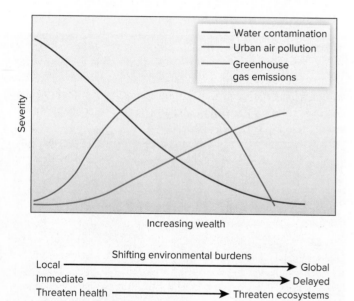

FIGURE 1.18 Environmental indicators show different patterns as incomes rise. Water contamination decreases as people can afford wastewater treatment and drinking water filtration. Local air pollution, on the other hand, often increases as more fuel is burned; eventually, development reaches a point at which people can afford clean air technology. Delayed, distant problems, such as greenhouse gas emissions that lead to global climate change, tend to rise steadily with income because people make decisions based on immediate needs and wants rather than long-term consequences. Thus, we tend to shift environmental burdens from local and immediate to distant and delayed if we can afford to do so.
Source: World Energy Assessment, UNDP.

Quantitative Reasoning

Examine figure 1.17. Describe in your own words how increasing wealth might affect water contamination, air pollution, and greenhouse gas emissions. Why might greenhouse gas emissions rise? Describe and explain the air pollution curve.

Is sustainable development possible?

Most poor nations, and most poor families, desperately want to become wealthier, to have more material goods, more food, and more cars, lights, computers, and other amenities. Can everyone have their share without destroying our shared environment? This depends on how development happens, how resources are shared, and whether developing regions find new paths to prosperity, different from the resource-intensive paths of the twentieth century.

Two basic principles of sustainable development, put forth by Herman Daly in the 1970s, are that we cannot consume natural resources faster than they are produced, and we cannot produce waste faster than nature can recycle it, if we intend to be here for the long term. Note, however, that the measures in the Human Development Index (table 1.1) are not resource-intensive factors. Providing basic access to health care and education, or providing clean drinking water and controlling pollution, improves people's lives with few natural resources. At the same time, they greatly advance human resources—our capacity to work and dream. Increasing wealth for the already-wealthy is more costly.

Sustainable development can mean many things. Many forms of sustainable farming, for example, produce food while protecting wildlife habitat, soil, and water resources. Often, sustainable development simply means distributing investment to small

average American each year. By contrast, the average person in China consumes about 2.1 gha per year. Providing the 1.3 billion Chinese with American standards of consumption would require about 10 billion gha, or almost another entire earth's worth of producers, who circulate money in their local communities. Fair trade organizations, for example, help people in developing countries to grow or make high-value products—often using traditional techniques and designs—that can be sold on world markets for good prices (fig. 1.19).

It is important to recognize that many innovations for climate action, along with sustainable development, are coming from the developing world. Already by 2015, Kenya produced more than half its electric power from geothermal, wind, and solar energy sources. The World Bank calculates that Sub-Saharan Africa could produce 170 gigawatts of low-carbon energy. This sustainable development potential exists across much of the world. In the Americas, Uruguay achieved 95 percent renewable electricity due to a decade of policy commitment. Costa Rica, with abundant geothermal energy, is approaching 100 percent renewable energy. Nicaragua is aiming for 90 percent renewable energy by 2020.

Growth can also occur without increased resource consumption. Markets in arts, entertainment, education, services, and leisure time can improve our lives with little environmental cost. These are growth industries whose primary resource is human ingenuity. This is an idea recognized by economists at least since John Stuart Mill in 1857: "It is scarcely necessary to remark that [resource limitation] implies no stationary state of human improvement. There would be just as much scope as ever for all kinds of mental culture and moral and social progress; as much room for improving the art of living and much more likelihood of its being improved when minds cease to be engrossed by the art of getting on."

The UN has identified 17 Sustainable Development Goals

Beginning in 2016, the United Nations initiated a program to promote 17 **Sustainable Development Goals** (SDGs). The goals are ambitious and global, and they include eliminating the most severe poverty and hunger, promoting health, education, and gender equality, providing safe water and clean energy, and preserving biodiversity (see fig. 1.20). This global effort, developed by representatives of the member states of the UN, seeks to coordinate data gathering and reporting, so that countries can monitor their progress, to share resources, and to promote sustainable investment in developing areas.

For each of the 17 goals, organizers identified targets, some quantifiable, some more general. For example, Goal 1, "End poverty," includes targets to eradicate extreme poverty everywhere, and to ensure that all people have rights to basic services, ownership and inheritance of property, and other economic needs. Goal 7, "Ensure access to affordable, sustainable energy," includes targets of doubling energy efficiency and enhancing international investment in clean energy. Goal 12, "Ensure sustainable consumption and production," calls for cutting per capita food waste in half, as well as phasing out fossil fuel subsidies that encourage wasteful consumption. The UN aims to meet these targets by 2030, in a span of 15 years.

The SDGs also include a number of targets for economic and social equity, and for better governance. To most economists and policymakers, it seems obvious that economic growth is the only way to bring about a long-range transformation to more advanced and productive societies and to provide resources to improve the lot of all people. As former U.S. president John F. Kennedy said, "A rising tide lifts all boats." But the Brundtland report (previous section) emphasized that equity is also essential: Political stability, democracy, and fair access to resources and opportunity are needed to ensure that the poor will get a fair share of the benefits of greater wealth in a society. According to a study released in 2006 by researchers at Yale and Columbia Universities, environmental sustainability tends to occur where there are open political systems and good government.

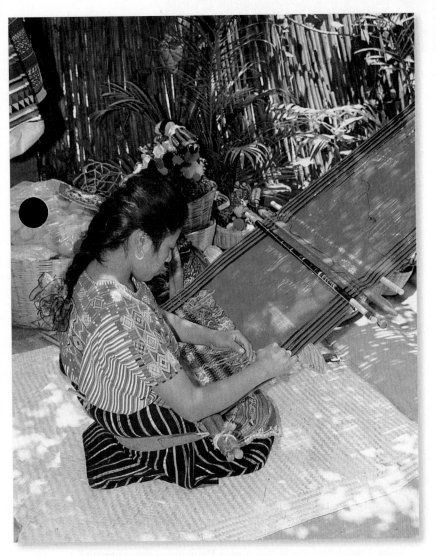

FIGURE 1.19 A Mayan woman from Guatemala weaves on a backstrap loom. A member of a women's weaving cooperative, she sells her work to nonprofit organizations in the United States at much higher prices than she would get at the local market.

Barry Barker/McGraw-Hill Education

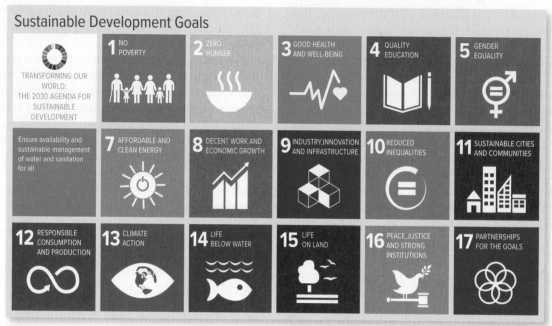

FIGURE 1.20 The United Nations Sustainable Development Goals are intended to improve well-being of the world's poorest people while also protecting biodiversity, natural resources, and climate. These goals follow the largely successful Millennium Development Goals.
Source: UN Development Programme.

The Millennium Development Goals were largely successful

These ambitious goals might appear unrealistic, but they build on the remarkable (though not complete) successes of the **Millennium Development Goals.** These eight goals were a 15-year effort, from 2000 to 2015, to improve literacy, health, access to safe water, child survival, and other goals. Targets included ending poverty and hunger, universal education, gender equity, child health, maternal health, combating HIV/AIDS, environmental sustainability, and global cooperation in development efforts. While only modest progress was achieved on some goals, UN Secretary General Ban Ki-Moon called that effort the most successful anti-poverty movement in history. Extreme poverty dropped from nearly half the population of developing countries to just 14 percent in only 15 years. The proportion of undernourished people dropped by almost half, from 23 percent to 13 percent. Primary school enrollment rates climbed from 83 percent to 91 percent in developing countries. Girls gained access to education, employment, and political representation in national parliaments.

At the same time, UN reports stressed that many goals were not met. Some 2.6 billion people gained access to safe drinking water, but over 40 percent of the world's population still lacks access to piped drinking water at home. The proportion of urban populations living in slums fell from 39 percent to 30 percent, but that still represents a large population with inadequate and unsafe housing.

The value of having clearly stated goals, especially with quantifiable targets, is that they help people agree on what to work for. With so many simultaneous problems in developing areas, it can be hard for leaders to know where to focus first. Agreed-upon targets, especially when they are shared and monitored by many countries, can strongly motivate action. International agreement on goals can also help motivate financial and planning assistance, both often badly needed in developing areas.

Development depends on how wealthy countries allocate spending

Economist Jeffery Sachs, director of the UN Millennium Development Project, says we could end extreme poverty worldwide by 2025 if the richer countries would donate just 0.7 percent of their national income for development aid in the poorest nations. These funds could be used for universal childhood vaccination against common infectious diseases, access to primary schools for everyone, family planning and maternal health services, safe drinking water and sanitation for all, food supplements for the hungry, and microcredit loans to promote self-employment.

The United States, the world's largest total donor, sets aside only 0.16 percent of its gross domestic product for development aid. That amounts to about 18 cents per citizen per day for both private and government aid to foreign nations. As former Canadian prime minister Jean Chrétien said, "Aid to developing countries isn't charity; it's an investment. It will make us safer, and when standards of living increase in those countries, they'll become customers who will buy tons of stuff from us."

The United Nations Development Programme has estimated that it would take about (U.S.) $135 billion per year to abolish extreme poverty and the worst infectious diseases over the next 20 years. That's a lot of money, but it's not much more than the $120 billion in subsidies and tax breaks the U.S. government gives

FIGURE 1.21 Every year, military spending equals the total income of half the world's people. The cost of a single large aircraft carrier equals 10 years of human development aid given by all the world's industrialized countries.
Source: Kyle Gahlau/DVIDS.

to oil companies each year. And it's far less than the $1 trillion of global military spending each year (fig. 1.21).

Many experts propose that if we were to shift one-tenth of that spending to development aid, we would be much safer than we are spending it on the military. The Worldwatch Institute warns that "poverty, disease and environmental decline are the true axis of evil." Terrorist attacks—and the responses they provoke—are symptoms of the underlying sources of global instability, including the dangerous interplay among poverty, hunger, disease, environmental degradation, and rising competition for resources. Unless the world takes action to promote sustainability and equity, Worldwatch suggests, we are unlikely to resolve persistent problems of wars, terrorism, and natural disasters. As writer H. L. Mencken once said, "If you want peace, work for justice."

Section Review

1. List any three quality of life indicators (table 1.1). How do they differ between wealthy and poor countries?
2. Why is affluence a liability? Give an example.
3. Explain the idea of sustainable development.

1.4 CORE CONCEPTS IN SUSTAINABLE DEVELOPMENT

- *"Ecosystem services" is a term for goods, services, and products we rely on; often these are invisible.*
- *Shared resources and ecosystem services can be described as "common property," or as a "commons." Managing common property is a key challenge.*
- *Indigenous peoples often protect biodiversity.*

Some general organizing ideas help us make sense of the problem of sustainable development and of how to think about prosperity in a world of limits. The ideas raised in this section will help you think about issues such as water resource management, ways to preserve water quality, how to encourage greenhouse gas reductions, questions of biodiversity, air quality, and many other natural resource considerations. These issues will also come up in later chapters on economics and policy.

How do we describe resource use?

The natural world supplies the water, food, metals, energy, and other resources we use. Because we use so many kinds of resources, one widely used measure for evaluating resource consumption is the ecological footprint (see What Do You Think?, p. 15, and the Data Analysis section, at the end of this chapter). Our ecological footprint is an index, a number representing a complex array of factors.

Another widely used concept for describing resource use is **throughput,** the amount of material or resources that flow through a system. A household that consumes abundant consumer goods, foods, and energy brings in a great deal of natural resource–based materials; that household also disposes of a great deal of materials. Conversely, a household that consumes very little also tends to produce little waste.

Ecosystem services refers to services or resources provided by environmental systems (fig. 1.22). Provisioning of resources, such as the fuels we burn, may be the most obvious service we require. Supporting services are less obvious until you start listing them: These include water purification, production of food and atmospheric oxygen by plants, or decomposition of waste by fungi and bacteria. Regulating services include maintenance of temperatures suitable for life by the earth's atmosphere, or carbon capture by green plants, which maintains a stable atmospheric composition. Cultural services include a diverse range of recreational, aesthetic, and other nonmaterial benefits. Usually we rely on these resources without thinking about them. They support all our economic activities in some way, but we don't put a price on them because nature doesn't force us to pay for them.

How can we protect these services over the long term? One of the answers to this basic question was given in an essay entitled **"Tragedy of the Commons,"** published in 1968 in the journal *Science* by ecologist Garret Hardin. In this classic framing of the problem, Hardin argued that population growth leads inevitably to overuse and then destruction of common resources—such as shared pastures, unregulated fisheries, fresh water, land, and clean air. Hardin's essay has influenced our ideas about resource management for decades.

Examples of destroyed commons abound. The North Atlantic cod fishery, once one of the world's greatest fish populations, was functionally destroyed by a free-for-all of unregulated fishing by fleets of many nations. The species is not extinct, but it may never recover to its historic abundance. Air pollution is another familiar example: Industries emit pollution from unregulated incinerators and burners, spilling soot, sulfur dioxide, and carbon dioxide into

FIGURE 1.22 Ecosystem services we depend on are countless and often invisible.

In the image, the following labels appear:

CO_2

O_2

Photosynthesis (provisioning, supporting)

Carbon capture, atmospheric regulation (regulating, supporting)

Food, fuel (provisioning)

Decomposition, nutrient cycling by decomposers (supporting)

Water purification by streams, soil bacteria (supporting)

Temperature regulation by water bodies (regulating)

the air, which is then contaminated for all users. Hardin proposed that there are only two ways to avoid this destruction: (a) a system of private property, in which owners protect resources because of self-interest, or (b) coercive regulation by the state.

An alternative perspective to Hardin's framework is strategies for **managing the commons;** that is, for collectively safe-guarding commonly used resources. The importance of common property management was publicized by Elinor Ostrom, who won the 2009 Nobel Prize in Economic Sciences. Ostrom and her colleagues demonstrated that ordinary people have often created rules and institutions for the sustainable management of shared resources. They examined cases of successful long-term management of sustainably managed fisheries, common forests, common grazing lands, and other resources in communities around the world. They emphasized that not all common properties are well managed, but a great many are.

What conditions can help communities manage their commons over the long term? Many strategies exist, but Ostrom and her colleagues found that some conditions occur frequently in successful cases. Among these are (1) effective and inexpensive monitoring of resource use; (2) an ability to exclude outsiders, who don't understand rules of use; and (3) frequent face-to-face communications and strong social networks among users, which reduce distrust and promote communication about the state of the resource.

What type of institution is best for managing a global commons, such as climate or biodiversity? Garret Hardin, with many others, has argued that local solutions to climate change are irrelevant as long as countries and international institutions fail to make policy changes. Ostrom and her colleagues argued that these large institutions are often incapable of taking strong or quick action for resource conservation. So it is also important to invest in smaller, local, even individual policy changes, whose effects and ideas may spread contagiously or inform broader improvements in resource management. Both positions are probably correct: Often mixtures of different types of institutions, large and small, may be needed, to contribute simultaneously to solutions at different scales.

Planetary boundaries define broad limits

Another way to think about environmental services is planetary boundaries, or thresholds of abrupt or irreversible environmental change. Studies by Johan Rockström and colleagues at the Stockholm Resilience Centre have identified nine major systems with these critical thresholds: climate change, biodiversity, land system change, freshwater use, biogeochemical flows (nitrogen and phosphorus), ocean acidification, atmospheric aerosols, stratospheric ozone loss, and "novel entities," including chemical pollution and other factors (fig. 1.23). Calculations are that we have already passed the planetary boundaries for some of these systems, and that we

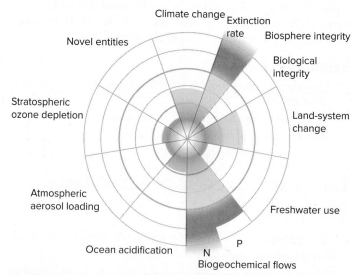

FIGURE 1.23 Calculated planetary boundaries, or thresholds beyond which irreversible change is likely. Green shading represents safe ranges; yellow represents a zone of increasing risk; red wedges represent factors exceeding boundaries.

are approaching limits for others. These overshoots are expected to cause rapid declines in ecosystem function and ecosystem services in many areas.

Ecosystem services are tightly coupled. Destruction of tropical forests in Southeast Asia, for example, can influence heat and drought in North America. Drought and fires in North America enhance climate warming and sea ice loss in the Arctic. A planetary perspective helps us see interconnections in global systems and their effects on human well-being. What it means to pass these boundaries remains uncertain.

Indigenous peoples often protect biodiversity

Development challenges are especially severe for indigenous peoples in both rich and poor countries. Typically descendants of the original inhabitants of an area taken over by more powerful outsiders, indigenous peoples are distinct from their country's dominant language, culture, religion, and racial communities. Consequently, these groups often generally are the least powerful, most neglected groups in an area. Of the world's nearly 6,000 recognized cultures, 5,000 are indigenous; and these account for only about 10 percent of the total world population. In many countries, traditional caste systems, discriminatory laws, economics, or prejudices repress indigenous people. At least half of the world's 6,000 distinct

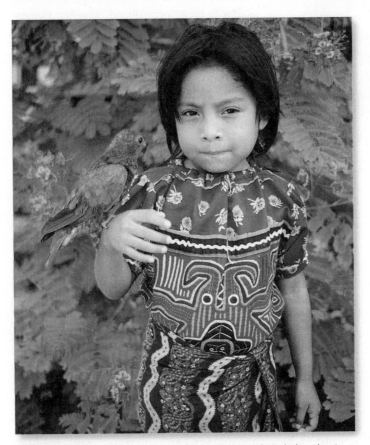

FIGURE 1.24 Do indigenous people have unique knowledge about nature and inalienable rights to traditional territories?
Barry Barker/McGraw-Hill Education

languages are dying because they are no longer taught to children. When the last elders who still speak the language die, so will the culture that was its origin. Lost with those cultures will be a rich repertoire of knowledge about nature and a keen understanding of a particular environment and way of life (fig. 1.24).

Nonetheless, the 500 million indigenous people who remain in traditional homelands still possess valuable ecological wisdom and are the guardians of little-disturbed habitats that are refuges for rare and endangered species and undamaged ecosystems. As we seek strategies for sustainable development and biodiversity conservation, this knowledge may be an essential resource.

Recognizing native land rights and promoting political pluralism can be among the best ways to safeguard ecological processes and endangered species. A few countries, such as Papua New Guinea, Fiji, Ecuador, Canada, and Australia, acknowledge indigenous title to extensive land areas. As the Kuna Indians of Panama say, "Where there are forests, there are native people, and where there are native people, there are forests."

Section Review

1. Think of five ecosystem services on which you rely.
2. What is the "tragedy of the commons"? What are two ways to avoid it?
3. List any two of the factors that can help communities manage a commons.
4. Why do indigenous people often have an interest in protecting biodiversity?

1.5 ENVIRONMENTAL ETHICS, FAITH, AND JUSTICE

- *Moral extensionism means extending value beyond ourselves.*
- *Many faiths encourage stewardship because they see divine value in our environment.*
- *Environmental justice involves human rights and environmental justice.*

The ways we interpret environmental issues, or our decisions about what we should or should not do with natural resources, depend partly on our basic worldviews. Perhaps you have a basic ethical assumption that you should be kind to your neighbors, or that you should try to contribute in positive ways to your community. Do you have similar responsibilities to take care of your environment? To conserve energy? To prevent the extinction of rare species? Why? Or why not?

Your position on these questions is partly a matter of **ethics**, or your sense of what is right and wrong. Some of these ideas you learn early in life; some might change over time. Ethical views in society also change over time. In ancient Greece, many philosophers who were concerned with ethics and morality owned slaves. A slave owner could mistreat or even kill a slave with little or no consequences. Today, few societies condone slavery. Most societies now believe it is unethical to treat other humans as property. On the other hand, most societies now consider land, water, forests, and other natural resources as private property. It is the owner's right to conserve or degrade those private resources as they like. Other people (or other organisms) have no legal right to restrict how private property is used or abused. Normally, if you have ancient trees on your property, it is your right to cut and sell them, regardless of your neighbors' opinions. Is this ethical? That depends on your perspective.

Often, our core beliefs are so deeply held that we have difficulty even identifying them. But they can influence how you act, how you spend money, or how you vote. Try to identify some of your core beliefs. What is a basic thing you simply should or should not do? Where does your understanding come from about those actions?

We can extend moral value to people and things

One of the reasons we don't accept slavery now, as the ancient Greeks did, is that most societies believe that all humans have basic rights. The Greeks granted **moral value** (value or worth, based on moral principles) only to adult male citizens within their own community. Women, slaves, and children had few rights and were essentially treated as property. Over time, we have gradually extended our sense of moral value to a wider and wider circle, an idea known as **moral extensionism;** that is, extending moral value to a larger circle of people, organisms, or objects (fig. 1.25). In most countries, women and minorities have basic civil rights, children cannot be treated as property, and even domestic pets have some legal protections against cruel treatment. For many people, moral value also extends to domestic livestock (cattle, hogs, poultry), which makes eating meat a fundamentally wrong thing to do. For others, this moral extension ends with pets, or with humans. Some people extend moral value to include forests, biodiversity, inanimate objects, or the earth as a whole.

How we treat other people, animals, or things can also depend on whether we believe they have **inherent value** (an intrinsic right to exist) or **instrumental value** (usefulness to someone). If I hurt you, I owe you an apology. If I borrow your car and crash it into a tree, I don't owe the car an apology, I owe *you* an apology—or reimbursement.

Does this apply to other species? Domestic animals clearly have an instrumental value because they are useful to (or give comfort to) their owners. But some philosophers would say they also have inherent values and interests. By living, breathing,

FIGURE 1.25 Moral extensionism describes an increasing consideration of moral value in other living things—or even nonliving things.

struggling to stay alive, the animal carries on its own life independent of its usefulness to someone else.

Some argue that even nonliving things also have inherent worth. Rocks, rivers, mountains, landscapes, and certainly the earth itself, have value. These things were in existence before we came along, and we couldn't re-create them if they are altered or destroyed. This philosophical debate became a legal dispute in a historic 1969 court case, when the Sierra Club sued the Disney Corporation on behalf of the trees, rocks, and wildlife of Mineral King Valley in the Sierra Nevada Mountains (fig. 1.26), where Disney wanted to build a ski resort. The Sierra Club argued that it represented the interests of beings that could not speak for themselves in court.

A legal brief entitled *Should Trees Have Standing?*, written for this case by Christopher D. Stone, proposed that organisms as well as ecological systems and processes should have standing (or rights) in court. After all, corporations—such as Disney—are treated as persons and given legal rights even though their "personhood" is only a figment of our imagination. Why shouldn't nature have similar standing? The case went all the way to the Supreme Court but was overturned on a technicality. In the meantime, Disney lost interest in the project and the ski resort was never built. What do you think? Where would you draw the line, regarding what deserves moral standing? Are there ethical limits on what we can do to nature?

Many faiths promote conservation and justice

Ethical and moral values are often rooted in religious traditions, which try to guide us in what is right and wrong to do. With growing public awareness of environmental problems, religious organizations have begun to take stands on environmental concerns. They recognize that some of our most pressing environmental problems don't need technological or scientific solutions; they're not so much a question of what we're able to do, but what we're willing to do. Are we willing to take the steps necessary to stop global climate change? Do our values and ethics require us to do so? In what ways might religious views influence our attitudes toward nature?

Environmental scientists have long been concerned about religious perspectives. In 1967, historian Lynn White Jr. published a widely influential paper, "The Historic Roots of Our Ecological Crisis." He argued that Christian societies have often exploited natural resources carelessly because the Bible says that God

FIGURE 1.26 Mineral King Valley at the southern border of Sequoia National Park was the focus of an important environmental law case in 1969. The Disney Corporation wanted to build a ski resort here, but the Sierra Club sued to protect the valley on behalf of the trees, rocks, and native wildlife.

Photo by Laurel Di Silvestro courtesy of the Mineral King Valley Preservation Society

Table 1.2	Principles and Actions for Stewardship in the Ohito Declaration

Spiritual Principles

1. Religious beliefs and traditions call us to care for the earth.
2. For people of faith, maintaining and sustaining environmental life systems is a religious responsibility.
3. Environmental understanding is enhanced when people learn from the example of prophets and of nature itself.
4. People of faith should give more emphasis to a higher quality of life, in preference to a higher standard of living, recognizing that greed and avarice are root causes of environmental degradation and human debasement.
5. People of faith should be involved in the conservation and development process.

Recommended Courses of Action

The Ohito Declaration calls upon religious leaders and communities to

1. emphasize environmental issues within religious teaching: faith should be taught and practiced as if nature mattered.
2. commit themselves to sustainable practices and encourage community use of their land.
3. promote environmental education, especially among youth and children.
4. pursue peacemaking as an essential component of conservation action.
5. take up the challenge of instituting fair trading practices devoid of financial, economic, and political exploitation.

commanded Adam and Eve to dominate nature: "Be fruitful, and multiply, and replenish the earth and subdue it: and have dominion over the fish of the sea, and over the fowl of the air, and over every living thing that moveth upon the earth" (Genesis 1:28). Since then, many religious scholars have pointed out that God also commanded Adam and Eve to care for the garden they were given, "to till it and keep it" (Genesis 2:15). Furthermore, Noah was commanded to preserve individuals of all living species, so that they would not perish in the great Flood. Passages such as these inspire many Christians to insist that it is our responsibility to act as stewards of nature, and to care for God's creations.

The idea of **stewardship,** or taking care of the resources we are given, inspires many religious leaders to promote conservation. "Creation care" is a term that has become prominent among evangelical Christians in the United States. In 1995, representatives of nine major religions met in Ohito, Japan, to discuss their various traditions' views of environmental stewardship. The resulting document, the Ohito Declaration, outlined common beliefs and responsibilities of these different faiths toward protecting the earth and its life (table 1.2). In recent years, religious organizations have played important roles in nature protection. A coalition of evangelical Christians has been instrumental in promoting stewardship of many aspects of our environment, from rare plants and animals to our global climate.

Religious concern extends beyond our treatment of plants and animals. Pope Francis has repeatedly called for Christians to care for God's creation: "let us become . . . channels through which God can water the earth, protect all creation and make justice and peace flourish." Pope John Paul II and Orthodox Patriarch Bartholomew called on countries bordering the Black Sea to stop pollution, saying that "to commit a crime against nature is a sin."

Many religious organizations are also working for integrated justice and environmental goals. Interfaith Power and Light is an organization promoting indigenous interests, clean energy, and greenhouse gas reduction. The Creation Care Network has also launched initiatives against energy inefficiency, mercury pollution, mountaintop removal mining, and destruction of endangered species, in addition to its campaign to combat global warming. For many people, religious beliefs provide the best justification for environmental protection.

Calls for both stewardship and human domination over nature can be found in the writings of most major faiths. The Koran teaches that "each being exists by virtue of the truth and is also owed its due according to nature," a view that extends moral rights and value to all other creatures. Hinduism and Buddhism teach *ahimsa,* or the practice of not harming other living creatures, because all living beings are divinely connected (fig. 1.27).

Environmental justice integrates civil rights and environmental protection

People of color in the United States and around the world are subjected to a disproportionately high level of environmental health risks in their neighborhoods and on their jobs. Minorities, who

FIGURE 1.27 Many religions emphasize the divine relationships among humans and the natural world. The Tibetan Buddhist goddess Tara represents compassion for all beings.
William P. Cunningham

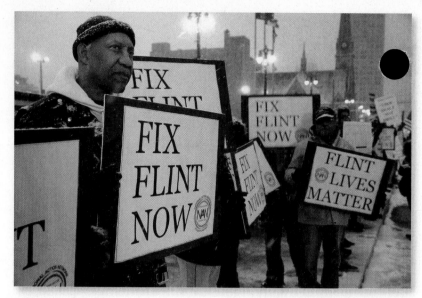

FIGURE 1.28 Poor people and minorities often live in the most dangerous and least desirable places. Here residents of Flint, Michigan, protest lead contamination in their drinking water.
Chip Somodevilla/Getty Images

tend to be poorer and more disadvantaged than other residents, work in the dirtiest jobs where they are exposed to toxic chemicals and other hazards. More often than not they also live in urban ghettos, barrios, reservations, and rural poverty pockets that have shockingly high pollution levels and are increasingly the site of unpopular industrial facilities, such as toxic waste dumps, landfills, smelters, refineries, and incinerators. **Environmental justice** combines civil rights with environmental protection to demand a safe, healthy, life-giving environment for everyone.

Among the evidence of environmental injustice is the fact that three out of five African Americans and Latino/as, and nearly half of all Native Americans, Asians, and Pacific Islanders, live in communities with one or more uncontrolled toxic waste sites, incinerators, or major landfills, while fewer than 10 percent of all whites live in these areas. Using zip codes or census tracts as a unit of measurement, researchers found that minorities make up twice as large a population share in communities with these locally unwanted land uses (**LULUs**) as in communities without them.

Poverty and race are often closely correlated, so environmental justice struggles often form around racial differences. Poor people in general are likely to live in polluted neighborhoods, but even in middle-class income ranges, people of color are more

likely than whites to be explosed to pollution. Where upper-class whites can "vote with their feet" and move out of polluted and dangerous neighborhoods, blacks and other minorities tend to be restricted by color barriers and prejudice (overt or covert) to the less desirable locations (fig. 1.28).

Racial prejudice is a belief that people are inferior merely because of their race. **Environmental racism** is inequitable distribution of environmental hazards based on race. Evidence of environmental racism can be seen in lead poisoning in children. The Federal Agency for Toxic Substances and Disease Registry considers lead poisoning to be the number one environmental health problem for children in the United States. Some 4 million children—many of whom are African American, Latino, Native American, or Asian, and most of whom live in inner-city areas—have dangerously high lead levels in their bodies. This lead is absorbed from old lead-based house paint, contaminated drinking water from lead pipes or lead solder, and soil polluted by industrial effluents and automobile exhaust. The evidence of racism is that at every income level, whether rich or poor, black children are two to three times more likely than whites to suffer from lead poisoning.

Native American communities also face high exposure to environmental contaminants. In addition to poverty, an important factor is that reservations are sovereign nations that are not covered by state environmental regulations. Court decisions holding that reservations are specifically exempt from hazardous waste storage and disposal regulations have resulted in a land rush of seductive offers from waste disposal companies to Native American reservations for onsite waste dumps, incinerators, and landfills. The short-term economic incentives can be overwhelming for communities with chronic poverty. Nearly every tribe in America

FIGURE 1.29 Environmental scientists increasingly try to address both public health and environmental quality. The poorest populations often suffer most from environmental degradation.
Kaetana/Shutterstock

has been approached with proposals for some dangerous industry or waste facility.

The practice of targeting poor communities of color in the developing nations for waste disposal or experimentation with risky technologies has been described as **toxic colonialism.** Internationally, the trade in toxic waste has mushroomed in recent years as wealthy countries have become aware of the risks of industrial refuse (fig. 1.29). Poor minority communities at home and abroad are being increasingly targeted as places to dump unwanted wastes. Although a treaty regulating international shipping of toxics was signed by 105 nations in 1989, millions of tons of toxic and hazardous materials continue to move—legally or illegally—from the richer countries to the poorer ones every year. This issue is discussed further in chapter 23.

Section Review

1. Explain the idea of moral extensionism, and give an example.
2. How does *inherent value* differ from *instrumental value*?
3. Why is stewardship important in many faiths?
4. What is environmental justice?

Connecting the Dots

One of our most important questions today is, How can we improve lives without damaging our shared environment? We face many serious environmental challenges, and many of these challenges are especially visible in the rapidly expanding urban areas of the developing world, where people struggle with air and water pollution, chronic hunger, water shortages, and other problems. These topics are among the major themes in environmental science, because they have to do with environmental systems, environmental quality, and our impacts on our environment.

Sustainable development—seeking to improve lives without further damaging environmental systems—and climate change are major topics that you will revisit in many chapters of this book. UN Sustainable Development Goals identify strategies and targets for improving human well-being. These goals build on the largely successful UN Millennium Development Goals, which produced substantial progress. Innovations in transportation, energy sources, food production, and international cooperation for environmental protection. Often these goals are complementary with environmental protections.

Environmental science is a discipline that uses an understanding of natural systems to seek solutions to environmental problems and to help find solutions—which can draw on knowledge from technological, biological, economic, political, social, and many other fields of study. Ideas in environmental science, and in development, also draw on a long history of approaches to conservation. Utilitarian conservation has been a common incentive; aesthetic preservation also motivates many people to work for conservation. Social progress and a concern for making sure that all people have access to a healthy environment have also been important motivating factors in environmental science and in environmental conservation.

Designation of ecosystem services also informs the ways we think about development and resources. Identifying and quantifying these services helps us prioritize resource protection. Narratives about the tragedy of the commons, or managing commons, also help explain how we treat resource. Faith-based perspectives are also influential, promoting ideas of stewardship and responsibility for others. All of these ideas can inform the goals and strategies of understanding and sustaining the natural environment that sustains us.

Reviewing Key Terms

Can you define the following terms in environmental science?

biocentric
  preservation 1.2
ecological footprint 1.1
ecosystem services 1.4

environment 1.1
environmental
  justice 1.5
environmental racism 1.5

environmental
  science 1.1
ethics 1.5
global environmentalism 1.2

inherent value 1.5
instrumental value 1.5
LULUs 1.5
managing the commons 1.4

Critical Thinking and Discussion Questions

1. Overall, do environmental and social conditions in Kibera give you hope or fear about the future?

2. What are the underlying assumptions and values of utilitarian conservation and altruistic preservation? Which do you favor?

3. Suppose a beautiful grove of trees near your house is scheduled to be cut down for a civic project such as a swimming pool. Would you support this? Why or why not?

4. What resource uses are most strongly represented in the ecological footprint? What are the advantages and disadvantages of using this assessment?

5. Are there enough resources in the world for 8 or 10 billion people to live decent, secure, happy lives? What do these terms mean to you? Try to imagine what they mean to residents of other countries.

6. Identify several types of ecosystem services. How might their value be accounted for?

Data Analysis

Working with Graphs

Graphs are one of the most common and important ways scientists communicate their results. Learning to understand the language of graphs will help you understand ideas in this book. Graphs help us identify trends and understand relationships. We could present a table of numbers, but most of us have difficulty seeing a pattern in a large field of numbers. In a graph, we can quickly and easily see trends and relationships.

Below is a graph showing ecological footprints. Often we pass quickly over graphs like these that appear in text, but it's rewarding to investigate them more closely, because their relationships can raise interesting questions. Go to Connect to answer questions about this graph and to demonstrate your understanding of graph reading.

For a full discussion of ecological footprints and the many components that contribute to its calculation, see the WWF Living Planet Report, http://wwf.panda.org/about_our_earth/all_publications/living_planet_report/. To calculate your own footprint, visit the Global Footprint Network http://footprintnetwork.org/.

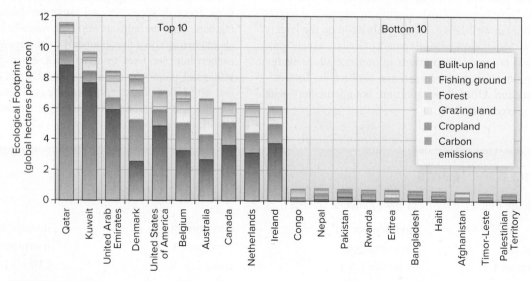

FIGURE 1 Ecological footprint for the 10 highest and 10 lowest countries, in terms of impact per person

2

Principles of Science and Systems

▲During the annual migration, 1.6 million wildebeests and zebras make a great circle across the Serengeti plains. New scientific techniques aid in monitoring this vast, complex ecosystem.
Source: Mary Ann Cunningham

Learning Outcomes

After studying this chapter, you should be able to:

2.1 Describe the scientific method and explain how it works.
2.2 Explain systems and how they're useful in science.
2.3 Evaluate the role of scientific consensus and conflict.

If only an elite minority of Americans understands science and technology, there is no hope for democracy.

– Ann Druyan, Cosmos Studios

Snapshot Serengeti

What would it take for you to become a researcher for one of Africa's most renowned wildlife refuges? Generally the answer would be a great deal of time, education, and funding. But an unusual project, Snapshot Serengeti, lets you participate in cutting-edge science—and you can do it from the comfort of your own couch.

The Serengeti is an amazing ecosystem. Every year more than 1.6 million wildebeest and zebra follow seasonal rainfall, and the lush grass it produces, on a circular route from Tanzania into Kenya and back. Lions, leopards, hyenas, and other carnivores shadow the vast herds of grazers. Despite 40 years of studies, though, much remains unknown about basic dynamics and interactions among species. The ecosystem is vast, and collecting data is difficult. How do predator and prey species shift over time? How are they affected by droughts, or by poachers? Where do animals go, and what do they do at night, when it's difficult and dangerous for biologists to be out on the plains?

Wildlife researchers often use motion sensor cameras to monitor wildlife and to answer questions like these. In the Snapshot Serengeti project, biologists have deployed 225 camera traps across 1,125 km^2 (about 500 mi^2) of open grassland and wooded savanna in Tanzania's Serengeti National Park (fig. 2.1). The cameras snap photos automatically, day or night whenever they detect motion or a change in heat. Operating continuously since 2010, they have collected millions of images (fig. 2.2).

These photos represent a terrific trove of data, but nobody has time to evaluate it. So project organizers turned to crowdsourcing. Every six months, the memory cards are collected, and images are uploaded to a server. Volunteers from around the world can view the images at www.snapshotserengeti.org, where they classify any species present, count individuals, record the presence of offspring, and describe basic behaviors in each photo. You could be one of those volunteers, even if you have only a few hours now and then to give. And you don't have to be an expert at wildlife identification. The web page has prompts and filters to help you recognize species and behaviors.

By the end of 2018, Snapshot Serengeti volunteers had identified 48 different mammal

FIGURE 2.1 The grid of camera traps in the Snapshot Serengeti project.

Source: Dr. Craig Packer, Department of Ecology, Evolution and Behavior, University of Minnesota, Saint Paul, MN 55108, USA.

species, from mongooses to elephants. More than 70,000 volunteers, citizen scientists from around the world, had participated, reviewing more than 3 million image sets to make more than 10 million classifications. Multiple volunteers view each image set to produce a consensus identification, and experts monitor accuracy by spot-checking random images. Fully 96.6 percent of the time, the experts have agreed with volunteer analysts. This process can make more progress in just a few days than the project team could make in years of work. Observations are then condensed into a summary report, providing data on each species. Images also are shared with younger students who can enjoy a virtual safari to Africa together with the thrill of discovering amazing wildlife living in the wild.

This rich data source also supports graduate student research, and other projects now mine the data set as they form new research questions on wildlife distributions, species interactions, and animal behavior. Citizen scientists have made this possible by effectively producing large quantities of data that would otherwise be unavailable. Their work makes it possible to understand more of the ecological structure of the system, test hypotheses of community ecology, and inform management decisions.

Increasingly, science intersects with many parts of our lives. A foundation of basic principles of science is essential to decisions about how we act, work, consume, and even vote. How do these principles shape the ways we ask questions, and the ways we seek answers? How do we gather and evaluate evidence (data)? In this chapter, we'll examine these principles, including how research questions are formulated and tested, how hypotheses are evaluated, and how data can be used to generate models and understand complex systems.

FIGURE 2.2 A curious lioness at one of the camera traps.
SNAPSHOTSAFARI/Sarah Huebner/Dr. Craig Packer

2.1 WHAT IS SCIENCE?

- *Science depends on careful, objective, logical analysis.*
- *Reproducible results are essential in science.*
- *Science depends on orderly testing of hypotheses.*

Science is a way of seeking answers to questions that are important to us. Are wildlife populations surviving in the Serengeti? How are they affected by human pressures? In general, science begins with curiosity about the world. More formally, **science** is a process for producing knowledge methodically and logically. Derived from *scire,* "to know" in Latin, science depends on making precise observations of natural phenomena. We develop or test theories (proposed explanations of how a process works) using these observations. "Science" also refers to the cumulative body of knowledge produced by many scientists. Science is valuable because it helps us understand the world and meet practical needs, such as new medicines, new energy sources, or new foods.

Science rests on the assumption that the world is knowable and that we can learn about the world by careful observation (table 2.1). For early philosophers of science, this assumption was a radical departure from religious and philosophical approaches. In the Middle Ages, the ultimate sources of knowledge about how crops grew, how diseases spread, or how the stars moved, were religious authorities or cultural traditions. These sources provided many useful insights,

but there was no way to test their explanations independently and objectively. The benefit of scientific thinking is that it searches for testable evidence. By testing our ideas with careful observations, we can evaluate whether our explanations are reasonable or not.

Science depends on skepticism and accuracy

Ideally, scientists are skeptical. They are cautious about accepting proposed explanations, including their own, until there is substantial evidence to support them. Even when observations support an idea, scientists try to remain cautious, and to treat explanations as provisionally true. There is always a possibility that some additional evidence may appear to disprove them. Scientists also aim to be methodical and unbiased. Because unintentional biases and errors are hard to avoid, scientific tests are subject to review by informed peers, who can evaluate results and conclusions (fig. 2.3). The peer review process is an essential part of ensuring that scientists maintain good standards in study design, data collection, and interpretation of results.

Scientists demand **reproducibility** because they are cautious about accepting conclusions. Making an observation or obtaining a result just once doesn't count for much. You have to produce the same result consistently to be sure that your first outcome wasn't a fluke. Even more important, you must be able to describe the conditions of your study so that someone else can reproduce your findings. Repeating studies or tests is known as **replication.**

Science also relies on accuracy and precision. Accuracy is correctness of measurements. Inaccurate data can produce sloppy and misleading conclusions (fig. 2.4). Precision means repeatability of

Table 2.1 Basic Principles of Science

1. *Empiricism:* We can learn about the world by careful observation of empirical (real, observable) phenomena; we can expect to understand fundamental processes and natural laws by observation.

2. *Uniformitarianism:* Basic patterns and processes are uniform across time and space; the forces at work today are the same as those that shaped the world in the past, and they will continue to do so in the future.

3. *Parsimony:* When two plausible explanations are equally reasonable, the simpler (more parsimonious) one is preferable. This rule is also known as Ockham's razor, after the English philosopher who proposed it.

4. *Uncertainty:* Knowledge changes as new evidence appears, and explanations (theories) change with new evidence. Theories based on current evidence should be tested on additional evidence, with the understanding that new data may disprove the best theories.

5. *Repeatability:* Tests and experiments should be repeatable; if the same results cannot be reproduced, then the conclusions are probably incorrect.

6. *Proof is elusive:* We rarely expect science to provide absolute proof that a theory is correct, because new evidence may always undermine our current understanding.

7. *Testable questions:* To find out whether a theory is correct, it must be tested; we formulate testable statements (hypotheses) to test theories.

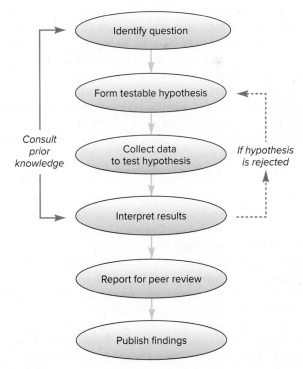

FIGURE 2.3 Ideally, scientific investigation follows a series of logical, orderly steps to formulate and test hypotheses.

FIGURE 2.4 Making careful, accurate measurements and keeping good records are essential in scientific research.
David L. Hansen, University of Minnesota Agricultural Experiment Station

results and level of detail. The classic analogy for repeatability is throwing darts at a dart board. You might throw ten darts and miss the center every time, but if all the darts hit nearly the same spot, they were very precise. Another way to think of precision is levels of detail. Suppose you want to measure the growth of a spruce tree (fig. 2.4). If you use a ruler marked in centimeters, you could report that the stem is about 2 cm in diameter. This measurement is accurate (true), but only to the nearest centimeter. You could get a more precise estimate with a ruler measured in millimeters. With this finer ruler, you could meaningfully report that the stem was 2.4 cm wide.

Suppose you took four measurements, just to be sure, and they averaged to 2.423 cm. This would increase your confidence that 2.4 cm was accurate. You would not, however, report all three decimal places. That would imply that you know more detail than you do. You don't know the stem width to 1/1000th of a centimeter. One decimal place (2.4 cm) would be a **significant figure,** or a level of detail you actually knew. If you reported 2.423 cm, you couldn't be certain you were telling the truth at that level of detail.

Deductive and inductive reasoning are both useful

Ideally, scientists deduce conclusions from general laws that they know to be true. For example, if we know that massive objects attract each other (because of gravity), then it follows that an apple will fall to the ground when it releases from the tree. This logical reasoning from general to specific is known as **deductive reasoning.** Often, however, we do not know general laws that guide natural systems. We observe, for example, that birds appear and disappear as a year goes by. Through many repeated observations in different places, we can infer that the birds move from place to place. We can develop a general rule that birds migrate seasonally. Reasoning from many observations to produce a general rule is **inductive reasoning.** Although deductive reasoning is more logically sound than inductive reasoning, it only works when

our general laws are correct. We often rely on inductive reasoning to understand the world, because we have few immutable laws.

Sometimes it is insight, as much as reasoning, that leads us to an answer. Many people fail to recognize the role that insight, creativity, aesthetics, and luck play in research. Some of our most important discoveries were made not because of superior scientific method and objective detachment, but because the investigators were passionately interested in their topics and pursued hunches that appeared unreasonable to fellow scientists. A good example is Barbara McClintock, the geneticist who discovered that genes in corn can move and recombine spontaneously. Where other corn geneticists saw random patterns of color and kernel size, McClintock's years of experience in corn breeding, and an uncanny ability to recognize patterns, led her to guess that genes could recombine in ways that no one had yet imagined. Her intuitive understanding led to a theory that took other investigators years to accept.

Testable hypotheses and theories are essential tools

Science also depends on orderly testing of hypotheses, a process known as the scientific method. You may already be using the scientific method without being aware of it. Suppose you have a flashlight that doesn't work. The flashlight has several components (switch, bulb, batteries) that could be faulty. If you change all the components at once, your flashlight might work, but a more methodical series of tests will tell you more about what was wrong with the system—knowledge that may be useful the next time you have a faulty flashlight. So you decide to follow the standard scientific steps:

1. *Observe* that your flashlight doesn't light; also, that there are three main components of the lighting system (batteries, bulb, and switch).

2. Propose a **hypothesis,** a testable explanation: "The flashlight doesn't work because the batteries are dead."

3. Develop a *test* of the hypothesis and *predict* the result that would indicate your hypothesis was correct: "I will replace the batteries; the light should then turn on."

4. Gather *data* from your test: After you replaced the batteries, did the light turn on?

5. *Interpret* your results: If the light works now, then your hypothesis was right; if not, then you should formulate a new hypothesis, perhaps that the bulb is faulty, and develop a new test for that hypothesis.

In systems more complex than a flashlight, it is almost always easier to prove a hypothesis wrong than to prove it unquestionably true. This is because we usually test our hypotheses with observations, but there is no way to make every possible observation. The philosopher Ludwig Wittgenstein illustrated this problem as follows: Suppose you saw hundreds of swans, and all were white. These observations might lead you to hypothesize that all swans were white. You could test your hypothesis by viewing thousands of swans, and each observation might support your hypothesis, but you could never be entirely sure that it was correct. On the other hand, if you saw just one black swan, you would know with certainty that your hypothesis was wrong.

As you'll read in later chapters, the elusiveness of absolute proof is a persistent problem in environmental policy and law. You can rarely prove absolutely that the toxic waste dump up the street is making you sick. The elusiveness of proof often decides environmental liability lawsuits.

When an explanation has been supported by a large number of tests, and when a majority of experts have reached a general consensus that it is a reliable description or explanation, we call it a **scientific theory.** Note that scientists' use of this term is very different from the way the public uses it. To many people, a theory is speculative and unsupported by facts. To a scientist, it means just the opposite: While all explanations are tentative and open to revision and correction, an explanation that counts as a scientific theory is supported by an overwhelming body of data and experience, and it is generally accepted by the scientific community, at least for the present (fig. 2.5).

Science communication is increasingly recognized as a fundamental aspect of scientific work. Research should inform public policy on issues from endangered species and land management to public health and air pollution. So it is essential that the public, as well as other researchers, have access to findings. Science communication is a specialized skill, just as research is. It requires an ability to translate the language and ideas to conventional language. It also requires an ability to give context, to explain why findings matter. Take some time to find a public media report on science or health topics, perhaps in a newspaper or online news service. Notice how the writer picks apart a study to explain its meaning to a general audience. This communication supports **science literacy,** an ability to understand basic scientific concepts, methods, implications. Science literacy in turn supports democracy, since it helps ensure that the public is well-informed about issues that affect their lives.

FIGURE 2.5 Data collection and repeatable tests support scientific theories. Here students use a radio antenna to locate radio-tagged fish.
David L. Hansen, University of Minnesota Agricultural Experiment Station

Understanding probability helps reduce uncertainty

One strategy to improve confidence in the face of uncertainty is to focus on probability. Probability is a measure of how likely something is to occur. Usually, probability estimates are based on a set of previous observations or on standard statistical measures. Probability does not tell you what *will* happen, but it tells you what *is likely* to happen. If you hear on the news that you have a 20 percent chance of catching a cold this winter, that means that 20 of every 100 people are likely to catch a cold. This doesn't mean that you will catch one. In fact, it's more likely that you won't catch a cold than that you will. If you hear that 80 out of every 100 people will catch a cold, you still don't know whether you'll get sick, but there's a much higher chance that you will.

Science often involves probability, so it is important to be familiar with the idea. Sometimes probability has to do with random chance: If you flip a coin, you have a random chance of getting heads or tails. Every time you flip, you have the same 50 percent probability of getting heads. The chance of getting ten heads in a row is small (in fact, the chance is 1 in 2^{10}, or 1 in 1,024), but on any individual flip you have exactly the same 50 percent chance, because this is a random test.

Sometimes probability is weighted by circumstances: Suppose that about 10 percent of the students in this class earn an A each semester. Your likelihood of being in that 10 percent depends a great deal on how much time you spend studying, how many questions you ask in class, and other factors. Sometimes there is a combination of chance and circumstances: The probability that you will catch a cold this winter depends partly on whether you encounter someone who is sick (largely random chance) and whether you take steps to stay healthy (get enough rest, wash your hands frequently, eat a healthy diet, and so on).

Scientists often increase their confidence in a study by comparing results to a random sample or a larger group. Suppose that 40 percent of the students in your class caught a cold last winter. This *seems* like a lot of colds, but is it? One way to decide is to compare to the cold rate in a larger group. You call your state epidemiologist, who took a random sample of the state population last year: She collected 200 names from the telephone book and called each to find out if each got a cold last year. A larger sample, say 2,000 people, would have been more likely to represent the actual statewide cold rate. But a sample of 200 is much better than a sample of 50 or 100.

Suppose the epidemiologist tells you that in your state as a whole, only 20 percent of people caught a cold. Now you know that the rate in your class (40 percent) was quite high, and you can investigate possible causes for the difference. Perhaps people in your class got sick because they were short on sleep, because they tended to stay up late studying. You could test whether studying late was a contributing factor by comparing the frequency of colds in two groups: those who study long and late, and those who don't. Suppose it turns out that among the 40 late-night studiers, 30 got colds (a rate of 75 percent). Among the 60 others, only 10 got colds (17 percent). This difference would

Why Do Scientists Answer Questions with a Number?

Statistics are numbers that let you evaluate and compare things. "Statistics" is also a field of study that has developed meaningful methods of comparing those numbers. By both definitions, statistics are widely used in environmental sciences, partly because they can give us a useful way to assess patterns in a large population, and partly because the numbers can give us a measure of confidence in our research or observations. Understanding the details of statistical tests can take years of study, but a few basic ideas will give you a good start toward interpreting statistics.

1. **Descriptive statistics help you assess the general state of a group.** In many towns and cities, the air contains dust, or particulate matter, as well as other pollutants. From personal experience you might know your air isn't as clean as you'd like, but you may not know how clean or dirty it is. You could start by collecting daily particulate measurements to find average levels. An averaged value is more useful than a single day's values, because daily values may vary a great deal, but general, long-term conditions affect your general health. Collect a sample every day for a year; then divide the sum by the number of days, to get a **mean** (average) dust level. Suppose you found a mean particulate level of 30 micrograms per cubic meter ($\mu g/m^3$) of air. Is this level high or low? In 1997, the EPA set a standard of 50 $\mu g/m^3$ as a limit for allowable levels of coarse particulates (2.5–10 micrometers in diameter). Higher levels tend to be associated with elevated rates of asthma and other respiratory diseases. Now you know that your town, with an annual average of 30 $\mu g/m^3$, has relatively safe air, after all.

2. **Statistical samples.** Although your town is clean by EPA standards, how does it compare with the rest of the cities in the country? Testing the air in *every* city is probably not possible. You could compare your town's air quality with a **sample,** or subset of cities, however. A large, random sample of cities should represent the general "population" of cities reasonably well. Taking a large sample reduces the effects of outliers (unusually high or low values) that might be included. A random sample minimizes the chance that you're getting only the worst sites, or only a

collection of sites that are close together, which might all have similar conditions. Suppose you get average annual particulate levels from a sample of 50 randomly selected cities. You can draw a frequency distribution, or histogram, to display your results (fig. 1). The mean value of this group is 36.8 $\mu g/m^3$, so by comparison your town (at 30 $\mu g/m^3$) is relatively clean.

Many statistical tests assume that the sample has a normal, or Gaussian, frequency distribution, often described as a bell-shaped curve (fig. 2). In this distribution, the mean is near the center of the range of values, and most values are fairly close to the mean. Large and random samples are more likely to fit this shape than are small and nonrandom samples.

3. **Confidence.** How do you know that the 50 cities you sampled really represent all the cities in the country? You can't ever be completely certain, but you can use estimates, such as confidence limits, to express the reliability of your mean statistic. Depending on the size of your sample (not 10, not 500, but 50) and the amount of variability in the sample data, you can calculate a confidence interval that the mean represents the whole population (all cities). Confidence limits, or confidence intervals, represent the likelihood that your statistics correctly represent the entire population. For the mean of your sample, a confidence interval tells you the probability that your sample is similar to other random samples of the population. A common convention is to compare values with

FIGURE 2 A normal distribution.

a 95 percent confidence level, or a probability of 5 percent or less that your conclusions are misleading. Using statistical software, we can calculate that, for our 50 cities, the mean is 36.8 $\mu g/m^3$, and the confidence interval is 35.0 to 38.6. This suggests that, if you take 1,000 samples from the entire population of cities, 95 percent of those samples ought to be within 2 $\mu g/m^3$ of your mean. This indicates that your mean is reliable and representative.

4. **Is your group unusual?** Once you have described your group of cities, you can compare it with other groups. For example, you might believe that Canadian cities have cleaner air than U.S. cities. You can compare mean air quality levels for the two groups. Then you can calculate confidence intervals for the difference between the means, to see if the difference is meaningful.

5. **Evaluating relationships between variables.** Are respiratory diseases correlated with air pollution? For each city in your sample, you could graph pollution and asthma rates (fig. 3). If the graph looks like a loose cloud of dots, there is no clear relationship. A tight, linear pattern of dots trending upward to the right indicates a strong and positive relationship. You can also use a statistical package to calculate an equation to describe the relationship and, again, confidence intervals for the equation. This is known as a regression equation.

6. **Lies, damned lies, and statistics.** Can you trust a number to represent a complex or large phenomenon? One of the devilish details of representing the world with numbers is that those numbers can be tabulated in many

FIGURE 1 Average annual airborne dust levels for 50 cities.

(header photo): David L. Hansen

Asthma Cases

FIGURE 3 A scatter plot shows relationships between variables.

ways. If we want to assess the greatest change in air quality statistics, do we report rates of change or the total amount of change? Do we look at change over five years? Twenty-five years? Do we accept numbers selected by the EPA, by the cities themselves, by industries, or by environmental groups? Do we trust that all the data were collected with a level of accuracy and precision that we would accept if we knew the hidden details in the data-gathering process? Like all information, statistics need to be interpreted in terms of who produced them, when, and why. Awareness of some of the standard assumptions behind statistics,

such as sampling, confidence, and probability, will help you interpret statistics that you see and hear.

Test your comprehension

1. What is a mean? How would you use one?
2. What is a Gaussian or normal distribution? What shape does it create in a graph?
3. What do statisticians mean by confidence limits?

give you a good deal of confidence that staying up late contributes to getting sick.

Statistics can indicate the probability that your results were random

Statistics can help in experimental design as well as in interpreting data (see Exploring Science, above). Many statistical tests focus on calculating the probability that observed results could have occurred by chance. Often, the degree of confidence we can assign to results depends on sample size as well as the amount of variability between groups.

Ecological tests are often considered significant if there is less than 5 percent probability that the results were achieved by random chance. A probability of less than 1 percent gives still greater confidence in the results.

As you read this book, you will encounter many statistics, including many measures of probability. When you see these numbers, stop and think: Is the probability high enough to worry about? How high is it compared to other risks or chances you've read about? What are the conditions that make probability higher or lower?

Experimental design can reduce bias

The study of large mammals in the Serengeti described in the case study for this chapter is an example of an observational experiment, one in which you observe natural systems and then test explanations for patterns and relationships you find. This kind of study can't prove cause and effect; it only provides correlations. But it may be the only practical way to get information about species in a natural setting (fig. 2.6). Despite uncertainty about the interpretation of some evidence, observational studies have provided much of the information we have about

FIGURE 2.6 Observational studies may be the only way to get large amounts of information about big, rare animals in a natural setting.
William P. Cunningham

large-scale natural systems. Having a very large data set (as in the case for the Snapshot Serengeti Project) makes inferences about questions such as species abundances and distributions, daily activity patterns, and seasonal movements more reliable. There are many other opportunities to participate in citizen science research projects. We'll discuss some of these projects in later chapters of this book.

This kind of study is also called a **natural experiment;** that is, a study of events that have already happened. Many scientists depend on natural experiments: A geologist, for instance, might want to study mountain building, or an ecologist might want to learn about how species coevolve, but neither scientist can spend millions of years watching the process happen. Similarly, a toxicologist can't give people a disease just to see how lethal it is.

Other scientists can use **manipulative experiments.** In one landmark manipulative study, ecologists Edward O. Wilson and Robert MacArthur were interested in how quickly species colonize small islands, depending on distance to the mainland. They fumigated several tiny islands in the Florida Keys, killing all resident insects, spiders, and other invertebrates. They then monitored the islands to test a theory that populations would reestablish faster on large islands and those near the mainland, compared to small or remote islands.

In manipulative experiments in real environments, like a set of Florida islands, controlling all variables is difficult and costly—and potentially ecologically damaging. So most manipulative experiments are done in the laboratory, where conditions can be carefully controlled. Suppose you were interested in studying whether lawn chemicals contribute to deformities in tadpoles. You might keep two groups of tadpoles in fish tanks and expose one to chemicals. In the lab, you could ensure that both tanks had identical temperatures, light, food, and oxygen. By comparing a treatment (exposed) group and a control (unexposed) group, you have also made this a **controlled study.**

Often there is a risk of experimenter bias. Suppose the researcher sees a tadpole with a small nub that looks like it might become an extra leg. Whether she calls this nub a deformity might depend on whether she knows that the tadpole is in the treatment group or the control group. To avoid this bias, **blind experiments** are often used, in which the researcher doesn't know which group is treated until after the data have been analyzed. In health studies, such as tests of new drugs, **double-blind experiments** are used, in which neither the subject (who receives a drug or a placebo) nor the researcher knows who is in the treatment group and who is in the control group.

In each of these studies there is one **dependent variable** and one, or perhaps more, **independent variables.** The dependent variable, also known as a response variable, is affected by the independent variables. Independent variables are rarely really independent (they are affected by the same environmental conditions as the dependent variable, for example). Many people prefer to call them explanatory variables, because we hope they will explain differences in the dependent variable.

Models are an important experimental strategy

Another way to gather information about environmental systems is to use **models.** A model is a simple representation of something. Perhaps you have built a model airplane. The model doesn't have

FIGURE 2.7 A model uses just the essential elements to represent a complex system. DNY59/Getty Images

Geographical pattern of surface warming

0 0.5 1 1.5 2 2.5 3 3.5 4 4.5 5 5.5 6 6.5 7 7.5
Temperature change (°C)

FIGURE 2.8 Numerical models, calculated from observed data, can project future scenarios. Here, temperature changes in 2090–2099 are modeled, relative to 1980–1999 temperatures.
Figure SPM.6 from Climate Change 2007: Synthesis Report. Contribution of Working Groups I, II and III to the Fourth Assessment Report of the Intergovernmental Panel on Climate Change [Core Writing Team, Pachauri, R.K. and Reisinger, A. (eds.)]. IPCC, Geneva, Switzerland. Used with permission.

all the elements of a real airplane, but it has the most important ones for your needs. A simple wood or plastic airplane has the proper shape, enough to allow a child to imagine it is flying (fig. 2.7). A more complicated model airplane might have a small gas engine, just enough to fly it around for short distances.

Similarly, scientific models vary greatly in complexity, depending on their purposes. Some models are physical models: Engineers test new cars and airplanes in wind tunnels to see how they perform, and biologists often test theories about evolution and genetics using "model organisms" such as fruit flies or rats as a surrogate for humans.

Most models are numeric, though. A model could be a mathematical equation, such as a simple population growth model ($N_t = rN_{(t-1)}$). Here the essential components are number (N) of individuals at time t (N_t), and the model proposes that N_t is equal to the growth rate (r) times the number in the previous time period ($N_{(t-1)}$). This model is a simplistic representation of population change, but it is useful because it precisely describes a relationship between population size and growth rate. Also, by converting the symbols to numbers, we can predict populations over time. For example, if last year's rabbit population was 100, and the growth rate is 1.6 per year, then this year's population should be 160. Next year's population should be 160×1.6, or 256. This is a simple model, then, but it can be useful. A more complicated model might account for deaths, immigration, emigration, and other factors (see chapter 6).

More complicated mathematical models can be used to describe and calculate more complex processes, such as climate change (fig. 2.8) or economic growth. These models are also useful because they allow the researcher to manipulate variables quickly and easily, without actually destroying anything. An economist can experiment with different interest rates to see how they affect economic growth. A climatologist can raise CO_2 levels and see how quickly temperatures respond. These models are often called simulation models, because they simulate a complex system. Of course, the results depend on the assumptions built into the models. One model might show temperature rising quickly in response to CO_2; another might show temperature rising more slowly, depending on how evaporation, cloud cover, and other variables are taken into account. Consequently, simulations can produce powerful but controversial results. If multiple models generally agree, though, as in the cases of climate models that agree on upward temperature trends, we can have confidence that the overall predictions are reliable. These models are also very useful in laying out and testing our ideas about how a system works.

Section Review

1. What is science? What are some of its basic principles?
2. Why are widely accepted, well-defended scientific explanations called "theories"?
3. Draw a diagram showing the steps of the scientific method, and explain why each is important.

2.2 SYSTEMS INVOLVE INTERACTIONS

- *A system is a network of interdependent components and processes that together have properties beyond those of individual parts.*
- *Feedbacks are self-regulating mechanisms in which the results of a process affect the process itself.*
- *Homeostasis (the ability to maintain stability) and resilience (ability to recover from disturbance) are system traits.*

The Serengeti ecosystem discussed in the opening case study is interesting because it is composed of many interdependent parts. By studying those parts, we can understand how similar ecosystems might function, and why. The idea of **systems,** including ecosystems, is central in environmental science. A system is a network of interdependent components and processes, with materials and energy flowing from one component of the system to another. For example, "ecosystem" is probably a familiar term for you. This simple word stands for a complex assemblage of animals, plants, and their environment, through which materials and energy move. A river is also a system, with its network of tributary streams, its forested or farmed watershed, fish, aquatic plants, and other interacting components.

The idea of systems is useful because it helps us organize our thoughts about the inconceivably complex phenomena around us. For example, an ecosystem might consist of countless animals and plants and their physical surroundings. You yourself are a system consisting of millions of cells, complex organs, and innumerable bits of energy and matter that move through your body. Keeping track of all the elements and their relationships in an ecosystem would probably be an impossible task. But if we step back and think about them in terms of broad functional categories, then we can start to comprehend how it works (fig. 2.9).

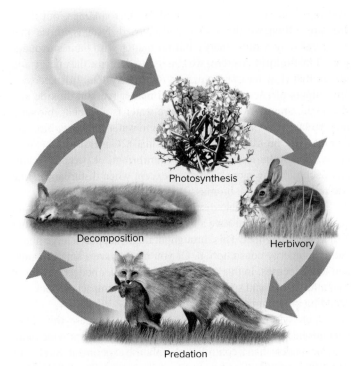

FIGURE 2.9 A system can be described in very simple terms.

We can use some general terms to describe the components of a system. A simple system consists of state variables (also called compartments), which store resources such as energy, matter, or water; and flows, or the pathways by which those resources move from one state variable to another. In figure 2.9, the plant and animals represent state variables. The plant represents many different plant types, all of which are things that store solar energy and create carbohydrates from carbon, water, and sunlight. The rabbit represents many kinds of herbivores, all of which consume plants, then store energy, water, and carbohydrates until they are used, transformed, or consumed by a carnivore. We can describe the flows in terms of herbivory, predation, or photosynthesis, all processes that transfer energy and matter from one state variable to another.

It may seem cold and analytical to describe a rabbit or a flower as a state variable, but it is also helpful to do so. When we start discussing natural complexity in the simple terms of systems, we can identify common characteristics. Understanding these characteristics can help us diagnose disturbances or changes in the system: For example, if rabbits become too numerous, herbivory can become too rapid for plants to sustain. Overgrazing can lead to widespread collapse of this system. Let's examine some of the common characteristics we can find in systems.

Systems can be described in terms of their characteristics

Open systems are those that receive inputs from their surroundings and produce outputs that leave the system. Almost all natural systems are open systems. In principle, a **closed system**

exchanges no energy or matter with its surroundings, but these are rare. Often we think of pseudo-closed systems, those that exchange only a little energy but no matter with their surroundings. **Throughput** is a term we can use to describe the energy and matter that flow into, through, and out of a system. In an ecosystem, inputs of energy (sunlight) and matter (carbon dioxide and water) are stored in biomass. Where water is abundant, biomass may be large (as in a rain forest). Where water is sparse, biomass storage may be slight (as in a grassland). Generally we expect ecosystems to be more or less in equilibrium, or perhaps gradually accumulating more biomass—except when disturbances like fire release or remove biomass.

In chapter 1, we also discussed throughput in households and economics. You might have a large income and high expenses, that is, a large throughput. This throughput might support a large house and high energy consumption. Or you might have a small income and low expenses. In this case you have a smaller budget, probably a smaller house, and smaller throughput. Your impact on your surrounding environment is also smaller.

Open systems can be difficult to study, because they have uncontrolled variables (fig. 2.10). However, they may provide more realistic models than a controlled greenhouse experiment. Studies of simplified, pseudo-closed systems are also useful: They allow more control of inputs, and thus relatively exact measurement of state variables and flows in the system (fig. 2.11).

Systems also experience positive and negative feedback mechanisms. A **positive feedback** is a self-perpetuating process. When a plant grows new leaves, and the more leaves it has, the more energy it can capture for producing more leaves. In other words, positive feedback amplifies a process (such as plant growth). If they proceed unchecked, positive feedbacks

(a) A simple system

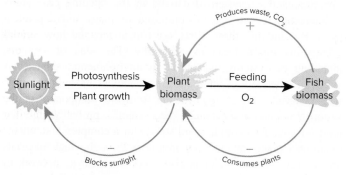

(b) A model of a system

FIGURE 2.11 Systems consist of compartments (also known as state variables) such as fish and plants, and flows of resources, such as nutrients or O_2 (a). Feedback loops (b) enhance or inhibit flows and the growth of compartments.

FIGURE 2.10 Environmental scientists often study open systems. Here students collect plant samples grown in open, uncontrolled moisture and temperature conditions, but with high CO_2 (pumped from white pipes) at the the Cedar Creek Long-Term Ecological Research Station.
William P. Cunningham

tend to destabilize a system (plants become overcrowded and then die back, for example). In contrast, a **negative feedback** is a process that suppresses change. If a plant grows very rapidly, it may produce more leaves than can be supported by available soil moisture. With insufficient moisture, the plant begins to die back. By dampening positive feedbacks, negative feedbacks often stabilize a system. Can you apply these ideas to a system you are familiar with?

In climate systems (chapter 15), positive and negative feedbacks are important ideas. For example, as warm summers melt ice in the Arctic, newly exposed water surfaces absorb heat, which leads to further melting, which leads to further heat absorption . . . This is positive feedback. In contrast, clouds can have a negative feedback effect (although there are debates on the net effect of clouds). A warming atmosphere can evaporate more water, producing clouds. Clouds block some solar heat, which reduces the evaporation. Thus, clouds can slow the warming process.

Positive and negative feedback mechanisms are also important in understanding population dynamics (chapter 6). For example, more individuals produce more young, which produces more individuals . . . (a positive feedback). But sometimes environmental limits reduce the number of young that survive to reproduce (a negative feedback). Your body is a system with active negative feedback mechanisms: For example, if you exercise, you become hot, and your skin sweats, which cools your body.

Systems may exhibit stability

Negative feedbacks tend to maintain stability in a system. We often think of systems exhibiting **homeostasis,** or a tendency to remain more or less stable and unchanging. Equilibrium is another term for stability in a system. Your body temperature stays remarkably constant despite dramatic changes in your environment and your activity levels. Changing by just a few degrees is extremely unusual—a fever just 4–6°F above normal is unusual and serious. Natural systems such as grasslands can be fairly stable, too. If the climate isn't too dry or too wet, a grassland tends to remain grassland, although the grass may be dense in some years and sparse in others. Cycles of wet and dry years may be part of the system's normal condition.

Disturbances, events that can destabilize or change the system, might also be normal for the system. There can be many kinds of disturbance in a grassland. Severe drought can set back the community, so that it takes some time to recover. Many grasslands also experience occasional fires, a disturbance that stimulates grass growth (by clearing accumulated litter and recycling nutrients) but destroys trees that might be encroaching on the grassland. Thus, disturbances are often a normal part of natural systems. Sometimes we consider this "dynamic equilibrium," or a tendency for a system to change and then return to normal.

Ecosystems can also show **resilience,** an ability to recover from disturbance. Grasslands regrow after a fire or drought, and wildlife populations can recover after hunting or disease reduces their numbers. Sometimes severe disturbance can lead to a **state shift,** in which conditions do not return to "normal." For example, a climate shift that drastically reduced rainfall could lead to a transition from grassland to desert. Plowing up grassland to plant crops is basically a state shift from a complex system to a single-species system.

FIGURE 2.12 Emergent properties of systems, including beautiful sights and sounds, make them exciting to study.
William P. Cunningham

Emergent properties are another interesting aspect of systems. Sometimes a system is more than the sum of its parts. For example, a tree is more than just a mass of stored carbon. It provides structure to a forest, habitat for other organisms, it shades and cools the ground, and it holds soil in place with its roots. All these functions are emergent properties, or characteristics of the system that are greater than the sum of its parts. An ecosystem can have emergent properties such as spatial structure, complexity, and diversity that individual components of the system could not have. It can also have emergent properties that we appreciate, such as beautiful sights and sounds, beyond its functioning as a system (fig. 2.12). In a similar way, you are a system made up of component parts, but you have many emergent properties, including your ability to think, share ideas with people around you, sing, and dance. These are properties that emerge because you function well as a system.

Section Review

1. Why are systems important in our environment?
2. What are feedback mechanisms?
3. Describe some emergent properties of ecosystems.

2.3 SCIENTIFIC CONSENSUS AND CONFLICT

- *Science is an incremental process in which many people gradually reach a consensus.*
- *Critical thinking helps us evaluate scientific evidence.*
- *Many people misunderstand the role of uncertainty in science.*

The scientific method outlined in figure 2.3 is the process used to carry out individual studies. Larger-scale accumulation of scientific knowledge involves cooperation and contributions from countless people. Good science is rarely carried out by a single individual working in isolation. Instead, a community of scientists collaborates in a cumulative, self-correcting process. You often hear about big breakthroughs and dramatic discoveries that change our understanding overnight, but in reality these changes are usually the culmination of the labor of many people, each working on different aspects of a common problem, each adding small insights to solve it. Ideas and information are exchanged, debated, tested, and retested to arrive at **scientific consensus,** or general agreement among informed scholars.

The idea of consensus is important. For those not deeply involved in a subject, the multitude of contradictory results can be bewildering: Are wildlife populations in the Serengeti under threat from human encroachment, or not? Is climate changing, and how much? Among those who have performed and read many studies, there tends to emerge a general agreement about the state of a problem. Scientific consensus now holds that many Serengeti species are declining, though opinions vary on how severe the problem is. Consensus is that global climates are changing, though models differ somewhat on how rapidly they will change under different policy scenarios.

Sometimes new ideas emerge that cause major shifts in scientific consensus. Two centuries ago, geologists explained many earth features in terms of Noah's flood. The best scientists held that the flood created beaches well above modern sea level, scattered boulders erratically across the landscape, and gouged enormous valleys where there is little water now (fig. 2.13). Then the Swiss glaciologist Louis Agassiz and others suggested that the earth had once been much colder and that glaciers had covered large areas. Periodic ice ages better explained changing sea levels, boulders transported far from their source rock, and the great, gouged valleys. This new idea completely altered the way geologists explained their subject. Similarly, the idea of tectonic plate movement, in which continents shift slowly around the earth's surface, revolutionized the ways geologists, biogeographers, ecologists, and others explained the development of the earth and its life-forms.

These great changes in explanatory frameworks were termed **paradigm shifts** by Thomas Kuhn, who studied revolutions in scientific thought. According to Kuhn, paradigm shifts occur when a majority of scientists accept that the old explanation no longer explains new observations very well. The shift is often contentious and political, because whole careers and worldviews, based on one sort of research and explanation, can be undermined by a new model. Sometimes a revolution happens rather quickly. Quantum mechanics and Einstein's theory of relativity, for example, overturned classical physics in only about 30 years. Sometimes a whole generation of scholars has to retire before new paradigms can be accepted.

As you study this book, try to identify some of the paradigms that guide our investigations, explanations, and actions today. This is one of the skills involved in critical thinking, discussed in the introductory chapter of this book.

Detecting pseudoscience relies on independent, critical thinking

Ideally, science should serve the needs of society. Deciding what those needs are, however, is often a matter of politics and economics. Should water be taken from a river for irrigation or left in the river for wildlife habitat? Should we force coal-burning power plants to reduce air pollution in order to lower health costs and respiratory illnesses, or are society and our economy better served by having cheap but dirty energy? These thorny questions are decided by a combination of scientific evidence, economic priorities, political positions, and ethical viewpoints.

On the other hand, in every political debate, lawyers and lobbyists can find scientists who will back either side. Politicians hold up favorable studies, proclaiming them "sound science," while they dismiss others as "junk science." Opposing sides dispute the scientific authority of the study they dislike. What is "sound" science, anyway? If science is often embroiled in politics, does this mean that science is always a political process?

Consider the case of climate change. If you judge only from reports in newspapers or on television, you'd probably conclude that scientific opinion is about equally divided on whether climate change is a threat or not. In fact, the vast majority of scientists

FIGURE 2.13 Paradigm shifts change the ways we explain our world. Geologists now attribute Yosemite's valleys to glaciers, where once they believed events like Noah's flood carved its walls.
John A. Karachewski/McGraw-Hill Education

working on this issue agree that the earth's climate is being affected by human activities, and that threats to the systems we depend on are serious. This agreement has long been established. In a 2004 study of 928 papers published in peer-reviewed scientific journals between 1993 and 2003, not one disagreed with the broad scientific consensus on global warming.

Why, then, is there so much confusion among the public about this issue? Why do politicians continue to assert that the dangers of climate change are uncertain at best, or "the greatest hoax ever perpetrated on the American people," as Senator James Inhofe, chair of the Senate Committee on Environment and Public Works, claimed, and as President Trump has repeated more recently? A part of the confusion lies in the fact that media often present the debate as if it's evenly balanced. And controversy makes headlines. The fact that an overwhelming majority of working scientists agree on the issue doesn't make good drama, so some media give equal time to minority viewpoints just to make an interesting fight.

Perhaps a more important source of misinformation comes from conservative foundations and political action funds that finance climate deniers. Between 2002 and 2010, a small group of billionaires donated at least $120 million through secret channels to fund more than 100 think tanks, media outlets, and other groups that promote skepticism about global warming. Some of these organizations sound like legitimate science or grassroots groups but are really only public relations operations. Others are run by individuals who find it rewarding to offer contrarian views. This tactic of spreading doubt and disbelief through innocuous-sounding organizations or seemingly authentic experts isn't limited to the climate change debate. This strategy was pioneered by the tobacco industry to mislead the public about the dangers of smoking. Notably, some of the same individuals, groups, and lobbying firms employed by tobacco companies have been hired to spread confusion about climate change.

Given this highly sophisticated battle of "experts," how do you interpret these disputes, and how do you decide whom to trust? The most important strategy is to apply critical thinking as you watch or read the news. What is the agenda of the person making the report? What is the source of their expertise? What economic or political interests do they serve? Do they appeal to your reason or to your emotions? Do they use inflammatory words (such as "junk"), or do they claim that scientific uncertainty makes their opponents' study meaningless? If they use statistics, what is the context for their numbers?

You can also find reputable organizations, if you want reliable answers. Science news outlets and international organizations such as the Intergovernmental Panel on Climate Change (IPCC) provide thorough documentation of evidence, which helps ensure their honesty and thoroughness. Major news organizations are also usually reliable, because their methods and findings are publicly visible and closely scrutinized. Watching for likely bias also helps you decide what sources are trustworthy. When you watch or read the news, you can look for places where reporting

Table 2.2 Questions for Baloney Detection	
1.	How reliable are the sources of this claim? Is there reason to believe that they might have an agenda to pursue in this case?
2.	Have the claims been verified by other sources? What data are presented in support of this opinion?
3.	What position does the majority of the scientific community hold in this issue?
4.	How does this claim fit with what we know about how the world works? Is this a reasonable assertion or does it contradict established theories?
5.	Are the arguments balanced and logical? Have proponents of a particular position considered alternate points of view or only selected supportive evidence for their particular beliefs?
6.	What do you know about the sources of funding for a particular position? Are they financed by groups with partisan goals?
7.	Where was evidence for competing theories published? Has it undergone impartial peer review or is it only in proprietary publication?

looks incomplete, you can consider sources and ask yourself what unspoken interests might lie behind the story.

Another strategy for deciphering the rhetoric is to remember that there are established standards of scientific work, and to investigate whether an "expert" follows these standards: Is the report peer-reviewed? Do a majority of scholars agree? Are the methods used to produce results well documented?

Harvard's Edward O. Wilson writes, "We will always have contrarians whose sallies are characterized by willful ignorance, selective quotations, disregard for communications with genuine experts, and destructive campaigns to attract the attention of the media rather than scientists. They are the parasite load on scholars who earn success through the slow process of peer review and approval." How can we identify misinformation and questionable claims? The astronomer Carl Sagan proposed a "Baloney Detection Kit" to help identify questionable sources and arguments (table 2.2).

Most scientists have an interest in providing knowledge that is useful, and our ideas of what is useful and important depend partly on our worldviews and priorities. Science is not necessarily political, but it is often used for political aims. The main task of educated citizens is to discern where it is being misused or disregarded for purposes that undermine public interests.

Section Review

1. Why do we say that proof is elusive in science?
2. How can we evaluate the validity of claims about science?
3. What is the role of consensus in science?

Connecting the Dots

Science is a process for producing knowledge methodically and logically, using observations to discern patterns. Many aspects of scientific approaches and principles can be seen in the Snapshot Serengeti project, as it seeks to understand a vast, complex, and open system. Public visibility and careful documentation in the study's publications allow a community of peer scientists to closely observe methods and analysis. This is important because they are cautious about conclusions—because they care strongly about understanding the viability of the ecosystem, and they don't want to reach hasty conclusions.

The very large data set in the Serengeti study allows researchers to test hypotheses about predator and prey movements, frequency of young, hours of activity, or other factors. It is not possible to watch all antelopes at all times, but a large number of observations supports statistical assessment of distributions and relationships. Quantitative measures also provide insights about system stability and resilience. Doing a manipulative experiment is difficult (and probably damaging) in this unique ecosystem. Instead, environmental studies often involve observational data and natural experiments as they assess system function and dynamics.

Debates about scientific consensus and policy are not always prominent in a basic study of wildlife in their native habitat. But when wildlife studies begin to inform questions of land use, development, or human population pressures, these questions can arise quickly. This is one of the reasons scientists try to document and share all these aspects of their work. Visibility helps the rigor of the work and confidence in answers.

Ultimately all these aspects of how science is conducted (ideally) are about ensuring reliable answers to questions we care about. In environmental science, many questions directly affect our lives. This is one of the reasons environmental science is both useful and enjoyable. And sometimes it leads you to learning more about amazing creatures like those in the Serengeti.

Reviewing Key Terms

Can you define the following terms in environmental science?

blind experiment 2.1	homeostasis 2.2	open system 2.2	science literacy 2.1
closed system 2.2	hypothesis 2.1	paradigm shift 2.3	scientific theory 2.1
confidence 2.1	independent variable 2.1	positive feedback 2.2	significant figure 2.1
controlled study 2.1	inductive reasoning 2.1	replication 2.1	state shift 2.2
deductive reasoning 2.1	manipulative experiment 2.1	reproducibility 2.1	statistics 2.1
dependent variable 2.1	mean 2.1	resilience 2.2	systems 2.2
disturbances 2.2	models 2.1	sample 2.1	throughput 2.2
double-blind experiment 2.1	natural experiment 2.1	science 2.1	
emergent properties 2.2	negative feedback 2.2	scientific consensus 2.3	

Critical Thinking and Discussion Questions

1. Explain why scientific issues are or are not influenced by politics. Can scientific questions ever be entirely free of political interest? If you say no, does that mean that all science is merely politics? Why or why not?

2. Review the questions for "baloney detection" in table 2.2, and apply them to an ad on TV. How many of the critiques in this list are easily detected in the commercial?

3. How important is scientific thinking for you, personally? How important do you think it should be? How important is it for society to have thoughtful scientists? How would your life be different without the scientific method?

4. Many people consider science too remote from everyday life and from nonscientists. Do you feel this way? Are there aspects of scientific methods (such as reasoning from observations) that you use?

5. Many scientific studies rely on models for experiments that cannot be done on real systems, such as climate, human health, or economic systems. If assumptions are built into models, then are model-based studies inherently weak? What would increase your confidence in a model-based study?

Data Analysis

Go to Snapshot Serengeti (https://www.zooniverse.org/projects/zooniverse/snapshot-serengeti/classify) and review the tutorial information. This site provides motion-sensor activated snapshots, including the one shown here.

How many species are in the collection?

Go through the tutorial. How many species did you see?

After the tutorial, try continuing and classifying more species. As you do it, remember that you are helping to collect primary data, creating a data set that many researchers can use to answer ecological questions. As you work, identify two or three potential questions of your own. How might you form them into testable hypotheses?

FIGURE 1 Snapshot Serengeti lets you contribute to current research and also to learn about wildlife species.
SNAPSHOTSAFARI/Sarah Huebner/Dr. Craig Packer

▲Fishing crews have too often come home with empty nets along much of the Louisiana coast in recent years, as low oxygen levels are killing or driving away marine life.
Earl_of_omaha/iStock/Getty Images

Learning Outcomes

After studying this chapter, you should be able to:

3.1 Describe matter, elements, and molecules and give simple examples of the role of four major kinds of organic compounds in living cells.

3.2 Define energy; explain how organisms and ecosystems capture and transform energy.

3.3 Define species, populations, communities, and ecosystems, and summarize the ecological significance of trophic levels.

3.4 Understand pathways in the water, carbon, nitrogen, sulfur, and phosphorus cycles.

"When one tugs at a single thing in nature, he finds it attached to the rest of the world."

– John Muir

CASE STUDY

Death by Fertilizer: Hypoxia in the Gulf of Mexico

In the 1980s, fishing crews began observing large areas in the Gulf of Mexico, near the Mississippi River mouth, that were nearly devoid of aquatic life in early summer. This region supports shrimp, fish, and oyster fisheries worth $250 to $450 million per year, so this "dead zone" was an economic disaster, as well as an ecological one. Marine biologists suspected that the Gulf ecosystem was collapsing because of oxygen deprivation.

To evaluate the problem, marine scientist Nancy Rabelais began mapping areas of low oxygen concentrations along the Louisiana coast in 1985. Every summer since then, she has found vast areas with oxygen concentrations of less than 2 parts per million (ppm). At 2 ppm, nearly all aquatic life, other than microorganisms and primitive worms, is eliminated. In 2017 the Gulf's hypoxic (oxygen-starved) area was a record 22,730 km^2 (8,776 mi^2), an area the size of New Jersey. The 2019 dead zone was nearly as large.

What causes this huge dead zone? The familiar process of eutrophication, visible in golf course ponds and city parks, is responsible. Eutrophication is the explosive growth of algae and phytoplankton (tiny, floating plants) that occurs when normally scarce nutrients become available. When nutrients become available, especially the elements nitrogen (N) and phosphorus (P), they allow explosive growth of aquatic plants and algae. Overabundant plants then die off, and decomposers (mainly bacteria) use up nearly all available oxygen. Oxygen depletion is especially severe near the sea bed where dead matter falls and collects.

Rabelais and her team observed that each year, following spring rains in the farmlands of the upper Mississippi watershed, oxygen concentrations in the Gulf would drop from 5 ppm to less than 2 ppm. Spring rains wash nutrient-rich soil, organic debris, and fertilizers from farm fields. Pulses of agricultural runoff from as far away as Montana and Pennsylvania contribute to eutrophic conditions in the Gulf (fig. 3.1). In addition to oxygen depletion, putrefying sediments also produce hydrogen sulfide, which further poisons the water near the seafloor.

In well-mixed water bodies, such as the open ocean, oxygen from the surface mixes down into lower layers. But warm, protected water bodies like the Gulf are often stratified: Abundant sunlight keeps the upper layers warmer and less dense than lower layers; cold, dense layers lie stable at depth, and fresh oxygen from the surface can't mix downward. Fish may be able to swim away from the hypoxic zone, but bottom dwellers often simply die. Widespread fish kills are also associated with hypoxia in enclosed waters. The dead zone in the Gulf is especially severe because the area receives runoff from the enormous Mississippi-Missouri watershed, where abundant agriculture makes nutrient runoff unusually high.

First observed in the 1970s, dead zones now occur along the coast of nearly every major populated region. The number increases almost every year, and more than 600 are now known (see chapter 18). They occur mainly in enclosed coastal waters, which tend to be stratified and are vulnerable to nutrient influxes, such as Chesapeake Bay, Long Island Sound, the Mediterranean Sea, the Black Sea, and China's Bohai Bay. But they have also been observed on open coastlines.

Can dead zones recover? Yes. If the influx of nitrogen stops, the system can return to normal. In 1996 in the Black Sea region, farmers in collapsing communist economies were forced to cut nitrogen fertilizer use in half, as fertilizer subsidies collapsed. The Black Sea dead zone disappeared, while farmers saw little decline in their crop yields. But in the Mississippi River watershed, farmers upstream are far from the Gulf and its fisheries. Midwestern policymakers have shown little interest in what happens to fisheries in Louisiana.

The flow of nitrogen reaching U.S. coastal waters has grown by eightfold since the 1950s. Phosphorus, another key nutrient, has tripled. Despite decades of efforts to control nutrients upstream, the dead zone has continued to grow, as intensification of agriculture upstream continues.

The movement of nutrients and energy determines how ecosystems function and how organisms and biological communities flourish or collapse. These topics set the stage for much of the rest of our study of environmental science. In this chapter we examine terms of matter and energy, key elements in living systems, and how they shape ecosystems and communities.

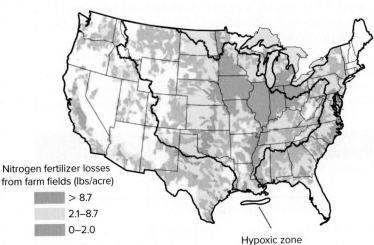

Nitrogen fertilizer losses
from farm fields (lbs/acre)

◼ > 8.7
◻ 2.1–8.7
◼ 0–2.0

Hypoxic zone

FIGURE 3.1 A hypoxic "dead zone" about the size of New Jersey forms in the Gulf of Mexico each summer, the result of nutrients from the Mississippi River.
Source: U.S EPA Water Atlas.

3.1 ELEMENTS OF LIFE

- *From living organisms to ecosystems, life can be understood in terms of the movement of matter and energy.*

- *To understand how matter and energy cycle through living systems, we must understand how atoms bond to form compounds.*

- *Carbon-based ("organic") compounds are the foundation of organisms.*

The accumulation and transfer of energy and nutrients allows living systems, including yourself, to exist. These processes tie together the parts of an ecosystem—or an organism. In this chapter, we'll introduce a number of concepts that are essential to understanding how living things function in their environment. These concepts include fundamental ideas of matter and energy, the ways organisms acquire and use energy, and the nature of chemical elements. We then apply these ideas to feeding relationships among organisms—the ways that energy and nutrients are passed from one living thing to another. In other words, we'll trace components from atoms to elements to compounds to cells to organisms to ecosystems.

Atoms, elements, and compounds

Everything that takes up space and has mass is **matter.** Matter exists in four distinct states, or phases—solid, liquid, gas, and plasma—which vary in energy intensity and the arrangement of particles that make up the substance. Water, for example, can exist as ice (solid), as liquid water, or as water vapor (gas). The fourth phase, plasma, occurs when matter is heated so intensely that electrons are released and particles become ionized (electrically charged). We can observe plasma in the sun, lightning, and very hot flames.

Under ordinary circumstances, matter is neither created nor destroyed; rather, it is recycled over and over again. This idea is known as the principle of **conservation of matter.** The molecules that make up your body may contain atoms that once were part of the body of a dinosaur. Most certainly you contain atoms that were part of many smaller prehistoric organisms. This is because chemical elements are used and reused by living organisms. Matter is transformed and combined in different ways, but it doesn't disappear; everything goes somewhere.

How does this principle apply to environmental science? It explains how components of environmental systems are intricately connected. In an agricultural watershed (opening case study) or in your own household, all matter comes from somewhere, and all

Table 3.1	Functions of Some Common Elements	
Function	**Elements**	**Comments**
Fertilizers	N nitrogen P phosphorus K potassium	Essential components of proteins, cells, other biological compounds; crucial fertilizers for plants
Organic compounds	C carbon O oxygen H hydrogen	Form the basic structure of cells and other components of living things, in combination with many other elements
Metals	Fe iron Al aluminum Au gold	Generally malleable; most (not all) react readily with other elements
Toxic elements	Pb lead Hg mercury As arsenic	Many are metals that can interfere with processes in nervous systems

waste goes somewhere. Pause to consider the source of nutrients causing hypoxia in the Gulf: How far back can you trace their origins? Also think about what you have consumed or acquired today. Where did it come from, and where will it go when you are done with it? You are intricately tied to both the sources and the destinations of everything you use. This is a useful idea for us as residents of a finite world. Ultimately, when we throw away our disposable goods, they don't really go "away," they just go somewhere else, to stay there for a while and then move on.

Matter consists of **elements** (basic substances that cannot be broken down into simpler forms by ordinary chemical reactions), such as carbon or oxygen. Each of the 122 known elements (92 natural, plus 30 created under special conditions) has distinct chemical characteristics. Just four elements—oxygen, carbon, hydrogen, and nitrogen—are responsible for more than 96 percent of the mass of most living organisms. A relatively short list makes up most of the rest (table 3.1). Shortages or excesses of these elements can dramatically destabilize ecosystems. See if you can find these elements in the periodic table of the elements.

Atoms are the smallest particles that exhibit the characteristics of an element. Atoms are composed of positively charged protons, negatively charged electrons, and electrically neutral neutrons. Protons and neutrons, which have approximately the same mass, are clustered in the nucleus in the center of the atom (fig. 3.2). Electrons, which are tiny in comparison to the other particles, orbit the nucleus at the speed of light.

Each element has a characteristic number of protons per atom, called its **atomic number.** Carbon, for example, has 6 protons (see fig. 3.2), so its atomic number is 6. Each element also has a characteristic atomic mass, which is the sum of protons and neutrons (each having a mass of about 1 atomic mass unit). Carbon normally has 6 neutrons and 6 protons, which sum to an atomic mass of 12. However, the number of neutrons can vary slightly. Forms of the same element that differ in number of neutrons (and therefore in mass) are called **isotopes.** For example, some carbon atoms have 7 or 8 neutrons. These atoms have a mass of 13 or 14, rather than the usual

Benchmark Data

Among the ideas and values in this chapter, the following are a few worth remembering.

7	pH of neutral substances
16, 18	Atomic mass of normal (16) oxygen and a heavy isotope (18)
1/10	Approximate energy remaining at successive levels of a trophic pyramid
>2:1	Nitrogen captured from atmosphere by human:natural processes

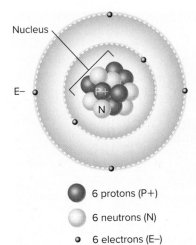

Nucleus

E−

6 protons (P+)

6 neutrons (N)

6 electrons (E−)

FIGURE 3.2 As difficult as it may be to imagine when you look at a solid object, all matter is composed of tiny, moving particles, separated by space and held together by energy. It is hard to capture these dynamic relationships in a drawing. This model represents carbon-12, with a nucleus of six protons and six neutrons; the six electrons are represented as a fuzzy cloud of potential locations rather than as individual particles.

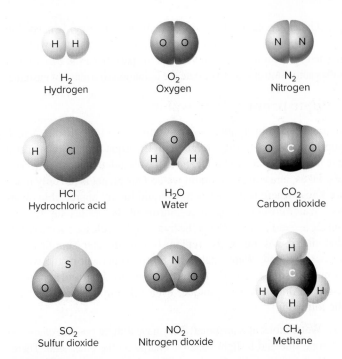

H_2 Hydrogen

O_2 Oxygen

N_2 Nitrogen

HCl Hydrochloric acid

H_2O Water

CO_2 Carbon dioxide

SO_2 Sulfur dioxide

NO_2 Nitrogen dioxide

CH_4 Methane

FIGURE 3.3 These common molecules, with atoms held together by covalent bonds, are important components of the atmosphere or are important pollutants.

12. Similarly, hydrogen (H) is the lightest element, and normally it has just one proton and one electron (and no neutrons) and an atomic mass of 1. A small percentage of hydrogen atoms also have a neutron in the nucleus, which gives those atoms an atomic mass of 2 (one proton + one neutron). We call this isotope deuterium (^{2}H). An even smaller percentage of natural hydrogen called tritium (^{3}H) has one proton plus two neutrons. Oxygen atoms can also have one or two extra neutrons, making them the isotopes ^{17}O or ^{18}O, instead of the normal ^{16}O.

This difference is important in environmental science. Oxygen isotopes, for example, tell us about ancient climates. Water (H_2O) containing ^{18}O is slightly more massive than water containing the normal ^{16}O. The higher-mass H_2O evaporates into the air more easily in hot climates, where there is plenty of evaporative energy, than in cold climates. When we examine bubbles of ancient air trapped in ice cores, abundant ^{18}O indicates a relatively warm ancient climate. Lower amounts of ^{18}O indicate a colder climate (chapter 15). Some isotopes are unstable—that is, they spontaneously emit electromagnetic energy or subatomic particles, or both. Radioactive waste and nuclear energy are both environmentally hazardous because they involve unstable isotopes of elements such as uranium and plutonium (chapters 19 and 21).

Chemical bonds hold molecules together

Atoms often join to form **compounds,** or substances composed of different kinds of atoms (fig. 3.3). If a pair or group of atoms can exist as a single unit, it is known as a **molecule.** Some elements commonly occur as molecules, such as molecular oxygen (O_2) or molecular nitrogen (N_2), and some compounds can exist as molecules, such as glucose ($C_6H_{12}O_6$). In contrast to these molecules, sodium chloride (NaCl, table salt) is a compound that cannot exist as a single pair of atoms. Instead it occurs in a solid mass of Na and Cl atoms or as two ions, Na$^+$ and Cl$^-$, dissolved in solution. Most molecules consist of only a few atoms. But many, such as proteins and nucleic acids, discussed below, can include millions or even billions of atoms.

Electrical attraction holds atoms together to form compounds. When *ions* with opposite charges (such as Na$^+$ and Cl$^-$) form a compound, the electrical attraction holding them together is an *ionic*

bond. Sometimes neither atom readily gives up an electron to the other, as when two hydrogen atoms meet. In this case, atoms form bonds by *sharing* electrons. Two hydrogen atoms can bond by sharing a pair of electrons—they orbit the two hydrogen nuclei equally and hold the atoms together. Such electron-sharing bonds are known as *covalent bonds.* Carbon (C) can form covalent bonds simultaneously with four other atoms, so carbon can create complex structures such as sugars and proteins, or hydrocarbons like methane (fig. 3.3).

Atoms in covalent bonds do not always share electrons evenly. An important example in environmental science is the covalent bonds in water (H_2O). The oxygen atom attracts the shared electrons more strongly than do the two hydrogen atoms. Consequently, the hydrogen portion of the molecule has a slight positive charge, while the oxygen has a slight negative charge. These charges create a mild attraction between water molecules, making water somewhat cohesive. This fact helps explain some of the remarkable properties of water that we'll discuss in the next section.

When an atom gives up one or more electrons, we say it is *oxidized* (because it is very often oxygen, an abundant and highly reactive element, that takes the electron). When an atom gains electrons, we say it is *reduced.* Oxidation and reduction reactions are necessary for life: Oxidation of sugar and starch molecules, for example, is an important part of how you gain energy from food.

Forming bonds usually releases energy. Breaking bonds generally requires energy. Think of this in burning wood: carbon-rich organic compounds such as cellulose are *broken,* which requires some activation energy (from a match, for example). Carbon atoms released from the wood then *form* bonds with oxygen from the air, making CO_2 and releasing heat. In a fire, more energy is produced than is consumed, and the net effect is that it feels hot to us.

Generally, some energy input (activation energy) is needed to start these reactions. In your fireplace, a match might provide the needed activation energy. In your car, a spark from the battery provides activation energy to initiate the oxidation (burning) of gasoline.

Unique properties of water

If travelers from another solar system were to visit our lovely, cool, blue planet, they might call it Aqua rather than Terra because of its outstanding feature: the abundance of streams, rivers, lakes, and oceans of liquid water. Our planet is the only place we know where water exists as a liquid in any appreciable quantity. Water covers nearly three-fourths of the earth's surface and moves around constantly via the hydrologic cycle (see section 3.4) that distributes nutrients, replenishes freshwater supplies, and shapes the land. Water makes up 60 to 70 percent of the weight of most living organisms. It fills cells, giving form and support to tissues. Among water's unique, even serendipitous qualities are the following:

1. Water molecules are polar: They have a slight positive charge on one side and a slight negative charge on the other side. Therefore, water readily dissolves polar or ionic substances, including sugars and nutrients, and carries materials to and from cells.

2. Water is the only inorganic substance that normally exists as a liquid at temperatures suitable for life. Most substances exist as either a solid or a gas; they are liquid only in a very narrow temperature range. Organisms synthesize organic compounds such as oils and alcohols that remain liquid at ambient temperatures and are therefore extremely valuable to life, but the original and predominant liquid in nature is water.

3. Water molecules are cohesive: They hold together tenaciously and create high surface tension (fig 3.4). You have experienced this property if you have ever done a belly flop off a diving board. Water has the highest surface tension of any common, natural liquid. Water also adheres to surfaces. As a result, water is subject to *capillary action:* It can be drawn into small channels. Without capillary action, movement of water and nutrients into groundwater reservoirs and through living organisms might not be possible.

4. Water is unique in that it expands when it crystallizes. Most substances shrink as they change from liquid to solid. Ice floats because it is less dense than liquid water. When temperatures fall below freezing, the surface layers of lakes, rivers, and oceans cool faster and freeze before deeper water. Floating ice then insulates underlying layers, keeping most water bodies liquid (and aquatic organisms alive) throughout the winter in most places. Without this feature, many aquatic systems would freeze solid in winter.

5. Water has a high heat of vaporization: It takes a great deal of heat to convert from liquid to vapor. Consequently, evaporating water is an effective way for organisms to shed excess heat. Many animals pant or sweat to moisten evaporative cooling surfaces. Why do you feel less comfortable on a hot, humid day than on a hot, dry day? Because the water vapor–laden

FIGURE 3.4 Surface tension is demonstrated by the resistance of a water surface to penetration, as when it is walked upon by a water strider. *Matti Suopajarvi/Getty Images*

air inhibits the rate of evaporation from your skin, thereby impairing your ability to shed heat.

6. Water also has a high specific heat: A great deal of heat is absorbed before it changes temperature. The slow response of water to temperature change helps moderate global temperatures, keeping the environment warm in winter and cool in summer. This effect is especially noticeable near the ocean, but it is important globally.

All these properties make water a unique and vitally important component of the ecological cycles that transfer matter and energy and that make life on earth possible.

Ions react and bond to form compounds

Atoms frequently gain or lose electrons, acquiring a negative or positive electrical charge. Charged atoms (or combinations of atoms) are called **ions.** Negatively charged ions (with one or more extra electrons) are *anions.* Positively charged ions are *cations.* A hydrogen (H) atom, for example, can give up its sole electron to become a hydrogen ion (H^+). Chlorine (Cl) readily gains electrons, forming chlorine ions (Cl^-).

Substances that readily give up hydrogen ions in water are known as **acids.** Hydrochloric acid, for example, dissociates in water to form H^+ and Cl^- ions. In later chapters, you may read about acid rain (which has an abundance of H^+ ions), ocean acidification, acid mine drainage, and other environmental concerns. In general, acids can cause environmental damage because the H^+ ions react readily with living tissues (such as tissues of fish larvae) and with nonliving substances (such as the limestone on buildings, or shells of marine organisms, which dissolve in acidic conditions).

Substances that readily bond with H^+ ions are called **bases** or alkaline substances. Sodium hydroxide (NaOH), for example, releases hydroxide ions (OH^-) that bond with H^+ ions in water. Bases can be highly reactive, so they also cause significant damage to living organisms. But acids and bases also are essential to living

FIGURE 3.5 The pH scale. The numbers represent the negative logarithm of the hydrogen ion concentration in water. Alkaline (basic) solutions have a pH greater than 7. Acids (pH less than 7) have high concentrations of reactive H^+ ions.

things: The acids in your stomach dissolve food, for example, and acids in soil help make nutrients available to growing plants.

We describe the strength of an acid and base by its **pH,** the negative logarithm of its concentration of H^+ ions (fig. 3.5). Acids have a pH below 7; bases have a pH greater than 7. A solution of exactly pH 7 is "neutral." Because the pH scale is logarithmic, pH 6 represents *ten times* more hydrogen ions in solution than pH 7.

A solution can be neutralized by adding buffers—substances that accept or release hydrogen ions. In the environment, for example, alkaline rock can buffer acidic precipitation, decreasing its acidity. Lakes with acidic bedrock, such as granite, are especially vulnerable to acid rain because they have little buffering capacity.

Quantitative Reasoning

The pH scale shows the availability of reactive hydrogen ions (H^+) in a liquid. The scale is logarithmic, so milk has 10 times as many H^+ ions as pure water, for a given volume. How many more H^+ ions does normal rain have compared to pure water? Soda pop? Vinegar? Is sea water more acidic or more basic than pure water?

Organic compounds have a carbon backbone

Organisms use some elements in abundance, others in trace amounts, and others not at all. Certain vital substances are concentrated within cells, while others are actively excluded. Carbon is a particularly important element because chains and rings of carbon atoms form the skeletons of **organic compounds,** the material of which biomolecules, and therefore living organisms, are made.

There are four major categories of organic compounds in living things ("bio-organic compounds"): lipids, carbohydrates, proteins,

FIGURE 3.6 The four major groups of biologically important organic molecules are based on repeating subunits of these carbon-based structures. Basic structures are shown for (a) butyric acid (a building block of lipids) and a hydrocarbon, (b) a simple carbohydrate, (c) a protein, and (d) a nucleotide (a component of nucleic acids).

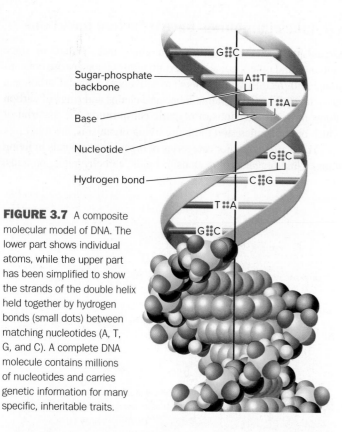

FIGURE 3.7 A composite molecular model of DNA. The lower part shows individual atoms, while the upper part has been simplified to show the strands of the double helix held together by hydrogen bonds (small dots) between matching nucleotides (A, T, G, and C). A complete DNA molecule contains millions of nucleotides and carries genetic information for many specific, inheritable traits.

and nucleic acids. Lipids (including fats and oils) store energy for cells, and they provide the core of cell membranes and other structures. Lipids do not readily dissolve in water, and their basic structure is a chain of carbon atoms with attached hydrogen atoms. A general term for compounds with this structure is hydrocarbons (fig. 3.6a). Fossil fuels, such as oil and propane, are hydrocarbons derived from living organisms (chapters 14 and 19). Carbohydrates (including sugars, starches, and cellulose) also have a basic structure of carbon atoms, but hydroxyl (OH) groups replace half the hydrogen atoms (fig. 3.6b). Like lipids, carbohydrates store energy and provide structure to cells. Glucose (fig. 3.6b) is an example of a very simple sugar. Most carbohydrates are much larger and more complex.

Proteins are composed of chains of subunits called amino acids (fig. 3.6c). Folded into complex three-dimensional shapes, proteins provide structure to cells and are used for countless cell functions. Most enzymes, such as those that release energy from lipids and carbohydrates, are proteins. Proteins also help identify disease-causing microbes, make muscles move, transport oxygen to cells, and regulate cell activity.

Nucleotides are complex molecules made of a five-carbon sugar (ribose or deoxyribose), one or more phosphate groups, and an organic nitrogen-containing base called either a purine or pyrimidine (fig. 3.6d). Nucleotides serve many functions. They carry information between cells, tissues, and organs. They are sources of energy for cells. They also form long chains called *ribo*nucleic *acid* (RNA) or **deoxyribonucleic acid (DNA)** that are essential for storing and expressing genetic information.

Just four kinds of nucleotides make up all DNA (these are adenine, guanine, cytosine, and thymine), but there can be billions of these molecules lined up in a very specific sequence. Long chains of DNA bind together to form a stable double helix (fig. 3.7). These chains separate and are duplicated when cells divide, so that genetic information is replicated. DNA molecules are responsible for differences among species and among individuals. Because they determine traits like growth rates and disease tolerance, DNA modification is now a dominant type of research in biology (see Exploring Science, p. 55).

Cells are the fundamental units of life

All living organisms are composed of **cells,** minute compartments within which the processes of life are carried out (fig. 3.8). Microscopic organisms such as bacteria, some algae, and protozoa are composed of single cells. Higher organisms have many cells, usually with many different cell varieties. Your

FIGURE 3.8 Plant tissues and a single cell's interior. Cell components include a cellulose cell wall, a nucleus, a large empty vacuole, and several chloroplasts, which carry out photosynthesis.

Gene Editing

Selective breeding has long been used to improve the characteristics of domestic plants and animals. But the process has been slow and rather unpredictable. With the development of molecular genetics, our ability to tailor organisms has improved considerably, but modifying specific genes has remained difficult and prone to errors. The discovery of a bacterial system for editing genes, however, may unleash a gold rush in genetic engineering.

This bacterial gene editing system is called CRISPR (pronounced crisper), which stands for "Clustered regularly-interspaced short palindromic repeats." CRISPR uses short sequences (palindromic repeats) of genetic material to attach to specific sections of DNA in a cell; it then uses an enzyme to cut the DNA in specific places. Bacteria use this process to cut and disable the DNA of invading viruses. Now geneticists have realized that this bacterial process could be used to recognize and modify, or "edit," genes in higher plants and animals. Just as bacteria target the DNA of an invading virus, CRISPR uses molecules synthesized to bind to and cut any gene we want to edit. When a target DNA sequence is identified and cut, it can inactivate the gene. As the cell tries to repair broken DNA, CRISPR can supply a template for new versions of the target gene, to replace the original sequence. Think of this as a molecular version of the search and replace function in your word processor.

The tool is being used in experiments to make human cells impervious to HIV, to correct a mutation that leads to blindness, and to cure mice of muscular dystrophy, cataracts, and a hereditary liver disease (fig. 1). It has been used to improve wheat, rice, soybeans, tomatoes, and oranges. Libraries of tens of thousands of DNA sequences are now available to target and activate, or silence, specific genes.

An exciting feature of CRISPR is that it can modify multiple genes at the same time. Eventually this may help us treat complex diseases, such as Alzheimer's or Parkinson's, that are regulated by many genes.

(header photo): David L. Hansen

FIGURE 1 Experiments with CRISPR have modified genes and edited inherited traits in lab mice and other organisms.
Nick Lander/123RF

CRISPR has also become the latest stage in the debate about genetically modified organisms (GMOs). On one hand, CRISPR makes it easier, quicker, and less expensive to create modified genomes. This will probably make GMOs more abundant. CRISPR edits genes much more precisely than other tools. Previous methods of creating GMOs generally involved blasting cells with chunks of DNA, often from unrelated species, and the outcome was often uncertain. Gene editing with CRISPR is far more precise and predictable. The implications of modified organisms for the ecosystems or social systems they inhabit, however, remain harder to predict.

In 2018, researchers in China shocked the medical and scientific community when they reported the birth of twin girls with DNA tweaked using CRISPR to make them resistant to HIV. The scientific community has long observed an unofficial moratorium on modifying viable human embryos. The ethical implications of modifying humans at will are enormous and complex. They are the stuff of science fiction: Will the rich produce designer super-babies? Will we generate drone classes, or create super-humans with physical perfection but lacking social traits that cannot be biologically engineered, such as empathy? Thus far, most genetic engineering has flown the flag of social progress and altruism—promising nutrient enriched rice or edible vaccines—but the industry focuses on products with the highest economic return, such as crops engineered to tolerate pesticides. These crops are worth billions of dollars per year and now dominate agricultural production. They have greatly increased use of pesticides and helped drive smaller farmers out of business (see chapter 10). We may have the technology, but do we have the wisdom, or the self-restraint, to use it for the betterment of society?

Why is this important in environmental science? Observers have worried that genetic engineering could produce superbugs, causing new epidemic diseases or disrupting ecosystems. On the other hand, CRISPR has been used in efforts to control invasive *Aedes aegypti* mosquitoes, which carry Zika virus and Dengue fever, by producing and releasing modified mosquitoes that mate with wild relatives and produce nonviable larvae. In Brazilian tests the number of observed *Aedes aegypti* larvae dropped by 82 percent in only 8 months. This approach could aid disease control without blasting ecosystems with powerful pesticides.

What do you think? If you were appointed to a regulatory panel commissioned to oversee gene editing, what limits—if any—would you impose on this new technology?

---

body, for instance, is composed of several trillion cells of about two hundred distinct types. Every cell is surrounded by a thin but dynamic membrane of lipid and protein that receives information about the exterior world and regulates the flow of materials between the cell and its environment. Inside, cells are subdivided into tiny organelles and subcellular particles that provide the machinery for life. Some of these organelles store and release energy. Others manage and distribute information. Still others create the internal structure that gives the cell its shape and allows it to fulfill its role.

A special class of proteins called **enzymes** carry out all the chemical reactions required to create these various structures. Enzymes also provide energy and materials to carry out cell functions, dispose of wastes, and perform other functions of life at the cellular level. Enzymes are molecular catalysts: They regulate chemical reactions without being used up or inactivated in the process. Like hammers or wrenches, they do their jobs without being consumed or damaged as they work. There are generally thousands of different kinds of enzymes in every cell, which carry out the many processes on which life depends. Altogether, the multitude of enzymatic reactions performed by an organism is called its **metabolism.**

Section Review

1. Define *atom* and *element*. Are these terms interchangeable?
2. What differentiates acids from bases? Give an example of each.
3. What are six characteristics of water that make it so valuable for living organisms and their environment?

3.2 ENERGY FOR LIFE

- *Energy occurs in different forms, such as kinetic energy, potential energy, chemical energy, or heat.*
- *The laws of thermodynamics state that energy is neither created nor destroyed, but that energy degrades to lower-intensity forms when used.*
- *Nearly all energy in ecosystems is captured from sunlight by photosynthesis; it is released through cellular respiration.*

Energy is the ability to do work, such as moving matter over a distance or causing a heat transfer between two objects at different temperatures. Energy can take many different forms. Heat, light, electricity, and chemical energy are examples that we all experience. Here we examine differences between forms of energy.

FIGURE 3.9 Water stored behind this dam represents potential energy. Water flowing over the dam has kinetic energy, some of which is converted to heat.
William P. Cunningham

Energy varies in intensity

The energy contained in moving objects is called **kinetic energy.** A rock rolling down a hill, the wind blowing through the trees, water flowing over a dam (fig. 3.9), or electrons speeding around the nucleus of an atom are all examples of kinetic energy. **Potential energy** is stored energy that is latent but available for use. A rock poised at the top of a hill and water stored behind a dam are examples of potential energy. **Chemical energy** is potential energy stored in the chemical bonds of molecules. The energy provided by the food you eat and the gasoline you put into your car are also examples of chemical energy that can be released to do useful work. Energy is often measured in units of heat (calories) or work (joules). One joule (J) is the work done when one kilogram is accelerated at one meter per second per second. One calorie is the amount of energy needed to heat one gram of pure water one degree Celsius. A calorie can also be measured as 4.184 J.

Heat is the energy that can be transferred between objects due to their difference in temperature. When a substance absorbs heat, the kinetic energy of its molecules increases. We sense this as a change in temperature. If enough heat is added, the substance may change to a different phase: A solid may become a liquid, or a liquid may become a gas. Note that when a substance changes phase, the added energy is used in the phase change, not in raising the temperature we sense. So as water evaporates, the process of evaporation takes heat from the environment, but the water vapor (a gas) is not warmer than the liquid water.

An object can have a high heat content but a low temperature, such as a lake that freezes slowly in the fall. Other objects, like a burning match, have a high temperature but little heat content. Heat storage in lakes and oceans is essential to moderating climates and maintaining biological communities. Heat absorbed in phase changes is also critical. As you will read in chapter 15, evaporation and condensation of water in the atmosphere help distribute heat around the globe.

Energy that is diffused, dispersed, and low in temperature is considered low-quality energy because it is difficult to gather and use for productive purposes. The heat stored in the oceans, for instance, is immense but hard to capture and use, so it is low quality. Conversely, energy that is intense, concentrated, and high in temperature is high-quality energy because of its usefulness in carrying out work. The intense flames of a very hot fire or high-voltage electrical energy are examples of high-quality forms that are valuable to humans. Many of our alternative energy sources (such as wind) are diffuse compared to the higher-quality, more concentrated chemical energy in oil, coal, or gas. This can mean that alternative energy sources are less intense—and less easy to use for work—than oil or gas.

Thermodynamics regulates energy transfers

Atoms and molecules cycle endlessly through organisms and their environment, but energy flows in a one-way path. A constant supply of energy—nearly all of it from the sun—is needed to keep biological processes running. Energy can be used repeatedly as it flows through the system, and it can be stored temporarily in the chemical bonds of organic molecules, but eventually it is released and dissipated.

The study of thermodynamics deals with the rates of flow and the transformation of energy from one form or quality to another. Thermodynamics is a complex, quantitative discipline, but you don't need a great deal of math to understand some of the broad principles that shape our world and our lives.

The **first law of thermodynamics** states that energy is *conserved;* that is, it is neither created nor destroyed under normal conditions. Energy may be transformed, for example, from the energy in a chemical bond to heat energy, or from kinetic energy to an electrical current, but the total amount does not change.

The **second law of thermodynamics** states that, with each successive energy transfer or transformation in a system, less energy is available to do work. That is, energy is degraded to lower-quality forms, or it dissipates and is lost, as it is used. When you drive a car, for example, the chemical energy of the gas is degraded to kinetic energy and heat, which dissipates, eventually, to space. The second law recognizes that disorder, or **entropy,** tends to increase in all natural systems. Consequently, there is always less *useful* energy available when you finish a process than there was before you started. Because of this loss, everything in the universe tends to fall apart, slow down, and get more disorganized.

How does the second law of thermodynamics apply to organisms and biological systems? Organisms are highly organized, both structurally and metabolically. Constant care and maintenance is required to keep up this organization, and a constant supply of energy is required to maintain these processes. Every time some energy is used by a cell to do work, some of that energy is dissipated or lost as heat. If cellular energy supplies are interrupted or depleted, the result—sooner or later—is death.

Ecosystems run on energy

All ecosystems require constant inputs of energy to function. Nearly all ecosystems rely on the sun for this energy. A few types of ecosystems, though, survive without sunlight by capturing other energy sources. Deep sea hydrothermal systems are an example.

Until recently, the deep ocean floor was believed to be essentially lifeless. Cold, dark, subject to crushing pressures, and without any known energy supply, it was a place where scientists thought nothing could survive. Undersea explorations in the 1970s, however, revealed dense colonies of animals—blind shrimp, giant tubeworms, strange crabs, and bizarre clams—clustered around vents called black chimneys, where boiling hot, mineral-laden water bubbles up through cracks in the earth's crust. How do these sunless ecosystems get energy? The answer is **chemosynthesis,** the process in which bacteria use chemical bonds between inorganic elements, such as hydrogen sulfide (H_2S) or hydrogen gas (H_2), to provide energy for synthesis of organic molecules.

Discovering organisms living under the severe conditions of deep-sea hydrothermal vents led to exploration of other sites that

FIGURE 3.10 A colony of tube worms and mussels cluster over a cool, deep-sea methane seep in the Gulf of Mexico.
Source: Image courtesy of Gulf of Mexico 2002, NOAA/OER.

seem exceptionally harsh to us. Fascinating organisms have been discovered in hot springs, such as in Yellowstone National Park, in intensely salty lakes, and even in deep rock formations, up to 1,500 m (nearly a mile) deep in Columbia River basalts. Some species are amazingly hardy. The single-celled organism *Pyrolobus fumarii* can withstand temperatures up to 113°C (235°F). Most of these extremophiles are archaea, single-celled organisms that are thought to be the most primitive of all living organisms, and the conditions under which they live are thought to be similar to those in which life first evolved.

Deep-sea exploration of areas without thermal vents also has found abundant life (fig. 3.10). We now know that archaea live in oceanic sediments in astonishing numbers. The deepest of these species (they can be 800 m or more below the ocean floor) make methane from gaseous hydrogen (H_2) and carbon dioxide (CO_2), derived from rocks. Other species oxidize methane using sulfur to create hydrogen sulfide (H_2S), which is consumed by bacteria, that serves as a food source for more complex organisms such as tubeworms, crabs, and shrimp. Methane-using bacteria can also help clean up pollution. After the Deepwater Horizon oil spill in the Gulf of Mexico in 2010, a deep-sea bloom of methane-metabolizing bacteria apparently consumed most of the methane (natural gas) escaping the spill.

In more familiar ecosystems that surround us, energy comes from the sun. Our sun is a star—a fiery ball of exploding hydrogen gas. Its energy comes from fission of hydrogen atoms, an intense and searing energy source that produces an abundant flow of radiative energy. Radiation occurs in a wide range of wavelengths (fig. 3.11). Most solar energy is emitted in a relatively narrow range of wavelengths, which we perceive as light.

Solar energy is essential to life for two main reasons. First, the sun provides warmth. Most organisms survive within a relatively narrow temperature range: above 40°C, most biomolecules

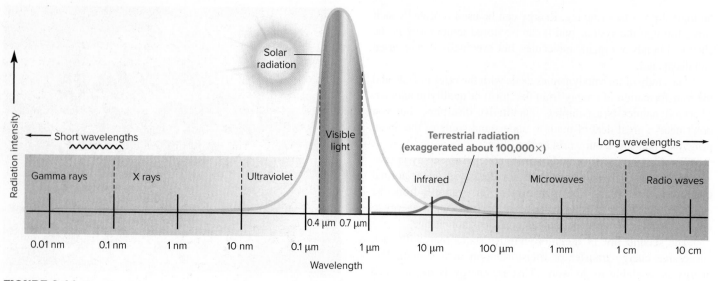

FIGURE 3.11 The electromagnetic spectrum. Our eyes are sensitive to light wavelengths, which make up nearly half the energy that reaches the earth's surface (represented by the area under the curve). Photosynthesizing plants also use the most abundant solar wavelengths. The earth re-emits lower-energy, longer wavelengths, mainly the infrared part of the spectrum.

begin to break down or become distorted and nonfunctional. At low temperatures (near 0°C), some chemical reactions of metabolism occur too slowly to enable organisms to grow and reproduce. Other planets in our solar system are either too hot or too cold to support life as we know it. The earth's water and atmosphere help to moderate, maintain, and distribute the sun's heat.

Second, as we have noted, nearly all organisms on the earth's surface depend on solar radiation for life-sustaining energy, which is captured by green plants, algae, and some bacteria through **photosynthesis.** Photosynthesis converts radiant energy into high-quality chemical energy in the bonds that hold together organic molecules. Photosynthetic organisms (plants, algae, and bacteria) capture roughly 105 billion metric tons of carbon every year and store it as biomass. About half of this carbon capture is on land; about half is in the ocean.

Plants use the abundant "visible" range of the spectrum, just as our eyes do, as well as near infrared. These wavelengths pass easily through our earth's atmosphere and reach the surface. About 45 percent of the radiation at the surface is visible, another 45 percent is infrared, and 10 percent is ultraviolet. Of the visible wavelengths, photosynthesis uses mainly red and blue light. Most plants reflect green wavelengths, so they appear to us. Half of the energy plants absorb is used in evaporating water. In the end, only 1 to 2 percent of the sunlight falling on plants is captured by photosynthesis. This small percentage is the energy base for virtually all life in the biosphere.

Photosynthesis captures energy; respiration releases that energy

Photosynthesis occurs in tiny organelles called chloroplasts that reside within plant cells (fig. 3.8). The main key to this process is chlorophyll, a green molecule that can absorb light energy and use the energy to create high-energy chemicals in compounds that serve as the fuel for all subsequent cellular metabolism. Chlorophyll

doesn't do this important job all alone, however. It is assisted by a large group of other lipid, sugar, protein, and nucleotide molecules. Together, these components carry out two interconnected cyclic sets of reactions (fig. 3.12).

Photosynthesis involves a sequence of reactions. Some photosynthetic reactions require energy from sunlight, and some don't. The process begins with a series of *light-dependent reactions*. These use solar energy directly to split water molecules into oxygen (O_2), which is released to the atmosphere, and hydrogen (H). This is the source of all the oxygen in the atmosphere on which all animals, including you, depend for life. Separating the hydrogen atom from its electron produces H^+ and an electron, both of which are used to form mobile, high-energy molecules called adenosine triphosphate (ATP) and nicotinamide adenine dinucleotide phosphate (NADPH).

Light-independent reactions then use the energy stored in ATP and NADPH molecules to create simple carbohydrates and sugar molecules (glucose, $C_6H_{12}O_6$) from carbon atoms (from CO_2) and water (H_2O). Glucose provides the energy and the building blocks for larger, more complex organic molecules. As ATP and NADPH give up some of their chemical energy, they are transformed to adenosine *di*phosphate (ADP) and NADP. These molecules are then reused in another round of light-dependent reactions. In most temperate-zone plants, photosynthesis can be summarized in the following equation:

$$6H_2O + 6CO_2 + \text{solar energy} \xrightarrow[\text{chlorophyll}]{} C_6H_{12}O_6 \text{ (sugar)} + 6O_2$$

We read this equation as "water plus carbon dioxide plus energy produces sugar plus oxygen." The reason the equation uses six water and six carbon dioxide molecules is that it takes six carbon atoms to make glucose (a sugar). The CO_2 in this equation is captured from the air by plant tissues. Thus, you could say that a plant is made primarily from air and water.

FIGURE 3.12 Photosynthesis involves a series of reactions in which chlorophyll captures light energy and forms high-energy molecules, ATP and NADPH. Light-independent reactions then use energy from ATP and NADPH (converting them to ADP and NADP) to fix carbon from air in organic molecules.

(Figure 3.12 labels: Light energy, Energized chlorophyll, Chlorophyll, Light-dependent reactions, Water H₂O, High-energy molecules, Carbon dioxide CO₂, Oxygen O₂, H⁺, Light-independent reactions, Carbohydrates (CH₂O))

What does the plant do with glucose? Because glucose is an energy-rich compound, it serves as the central, primary fuel for all cell growth and synthesis. The energy in its chemical bonds—created by photosynthesis—can be released by other enzymes and used to make other molecules (lipids, proteins, nucleic acids, or other carbohydrates), or it can drive kinetic processes such as the movement of ions across membranes, the transmission of messages, changes in cellular shape or structure, or movement of the cell itself in some cases.

This process of releasing chemical energy, called **cellular respiration,** involves splitting carbon and hydrogen atoms from the sugar molecule and recombining them with oxygen to re-create carbon dioxide and water. The net chemical reaction, then, is the reverse of photosynthesis:

$$C_6H_{12}O_6 + 6O_2 \longrightarrow 6H_2O + 6CO_2 + \text{released energy}$$

Note that in photosynthesis, energy is *captured,* while in respiration, energy is *released.* Similarly, photosynthesis *consumes* water and carbon dioxide to *produce* sugar and oxygen, while respiration does just the opposite. In both sets of reactions, chemical bonds are used to capture, store, and deliver energy within a cell. Plants carry out both photosynthesis and respiration, but during the day, if light, water, and CO_2 are available, they have a net production of O_2 and carbohydrates. "Net primary productivity" (NPP) is the rate of biomass accumulation—that is, the sum of

plant photosynthesis and respiration. We often describe ecosystems in terms of high NPP (for example, a tropical rainforest) or low NPP (a desert). Rates of NPP depend largely on availability of suitable temperatures and sufficient water for photosynthesis (see chapter 4, section 4.3).

We animals don't have chlorophyll and can't carry out photosynthetic food production. We do perform cellular respiration, however. We eat plants—or other animals that have eaten plants—and break down the organic molecules in our food through cellular respiration to obtain energy (fig. 3.13). In the process, we also consume oxygen and release carbon dioxide, thus completing the cycle of photosynthesis and respiration.

Section Review

1. Restate the first and second laws of thermodynamics.
2. Explain *kinetic energy, potential energy, high-quality energy,* and *low-quality energy.*
3. What are primary producers? Consumers?
4. What is the source of carbon for green plants? What is one product of photosynthesis?
5. How are photosynthesis and respiration related?

(Figure 3.13 labels: Sun, PHOTOSYNTHESIS, Oxygen (O₂), Light (diffuse energy), Sugars (high-quality energy), CO₂, H₂O, Producers, Oxygen (O₂), Consumers and decomposers, Carbon dioxide (CO₂), Water (H₂O), RESPIRATION, Heat (low-quality energy), High-quality energy for work: Biosynthesis, Movement, Membrane transport, Bioluminescence)

FIGURE 3.13 Energy exchange in ecosystems. Plants use sunlight, water, and carbon dioxide to produce sugars and other organic molecules. Consumers use oxygen and break down sugars during cellular respiration. Plants also carry out respiration, but during the day, if light, water, and CO_2 are available, they have a net production of O_2 and carbohydrates.

3.3 FROM SPECIES TO ECOSYSTEMS

- *Ecosystems consist of organisms and the systems they depend on.*
- *Food webs are structured by trophic levels.*
- *Higher trophic levels have fewer organisms and less mass than lower trophic levels.*

The Gulf of Mexico (opening case study) is part of a vast and complex system. The Mississippi River and land uses in its watershed influence the Gulf, as do currents and weather systems. The size and complexity of this system shows how difficult it can be to define boundaries around ecosystems. So we often define boundaries that make sense for the purposes at hand: We might discuss the Gulf as an ecosystem—with its primary producers and other trophic levels—even though we know is integrally linked to other systems. In this section, we consider the components and definitions of ecosystems. Understanding how nutrients and energy function in a system, and where they come from, and where they go, are essential to understanding **ecology,** the scientific study of relationships between organisms and their environment.

Ecosystems include living and nonliving parts

To biologists, terms like *species*, *population*, and *community* have very particular meanings. In Latin, *species* literally means "kind." In biology, **species** generally refers to all organisms of the same kind that are genetically similar enough to breed in nature and produce live, fertile offspring. There are important exceptions to this definition, and taxonomists increasingly use genetic differences to define species, but for our purposes this is a useful working definition.

A **population** consists of all the members of a species living in a given area at the same time. All the populations living and interacting in a particular area make up a **biological community.** What populations make up the biological community of which you are a part? If you consider all the populations of animals, plants, fungi, and microorganisms in your area, your community is probably large and complex. We'll explore the dynamics of populations and communities more in chapters 4 and 6.

As discussed in chapter 2, systems are networks of interaction among many interdependent factors. Your body, for example, is a very complex, self-regulating system. An ecological system, or **ecosystem,** is composed of a biological community and its physical environment. The environment includes abiotic factors (nonliving components), such as climate, water, minerals, and sunlight, as well as biotic factors, such as organisms and their products (secretions, wastes, and remains) and effects in a given area. It is useful to think about the biological community and its environment together, because energy and matter flow through both. Understanding how those flows work is a major theme in ecology.

For simplicity, we think of ecosystems as distinct ecological units with fairly clear boundaries. If you look at a patch of woods surrounded by farm fields, for instance, a relatively sharp line might separate the two areas, and conditions such as light levels, wind, moisture, and shelter are quite different in the woods than in the fields around them. Because of these variations, distinct populations of plants and animals live in each place. By studying each of these areas, we can make important and interesting discoveries about who lives where and why and about how conditions are established and maintained there.

The division between the fields and woods is not always clear, however. Air, of course, moves freely from one to another, and the runoff after a rainfall may carry soil, leaf litter, and live organisms between the areas. Birds may feed in the field during the day but roost in the woods at night, giving them roles in both places. Are they members of the woodland community or the field community? Is the edge of the woodland ecosystem where the last tree grows, or does it extend to every place that has an influence on the woods?

As you can see, it may be difficult to draw clear boundaries around communities and ecosystems. To some extent, we define these units by what we want to study and how much information we can handle. Thus, an ecosystem might be as large as a whole watershed or as small as a pond or even your own body. The thousands of species of bacteria, fungi, protozoans, and other organisms that live in and on your body make up a complex, interdependent community called the **microbiome.** You keep the other species warm and fed; they help you with digestion, nutrition, and other bodily functions. Some members of your community are harmful, but many are beneficial. You couldn't survive easily without them. We'll discuss your microbiome further in chapter 4.

You, as an ecosystem, have clear boundaries, but you are open in the sense that you take in food, water, energy, and oxygen from your surrounding environment, and you excrete wastes. This is true of most ecosystems, but some are relatively closed; that is, they import and export comparatively little from outside. Others, such as a stream, are in a constant state of flux with materials and even whole organisms coming and going. Because of the second law of thermodynamics, however, every ecosystem must have a constant inflow of energy and a way to dispose of heat. Thus, with regard to energy flow, every ecosystem is open.

Many ecosystems have feedback mechanisms that maintain generally stable structures and functions. A forest tends to remain a forest, for the most part, and to have forest-like conditions if it isn't disturbed by outside forces. Some ecologists suggest that ecosystems—or perhaps all life on the earth—may function as superorganisms, because they maintain stable conditions and can be resilient to change.

Food webs link species of different trophic levels

All ecosystems are based on photosynthesis (or, rarely, chemosynthesis). Organisms that photosynthesize, mainly green plants and algae, are therefore known as **producers.** One of the major properties of an ecosystem is its **productivity,** the amount of **biomass** (biological material) produced in a given area during a given period of time. Photosynthesis is described as *primary productivity* because it is the basis for almost all other growth in an ecosystem. Manufacture of biomass by organisms that eat plants is termed *secondary productivity*. A given ecosystem may have very high total productivity, but if decomposers decompose organic material as rapidly as it is formed, the *net primary productivity* will be low.

Think about what you have eaten today and trace it back to its photosynthetic source. If you have eaten an egg, you can trace it back to a chicken, which probably ate grain. This is an example of a **food chain,** a linked feeding series. Now think about a more complex food chain involving you, a chicken, a corn plant, and a grasshopper. The chicken could eat grasshoppers that had eaten leaves of the corn plant. You also could eat the grasshopper directly—some humans do. Or you could eat corn yourself, making the shortest possible food chain. Humans have several options of where we fit into food chains.

In ecosystems, some consumers feed on a single species, but most consumers have multiple food sources. Similarly, some species are prey to a single kind of predator, but many species in an ecosystem are beset by several types of predators and parasites. In this way, individual food chains become interconnected to form a **food web.** Figure 3.14 shows feeding relationships among some of the larger organisms in a woodland and lake community. If we were to add all the insects, worms, and microscopic organisms that belong in this picture, however, we would have overwhelming complexity. Perhaps you can imagine the challenge ecologists face in trying to quantify and interpret the precise matter and energy transfers that occur in a natural ecosystem!

An organism's feeding status in an ecosystem can be expressed as its **trophic level** (from the Greek *trophe*, "food"). In our first example, the corn plant is at the producer level; it transforms solar energy into chemical energy, producing food molecules. Other organisms in the ecosystem are **consumers** of the chemical energy harnessed by the producers. An organism that eats producers is a primary consumer. An organism that eats primary consumers is a secondary consumer, which may, in turn, be eaten by a tertiary consumer, and so on. Most terrestrial food chains are relatively short (seeds → mouse → owl), but aquatic food chains may be quite long (microscopic algae → copepod → minnow → crayfish → bass → osprey). The length of a food chain also may reflect the physical characteristics of a particular ecosystem. A harsh arctic landscape, with relatively low species diversity, can have a much shorter food chain than a temperate or tropical one (fig. 3.15).

Organisms can be identified both by the trophic level at which they feed and by the *kinds* of food they eat (fig. 3.16). **Herbivores** are plant eaters, **carnivores** are flesh eaters, and **omnivores** eat both plant and animal matter. What are humans? We are natural omnivores, by history and by habit. Tooth structure is an important clue to understanding animal food preferences, and humans are no exception. Our teeth are suited for an omnivorous diet, with a combination of cutting

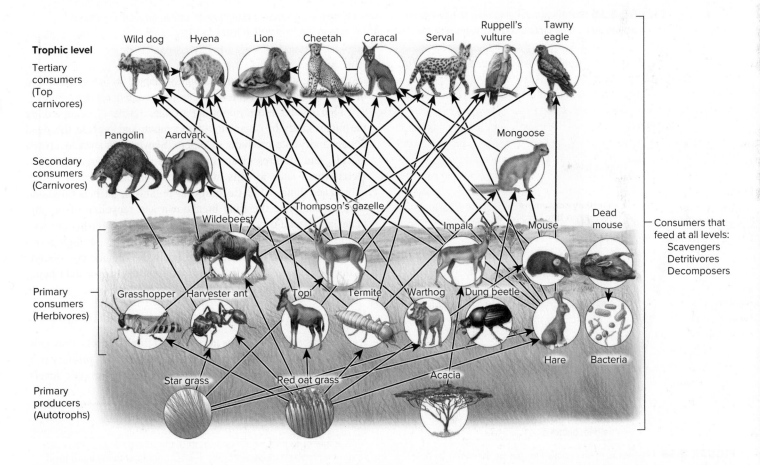

FIGURE 3.14 Each time an organism feeds, it becomes a link in a food chain. In an ecosystem, food chains become interconnected when predators feed on more than one kind of prey, thus forming a food web. The arrows in this diagram and in figure 3.15 indicate the direction in which matter and energy are transferred through feeding relationships. Only a few representative relationships are shown here. What others might you add?

FIGURE 3.15 Harsh environments tend to have shorter food chains than environments with more favorable physical conditions. Compare the arctic food chains depicted here with the longer food chains in the food web in figure 3.14.

FIGURE 3.16 Organisms in an ecosystem may be identified by how they obtain food for their life processes (producer, herbivore, carnivore, omnivore, scavenger, decomposer, reducer) or by consumer level (producer; primary, secondary, or tertiary consumer) or by trophic level (1st, 2nd, 3rd, 4th).

and crushing surfaces that are not highly adapted for one specific kind of food, as are the teeth of a wolf (carnivore) or a horse (herbivore).

One of the most important trophic levels is occupied by the many kinds of organisms that remove and recycle the dead bodies and waste products of others. **Scavengers** such as crows, jackals, and vultures clean up dead carcasses of larger animals. **Detritivores** such as ants and beetles consume detritus—litter, debris, and dung, while **decomposer** organisms such as fungi and bacteria complete the final breakdown and recycling of organic materials. It could be argued that these microorganisms are second in importance only to producers, because without their activity, nutrients would remain locked up in the organic compounds of dead organisms and discarded body wastes, rather than being made available to successive generations of organisms.

Ecological pyramids describe trophic levels

If we arrange the organisms according to trophic levels, they generally form a pyramid with a broad base representing primary producers and only a few individuals in the highest trophic levels.

Quantitative Reasoning

As a rule of thumb, about one-tenth of the energy or biomass consumed is stored at each trophic level. About how many kg of feed should it take to produce 1 kg of chicken meat that we eat? How much more energy should it take to provide you a meal of meat compared to vegetables?

	0.1%	Top carnivores
Detritivores and decomposers	1.8%	Primary carnivores
24.2%	16.1%	Herbivores
100%		Producers

FIGURE 3.17 A classic example of an energy pyramid from Silver Springs, Florida. The numbers in each bar show the percentage of the energy captured in the primary producer level that is incorporated into the biomass of each succeeding level. Detritivores and decomposers feed at every level but are shown attached to the producer bar because this level provides most of their energy.

This pyramid arrangement is especially true if we look at the energy content of an ecosystem (fig. 3.17).

Why is there so much less energy in each successive level in figure 3.17? Because of the second law of thermodynamics, which says that energy dissipates and degrades as it is reused. Thus, a rabbit consumes a great deal of chemical energy stored in carbohydrates in grass, and much of that energy is transformed to kinetic energy, when the rabbit moves, or to heat, which dissipates to the environment. A fox eats the rabbit, and the same degradation and dissipation happen again. As the fox uses the energy to live, some is lost in heat or movement. A little of the energy it has eaten is stored in chemical bonds in the fox's tissues. From an ecosystem energy perspective, there will always be smaller amounts of energy at successively higher trophic levels. Large top carnivores need a very large pyramid, and a large home range, to support them. A tiger, for example, may require a home range of several hundred square kilometers to survive.

As a broad generalization, only about one-tenth of the energy in one trophic level is represented in the next higher level (fig. 3.18). The amount of energy available is often expressed in biomass. For example, it generally takes about 100 kg of clover to produce 10 kg of rabbit, and 10 kg of rabbit to make 1 kg of fox.

The total number of organisms and the total amount of biomass in each successive trophic level of an ecosystem also may

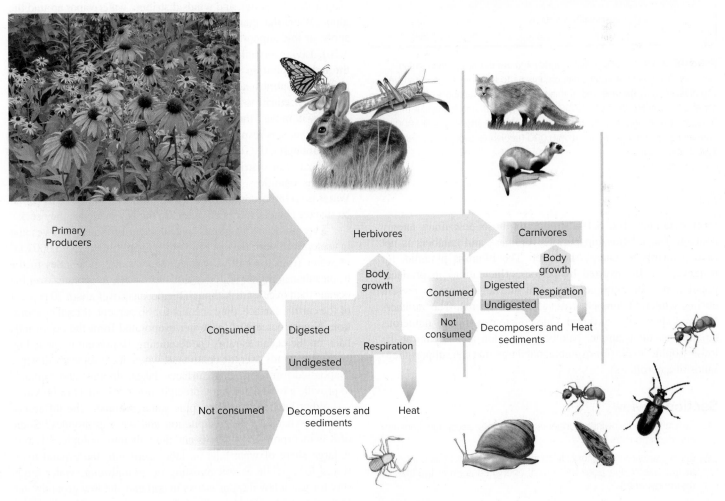

FIGURE 3.18 A biomass pyramid shows that, like energy, biomass storage decreases at higher trophic levels.

My Rubio Garden/Shutterstock

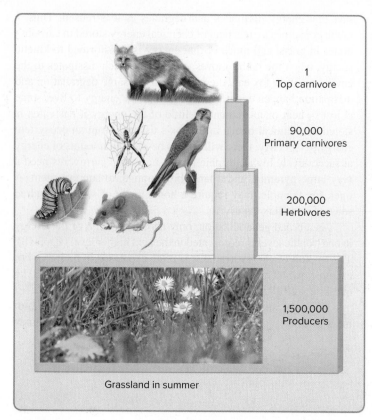

FIGURE 3.19 Usually, smaller organisms are eaten by larger organisms and it takes numerous small organisms to feed one large organism. The classic study represented in this pyramid shows numbers of individuals at each trophic level per 1,000 m² of grassland, and reads like this: To support one individual at the top carnivore level, there were 90,000 primary carnivores feeding upon 200,000 herbivores that in turn fed upon 1,500,000 producers.
Mahey/Shutterstock

form pyramids (fig. 3.19) similar to those describing energy content. The relationship between biomass and numbers is not as dependable as energy, however. The biomass pyramid, for instance, can be inverted by periodic fluctuations in producer populations (for example, low plant and algal biomass present during winter in temperate aquatic ecosystems). The numbers pyramid also can be inverted. One coyote can support numerous tapeworms, for example. Numbers inversion also occurs at the lower trophic levels (for example, one large tree can support thousands of caterpillars).

Section Review

1. Describe the following: producers; consumers; secondary consumers; decomposers.

2. Ecosystems require energy to function. Where does this energy go as it is used? How does the flow of energy conform to the laws of thermodynamics?

3. Why are there generally fewer organisms at the top of the food pyramid than at the bottom?

3.4 MATERIAL CYCLES

- *The water cycle distributes water among atmosphere, biosphere, surface, and groundwater.*
- *Carbon, nitrogen, and phosphorous are among the essential elements that also move through biological, atmospheric, and earth systems (biogeochemical cycles).*

Earth is the only planet in our solar system that provides a suitable environment for life as we know it. Even our nearest planetary neighbors, Mars and Venus, do not meet these requirements. Maintenance of these conditions requires a constant recycling of materials between the biotic (living) and abiotic (nonliving) components of ecosystems.

The hydrologic cycle redistributes water

The path of water through our environment, known as the **hydrologic cycle,** is perhaps the most familiar material cycle, and it is discussed in greater detail in chapter 17. Most of the earth's water is stored in the oceans, but solar energy continually evaporates this water, and winds distribute water vapor around the globe. Water that condenses over land surfaces, in the form of rain, snow, or fog, supports all terrestrial (land-based) ecosystems (fig. 3.20). Living organisms emit the moisture they have consumed through respiration and perspiration. Eventually this moisture reenters the atmosphere or enters lakes and streams, from which it ultimately returns to the ocean again.

As it moves through living things and through the atmosphere, water is responsible for metabolic processes within cells, for maintaining the flows of key nutrients through ecosystems, and for global-scale distribution of heat and energy (chapter 15). Water performs countless services because of its unusual properties. Water is so important that when astronomers look for signs of life on distant planets, traces of water are the key evidence they seek.

Everything about global hydrological processes is awesome in scale. Each year, the sun evaporates approximately 496,000 km³ of water from the earth's surface. More water evaporates in the tropics than at higher latitudes, and more water evaporates over the oceans than over land. Although the oceans cover about 70 percent of the earth's surface, they account for 86 percent of total evaporation. Ninety percent of the water evaporated from the ocean falls back on the ocean as rain. The remaining 10 percent is carried by prevailing winds over the continents, where it combines with water evaporated from soil, plant surfaces, lakes, streams, and wetlands to provide a total continental precipitation of about 111,000 km³.

What happens to the surplus water on land—the difference between what falls as precipitation and what evaporates? Some of it is incorporated by plants and animals into biological tissues. A large share of what falls on land seeps into the ground to be stored for a while as soil moisture or groundwater. Water might stay for just a few days or weeks in soil and shallow groundwater. In deep aquifers, water can reside for centuries or millennia. Eventually, all the water makes its way back downhill to the oceans.

FIGURE 3.20 The hydrologic cycle. Most exchange occurs with evaporation from oceans and precipitation back to oceans. About one-tenth of water evaporated from oceans falls over land, is recycled through terrestrial systems, and eventually drains back to oceans in rivers.

The 40,000 km³ carried back to the ocean each year by surface runoff or underground flow is the renewable supply available for human uses and sustaining freshwater-dependent ecosystems.

Carbon cycles through earth, air, water, and life

Carbon serves a dual purpose for organisms: (1) it is a structural component of organic molecules, and (2) the energy-holding chemical bonds it forms represent energy "storage." The **carbon cycle** begins with the intake of carbon dioxide (CO_2) by photosynthetic organisms (fig. 3.21). Photosynthesis produces plant cells from carbon in the air (plus hydrogen and oxygen in water). Carbon dioxide is eventually released during respiration, closing the cycle. The carbon cycle is of special interest because biological accumulation and release of carbon is a major factor in climate regulation.

The path followed by an individual carbon atom in this cycle may be quite direct and rapid, depending on how it is used in an organism's body. Imagine for a moment what happens to a simple sugar molecule you swallow in a glass of fruit juice. The sugar molecule is absorbed into your bloodstream, where it is made

available to your cells for cellular respiration or for making more complex biomolecules. If it is used in respiration, you may exhale the same carbon atom as CO_2 the same day.

Some forms of carbon cycle extremely slowly. Coal and oil are the compressed, chemically altered remains of plants or microorganisms that lived millions of years ago. Their carbon atoms (and hydrogen, oxygen, nitrogen, sulfur, etc.) are not released until the coal and oil are burned. We dramatically accelerate the movement of this carbon when we use fossil fuels. Enormous amounts of carbon also are locked up as calcium carbonate ($CaCO_3$), used to build shells and skeletons of marine organisms from tiny protozoans to corals. Most of these deposits are at the bottom of the oceans. The world's extensive surface limestone deposits are biologically formed calcium carbonate from ancient oceans, exposed by geological events. The carbon in limestone has been

FIGURE 3.21 The carbon cycle. Numbers indicate the approximate exchange of carbon in gigatons (Gt) per year. Natural exchanges are balanced, but human sources produce a net increase of CO_2 in the atmosphere.

locked away for millennia, which is probably the fate of carbon currently being deposited in ocean sediments. Eventually, even the deep ocean deposits are recycled, as they are drawn into deep molten layers and released via volcanic activity. Geologists estimate that every carbon atom on the earth has made about 30 such round trips over the last 4 billion years.

How does tying up so much carbon in the bodies and by-products of organisms affect the biosphere? Favorably. It helps balance CO_2 generation and utilization. Carbon dioxide is one of the so-called greenhouse gases because it absorbs heat radiated from the earth's surface, retaining it instead in the atmosphere (see chapter 15). Photosynthesis, accumulation of organic matter in soils and wetlands, and deposition of $CaCO_3$ remove atmospheric carbon dioxide; therefore, expansive forested areas, such as the boreal forests, and the oceans are very important **carbon sinks** (storage deposits). Cellular respiration and combustion both release CO_2, so they are referred to as carbon sources in the cycle.

Presently, combustion of organic fuels (mainly wood, coal, and petroleum products), removal of standing forests, and soil degradation are releasing huge quantities of CO_2 at rates that surpass the pace of CO_2 removal, a problem discussed in chapters 15 and 16.

Nitrogen occurs in many forms

Nitrogen is essential for living organisms, and it is often a key limiting factor in ecosystems. Nitrogen shortages slow biological productivity—a primary concern for farmers, for example. A surplus of nitrogen, as in the Gulf of Mexico's dead zone and in many other marine ecosystems, causes eutrophication, a common and often severe problem in aquatic systems.

Nitrogen is critical because it is an essential component of proteins, enzymes, DNA, and countless other organic compounds (see fig. 3.6). Primary producers incorporate nitrogen into these compounds, and consumers, such as yourself, acquire nitrogen by consuming plants or animals that have eaten plants.

Nitrogen is abundant in our environment—nitrogen gas molecules (N_2) make up 78 percent of our atmosphere. But these nitrogen molecules have strong bonds, and green plants can't use nitrogen from N_2 directly. Plants rely on bacteria living in soils, in plant tissues, or in aquatic systems to capture N_2. These "nitrogen fixing" bacteria have proteins that can break the N_2 bonds. The N is then available to be combined with hydrogen to form NH_3 (ammonia) or NH_4^+ (ammonium; the "+" indicates that the molecule has a positive ionic charge). Bacteria can gain energy by recombining these molecules with oxygen to form NO_2^- (nitrite) or NO_3^- (nitrate).

Of the many nitrogen compounds, only two, nitrate (NO_3^-) and ammonium (NH_4^+) can be used directly by plants. These are the sources of nitrogen for forming amino acids, the building blocks for complex organic compounds such as proteins.

Some plants, such as those in the bean and clover families, have nitrogen-fixing bacteria that reside in their roots. In beans, these bacteria live in nodules on plant roots (fig. 3.22). Having resident bacteria gives these plants built-in fertilizer. In return, the plant gives carbohydrates to the bacteria. This symbiotic relationship also enriches soil, so farmers often rotate legumes, such as

FIGURE 3.22 The roots of this bean plant are covered with bumps called nodules. Each nodule is a mass of root tissue containing many bacteria that help to convert nitrogen in the soil to a form the bean plants can assimilate and use to manufacture amino acids.
Kelly Marken/Shutterstock

beans, with nitrogen-hungry crops such as corn. Using natural fertilizers this way can improve both soil and productivity.

Bacteria in soil or water provide chemical reactions that drive many parts of the complex **nitrogen cycle** (fig. 3.23). Where oxygen is available, bacteria may combine ammonia (NH_3) with oxygen to form nitrous oxide (N_2O), nitric oxide (NO), nitrite (NO_2^-), or nitrate (NO_3^-).

In oxygen-poor conditions, such as in streambed sediments, saturated soils, or wetlands, denitrifying bacteria may remove oxygen from nitrate to form gaseous compounds, especially nitrous oxide or nitrogen gas (table 3.1). Conversion to these gaseous forms is known as denitrification. Denitrification can be important in removing nitrogen from aquatic systems that suffer from eutrophication.

Nitrogen moves through the food web as organisms die, decompose, or are consumed. Decomposers (fungi and bacteria) release ammonia and ammonium ions, which then are available for nitrate formation. Organisms also release proteins when plants shed their leaves, needles, flowers, fruits, and cones; or when animals shed hair, feathers, skin, exoskeletons, pupal cases, and silk, excrement, or urine, all of which are rich in nitrogen. Urinary wastes are especially high in nitrogen because they contain the detoxified wastes of protein metabolism.

Humans now capture a great deal of atmospheric nitrogen, and our activities have more than doubled the amount of nitrogen in circulation. Synthetic ammonia (NH_3) fertilizers are largely responsible. This form of nitrogen is produced from atmospheric nitrogen and natural gas under extreme heat and pressure. The spread of nitrogen-fixing soybeans, now one of the world's dominant crops, has also increased nitrogen capture. Even more important is the burning of fossil fuels in car engines and industrial furnaces. The high heat and pressure of combustion fix N_2 to form ammonia (NH_3) or nitrous oxide (N_2O). Rainfall washes these compounds out of the air and into soils and aquatic ecosystems. Excessive nitrogen fertilization is a dominant pollutant in many destabilized rivers, lakes, and

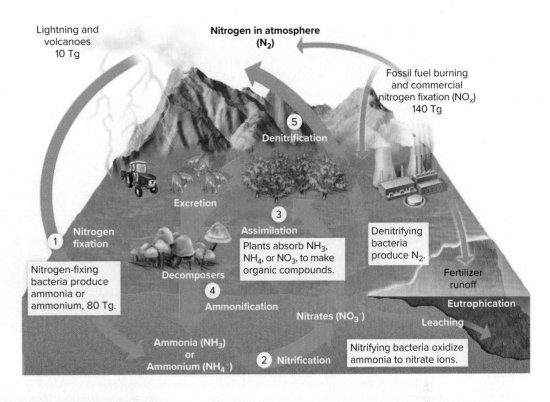

FIGURE 3.23 The nitrogen cycle. Nitrogen occurs in many different forms. Human activities have doubled the amount of nitrogen circulating, mainly because of industry, gasoline combustion in transportation, and agricultural fertilizers. Units are teragrams (Tg, equals one million tons) of nitrogen per year.

Lightning and volcanoes
10 Tg

Nitrogen in atmosphere
(N_2)

Fossil fuel burning and commercial nitrogen fixation (NO_x)
140 Tg

5 Denitrification

Excretion

3 Assimilation
Plants absorb NH_3, NH_4, or NO_3, to make organic compounds.

Denitrifying bacteria produce N_2.

1 Nitrogen fixation

Nitrogen-fixing bacteria produce ammonia or ammonium, 80 Tg.

Decomposers

4 Ammonification

Fertilizer runoff

Eutrophication

Nitrates (NO_3^-)

Leaching

Ammonia (NH_3) or Ammonium (NH_4^+)

2 Nitrification

Nitrifying bacteria oxidize ammonia to nitrate ions.

Table 3.2 Common Forms of Nitrogen

Name	Formula	Characteristics
Nitrogen gas	N_2	Makes up 78% of atmosphere; can be fixed by bacteria, cyanobacteria, some marine algae.
Ammonia	NH_3	Gaseous form; corrosive; common agricultural fertilizer that bonds with H^+ to form NH_4^+.
Ammonium	NH_4^+	Directly usable by plants; positive charge helps adhere to clays in soils.
Nitrite	NO_2^-	Ion with negative electrical charge; a step in *nitrification* process (conversion of NH_4^+ to NO_3^-); toxic to plants; usually present temporarily or in low quantities.
Nitrate	NO_3^-	Directly usable by plants; product of *nitrification*.
Organic nitrogen		Diverse compounds, such as proteins; must be converted to NH_4^+ for use by plants. Organic to inorganic transition is *mineralization*.
Nitrogen oxides	NO_x	Various combinations such as NO_2 (nitrogen dioxide), NO (nitric oxide), and N_2O (nitrous oxide); fuel combustion in vehicles and industry produces most NO_x; rainfall washes NO_x into soils and waterways. NO_2 and N_2 also result from *denitrification* by bacteria.

estuaries. In terrestrial systems, nitrogen enrichment encourages the spread of weeds into ecosystems that historically had low N inputs. In addition, nitrous oxide (N_2O) is an important greenhouse gas.

Phosphorus follows a one-way path

Phosphorus is an essential component of all cells. Compounds containing this element store and release a great deal of energy, so phosphorus-containing compounds, such as ATP, are primary participants in energy-transfer reactions in cells. Phosphorus is also a key component of proteins, enzymes, and tissues. The amount of available phosphorus in an ecosystem can have a dramatic effect on productivity. Low levels of phosphorus can limit plant growth. Abundant phosphorus stimulates lush plant and algal growth, making it a major water pollutant.

Most plants acquire phosphorus from soil, and animals acquire phosphorus by consuming plants or other animals. The ultimate source of phosphorus, though, is rocks. On the scale of centuries, the **phosphorus cycle** (fig. 3.24) is really a one-way path. This is because phosphorus has no atmospheric form in which it can quickly recirculate. Instead, phosphorus travels gradually downstream, as it is leached from rocks and minerals, taken up by the food web, and eventually released into water bodies that deliver it to the ocean. Phosphorus cycles repeatedly through the food web, as inorganic phosphorus is taken up by plants, incorporated into organic molecules, and passed on to consumers. Eventually, though, phosphorus washes down river to the ocean, where it accumulates in ocean sediments. Over geologic time, these deposits can be uplifted and exposed, so they become available to terrestrial life again. The phosphate used for detergents and fertilizers today are mined from

Geological uplift

Weathering and mining of phosphate rocks

Animals

Decomposers

Fertilizer

Farming

Plants

Runoff

21 Tg

200 Tg

Phosphate in soil $(HPO_4^{2-}$ $H_2PO_4^-)$

200 Tg

Dissolved phosphate

1,000 Tg

1,000 Tg

8 Tg

Leaching

Detritus

Marine organisms

Sedimentation forms new rocks 1.9 Tg

FIGURE 3.24 The phosphorus cycle. Natural movement of phosphorus is slight, involving recycling within ecosystems and some erosion and sedimentation of phosphorus-bearing rock. Use of phosphate (PO_4^{-3}) fertilizers and cleaning agents increases phosphorus in aquatic systems, causing eutrophication. Units are teragrams (Tg, equals one million tons) phosphorus per year.

exposed ocean sediments millions of years old. As with nitrogen, we have dramatically accelerated the movement of phosphorus in our environment. The effects of this increased phosphorus on aquatic ecosystems, as in the Gulf of Mexico, can include explosive growth of algae and photosynthetic bacteria, or eutrophication.

Phosphorus is essential for agriculture. There are worries about how long global phosphorus supplies will last. Some authorities estimate that we have several hundred years' worth of this essential mineral. Others worry that we may run out much sooner than that.

Section Review

1. What do we mean by carbon-fixation or nitrogen-fixation? Why is it important to humans that carbon and nitrogen be "fixed"?

2. Why is nitrogen important to living things? What type of organism is responsible for making nitrogen available to plants?

3. Describe the relative residence time of water in the ocean, the atmosphere, and in rivers and lakes.

4. In what forms is carbon stored in oceans, the atmosphere, living things, and rocks?

Connecting the Dots

Ecosystems, and all living things, persist by capturing, using, and recycling energy and matter. Understanding the different states of energy, and the different kinds of elements, isotopes, and molecules used by living things helps us understand how ecosystems function. These ideas help us piece together the problem of eutrophication in the Gulf of Mexico. The annual hypoxic zone, or dead zone, in the Gulf results from an influx of nutrients, especially nitrogen, from fertilizers upstream. Hypoxic zones are increasing in number and frequency worldwide. Human activities that capture nitrogen have more than doubled the flow of nitrogen through ecosystems, creating these imbalances.

Nitrogen is just one of a number of elements necessary for living cells, but it is a key element both in cellular energy transfer and in the formation of proteins and cell components. Yet it is only available in nature because bacteria (especially those living on the roots of legumes) capture it from the atmosphere. Carbon and oxygen, also necessary elements of cells and energy transfer, are not limiting factors in ecosystems because all green plants can capture CO_2 from the air to build cells and capture energy in sugars and starches.

The first and second laws of thermodynamics, which state that energy is conserved but also degraded as it is used, also helps us understand the structure of ecosystems. Of the energy an organism captures (by photosynthesis or eating), most is dissipated, for example, as heat or kinetic energy. Only about one-tenth as much energy is stored at successively higher trophic levels. This explains why an abundance of plant supports only a few herbivores, and even fewer predators.

Keep these principles in mind as you read later chapters in this book. These principles help us understand the function and organization of environmental systems all around us.

Reviewing Key Terms

Can you define the following terms in environmental science?

acids 3.1
atomic number 3.1
atoms 3.1
bases 3.1
biological community 3.3
biomass 3.3
carbon cycle 3.4
carbon sink 3.4
carnivore 3.3
cell 3.1
cellular respiration 3.2
chemical energy 3.2
chemosynthesis 3.2
compound 3.1

conservation of matter 3.1
consumer 3.3
decomposer 3.3
deoxyribonucleic acid
    (DNA) 3.1
detritivore 3.3
ecology 3.3
ecosystem 3.3
elements 3.1
energy 3.2
entropy 3.2
enzymes 3.1
first law of
    thermodynamics 3.2

food chain 3.3
food web 3.3
heat 3.2
herbivore 3.3
hydrologic cycle 3.4
ions 3.1
isotopes 3.1
kinetic energy 3.2
matter 3.1
metabolism 3.1
microbiome 3.3
molecule 3.1
nitrogen cycle 3.4
omnivore 3.3

organic compounds 3.1
pH 3.1
phosphorus cycle 3.4
photosynthesis 3.2
population 3.3
potential energy 3.2
producers 3.3
productivity 3.3
scavenger 3.3
second law of
    thermodynamics 3.2
species 3.3
trophic level 3.3

Critical Thinking and Discussion Questions

1. If your dishwasher detergents contained phosphorus, would you change brands? Would you encourage others to change? Why or why not?

2. Describe one or two practical examples of the laws of thermodynamics in your own life. Do they help explain why you can recycle cans and bottles but not energy? Which law is responsible for the fact that you get hot and sweaty when you exercise?

3. The ecosystem concept revolutionized ecology by introducing holistic systems thinking as opposed to individualistic life history studies. Why was this a conceptual breakthrough?

4. If ecosystems are so difficult to delimit, why is this such a persistent concept? Can you imagine any other ways to define or delimit environmental investigation?

5. Choose one of the material cycles (carbon, nitrogen, phosphorus, or sulfur) and identify the components of the cycle in which you participate. For which of these components would it be easiest to reduce your impacts?

Data Analysis

Inspect the Chesapeake's Report Card

As you know, nutrient cycling and nutrient supplies, or excesses, are key aspects of marine ecosystems in the Gulf of Mexico and many other areas. One system for which you can readily find recent, detailed insights on nutrient impacts is Chesapeake Bay. This system has been studied in detail because improving water quality is critical for economic health in the region.

Go to http://ecoreportcard.org/report-cards/chesapeake-bay/health and click on "Learn More" to see recent data. This site is maintained by the University of Maryland and the National Oceanic and Atmospheric Administration (NOAA), with the support of many collaborators and data providers. Use this site to examine recent data on chlorophyll, nutrients, and other measures of the bay's health.

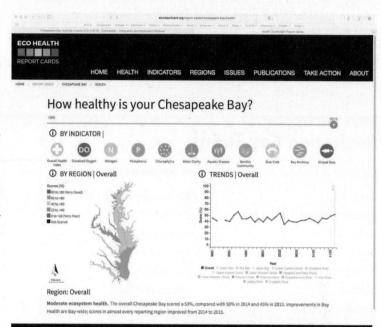

FIGURE 1 The EcoCheck website provides a wealth of water quality data.

Source: Integration and Application Network, University of Maryland Center for Environmental Science (http://ecoreportcard.org/report-cards/chesapeake-bay/health/). Used by permission.

4

Evolution, Biological Communities, and Species Interactions

▲Seagrass meadows provide numerous ecosystem services, including supporting many marine species.
Douglas Klug/Getty Images

Learning Outcomes

After studying this chapter, you should be able to:

4.1 Describe how evolution produces species diversity.

4.2 Discuss how species interactions shape biological communities.

4.3 Explain the dynamic nature of biological communities and some ecosystem services they provide.

4.4 Discuss the role of disturbance and resilience in ecosystems.

"When I view all beings not as special creations, but as lineal descendents of some few beings which have lived long before the first bed of the Cambrian system was deposited, they seem to me to become ennobled."

– Charles Darwin

Seagrass Meadows, the Planet's Hidden Productivity Powerhouse

Have you ever seen someone standing in shallow ocean water, peering down into their mobile phone? You may have spotted an intrepid citizen scientist contributing to marine conservation using the SeagrassSpotter app. Seagrass meadows are in trouble worldwide, disappearing at the same rate as rainforests and coral reefs, but few people even know they exist. The SeagrassSpotter project lets people around the world identify seagrass species and upload an image of the plant to an online map. Their input provides data for a growing, crowd-sourced planetary inventory of seagrass ecosystems. Online, we can see that people everywhere—vacationers at the ancient ruins in Lipsi, Greece; residents in Rameswaram, Sri Lanka; snorkelers in Changi Beach, Singapore; and couples on an after-dinner stroll in Mapua, New Zealand, have all contributed to conservation science with just a tap on their phone.

Why does the SeagrassSpotter project enlist the public's help to find seagrass meadows? Because these hidden ecosystems are powerhouses of biodiversity and productivity, and they deliver many benefits to us directly and indirectly.

Seagrasses are submerged, often deep-rooted, plants that grow along some 600,000 km^2 of the world's tropical and temperate coastlines. Because they photosynthesize, the plants typically occur in shallow ocean waters, between 2 and 12 m deep, but they can occur down to 50–60 m if water is extremely clear. The plants form extensive underwater ecosystems that can contain almost 1,000 different species of plants, invertebrates, mollusks, fish, and marine mammals (fig. 4.1).

Seagrass systems are among the most productive ecosystems in the world, more productive even than coral reefs (although reefs may have greater species diversity). While plant productivity and ecosystem services vary with geographic location and the type of seagrass present, seagrass communities form the primary production base of many coastal food webs. Even small areas of seagrass can yield big benefits: In the Mediterranean, seagrass covers less than 2 percent of the seafloor, but seagrass-associated fish and invertebrate species comprise 30 to 40 percent of the total value of the commercial fish catch.

Seagrass meadows are especially important on a planetary level because they deliver essential ecosystem services that benefit our lives, both economically and environmentally. They act as an enormous oxygen pump—one square meter can generate 10 l of oxygen every day. Their extensive network of roots and rhizomes trap and stabilize sediments on the seafloor. This helps control erosion, improve water clarity, and provide protection against floods.

Seagrass meadows perform a particularly important ecosystem service by capturing carbon dioxide through photosynthesis and storing it as organic carbon in the seafloor soil matrix. This "blue carbon" (organic carbon sequestered by coastal ecosystems like seagrass meadows, tidal marshes, and mangroves) is crucial for mitigating climate change. Seagrass meadows can store twice as much carbon as forests do; coastal mangroves can store four times as much. The carbon storage capabilities of seagrass meadows are tremendous for their size. Although though they cover only 0.1 percent of the seafloor, they provide 10 percent of ocean carbon storage.

Seagrasses are essential to the structure and functioning of coastal ecosystems and to the people that depend on them, but we are losing them at an alarming rate. In 2018, 140 ecologists from 35 countries sent a distress signal by declaring an International Call to Action, calling for increased conservation action, including management, monitoring, and protection. They warned that the loss of seagrass meadows threatens hundreds of millions of people with increased poverty, hunger, and disruption.

What threatens seagrass meadows? Runoff of fertilizers, pollution, and sediment is common. Dredging for ship channels, overfishing, and disease are widespread. Warmer ocean temperatures exacerbate these stressors.

But we can help these hidden gardens stay healthy and resilient. Everyone who visits sandy shorelines can help find and identify the grasses through programs like SeagrassSpotter, and we can promote their conservation through the media. Boaters can use low-impact anchoring systems to minimize damage.

FIGURE 4.1 Seagrass meadows are easily overlooked environments that are critical in supporting global biodiversity, including turtles, fish, invertebrates, and even humans.
M Swiet Productions/Moment/Getty Images

Communities and coastal managers can minimize impacts of coastal development, ban destructive fishing practices, and reduce erosion and surface run-off.

There are good reasons to be optimistic about seagrass meadows. Increasing evidence shows that they can be restored, and governments are showing interest in seagrass conservation as a tool for climate mitigation. You can be part of the solution by understanding and embracing the positive and powerful role of these secret gardens.

In this chapter, we consider some of the ways species adapt to their environments, interact, and shape their biological communities. We also examine effects of disturbance on those communities and properties that contribute to resilience in the face of environmental change.

4.1 EVOLUTION PRODUCES SPECIES DIVERSITY

- *Natural selection acting on spontaneous mutations results in evolution.*
- *Natural diversity develops as species allocate resources.*
- *All species have environmental tolerance limits; some narrower than others.*
- *Taxonomy describes relationships between species.*

What mechanisms promote the great variety of species on earth, and what determines which species will survive in one environment but not another? In this section, we examine (1) the process of evolution, which includes reproduction, variation, natural selection, and adaptation; (2) the limitations species face in their environments and implications for their survival; (3) the characteristics of species that make some of them ubiquitous and others rare;

and (4) the use of taxonomy to describe relationships between the vast complexity of organisms on earth. We'll start with the basics: How do species arise?

The basic idea of **evolution** is that species change over generations because individuals compete for scarce resources. The theory of evolution through natural selection was established most notably by Charles Darwin (1809–1882). Darwin's ideas on how species originate are generally traced to his visit to the Galápagos Islands in 1835, when he was serving as the ship naturalist on the *Beagle,* a ship sent to survey the coast of South America. Darwin's job was to collect specimens and record observations of the natural world, and in the Galápagos islands in particular, he was fascinated by the variety of unusual creatures, most occurring only in the Galápagos. Like other naturalists, Darwin was intrigued by the question of how such variety came to be. Most Europeans at the time believed that all living things had existed, unchanged, since a moment of divine creation, just a few thousand years ago. But Darwin had observed fossils of vanished creatures in South America. And he had read the new theories of geologist Charles Lyell (1797–1875), who argued that fossils showed that the world was much older than previously thought, and that species could undergo gradual but profound change over time. He also knew Thomas Malthus's (1766–1834) theory that competition for scarce resources can determine survival.

Observing that dog breeders created new varieties of dogs, from dachshunds to Great Danes, by selecting for certain traits, Darwin proposed that "natural selection" could explain the origins of species. Just as dog breeders favored individuals with particular characteristics, such as long legs or short noses, environmental conditions in an area could favor certain characteristics. Darwin proposed that individuals with traits suitable for their environment are the best competitors for scarce resources. As those individuals survive and reproduce, their traits eventually become common in the population. The evolution of species through natural selection is now a cornerstone of biology and its many subfields, from ecology to medicine and health care. Let's take a closer look at how this process works.

Benchmark Data	
Among the ideas and values in this chapter, the following are a few worth remembering.	
10 liters	Amount of oxygen one m^2 of seagrass generates each day
1,000	Number of species that may occur in a seagrass meadow
5:1	Ratio of carbon capture and storage in a coastal wetland: tropical forest
100 trillion	Approximate number of cells in your body
10%	Proportion of those cells that are mammalian
1/10th	Biological productivity of a desert compared to a rainforest
6	Number of "kingdoms" representing all life on earth

Evolution occurs through reproduction, variation, and natural selection

Evolution is a gradual process that results from three observable phenomena. The first is replication, which occurs through reproduction; the second is the variation in traits that exists in a population; and the third is the process of selection (nonrandom survival of offspring). Evolution is the unintended consequence of these three somewhat independent, observable facts of nature.

A logical sequence of events occurs in the process of evolution, so let's start at the beginning. The building blocks of evolution are formed during procreation, when parents produce offspring through sexual reproduction. This can happen through an asexual process, but it largely occurs when sperm and egg cells fuse together. The parental DNA builds offspring organisms that are near copies of the parents, but sometimes a gene of an individual plant or animal mutates, or changes by accident. **Mutations** in reproductive cells (gametes) can be passed on from generation to generation. These random mutations occur occasionally in any population. Arbitrary errors in DNA replication and exposure to radiation or toxic materials are the main causes of mutations. Most mutations have no effect on fitness.

When genes mutate, they often yield a slight variation (for example, a difference in visual acuity, length of limbs, or, feather coloring). Consider the evolution of giraffes, for example. Among herbivores in a savannah, where trees are an important year-round source of vegetation, those able to reach higher have access to more food. Individuals that happen to have longer necks are likely to be more vigorous, and more likely to survive and reproduce, compared to those that can't reach quite as high. When accidental

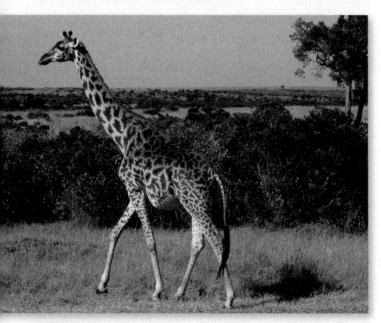

FIGURE 4.2 Giraffes don't have long necks because they stretch to reach tree-top leaves, but those giraffes that happened to have longer necks got more food and had more offspring, so the trait became prevalent in the population.
Tom Finkle

mutations produce individuals with longer necks, those traits are likely to become dominant in a population (fig. 4.2).

Adaptation, the acquisition of traits that allow a species to survive in its environment, is one of the most important concepts in biology. In evolutionary terms, adaptation affects populations, not individuals. Adaptation in a species happens over generations, as individual organisms that inherit advantageous traits persist, and other organisms which do not carry the advantageous trait die out.

The process of better-selected individuals passing their traits to the next generation is called **natural selection.** Natural selection helps explain how environmental conditions in a particular area can favor certain traits or characteristics. It occurs when the conditions of a particular environmental setting exert selective pressures on species, and over time the characteristics that help the species survive and reproduce persist. This is neither a random process, nor is it an "intentional" process that acts with a particular goal in mind. It's just that some trait variations worked better than others at a certain time, and their carriers were more likely to survive and leave offspring with the same variations. There is no ultimate goal, intended outcome, or "ladder of progress" inherent in evolution.

All species live within limits

No individual, population, or species can increase without limits. Under normal conditions, growth is limited by the availability of resources within the system, or by external factors in the environment. A **limiting factor** can be anything that constrains the growth of an organism or a population, such as *competition* (for space, food, nesting sites, etc.); *predation* (when either predator or prey populations increase or decrease); *parasitism or disease* (including the numbers of parasites or the number of disease-carrying organisms); *climate and weather changes* (like drought or heat waves); or human disturbance (pollution, habitat destruction).

In 1840, the chemist Justus von Liebig proposed that growth is controlled not by the total amount of resources available, but by the scarcest resource. He asserted that there is always a single limiting factor, or a critical factor that determines where a species lives. The giant saguaro cactus *(Carnegiea gigantea),* which grows in the dry, hot Sonoran Desert of southern Arizona and northern Mexico, offers an example (fig 4.3). Saguaros are tolerant of extreme heat and drought, intense sunlight, and nutrient-poor soils, but are extremely sensitive to freezing temperatures. A single winter night with temperatures below freezing for 12 or more hours kills growing tips on the branches, preventing further development. Thus, the northern edge of the saguaro's range corresponds to a zone where freezing temperatures last less than half a day at any time.

Ecologist Victor Shelford (1877–1968) later expanded Liebig's principle by stating that for each species, the environmental factor has a range of specific minimum and maximum levels that determine the species success. He called these **tolerance limits,** beyond which a particular species cannot survive or is unable to reproduce (fig 4.4). The single factor closest to these survival limits, Shelford postulated, is the critical factor that dictates where a particular

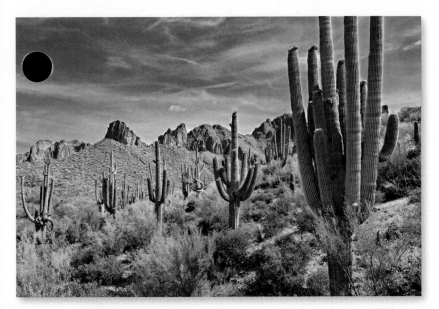

FIGURE 4.3 Saguaro cacti, symbolic of the Sonoran desert, are an excellent example of distribution controlled by a critical environmental factor. Extremely sensitive to low temperatures, saguaros are found only where minimum temperatures never dip below freezing for more than a few hours at a time.
Anton Foltin/Shutterstock

organism can live. At one time, ecologists tried to identify unique factors limiting the growth of every plant and animal population. We now know that several factors working together usually determine a species' distribution. If you ever explore the rocky coasts of New England or the Pacific Northwest, you might notice that mussels and barnacles grow thickly in the intertidal zone, the place between high and low tides. No one factor decides this pattern. Instead, the distribution of these animals is determined by a combination of temperature extremes, drying time between tides, salt concentrations, competitors, and food availability.

In some species, tolerance limits affect the distribution of young organisms differently than they affect adults. The desert pupfish, for instance, lives in small, isolated populations in warm springs in the northern Sonoran Desert. Adult pupfish can survive temperatures between 0° and 42°C (a remarkably high temperature for a fish) and tolerate a wide range of salt concentrations. Eggs and juvenile fish, however, can survive only between 20° and 36°C and are killed by high salt levels. Reproduction, therefore, is limited to a small part of the range of the adult fish.

Sometimes the sensitivity of a particular species makes it a useful **indicator species,** one that indicates the presence of particular environmental conditions. The presence or absence of an indicator species can reveal something about the community and the ecosystem as a whole. Lichens and eastern white pine, for example, are indicators of air pollution, because they are extremely sensitive to sulfur dioxide and ozone, respectively. Similarly, anglers know that trout species require cool, clean, oxygen-rich water; the presence or absence of trout is used as an indicator of water quality.

An ecological niche is a species' environment and its ecological role

A key habitat is the place or set of environmental conditions in which a particular organism lives. The term **ecological niche** is a more functional term that refers to a species' role in an ecosystem—how it uses its habitat, and how other organisms in the habitat use that species—as well as the habitat it occupies (fig. 4.5). The concept of an ecological niche was first defined in 1927 by the British ecologist Charles Elton (1900–1991). To Elton, each species had a role in a biological community, and the niche defined that species' way of obtaining food, the relationships it had with other species, and the services it provided to its community.

Thirty years later, the American limnologist G. E. Hutchinson (1903–1991) refined the definition of a niche to include physical and chemical conditions, such as temperature, light levels, acidity,

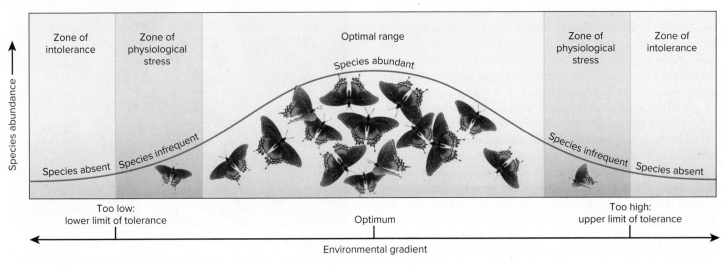

FIGURE 4.4 Tolerance limits affect species distributions. For every environmental factor, there is an optimal range within which a species lives or reproduces most easily, so abundance is high. The horizontal axis here could represent a factor such as temperature, rainfall, vegetation height, or the availability of some critical resource.

FIGURE 4.5 Each of the species in this African savanna has its own ecological niche that determines where and how it lives.
William P. Cunningham

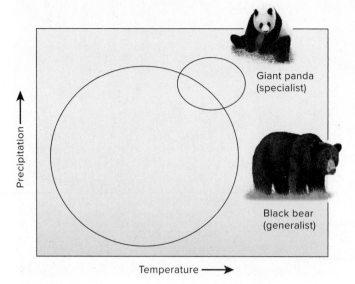

FIGURE 4.6 Generalists, such as the American black bear, tolerate a wide range of environmental conditions. Specialists, in contrast, have a narrower tolerance of environmental conditions.

humidity, or salinity. Hutchinson emphasized that these physical conditions occurred within a set of biological interactions, such as the presence of predators or prey, or the availability of nutritional resources.

Some species may be more flexible in their niche than others. Those that have greater niche flexibility can tolerate a wide range of habitat conditions or exploit a wide range of resources. These species are known as **generalists** (fig 4.6). A species can be a dietary generalist, like the racoon *(Procyon lotor),* who are omnivores and eat many different types of food, including human garbage; or they can be a habitat generalist, with the ability to live in many different habitat types, like the red fox *(Vulpes vulpes),* the most widely distributed wild carnivore in the world. Many generalists have large geographic ranges, as in the case of the black bear *(Ursus americana),* which is omnivorous, abundant, and ranges across most of North America's forested regions. We consider some generalists to be a nuisance or a "pest," such as rats, cockroaches, or dandelions. Other generalist species, like deer, aspen trees, or wild turkeys are looked on more favorably.

While generalists can exploit a wide range of resources or habitat, some species are **specialists** that occupy a narrow ecological niche. Over long periods of time, through the process of adaptation, natural selection, and evolution, these species developed features that allow them to live in a very limited set of conditions. Some specialists eat plants or animals that are toxic or undesirable to other species. Others live in very dry or nutrient-poor areas that few other species can tolerate, filling a niche opportunity.

As we saw with generalists, a species can be a *dietary specialist,* like the koala *(Phascolarctos cinereus),* which only eats leaves from eucalyptus trees; or a habitat specialist like the Kirtland's warbler *(Setophaga kirtlandii)* that requires a very specific forest type and stage of forest growth (young Jack Pine, *Pinus banksiana).*

The giant panda *(Ailuropoda melanoleuca)* is highly specialized in its diet. Pandas have evolved from carnivorous ancestors to subsist almost entirely on bamboo, a large but low-nutrient grass (fig 4.7). To acquire enough nutrients, pandas must spend as much as 16 hours a day eating. The panda's slow metabolism, slow movements, and low reproductive rate help it survive on this very specialized diet. Its narrow niche now endangers the giant panda, though, as recent destruction of most of its native bamboo forest has reduced its range and its population to the margin of survival. The giant saguaro is also a specialist. It is slow-growing and finely adapted to certain climatic conditions, but it cannot persist in wetter or cooler environments.

FIGURE 4.7 The giant panda has evolved from carnivorous ancestors to live on a diet composed almost exclusively of bamboo. Adaptions include "thumbs" that help it grasp bamboo leaves, and teeth that help it chew the grass.
Leungchopan/Getty Images

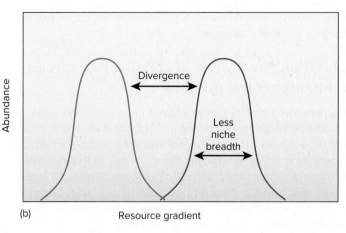

(a) Resource gradient

(b) Resource gradient

FIGURE 4.8 Competition causes resources partitioning and niche specialization. (a) Where niches of two species overlap along a resource gradient, competition occurs (shaded area). Individuals in this part of the niche have less success producing young. (b) Over time, the traits of the populations diverge, leading to specialization, narrower niche breadth, and less competition between species.

Because of their exacting habitat requirements, specialists do not tolerate environmental change well and may adjust poorly to changing conditions. In contrast, generalists can easily switch their food source or relocate to a different type of habitat if necessary. Habitat specialists are therefore more susceptible to extinction when habitat is lost or fragmented.

Often, specialists are **endemic species**—they occur only in one area (or one type of environment). The giant panda, for example, is endemic to the mountainous bamboo forests of southwestern China; the saguaro is endemic to the Sonoran Desert.

In most organisms, genetic traits and instinctive behaviors restrict the ecological niche. But some species have complex social structures that help them expand the range of resources or environments they can use. Elephants, chimpanzees, and dolphins, for example, learn from their social group how to behave and can invent new ways of doing things when presented with novel opportunities or challenges. In effect, they expand their ecological niche by transmitting cultural behavior from one generation to the next.

Resource partitioning can reduce competition

When two species compete for limited resources, one eventually gains the larger share, while the other finds a different habitat, dies out, or experiences a change in its behavior or physiology that reduces competition. Consequently, as explained by the Russian biologist G. F. Gause (1910–1986), "complete competitors cannot coexist." The general term for this idea is the **principle of competitive exclusion:** No two coexisting species can have identical niches for long. The species that is more efficient in using an available resource will begin to exclude the other. Eventually, the other species will disappear or it will develop a new niche, exploiting the resource in a slightly different way.

When species with similar niches divide up a resource or segregate it in some way, it is known as **resource partitioning** (fig. 4.8). Partitioning can allow several species to utilize different parts of the same resource and coexist within a single habitat. Species can partition a variety of resources, like food or nesting

sites. A classic example of resource partitioning is that of woodland warblers, studied by ecologist Robert MacArthur. Although several similar warblers species foraged in the same trees, they avoided competition by specializing in different levels of the forest canopy, or in inner and outer branches (fig 4.9).

Resources can be partitioned in time as well as space. Swallows and insect-eating bats both live by capturing flying insects, but bats hunt for night-flying insects, while swallows hunt

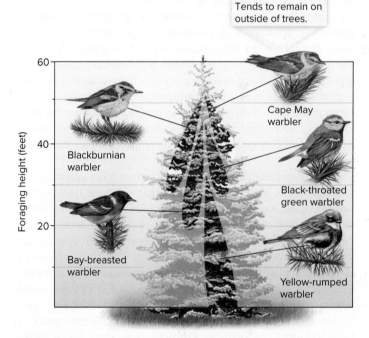

FIGURE 4.9 Several species of insect-eating wood warblers occupy the same forests in eastern North America. The competitive exclusion principle predicts that the warblers should partition the resource—insect food—in order to reduce competition. And, in fact, the warblers feed in different parts of the forest.
Source: Original observations by R. H. MacArthur (1958).

during the day. Thus, the two groups have noncompetitive feeding strategies for similar insect prey.

Speciation, the process of creating new species, maintains natural diversity

As a population becomes more adapted to its ecological niche, it may develop distinctive traits that differentiate it entirely from its biological cousins. The splitting of one species into two different species is called **speciation.** There are two well-documented mechanisms of speciation: allopatric ("different homelands") speciation, which usually involves a physical barrier and is common in nature; and sympatric ("same homeland") speciation, which most often involves genetic mutations in plants.

Allopatric speciation can occur when some kind of barrier isolates part of a population. Geographic isolation may develop, for example, as climate change alters habitat, as earthquakes or continental shifts separate land areas, or with sea level rise or island formation. We can see a recent example of reproductive isolation in populations of red squirrels in Arizona in just the past 10,000 years or so, as expanding lowland deserts have isolated populations on wooded mountains (fig 4.10).

The barriers that divide subpopulations aren't always physical. For example, two virtually identical tree frogs (*Hyla versicolor* and *H. chrysoscelis*) live in similar habitats of eastern North America but have different mating calls. This difference, which is enough to prevent interbreeding, is known as behavioral isolation. Behavioral isolation allows the two different species to evolve in the same geographic area at the same time.

Speciation that occurs without a geographical barrier, **sympatric speciation,** is common in plant populations. It is made possible when chromosome separation errors occur during the cell division of plants. When plant chromosomes do not separate fully into two cells, there can be more than two full sets of chromosomes in a single plant (polyploidy). Fern species and other plants sometimes undergo sympatric speciation by doubling or quadrupling the chromosome number of their ancestors, which makes them reproductively incompatible with the parent population.

Environmental change can make certain traits advantageous in a population. In the case of the Galapagos Islands finches studied by Charles Darwin, for example, an extended drought reduced the availability of small seeds on some islands. Large-billed individuals in the finch population had better success in opening the remaining larger seeds. Within two years, the large-billed trait came to dominate the population. This shift toward one extreme of a trait is known as directional selection (fig. 4.11). Environmental conditions may also make intermediate traits advantageous, producing stabilizing selection. Competition can cause divergence of traits and niche specialization (disruptive selection), which allows for better partitioning of resources.

Directional selection can be observed in many cases. Arctic foxes (*Vulpes Lagopus*), for example, occupy a landscape that is snowy in winter but grassy in summer, with very little cover to hide in. This environment favored individuals whose winter fur

1 Single population

2 Geographically isolated populations

FIGURE 4.10 Geographic barriers can result in allopatric speciation. During cool, moist glacial periods, what is now Arizona was forest-covered, and squirrels could travel and interbreed freely. As the climate warmed and dried, desert replaced forest on the plains. Squirrels were confined to cooler mountaintops, which acted as island refugia, where new, reproductively isolated species gradually evolved.

molted to white and whose summer fur was brown, which concealed them from predators as well as from prey. In a process of directional selection, foxes that lacked this ability to change colors died out, leaving the more extreme trait (contrasting seasonal colors) dominant in the population. We also see directional selection in the emergence of antibiotic-resistant bacteria (chapter 8). In this case, an environmental stressor (antibiotics) kills off most of a bacteria population. But a few individuals that happen to have better-than-average tolerance of antibiotics may survive. Resistant survivors produce new generations, leading to a population with resistance to the antibiotics.

Evolutionary change is typically slow

We generally believe that species arise slowly, and in fact many have existed unchanged for tens of millions of years. The American alligator (*Alligator mississippiensis*), for example, has existed

FIGURE 4.11 A species trait, such as beak shape, changes in response to selection pressure. Suppose a population starts with wide variation in beak shape (a). Sometimes selection pressures advantage just one beak type, which becomes dominant (b). Sometimes an intermediate or average type is advantageous (c). Disruptive selection moves characteristics to the extremes of the trait (d).

FIGURE 4.12 Female African bush elephants (*Loxodonta africana*) live to be about 70 years old, and they reach their highest fertility between 25 and 45 years old. Because of this slow growth and late maturity, the rate of evolutionary change through natural selection is much slower than for a species like a fruit fly, which becomes sexually mature at the age of one week.
William P. Cunningham

unchanged for over 150 million years. Some organisms, however, evolve swiftly: New flu viruses evolve every season and fruit flies in Hawaii mutate frequently and can give rise to new species in just a few years.

In 2018, three scientists won the Nobel Prize in chemistry for "speeding up" evolution. All three scientists accelerated the selection process to create brand new, "novel" molecules in a period of just months or weeks. The three chemists all identified and refined protein or enzyme molecules that attach to other molecules, speed up reactions, and accelerate change.

Given these examples of rapid evolution, why can't organisms adapt rapidly to the changes we are making to our environment, such as climate change, pollution, and habitat fragmentation? Species have evolved before. Why can't they just do it now?

There are several answers to this question. First, the frequency and intensity of modern environmental disturbances is unprecedented in geologic history, and the natural frequency of random mutations (required for evolution) is low. Second, genetic mutations cause change over generations, and many species have a long lifespan or age of reproductive maturity (fig 4.12). For example, each female lake sturgeon (*Acipenser fulvescens*) takes 14–33 years to reach sexual maturity, and then it only spawns once every 4–9 years; orange roughy (*Hoplostethus atlanticus*) grow very slowly and can live for 150 years; they don't begin to breed until they are at least 20 years old.

A third reason that species cannot easily or rapidly adapt to changing conditions is because species do not evolve alone, in isolation. Instead, species evolve over millennia as part of a biological community, and they are embedded in a complex web of interactions that includes competition, predator-prey relationships, and symbiosis. These biological and ecological interactions are discussed in sections 4.2 and 4.3.

Taxonomy describes relationships among species

Evolution has produced a bewildering variety of organisms. How can we make sense of them all? Taxonomy, the study of types of organisms and their relationships, helps biologists organize and understand biological diversity. Taxonomy identifies similar traits to trace diverse organisms back to common ancestors, as in a family tree.

We often use the most specific levels of the tree, genus and species, to compose names called **binomials.** Also called scientific or Latin names, binomial names identify and describe species using Latin, or Latinized nouns and adjectives, or names of people or places. The first word in a bionomial indicates the genus of the organism, while the second word designates the species to which it belongs.

Scientists communicate about species using bionomial names instead of common names (such as lion, dandelion, or ant), to avoid confusion. A common name can refer to a number of species in different places, and a single species might have many common names. On the other hand, the binomial *Pinus resinosa* refers to the same type of tree, whether you call it a red pine, Norway pine, or just pine.

Taxonomy defines relationships from broadest to most specific levels. The bionomial *Pinus resinosa* indicates the detailed levels of genus (pine) and the species (the type of pine). Similarly, you are *Homo sapiens* (genus *Homo*, and species *sapiens*, meaning "wise"). You and the pine represent two well-known examples of the six "kingdoms" into which taxonomists organize all living organisms (fig 4.13): animals, plants, fungi (molds and mushrooms), protists (algae, protozoans, slime molds), bacteria (or eubacteria), and archaea (or archaebacteria, ancient, single-celled organisms

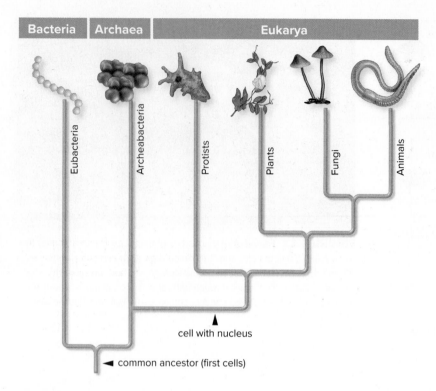

FIGURE 4.13 The six great kingdoms representing all life on earth. The kingdoms are grouped in domains indicating common origins.

FIGURE 4.14 Coevolution has led to close evolutionary relationships between many species, as in this star orchid and the specially adapted hawk moth that pollinates it. When Charles Darwin saw this orchid, he predicted there must be a moth with a 30-cm proboscis to reach the bottom of the nectary. Nearly a century later, the moth was discovered.

that live in harsh environments, such as hot springs). Within these kingdoms are millions of different species, which you will explore more in chapters 5 and 11.

Section Review

1. Is the evolutionary process random, does it occur with an ultimate end in mind, or is neither description an accurate portrayal of evolution?

2. Explain how tolerance limits to environmental factors determine distribution of a highly specialized species such as the saguaro cactus.

3. Describe why a habitat specialist is more vulnerable to extinction than a habitat generalist.

4. Describe how evolution produces species diversity.

4.2 SPECIES INTERACTIONS AND THE EVOLUTIONARY PROCESS

- *All complex organisms have evolutionary relationships with other species.*
- *Predation is an important type of selective pressure.*
- *Symbiosis involves long-term biological interactions that benefit one or both species.*
- *Keystone species serve a fundamentally critical role in an ecosystem.*

Over many millions of years, species evolve to better fit their environment, and part of their environment includes the presence, behavior, and actions of other species. In fact, much of the evolutionary process is shaped by species interactions, and these interactions create the web of life. When two (or more) species affect each other's evolutionary process, it is called **coevolution** (fig. 4.14). Coevolved relationships can involve antagonistic or mutually beneficial cases. Because no species lives in isolation, all complex organisms demonstrate the influence of coevolution. In fact, one of the most fundamental facts of biology is that specialized, coevolved relationships form the base of all natural ecosystems.

Predator-prey dynamics assert selective pressure

Almost everyone is familiar with predator-prey species interactions. To understand its pervasiveness, we need to revisit our discussion about trophic food webs. As we saw in chapter 3, food webs begin with a simple fact: All organisms need food. Producers like plants and algae capture energy to create organic molecules through the process of photosynthesis, while consumers eat organic matter created by other organisms. Consumers include herbivores, carnivores, omnivores, parasites, scavengers, detritivores, and decomposers.

We often think only of carnivores as predators, but in an ecological sense, a predator is a consumer that feeds directly on another living organism, whether or not the predator kills the prey (fig. 4.15). Herbivores, carnivores, and omnivores, which feed on live prey, are predators, but scavengers, detritivores, and decomposers, which feed on dead organisms, are not. In this sense, parasites (organisms that feed on a host organism or steal resources from it without necessarily killing it) and even pathogens (disease-causing organisms) can be considered predator organisms. Herbivory is the type of predation practiced by animals that graze and browse on plants.

Predation has a powerful but complex influence on species populations in communities. It affects (1) all stages in the life cycles of predator and prey species; (2) many specialized food-obtaining mechanisms; and (3) the evolutionary adjustments in behavior and body characteristics that help prey avoid being eaten and help predators more efficiently catch their prey.

Let's first examine predation through the lifecycle of an organism. An organism's predators, and its position in the food web, can change as individuals mature. In marine ecosystems, crustaceans, mollusks, and worms release eggs directly into the water, where they and hatchling larvae join the floating plankton community of small and microscopic organisms. Planktonic animals eat each other and are food for larger carnivores, including fish. As prey species mature, their predators change. Barnacle larvae are planktonic and are eaten by small fish. As adults their hard shells protect them from fish, but not from starfish and predatory snails. Predators often switch prey over the course of their lives. Carnivorous adult frogs usually begin their lives as herbivorous tadpoles. Predators also switch prey when their original prey becomes rare or something else becomes abundant. Many predators have morphologies and behaviors that make them highly adaptable to a changing prey base, but some, like the polar bear (which eats seals), are highly specialized in their prey preferences.

Predator–prey relationships exert selection pressures that favor evolutionary adaptation. This occurs when predators become more efficient at searching and feeding on prey, and the prey become more effective at escape and avoidance. An "evolutionary arms race" can occur between predator and prey as each party enhances its predatory or defensive traits in response to the other. Toxic chemicals, body armor, extraordinary speed, and the ability to hide are a few strategies organisms use to protect themselves. Plants have thick bark, spines, thorns, or distasteful and even harmful chemicals in tissues—poison ivy and stinging nettle are examples. Arthropods, amphibians, snakes, and some mammals produce noxious odors or poisonous secretions that cause other species to leave them alone.

A classic example of an evolutionary arms race is the relationship between the rough-skinned newt (*Taricha granulosa*) and its predator, the common garter snake (*Thamnophis sirtalis*, fig. 4.16). Biologists in Oregon discovered rough-skinned newts carrying a neurotoxin called tetrodotoxin, TTX, that was two to three times stronger than the venom of the world's deadliest

FIGURE 4.15 Insect herbivores are predators as much as are lions and tigers. In fact, insects consume the vast majority of biomass in the world. Complex patterns of predation and defense have often evolved between insect predators and their plant prey.
Brad Boland/Flickr Flash/Getty Images

snake (the Inland Taipan, *Oxyuranus microlepidotus*). Some rough-skinned newts were so poisonous that the amount of toxin in a single individual could kill 50,000 mice or over 100 people. Why were the newts producing such extreme levels of TTX? The researchers hypothesized that a newt-eating predator must be resistant to this amount of TTX. They found that the common garter snake could eat the newts with relative ease. Note that the evolutionary process is evident in the newt–snake relationship: Because of random genetic variation, some newts produced more toxin than others, and the more toxic newts escaped predation, lived longer, and produced more offspring through the process

FIGURE 4.16 Some garter snakes have a trait for tolerating poison in their prey. This advantage also has trade-offs in other traits.
PhotoLink/Getty Images

of natural selection. Over time, the newt offspring with higher levels of TTX became more prevalent in the population. At the same time, garter snakes that tolerated TTX venom survived and reproduced better than others.

To show that the snakes and the newts coevolved in response to each other, scientists mapped different populations of newts and snakes, revealing that newt toxicity and snake resistance "matched up" by location. In areas where newts were extremely toxic, the snakes were extremely resistant, but in areas where the newts had low toxicity, the snakes had low resistance.

Why wouldn't all of the snakes in the Pacific Northwest be resistant to TTX, no matter the toxicity of their prey? Because toxin resistance comes at a price for snakes. As the snake's toxin resistance increases, their muscle performance declines, slowing their crawling speed and making them more vulnerable to their own predators. Likewise, it takes a lot of energy for a newt to produce TTX, and that's energy it could spend on finding food or reproducing. For many organisms, optimizing a single trait means downgrading another trait; these compromises are called evolutionary trade-offs.

Many species are able to capitalize on the toxic defenses of other species because species with chemical defenses often display distinct coloration and patterns to warn away enemies. In a neat evolutionary twist, certain species that are harmless resemble poisonous or distasteful ones, gaining protection against predators who remember a bad experience with the actual toxic organism. This is called **Batesian mimicry,** after the English naturalist H. W. Bates (1825–1892). Many wasps, for example, have bold patterns of black and yellow stripes to warn potential predators (fig. 4.17a). The much rarer longhorn beetle has no stinger but looks and acts much like a wasp, tricking predators into avoiding it (fig. 4.17b). The distasteful monarch and benign viceroy butterflies are a classic case of Batesian mimicry. Another form of mimicry, **Müllerian mimicry** (after the biologist Fritz Müller), involves *two unpalatable* or dangerous species that look alike. When predators learn to avoid either species, both benefit.

Competition occurs between and within species

Competition, an adversarial relationship that occurs when species or organisms compete for a resource that is in limited supply, is another type of ecological interaction. Competition includes both **interspecific competition,** which occurs between different species, and **intraspecific competition,** which occurs within a single species. When members of the same species compete directly with each other for a resource, it is called intraspecific competition (fig. 4.18a). Intraspecific competition tends to be a more powerful coevolutionary force than interspecific competition. Recall the competitive exclusion principle as it applies to intraspecific competition. Competition shapes a species' population and biological community by causing individuals and species to shift their focus from one segment of a resource type to another. Thus, warblers competing with each other for insect food in New England have been observed to specialize on different areas of the forest's trees, reducing or avoiding competition (see fig. 4.9).

Several avenues exist to reduce intraspecific conflict within a species' population. First, offspring tend to disperse geographically. Even plants practice dispersal: Seeds are carried by wind, water, and passing animals to less crowded conditions away from the parent plants. Second, by exhibiting strong territoriality, many animals force their own offspring or trespassing adults out of their vicinity. In this way, territorial species (such as bears, songbirds, ungulates, and even fish) minimize competition for scarce resources. A third way to reduce intraspecific competition is resource partitioning between generations. This happens when adults and juveniles of these species occupy different ecological

(a)

(b)

FIGURE 4.17 An example of Batesian mimicry. The dangerous wasp (a) has bold yellow and black bands to warn predators away. The longhorn beetle (b) has no poisonous stinger, but it has a similar color and pattern that helps it avoid predators as well.
Martin Fowler/Alamy; Ger Bosma/Moment Open/Getty Images

niches. For instance, monarch caterpillars munch on milkweed leaves, while adult butterflies sip nectar.

Interspecific competition involves different species that use the same resource. In open savannah environments, for example, many competitors depend on carcasses of large animals, such as wildebeest (fig. 4.18b), and there are often intense battles over kill sites. Spotted hyenas, vultures, and other scavengers compete, sometimes violently, over this resource.

Often, the superior competitors eliminate other species from the habitat, but predation can influence, or even out, competition between species. Predator-meditated competition is a general term for this. If one species is an especially good competitor, for example, its population may expand, making it an easier target for predators. Increased predation can then reduce its abundance, allowing other species to increase in number. In a classic test of this idea, biologists removed the ochre starfish (Pisaster ochraceus) from Pacific tidal zones. Its main prey, the common mussel (Mytilus californicus), exploded in numbers and crowded out other intertidal species.

Symbiosis involves long-term interaction between species

Many relationships in nature are neither antagonistic nor competitive; instead, they are cooperative, and sometimes even beneficial. Long-term biological interactions, called **symbiosis,** exist when two or more species live together intimately, with their fates linked. Symbiotic relationships enhance the survival odds of one or both partners. Symbiotic interactions include *mutualistic, commensalistic,* and *parasitic* relationships.

Some ecologists assert that cooperative evolutionary relationships deserve more attention. Biologist Lynn Margulis, who has shown the widespread importance of symbiotic relationships, has noted that "Life did not take over the globe by combat, but by networking." Recent research supports this idea, asserting that symbiotic cooperation, not competition, is what drives most of the expansion of living diversity.

Mutualism is a term for relationships in which both species benefit from the interaction. Pollination is a classic example of mutualism. The pollinator gets a food source like nectar or pollen, and the plant gets help spreading its reproductive genes. Another mutualistic relationship is the evolutionary cooperation between Goby fish and pistol shrimp. When they are both juveniles, the Goby and the shrimp develop a lifelong partnership. The shrimp builds and maintains a burrow that is used by both animals. When they are outside of the tunnel, the Goby fish sends a signal to the shrimp, which is mostly blind, that a predator is nearby and the shrimp darts into the tunnel. The Goby contributes its superior eyesight; in return, the shrimp provide a safe home for the Goby.

Sometimes species develop mutualistic relationships for defense. We can see this in the relationship between aphids and several species of ants. The aphids secrete honeydew when they feed on plant sap, and the honeydew serves as a source of food for the ants. In return, the ants take the aphids into their nests at night to protect them from predators, and then move the aphids back outside in the morning.

Sometimes, a mutualistic relationship can evolve to the point where a pair of interacting species literally cannot survive without each other. In the depths of the ocean, giant tube worms live on the

FIGURE 4.18 (a) Intraspecific competition occurs among individuals of a species, such as male elephant seals fighting over a harem of female seals (b) Interspecific (between species) competition can be seen as spotted hyenas battle white-backed vultures for a wildebeest carcass.
eoind/Getty Images; Peter Miller/iStock/Getty Images

seafloor next to hydrothermal vents that spew superheated (250° to 400°C/ 482° to 752°F) water and compounds like hydrogen sulfide. Tiny bacteria enter the tube worms' skin and establish a safe refuge inside the worms, where they oxidize hydrogen sulfide to create energy for their host (fig. 4.19).

Commensalism is a type of symbiosis that clearly benefits one species while the other neither benefits nor is harmed. The species that benefits from the relationship may gain shelter, nutrition, or transportation from its interaction. Many mosses, bromeliads, and other plants growing on trees in the moist tropics are considered commensals. These epiphytes (plants that grow on other plants) obtain watered from rain and nutrients from leaf litter and falling dust. Generally they neither help nor hurt the trees on which they grow (unless they become so heavy they break limbs).

There are numerous examples of commensalism in marine environments (fig. 4.20). For example, Baleen whales often carry hitchhiking barnacles. The whale does not benefit, nor does it suffer any ill effects from having the barnacle attached, but the

FIGURE 4.20 Barnacles attached to a humpback whale are an example of commensalism, which benefits the barnacle while leaving the whale seemingly unaffected.
Source: Ken Balcomb, National Marine Sanctuaries Media Library, NOAA.

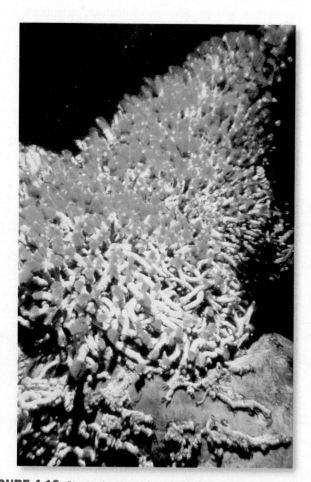

FIGURE 4.19 Giant tube worms at the base of a deep sea hydrothermal vent. Remarkably, the tube worms have no mouth, digestive tract, or anus. Instead, a symbiotic bacteria lives inside the worm's body where it oxidizes hydrogen sulfide to create energy. The bacteria get a safe place to live, while the tube worms get a steady supply of organic carbon.
Source: OAR/National Undersea Research Program (NURP)/NOAA.

barnacle benefits by increasing access to food sources and habitat. When the whale swims into schools of plankton to feed, the barnacle is also able to feed on the plankton.

Parasitism can be considered a form of predation, in which the parasite benefits, while the host is harmed. Parasitism is a common life strategy; some ecologists have even argued that parasitism is the most widespread life-history strategy in nature, and that food webs often contain more connections between parasites and their hosts than between predators and prey. In fact, the more intact and healthy an ecosystem, the more parasites are present there.

Since parasites are typically small and inconspicuous, it can be easy to assume that they play an unassuming role in community ecology. But parasites are ecologically important, and they can alter ecosystems. For example, as parasites weaken their hosts, they aid predators and keep host populations in check. Parasites can contribute a considerable amount of biomass to an ecosystem: A study of California estuaries revealed that the yearly productivity of trematode parasites was higher than the biomass of birds. In a similar case, the biomass of plant parasites at the Cedar Creek Ecosystem Science Reserve in Minnesota was equal to the biomass of herbivores in the study area.

Ecologists now hypothesize that parasites may stabilize food webs by increasing connections between elements. The estuary study in California found that parasites were present in 78 percent of all links in the food web. "Parasites may well be the thread that holds the structure of ecological communities together," says Andrew Dobson of Princeton University.

Parasites may play a crucial role in our ecosystems, but often they inflict considerable suffering. Parasitic flatworms (fig. 4.21), for example, cause schistosomiasis, a devastating disease in some tropical regions. The parasitic plasmodium that causes malaria is chronically debilitating and often deadly for its human hosts.

Some parasites even eat the entirety of their host. Many parasitic wasps, for example, lay their eggs on or inside caterpillars. As soon as the eggs hatch, the wasp larvae begin eating their host, but

Say Hello to Your 90 Trillion Little Friends

Have you ever thought of yourself as a biological community or an ecosystem? Researchers estimate that each of us has about 90 trillion bacteria, fungi, protozoans, and other organisms living in or on our bodies. And the viruses inside those commensal species increase our biodiversity by another order of magnitude. The largest group—around 2 kg worth—inhabit your gut, but there are thousands of species living in every orifice, gland, pore, and crevice of your anatomy. Although the 10 trillion or so mammalian cells make up more than 95 percent of the volume of your body, they represent less than 10 percent of all the cell types that occupy that space.

Because most of the other species with which we coexist are microorganisms, we call the collection of cells that inhabit us our microbiome. The species composition of your own microbial community will be very similar to that of other people and pets with whom you live, but each of us has a unique collection of species that may be as distinctive as our fingerprints.

As is the case in other species interactions, these relationships can be mutalistic, symbiotic, commensal, or predatory. We used to think of all microorganisms as germs to be eliminated as quickly and thoroughly as possible. Current research suggests, however, that many of our fellow travelers are beneficial, perhaps even indispensable, to our good health and survival.

Your microbiome is essential, for example, in the digestion and absorption of nutrients. Symbiotic bacteria in your gut supply essential nutrients (important amino acids and short-chain fatty acids), vitamins (such as K and some B varieties), hormones and neurotransmitters (such as serotonin), and a host of other signaling molecules that

FIGURE 1 *Lactobacillus* bacteria are part of the normal flora of human intestine and are often used as probiotic supplements. These intestinal bacteria help crowd out pathogens, aid in digestion, and supply your body with essential nutrients.

Kateryna Kon/Shutterstock

communicate with, and modulate, your immune and metabolic systems. They help exclude pathogens by competing with them for living space, or by creating an environment in which harmful species can't grow or prosper.

The inhabitants of different organs can have important roles in specific diseases. Oral bacteria, for example, have been implicated in cardiovascular disease, pancreatic cancer, rheumatoid arthritis, and preterm birth, among other things. Symbionts in the lung have been linked to cystic fibrosis and chronic obstructive pulmonary disease (COPD). And the gut community seems to play a role in obesity, diabetes, colitis, susceptibility to infections, allergies, and other chronic problems. A healthy biome seems to be critical in controlling chronic inflammation that triggers many important long-term diseases.

As is true in many ecosystems, the diversity of your microbiome may play an important role in its stability and resilience. Having

a community rich in good microbes will not only help you resist infection by pathogens but will allow faster recovery after a catastrophic event. People in primitive or rustic societies who eat a wide variety of whole grains, raw fruits and vegetables, and unprocessed meat or dairy products tend to have a much greater species variety than those of us who have a diet full of simple sugars and highly processed foods. Widespread use of antibiotics to treat illnesses, as well as chronic low levels of antimicrobials, preservatives, and stabilizers in our food, toothpaste, soap, and many other consumer products also limits diversity in our symbiotic community.

A growing problem in many places is antibiotic-resistant, hospital-acquired infections. One of the most intractable of these is *Clostridium difficile,* or *C. diff.,* which infects 250,000 and kills 14,000 Americans every year. An effective treatment for this superpathogen is a fecal transplant. A sample of the microbiome from a healthy person is implanted either directly through a feeding tube into the patient's stomach or in frozen, encapsulated pellets of feces that are delivered orally. In one trial, 18 of 20 patients who received fecal transplants recovered from *C. diff.*

Similarly, obese mice given fecal transplants from lean mice lose weight, while lean mice that receive samples of gut bacteria from obese mice gain weight. The microbiome may even regulate mood and behavior. When microbes from easygoing, adventurous mice are transplanted into the gut of anxious, timid mice, they become bolder and more adventurous.

So, it may pay to take care of your garden of microbes. If you keep them happy, they may help keep you happy as well.

(header photo): David L. Hansen

they do so strategically. The larvae focus on the less fundamental components of the caterpillar, keeping its essential organs intact. Eventually, only the vital organs are left, at which point the larvae eat the vital organs and emerge out of the caterpillar's now-dead body.

You have many examples of mutualism, commensalism, and parasitism in and on your own body. Thousands of species live in and on your body. We call this highly diverse community your microbiome. You couldn't survive without it (see Exploring Science, above).

FIGURE 4.21 Many people are intimately familiar with parasites. Parasitic flatworms (*Schistosoma*) are responsible for infecting more than 250 million people worldwide. The World Health Organization reports that Schistosomiasis (also known as snail fever and bilharzia) is the second most socioeconomically devastating parasitic disease after malaria.
Source: Bruce Wetzel and Harry Schaefer/National Cancer Institute (NCI).

Keystone species have disproportionate influence

A **keystone species** plays a critical role in a biological community that is out of proportion to its abundance. Ecologists long considered keystone species to be mainly top predators—lions, wolves, tigers—that limited herbivore abundance and thus reduced the herbivorous depletion of plants in an ecosystem. Scientists now recognize that less-conspicuous species also play keystone roles. Some tropical figs, for example, bear fruit year-round at a low but steady rate. If figs are removed from a forest, many fruit-eating animals (frugivores) would starve in the dry season when other fruits are scarce. In turn, the disappearance of frugivores would affect plants that depend on them for pollination and seed dispersal. In this way, the effect of a keystone species on communities often ripples across trophic levels.

Keystone functions have been documented for vegetation-clearing elephants, the predatory ochre sea star, and frog-eating salamanders in coastal North Carolina. Even microorganisms can play keystone roles: In many temperate forest ecosystems, groups of fungi that are associated with tree roots (mycorrhizae) facilitate the uptake of essential minerals much like the microbiome in mammals. When fungi are absent, trees grow poorly or not at all. Overall, keystone species seem to be more common in aquatic habitats than in terrestrial ones, but they always play an important role in food webs.

A concept related to keystone species is that of **ecosystem engineers.** These organisms affect the availability of resources by causing physical changes in biotic or abiotic materials. Ecosystem engineers are able to modify, maintain, and/or create *habitats* either by their own physical bodies (living or dead) or by moving living or nonliving materials from one area to another. Engineering effects have been described for many species, including beavers, earthworms, ants, and whales (fig 4.22). Earthworms, for example, alter the physical, chemical, and biological properties of soil by creating horizontal and vertical burrows in the soil structure. This activity disperses nutrients, making them available to plants, bacteria, and fungi. It can also accelerate decomposition and loss of organic matter from the

FIGURE 4.22 The lowly earthworm (*Lumbricus* sp.) and the majestic humpback whale (*Megaptera novaeangliae*) both act as ecosystem engineers in their biological communities by creating, maintaining, and modifying habitats.
Hemera/Getty Images; Chase Dekker Wild-Life Images/Getty Images

soil. Whales distribute resources by depositing large amounts of nutrients vertically through the water column and horizontally across large expanses of ocean habitat with their feces.

The concepts of ecosystem engineers and keystone species partially overlap. An ecosystem engineer can also be considered a keystone species, because while a keystone species has disproportionate importance in its ecosystem, that importance sometimes has to do with its modification of the environment.

Section Review

1. Explain how predators affect the adaptations of their prey.
2. What concept explains that the most intense species interactions occur between individuals of the same species?
3. List and explain three types of symbiotic interactions and give an example of each.
4. Give an example of a keystone species.

4.3 COMMUNITY PROPERTIES AFFECT SPECIES, POPULATIONS, AND PRODUCTIVITY

- *Community structure and function depend on species diversity, abundance, and distribution.*
- *Complexity and connectedness describe the extent of species interrelationships.*
- *Biological communities provide ecosystem services.*

So far, we have examined how species evolve and change over time, and how species interact through predation, competition, or symbiosis. Next we will consider the larger picture, the dynamic nature of biological communities and how they contribute to humans through ecosystem services. Ecosystems are woven together through a complex web of interactions, and when we understand the structure and function of these webs, we can learn how to support healthy ecosystem functioning both now and in the future.

Community dynamics involves diversity, abundance, and distribution of species

To understand how a biological system works, we will need to understand a few basics about how the community fits together. The first step is to record the number of species that live within a community. This is species **diversity.** In general, we expect a more diverse community to have a more complex web of relationships among organisms and more specialized and unusual species.

But this count only tells us about one dimension of the biological community. We would also need to know about the **abundance** of different species. Species abundance refers to the number of individuals of a species; *relative abundance* expresses how common or rare a species is relative to other species in a given location. Some communities have a great variety of organisms but only a few members of any given species; here relative abundance is fairly even among species. Tropical rainforest communities, for example, might have dozens of tree species but only a few of each. Other communities have few species that occur in great abundance. A boreal spruce forest, on the other hand, might have thousands of trees, but just two or three tree species.

A third component of community structure is the spatial distribution of species and populations across the landscape. Individuals in a population can be distributed randomly with no predictable pattern, uniformly in a regular pattern, or clustered together. In a randomly arranged population, individuals live wherever resources are available and the environment is favorable (fig. 4.23a). Ordered patterns often result from biological competition. In seabird colonies, for example, competition for nesting space is often fierce. Each nest tends to be just out of reach of the neighbors sitting on their own nests. Constant squabbling produces a highly regular pattern (fig. 4.23b).

Some other species cluster together for protection, mutual assistance, reproduction, or access to a resource. Dense schools of fish, for instance, cluster together in the ocean, increasing their chances of detecting and escaping predators (fig. 4.23c). Similarly, predators, whether sharks, wolves, or humans, often hunt in packs

(a) Random

(b) Uniform

(c) Clustered

FIGURE 4.23 Distribution of members of a population in a given space can be (a) random, (b) uniform, or (c) clustered. The physical environment and biological interactions determine these patterns. The patterns may produce a graininess or patchiness in community structure.
Jim Zuckerman/Getty Images; Eric and David Hosking/Getty Images; image100/PunchStock

Quantitative Reasoning

In figure 4.21, how much *primary productivity* occurs in a desert? How much more is there in a tropical rainforest? Why are the open oceans called the deserts of the sea?

to catch their prey. Birds often flock together, both to avoid predators and to find food more efficiently.

Complexity and connectedness are important ecological indicators

Community complexity and connectedness are related to species diversity, abundance, and distribution. In ecological terms, **complexity** refers to the variety of biological interactions, including the links or connections between species and trophic levels. A diverse community may not be very complex if all its species are clustered in only a few trophic levels and form a relatively simple food chain.

A complex, highly interconnected community like a tropical forest might have many trophic levels, some of which can be compartmentalized into subdivisions. In tropical rainforests, the herbivores can be described in terms of the specialized ways they feed on plants. There may be fruit eaters, leaf nibblers, root borers, seed gnawers, and sap suckers. Each of these groups, or "guilds," includes diverse species, of different sizes, shapes, and even biological kingdoms, that feed in similar ways. In a seagrass meadow ecosystem, we would likely find guilds of suspension feeders such as bay scallops, clams, mussels, and amphipods; epiphyte-grazing guilds including shrimp, amphipods, and spider crabs; and deposit-feeding detritivore guilds containing ostracods, gastropods, and polychaetas.

Biological systems also have vertical structure. A tropical forest may include an emergent layer, canopy, understory, shrub layer, herb layer, and ground layer, each with different environmental conditions and combinations of species. Distinct communities of smaller plants, animals, and microbes live at different levels (fig 4.24). Many aquatic communities are stratified into layers based on light penetration in the water, temperature, salinity, pressure, or other factors.

A tropical forest usually has high species diversity. Because so many species are present, various species may play similar roles in ecosystem functions like pollination, pest control, decomposition, soil formation, seed dispersal, and the maintenance of water and nutrient cycles. For instance, seed dispersal in Ugandan forests is performed by different-sized mammals, from small red-tailed monkeys *(Cercopithecus Ascanius)* and cane rats *(Thryonomys sp.)* to larger species like Mangabey monkeys *(Cercocebus albigena),* and chimpanzees *(Pan troglodytes).* When several species in a system perform a function in a similar way, ecologists call it **redundancy.** In this case, mammals of different sizes and mobility levels provide seed dispersal services for fleshy-fruited plant species in the forest.

FIGURE 4.24 Tropical rainforests are complex structurally and ecologically. Trees form layers, each with a different amount of light and a unique combination of flora and fauna. Many insects, arthropods, birds, and mammals spend their entire life in the canopy. In Brazil's Atlantic Rainforest, a single hectare had 450 tree species and many times that many insects. With so many species, the ecological relationships are complex and highly interconnected.

Redundancy is often useful repetition, providing "backup" if a particular species undergoes a sharp decline as a result of disease, overharvest, or a changing climate. In our forest example, the loss of a particular mammal may not have a large effect on ecosystem functioning because other species that perform the seed-dispersal service would step in and compensate for the loss.

While the relationship between redundancy and ecosystem stability is not simple, the importance of redundancy is particularly clear once it is lost. Consider the example of a coral reef system where algae is controlled by plant-eating fish and urchins. If the plant-eating fish are removed through overfishing, the algae-grazing function is now dependent the urchin. If disease strikes the

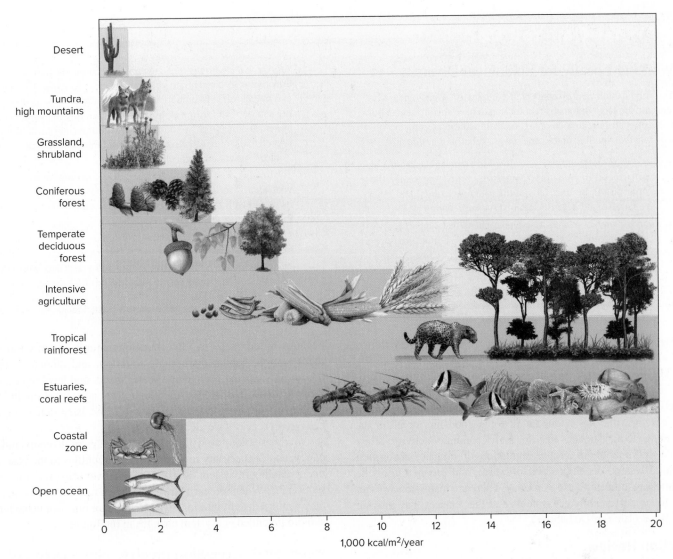

FIGURE 4.25 Relative biomass accumulation of major world ecosystems. Only plants and some bacteria capture solar energy. Animals consume biomass to build their own bodies.

urchin, algae may overtake the system because nothing is left to keep the algae in check.

Biological communities vary in productivity

One of the ways biological communities differ is in their rates of energy capture and biomass accumulation. A community's **primary productivity** is the rate of biomass production through photosynthesis. Primary productivity can be reported either as in mass (grams) or energy (calories) per unit area, because photosynthesis captures both energy and mass. Living organisms also release energy and biomass (carbon) from the system through respiration. The net accumulation in a community—primary productivity minus respiration—is known as net primary productivity.

The rate of photosynthesis depends on temperature and moisture, as well as nutrient availability and light levels, so primary productivity varies dramatically among environments (fig 4.25).

Tropical forests, coral reefs, and estuaries (where rivers meet the ocean), for example, have high biological productivity, capturing in biomass around 20,000 kcal of energy per m^2 per year. These environments have abundant light, moisture, nutrients, and warm temperatures. In deserts, a lack of water limits photosynthesis, and biological productivity may be less than 1/10th that of a tropical forest. On the arctic tundra or in high mountains, low temperatures inhibit plant growth. In the open ocean, a lack of nutrients reduces the ability of algae to make use of plentiful sunshine and water. Productivity is related to abundance and diversity, both of which are dependent on the total resource availability in an ecosystem, as well as the reliability of resources, the adaptations of the member species, and the interactions between species.

Some agricultural crops such as corn (maize) and sugar cane grown under ideal conditions in the tropics approach the productivity levels of tropical forests. Shallow water ecosystems such as coral reefs, seagrass beds, and mangroves are highly productive. The vast extent of open oceans, on the other hand, often are called

What Can You Do?

Working Locally for Ecological Diversity

You might think that diversity and complexity of ecological systems are too large or too abstract for you to have any influence. But you can contribute to a complex, resilient, and interesting ecosystem, whether you live in the inner city, a suburb, or a rural area.

- *Keep your cat indoors.* Our lovable domestic cats are also very successful predators. Migratory birds, especially those nesting on the ground, as well as rodents, lizards, and others, have not evolved defenses against these predators.

- *Plant a butterfly garden.* Use native plants that support a diverse insect population. Native trees with berries or fruit also support birds. (Be sure to avoid non-native invasive species; see chapter 11.) Allow structural diversity (open areas, shrubs, and trees) to support a range of species.

- *Join a local environmental organization.* Often, the best way to be effective is to concentrate your efforts close to home. City parks and neighborhoods support ecological communities, as do farming and rural areas. Join an organization working to maintain ecosystem health; start by looking for environmental clubs at your school, park organizations, a local Audubon chapter, or a local Nature Conservancy branch.

- *Take walks.* The best way to learn about ecological systems in your area is to take walks and practice observing your environment. Go with friends and try to identify some of the species and trophic relationships in your area.

- *Live in town.* Suburban sprawl consumes wildlife habitat and reduces ecosystem complexity by removing many specialized plants and animals. Replacing forests and grasslands with lawns and streets is the surest way to simplify, or eliminate, ecosystems.

biological deserts, with low productivity resulting largely from a lack of nutrients.

Highly productive and diverse communities also tend to provide abundant **ecosystem services,** a general term for the diverse benefits we receive from natural systems (chapter 1). Ecosystem services form the basic foundation for a good life for all of us, including food, shelter, health, and economic activity. They include pollination of crops, regulation of climate, absorption of flood waters, decomposition of waste, and production of the countless resources we use every day. There are many steps we can take to support healthy ecosystems and the services they provide (see What Can You Do? above).

Section Review

1. If you were to study a biological community, list and describe three characteristics you would first measure.
2. Why would a system benefit from having multiple species that basically "do the same thing"?
3. Which ecosystems have the greatest biological productivity?
4. How do ecosystems benefit people? What types of services do they perform?

4.4 SYSTEM CHANGE AND RESILIENCE

- *The adaptive cycle explains disturbance and change in a complex system.*

- *Large, abrupt, nonlinear change results in a fundamental restructuring of the biological community.*

- *The risk of ecological systems tipping into undesirable states is significant for people and their well-being.*

How much can you change a system and have it still remain recognizable or deliver something you have come to rely on? This is a question of *resilience*. It's of the utmost importance right now because humans are rapidly changing natural systems through deforestation, habitat disruption, pollution, and climate change. We are introducing change at a scale, speed, and level of connectivity that was unthinkable a hundred years ago. What might happen? Can natural systems remain stable and keep delivering the ecosystem services we all depend on?

In this section, we will use our knowledge of diversity, redundancy, and connectivity in biological communities to increase our understanding of system dynamics, which were first introduced in section 2.2. When we learn how ecosystems respond to change, we may acquire insight into how we might adapt our own behavior to avoid unintended ecosystem changes in the future.

Ecological succession involves changes in community composition

Ecologists have long debated the basic nature of communities. Many of the early disputes still influence how we understand communities, and how and why they change over time. Two ecologists, J. E. B. Warming (1841–1924) in Denmark and Henry Chandler Cowles (1869–1939) in the United States, established the idea that biological communities develop in a sequence of stages, starting either from bare rock or after a severe disturbance. They worked in sand dunes and watched the changes as plants first took root in bare sand and, with further development, created forest. The community that developed last and lasted the longest was called the **climax community.**

The idea of a climax community was championed by the biogeographer F. E. Clements (1874–1945). He made the analogy that the development of a climax community resembled the maturation of an organism. Like organisms, Clements argued, communities begin simply, and then mature through predictable stages until reaching the highest stage of development allowed by genetics or environmental conditions. Clements and others developed the concept of ecological succession to describe the transition from one community composition to another. Once you understand likely transitions, you can see the history of an area by observing what is growing there. For example, in humid environments, abandoned farm fields are gradually overtaken by shrubs, then by sun-loving trees, then by mature forest canopy trees.

This organism-like theory of community growth and maturation was disputed by H. A. Gleason (1882–1975), who saw community composition as less predictable. He argued that species are individualistic, each establishing in an environment according to its ability to colonize, tolerate the environmental conditions, and reproduce in a place. This idea allows for myriad associations of plants and animals to emerge in slightly different forms, depending on environmental conditions and the species in the neighborhood. Imagine a time-lapse movie of a busy airport terminal. Passengers come and go; groups form and dissipate. Patterns and assemblages that seem significant may not mean much at a later time. Where Clements envisioned a predictable climax community composition, Gleason's individualistic concept left open the possibility of different, unpredictable compositions of species.

Both Clements and Gleason agreed, though, that communities change through succession. During succession, different types of organisms occupy a site and change the environmental conditions. In **primary succession,** land that is bare of soil—a sandbar, mudslide, rock face, volcanic flow—becomes inhabited by living organisms where none lived before (fig 4.26). When an existing community is disturbed, a new one develops from the biological legacy of the old in a process called **secondary succession.** In these stages of succession, organisms change the environment by modifying soil, light levels, food supplies, and microclimate. This change permits new species to colonize and eventually replace the previous species, a process known as ecological development or facilitation.

In primary succession on land, the first colonists are hardy **pioneer species** carried by the wind. Often these are microbes, mosses, algae, fungi, and lichens that can withstand a harsh environment with few resources. These pioneers capture moisture and organic material, which then allows the seeds of slightly larger plants to germinate and grow. These first plants might be short-lived annuals, surviving just a year or two. Their light seeds travel far on the wind, and their seedlings tolerate full sun and extreme heat. These plants capture, or deposit, still more organic material that improves the soil's fertility and shelters other seedlings. Over time, long-lived and deep-rooted perennial grasses, herbs, shrubs, and trees take hold, building up the soil's organic matter and increasing its ability to store moisture.

Generalists figure prominently in early succession. Over time, more diverse niches develop, more long-lived species establish, and a stable, persistent community is likely to develop. In this climax community, ecologists generally expect greater community complexity, high nutrient conservation and recycling, and stable productivity.

Biological communities may be adapted to disturbance

Disturbances occur in most environments. Landslides, mudslides, hailstorms, earthquakes, hurricanes, tornadoes, tidal waves, wildfires, and volcanoes all disrupt ecosystems. Animals can cause disturbance, as when African elephants rip out small trees, trample shrubs, and tear down tree limbs as they forage. People also cause disturbances with agriculture, forestry, and construction. A **disturbance** is any force that disrupts a community or ecosystem, altering its patterns of species diversity and abundance, community structure, or community properties.

Ecologists often find that disturbance can benefit diversity by setting back the strongest competitors and allowing less-competitive species to persist. In northern temperate forests, maples (especially the sugar maple) compete especially well because they produce extraordinary numbers of seeds and shade-tolerant seedlings. After decades of stable growth, maples can outcompete other trees and forest understory plants, starving them for light. Oak, hickory, and other species that require more sunlight become uncommon. Occasionally, though, storms or old age might cause a large maple to fall. In the gap in the canopy, hickory, and other light-requiring trees diminish in abundance. Light again reaches the forest floor

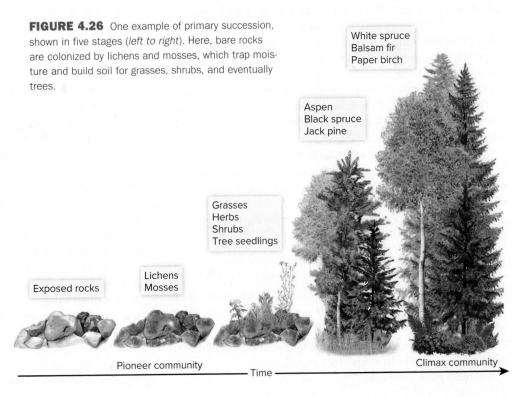

FIGURE 4.26 One example of primary succession, shown in five stages (*left to right*). Here, bare rocks are colonized by lichens and mosses, which trap moisture and build soil for grasses, shrubs, and eventually trees.

White spruce
Balsam fir
Paper birch

Aspen
Black spruce
Jack pine

Grasses
Herbs
Shrubs
Tree seedlings

Lichens
Mosses

Exposed rocks

Pioneer community

Time

Climax community

and stimulates seedlings of oaks and hickories and understory plants. Breaking the grip of a supercompetitor is the helpful role disturbances often play.

Many species tolerate or even require occasional disturbances; these are **disturbance-adapted species.** For instance, grasslands, savannas, the chaparral scrubland of California and the Mediterranean region, and some kinds of coniferous forests are shaped and maintained by periodic fires that have long been a part of their history. A variety of traits help fire-adapted species survive, such as deep roots, thick and fire-resistant bark, or seeds that sprout quickly in the full sunshine after a fire. Frequently, the dominant plant species in these communities depend on fire to suppress competitors, to prepare the ground for seeds to germinate, or to pop open protective resin-sealed pine cones or split thick seed coats and release seeds. Without fire, the ecological community structure would be quite different.

Ecologists often distinguish between natural disturbances and human-caused (or anthropogenic) disturbances. Natural disturbances tend to be considered a normal part of an ecosystem, while anthropogenic disturbances, especially mechanized activities such as clear-cut logging or industrial scale farming, are often considered new and disruptive. But smaller-scale anthropogenic disturbances have probably been part of many ecosystems for thousands of years. Indigenous cultures have often introduced disturbances, such as fire to improve grazing lands or open hunting grounds, or clearing forest patches for agriculture. In general, impacts are considered modest when populations are small, and when disturbances are patchy and limited in scale. Frequently, these disturbances are restricted to quickly passing wildfires in grasslands, savannas, or woodlands, which are composed of species already adapted to fire.

However, historical disturbances have sometimes had profound, long-lasting effects. When aboriginal people reached Australia, 50,000 years ago, it's thought they set fires and eliminated plant and animal species to produce the dry, desert landscape we see there now. More modern disturbances have also caused rapid and fundamental ecosystem shifts. In the Kingston Plains of Michigan's Upper Peninsula, clear-cut logging followed by repeated human-set fires from 1880 to 1900 caused a change in basic ecological conditions such that the white pine forest has never regenerated.

Although disturbance is normal in many ecosystems, the disturbances created by the production and development of modern human life can be more challenging for ecosystems than in the more distant past.

A number of factors make current changes more extreme and persistent:

- *Change in rate and scale of disturbances.* Natural disasters or disturbances are happening with increasing frequency or intensity.
- *Habitat fragmentation.* When human activities fragment habitats into smaller patches, subpopulations become isolated and systems become disconnected.
- *Habitat degradation.* Pollution or simplification of systems can weaken the productive capacity or resilience of systems, making them more vulnerable.

- *Invasive species.* The spread of invasive species may destabilize a system, for example by eliminating a keystone species or an important predator or prey species (see chapter 13).
- *Multiple stressors.* Multiple human-induced disturbances may occur simultaneously. For instance, a lake ecosystem may be affected simultaneously by sewage, industrial discharge, surface sediment runoff, airborne pollutants, introduced species, and warmer water temperatures due to a changing climate.
- *Climate change.* Stronger storms, longer droughts, or other aspects of climate change often amplify other disturbances (see chapter 15).

The adaptive cycle explains a system's response to disturbance

In 1973, the Canadian ecologist C.S. "Buzz" Holling introduced a model that gave ecologists a more complete picture of change and disturbance. Through his observations of natural systems, Holling confirmed that the early species of primary succession and the sustaining species of secondary succession were indeed important, but he didn't think they told the whole story.

At the time, natural and social scientists across disciplines were beginning to distinguish biological and social systems from other types of systems. Specifically, they defined Complex Adaptive Systems (CAS) as networks of elements connected to each other through feedback loops (see chapter 2). CAS learn from experience and "adapt" to changes in the environment through their feedback cycles. Human beings, immune systems, ecosystems, and social institutions are all examples of CAS. Because each CAS is unique, it is difficult to predict exactly how it will respond to change or disturbance. In contrast, some systems may be complicated (composed of many parts) but not adaptive. These systems are predictable: We can use formulas to design machines to perform a certain way, and there is a high degree of certainty that each machine will deliver the same outcome.

Holling proposed that most complex systems move through cycles of change according to a cyclical pattern, the **adaptive cycle** (fig. 4.27). The iconic representation of the adaptive cycle is a "lazy 8" with four phases. The progression typically begins with rapid growth (primary succession), followed by a conservation phase (secondary succession). Ecologists have long recognized these phases. Holling's innovation was the idea that a release phase, brought about by a disturbance, leads to a reorganization phase, in which the system recovers. This framework is useful because it helps us describe system transformations in human communities, economic systesms, ecosystems, businesses, nations, global systems, and paradigms of thought.

The first phase of the adaptive cycle resembles primary succession. This is the *rapid growth phase* where hardy species rush in to exploit newly available resources. In natural systems, this may consist of fast-growing grasses and shrubs on newly cleared land; in economic systems, it may be entrepreneurs and innovators who rush in to seize on opportunity.

The system then moves to a *conservation phase,* which resembles secondary succession. In this phase, the community

4. Reorganization phase
Renewal, growth, invention, innovation

2. Conservation phase
Secondary succession, specialization, stability

1. Rapid growth phase
Early pioneers expand and colonize territory

3. Release phase
Disturbance, rapid change, disruption

FIGURE 4.27 The adaptive cycle model of ecosystem change proposes that disturbance resets a system, creating opportunities for change, or for a possible "exit" to a new state or system.

switches from opportunists to specialists. A small steel-making business that grows from a local producer to a national, and then global competitor serves as an example. In the conservation phase, stability makes the future seem more certain, but there is a loss of flexibility as different ways of doing things (redundancy) are eliminated for efficiency.

A *release phase* results from disturbance or a shock to the system, and it happens quickly. In ecosystems, the disturbance might take the form of fire, drought, insects, pests, or disease. In an economy, it might be a "disruptive" technology (which changes how

we do things) or a market shock. Things change quickly in the release phase; dynamics are chaotic, and historic connections are broken.

The *reorganization phase* is one of renewal and growth in which novelty can thrive. Invention, experimentation, and creative ideas are the order of the day. In ecosystems, we might see buried seeds germinating or the surviving trees of a forest fire begin to regenerate. In economic and social systems, we would see many new groups appear to try and take control of the system. This phase begins to sort out and capitalize on the chaos of the release phase.

Holling observed that when a disturbance appears in an ecosystem, the system could stay in its prior path (continuing again into box 1, the rapid growth phase) or it can leave its historical path and transform into something fundamentally different ("possible exit" in fig. 4.24). For example, a clear-water lake could survive a disturbance like flooding or an invasive species and recover to a state that resembled its historical clear-water structure. Alternatively, the clear-water system could suddenly shift into a turbid-water system that behaves in an entirely different way.

Systems can shift abruptly

There are numerous examples of unexpected and abrupt shifts in nature, including kelp forests that transform into systems dominated by sea urchins, or coral-dominated reefs that transition into algae-dominated reefs (fig. 4.28). In ecology, we refer to the movement from one ecologically stable state to another as a **regime shift**. Regime shifts are usually sudden, abrupt, and nonlinear. They occur over a period of weeks to decades, which are relatively rapid when compared to the long time frame of the

(a)

(b)

FIGURE 4.28 Regime shifts happen when a system suddenly transforms into a different type of system, as when a coral–dominated reef (a) shifts to an algae-dominated reef (b). These shifts may take people by surprise, impacting fisheries, livelihoods, recreation, economic activities, or food availability.
Csaba Tökölyi/Getty Images; Rainer von Brandis/iStockphoto/Getty Images

conservation phase, which may last hundreds of years or more. These transitions have significant implications for human well-being because they bring surprising, expansive, long-lasting, and frequently negative changes to the benefits and services we receive from nature. You may be familiar with this idea through the popular notion of "tipping points," which have been documented in financial markets, climate systems, the brain, and social networks.

Ecologists study regime shifts in nature by using models, paleoenvironmental records, contemporary observation, and experiments to compare and learn from marine, terrestrial, and polar systems worldwide. They assert that regime shifts occur because of two changes working in combination: (1) a significant disturbance like a storm, fire, or pest outbreak and (2) a background change in the biological community, like overfishing, disease, or the introduction of an invasive species that has left the system vulnerable.

Often, the new regime is less ecologically complex and diverse than the original regime, and it delivers fewer ecosystem benefits. For instance, seagrass meadows can undergo sudden and abrupt change when a background variable like overfishing combines with a shock or disturbance like a heat wave. The seagrass meadow regime can abruptly transform into an algae-dominated regime or a barren-sediment regime. The new degraded system can be persistent and difficult, if not impossible, to reverse.

Resilience is the ability of a system to absorb disturbance and maintain its historic identity

Knowledge of tipping points and regime shifts is key to our understanding of resilience. **Resilience** is the capacity of a system to absorb a spectrum of shocks or disturbances and retain its fundamental function, structure, and identity. Even when shocks and disturbances occur, a resilient system absorbs these shocks and reorganizes itself to continue along in its predetermined or traditional cycle that occurred before the disturbance (it "stays in its loop.") In a resilient system, things may seem chaotic after a disturbance, but over time community structure and productivity get back to normal, species diversity is preserved, and nature seems to reach a dynamic balance.

Many biological communities are resilient and remain relatively stable and constant over time. An oak forest tends to remain an oak forest, for example, because the species that compose have self-perpetuating mechanisms. Through coevolution, the system developed ways to adjust to historic changes through mechanisms called feedback loops. Feedback loops can reinforce or amplify change (a positive feedback loop), or they can reduce or stabilize the force of a change (a negative feedback loop, see section 2.2).

What helps a system be resilient? Our study of the structure of biological communities (section 4.3) may be helpful here. While the relationship between diversity and resilience is complex, research indicates that the simplification of systems (the reduction of diversity at all scales) often contributes to a loss of resilience. To illustrate, we will return to our example of a tropical rainforest (see fig 4.24) with several structural layers, including an emergent layer, canopy, understory, shrub layer, herbaceous layer, and ground layer, each consisting of multiple species. In a complex, highly connected forest like this, each layer is working in its own way to trap moisture in the system and to maintain a hydrological balance. If the forest is simplified to include only a canopy, shrub, and ground layer, it may be more susceptible to disturbance by drought because there are fewer subsystems working to trap moisture, and the water more easily evaporates.

Resilience analysis is bringing new insight to how we manage natural systems. Traditionally, a focus on individual organisms or species led scientists to view disturbances as generally harmful. Land managers worked to suppress naturally occurring forest fires, in theory to protect forest communities as well as commercial timber supplies. Seasonal floods were also seen threat to farmlands and floodplains, rather than a natural part of river systems and a means of replenishing soil with fresh sediment. Fire suppression and flood control became the central policies in American natural resource management for most of the twentieth century.

Recent concepts like the adaptive cycle are changing our understanding of disturbance and are informing our land management policies. Resilience thinking brings us the insight that change can be a normal and natural process, and not something to prevent at all costs. Grasslands and some forests are now considered "fire-adapted," and managers allow fires to burn if weather conditions are safe. Floods also are seen as crucial for maintaining floodplain and river health. Preserving species diversity by allowing in natural disturbances (or judiciously applied human disturbances) may ensure stability over the long run, just as diverse prairies managed with fire recover after drought.

This view is especially important today because many places on earth are experiencing an increase in wildfires, flooding, droughts, storms, and pest outbreaks that are made worse by climate change. Managers can use their knowledge of resilience to help prevent systems from moving into undesirable configurations (fig 4.29). For example, administrators can work to foster

FIGURE 4.29 Ecological resilience is a guiding principle in managing wildlife at South Africa's Kruger National Park, including these impala.
Shawn Mahoney

resilience by maintaining diversity and redundancy, managing connections between elements, and enhancing (or reducing) the strength of feedback loops. Large regime shifts are frequently preceded by changes in early-warning indicators, and the monitoring of these signals can provide cost-effective early action options for managers. Early action is a more practical, affordable, and effective strategy than trying to reverse a tipping point shift.

Section Review

1. Describe the process of succession that occurs after a forest fire.
2. Why can disturbance can be good for an ecosystem?
3. What is a regime shift, and how can it be important for natural or human systems?
4. How does resilience thinking challenge our traditional thinking about change and disturbance?

Connecting the Dots

The study of biological communities begins with the principle that genetic variations and evolution produce diverse traits in a population, and diverse species in a community. When individual organisms inherit advantageous traits, they persist, live longer, and reproduce more; the result is natural selection, which alters species characteristics over time. Species' traits are also heavily influenced by close relationships with other species, which typically evolve over millennia.

As in the case of seagrass meadows, many ecosystems involve complex and connected networks. Species evolve to partition resources and occupy different niches. They also evolve intimate symbiotic relationships. These systems generate significant productivity and biomass, as well as providing benefits like clean air, healthy soils, and nutrient cycling. All of us depend on these essential ecosystem services.

Disturbances are often normal parts of an ecosystem, and generally, resilient ecosystems can absorb disturbances. But human activity creates disturbance at increasing rates and scales that may reduce the capacity of ecosystems to cope with change. When pushed past some threshold of resilience, ecosystems can shift into a new form. Often, the new ecosystem state is less complex, and it delivers fewer benefits, such as production of fresh water, carbon and nutrient cycling, and the ability to grow nutritious food.

A whole-system perspective, with appreciation of cycles of change, species relationships, and the structure and complexity of biological communities, is crucial to our survival, as well as to the survival of ecosystems. Ideally, seeing the big picture and understanding the nature of regime shifts can help us avoid surprising, or destabilizing changes.

Reviewing Key Terms

Can you define the following terms in environmental science?

abundance 4.3	disturbance 4.4	generalists 4.1	natural selection 4.1	resource
adaptation 4.1	disturbance-adapted	indicator species 4.1	parasitism 4.2	partitioning 4.1
adaptive cycle 4.4	species 4.4	interspecific	pioneer species 4.4	secondary
allopatric	diversity 4.3	competition 4.2	primary	succession 4.4
speciation 4.1	ecological niche 4.1	intraspecific	productivity 4.3	selection pressures 4.1
Batesian mimicry 4.2	ecosystem	competition 4.2	primary	specialists 4.1
binomials 4.1	engineers 4.2	keystone species 4.2	succession 4.4	speciation 4.1
climax community 4.4	ecosysem	limiting factor 4.1	principle of competitive	symbiosis 4.2
coevolution 4.2	services 4.3	Müllerian mimicry 4.2	exclusion 4.1	sympatric
commensalism 4.2	endemic species 4.1	mutation 4.1	redundancy 4.3	speciation 4.1
complexity 4.3	evolution 4.1	mutualism 4.2	regime shift 4.4	tolerance limits 4.1

Critical Thinking and Discussion Questions

1. The concepts of natural selection and evolution are central to how most biologists understand and interpret the world, and yet the theory of evolution is contrary to the beliefs of many religious groups. Why do you think this theory is so important to science and so strongly opposed by others? What evidence would be required to convince opponents of evolution?

2. What is the difference between saying that a duck has webbed feet because it needs them to swim and saying that a duck is able to swim because it has webbed feet?

3. The concept of keystone species is controversial among ecologists because most organisms are highly interdependent. If each of the trophic levels is dependent on all the others, how can we say one is most important? Choose an ecosystem with which you are familiar and decide whether it has a keystone species or keystone set.

4. Some scientists look at the boundary between two biological communities and see a sharp dividing line. Others looking at the same boundary see a gradual transition with much intermixing of species and many interactions between communities. Why are there such different interpretations of the same landscape?

5. The absence of certain lichens is used as an indicator of air pollution in remote areas such as national parks. How can we be sure that air pollution is really responsible? What evidence would be convincing?

6. We tend to regard generalists or "weedy" species as less interesting and less valuable than rare and highly specialized endemic species. What values or assumptions underlie this attitude?

Data Analysis

SeagrassSpotter

In the opening case study you read about the SeagrassSpotter project. If you live or vacation near a coastal area, this is something you can try contributing to. While you are in the classroom or the library, though, you can still learn more about the project and about the species it is trying to save. Try answering these questions with a friend or another student:

Find the SeagrassSpotter.org web site. To begin, go to the *Map* section. The map initially focuses on Europe, because the project started there.

1. Which region of Europe has the most observations?

2. Zoom in to an area that you are interested in or know something about that has some observations. If you zoom in closely, what kind of coastline do you see? (flat and calm, rocky, deep, shallow?)

3. Make note of the seagrass species observed in your location. Then go a location in the Americas or Asia. Do you see the same species there?

4. Go to the *Explore* section. Describe the growth forms (leaf shapes, branching patterns) of the different species you see there.

FIGURE 1 Seagrass meadows are easily overlooked environments that are critical in supporting global biodiversity, including turtles, fish, invertebrates, and even humans.
M Swiet Productions/Moment/Getty Images

5. Finally, go to the *Identify* section. How many traits differentiate the species in your area? What are the names of some of the species that can be found in your area?

5

Biomes
Global Patterns of Life

▲Many regions are identified by their iconic trees. Sugar maples, whose sweet sap makes maple syrup, are integral to lives and identities of the Northeast.
API/Alamy

Learning Outcomes

After studying this chapter, you should be able to:

5.1 Recognize the characteristics of major terrestrial biomes and how temperature and precipitation influence their distribution.

5.2 Understand how and why marine environments vary with temperature and light.

5.3 Compare the characteristics and biological importance of major freshwater ecosystems.

5.4 Describe important factors that make these systems vulnerable to human disturbance.

"What is the use of a house if you haven't got a tolerable planet to put it on?"

– Henry David Thoreau

Shifting Biomes, Shifting Ways of Life?

Every year, as winter wanes and the spring sun grows stronger, a ritual begins that is hundreds, perhaps thousands of years old in the hills of Vermont: sugar season. While snow still covers the ground, sweet sap starts flowing up to fattening buds, and syrup makers capture some of it on the way. Maple sap appears in legends of indigenous peoples all over the Northeast, and farm families recall long tradition as they boil syrup in the sugar shack. Maple syrup is produced across northeastern North America, but some of the best concentrations of sugar maples can be found in the cold forests of Vermont and southern Quebec. The small state of Vermont produces nearly half of all U.S. maple syrup, and southern Quebec makes 70 percent of the global supply. Sugar maples are integral to the identity of this region. Rural landowners rely on maple syrup for income. Towns and villages depend on tourists who travel in late winter to visit the sugar houses and again in fall to see the stunning colors unique to the sugar maple and its close cousin, the red maple.

Maple syrup is also a distinctive industry in which people make a living from the forest without cutting down the trees. It encourages stewardship of standing forests. But sugar producers also know that the species and its traditions are going to disappear. The forest their children inherit will not be the forest they grew up in.

Ecological communities are strongly defined by the characteristic temperature and precipitation ranges found in any environment. This means that if temperature and precipitation change, the community changes, too. The deciduous northeastern forest consist largely of maple, beech, and birch species, adapted to cold winters and temperate, humid summers. The sugar maple thrives in the nutrient-rich forest soils that develop here. Winter snow cover slows decomposition of fallen leaves, so that nutrients accumulate in the ground to fertilize new seedlings, while mild summers allow abundant growth. As a changing climate makes winters shorter and snow more unpredictable, the timing and volume of sap is becoming less reliable. As summers get hotter, trees are more likely to suffer from drought stress, insects, or competition. As decomposition in the soil accelerates, nutrients to support young seedlings become less available. Older trees eventually die or fall in storms, but conditions are becoming less suitable for regeneration of young trees.

Forest ecologists have modeled current and future potential ranges for many species, and they show that climate conditions suitable for regeneration of sugar maples are currently widespread (the year 2000 is shown, fig. 5.1) but will shift by 2050 and will decline dramatically by 2100. Note that these are potential, or suitable ranges: The species doesn't actually occur in all places with suitable climate and soils. Species of beech, birch, and northern red oaks will shift north and decline at similar rates. Of course, trees and their ecosystems have advanced and retreated repeatedly with ice age ages, over millennia. But historical shifts were on the order of 50 km in a century. Near-future shifts will be hundreds of kilometers in decades.

What will replace the northeastern maple-beech-birch forest community? Most likely warm-temperate oaks and hickories now characteristic of Tennessee and Missouri. These forests also have great ecological importance, but the brilliant fall colors that identify the Northeast will change. And they won't provide the maple syrup that we enjoy, and that supports rural families. The iconic sugar maple leaf on Canada's flag, and Vermont's state tree, may become hard to find.

In this chapter we explore the character and distribution of biomes—general categories of ecosystem types—as well as the characteristics of wetland and marine ecosystems. We have long considered the map of these systems generally stable on human time scales. As we change our climate and our oceans, though, this map may be changing. What that means for the ways we live in these regions remains to be seen.

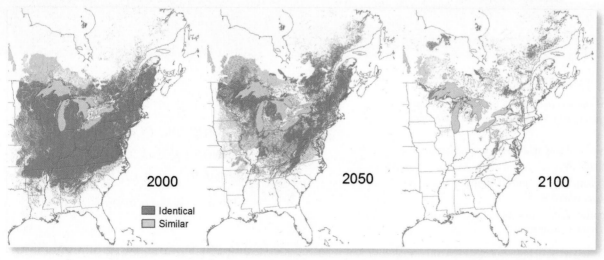

2000

Identical
Similar

2050

2100

FIGURE 5.1 Modeled area suitable for sugar maple with extreme climate change (the path we are now on). Maps show areas with climate and soil conditions statistically identical or similar to forest inventory sites where sugar maples occurred.

W. Hargrove et al. ForeCASTSProject. https://www.geobabble.org/ForeCASTS/atlas.html

5.1 TERRESTRIAL BIOMES

- *Characteristics of biological communities vary with temperature, precipitation, and latitude.*
- *Hot, humid regions generally have greater biological productivity than cold or dry regions.*
- *Climographs help us describe and compare precipitation and temperature in different biomes.*

Ecological communities are shaped by many factors, but the broad categories known as **biomes** are strongly shaped by temperature and precipitation patterns (fig. 5.2). Often temperature varies with latitude, but it also varies with elevation. When we describe biomes, we usually note mainly vegetation types, as in figure 5.2. It is important to note, though, that biomes also have characteristic types of animals and soils.

Understanding the global distribution of biomes, and knowing what grows where and why, is essential to the study of global environmental science. Biological productivity—and ecosystem resilience—varies greatly from one biome to another. Human use of biomes depends largely on those levels of productivity. Our ability to restore ecosystems and nature's ability to restore itself depend largely on biome conditions. Clear-cut forests can regrow relatively quickly in Brazil, but very slowly in Siberia, both areas now seeing rapid forest losses. Some grasslands rejuvenate rapidly after grazing, and some are very slow to recover. Why these differences? The sections that follow seek to answer this question.

Temperature and precipitation change dramatically with elevation. In mountainous regions, temperatures are cooler and precipitation is usually greater at high elevations. **Vertical zonation** occurs as vegetation types change rapidly from warm and dry to cold and wet as you go up a mountain. A 100-km transect from California's Central Valley up to Mt. Whitney, for example, crosses as many vegetation zones as you would find on a journey from southern California to northern Canada (fig. 5.3).

Because temperatures are cooler at high latitudes (away from the equator), temperature-controlled biomes often occur in latitudinal bands. For example, a band of boreal (northern) forests crosses Canada and Siberia, tropical forests occur near the equator, and expansive grasslands lie near—or just beyond—the tropics

Benchmark Data

Among the ideas and values in this chapter, the following are a few worth remembering.

>200 cm	Rainfall in a tropical rainforest, annually
<30 cm	Rainfall in temperate deserts (approximately)
23° N, S	Latitudes of the Tropics; the world's year-round rainy climates occur mainly between these latitudes
30° N, S	Approximate latitude of many global deserts
>72%	Percentage of temperate forests under severe human pressure

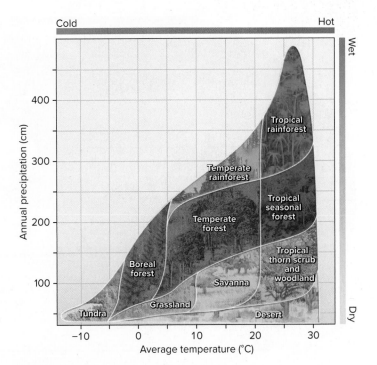

FIGURE 5.2 Temperature and precipitation conditions predict the general type of biome most likely to occur in a place. (Human disturbance, soil types, and other factors can also help determine the biological community that occurs in a local area.)

FIGURE 5.3 Vegetation changes with elevation because temperatures are lower and precipitation is greater high on a mountainside. A 100-km transect from Fresno, California, to Mt. Whitney (California's highest point) crosses vegetation zones similar to about seven different biome types.

■ Tropical rainforest, subtropical moist forest	Temperate rainforest	■ Boreal forests
■ Tropical and subtropical seasonal forests	Temperate conifer forests	■ Tundra
■ Tropical grasslands and savannas	Temperate broadleaf and mixed forests	Rock and ice
Deserts and dry shrublands	Mediterranean woodlands and scrub	Montane grasslands and shrublands
	Temperate grasslands and savannas	

FIGURE 5.4 Major world biomes. Compare this map to figure 5.2 for generalized temperature and moisture conditions that control biome distribution. Also compare it to the satellite image of biological productivity (fig. 5.13).

(fig. 5.4). Many biomes are even named for their latitudes. Tropical rainforests occur between the Tropic of Cancer (23° north) and the Tropic of Capricorn (23° south); arctic tundra lies near or above the Arctic Circle (66.6° north).

In this chapter, we'll examine the major terrestrial biomes. We give special attention to climate graphs that explain their distribution (Exploring Science, p. 101, and the Data Analysis exercise at the end of this chapter.). We then examine major ocean and freshwater communities. Ocean environments cover two-thirds of the earth's surface, provide food for much of humanity, and help regulate our climate through photosynthesis. Freshwater systems have tremendous influence on environmental health, biodiversity, and water quality.

Tropical moist forests have rain year-round

The humid tropical regions support one of the most complex and biologically rich biome types in the world (fig. 5.5). There are several kinds of moist tropical forests; all have abundant moisture and uniform temperatures. Cool **cloud forests** are found high in the mountains where fog and mist keep vegetation wet all the time. **Tropical rainforests** occur where rainfall is abundant—more

than 200 cm (80 in.) per year—and temperatures are warm to hot year-round.

The soil of both of these tropical moist forest types tends to be old, thin, acidic, and nutrient-poor, yet an enormous number of species can be present. For example, it has been estimated that there are millions of insect species in the canopy of tropical rainforests! And it is believed that one-half to two-thirds of all species of terrestrial plants and insects live in tropical forests.

The nutrient cycles of these forests also are distinctive. Almost all (90 percent) of the nutrients in the system are contained in the bodies of the living organisms. This is a striking contrast to temperate forests, where nutrients are held within the soil and made available for new plant growth. The luxuriant growth in tropical rainforests depends on rapid decomposition and recycling of dead organic material. Leaves and branches that fall to the forest floor decay and are almost immediately incorporated back into living biomass.

When the forest is removed for logging, agriculture, and mineral extraction, the thin soil generally cannot support continued cropping or resist erosion from the abundant rains. And if the cleared area is too extensive, it cannot be repopulated by the rainforest community. Rapid deforestation is occurring in many

How Do We Describe Climate Regions?

Differences between climate regions shape our experience of the world. It makes a difference in your daily life if you live in a hot, humid area, where your air–conditioning costs are high, or if you live in a cold, dry region where you pay for heating much of the year. Perhaps you have to follow special water conservation rules because demand for water is greater than rainfall can supply. Climate regions also shape the plants and animals that live around you. Does the climate in your area produce abundant, dense vegetation? Vegetation that is low or sparse? Do plants and animals have special adaptations to survive in local conditions? You probably observe biome conditions every day, whether or not you realize it.

Comparing regions is a more specialized task. Climatologists, geographers, biologists, and others often want to explain differences in biodiversity, or in water resources, for example. To describe and compare regional climate

conditions, they use climographs, graphs that plot precipitation and temperature variation over a year (fig. 1). If you take a few minutes to make sure you understand these graphs, they will help you remember and explain differences in the biomes discussed in this chapter.

Each climograph shows long-term averages for each month. Plotting 12 months gives a picture of the year. Temperature (°C) is plotted on the left axis. Precipitation (mm of rain or snow) uses the right axis. Yellow areas indicate how much of the year evaporation exceeds precipitation—that is, when conditions are dry. Dry conditions could occur because it's hot and evaporation is rapid or simply because there's little precipitation. In yellow months, little moisture is available for plant growth. Blue areas show when precipitation exceeds evaporation, leaving moisture available for plants to thrive. Examine these graphs to answer these questions.

1. What are the maximum and minimum temperatures in each location?
2. What do these temperatures correspond to in Fahrenheit? (Hint: Look at the conversion table in the back of your book.)
3. Which area has the wettest climate? Which is driest?
4. How do the maximum and minimum monthly rainfalls in San Diego and Belém compare? Note that the axis scale changes for Belém, because there's not enough room to plot such high numbers.
5. What would a climate graph look like where you live? Try sketching one out, then compare it to a graph for a biome similar to yours in this chapter.
6. Which of these graphs is probably most similar to that in Kenya, the subject of the opening case study?

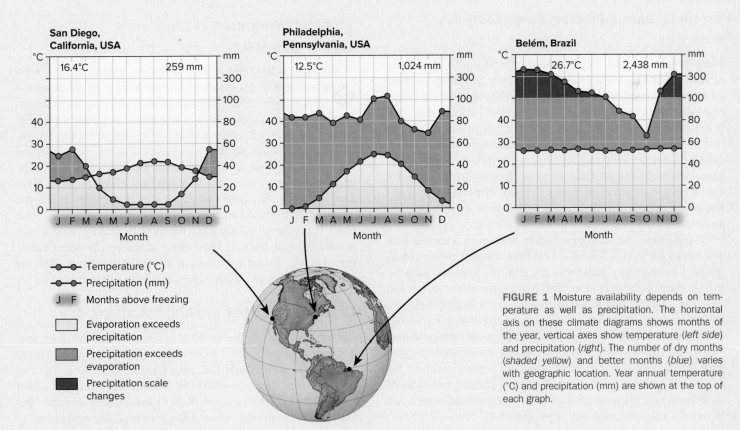

FIGURE 1 Moisture availability depends on temperature as well as precipitation. The horizontal axis on these climate diagrams shows months of the year, vertical axes show temperature (*left side*) and precipitation (*right*). The number of dry months (*shaded yellow*) and better months (*blue*) varies with geographic location. Year annual temperature (°C) and precipitation (mm) are shown at the top of each graph.

(header photo): David L. Hansen

Annual mean temperature and precipitation

Monthly precipitation (mm)

27.5°C   2,685 mm

Moisture surplus (blue)

Monthly temperature (°C)

Shaded months are above freezing

Month

FIGURE 5.5 Tropical rainforests have luxuriant and diverse plant growth. Heavy rainfall in most months supports this growth.
Adalberto Rios Szalay/Sexto Sol/Getty Images

tropical areas as agribusiness, logging firms, and settlers move in and displace ancient forests.

Tropical seasonal forests have yearly dry seasons

Many tropical regions are characterized by distinct wet and dry seasons, although temperatures remain hot year-round. These areas support **tropical seasonal forests:** Drought-tolerant forests that look brown and dormant in the dry season but burst into vivid green during rainy months. These forests are often called dry tropical forests because they are dry much of the year; however, there must be some periodic rain to support tree growth. Many of the trees and shrubs in a seasonal forest are drought-deciduous: They lose their leaves and cease growing when no water is available. Seasonal forests are often open woodlands that grade into savannas.

Tropical dry forests have typically been more attractive than wet forests for human habitation and have suffered greater degradation. Clearing a dry forest with fire is relatively easy during the dry season. Soils of dry forests often have higher nutrient levels and are more agriculturally productive than those of a rainforest. Finally, having fewer insects, parasites, and fungal diseases than a wet forest makes a dry or seasonal forest a healthier place for humans to live. Consequently, these forests are highly endangered in many places. Less than 1 percent of the dry tropical forests of the Pacific coast of Central America or the Atlantic coast of South America, for instance, remain in an undisturbed state. In some areas, however, forest restoration efforts are under way.

28.6°C   386 mm

Month

FIGURE 5.6 Tropical savannas and grasslands experience annual drought and rainy seasons and year-round warm temperatures. Thorny acacias and abundant grazers thrive in this savanna. Yellow areas show moisture deficit.
Tom Finkle

Tropical savannas and grasslands support few trees

Where there is too little rainfall to support forests, we find open **grasslands** or grasslands with sparse tree cover, which we call **savannas** (fig. 5.6). Like tropical seasonal forests, most tropical savannas and grasslands have a rainy season, but generally the rains are less abundant or less dependable than in a forest. During dry seasons, fires can sweep across a grassland, killing off young trees and keeping the landscape open. Savanna and grassland plants have many adaptations to survive drought, heat, and fires. Many have deep, long-lived roots that seek groundwater and that persist when leaves and stems above the ground die back. After a fire, or after a drought, fresh green shoots grow quickly from the roots. Migratory grazers such as wildebeest, antelope, or bison thrive on this new growth. Grazing pressure from domestic livestock is an important threat to both the plants and the animals of tropical grasslands and savannas.

Deserts can be hot or cold, but all are dry

Deserts occur where precipitation is rare and unpredictable, usually with less than 30 cm of rain per year. Deserts can appear barren and biologically impoverished, but while their vegetation is sparse, it can be surprisingly diverse. Most desert plants and animals have special adaptations that allow them to survive long droughts and extreme heat, and many can survive extreme cold. Adaptations to these conditions include water-storing leaves and stems, thick epidermal layers to reduce water loss, and salt tolerance. As in other dry environments, many desert plants are drought-deciduous. Most also bloom and set seed quickly when a spring rain does fall.

Warm, dry, high-pressure climate conditions create desert regions at about 30° latitude north and south (see fig. 5.4 and chapter 15). Extensive deserts occur in the continental interiors (far from oceans, which evaporate the moisture for most precipitation) of North America, Central Asia, Africa, and Australia (fig. 5.7). The rain shadow of the Andes produces the world's driest desert in coastal Chile. Deserts can also be cold, as in many parts of the Canadian Arctic. Antarctica is a desert, despite its accumulated snow and ice. Its Dry Valleys get no precipitation at all.

Like plants, animals in deserts are specially adapted. Many are nocturnal, spending their days in burrows to avoid the sun's heat and desiccation. Pocket mice, kangaroo rats, and gerbils can get most of their moisture from seeds and plants. Desert rodents also have highly concentrated urine and nearly dry feces that allow them to eliminate body waste without losing precious moisture.

Because biological activity is limited in deserts, they are very vulnerable to disturbance. Sparse, slow-growing vegetation is quickly damaged by off-road vehicles. Desert soils recover slowly. Tracks left by army tanks practicing in California deserts during World War II can still be seen today.

Deserts are also vulnerable to overgrazing. In Africa's vast Sahel (the southern edge of the Sahara Desert), livestock are destroying much of the plant cover. Bare, dry soil becomes drifting sand, and restabilization is extremely difficult. Without plant roots and organic matter, the soil loses its ability to retain what rain does fall, and the land becomes progressively drier and more bare. Similar depletion of dryland vegetation is happening in many desert areas, including Central Asia, India, and the American Southwest and Plains states.

Temperate grasslands have rich soils

As in tropical latitudes, temperate (midlatitude) grasslands occur where there is enough rain to support abundant grass but not enough for forests (fig. 5.8). Most grasslands are a complex, diverse mix of grasses and flowering herbaceous plants, generally known as forbs. Myriad flowering forbs make a grassland colorful and lovely in summer. In dry grasslands, vegetation may be less than a meter tall. In more humid areas, grasses can exceed 2 m. Where scattered trees occur in a grassland, we call it a savanna. Grasslands historically supported the world's greatest herds of migrant grazers, such as bison and antelope. Vast herds of wildebeest in East Africa are a remaining example.

Deep roots help plants in temperate grasslands and savannas survive drought, fire, and extreme heat and cold. Often, roots reach 2 m or more into the ground, storing nutrients, and delivering moisture deep into the soil. Cold winters kill off most vegetation above the surface, but roots put up new growth rapidly in spring. These roots, together with an annual winter accumulation

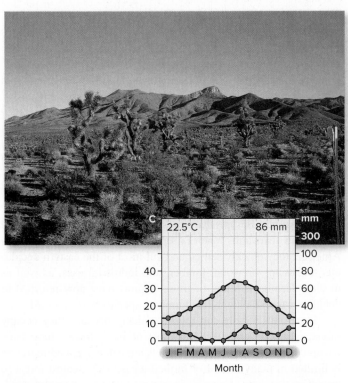

FIGURE 5.7 Deserts generally receive less than 300 mm (30 cm) of precipitation per year. Hot deserts, as in the American Southwest, endure year-round drought and extreme heat in summer.
William P. Cunningham

FIGURE 5.8 Grasslands occur at midlatitudes on all continents. Kept open by extreme temperatures, dry conditions, and periodic fires, grasslands can have surprisingly high plant and animal diversity.
Mary Ann Cunningham

of dead leaves on the surface, produce thick, organic-rich soils in temperate grasslands. Because of this rich soil, many grasslands have been converted to farmland. The legendary tallgrass prairies of the central United States and Canada have been almost completely replaced by corn, soybeans, wheat, and other crops. Most remaining grasslands in this region are too dry to support agriculture, and the greatest threat to them is overgrazing.

Temperate shrublands have summer drought

Many dry environments support sparse shrubs and trees that are adapted to survive drought, in addition to grasses. These mixed environments can be highly variable. They can also be very rich biologically. Such conditions are often described as Mediterranean (where the hot season coincides with the dry season, producing hot, dry summers and cool, moist winters). Evergreen shrubs with small, leathery, sclerophyllous (hard, waxy) leaves form dense thickets. Scrub oaks, drought-resistant pines, or other small trees often cluster in moist and sheltered valleys. Periodic fires burn fiercely in this fuel-rich plant assemblage and are a major factor in plant succession. Annual spring flowers often bloom profusely, especially after fires. In California, this landscape is called **chaparral,** Spanish for "thicket." These areas are inhabited by drought-tolerant animals such as jackrabbits, kangaroo rats, mule deer, chipmunks, lizards, and many bird species. Very similar landscapes are found along the Mediterranean coast, as well as southwestern Australia, central Chile, and South Africa. Although this biome doesn't cover a very large total area, it contains a high number of unique species and is often considered a "hot spot" for biodiversity. It also is highly desired for human habitation, often leading to conflicts with rare and endangered plant and animal species.

Areas that are drier year-round, such as the African Sahel, northern Mexico, or the American Intermountain West (or Great Basin), tend to have a more sparse, open shrubland, characterized by sagebrush (*Artemisia* sp.), chamiso (*Adenostoma* sp.), or salt-bush (*Atriplex* sp.).

Temperate forests can be evergreen or deciduous

Temperate, or midlatitude, forests occupy a wide range of precipitation conditions but occur mainly between about 30° and 55° latitude (see fig. 5.3). In general, we can group these forests by tree type, which can be broadleaf **deciduous** (losing leaves seasonally) or evergreen **coniferous** (cone-bearing). We usually treat these ideas as overlapping: Most broadleaf species, such as maples or oaks, lose their leaves seasonally, while most needle-leafed trees, such as pine or spruce, keep their needles year-round. It's important to note that there are exceptions, though: Southern live oaks are evergreen broad-leafed oaks, while cyprus and tamarack trees are needle-leafed conifers that lose their needles in winter. And some trees are drought-deciduous, losing their leaves in dry seasons.

Deciduous Forests

Broadleaf forests occur throughout the world where rainfall is plentiful. In midlatitudes, these forests are deciduous and lose

FIGURE 5.9 Temperate deciduous forests have year-round precipitation and winters near or below freezing.
William P. Cunningham

their leaves in winter. The loss of green chlorophyll pigments can produce brilliant colors in these forests in autumn (fig. 5.9).

Although these forests have a dense canopy in summer, they have a diverse understory that blooms in spring, before the trees leaf out. Spring ephemeral (short-lived) plants produce lovely flowers, and vernal (springtime) pools support amphibians and insects. These forests also shelter a great diversity of songbirds.

North American deciduous forests once covered most of what is now the eastern half of the United States and southern Canada. Most of western Europe was once deciduous forest but was cleared a thousand years ago. When European settlers first came to North America, they quickly settled and cut most of the eastern deciduous forests for firewood, lumber, and industrial uses, as well as to clear farmland. Many of those regions have now returned to deciduous forest, though the dominant species have changed.

Deciduous forests can regrow quickly because they occupy moist, moderate climates. But most of these forests have been occupied so long that human impacts are extensive, and regrowth is limited in many settled or farmed areas. The greatest threat to broadleaf deciduous forests is in eastern Siberia, where deforestation is proceeding rapidly. Siberia has one of the highest deforestation rates in the world. As forests disappear, so do Siberian tigers, bears, cranes, and a host of other endangered species.

Boreal forests occur at high latitudes

Because conifers can survive winter cold, they tend to dominate the **boreal forest,** or northern forests, that lie between about 50° and 60° north (fig. 5.11). Mountainous areas at lower latitudes may also have many characteristics and species of the boreal forest. Dominant trees are pines, hemlocks, spruce, cedar, and fir. Some deciduous trees are also present, such as maples, birch, aspen, and alder. These forests are slow-growing because of the cold temperatures and a short frost-free growing season, but they are still an expansive resource. In Siberia, Canada, and the western United States, large regional economies depend on boreal forests.

The extreme, ragged edge of the boreal forest, where forest gradually gives way to open tundra, is known by its Russian name, **taiga.** Here the extreme cold and short summers limit the growth rate of trees. A 10–cm diameter tree may be over 200 years old in the far north.

Tundra can freeze in any month

Where temperatures are below freezing most of the year, only small, hardy vegetation can survive. **Tundra,** a treeless landscape that occurs at high latitudes or on mountaintops, has a growing season of only two to three months and may have frost any month of the year.

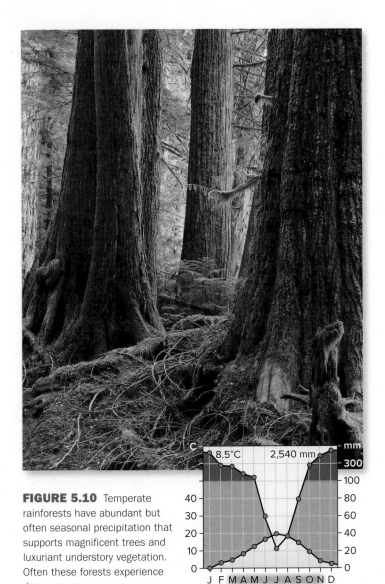

FIGURE 5.10 Temperate rainforests have abundant but often seasonal precipitation that supports magnificent trees and luxuriant understory vegetation. Often these forests experience dry summers.
Timothy Epp/Shutterstock

Coniferous Forests

Coniferous forests grow in a wide range of temperature and moisture conditions. They often occur where moisture is limited: In cold climates, moisture is unavailable (frozen) in winter; hot climates may have seasonal drought; sandy soils hold little moisture, and they are often occupied by conifers. Thin, waxy leaves (needles) help these trees reduce moisture loss. Coniferous forests provide most wood products in North America. Dominant wood production regions include the southern Atlantic and Gulf Coast states, the mountain West, and the Pacific Northwest (northern California to Alaska), but coniferous forests support forestry in many regions.

Remaining fragments of ancient temperate rainforests are important areas of biodiversity (fig. 5.10). Recent battles over old-growth conservation (chapter 12) focus mainly on these areas. As with deciduous forests, Siberian forests are especially vulnerable to old-growth logging. The rate of this clearing, and its environmental effects, remain largely unknown.

FIGURE 5.11 Boreal forests have moderate precipitation but are often moist because temperatures are cold most of the year. Cold-tolerant and drought-tolerant conifers dominate boreal forests and taiga, the forest fringe.
William P. Cunningham

Arctic tundra covers vast areas of the far north, as in northern Canada and Siberia. This biome has low productivity because it has a short growing season (fig. 5.12). During midsummer, however, 24-hour sunshine supports a burst of plant growth and an explosion of insect life. Tens of millions of waterfowl, shorebirds, terns, and songbirds migrate to the Arctic every year to feast on the abundant invertebrate and plant life and to raise their young on the brief bounty. These birds then migrate to wintering grounds, where they may be eaten by local predators—effectively they carry energy and protein from high latitudes to low latitudes. Arctic tundra is essential for global biodiversity, especially for birds.

Alpine tundra, occurring on or near mountaintops, has environmental conditions and vegetation similar to arctic tundra. These areas have a short, intense growing season. Often one sees a splendid profusion of flowers in alpine tundra; everything must flower at once in order to produce seeds in a few weeks before the snow comes again. Many alpine tundra plants also have deep pigmentation and leathery leaves to protect against the strong ultraviolet light in the thin mountain atmosphere.

Compared to other biomes, tundra has relatively low diversity of plants and year-round animal residents. But there is tremendous abundance of the relatively few species that survive the climate. Dwarf shrubs such as willows, sedges, grasses, mosses, and lichens tend to dominate the vegetation. Musk oxen, caribou, or alpine mountain sheep and mountain goats can live on the vegetation because they move frequently to new pastures.

Because these environments are too cold for most human activities, especially agriculture, they are not as badly threatened as other biomes. There are important problems, however. Global climate change is changing the composition of tundra ecosystems, melting frozen soils, and eroding coastlines. In addition, air pollution from distant cities tends to accumulate at high latitudes (chapter 16). In eastern Canada, coastal tundra is being badly overgrazed and degraded by overabundant populations of snow geese, whose numbers have exploded due to winter grazing on the rice fields of Arkansas and Louisiana. Oil and gas drilling—and associated truck traffic—threatens tundra in Alaska and Siberia. Clearly, this remote biome is not independent of human activities at lower latitudes.

Section Review

1. Grassland historically occupied much of the center of North America. Why is this, in terms of environmental factors?
2. What is taiga and where is it found? Why might taiga be slower to recover from logging than southern forests?
3. How does the amount of nutrients in soils compare in tropical rainforests and temperate grasslands?
4. Find out the annual temperature and precipitation conditions where you live (fig. 5.2). Which biome type do you occupy?

5.2 MARINE ECOSYSTEMS

- *Marine ecosystems vary with depth, temperature, and salinity.*
- *Coral reefs and estuaries are among the world's most productive and diverse ecosystems.*

The biological communities in oceans and seas are less understood than terrestrial biomes, but they can be just as diverse and complex. Oceans cover nearly three-fourths of the earth's surface, and they contribute in important, although often unrecognized, ways to terrestrial ecosystems. Like land-based systems, most marine communities depend on photosynthetic organisms. Often, it is algae or tiny, free-floating photosynthetic plants (**phytoplankton**) that support a marine food web, rather than the trees and grasses we see on land. In oceans, photosynthetic activity tends to be greatest near coastlines, where nitrogen, phosphorus, and other nutrients wash offshore and fertilize primary producers. Ocean currents also contribute to the distribution of biological productivity, because they transport nutrients and phytoplankton far from shore (fig. 5.13).

As plankton, algae, fish, and other organisms die, they sink toward the ocean floor. Deep-ocean ecosystems, consisting of crabs, filter-feeding organisms, strange phosphorescent fish, and many other life-forms, often rely on this "marine snow" as a

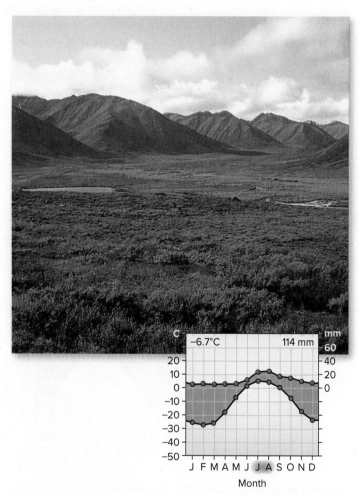

FIGURE 5.12 This landscape in Canada's Northwest Territories has both alpine and arctic tundra. Plant diversity is relatively low, and frost can occur even in summer.
Mary Ann Cunningham

FIGURE 5.13 Satellite measurements of chlorophyll levels in the oceans and on land. Dark green to blue land areas have high biological productivity. Dark blue oceans have little chlorophyll and are biologically impoverished. Light green to yellow ocean zones are biologically rich.
Source: Courtesy of SeaWiFS/NASA.

primary nutrient source. Surface communities also depend on this material. Upwelling currents circulate nutrients from the ocean floor back to the surface. Along the coasts of South America, Africa, and Europe, these currents support rich fisheries.

Vertical stratification (differing conditions from upper to lower layers) is a key feature of aquatic ecosystems, mainly because light and temperatures decrease rapidly with depth. Photosynthesis can only occur in the upper sunlit layers (the photic zone, ususally up to about 20 m deep). Well-lit, near-surface communities such as coral reefs and estuaries are among the world's most biologically productive environments, supported by photosynthetic algae growing within the corals, and phytoplankton. Deep-ocean communities, in contrast, have little primary productivity, and lower diversity. Deep species often grow slowly, in part because metabolism slows in very cold conditions. On the other hand, temperature also affects the amount of oxygen that can be absorbed in water. Cold water holds abundant oxygen, so productivity is often high in cold oceans, as in the North Atlantic, North Pacific, and Antarctic, especially for fish and other mobile or migratory species that can travel to find food resources.

Ocean systems can be described by depth and proximity to shore (fig. 5.14). In general, **benthic** communities occur on the bottom, and **pelagic** (from "sea" in Greek) zones comprise the water column. The epipelagic zone (*epi* = on top) has photosynthetic organisms. Below this are the mesopelagic (*meso* = medium), and bathypelagic (*bathos* = deep) zones. The deepest layers are the abyssal zone (at 4,000 m) and hadal zone (deeper than 6,000 m). Shorelines are known as littoral zones, and the area exposed by low tides is known as the intertidal zone. Often, there is a broad, relatively shallow coastal plain along a continent's coast, which may reach a few kilometers or hundreds of kilometers from shore. This undersea area is the continental shelf.

Depth controls light penetration and temperature

The open ocean is often referred to as a biological desert because it has relatively low productivity. But like terrestrial deserts, the open ocean has areas of rich productivity and diversity. Fish and plankton abound in regions such as the equatorial Pacific

FIGURE 5.14 Light penetrates only the top 10–20 m of the ocean. Below this level, temperatures drop and pressure increases. Nearshore environments include the intertidal zone and estuaries.

FIGURE 5.15 Deep-ocean thermal vent communities have great diversity and are unusual because they rely on chemosynthesis, not photosynthesis, for energy.
Source: C. Van Dover/NOAA.

and Antarctic oceans, where nutrients are distributed by currents. Another notable exception, the Sargasso Sea in the western Atlantic, is known for its free-floating mats of brown algae. These algae mats support a phenomenal diversity of animals, including sea turtles, fish, and even eels that hatch amid the algae, then eventually migrate up rivers along the Atlantic coasts of North America and Europe.

Deep-sea thermal vent communities are another remarkable type of marine system (fig. 5.15) that was completely unknown until 1977 explorations with the deep-sea submarine *Alvin*. These communities are based on microbes that capture chemical energy, mainly by metabolizing sulfur compounds released from thermal vents—jets of hot water and minerals on the ocean floor. Magma below the ocean crust heats these vents. Tube worms, mussels, and microbes on these vents are adapted to survive both extreme temperatures, often above 350°C (700°F), and the intense water pressure at depths of 7,000 m (20,000 ft) or more. Oceanographers have discovered thousands of different types of organisms, most of them microscopic, in these communities (chapter 3).

Coastal zones support rich, diverse communities

As in the open ocean, shoreline communities vary with depth, light, nutrient concentrations, and temperature. Some shoreline communities, such as estuaries, have high biological productivity and diversity because they are enriched by nutrients washing off from the land. But nutrient loading can be excessive. Around the world, more than 200 "dead zones" occur in coastal areas where excess nutrients stimulate bacterial growth that consumes almost all oxygen in the water and excludes most other life (see opening case study, chapter 3).

Coral reefs are among the best-known marine ecosystems because of their extraordinary biological productivity and their diverse and beautiful organisms. Reefs are aggregations of minute colonial animals (coral polyps) that live symbiotically with photosynthetic algae. Calcium-rich coral skeletons build up to make reefs, atolls, and islands (fig. 5.16a). Reefs protect shorelines and shelter countless species of fish, worms, crustaceans, and other life-forms. Reef-building corals live where water is shallow and clear enough for sunlight to reach the photosynthetic algae. They need warm (but not too warm) water, and can't survive where high nutrient concentrations or runoff from the land create dense layers of algae, fungi, or sediment. Coral reefs also are among the most endangered biomes in the world. Destructive fishing practices can damage or destroy coral communities. In addition, polluted urban runoff, trash, sewage and industrial effluent, sediment from agriculture, and unsustainable forestry are smothering coral reefs along coastlines that have high human populations. Introduced pathogens and predators also threaten many reefs. But the greatest global threat to reefs is climate warming. Elevated water temperatures cause **coral bleaching,** in which corals expel their algal partner and then die. At least three-fourths of reefs globally are threatened by bleaching, pollution, or other factors.

The value of an intact reef in a tourist economy can be over $1 million per square kilometer. The costs of conserving these same reefs in a marine-protected area would be just $775 per square kilometer per year, the UN Environment Programme estimates. Of the approximately 30 million small-scale fishers in the developing world, most are dependent to a greater or lesser extent on coral reefs. In the Philippines, the UN estimates that more than 1 million fishers depend directly on coral reefs for their livelihoods. We'll discuss reef restoration efforts further in chapter 13.

Sea-grass beds, or eel-grass beds, are another critically important ecosystem that occurs in shallow, warm, sandy areas near coral reefs. Like reefs, these communities support a rich diversity of grazers, from snails and worms to turtles and manatees. Also like reefs, these environments are easily smothered by sediment originating from onshore agriculture and development.

Mangroves are a special variety of trees that grow in salt water (fig. 5.16b). They occur along calm, shallow, tropical coastlines around the world. Mangrove forests or swamps help stabilize shorelines, and they are also critical nurseries for fish, shrimp, and other commercial species. Like coral reefs, mangroves line tropical

(a) Coral reefs

(b) Mangroves

(c) Estuary and salt marsh

(d) Tide pool

FIGURE 5.16 Coastal environments support incredible diversity and help stabilize shorelines. Coral reefs (a), mangroves (b), and estuaries (c) also provide critical nurseries for marine ecosystems. Tide pools (d) also shelter highly specialized organisms.
Glen Allison/Photodisc/Getty Images; shakzu/iStock/360/Getty Images; andykazie/123RF; Cary Kalscheuer/Shutterstock.com

and subtropical coastlines, where they are vulnerable to development, sedimentation, and overuse. Unlike reefs, mangroves provide commercial timber, and they can be clear-cut to make room for aquaculture (fish farming) and other activities. Ironically, mangroves provide the protected spawning beds for most of the fish and shrimp farmed in these ponds. As mangroves become increasingly threatened in tropical countries, villages relying on fishing for income and sustenance are seeing reduced catches and falling income.

Estuaries are bays where rivers empty into the sea, mixing fresh water with salt water. **Salt marshes,** shallow wetlands flooded regularly or occasionally with seawater, occur on shallow coastlines, including estuaries (fig. 5.16c). Usually calm, warm, and nutrient-rich, estuaries and salt marshes are biologically diverse and productive. Rivers provide nutrients and sediments; and a muddy bottom supports emergent plants (whose

leaves emerge above the water surface), as well as the young forms of crustaceans, such as crabs and shrimp, and mollusks, such as clams and oysters. Nearly two-thirds of all marine fish and shellfish rely on estuaries and saline wetlands for spawning and juvenile development.

Estuaries near major American cities once supported an enormous wealth of seafood. Oyster beds and clam banks in the waters adjacent to New York, Boston, and Baltimore provided free and easy food to early residents. Sewage and other contaminants long ago eliminated most of these resources. As eastern rivers have become cleaner, since the Clean Water Act of 1972, interest has grown in reviving these oyster beds. New York is among the leaders in restoring oyster reefs, with the added benefit of further filtering and cleaning its estuary.

In contrast to the shallow, calm conditions of estuaries, coral reefs, and mangroves, there are violent, wave-blasted shorelines that support fascinating life-forms in **tide pools.** Tide pools are depressions in a rocky shoreline that are flooded at high tide but retain some water at low tide. These areas remain rocky where wave action prevents most plant growth or sediment (mud) accumulation. Extreme conditions—with frigid flooding at high tide, and hot desiccating sunshine at low tide—make life impossible for most species. But the specialized animals and plants that do occur in this rocky intertidal zone are astonishingly diverse and beautiful (fig. 5.16*d*).

Barrier islands are low, narrow, sandy islands that form parallel to a coastline (fig. 5.17). They occur where the continental shelf is shallow and rivers or coastal currents provide a steady source of sediments. They protect brackish (moderately salty), inshore lagoons and salt marshes from storms, waves, and tides. One of the world's most extensive sets of barrier islands lines the Atlantic coast from New England to Florida, as well as along the Gulf Coast of Texas. Composed of sand that is constantly reshaped by wind and waves, these islands can be formed or removed by a single violent storm. Because they are mostly beach, barrier islands are also popular places for real estate development. About 20 percent of the barrier island surface in the United States has been developed. Barrier islands are also critical to preserving coastal shorelines, settlements, estuaries, and wetlands.

Human occupation often destroys the value that attracts us there in the first place. Barrier islands and beaches are dynamic environments, and sand is hard to keep in place. Wind and wave erosion is a constant threat to beach developments. Walking or driving vehicles over dune grass destroys the stabilizing vegetative cover and accelerates, or triggers, erosion. Cutting roads through the dunes further destabilizes these islands, making them increasingly vulnerable to storm damage. When Hurricane Katrina hit the U.S. Gulf Coast in 2005, it caused at least $200 billion in property damage and displaced 4 million people. Thousands of homes were destroyed (fig. 5.18), particularly on low-lying barrier islands.

FIGURE 5.18 Coastal beaches and barrier islands are extremely vulnerable to erosion, especially when shoreline plant communities are damaged. Protecting property on naturally shifting coastlines is difficult and costly.
David L. Ryan/The Boston Glove via Getty Images

Similarly, when Superstorm Sandy washed over New York and New Jersey in 2012, most of the $50 billion in damage occurred on the heavily built-up but low-lying beaches and barrier islands of the region.

Because of these problems, we spend billions of dollars each year building protective walls and barriers, pumping sand onto beaches from offshore, and moving sand from one beach area to another. Some planners question whether we should allow rebuilding on barrier islands, especially after they've been destroyed multiple times.

Section Review

1. How do physical conditions change with depth in marine environments?
2. Describe four different coastal ecosystems.
3. What is coral bleaching?

5.3 FRESHWATER ECOSYSTEMS

- *Freshwater systems vary according to depth and light penetration, which control size and types of vegetation.*
- *Shallow wetlands can be highly productive and diverse.*

Freshwater environments are often small, but they are disproportionately important in biodiversity. Wetlands support many species that can live nowhere else, including amphibians,

FIGURE 5.17 A barrier island, Assateague, along the Maryland–Virginia coast. Grasses cover and protect dunes, which keep ocean waves from disturbing the bay, salt marshes, and coast at right. Roads cut through the dunes expose them to erosion.
Source: U.S. Geological Survey.

fish, aquatic plants, and many insects and birds. Upland ecosystems (drier areas around wetlands) often rely on these wetland communities. Insects, birds, and frogs, for example, are eaten by predators in grassland and forest ecosystems. In deserts or dry grasslands, isolated pools and streams support astonishing concentrations of biodiversity, including streamside shrubs and trees that support birds, insects, and small mammals.

Temperature and light vary with depth in lakes

Freshwater lakes, like marine environments, have distinct vertical zones (fig. 5.19). Near the surface, a subcommunity of plankton—mainly microscopic plants, animals, and protists (single-celled organisms such as amoebae)—floats freely in the water column. Insects such as water striders and mosquitoes also live at the air-water interface. Fish move through the water column, sometimes near the surface and sometimes at depth.

Finally, the bottom, or *benthos,* is occupied by a variety of snails, burrowing worms, fish, and other organisms. These make up the benthic community. Oxygen levels are lowest in the benthic environment, mainly because there is little mixing to introduce oxygen to this zone. Anaerobic bacteria (not using oxygen) may live in low-oxygen sediments. In the littoral zone, emergent plants such as cattails and rushes grow in the bottom sediment. These plants create important functional links between layers of an aquatic ecosystem, and they may provide the greatest primary productivity to the system.

Lakes, unless they are shallow, have a warmer upper layer that is mixed by wind and warmed by the sun. This layer is the *epilimnion.* Below the epilimnion is the *hypolimnion (hypo = below),* a colder, deeper layer that is not mixed. If you have gone swimming in a moderately deep lake, you may have discovered the sharp temperature boundary, known as the **thermocline,** between these layers. Below this boundary, the water is much colder. This boundary is also called the *mesolimnion.*

FIGURE 5.19 The layers of a deep lake are determined mainly by gradients of light, oxygen, and temperature. The epilimnion is affected by surface mixing from wind and thermal convections, while mixing between the hypolimnion and epilimnion is inhibited by a sharp temperature and density difference at the thermocline.

FIGURE 5.20 The character of freshwater ecosystems is greatly influenced by the immediately surrounding terrestrial ecosystems, and even by ecosystems far upstream or far uphill from a particular site.

Local conditions that affect the characteristics of an aquatic community include (1) availability (or excess) of nutrients, such as nitrates and phosphates; (2) suspended matter, such as silt, that affects light penetration; (3) depth; (4) temperature; (5) currents; (6) bottom characteristics, such as muddy, sandy, or rocky floor; (7) internal currents; and (8) connections to, or isolation from, other aquatic and terrestrial systems (fig. 5.20).

Wetlands are shallow and productive

Wetlands are shallow ecosystems in which the land surface is saturated or submerged at least part of the year. Wetlands have vegetation that is adapted to grow under saturated conditions. These legal definitions are important, because although wetlands make up only a small part of most countries, they are disproportionately important in conservation debates and are the focus of continual legal disputes around the world and in North America. Beyond these basic descriptions, defining wetlands is a matter of hot debate. How often must a wetland be saturated, and for how long? How large must it be to deserve legal protection? Answers can vary, depending on political, as well as ecological, concerns.

These relatively small systems support rich biodiversity, and they are essential for both breeding and migrating birds. Although wetlands occupy less than 5 percent of the land in the United States, the Fish and Wildlife Service estimates that one-third of all endangered species spend at least part of their lives in wetlands. Wetlands retain storm water and reduce flooding by slowing the rate at which rainfall reaches river systems. Floodwater storage is worth $3 billion to $4 billion per year in the United States. As

(a) Swamp, or wooded wetland

(b) Marsh

(c) Coastal saltmarsh

FIGURE 5.21 Wetlands provide irreplaceable ecological services, including water filtration, water storage and flood reduction, and habitat. Forested wetlands (a) are often called swamps; marshes (b) have no trees; coastal saltmarshes (c) are tidal and have rich diversity.
Warren Price Photography/Shutterstock; William P. Cunningham;TerryJ/iStock/Getty Images

water stands in wetlands, it also seeps into the ground, replenishing groundwater supplies. Wetlands filter, and even purify, urban and farm runoff, as bacteria and plants take up nutrients and contaminants in water. They are also in great demand for filling and development. They are often near cities or farms, where land is valuable, and once drained, wetlands are easily converted to more lucrative uses.

Wetlands are generally described by their vegetation. **Swamps,** also known as wooded wetlands, are wetlands with trees (fig. 5.21*a*). **Marshes** are wetlands without trees (fig. 5.21*b*). **Bogs** are areas of saturated ground, and usually the ground is composed of deep layers of accumulated, undecayed vegetation known as peat. **Fens** are similar to bogs except that they are mainly fed by groundwater, so that they have mineral-rich water and specially adapted plant species. Many bogs are fed mainly by precipitation. Swamps and marshes have high biological productivity. Bogs and fens, which are often nutrient-poor, have low biological productivity. They may have unusual and interesting species, though, such as sundews and pitcher plants, which are adapted to capture nutrients from insects rather than from soil.

The water in marshes and swamps often is shallow enough to allow full penetration of sunlight and seasonal warming (fig. 5.21*c*). These mild conditions favor great photosynthetic activity, resulting in high productivity at all trophic levels. In short, life is abundant and varied. Wetlands are major breeding, nesting, and migration staging areas for waterfowl and shorebirds.

Wetlands may gradually convert to terrestrial communities as they fill with sediment, and as vegetation gradually fills in toward the center. Often, this process is accelerated by increased sediment loads from urban development, farms, and roads. Wetland losses are one of the areas of greatest concern to biologists.

Section Review

1. Describe four different kinds of wetlands.
2. Why are wetlands sites of high biodiversity and productivity?

5.4 Human Disturbance

Understanding characteristic biomes and communities that occur in different environments is an essential part of understanding environmental science. Living systems have invented countless ways to occupy the earth's diverse environments; it is important, and enjoyable, to understand some of the diversity that surrounds us, and also to define categories that help us make order of natural variety.

Understanding these different systems also helps us make sense of the different kinds of impacts that result from human activities. We have noted that some biomes can rebound relatively readily after disturbance (such as temperate forests), while some recover very slowly (such as deserts). Some biomes are vulnerable to disturbance because their high productivity makes them valuable to us (such as estuaries and grasslands) and some because we find them highly habitable (such as temperate forests). Some are vulnerable because they are rare (including many temperate shrublands).

Agriculture is responsible for most land conversion

Conversion of natural environments to human uses is the single largest cause of biodiversity loss in terrestrial systems. (Pollution and habitat destruction are dominant causes in marine and wetland systems). Climate change promises to be one of the most important future causes of biome and ecosystem shifts. But centuries of more conventional land conversion remain, thus far, more dramatic.

Among causes of land conversion, agriculture is the leading factor. Most of the world's great grasslands have been converted to farmland because of their level ground and rich soils. In Europe and much of the eastern United States, temperate forests were long ago converted to farmland. Temperate broadleaf forests are among the most completely human-dominated of terrestrial biomes (table 5.1).

The most extensive and rapid ongoing changes are conversion of the tropical forests to industrial-scale agriculture, especially soybean and oil palm production. Across the world's great

Table 5.1 Proportion of major biomes under high pressure from human activities		
Biome description[1]	High pressure (%)	Land area (%)[2]
Tropical Dry Forests	76.1	2.3
Temperate Broadleaf, Mixed Forests	72.7	9.5
Mangroves	65.4	0.2
Temperate Grasslands & Shrublands	52.5	7.6
Tropical Moist Forests	45.3	14.8
Tropical Grasslands & Shrublands	43.1	15.2
Temperate Conifer Forests	31.7	3.0
Deserts & Dry Shrublands	23.0	21.0
Boreal Forests/Taiga	7.0	11.3
Tundra	1.6	6.0

[1] Selected biome types shown.

[2] Percentage of total land area

O. Venter, EW Sanderson, et al. 2016. Global terrestrial Human Footprint maps for 1993 and 2009. Scientific Data 3: 160067. https://doi.org/10.1038/sdata.2016.67

FIGURE 5.23 A restored wetland, previously drained for farmland. Restoring and protecting ecosystems has both environmental and economic benefits.
U.S. Fish & Wildlife Service

tropical forests, these two crops are expanding rapidly. They provide biofuels and edible oils to industrialized nations, as well as livestock feed for rapidly expanding meat production, especially in China (fig. 5.22).

Small systems are most at risk

Both terrestrial and aquatic systems are most at risk when they are small in extent. Among terrestrial biomes, temperate shrublands and temperate rainforests are severely depleted, in part because their original extent was very limited (see fig. 5.4). Island ecosystems are notoriously vulnerable, with small populations and small extent, easily overrun by settlement or by invasive species.

FIGURE 5.22 Deforestation in Brazil's Amazon River basin is one of the most dramatic frontiers of biome change today. Here red soils and distant soy fields are replacing ancient rainforest.
Frontpage/Shutterstock

Freshwater wetland systems, similarly, are very small in extent relative to other ecosystems. Their biological importance is extraordinary, however, as they support tremendous diversity of both wetland and upland species. Freshwater wetlands are also largely lost in many countries, from the United States and Canada to China. In some Midwestern farming states, 99 percent of wetlands are gone, as well as 99 percent of native grasslands, both converted to farmland. California has lost 90 percent of the extensive marshes and deltas that once stretched across its central valley. Wildlife officials have worked to protect remaining wetlands for migratory birds and for fish, and in some cases they have worked with farmers to provide flooded fields for seasonal bird habitat. Overall, U.S. wetlands are less than half as extensive as they were a century ago. As sea level rise depletes coastal wetlands in Louisiana, Texas, and the Atlantic coast, U.S. wetland area stands to diminish even further.

Mangroves are also limited in extent and largely depleted. In Indonesia, almost all mangroves have been converted to shrimp farms and other uses. In the Philippines and Thailand, a majority have been cleared for timber or converted to shrimp and other aquaculture.

Slowing the loss of these systems, or even reversing it and restoring ecosystems, is an increasingly important response to these changes. There are many strategies for protecting critical areas, and environmental scientists and conservationists are increasingly developing strategies for progress (fig. 5.23). We discuss these efforts further in the context of parks and preserves (chapter 12) and restoration ecology (chapter 13).

Section Review

1. What percentage of temperate grasslands and forests face high pressure from human activity (table 5.1)? Why?
2. What is one reason for the clearing of tropical coastal mangroves?
3. How have temperate wetlands in the United States been lost?

Connecting the Dots

While we don't always think about it closely, we depend directly on the familiar biomes we inhabit, which in turn depend on generally stable climate conditions. Maple syrup production in Vermont and Quebec is an obvious case where shifting biomes will alter livelihoods, but all of us rely to some extent on these climate-determined systems. As climates and biomes shift, how will changes affect the places where we live?

Understanding the global distribution of biomes, and knowing the differences in what lives where and why, are essential to the study of global environmental science. Plants and animals have evolved characteristics that allow them to live in particular biomes, such as seasonal tropical forests, alpine tundra, or chaparral. Recognizing these adaptations helps you understand limiting factors for survival in those biomes. Human occupation and use of natural resources are strongly dependent on the biomes found in particular locations. Humans tend to prefer mild climates and the highly productive biological communities found in temperate zones. These biomes also suffer the highest rates of degradation and overuse.

While many of us pay most attention to terrestrial systems, oceans cover over 70 percent of the earth's surface. Marine biomes, such as coral reefs or mangroves, can be as biologically diverse and productive as any terrestrial biome. Freshwater ecosystems, too, are critically important, even though their extent is small overall. People have always depended on rich, complex ecosystems. In recent times, the rapid growth of human populations, coupled with more powerful ways to harvest resources, has led to extensive destruction of these environments. Awareness of emerging threats like climate change may help inspire more action to protect these living systems.

Reviewing Key Terms

Can you define the following terms in environmental science?

barrier islands 5.2	coniferous 5.1	grasslands 5.1	swamp 5.3	tundra 5.1
benthic 5.2	coral bleaching 5.2	mangroves 5.2	taiga 5.1	vertical zonation 5.1
biome 5.1	coral reefs 5.2	marsh 5.3	thermocline 5.3	wetlands 5.3
bog 5.3	deciduous 5.1	pelagic 5.2	tide pool 5.2	
boreal forest 5.1	desert 5.1	phytoplankton 5.2	tropical rainforests 5.1	
chaparral 5.1	estuary 5.2	salt marsh 5.2	tropical seasonal	
cloud forests 5.1	fen 5.3	savannas 5.1	forest 5.1	

Critical Thinking and Discussion Questions

1. What physical and biological factors are most important in shaping the biome in which you live? How have those factors changed in the past 100 or 1,000 years?

2. Fire is a common component of forest biomes. As more of us build homes in these areas, can we protect people from natural disturbances? How?

3. Often, humans work to preserve biomes that are visually attractive. Are there biomes that might be lost because they are not attractive? Is this a problem?

4. Disney World in Florida wants to expand onto a wetland. It has offered to buy and preserve a large nature preserve in a different area to make up for the wetland it is destroying. Is that reasonable? Why or why not?

5. Suppose further that the wetland being destroyed in question 4 and its replacement area both contain several endangered species (but different ones). How would you compare different species against each other? Should we preserve animals before we preserve plant or insect species?

6. Historically, barrier islands have been hard to protect because links between them and inshore ecosystems are poorly recognized. What kinds of information would help a community distant from the coast commit to preserving a barrier island?

Data Analysis

Reading Climate Graphs

As climate conditions change, interest in drought, rainfall, and temperature conditions are becoming more important to farmers, producers and users of hydroelectric power, policymakers, and others. The National Drought Mitigation Center at the University of Nebraska is a resource for monitoring climate conditions for the United States. Examine its website (https://drought.unl.edu/Climographs.aspx) and go to Connect to find questions and demonstrate your understanding of these graphs.

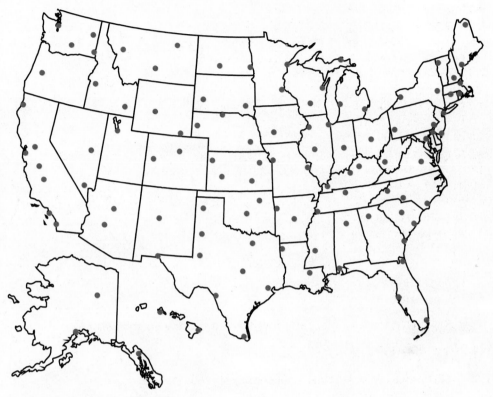

Climographs for Selected U.S. Cities

▲Asian silver carp, which leap from the water when startled by passing boats, are spreading rapidly throughout the Mississippi River Basin and displacing native species.
Source: Ryan Hagerty/USFWS

Learning Outcomes

After studying this chapter, you should be able to:

6.1 Describe the dynamics of population growth.

6.2 Compare and contrast the factors that regulate population growth.

6.3 Identify some applications of population dynamics in conservation biology.

"Nature teaches more than she preaches."

– John Burroughs

Flying Fish

Explosive populations of invasive species are a familiar problem in many ecosystems, but the leaping carp now spreading through the Mississippi River watershed are a whole new ballgame. Silver carp, one of several introduced Asian carp species, have become an Internet sensation and a notorious menace, launching themselves through the air when disturbed by boat engines. Dense populations of these large fish—often 10 kg (22 pounds) or more, produce barrages of flying fish cannonading from the water, startling and even injuring boaters. This invasive species is a concern not just to ecologists but also to the many people who fish, work, live, and play in this vast river system. Among the most notable, and fun, responses has been the Redneck Fishing Tournament, a boisterous event with the serious aim of catching as many of the invasive fish as possible. Competitors in hockey masks with nets and garbage cans can be seen scooping up fish, getting slimed and bruised, and helping to protect the watershed and its native fish.

How did the leaping silver carp become such a problem in just a few years? The silver carp is really just the most visible of four new carp species in the watershed—along with the bighead, black, and grass carp (fig 2.1). Introduced in the 1960s to control algae in wastewater treatment ponds and catfish farms, they probably escaped during floods (as introduced fish often do). Once they entered the river system, they had access to a watershed covering two-thirds of the United States. They have now been found in 45 states.

This is a remarkable case study in population biology. Populations grow or decline according to fertility rates, predation, and resource availability. In the case of introduced Asian carp species, fertility rates are astonishing. These carp are prolific breeders. A single female, reaching maturity in just two years, can produce 2 million eggs a year. They can spawn year round in running water, and in some stretches of river, they have been observed to double in number in a single year.

They also lack natural predators that might keep numbers in check. And although their white, firm flesh makes them a very popular dish throughout Asia, Americans have not developed a taste or a market for them. They also excel at using resources, which are partitioned among species. Bighead and silver carp are filter feeders, eating tiny floating plankton. Black carp, which have powerful molars capable of crunching shells, feed on native clams and snails. Grass carp eat aquatic plants and submerged terrestrial vegetation but may also feed on detritus, insects, and other invertebrates. Feeding voraciously, these fish can eat up to 40 percent of their body weight per day and can grow to 2 m (6.6 ft) and 45 kg (99 lbs) and live 20 years. Carp compete directly with native fish, depleting nutrients and modifying in-stream habitats, and there is concern that some native species may be driven to extinction. The large, hardy carp easily reach population densities that disrupt ecosystems and biological communities. The U.S.

Fish and Wildlife Service reports that in some rivers, Asian carp may account for 90 percent of all animal biomass.

A major concern has been the high likelihood that these carp will invade the Great Lakes, where commercial fisheries and ecosystems could be devastated. In 2007, the U.S. Department of the Interior declared silver carp an injurious species, and in 2012, Congress passed the "Stop Invasive Species Act," which requires action to protect the Great Lakes and the upper Mississippi River from the spread of Asian carp. Interstate transportation of live carp is forbidden. Several dams have been built on tributary streams, and existing locks have been closed to block upstream migration. The most likely route of entry to the Great Lakes is a canal in Chicago that connects Lake Michigan to the Mississippi watershed. A $30 million electric barrier was built that constantly electrifies the waterway, to try to block the fish. However, carp DNA has been found in all the Great Lakes except Lake Superior.

Population biology provides analytical tools and insights needed to understand the rapid spread of invasive species, such as Asian carp. In this chapter, we'll look at causes and consequences of population growth as well as the factors that regulate reproduction and survival of different species. And we'll use these ideas to understand conservation of endangered species.

FIGURE 6.1 Grass carp grow extremely rapidly, and populations can double in a single year.
Ryan Hagerty/USFWS

6.1 DYNAMICS OF POPULATION GROWTH

- *Unrestrained exponential growth can lead to boom and bust cycles.*
- *Logistic growth slows as populations increase.*
- *Carrying capacity is the number of individuals that live in a given area for an extended period of time.*

Population dynamics are one of the most important factors in questions of environmental quality and biodiversity. Living organisms modify the environments they—and we—occupy, affecting water quality, the stability of a river system, the viability of economically important species we rely on, and many other factors. So we want to understand how their numbers change, and how one species displaces another, or when rare species are at risk of extinction. In this chapter we explore the central concepts involved in understanding population dynamics, that is, what makes a species' numbers grow or decline.

Without constraints, population growth can be rapid. As in the case of invasive carp, many organisms can produce unbelievable numbers of offspring if environmental conditions are favorable, and if predators don't control them. Consider the common housefly (*Musca domestica*). A single female fly can lay 120 eggs. In 56 days those eggs become mature adults, about half of them reproductive females. In one year, after seven generations of flies being born and reproducing, that original fly would be the proud grandparent of 5.6 trillion offspring. If this rate of reproduction continued for ten years, the entire earth would be covered in several meters of housefly bodies. Luckily, housefly reproduction is constrained, as it is for most organisms— by predation, competition, scarcity of resources, disease, or accidents. Both the housefly and carp demonstrate the remarkable amplification—the **biotic potential**—of unrestrained biological reproduction.

FIGURE 6.2 Reproduction gives many organisms the potential to expand populations explosively. A single female cockroach can produce up to 80 eggs every six months.
Polovynkin Oleksandr/Shutterstock

Benchmark Data	
Among the ideas and values in this chapter, the following are a few worth remembering.	
70/n	Doubling time (n = percentage growth rate)
1 year	Population doubling time for Asian carp under ideal conditions
2 million	Maximum number of eggs in single female Asian carp
30 million	Current number of white-tailed deer in U.S.
0.5 million	Number of white-tailed deer in U.S. in 1920
100 billion	Estimated number of locusts in a single swarm

We can describe growth symbolically

Describing population growth is easiest if we can reduce it to a few general factors. Ecologists find it most efficient and simplest to use symbolic terms such as N, r, and t to refer to these factors. At first, this symbolic form might seem hard to interpret, but as you become familiar with the terms, you'll probably find them quicker to follow than longer text versions.

Here are some examples to show how you can describe population change. Consider cockroaches, a species capable of reproducing very rapidly (fig. 6.2). If there are no predators and food is abundant, then its population depends mainly on two factors: the number you start with, and the rate of reproduction. Start with 2 cockroaches, one male and one female, and suppose they can lay eggs and increase to about 20 cockroaches in the course of three months. You can describe the rate of growth (r) per adult in one three-month period like this: $r = 20$ per 2 adults, or 10/adult, or "$r = 10$." If nothing limits population growth, numbers will continue to increase at this rate of $r = 10$ for each three-month time step. You can call each of these time steps (t). The starting point, before population growth begins, is "time 0" (t_0). The first time step is called t_1, the second time step is t_2, and so on. If $r = 10$, and the population (N) starts at 2 cockroaches, then the numbers will increase like this:

time	N	rate (r)	$r \times N$
t_1	2	10	$10 \times 2 = 20$
t_2	20	10	$10 \times 20 = 200$
t_3	200	10	$10 \times 200 = 2,000$
t_4	2,000	10	$10 \times 2,000 = 20,000$

This is a very rapid rate of increase, from 2 to 20,000 in four time steps (fig. 6.3). It's a very simplified explanation of growth, but it's a fair approximation of the growth rate. This rate is described as a "geometric" rate of increase, in which each time step increases by the same multiple (in this case, multiplied by 10). Look carefully at the numbers above, and you might notice that the population at t_2 is $2 \times 10 \times 10$, and the population at t_3 is $2 \times 10 \times 10 \times 10$. Another way to say this is that the population at t_2 is 2×10^2, and at t_3 the population is 2×10^3. In fact, the population at any given time is equal to the starting number (2) times the rate (10) raised to the exponent of the number of time steps (10^t). The short way to express the geometric rate of increase is below. This growth rate is also known as an **exponential growth** rate, because the number at time t is a function of the growth rate raised to the exponent t. Note that geometric. Stop here and make sure you understand the terms N, r, and t:

$$N_t = N_0 r^t$$

Exponential growth involves continuous change

Thus far we have considered growth one time step at a time, but usually we think of populations increasing continuously. A standard way to describe continuous change is by using the same terms, r, N, and t, plus the added term delta (d), for *change* (fig. 6.4).

You can read the equation in fig. 6.4 like this: The change in N (dN) per change in time (dt) equals rate of increase (r) times the population size (N). This equation is a model, a very simplified description of the dynamic process of population growth. Models like this are convenient because you can use them to describe many different growth trends, just by changing the "r" term. If $r > 0$, then dN increases over time. If $r < 0$, then dN is negative, and the population is declining. If $r = 0$, then dN is 0 (no change), and the population is stable.

The equation for exponential growth can be written in different forms, but they all describe the same accelerating growth pattern: a J-shaped curve, as shown in fig. 6.3. This curve is characteristic of many species that grow rapidly when food is abundant.

Doubling times and the rule of 70

Often, the easiest way to grasp the effect of rapidly growing populations is to calculate the doubling time (the time it takes to reach twice the previous number). There's a shortcut called the rule of 70 for this computation. If you divide 70 by the rate of growth in percentage, you'll get the approximate doubling time.

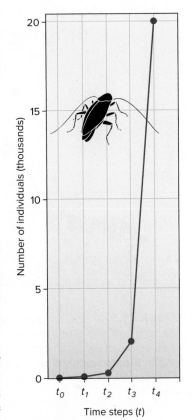

FIGURE 6.3 Population increase with a constant growth rate.

FIGURE 6.4 Exponential growth.

For example, our cockroach population above increased at a rate of 10 times (that is, 1,000 percent) in each time interval. If we divide 70/1,000 we calculate that the population doubles every 0.07 time interval. (And it increases 10-fold in one interval). A larger species, say moose, might increase at 2 percent each year, doubling in 70/2 = 35 years. Or a population growing at 10 percent per year will double in 7 years. This rule makes it easy to convert one common measure of growth (percent per year) to another (time to doubling).

We frequently use this approach for describing human population growth. The United States, for example, has a growth rate of about 0.7 percent per year. We can calculate a doubling time of 70/0.7 = 100 years to double (from about 325 million to 650 million), if the rate doesn't change. Nigeria, meanwhile, has a growth rate of about 2.5 percent per year. At this rate, its population will double (from 60 to 120 million) in 70/2.5 = 28 years. In many countries, such a rapid increase would have important social and environmental consequences (see chapter 7).

Exponential growth leads to crashes

A population can grow very rapidly only if nothing limits its growth. Usually there are many factors that reduce the rate of increase. Individuals die, they might mature slowly, they may fail to reproduce. A population that has few or no predators, though, can grow at an exponential rate, at least for a while.

But all environments have a limited capacity to provide food and other resources for a particular species. **Carrying capacity** is the term for the number (or the biomass) of a species that can be supported in a certain area without depleting resources. Eventually, a rapidly growing population reaches and overshoots this carrying capacity (fig. 6.5). Shortages of food or other resources eventually lead to a **population crash,** or rapid dieback. Once below the carrying capacity, the population may rise again, leading to boom and bust cycles. These oscillations can eventually lower the environmental carrying capacity for an entire food web.

A classic long-term study of moose on Isle Royale, a large island in Lake Superior, demonstrates this pattern of growth and dieback (fig. 6.5). Because the island has a limited carrying capacity, and it is isolated from the mainland, moose can quickly overshoot the island's resources, after which they die back. As vegetation recovers, the moose population can regrow. Over time, however, repeated overshoots may be degrading vegetation and reducing the island's carrying capacity.

Invasive species, like the carp discussed in the opening case study, often lack predators and thus follow a steep J-shaped pattern of population growth. Because successful invasive species are usually generalists, able to exploit a wide range of food resources, they may displace many other species before experiencing a dieback.

Logistic growth slows with population increase

Sometimes growth rates slow down as the population approaches carrying capacity—as resources become scarce, for example. In symbolic terms, the rate of change (dN/dt) depends on how close population size (N) is to the carrying capacity (K).

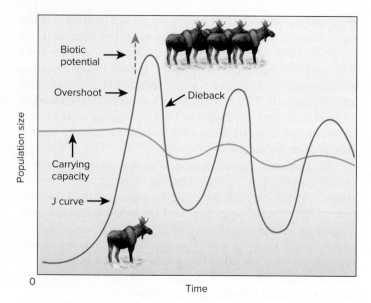

FIGURE 6.5 J curve, or exponential growth curve, with overshoot of carrying capacity. Exponential growth in an unrestrained population (*left side of curve*) leads to a population crash and oscillations below former levels. After the overshoot, carrying capacity may be reduced because of damage to the resources of the habitat. Moose on Isle Royale in Lake Superior may be exhibiting this growth pattern in response to their changing environment.

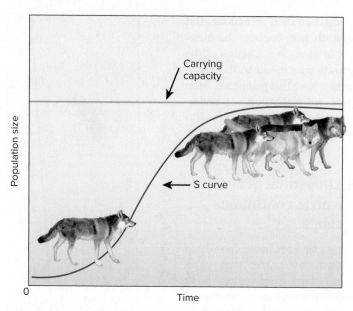

FIGURE 6.6 S curve, or logistic growth curve, describes a population's changing number over time in response to feedback from the environment or its own population density. Over the long run, a conservative and predictable population dynamic may win the race over an exponential population dynamic. Species with this growth pattern tend to be *K*-selected.

For example, suppose you have an area that can support 100 wolves. Let's say that 20 years ago, there were only 50 wolves, so there was abundant space and prey. The 50 wolves were healthy, many pups survived each year, and the population grew rapidly. Now the population has risen to 90. This number is close to the maximum 100 that the environment can support before the wolves begin to deplete their prey. Now, with less food per wolf, fewer cubs are surviving to adulthood, and the rate of increase has slowed. This slowing rate of growth makes an S-shaped curve (fig. 6.6). This S-shaped growth pattern is also called **logistic growth,** because the curve is shaped like a logistic function used in math.

You can describe the general case of this growth by modifying the basic exponential equation with a feedback term—a term that can dampen the exponential growth of *N* (fig. 6.7).

If you are patient, you can see interesting patterns in this equation. Look first at the $\frac{N}{K}$ part. For the wolf example, *K* is 100 wolves, the maximum that can be supported. If *N* is 100, then $\frac{N}{K} = \frac{100}{100}$, which is 1. So $1 - \frac{N}{K} = 1 - 1$, which is 0. As a consequence, the right side of the equation is equal to 0 ($rN \times 0 = 0$), so $\frac{dN}{dt} = 0$. So there is no change in *N* if *N* is equal to the carrying capacity. Try working out the following examples on paper as you read, so you can see how *N* changes the equation.

What if *N* is only 50? Then $\frac{N}{K} = \frac{50}{100} = \frac{1}{2}$. So $1 - \frac{N}{K} = 1 - \frac{1}{2} = \frac{1}{2}$. In this case, the rate of increase is ½*rN*, or half of the maximum possible reproductive rate. If $N = 10$, then $1 - \frac{N}{K} = 1 - \frac{10}{100}$, which is 0.90. So $\frac{dN}{dt}$ is increasing at a rate 90 percent as fast as the maximum possible reproductive rate for that species.

What if the population grows to 120? Overpopulation will likely lead to starvation or low birth rates, and the population will decline to something below 100 again. In terms of the model, now $1 - \frac{N}{K} = 1 - \frac{120}{100} = 0.2$. Now the rate of change, $\frac{dN}{dt}$, is *declining* at a rate of $-0.2rN$.

Logistic growth is **density-dependent,** meaning that the growth rate slows as population density increases. Many factors can cause growth rates to slow as density increases: Overcrowding can increase disease rates, stress, and predation, for example. Reduced food availability or disease can lead to smaller body size and lower fertility rates. **Density-independent** factors also affect populations. These are factors that influence population growth but that do not vary with population density. Often, these are abiotic (nonliving) disturbances, such as drought or fire or habitat destruction, which disrupt an ecosystem.

These values help predict sustainable yield

Natural resource managers are often very interested in predicting a "sustainable yield" for catching or hunting fish, deer, or other economically valuable species. The aim is to maintain both economic use of a species and population survival. Understanding rates of reproduction and population size, that is, *r*, *N*, and *K*, is critical for calculating this balance. A "maximum sustained yield" is the highest number that can be regularly captured while maintaining a reproductive population.

$$\frac{dN}{dt} = rN\left(1 - \frac{N}{K}\right)$$

> Population size as a proportion of carrying capacity

FIGURE 6.7 Logistic growth.

Population biologists have often been very successful in using growth rates to identify a sustainable yield. In North America, game laws restrict hunting of ducks, deer, fish, and other game species, and most hunters and fishers now understand and defend those limits. Acceptable harvest levels are set by population biologists, who have studied reproductive rates, carrying capacities, and population size of each species, in order to determine a sustainable yield. (In some cases, r is now too rapid for K: see What Do You Think? p. 122.) Similarly, the Pacific salmon fishery, from California to Alaska, is carefully monitored and is considered a healthy and sustainable fishery.

Species respond to limits differently: r- and K-selected species

Which is more successful for increasing a population, rapid reproduction or long-term survival within the carrying capacity? Different species place their bets on different strategies. Some organisms, such as dandelions, depend on a high rate of reproduction and growth (rN) to secure a place in the environment. In 1970, Robert MacArthur and E. O. Wilson described these organisms as **r-selected species** because they have a high reproductive rate (r) but give little or no care to offspring, which have high mortality. Seeds or larvae are cast far and wide, and there is always a chance that some will survive and prosper. If there are no predators or diseases to control their population, these abundantly reproducing species can overshoot carrying capacity and experience population crashes, but as long as vast quantities of young are produced, a few will survive. Other organisms reproduce more conservatively—with longer generation times, late sexual maturity, and fewer young. MacArthur and Wilson called these **K- selected species,** because their growth slows as the carrying capacity (K) of their environment is approached.

Many species blend exponential and logistic reproductive patterns, and ecologists often use the broader term **life history strategies** to describe the age-and stage-specific patterns and events that make up an organism's life. Still, it's useful to contrast the advantages and disadvantages of organisms at the extremes of the evolutionary continuum (Table 6.1).

Organisms with r-selected, or exponential, growth patterns tend to occupy low trophic levels in their ecosystems (see chapter 3) or they are successional pioneers. These species, which generally have wide tolerance limits for environmental factors, and thus can occupy many different niches and habitats, are the ones we often describe as "weedy." They tend to occupy disturbed or new environments, grow rapidly, mature early, and produce many offspring with excellent dispersal abilities. As individual parents, they do little to care for their offspring or protect them from predation. They invest their energy in producing huge numbers of young and count on some surviving to adulthood.

A female clam, for example, can release up to 1 million eggs in her lifetime. The vast majority of young clams die before reaching maturity, but a few survive and the species persists. Many marine invertebrates, parasites, insects, rodents, and annual plants follow this life history strategy. Also included in this group are most invasive and pioneer organisms, weeds, and pests.

Table 6.1 Life History Strategies

r-Selected Species	K-Selected Species
1. Short life	1. Long life
2. Rapid growth	2. Slower growth
3. Early maturity	3. Late maturity
4. Many small offspring	4. Few, large offspring
5. Little parental care or protection	5. High parental care or protection
6. Little investment in individual offspring	6. High investment in individual offspring
7. Adapted to unstable environment	7. Adapted to stable environment
8. Pioneers, colonizers	8. Later stages of succession
9. Niche generalists	9. Niche specialists
10. Prey	10. Predators
11. Regulated mainly by extrinsic factors	11. Regulated mainly by intrinsic factors
12. Low trophic level	12. High trophic level

So-called K-selected organisms are usually larger, live long lives, mature slowly, produce few offspring in each generation, and have few natural predators. Elephants, for example, are not reproductively mature until they are 18 to 20 years old. In youth and adolescence, a young elephant belongs to an extended family that cares for it, protects it, and teaches it how to behave. A female elephant normally conceives only once every 4 or 5 years. The gestation period is about 18 months; thus, an elephant herd doesn't produce many babies in any year. Because elephants have few enemies and live a long life (60 or 70 years), this low reproductive rate produces enough elephants to keep the population stable, given good environmental conditions and no poachers.

When you consider the life histories of species you recognize from around the world, can you designate them as r- or K-selected species? What strategies seem to be operating for ants, bald eagles, cheetahs, clams, dandelions, giraffes, or sharks?

Section Review

1. What factors influenced the spread of Asian carp?
2. Define *exponential growth* and *logistic growth*.
3. Explain these terms: r, N, t, dN/dt.

Quantitative Reasoning

The logistic population model (fig. 6.7) is an important idea for explaining how populations grow. To what do the terms dN, dt, r, and K refer? What factors influence changes in population numbers over time? What is the growth rate if $N = K$? Is growth positive or negative if $N > K$? If $N < K$? How does this relate to the boom and bust cycles in figure 6.5?

Too Many Deer?

A century ago, few Americans had ever seen a wild deer. Uncontrolled hunting and habitat destruction had reduced the deer population to about 500,000 animals nationwide. Some states had no deer at all. To protect the remaining deer, laws were passed in the 1920s and 1930s to restrict hunting, and the main deer predators—wolves and mountain lions—were exterminated throughout most of their former range.

As Americans have moved from rural areas to urban centers, forests have regrown, and with no natural predators, deer populations have undergone explosive growth. Maturing at age two, a female deer can give birth to twin fawns every year for a decade or more. Increasing more than 20 percent annually, a deer population can double in just three years, an excellent example of irruptive, exponential growth.

Wildlife biologists estimate that the contiguous 48 states now have a population of more than 30 million white-tailed deer (*Odocoileus virginianus*), probably triple the number present in pre-Columbian times. Some areas have as many as 200 deer per square mile (80/km^2). At this density, woodland plant diversity is generally reduced to a few species that deer won't eat. Most deer, in such conditions, suffer from malnourishment, and many die every year of disease and starvation.

White-tailed deer (*odocoileus virginianus*) can become unhealthy and also reduce other populations when they exceed their environment's carrying capacity.

Howard Sandler/Shutterstock

Overpopulation of deer affects other species as well. Many small mammals and ground-dwelling birds begin to disappear when deer populations reach just 25 animals per square mile. At 50 deer per square mile, most ecosystems are seriously impoverished.

The social costs of large deer populations are high. In Pennsylvania alone, where deer numbers are now about 500 times greater than a century ago, deer destroy about $70 million worth of crops and $75 million worth of trees annually. Nationally, there are over 1.3 million automobile collisions with deer each year. (With an average insurance claim of over $4,300, these amount to over $5.6 billion in damages each year.) Deer help spread Lyme disease, and in some states chronic wasting disease is found in wild deer herds. Some of the most heated criticisms of current deer management policies are in the suburbs. Deer love to browse on the flowers, young trees, and ornamental bushes in suburban yards, arousing the ire of gardeners and home owners.

At the same time, many people are fond of deer. We enjoy watching them—it helps that deer are much easier to spot than rare or nocturnal wildlife—and many people feel sympathy for deer as fellow creatures. Many people feel more connected to nature when they see deer in their neighborhoods.

In remote forest areas, many states have extended hunting seasons, increased the bag limit to four or more animals, and encouraged hunters to shoot does (females) as well as bucks (males). Some hunters criticize these changes because they believe that fewer deer will make it harder to hunt successfully and less likely that they'll find a trophy buck. Others, however, argue that a healthier herd and a more diverse ecosystem is better for all concerned.

In urban areas, increased sport hunting usually isn't acceptable. Wildlife biologists argue that the only practical way to reduce deer herds is culling by professional sharpshooters. Animal rights activists protest lethal control methods as cruel and inhumane. They call instead for fertility controls, the reintroduction of predators, such as wolves and mountain lions, or trap and transfer programs. Birth control works in captive populations but is expensive and impractical with wild animals because of the need for frequent treatments. Trapping is expensive, and few places are willing to take surplus animals, which usually die after relocation, having lost their home territory, resources, and social group.

This case shows that carrying capacity can be more complex than simply the maximum number of organisms an ecosystem can support. While it may be possible for 200 deer to survive in a square mile, the ecological carrying capacity—the population that can be sustained without damage to the ecosystem and to other species—is usually considerably lower. There's also an ethical carrying capacity, if we don't want to see animals suffer from malnutrition, disease, or starvation. There may also be a cultural carrying capacity, if we consider the tolerable rate of depredation on crops and lawns or an acceptable number of motor vehicle collisions.

Try debating this issue with your fellow students. Suppose that some of you are wildlife biologists, charged with managing the deer herd in your state, while others are deer defenders. How would you reconcile the different interests in this issue? What sources of information or ideas shape views for and against population control in deer? What methods would you suggest to reach the optimal population size? What social or ecological indicators would you look for to gauge whether deer populations are excessive or have reached an appropriate level?

6.2 FACTORS THAT REGULATE POPULATION GROWTH

- *Births, immigration, deaths, and emigration impact population size.*
- *Biotic factors are often intrinsic to an organism, while abiotic factors are always extrinsic.*
- *Density-dependent effects can be dramatic.*

By adding carrying capacity, we complicated our first simple population model, and we made it more realistic. To complicate it still further, we can consider the four factors that contribute to *r*, or rate of growth. These factors are **B**irths, **I**mmigration from other areas, **D**eaths, and **E**migration to other areas. More specifically, rate of growth is equal to Births + Immigration – Deaths – Emigration. In a detailed population model, populations receive immigrants and lose individuals to emigration. Number of births might rise more rapidly than number of deaths. Models of human populations (see chapter 7), as well as animal populations, involve detailed calculations of the four BIDE factors.

The two terms that make population grow—births and immigration—should be relatively easy to imagine. Birth rates are different for different species (house flies versus elephants, for example), and a birth rate can decline if there are food shortages or if crowding leads to stress, as noted earlier. Of the two negative terms, deaths and emigration, the emigration idea simply means that sometimes individuals leave the population. Deaths, on the other hand, can have some interesting patterns.

Mortality, or death rate, is the portion of the population that dies in any given time period. Some of mortality is determined by environmental factors, and some of it is determined by an organism's physiology, or its natural life span. Life spans vary enormously. Some microorganisms live whole life cycles in a few hours or even minutes. Bristlecone pine trees in the mountains of California, on the other hand, have life spans up to 4,600 years.

Survivorship curves show life histories

Different organisms have different characteristic rates of growth, maturity, and survival over time. Some organisms have just a few offspring, most of which survive to reproductive age (fig. 6.8, Type I). Survivorship declines sharply in the older, postreproductive phase. This pattern is followed by many larger mammals, such as whales, bears, and elephants (and many human populations). Juvenile survival tends to be fairly high, in part because

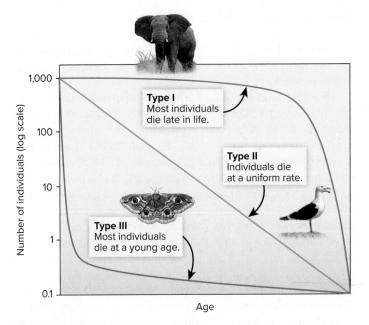

FIGURE 6.8 Idealized survivorship curves include high survival rates for juveniles (Type I), uniform mortality risk throughout life (Type II), and high mortality for juveniles, with long survival for a few individuals that reach maturity (Type III).
(elephant): Digital Vision/Getty Images; (sea gull): Stockbyte; (moth): simon22/123RF

parents invest considerable energy in tending to one or two young at a time. Adult survival is generally high because these organisms have few predators. Some very small organisms also have this type of survivorship curve. Among predatory protozoa, for example, a large proportion survives to what, for a microorganism, is a mature age.

Other organisms are about equally likely to die at any time of their lifespan (fig. 6.8, Type II). Here, the probability of death is unrelated to age, once infancy is past. Sea gulls, mice, rabbits, and other organisms face risks that affect all ages, such as predation, disease, or accidents. Mortality rates can be more or less constant with age, and the survivorship curve can be described as a straight line.

Many organisms produce tremendous numbers of offspring, only a few of which survive to reproductive age (fig. 6.8, Type III). Most species of fish, including Asian carp (opening case study), as well as trees, clams, crabs, and other invertebrate species produce thousands, or even millions, of seeds or eggs. The tremendous number ensures that predators miss at least a few individuals, which are then able to survive to maturity. Those that do survive to adulthood, however, have a very high chance of living nearly the maximum life span for the species.

Intrinsic and extrinsic factors affect births and deaths

So far, we have seen that differing patterns of natality, mortality, life span, and longevity can produce quite different rates of population growth. The patterns of survivorship and age structure

Quantitative Reasoning

In figure 6.8, Type I represents animals such as elephants. At what point in the curve does the number of elephants start to decline? What factors might explain the shape of this curve? (Hint: Refer to table 6.1.) For moths, Type III, at what point does the frequency of survivors change sharply? Explain this survivorship curve in words. Which of these survivorship patterns best describes humans?

created by these interacting factors not only show us how a population is growing but also can indicate what general role that species plays in its ecosystem. They also reveal a good deal about how that species is likely to respond to disasters or resource bonanzas in its environment.

We can describe factors affecting population dynamics in a number of ways. They can be *intrinsic* (operating within individual organisms or between organisms in the same species) or *extrinsic* (imposed from outside the population). Factors can also be **biotic** (caused by living organisms) or **abiotic** (caused by nonliving components of the environment). Biotic factors might include the presence of predators, competitors, or prey species; the abundance of these can strongly influence how many of a given species are born or survive. Abiotic factors include environmental conditions, such as climate, soil types, or moisture availability, which might support an abundance of a species, or make reproduction difficult and mortality high. Occasional events or disturbances, such as hurricanes, droughts, floods, landslides, or fire are also important abiotic factors. These can have devastating impacts on particular populations. They can also have beneficial effects: A spring rain can make the desert burst into flower, and fire can rejuvenate understory plants; some trees, such as the jack pine, even require fire to release seeds from cones, so in this case, an abiotic disturbance is critical to survival.

Finally, some regulatory factors are *density-dependent*. For example, predation or competition may increase as a population grows. Spread of disease in dense populations, or starvation due to population overshoots, are density-dependent factors that can change a population. In laboratory experiments, stress and crowding can also reduce reproduction and increase mortality, although the mechanisms are not well understood.

Others factors are *density-independent, unaffected by population size.* In general, biotic regulatory factors tend to be density-dependent, while abiotic factors tend to be density-independent: Incidence of a fire or drought affects a population regardless of its size.

Interspecific interactions are between species; intraspecific interactions are within a species

The biotic factors noted above often involve interactions among different species, such as predation or competition. Effects of these interactions can be complex: For example, predation may harm a population, or it may benefit the population by keeping numbers in check. While a moose eaten by wolves is harmed individually, the moose *population* is strengthened because the wolves tend to kill old or sick members of the herd. Thus, predation helps prevent population overshoot, so the remaining moose are stronger and healthier.

Sometimes predator and prey populations oscillate in a sort of synchrony with each other. Figure 6.9 represents a classic study, which plotted the number of furs brought into Hudson Bay Company trading posts in Canada between 1840 and 1930. As you can see in their graph, the numbers of Canada lynx fluctuate on about

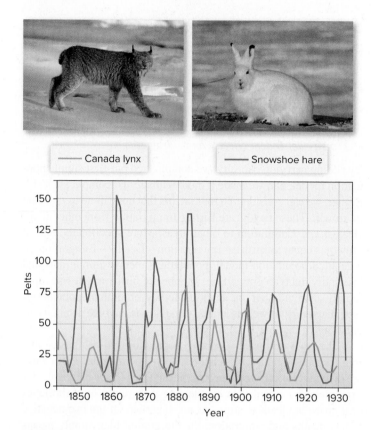

FIGURE 6.9 Ten-year oscillations in the populations of snowshoe hare and lynx in Canada suggest a close linkage of predator and prey, but that might not tell the whole story. These data are based on the number of pelts received by the Hudson Bay Company each year, making fur traders unwitting accomplices in later scientific research.
(a): Erni/Shutterstock; (b): Sophia Granchinho/Shutterstock

a ten-year cycle that is similar to, but slightly out of phase with, the population peaks of snowshoe hares. When prey populations (hares) are abundant, predators (lynx) reproduce more successfully and their population grows. When hare populations crash, so do the lynx. This predator–prey oscillation is known as the Lotka–Volterra model, after the two scientists who first described it mathematically. Subsequent studies have raised doubts about how and where these data were collected, but this remains a classic example of population dynamics.

Intraspecific ("within a species") competition is also a key factor in population dynamics. Individuals of a species often compete for food, territory, mates, and other resources. The stronger, quicker, more aggressive, or luckier members get a larger share, while others get less. If the population exceeds the limits for key resources (if it exceeds the carrying capacity of its environment) it is likely to decline.

Establishing separate territories is one principal way many animal species control access to environmental resources—and to limit competition with neighbors (fig. 6.10). The individual, pair, or group that holds the territory will drive off rivals, usually by threats or displays of superiority (colors, size, dancing ability, bigger antlers or teeth). Among birds, for example, males compete by displaying better colors or songs. Females then select the partners with the best attributes, or with the best territories. From a selective point of view, these successful individuals presumably are superior members of the population and the ones best able to produce offspring that will survive.

Stress and crowding can affect reproduction

Stress and crowding also are density-dependent population control factors. When population densities get very high, organisms often exhibit symptoms of what is called stress shock or **stress-related diseases**—physical, psychological, and/or behavioral changes that are thought to result from the stress of crowding and competition. Crowding-related changes in reproduction are not well understood. Lower reproduction could result from stress. Alternatively, they may simply result from malnutrition, infectious disease, or other more obvious mechanisms.

Some of the best evidence for the existence of stress-related disease comes from experiments in which laboratory animals, usually rats or mice, are grown in very high densities with plenty of food and water but very little living space. A variety of symptoms are reported, including reduced fertility, low resistance to infectious diseases, and pathological behavior. Dominant animals seem to be affected least by crowding, while subordinate animals—the ones presumably subjected to the most stress in intraspecific interactions—seem to be the most severely affected.

Density-dependent effects can be dramatic

The desert locust, *Schistocerca gregarius*, has been called the world's most destructive insect. Throughout recorded human history, locust plagues have periodically swarmed out of deserts and into settled areas. Their impact on human lives has often been so disruptive that records of plagues have taken on religious significance and made their way into sacred and historical texts.

Locusts usually are solitary creatures resembling ordinary grasshoppers. Every few decades, however, when rain comes to the desert and vegetation flourishes, locusts reproduce rapidly until the ground seems to be crawling with bugs. High population densities and stress bring ominous changes in these normally innocuous insects. They stop reproducing, grow longer wings, group together in enormous swarms, and begin to move across the desert. Dense clouds of insects darken the sky, moving as much as 100 km per day. Locusts may be small, but they can eat their own body weight in vegetation every day. A single swarm can cover 1,200 km^2 and contain 50 to 100 billion individuals. The swarm can strip pastures, denude trees, and destroy crops in a matter of hours, consuming as much food in a day as 500,000 people would need for a year. Eventually, having exhausted their food supply and migrated far from the desert where conditions favor reproduction, the locusts die and aren't seen again for decades.

FIGURE 6.10 Individuals within a species often compete for territory and resources. This is called intraspecific competition. Resource scarcity can induce stress and affect reproductive success.
Art Wolfe/Iconica/Getty Images

Huge areas of crops and rangeland in northern Africa, the Middle East, and Asia are within the reach of the desert locust. This small insect, with its voracious appetite, can affect the livelihood of at least one-tenth of the world's population. During quiet periods, called recessions, African locusts are confined to the Sahara Desert, but when conditions are right, swarms invade countries as far away as Spain, Russia, and India. Swarms are even reported to have crossed the Atlantic Ocean from Africa to the Caribbean.

Unusually heavy rains in the Sahara in 2004 created the conditions for a locust explosion. Four generations bred in rapid succession, and swarms of insects moved out of the desert. Twenty-eight countries in Africa and the Mediterranean area were afflicted. Crop losses reached 100 percent in some places, and food supplies for millions of people were threatened. Officials at the United Nations warned that we could be headed toward another great plague. Hundreds of thousands of hectares of land were treated with pesticides, but millions of dollars of crop damage were reported anyway.

This case illustrates the power of exponential growth and the disruptive potential of a boom-and-bust life cycle. Stress, population density, migration, and intraspecific interactions all play a role in this story. Although desert conditions usually keep locust numbers under control, the locust's biotic potential for reproduction is a serious worry for residents of many countries.

Section Review

1. Describe three major types of survivorship patterns, and explain what they show about a species' role in its ecosystem.
2. Explain how biotic and abiotic factors affect population growth.
3. What are the main interspecific population regulatory interactions? How do they work?

How do you measure populations?

To study populations dynamics, you need to be able to count them. Monitoring fish, as in the opening case study for this chapter, is especially tricky. Invasive fish in the muddy Mississippi, particularly bottom-dwelling ones, are invisible and silent. Even large fish are hard to spot and harder still to count. Young fish are still more elusive. How do population biologists get an estimate of numbers?

You can't hope to get a complete count of many populations, so most studies use *sampling* to estimate population size. For example, you can count individuals in a small area, then multiply the number by the total area, assuming the organisms are fairly evenly distributed. For fish, a standard way to get a snapshot of population size and distribution is electroshocking. Researchers insert electrodes into a stream, with a small amount of electric current running between the electrodes (fig. 1). Any fish in the vicinity are temporarily immobilized, so they can be scooped up in nets to be weighed, measured, and counted. Generally, fish recover in a couple of minutes. In a stream, you might collect all the fish along a 500-m distance, then multiply the number caught by the length of the stream.

FIGURE 1. A research crew uses non-lethal electroshocking to collect fish in a stream for census purposes.
David L. Hansen, University of Minnesota Agricultural Experiment Station

For a larger stream or lake, a *mark and recapture* approach is often used to estimate population size. In this approach, you capture and mark individuals (perhaps with a small nick in a fin, or with a small tag), then release them, keeping track of the number marked (M). Marked individuals are then released, to mix with unmarked population. At a later date, the same area is sampled again, and the total number captured is recorded again (C), as well as the number of marked individuals recaptured (R). From these numbers, the total population (N) can be estimated as $N = (M \times C)/R$.

What about highly mobile fish, invasive species, or uncommon species that may not be present when sampling is done? For these, environmental DNA markers are increasingly important in monitoring aquatic systems. A water sample contains traces of genetic material from all the organisms present, from feces, fluids, skin, scales, and so on. The mix of genes present can be compared to a DNA reference library, to check for a species of interest. If the genes are present, there's a good chance the species has been, as well.

Environmental DNA sampling has been a key strategy in monitoring fish, especially invasive carp, in the Mississippi and other water systems. This approach generally indicates presence of a species, rather than abundance, but it has been essential for monitoring bacteria, algae, plankton, and other populations quickly and efficiently.

These are a few of the approaches ecologists use to gather basic data on populations and ecosystem stability.

(header photo): David L. Hansen

6.3 POPULATION SIZE AND CONSERVATION

- *Small and isolated populations are prone to extinction, as predicted by the theory of island biogeography.*
- *Loss of genetic diversity can reduce a population's chances of survival.*
- *Gene flow within a metapopulation can maintain genetic diversity.*

Biologists use all the principles discussed so far in this chapter when trying to manage populations to maintain biodiversity and prevent extinctions. A species' survival depends on its rate of reproduction, growth, and mortality, and the size of the population. It also depends on other factors, such as how widely distributed or isolated a population is. Small or isolated populations of any species can undergo catastrophic declines due to environmental change, loss of genetic diversity, or random accidents. Below a critical size, a small population may not be viable in the long term. This is why understanding population numbers has been an important goal for population biologists (Exploring Science, above) or any other species with declining numbers. In this section, we examine how population size can influence survival.

Small, isolated populations are vulnerable

It has often been observed that small, remote islands have lower biodiversity than larger islands. Why is this? In a classic 1967 study, R. H. MacArthur and E. O. Wilson proposed an explanation they described as the equilibrium theory of **island biogeography.**

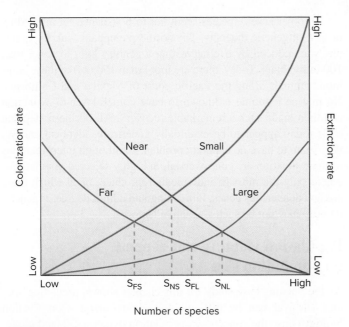

FIGURE 6.11 The theory of island biogeography predicts that the number of species in an island depends on colonization and extinction rates. An island that is small and far from other land masses, for example, should have fewer species (S_{FS}) than a large and near island (S_{NL}).

Source: Modified from The Theory of Island Biogeography, 1967, Princeton University Press.

FIGURE 6.12 Extinction rates of bird species on the California Channel Islands as a function of population size over 80 years.

Source: H. L. Jones and J. Diamond, "Short-term-base studies of turnover in breeding bird populations on the California coast island," in Condor, vol. 78:526–549, 1976.

Their explanation was that diversity (the number of species) in isolated islands depends on rates of colonization and extinction. The rate of colonization depends on proximity of the island to another source of species, such as another island or a continent. If new species can readily reach the island, they may establish new populations there (fig. 6.11). Extinction rates depend on the size of the island and of the population that it can support. Small islands can support only small populations, which are more likely to go extinct due to natural disasters, diseases, or demographic factors such as imbalance between sexes. Larger islands are more likely to sustain larger populations, and thus extinction rates tend to be lower on large islands. The loss of island species, such as birds in Hawai'i, is one of the greatest concerns for population biologists today.

Island biogeographical patterns have been observed in many places. In the Caribbean, for instance, Cuba is 100 times as large as Montserrat and has about 10 times as many amphibian species. Similarly, in a study on the California Channel Islands, Jared Diamond observed that for island bird species with fewer than 10 breeding pairs, 39 percent of the populations went extinct over an 80-year period. In contrast, for populations with 10–100 breeding pairs, only 10 percent went extinct. No species with more than 1,000 pairs disappeared over this time (fig. 6.12). Island biogeographical patterns, with more species in larger and less remote islands, have since been observed in many different types of "islands," such as isolated patches of forest, or isolated wetlands. Often, greater habitat diversity on larger "islands" helps explain this pattern.

Genetic diversity may help a population survive

Genetic diversity plays an important role in the survival or extinction of small, isolated populations. In large populations, genetic variation tends to remain stable, provided environmental conditions remain stable. The ratio of genetic traits tends to remain the same from one generation to the next. This stability in genetic diversity is known as the Hardy–Weinberg principle, named after the two scientists who first described it. A shift in traits could occur, though, if there is environmental change. For example, consider the classic case of finches in the Galapagos Islands, studied by Charles Darwin (chapter 3). These finches originated from larger mainland populations with considerable genetic variation. When they arrived on the small, isolated Galapagos Islands, though, environmental conditions and food resources were constrained. On an island where heavy-shelled seeds were the main food resource, the individuals with slightly stronger beaks tended to survive best, and these came to dominate the local population. The dominant traits for the whole local population shifted. We call the gradual changes in gene frequencies due to random events **genetic drift.** The presence of genetic variety in a population can be important: If environmental conditions shift, suitable traits are likely to be present that help the species survive.

In small or isolated populations, the pool of traits is smaller. Genetic drift can occur because unusual traits in a few individuals can become dominant in a small population.

The reduction in genetic diversity in a small population is sometimes called a **founder effect,** when it results from a small colonizing population. When a population, and its genetic diversity, are dramatically reduced, for example, by a catastrophic event, we call it a **demographic bottleneck** (fig. 6.13).

Sometimes the loss of genetic diversity can limit adaptability, reproduction, and species survival. Any deleterious genes present in the founders will be overrepresented in subsequent generations. Inbreeding, the mating of closely related individuals, also makes

Parent population | Bottleneck (drastic reduction in population) | Surviving individuals | Subsequent generations

FIGURE 6.13 Genetic drift resulting from a bottleneck effect. The parent population contains roughly equal numbers of blue and yellow individuals. Only a few individuals make it through a "bottleneck" that reduces the population, such as an environmental disturbance. If those few individuals are blue, then subsequent generations will be mostly blue.

expression of rare or recessive genes more likely. Cheetahs appear to have undergone a demographic bottleneck sometime in the not-too-distant past. All the male cheetahs alive today appear to be nearly genetically identical, suggesting that they all share a single male ancestor (fig. 6.14). This lack of diversity is thought to be responsible for an extremely low fertility rate, a high abundance of abnormal sperm, and a low survival rate for offspring. All these factors threaten the survival of the species.

FIGURE 6.14 Sometime in the past, cheetahs underwent a severe population crash. Now, all male cheetahs alive today are nearly genetically identical, and deformed sperm, low fertility levels, and low infant survival are common in the species.
Barry Barker/McGraw-Hill Education

In other cases, species seem not to be harmed by inbreeding or lack of genetic diversity. The northern elephant seal, for example, was reduced by overharvesting a century ago, to fewer than 100 individuals. Today, there are more than 150,000 of these enormous animals along the Pacific coast of Mexico and California. No marine mammal is known to have come closer to extinction and then made such a remarkable recovery. All northern elephant seals today appear to be essentially genetically identical and yet they seem to have no apparent problems. Although interpretations of their situation are controversial, in highly selected populations, where only the most fit individuals reproduce, or in which there are few deleterious genes, inbreeding and a high degree of genetic identity may not be such a negative factor.

Population viability can depend on population size

As the opening case study for this chapter shows, predicting species survival can be difficult and controversial. Conservation biologists use the concepts of population dynamics, island biogeography, genetic drift, and founder effects to determine **minimum viable population size,** or the number of individuals needed for long-term survival of rare and endangered species. A classic example is that of the grizzly bear in North America. Before European settlement, grizzlies roamed from the Great Plains west to California and north to Alaska. Hunting and habitat destruction reduced the number of grizzlies from an estimated 100,000 in 1800 to fewer than 1,200 animals in six separate subpopulations, which now occupy less than 1 percent of the historic range. Biologists have calculated that for some subpopulations, the environmental carrying capacity is fewer than 100 bears. But genetic models predicts that an isolated population of 100 bears is unlikely to persist for more than a few generations. Even the 600 bears now in the Greater Yellowstone Ecosystm in Wyoming and Montana will be susceptible to genetic problems if completely isolated. Interestingly, computer models suggest that translocating only two unrelated bears into small populations every generation (about ten years) could greatly increase population viability.

For mobile organisms, separated populations can have gene exchange if suitable corridors or migration routes exist. A **metapopulation** is a collection of populations that have regular or occasional gene flow between geographically separate units (fig. 6.15). For example, the northern spotted owl is an endangered species occupying remnant patches of the once-continuous forest of the Pacific Northwest. Historically, there has been interaction among forest areas, so that the many subpopulations are considered a metapopulation.

In a metapopulation, some habitat areas may be better than others at supporting reproductive subpopulations. A "source" habitat, where birth rates are higher than death rates, produces surplus individuals that can migrate to new locations within a metapopulation. In "sink" habitats, on the other hand, mortality exceeds birth rates. For spotted owls, sink habitats might be those too small to support sufficient prey (various small rodents) to raise young successfully, or it might be fragmented habitats where spotted owls

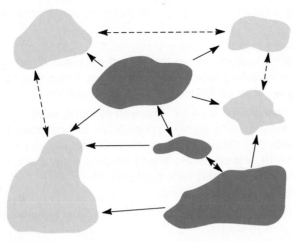

FIGURE 6.15 A metapopulation is composed of several local populations linked by regular (*solid arrows*) or occasional (*dashed lines*) gene flows. Source populations (*dark*) provide excess individuals, which emigrate to and colonize sink habitats (*light*).

must compete with barred owls, which thrive in mixed environments. The species would disappear from sink habitats if it were not periodically replenished from a source population. Recent studies of a metapopulation model for spotted owls predict just such a problem for this species in the Pacific Northwest. If many of these endangered owls end up in habitat patches too small to support reproduction, the entire population may decline. In this case and many others, understanding population biology and population dynamics is critical to conserving the species.

Section Review

1. What is island biogeography and how does it explain population survival?
2. Why does genetic diversity tend to persist in large populations, but gradually drift or shift in small populations?
3. Define the following: *metapopulation, genetic drift, demographic bottleneck.*

Connecting the Dots

Biological populations exploit diverse niches, and they generally increase as rapidly as resources will support. Invasive carp species demonstrate the power of exponential growth—in which a species can multiply to extraordinary numbers in a few years. Like many species, carp produce prodigious numbers of eggs, and their growth rate is limited mainly by their rate of reproduction (r). Large or predatory species are more likely to grow slowly up to their environment's carrying capacity (K). But most populations are likely to crash if they exceed their carrying capacity. Asian carp, which have been expanding into new territories, have evidently not reached the point of a dieback yet.

In addition to births and deaths, immigration to and emigration from an area help determine population size. Some population-limiting factors are density-dependent, for example limits to food or habitat resources, or intraspecific competition. Factors that are not density-dependent tend to be abiotic.

One of the reasons carp have been successful is that they occupy expansive and well-connected river systems. Small and isolated populations are more likely to die out—thus many island populations are at risk—but large and well-connected populations tend to persist. How the interaction of carp species ultimately changes the distribution of other species in the system is one of the great questions of fish management, and of population biology, for the Mississippi region.

Reviewing Key Terms

Can you define the following terms in environmental science?

abiotic 6.2
biotic 6.2
biotic potential 6.1
carrying capacity 6.1
demographic bottleneck 6.3

density-dependent 6.1
density-independent 6.1
exponential growth 6.1
founder effect 6.3
genetic drift 6.3

island biogeography 6.3
K-selected species 6.1
life history strategies 6.1
logistic growth 6.1
metapopulation 6.3

minimum viable population size 6.3
population crash 6.1
r-selected species 6.1
stress-related diseases 6.2

Critical Thinking and Discussion Questions

1. Compare the advantages and disadvantages to a species that result from exponential or logistic growth. Why do you think hares have evolved to reproduce as rapidly as possible, while lynx appear to have intrinsic or social growth limits?

2. Are humans subject to environmental resistance in the same sense that other organisms are? How would you decide whether a particular factor that limits human population growth is ecological or social?

3. What are advantages and disadvantages in living longer or reproducing more quickly? Why hasn't evolution selected for the most advantageous combination of characteristics so that all organisms will be more or less alike?

4. Why do abiotic factors that influence population growth tend to be density-independent, while biotic factors that regulate population growth tend to be density-dependent?

5. Some people consider stress and crowding studies of laboratory animals highly applicable in understanding human behavior. Other people question the cross-species transferability of these results. What considerations would be important in interpreting these experiments?

6. What implications (if any) for human population control might we draw from our knowledge of basic biological population dynamics?

Data Analysis

Experimenting with Population Growth

The previous data analysis lets you work through an example of population growth by hand, which is an important strategy for understanding the equations you've seen in this chapter.

Now try experimenting with more growth rates in an Excel "model." What value of r makes the graph extremely steep? What value makes it flat? Can you model a declining population?

Go to Connect to find the Excel spreadsheet that lets you experiment with different growth rates, as well as questions to demonstrate your understanding.

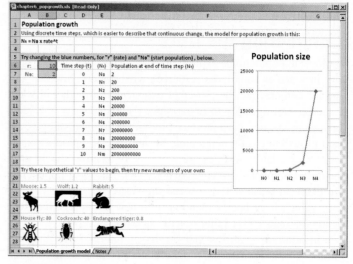

Source: Microsoft

7

Human Populations

▲Senior citizens gather in Xingyang parks every morning to practice ballroom dancing, do tai chi, and participate in exercise programs.
William P. Cunningham

Learning Outcomes

After studying this chapter, you should be able to:

7.1 Trace the history of human population growth.

7.2 Summarize different perspectives on population growth.

7.3 Analyze some of the factors that determine population growth.

7.4 Describe how a demographic transition can lead to stable population size.

"For every complex problem there is an answer that is clear, simple, and wrong."

– *H. L. Mencken*

China Is Aging

As dawn breaks over Xingyang, a medium-size city of about 600,000 people in east central China, elderly people appear in parks to practice ballroom dancing, do tai chi, and exercise while the morning is cool. In coming decades, the parks are expected to get more crowded. Retired people make up an ever-growing segment of Xingyang's population, like that of the rest of China. In 1980, people over 60 made up less than 5 percent of China's population. In 2018, they were 17 percent, and by 2030, they will constitute 35 percent, or nearly 500 million seniors. This will be the largest group of elderly on the planet. There are several reasons for this rapid growth: (1) life expectancy has risen rapidly, from 44 years in 1950 to 76 years now; (2) the large generation of baby boomers born in the 1960s and 1970s is now reaching retirement age; and (3) births have declined sharply. This decline is often attributed to an aggressive one-child-per-family policy from 1980–2015, although it began in the 1960s (fig. 7.1). We'll discuss this policy later in this chapter.

In fact, Xingyang may be an epicenter of population decline in China. The fertility rate here is only 0.4 children per woman, less than one-fifth of the replacement rate of 2.1 children per woman (the number of births to offset deaths). As a consequence, young people make up a smaller and smaller segment of society, and fewer people are entering the work force, a serious threat to the Chinese economy (China's overall fertility rate is 1.6 children per woman). Furthermore, the imbalance between youth and seniors means there are more people to care for even as number of caregivers declines. This problem is especially serious in rural areas, which young people abandon for the city in search of education and jobs. Many rural villages are now inhabited almost entirely by grandparents and grandchildren.

As a developing society where grandparents traditionally lived with their extended families, China lacks institutions and facilities to care for the elderly. But now about one-quarter of Chinese seniors need care and have no family nearby to help them. The few nursing homes or elder care services can only accommodate about one percent of the senior population. And only one-third of rural residents are covered by pensions, averaging about $12 per month. Leaders worry that growing demands for a better social security system, coupled with a shrinking economy, could cause political instability.

China is extreme but not alone in these trends. Nearly half the world's population lives in nations that are below the replacement fertility rate of 2.1 children, so aging populations are a common concern. Richer countries can depend on immigrants to replenish their work force and care for the elderly, but anti-immigrant sentiments make many countries reluctant to do this. Many people are wondering what the future of care looks like.

In environmental science, the main demographic concern has been rapid growth, ever since the 1968 publication of Paul Ehrlich's book, *The Population Bomb*. Published when growth was at a historic high, the book sounded alarms about environmental impacts of population expansion, especially in developing countries. Many have considered the book xenophobic, but it may have also contributed to global family planning efforts.

Population momentum means that our numbers are still growing. But now we have new questions like those in Xingyang. What do demographic trends mean for quality of life? How does an aging population care for itself? What factors most influence rates of growth? As evident in figure 7.1, state policy is one of multiple factors. Does slowing growth mean that we will stay within the earth's carrying capacity, rather than overshoot it, as described in chapter 6? What might that carrying capacity be? And does slowing growth mean we can lessen our impacts on environmental systems and biodiversity? All of these questions intersect with major themes in environmental science. Try to decide your answers to them as you read this chapter.

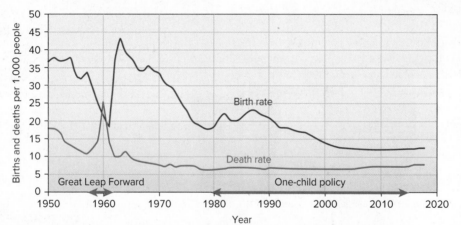

FIGURE 7.1 Birth and death rates in China per 1,000 people.
UN Population Division

7.1 PERSPECTIVES ON POPULATION

- *Human populations have grown rapidly in the past two centuries.*
- *Rapid growth has important social and environmental costs.*
- *Poverty contributes to population growth.*
- *Technology can both change carrying capacity and cause further damage.*

Every second, on average, 4.3 children are born somewhere on the earth. In that same second, 4 people die. This difference between births and deaths means a net gain of roughly 0.3 more humans per second in the world's population. In 2019, the total world population was 7.7 billion people and was growing at 1.08 percent per year. This means we are now adding nearly 83 million more people per year, and if this rate persists, our global population will double in about 65 years. Already humans are among the most numerous and widely distributed of vertebrate species on the earth. We also have a greater global environmental impact than any other species. For families, the birth of a child is a joyous event (fig. 7.2). But collectively, what is the effect of our growing population? The answers to this question depend on our estimates of the causes of change, rates of change, and ideas about resource consumption per person.

Many people worry that overpopulation is causing resource depletion and environmental degradation that threaten the ecological life-support systems on which we all depend. Recall from chapter 6 that carrying capacity is the maximum population size that can be supported for a particular species in a specific environment without depleting resources. Are there enough resources for 10 or 11 billion people to live a good and healthy life? Are we on track to overshoot the Earth's carrying capacity?

Alternatively, can human ingenuity, technology, and enterprise extend the world carrying capacity and allow us to live within environmental limits? A larger population means a larger

FIGURE 7.2 The number of children a family chooses to have is determined by many factors. In urban industrialized countries like Brazil, most families now want only one or two children.
EyeWire/Photodisc Collection/Getty Images

workforce, more inventions, more ideas about better solutions. Along with every new mouth comes a pair of hands. Can continued economic and technological growth provide the stability to reduce population growth voluntarily?

Another question is whether we can address social justice concerns. As the opening case study for this chapter shows, rapid population stabilization can cause stresses on societies, economies, and political systems. Countries with the highest current growth rates are also some of the poorest in the world. But the richest countries use the most resources. Many argue that there are sufficient resources for everyone, and the root cause of environmental degradation is inequitable distribution of wealth and power, rather than simple population size. Fostering democracy, empowering women and minorities, and improving the standard of living of the world's poorest people are what are really needed, from this point of view. A narrow focus on population growth can foster racism and blame the poor for their own poverty, while ignoring the deeper social and economic forces at work.

Will human populations continue to grow at present rates? The United Nations Population Division reports that the era of high fertility and rapid population growth is ending. Most of the world is now changing to stable or even declining population. What will this imply for environmental quality and for human life? These are among the most central and pressing questions in environmental science.

In this chapter we examine human population dynamics. A general term for population characteristics is **demographics**, from the Greek *demos* (people) and *graphein* (to write or measure). Demographic measures include statistics such as birth and death rates, age and gender distributions, and total population size. We examine how and why these factors are changing, and how that matters for population size and for our environmental impacts.

How many of us are there?

There were about 7.7 billion people in the world in 2019. This is an approximation, since it's not possible to do a precise count of that many people, with constant births and deaths and migrations. All national census counts have errors, and some countries have no censuses at all. Despite these uncertainties, demographers have developed confidence that their estimates are within reasonably small margins of error.

Benchmark Data	
Among the ideas and values in this chapter, the following are a few worth remembering.	
7.7 billion	Global population, 2019
70/*n*	Doubling time (*n* = percentage growth rate)
1.09%	Global percentage growth rate (2018)
2.43	Global average fertility rate (children/woman)
2.1	Replacement fertility rate
69%	Percentage of population in countries < 2.1 (2050)

Table 7.1	World Population Growth and Doubling Times	
Date	Population	Doubling Time
5000 B.C.E.	50 million	?
800 B.C.E.	100 million	4,200 years
200 B.C.E.	200 million	600 years
1200 C.E.	400 million	1,400 years
1700 C.E.	800 million	500 years
1900 C.E.	1,600 million	200 years
1965 C.E.	3,200 million	33 years
2004 C.E.	6,400 million	58 years
2050 C.E. (estimate)	9,710 million	155 years

United Nations Population Division, U.S. Census Bureau

This 7.7 billion is twice the number that existed 50 years ago. That magnitude of this change helps explain why many ecologists worry about our effects on other species and resources. Demographers predict growth in the near future with some confidence. Of course, dramatic changes can happen, as with birth rates in fig. 7.1, but demographers anticipate that the number by 2050 will be about 10 billion.

Human populations grew slowly until relatively recently

For most of our history, humans lived by hunting and gathering, and our population was probably not more than a few million. The invention of agriculture and the domestication of animals around 10,000 years ago allowed our numbers to grow. With a larger and more secure food supply, the population reached perhaps 50 million people by 5000 B.C.E Still, for thousands of years, the number of humans increased slowly. Archaeological evidence and historical descriptions suggest that only about 300 million people were living 200 years ago. It took all of human history to reach 1 billion people in 1804, but little more than 150 years to reach 3 billion in 1960. To go from 5 to 6 billion took only 12 years (fig. 7.3).

Demographers often describe this kind of growth in terms of doubling times (table 7.1). Over time, the growth rate has accelerated, and the time needed to double the population size has decreased from several hundred years to just a few decades. Population growth was most rapid in the early 1960s, when the global population doubling time fell to 33 years or less. The doubling time is now lengthening again, as birth rates slow around the world. Demographers expect that by 2050 the global population will reach about 9.7 billion, with a doubling time of 155 years or more. However, most analysts hope that the population will not actually double again. While we use doubling time as a relative measure of growth rates, the population is expected to stabilize by about 2100 at around 9 to 11 billion.

Until the Middle Ages, human population growth was limited by diseases and poor nutrition, which caused high child mortality and short life expectancy for those who reached adulthood. There is evidence that many early societies regulated their population size through cultural taboos and practices such as abstinence and infanticide. War and famine also caused episodes of population decline. Among the most destructive of natural population controls were bubonic plagues (or Black Death) that periodically swept across Europe between 1348 and 1650 (see fig. 7.3). It has been estimated that during the worst plague years (1348–1350) at least

FIGURE 7.3 Human population levels through history. Since about 1000 C.E., our population curve has assumed a J shape. Are we on the upward slope of a population overshoot? Will we be able to adjust our population growth to an S curve? Or can we just continue the present trend indefinitely?

one-third of the European population perished. Notice, however, that this didn't retard population growth for very long. In 1650, at the end of the last great plague, there were about 600 million people in the world.

As you can see in figure 7.3, human populations began to increase rapidly after 1600 C.E. Many factors contributed to this rapid growth. Increased sailing and navigating skills stimulated commerce and communication between nations. Trade increased access to resources, including more food. Agricultural developments, better sources of power, and better health care and hygiene also played a role.

We have had an exponential or J curve pattern of growth for centuries. Recently, however, the growth rate has begun flattening, and turning into an S-shaped curve (see section 7.3). If we do not shift to an S-shaped curve, will we overshoot the carrying capacity of our environment and experience a catastrophic dieback similar to those described in chapter 6? Or we will reach equilibrium soon enough and at a size that can be sustained over the long term? No one knows, of course, but in the sections that follow we analyze factors that influence growth rates.

Do large families cause poverty, or does poverty cause large families?

As with many topics in environmental science, people have widely differing opinions about population and resources. Some believe that population growth is the ultimate cause of poverty and environmental degradation. Others argue that poverty, environmental degradation, and overpopulation are all merely symptoms of deeper social and political factors. The worldview we choose to believe will profoundly affect our approach to population issues. In this section, we will examine some of the major figures and their arguments in this debate.

Since the time of the Industrial Revolution, when the population growth began accelerating rapidly, scholars have argued about the causes and consequences of population growth. One of the most influential explanations today was proposed in 1798 by Thomas Malthus (1766–1834), in *An Essay on the Principle of Population*. Malthus argued that populations tend to increase exponentially while food supplies tend to increase linearly. Eventually, human populations should therefore outstrip their food supply and collapse into starvation, crime, and misery. Population growth, Malthus said, slows only when disease or famine reduces population size, or when constraining social conditions—late marriage, insufficient resources, celibacy, and "moral restraint"—compel a population to reduce birth rates.

In Malthusian terms, then, population growth causes resource depletion and poverty (fig. 7.4*a*). At that time, most economists believed that high fertility was good for the economy, increasing gross domestic output. Malthus converted them to believing that population growth was an economic danger, because per capita output actually fell with rapidly increasing population, and because large populations deplete resources rapidly.

Malthus was observing population shifts at a time of dramatic social change and industrialization in Europe. Fast-growing cities

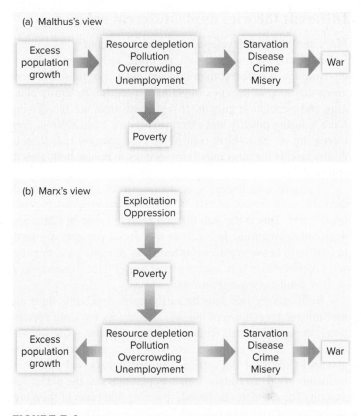

FIGURE 7.4 (a) Thomas Malthus argued that excess population growth is the ultimate cause of many other social and environmental problems. (b) Karl Marx argued that oppression and exploitation are the real causes of poverty and environmental degradation. Population growth in this view is a symptom or result of other problems, not the source.

teemed with poor laborers in the new factories, powered by newly invented steam engines. Meanwhile, the countryside was cleared of peasants to make way for expanding production of beef, wool, and other goods. Population was increasing and shifting, and hunger, poverty, and illness were widespread.

Later writers had different explanations for these conditions. Where Malthus understood poverty to result from population growth, Karl Marx (1818–1883) understood population growth to result from poverty. He argued that exploitation of workers, especially low wages and job insecurity, causes a variety of social ills. Poverty can lead to high birth rates, for example, when people in insecure or desperate conditions lack either opportunities (employment or education) or social stability that discourage early and frequent childbearing. In these conditions, large families can exacerbate insecurity.

Others since Marx have argued that the wealthy consume more resources, and cause more resource depletion, than the poor do. Both population growth and exploitation, then, cause resource depletion (fig. 7.4*b*). Reversing poverty, population growth, and environmental degradation, according to this perspective, requires the elimination of exploitation and oppression of the poor. As Mohandas Gandhi stated, "There is enough for everyone's need, but not enough for anyone's greed."

Different theories imply different solutions

Marx and Malthus developed their theories about human population growth when views about the world, technology, and society were much different than they are today. But these different views continue to inform competing approaches to family planning and population growth. If high birth rates are the driving force, causing poverty and extreme resource consumption, then increasing access to birth control, or even coercive reduction in family size, is the most important strategy to reduce both growth and poverty.

If poverty contributes to population growth, on the other hand, then the most important task is to reduce poverty and improve opportunity. This is the path that has been followed in China and many other countries. In much of the world, population growth has fallen to below replacement levels, as education and employment opportunities have improved and people have chosen to have smaller families (see section 7.4).

Both perspectives may be partly right. And while there are multiple interpretations of the best strategy for reducing populations, the question still remains whether, or how soon, we will exceed the earth's carrying capacity. The mathematical biologist Joel Cohen, of Rockefeller University, reviewed published estimates of the maximum human population size the planet can sustain. The estimates he found, spanning 300 years of thinking, converged on a median value of 10–12 billion. We are more than 7.7 billion strong today, and growing, an alarming prospect for some (fig. 7.5). Cornell University entomologist David Pimentel, for example, has said: "By 2100, if current trends continue, twelve billion miserable humans will suffer a difficult life on Earth."

Would you agree with Pimentel or with Ghandi? If environmental quality and resource consumption are the ultimate concern, then is population growth the most important problem to address, or are there other drivers of resource consumption that deserve attention?

Technology can change carrying capacity

Still another view is that humans will not exceed our environmental carrying capacity because we are innovative. **Technological optimists,** those who argue that innovation will allow us to overcome limits, argue that resources do not necessarily constrain human populations as tightly as other species. As resources become scarce, we tend to invent alternative resources or strategies. They note that Malthus was wrong 200 years ago in his predictions of famine and disaster because he failed to account for scientific and technical progress.

For example, even though urban populations have grown dramatically since Malthus's time, we have invented systems of trade, transportation, sanitation, and building that make cities healthier and safer than in the 1790s. And despite having nearly ten times the global population of Malthus's time, we now have have more and better food resources per capita than ever

FIGURE 7.5 Is the world overcrowded already, or are people a resource? In large part, the answer depends on the kinds of resources we use and how we use them. It also depends on democracy, equity, and justice in our social systems.
William P. Cunningham

before. According to the UN Food and Agriculture Organization (FAO), the average per capita daily food consumption across the whole world has grown from 2,400 calories per day in 1970, to 2,900 calories in 2018. This is well above the 2,200 calories considered necessary for an average adult. Even poorer developing countries saw a rise, from an average of 2,100 calories per day in 1970 to 2,700 now. In that same period, the world population went from 3.7 to more than 7 billion people. Of course, food resources are not evenly distributed. Over 800 million people still suffer from persistent hunger. But conflict and economics explain most of this hunger.

The burst of world population growth that began 200 years ago was stimulated by scientific and industrial revolutions. Progress in agricultural productivity, engineering, information technology, commerce, medicine, sanitation, and other achievements of modern life have made it possible to support thousands of times as many people per unit area as was possible 10,000 years ago.

Environmental Impact (I) = PAT

Technology tends to increase our environmental impacts as well as improving our survival rate. Our environmental impacts result from a combination of population size and how much we consume per person. A summary for this explanation of impacts is **I = PAT:** Environmental impacts (I) equal the product of our population size (P) times affluence (A) and the technology (T) used to produce the goods and services we consume (fig. 7.6).

While increased standards of living in China for example, have helped stabilize population, they also bring about higher technological impacts. A family living an affluent lifestyle that

Impact = Population × Affluence × Technology

FIGURE 7.6 Environmental impacts of population growth (I) are the product of population size (P), times affluence (A), times the technology (T) used to create wealth.

depends on high levels of energy and material consumption, and that produces excessive amounts of pollution, could cause greater environmental damage than a whole village of hunters and gatherers or subsistence farmers.

If the billions of people in Asia, Africa, and Latin America develop their economies with the same resource-intensive technology as that used in North America or Europe, the environmental effects will be disastrous. Already, growing wealth in China, whose middle class is expected to reach 700 million by 2020, or approximately twice the total population of the United States, is straining global resources and has made China the world's largest emitter of CO_2. There are now more millionaires in China than in all of Europe, and China has passed the United States in annual automobile production. As noted in the discussion of ecological footprints in chapter 1, it would take about four additional Earths to support everyone if the entire world were living an American lifestyle—using current technology.

But China has also become a global leader in renewable energy, public transportation, and electric vehicles. Ideally, all of us will soon be using nonpolluting, renewable energy and material sources. If we can extend the benefits of environmentally friendly technology to the poorer people of the world, then everyone can enjoy the benefits of a better standard of living without the environmental costs historically associated with development and growth.

Population growth can power innovation

Large populations of United States, China, and other regions consume tremendous resources, but they also represent gigantic economic engines and sources of innovation. More people mean larger markets, more workers, and efficiencies of scale in mass production of goods. If future economies are based largely on

services, such as education, information, or data management, then larger populations mean larger markets for those services. Moreover, adding people boosts human ingenuity and intelligence that will create new resources by finding new materials and discovering new ways of doing things.

Economist Julian Simon (1932–1998), a champion of this rosy view of human history, believed that people are the "ultimate resource" and that no evidence suggests that pollution, crime, unemployment, crowding, the loss of species, or any other resource limitations will worsen with population growth. In a famous bet in 1980, Simon challenged Paul Ehrlich, author of *The Population Bomb,* to pick five commodities that would become more expensive by the end of the decade. The resources Ehrlich chose— copper, chromium, nickel, tin, and tungsten—actually declined in price by an average of 50 percent. Simon, the technological optimist, won the bet. It has been argued that the 10 year span was too short, or that different resources might have performed differently, but it remains true that resource substitution has made many formerly critical resources cheap or obsolete. Leaders of developing countries insist that, instead of being obsessed with population growth, we should focus on reducing consumption of the world's resources by people in richer countries.

Section Review

1. At what point in history did the world population pass its first billion? What is the total population now?
2. How did Malthus and Marx differ in their explanations of growth?
3. What is the I = PAT formula, and what does it mean?

7.2 WAYS WE DESCRIBE GROWTH

- *Births, deaths, immigration, and emigration (BIDE) determine growth.*
- *Fertility rates have fallen in most countries.*
- *Life expectancies have risen in most countries.*

Governments like that in China often seek to influence population size, either because they are worried about providing resources for growing numbers, or because they want larger populations. But as we have noted, there have been conflicting theories about what factors most influence population growth. In this section, we review some of these factors and the terms used to describe growth rates.

In broad terms, the factors affecting changes in population size are birth and death rates together with immigration and emigration (fig. 7.7). As discussed in chapter 6, this relationship is often summarized as BIDE, or Births + Immigration − Deaths − Emigration.

We describe growth rates in several ways

We often compare countries in terms of **total growth rate,** the net change in population per year, expressed as a percentage. Low-income countries often have a high growth rate. For example,

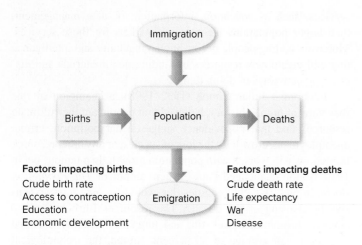

$$Population_{(t)} = Population_{(t+1)} + B + I - D - E$$

FIGURE 7.7 We can describe population at a given time (t) as a function of the population size at a previous time (t-1) + births + immigration − deaths − emigration.

Niger, a poor desert country in West Africa, has one of the world's highest growth rates at 2.7 percent per year (table 7.2). Portugal, on the other hand, has a developed economy characteristic of much of Europe and North America. Its population growth rate is −0.4 percent, a slight decline. China is intermediate, at 0.5 percent per year, but note that 0.5 percent of 1.4 billion people is a large number of people added per year.

We also describe growth in terms of doubling times (see table 7.1). You can easily use the "**rule of 70**" to calculate population doubling times from growth rate. If you divide 70 by the annual percentage growth, you will get the approximate doubling time in years. Niger, for example, with a growth rate of 2.7 percent per year, will double its population in just 26 years.

TABLE 7.2	Demographic Metrics of Selected Countries		
Metric	Niger	China	Portugal
Growth rate %	2.7	0.5	−0.4
Births/1,000	49	13	8
Deaths/1,000	11	7	10
Total fertility[1]	7.4	1.6	1.3
Life expectancy[2]	59	76	81
Under 5 mortality[3]	119	14	4
% less than 15	50	18	14
% over 60	4	16	28
Annual income	$917	$7,329	$27,124

[1]average per woman in lifetime
[2]at birth
[3]per 1,000 births
UN Population Division 2019

The world's total growth rate peaked in 1960 at 2.1 percent (with a doubling time of about 33 years), and it is now about 1.1 percent (about 64 years to double, if the rate did not change). Note that while growth *rates* can change fairly rapidly, total population *size* changes gradually. Children born during times of high growth rates will mature and have children decades later, and so population size stabilizes many years after growth rate declines (fig. 7.8). Demographers call this **population momentum.**

Demographers also report **crude birth rate,** the number of births in a year per thousand people in any given year. It is statistically "crude" in the sense that it is not adjusted for population characteristics such as the number of women of reproductive age. In low-income countries, birth rates tend to be high, around 35 to 40 births per 1,000 people in a given year. **Crude death rate** (or crude mortality rate) is the number of deaths per thousand persons. Countries with limited health care and sanitation may have mortality rates of 20 or more per 1,000 people. The number of deaths in a population is sensitive to the age structure of the population. Rapidly growing, developing countries such as Qatar, have lower crude death rates (1 per 1,000) than do the more-developed, slowly growing countries, such as Portugal (10 per 1,000). This is because there are proportionately more children and fewer elderly people in a rapidly growing country than in a more slowly growing one.

Mortality rates influence birth rates. In a country like Niger, very high mortality of children under 5 and short life expectancy encourage families to have more children (table 7.2). Child mortality, in turn, is associated with poverty: In Niger, 43.6 percent of the population is below the international poverty line of US $1.90 per day, and the adult female literacy rate is around 10 percent. Portugal, in contrast, has low birth and death rates, low fertility, and low child mortality. Portugal is an aging country, with twice as many old people as youngsters. While not among the richest countries in the world, Portugal has 30 times as much income per person per year as Niger. China, while still growing, has transitioned from a profile like Niger's to one more like Portugal's.

Fertility rate is the number of children per woman

The **total fertility rate** is the number of children born to an average woman in a population during her entire reproductive life. In most tribal or traditional societies, food shortages, illness, and cultural practices limit total fertility to about six or seven children per woman even without modern birth control methods. On the other hand, male pride, religious dogma, or a lack of women's control of their own bodies can result in very high fertility.

Upper-class women in seventeenth- and eighteenth-century England, whose babies were given to wet nurses immediately after birth and who were expected to produce as many children as possible, sometimes had 25 or 30 pregnancies. The highest recorded total fertility rates for working-class people is among some Anabaptist agricultural groups in North America, who have averaged up to 12 children per woman.

The **replacement fertility rate** (sometimes called zero population growth) occurs when birth rates just compensate for deaths. In general, this number is 2.1 children per woman. If there are roughly equal numbers of men and women in a population, then each woman would need to produce about two children, with a few extra to account for childless couples and early deaths. Where infant mortality rates are high, the replacement level may actually be higher, perhaps five or more per couple to compensate for deaths. In general, though, 2.1 is the benchmark value for the replacement rate. It takes several generations of replacement level fertility (where people just replace themselves) to reach a stable population size.

Fertility rates are falling globally

Fertility rates have declined in every region of the world except sub-Saharan Africa over the past 50 years (fig. 7.9). The map in fig. 7.8 corresponds almost exactly to a map of world poverty levels. In the 1960s, total fertility rates above 6 were common in many countries. The average family in Mexico in 1975, for instance, had seven children. By 2010, however, the average Mexican woman had only 2.3 children. Similarly, in Iran, total fertility fell from 6.5 in 1975 to 2.04 in 2010. According to the World Health Organization, 100 out of the world's 220 countries are now at or below a replacement fertility, and by 2050 all but a few of the least-developed countries are expected to have reached that milestone. For many of these countries, population growth will continue for a generation because they have such a large number of young people. China, for example, now has a fertility rate of only 1.6 children per woman. But 26 percent of its population is under 14 years old. Those children will mature and start to have families before their parents and grandparents die, so the population will continue to grow for some decades.

Some of the greatest fertility reduction in the twentieth century occurred in China (What Do You Think? p. 149). But as a result of selective abortions for girls (who are less valued in Chinese society), China now reports that 119 boys are being born for every 100 girls. Normal ratios would be about 105 boys to 100 girls. This imbalance is causing a shortage of brides, especially in rural areas.

Although the world as a whole still has an average total fertility rate of 2.4, growth rates are now lower than at any time since World War II. Most of Eastern Europe now has

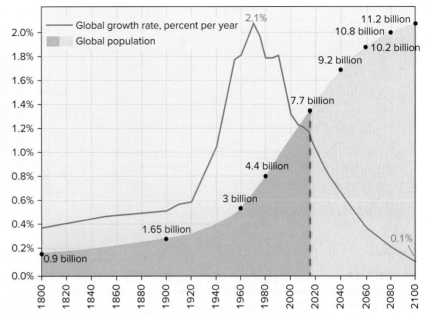

FIGURE 7.8 Population growth rates (percentage per year) peaked in the 1960s, but population momentum means that our total numbers will continue to grow for decades more.

United Nations, Department of Economic and Social Affairs, Population Division (2017). World Population Prospects: The 2017 Revision. New York: United Nations. Fig 5, pg 7

fertility levels of 1.2 children per woman. The United Nations Population Division reports that the era of high fertility is declining nearly everywhere. By 2050, more than two-thirds of the world population is expected to be in countries with below-replacement fertility (fig. 7.10).

Section Review

1. Explain the terms of BIDE and how they influence growth.
2. What are crude birth and death rates?
3. How does percentage growth rate relate to doubling time?

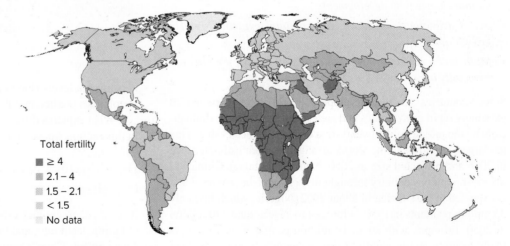

FIGURE 7.9 Population growth rates remain high in sub-Saharan Africa and a few other regions affected by poverty and/or conflict.

World Fertility Patterns 2015 Data Booklet, United Nations p 5 http://www.un.org/en/development/desa/population/publications/pdf/fertility/world-fertility-patterns-2015.pdf

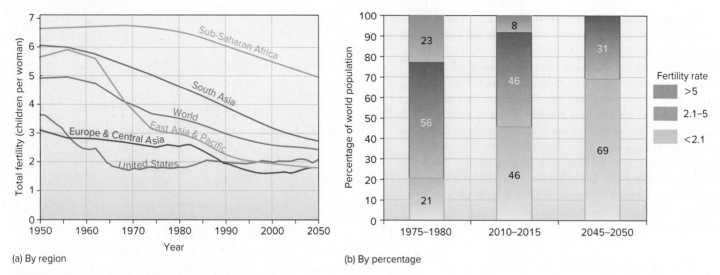

(a) By region

(b) By percentage

FIGURE 7.10 Two views of changing fertility rates. Most regions have declined toward the replacement rate of 2.1 (a). As a result, about half the world's people live in countries with this rate, and two-thirds are expected to by 2050 (b).
Figure 3 p 3. World Population Prospects: The 2017 Revision.

Quantitative Reasoning

Examine the graph in figure 7.10. What was the total world fertility rate in 2000? When will the whole world likely reach the replacement fertility rate of 2.1 children per woman, on average? When did the major change in world fertility rates occur worldwide?

7.3 WHAT FACTORS AFFECT POPULATION GROWTH?

- *The demographic transition describes a general pattern of population change with development.*
- *Age distributions, as shown in population pyramids, project growth rates*
- *Many factors affect growth; education and rights for girls are especially important.*

Some countries in the developing world have experienced extremely rapid growth rates and are expected to increase dramatically by the middle of the twenty-first century. Table 7.3 shows the ten largest countries in the world, arranged by their estimated size in 2018 and projected size in 2050. Note that although China was the most populous country throughout the twentieth century, India is expected to pass China in about 2022. Nigeria, which had only 33 million residents in 1950, is forecast to have around 390 million in 2050. Ethiopia, with about 18 million people 50 years ago, is likely to grow nearly eightfold over a century. In many of these countries, rapid population growth is a serious problem. Low-income countries cannot expand housing, employment, clean water, and services as fast as the population grows. Bangladesh, about the size of Iowa, is already crowded at 162 million people.

TABLE 7.3	The World's Largest Countries		
2018		**2050 (estimate)**	
Country	**Population (millions)**	**Country**	**Population (millions)**
China	1,409	India	1,692
India	1,339	China	1,295
United States	324	United States	403
Indonesia	264	Nigeria	390
Brazil	209	Indonesia	293
Pakistan	197	Pakistan	275
Nigeria	191	Brazil	223
Bangladesh	167	Bangladesh	194
Russia	144	Philippines	155
Japan	127	Dem Rep. Congo	149

UN Population Division 2019

If rising sea levels flood one-third of the country by 2050, as climatologists predict, population density will increase dramatically.

What explains differences in growth rates for these countries? In this section we examine some of the influential factors.

Development promotes a demographic transition

In 1945, demographer Frank Notestein pointed out that a pattern of falling death rates and birth rates tends to accompany economic development. China's population, for example, has seen a transition like this (fig. 7.1, table 7.1). Notestein called this pattern the **demographic transition** from a state of high birth and death rates to one of lower birth and death rates. Figure 7.11 shows an idealized model of this phenomenon.

The model has several stages. Stage I in figure 7.11 represents the conditions in a premodern society. Food shortages, malnutrition, lack of sanitation and medicine, accidents, and other hazards generally keep deaths around 35-40 per 1,000 people. Birth rates are correspondingly high, keeping population size relatively constant. In Stage II, death rates fall rapidly, as development brings better health care, sanitation, and nutrition. Birth rates may rise at first as more prosperity allows people to have the children they always wanted. Eventually, in a mature industrial economy (Stage III), birth rates fall as people see that their children are likely to survive, and families concentrate their resources on fewer children. The population continues to grow rapidly during this stage because of population momentum. Numbers may double more than once, as population stabilizes.

Stage IV represents post-industrial countries, with low birth and death rates. The population reaches a new equilibrium in this phase, but at a much larger size than before. Most countries of northern and western Europe and industrialized economies in Asia, such as Japan, and Korea, went through a demographic transition in the nineteenth or early twentieth century similar to the patterns in fig. 7.11.

Many of the most rapidly growing countries in the world are in Stage II of this demographic transition. Their death rates have fallen close to the rates of the fully developed countries, but birth rates have not fallen correspondingly. In fact, their birth rates are higher than those in most European countries when industrialization began 300 years ago. A large disparity between birth and death rates means that some sub-Saharan African countries now are growing between 3 to 4 percent per year, and Lebanon, Oman, and Qatar are all above 6 percent.

Urbanization contributes to falling fertility rates. Cities provide many more options for education, employment, and entertainment (fig. 7.12) than do rural areas. Small apartments make it difficult to raise children or have extended families. And when women work outside the home, child care becomes expensive. In most urbanized countries, women (and many men) prefer to have small families.

Long life expectancy increases populations

Life span is the oldest age to which a species is known to survive. Among humans, the oldest living person whose birth was documented was Jeanne Louise Calment of Arles, France, who was reported to be 122 years old at her death in 1997. The aging process is still a medical mystery, but it appears that cells in our bodies have a limited ability to repair damage and produce new components. At some point, they simply wear out, and we fall victim to disease, degeneration, accidents, or senility.

Because few people reach the maximum age, we usually examine **life expectancy,** the average age a newborn infant can expect to reach. For most of human history, the average life expectancy has probably been about 30 years. In the past century, improved medicine, sanitation, and nutrition have raised life expectancies dramatically. Longer life spans, not rising fertility, explain most population growth in this time. By 1960, average global life expectancy was 56 years, and today it is over 72 years.

The greatest recent progress has been in developing countries (table 7.4). In 1930 the average person in India could expect to live only 29 years. By 2000, although India's per capita income was only $2,500 (U.S.), average life expectancy was 63 years, approaching that of much wealthier countries. As elsewhere,

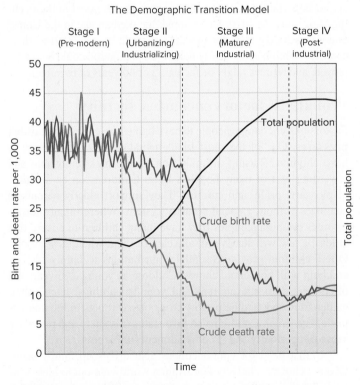

FIGURE 7.11 Theoretical birth, death, and population growth rates in a demographic transition accompanying economic and social development. In a predevelopment society, birth and death rates are both high, and total population remains relatively stable. During development, death rates tend to fall first, followed in a generation or two by falling birth rates. Total population grows rapidly until both birth and death rates stabilize in a fully developed society.

FIGURE 7.12 About 60 percent of China's population lives in urban areas. Urbanization, industrialization, and better job opportunities have played a crucial role in reducing population growth.
Yongyuan Dai/Getty Images

TABLE 7.4	Life Expectancy at Birth for Selected Countries, 1900 and 2018			
	1900		**2018**	
Country	**Males**	**Females**	**Males**	**Females**
India	23	23	67	71
China	39	42	75	78
Russia	31	33	66	77
United States	46	48	77	82
Sweden	57	60	81	84
Japan	42	44	81	87

Population Reference Bureau, 2019

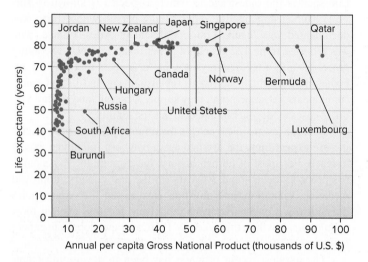

FIGURE 7.13 As incomes rise, so does life expectancy—up to about (U.S.) $10,000. Above that amount, the curve levels off. Some countries, such as South Africa and Russia, have far lower life expectancies than their GDP would suggest. Jordan, on the other hand, which has only one-fifth the per capita GDP of the United States, has the same life expectancy.

longer lives were due primarily to better nutrition, sanitation, and clean water, rather than to miracle drugs or high-tech medicine.

Life expectancy increases with income, but only up to a point (fig. 7.13). Above about $10,000 per person, enough for adequate food, shelter, and sanitation for most people, life expectancies level out at about 75 years for men and 85 for women. Women usually have higher life expectancies than men. Worldwide, the average difference between sexes is three years, but there are exceptions. In Russia the difference between men and women is 11 years.

Large discrepancies exist within countries. In the United States, for example, while the nationwide average life expectancy is 77.5 years, Asian American women in Bergen County, New Jersey, can expect to live 91 years, while Native American men on the Pine Ridge Reservation in South Dakota are reported to typically live only to 48. Two-thirds of African countries have life expectancies greater than that on Pine Ridge.

Will life expectancies continue to rise? Some demographers believe that life expectancy is approaching biological limits. Others

predict that medical advances might make it possible to live 150 years or more. Much longer life spans would profoundly affect society. We may need to find new ways to structure and finance our lives.

Age distributions determine future growth

One way to compare growth rates is with a population pyramid, which shows two back-to-back histograms of age classes, for males and females (fig. 7.14). In Niger, whose growth rate is 3.4 percent per year, 49 percent of the population is younger than age 15, so Niger can expect a baby boom when those children reach reproductive age. Even if fertility rates fall sharply, the population will continue to grow for some years as this younger generation starts families.

By contrast, a country with a stable population, like Sweden, has nearly the same number in each age cohort. Notice that there are more older females than males in Sweden because of differences in longevity between the sexes. A population that has recently entered a lower growth rate pattern, such as Singapore, has a bulge in the age classes for the last high-birth-rate generation.

A new challenge in understanding impacts of slowing population growth may be that policymakers don't yet know how to expand economies as populations stabilize or shrink. Even in China, economists worry about slowing population growth. This is partly because of the **dependency ratio,** the number of nonworking compared to working individuals in a population (fig. 7.15). Dependency ratios are an important question in many countries, as noted in the opening case study. In 1970, for example, the median age in the United States was 30. By 2100, the median age could be over 60. If workers continue to retire at 65, nearly half of the population could be unemployed, and retirees might be facing 35 or 40 years of retirement.

Pronatalist factors encourage fertility

Factors that encourage people to have babies are called **pronatalist pressures.** One of the most fundamental of these is a desire to have children. Raising a family can be the most enjoyable and rewarding part of many people's lives. Children can be a source of pleasure, pride, and comfort. They support elderly parents in countries without a social security system. Where infant mortality rates are high, couples may have many children to ensure that at least a few will survive to take care of them when they are old. Often, children provide essential contributions to family income and help with household chores. Children as young as 6 years old often tend domestic animals and younger siblings, fetch water, gather firewood, and help grow crops or sell things in the marketplace (fig. 7.16). Children also give status in society and provide a sense of continuity and accomplishment for parents.

Unmet need for contraceptives also drives up family sizes. Given a choice, relatively few women worldwide opt to have very large families, but in many areas they have little access to contraceptive methods. Global family planning agencies often focus on helping to make contraceptives available where they are wanted, and they find that this is also one of the best ways to protect the health of women and children.

Cultural and religious values often promote large families. Having many children can be a point of pride and a contribution

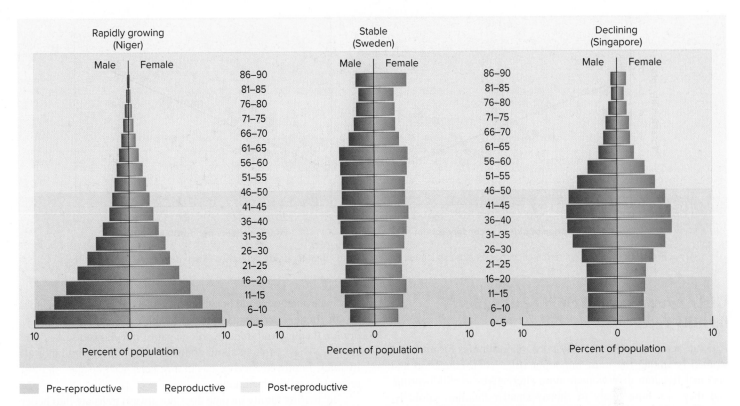

Rapidly growing (Niger)

Male | Female

Stable (Sweden)

Male | Female

Declining (Singapore)

Male | Female

Age classes: 86–90, 81–85, 76–80, 71–75, 66–70, 61–65, 56–60, 51–55, 46–50, 41–45, 36–40, 31–35, 26–30, 21–25, 16–20, 11–15, 6–10, 0–5

10   0   10
Percent of population

10   0   10
Percent of population

10   0   10
Percent of population

Pre-reproductive   Reproductive   Post-reproductive

FIGURE 7.14 The shape of each age-class histogram is distinctive for a population that is rapidly growing (Niger), stable (Sweden), or declining (Singapore). Horizontal bars represent the percentage of the country's population in consecutive age classes (0–5 yrs, 6–10 yrs, etc.).

to the community, and women with children—especially sons—have higher status than those who don't. In many cultures, boys are more valued than girls, so families may have more children in efforts to produce a son.

Male pride often is linked to having as many children as possible. In Niger and Cameroon, for example, men, on average, want 12.6 and 11.2 children, respectively. Women in these countries consider the ideal family size to be only about 5 or 6 children. Even though a woman might desire fewer children, however, she may have few choices and little control over her own fertility. In many societies, a woman has no status outside of her role as wife and mother. And without children, she has no source of support.

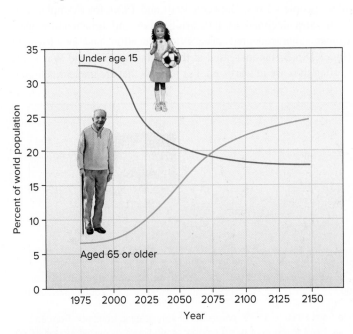

FIGURE 7.15 By the mid-twenty-first century, children under age 15 will make up a smaller percentage of world population, while people over age 65 will comprise an increasingly larger share of the population.

FIGURE 7.16 In rural areas with little mechanized agriculture, children are needed to tend livestock, care for younger children, and help parents with household chores.
William P. Cunningham

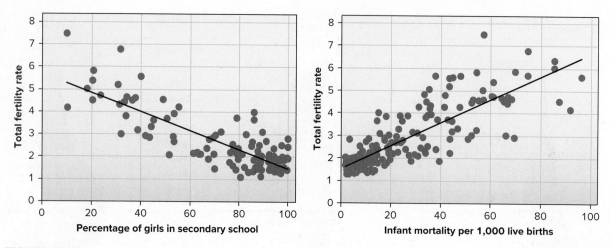

FIGURE 7.17 Total fertility declines as girls' education increases and infant mortality decreases. Each dot represents a country.
World Bank, 2015

Girls' education and child health affect fertility rates

Among factors reducing fertility rates, opportunities for girls and women is among the most important (fig. 7.17). Education and personal freedom give women more alternatives to childbearing, and they are more likely to choose smaller families. Globally, changes in birth rates show that when they have a choice, most women choose to have two or three children, rather than the historical average of six or more. Later marriage and delayed childbearing also occur when women have other opportunities. When women earn a salary, their income becomes an important part of the family budget. Thus, education and socioeconomic status are usually inversely related to fertility in most countries. Independence and choice also allow girls to push back against pronatalist pressures such as early marriage and large families.

As we have noted, high child mortality is tied to high birth rates. There has never been a sustained drop in birth rates that was not preceded by a sustained drop in infant and child mortality. This is why development and health agencies focus so strongly on child health. Better nutrition, improved health care, simple oral rehydration therapy, and immunization against infectious diseases (see chapter 8) have brought about dramatic reductions in child mortality rates. Improved survival has in turn been accompanied in most regions by falling birth rates.

Increasing family income does not always translate into better welfare for children, because in many cultures men control most financial assets. Often, the best way to improve child survival is to ensure the rights of their mothers.

Major events influence birth rates

Birth rates often change as social, economic, and political events affect people's lives. Between 1910 and 1930, for example, U.S. birth rates declined as industrialization and urbanization gave women unprecedented access to education and employment (fig. 7.18). Births increased at the beginning of World War II (as is often the case in wartime). For reasons that are unclear, a higher

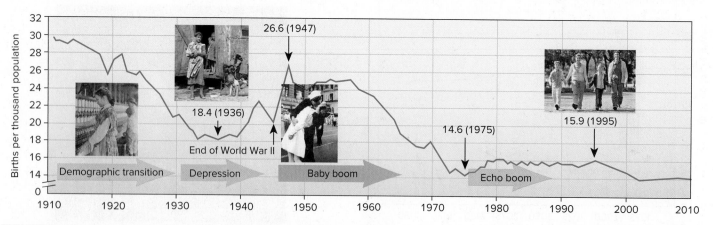

FIGURE 7.18 Birth rates in the United States, 1910–2000. The falling birth rate from 1910 to 1929 represents a demographic transition from an agricultural to an industrial society. The baby boom following World War II lasted from 1945 to 1965. A much smaller "echo boom" occurred around 1980 when the baby boomers started to reproduce.

Source: Data from the Population Reference Bureau and the U.S. Bureau of the Census (a): Library of Congress Prints and Photographs Division LC-DIG-nclc-01366; (b): U.S. Department of Agriculture Photograph Archives; (c): National Archives and Records Administration (NWDNS-80-G-377094); (d): Stockbyte/Getty Images percentage of boys are usually born during war years. At the end of the war, federal policy encouraged women to give up jobs for returning service men, and to stay home and be domestic. A baby boom resulted, producing a large generation that is currently approaching retirement age. Birth rates fell again in the 1960s, partly because of the small number of babies born in the 1930s (who were of reproductive age in the 1960s), and partly because of changes in the ideal family size. Probably even more important, birth control pills became available in the 1960s, revolutionizing women's control of their fertility and their life choices.

A small "echo boom" occurred in the 1980s as people born in the 1960s began to have babies, but changing economics and attitudes seem to have permanently altered our view of ideal family size in the United States.

Even more astonishing changes in economics and demography occurred in China over the past four decades. In 1980 about 85 percent of Chinese citizens lived in extreme poverty, one of the highest poverty rates in the world at the time. Between 1981 and 2017, more than 700 million Chinese moved out of poverty, the greatest economic development in world history.

The average income now is 15 times what it was in 1980, and less than 5 percent of the population lives in extreme poverty. Since 1980, about 400 million people moved from the countryside to cities, life expectancy more than doubled, health care improved dramatically, and hundreds of millions of children were able to attend school who would never have in the past, and a large fraction of women got education, jobs, and financial independence. While these dramatic developments were occurring, the total average fertility in China dropped from 5.8 in 1980 to 1.6 in 2018. All these social changes may have had as much to do with declining births as China's one-child policy (What Do You Think?, p. 146).

Family planning gives us choices

Family planning allows couples to determine the number and spacing of their children. It doesn't always mean fewer children, but it does allow conscious decisions about the timing and number of children they want. Family planning includes counseling and education as well as access to contraceptives. Access to these services varies dramatically around the world. In China, Thailand, and Vietnam, 80 percent or more of women in childbearing ages use some form of contraceptive. In North American and Europe, 70 percent or more do. In most countries of sub-Saharan Africa, prevalence is less than 20 percent.

One of the important strategies for progress in some of these countries is public education to reduce stigmas, and to increase awareness and access of options. Family planning programs also encourage both men and women to learn to communicate about family planning preferences, and to respect the views and preferences of women in household decisions. In many traditional societies, these changes are not easy, but often these lessons are received well, because people feel good about knowing they are promoting fairness and progress in their communities.

American policy in recent years has been not to fund global family planning. Because there is great unmet need for contraceptives—over 200 million women in developing countries want access and have none—this policy directly impacts families and women's health.

Reducing child marriage is an important step. Bearing children at a young age is very risky; in addition, the earlier childbearing begins, the more children a family is likely to have. Counseling on spacing children also slows birth rates. Spacing of children has been practiced in many traditional cultures. Studies of hunting and gathering people, such as the !Kung or San people of the Kalahari Desert in southwest Africa, for example, indicate that our early ancestors had stable population densities not because of starvation or violence but because they controlled fertility. For instance, San women breast-feed children for three or four years. When calories are limited, lactation depletes body fat stores and suppresses ovulation. Along with taboos against intercourse while breastfeeding, this is an effective way of spacing children.

Modern medicine gives us numerous options for controlling fertility, and many new methods are in development. Most involve hormones to prevent pregnancy. Established hormonal methods involve treatments for women, but a reversible contraceptive gel for men is now in clinical trials. It reduces sperm production when applied daily on the upper arms or shoulders. Preliminary results suggest that the gel is 90 percent effective in preventing conception. That's better than condoms. Another approach for men is an injection that reduces sperm production. This method has proven effective in mice and may be tested soon in humans.

Could we have a birth dearth?

While demographers celebrate stabilization of populations, and environmental scientists hope this means reduced impacts on resources and other species, economists and policymakers are often less enthusiastic about declining and aging populations. Modern economies depend on consumers to keep buying goods, and on workers to produce those goods. Fewer and older populations will not buy as many things as expanding young families (except for health care).

Most European countries now have birth rates of 1.5 or fewer children per woman, well below the replacement level of 2.1. Italy, Russia, Austria, Germany, Greece, and Spain are experiencing negative rates of natural population change. Families in Japan, Singapore, and Taiwan have fewer than 1.3 children and are also facing a "child shock" as fertility rates have fallen. If low birth rates persist or are not balanced by immigration, there are concerns in these countries that declining populations will be unable to provide enough soldiers to maintain military strength, enough workers and consumers to maintain economic power, and enough taxpayers to support social welfare systems. Some analysts now emphasize that the United States and Northern Europe are fortunate to have an influx of immigrants, who bring youth, energy, and consumers to shore up the economy.

Many countries, including China, are rethinking their population policies and beginning to encourage marriage and childbearing. Germany, Hungary, Denmark, and Russia offer incentives to encourage women to bear children. Japan offers financial support

China's One-Child Policy

When the People's Republic of China was founded in 1949, it had about 540 million residents, and official government policy encouraged large families. The Republic's First Chairman, Mao Zedong, proclaimed, "Of all things in the world, people are the most precious." He reasoned that more workers would mean greater output, increasing national wealth and international standing. This optimistic outlook was challenged, however, in the 1960s, when a series of disastrous government policies triggered massive famines and resulted in at least 30 million deaths (see fig. 7.1).

After Deng Xiaoping became chairman in 1978, he reversed many of Mao's policies by decollectivizing farms, encouraging private enterprise, and discouraging large families. China's population had already reached 975 million, and with annual growth of 2.5 percent, Deng recognized that number would double in only 28 years. China might have nearly 2 billion residents now if that growth had continued. Feeding, housing, educating, and employing all those people would put a severe strain on China's already limited resources.

Deng introduced a successful—but also controversial— one-child-per-family policy. Rural families and ethnic minorities were supposedly exempt from this rule, but local authorities often were capricious and tyrannical in applying sanctions. Ordinary families were punished harshly for having unauthorized children, while government officials and other privileged individuals could have more. There were many reports of bribery, forced abortions, coerced sterilizations, and even infanticide as a result of this policy. Because of a traditional bias for sons, China also has the world's most serious gender imbalance in the world. Under the one-child policy 118 boys were born for every 100 girls compared with the normal ratio of 103 boys for 107 girls. This has resulted in millions of men who may never find a bride.

Some analysts point out that the one-child policy does not fully explain falling birth rates. Declines started well before 1980, indicating that multiple social and economic factors contributed. Still, the policy may have helped establish the small family as an ideal standard.

A result of declining birth rates is the 4:2:1 problem. There are often four grandparents and two parents doting on a single child. Social scientists often refer to this highly spoiled generation as "little emperors" (fig. 1).

The Chinese government is now concerned about a "birth dearth." If the population declines, will there be enough consumers to keep the economy growing? Will there be enough workers, soldiers, farmers, scientists, inventors, and other productive individuals to keep society functioning and stable?

The one-child policy, which was intended to be temporary, ended in 2015. All couples are now allowed to have two children, and the state has begun valorizing the two-child family. But many demographers doubt that birth rates will change dramatically. Industrialization, urbanization, and economic growth encourage low growth. Many argue that these factors might have produced a demographic transition

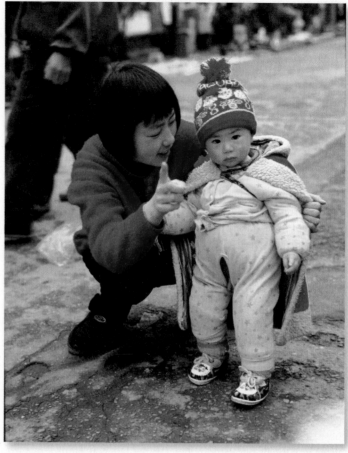

FIGURE 1 China's One-Child Policy is said to have produced a generation of privileged "little emperors."

William P. Cunningham

without such severe policies. Neighboring countries throughout East and Southeast Asia, including Singapore, Macao, Taiwan, South Korea, and Japan all have lower fertility rates than China without policies restricting child bearing. But these countries have long had higher average incomes than China.

Whatever the causes, it's important that China's growth rate has declined. China's population in 2018 was about 1.417 billion, or about half a billion less than it might be given its trajectory in 1978. What would the effects be if there were another half a billion Chinese today? What would you have done if you were Deng Xiaoping in 1980? What information would you have considered in planning the country's growth? What policies would you pursue today, now that the country is wealthier?

FIGURE 7.19 Some countries with a declining work force, such as Japan, are turning to robots to do many jobs that people once did.
Westend61/Getty Images

to new parents, and Singapore provides a dating service to encourage marriages among the upper classes as a way of increasing population. Japan, which is highly resistant to allowing immigration, is planning to cope with a declining work force and a rapidly aging population by developing robots to work both in factories and to provide personal care for those who need it (fig. 7.19).

Section Review

1. What are four stages of the demographic transition model?
2. Explain the difference between the three population pyramids in figure 7.14.
3. List several pronatalist factors and several that can slow growth.

7.4 WHAT IS THE FUTURE OF GROWTH?

- *Development usually slows population growth.*
- *The "demographic trap" describes the difficulty of transitioning from rapid growth.*
- *Social justice is an important factor in population growth.*

As we have seen, a wide array of factors affect population growth, from economic and technological change to social expectations, disruption, conflict, and educational access can influence population growth rates, at least over the short term. Governments often implement policies to affect growth, but those policies may or may not actually drive change, as we see in figure 7.1.

What does this imply about population change in coming decades? This depends on whether, and how rapidly, the world's developing countries stabilize their populations. Will wealthy nations help or hinder that stabilization? This question brings us back to the debates early in this chapter: Does poverty cause rapid growth, or does growth cause poverty? Do technological advances expand our environmental carrying capacity, or do they only increase our environmental impacts?

Development is seen as the main path to slower growth

Among development experts, consensus is now generally that poverty causes growth. This means that the best path to stabilizing growth is to reduce poverty, improve health care, and provide educational access. These are, not surprisingly, the main processes understood to drive the demographic transition.

As a reminder, four conditions are necessary for a demographic transition to occur: (1) improved standard of living, (2) increased confidence that children will survive to maturity, (3) improved social status of women, and (4) increased availability and use of birth control. As the example of China in the opening case study for this chapter shows, these conditions can be met, even countries with modest levels of income and development.

Statistically, demographers find that stabilization is occurring in most developing nations. They expect the world population should stabilize in most countries by the end of this century. There is, however, a long road ahead, and we don't know where stabilization will happen, or how it will resonate with demands for economic growth.

Migration is a growing concern

In general, birth and death rates have greater net impact on population change than immigration and emigration. However, increases in climate refugees and conflict-related migration are changing this rule in some areas. The United Nations High Commission on Refugees (UNHCR) reported in 2019 that more than 70 million people were displaced from their homes by economics, politics, persecution, or war. This is the highest number of refugees since World War II. About 41 million of these were internal refugees within their own country, while another 31 million moved (or attempted to move) to another country (fig. 7.20).

FIGURE 7.20 In 2019 at least 70 million people were forced from their homes by war, famine, persecution, or economics. In the future there could be hundreds of millions of climate refugees. What will be the social impacts of these migrations?
Anjo Kan/Shutterstock

This mass migration has provoked a shocking rise in nationalism, xenophobia, and authoritarian rule in many places. The United Nations High Commission on Refugees doesn't recognize environmental refugees, but it's likely that in coming years many millions more will be driven from their homes by climate change, water shortages, environmental degradation, and conflicts that arise from these stresses. Rising sea levels, droughts and famines, collapsing economies, and climates that are simply too hot to survive, are likely to trigger mass migrations never seen before in human history.

Consider the example of Syria, once the breadbasket of the Middle East. Starting in 2006 the worst drought in 500 years, together with poorly planned policies on irrigation subsidies, caused three-quarters of Syrian farms to fail and killed 85 percent of livestock. Millions of people migrated to urban centers, but they found there few jobs and poor living conditions. Angry at a lack of government assistance, residents rose up to oppose the Assad government. In the civil war that ensued, at least half a million civilians died, and more than 5.5 million fled the country. This flood of refugees destabilized neighboring countries and caused a backlash in Europe that continues to reverberate in world politics.

The demographic trap and lifeboat ethics describe challenges of poverty

Turning around a fast-growing population is not easy. Many development economists have warned that poorer countries can be caught in a "demographic trap" of unstable growth and poverty that prevents them from escaping from the middle phase of the demographic transition. Demands on resources exceed the sustainable yield of local forests, grasslands, croplands, or water resources, leading to further hunger and poverty. High resource demands threaten to destroy the resource base, causing environmental deterioration, economic decline, and political instability.

Many people have argued that the only way to break out of the demographic trap is to immediately and drastically reduce population growth by whatever means are necessary. They argue strongly for increasing access to birth control (a priority supported from many perspectives) and for bold policies encouraging lower birth rates. Historically, these policies have sometimes justified programs of forced sterilization, or of oppression of cultural minorities.

Some analysts agree with Malthus that helping the poor will simply increase their reproductive success and further threaten the resources on which we all depend. Author Garret Hardin described this view in 1968 as "lifeboat ethics." "Each rich nation," he wrote, "amounts to a lifeboat full of comparatively rich people. The poor of the world are in other much more crowded lifeboats. Continuously, so to speak, the poor fall out of their lifeboats and swim for a while, hoping to be admitted to a rich lifeboat, or in some other way to benefit from the goodies on board. . . . We cannot risk the safety of all the passengers by helping others in need. What happens if you share space in a lifeboat? The boat is swamped and everyone drowns. Complete justice, complete catastrophe."

How would you respond to this view of lifeboat ethics? Does it inform attitudes toward migrant populations today?

Social justice is an important consideration

Many analysts argue that because poverty contributes to population growth, **social justice** (ensuring a fair share of social benefits for everyone) is necessary for successful demographic transitions. The world has enough resources for everyone, but inequitable social, political, and economic systems cause uneven distribution of those resources. In this view, it is justice and access, not resources, that are limited.

Low income and developing nations often point out that per capita resource consumption is as important as, and sometimes more important than, total population count (fig. 7.21). They also note that many of today's wealthy countries are, or were, colonial powers, and many poor, fast-growing countries were badly destabilized by those colonial empires. The wealth that paid for progress and security for developed countries was often extracted from colonies,

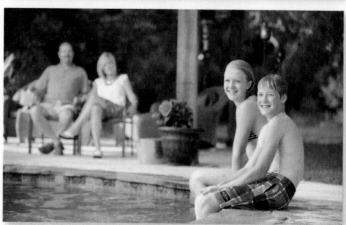

FIGURE 7.21 Population growth, resulting from high birth rates, has important impacts on resource consumption. But small, wealthy families consume far more resources per person, usually sourced from around the globe. Many people debate which impact is more important, but ultimately there are good reasons for reducing both family size and consumption rates.
Duncan Vere Green/Alamy; Andre Babiak/Alamy

which now suffer from exhausted resources, exploding populations, and chaotic political systems. Some of the world's poorest countries, such as India, Ethiopia, Mozambique, and Haiti, had rich resources and adequate food supplies before they were impoverished by colonialism. Those of us who now enjoy abundance may need to help the poorer countries not only as a matter of justice but because we all share the same environment.

Many also inquire about justice for other species. Should we be asking what is the maximum number of humans that the world can support, or should we ask how many can be supported while allowing other species to persist as well? How might we evaluate this balance?

Our choices now determine our future

Because there's often a lag between the time when a society reaches replacement birth rate and the end of population growth, we are deciding now what the world will look like in a hundred years. How many people will be in the world a century from now? Most demographers believe that population will stabilize sometime during the twenty-first century at a total of 9–12 billion. The United Nations Population Division projects four population scenarios. The optimistic (low) projection suggests that world population might stabilize by about 2030 and then drop back below current levels. This doesn't seem likely. The medium projection shows a population of about 9 billion by 2040 (fig. 7.22), while the high projection would reach 12 billion by midcentury.

Which of these scenarios will we follow? As you have seen, population growth is a complex system. Stabilizing or reducing human populations will require consistent attention, but experience now shows that people generally prefer smaller families if they have the option. Worldwide contraceptive use has increased steadily in recent years. Child marriages are declining in most countries, and the world average fertility rate of 2.6 births per woman is less than half what it was 50 years ago. If similar progress could be sustained for the next half century, populations should stabilize.

Already, nearly half the world population lives in countries where the total fertility rate is at or close to the replacement rate (fig. 7.23). The example of China shows that with rising standards of living, democracy, and social justice, population growth will slow without harsh government intervention. Whether we can control our per-capita consumption of environmental resources is another critical question. Furthermore, many countries still suffer from wars, corruption, colonial history, religious tensions, and other factors that continue to prevent economic and social development while perpetuating high population growth. Clearly these are places where humanitarian and environmental needs converge, and both will benefit as we work to create a more humane, sustainable world.

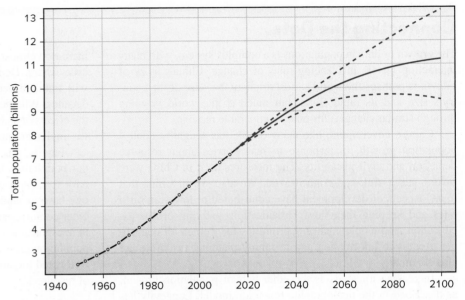

Estimates: — median   - - 95% confidence intervals

FIGURE 7.22 Population projections for different growth scenarios. Recent progress in family planning and economic development has led to significantly reduced estimates compared to a few years ago. The median projection is 8.9 billion in 2050, compared to previous estimates of over 10 billion for that date.

Source: World Population Prospects The 2015 Revision Key Findings and Advance Tables © 2015 United Nations.

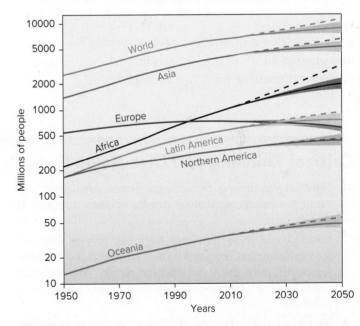

FIGURE 7.23 Estimates of population change in different continents between 1950 and 2050. The vertical axis is logarithmic.
United Nations 2018

Section Review

1. How can prosperity and urbanization affect birth rates?
2. Explain the ideas of a demographic trap and of lifeboat ethics.
3. Why does social justice matter in growth?

Connecting the Dots

The growth of human populations is a complex system, with many interacting factors influencing rates of change. China's story of growth is perhaps the world's most dramatic because of the country's size and its rapid shifts, but much of the world is going through similar changes, for many comparable reasons.

China's government, like many others, has sought to control population growth, in response to real worries about resource depletion and civil instability. The measures used in China were severe, successful, and controversial. We also see in China that many aspects of development have contributed to lowering birth rates and keeping them low. Urbanization, growing prosperity, education, and employment opportunities for women all contribute. Access to information and to family planning resources are also essential.

Increasing longevity is the other half of population growth, as described in the demographic transition model. Longevity has increased almost everywhere, just as birth rates have fallen almost everywhere. Demographers anticipate that the worlds population will stabilize between 9 and 11 billion sometime in the coming decades. Whether that happens, what number we reach, and what the effects of stabilization will be, depends on choices and policies we implement now.

Finally, the impacts of our resource use depend on not just our population, but also on our affluence and the technology we have to exploit our environment. China's affluence and technology have expanded at record paces, to match its population. And other countries are expanding similarly. Our task is to try to reign in irresponsible resource use at the same time as we keep slowing population growth, so as to ensure that all children are wanted and well cared for, and that we establish some socially just distribution of resources.

Reviewing Key Terms

Can you define the following terms in environmental science?

crude birth rate 7.2	dependency ratio 7.3	population momentum 7.2	social justice 7.4
crude death rate 7.2	family planning 7.3	pronatalist pressures 7.3	technological optimists 7.1
demographics 7.1	I = PAT 7.1	replacement fertility rate 7.2	total fertility rate 7.2
demographic transition 7.3	life expectancy 7.3	rule of 70 7.2	total growth rate 7.2

Critical Thinking and Discussion Questions

1. What do you think is the optimum human population? The maximum human population? Are the numbers different? If so, why?

2. Some people argue that technology can provide solutions for environmental problems; others believe that a "technological fix" will make our problems worse. What personal experiences or worldviews do you think might underlie these positions?

3. Karl Marx called Thomas Malthus a "shameless sycophant of the ruling classes." Why would the landed gentry of the eighteenth century be concerned about population growth of the lower classes? Are there comparable class struggles today?

4. Try to imagine yourself in the position of a person your age in a developing world country. What family planning choices and pressures would you face? How would you choose among your options?

5. Some demographers claim that population growth has already begun to slow; others dispute this claim. How would you evaluate the competing claims of these two camps? Is this an issue of uncertain facts or differing beliefs? What sources of evidence would you accept as valid?

6. What role do race, ethnicity, and culture play in our immigration and population policies? How can we distinguish between prejudice and selfishness, on one hand, and valid concerns about limits to growth, on the other?

Data Analysis

Population Change over Time

China's population trends have shifted dramatically in recent years. Is China unusual in this change? Take a look at population size and trends in different regions, and at some of the factors that influence growth rates. Gapminder.org is a rich source of data on global population, health, and development, including animated graphs showing change over time. Go to Connect to find a link to Gapminder graphs, and answer questions about what they tell you.

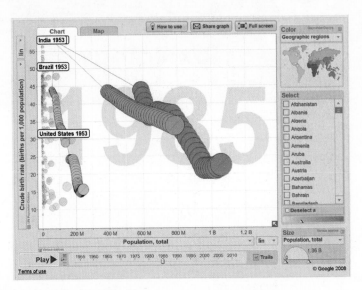

http://www.gapminder.org/world

Environmental Health and Toxicology

▲ How dangerous are new, synthetic chemicals like those in non-stick pans? Do we need to worry?
Paul Bradbury/OJO images/Getty Images

Learning Outcomes

After studying this chapter, you should be able to:

8.1 Describe health and disease and how the global disease burden is now changing.

8.2 Summarize the principles of toxicology.

8.3 Discuss the movement, distribution, and fate of toxic substances in the environment.

8.4 Explain ways we evaluate toxicity and risk.

8.5 Relate how we establish health policy.

"To wish to become well is a part of becoming well."

– Seneca

PFCs: Miracle or Menace?

Perfluorocarbons (PFCs) have been called miracle compounds. Recalling DuPont's slogan of "better living through chemistry," these slippery, heat-resistant, durable compounds occur in countless products (fig. 8.1), refrigerants, propellants, water- and stain-repellent coatings, non-stick cookware, electrical insulation, airplane and computer parts, fire-fighting foam, cosmetics, cleaning solvents, and hydraulic fluids.

There's also a dark side to these technological wonders. Because they're so persistent, PFCs are now found throughout the world, even in the most remote and seemingly pristine sites. They accumulate in birds, mammals, fish, and humans. Perfluorooctanoic acid (PFOA), the starting compound for most PFCs, now occurs in the blood of most people in industrialized nations at 2 to 8 parts per billion (ppb). In Korea, average levels are about 60 ppb. In the United States, nearly everyone is thought to have PFOA levels in their blood about ten times higher than the EPA (Environmental Protection Agency) limits suggested for drinking water.

These compounds were not regulated in the United States until recently because they were not legally designated as hazardous. There's a common misperception that if a chemical is dangerous, it will be regulated to protect public health. Under the 1976 Toxic Substances Control Act, however, the EPA can investigate a chemical only when it is proven harmful. Because long-lasting environmental exposure is hard to pin to any particular compound, proving cause and effect is extraordinarily difficult. This standard, therefore, largely eliminates public oversight of new chemicals. The EPA has restricted only five synthetic chemicals in the last 40 years, out of an estimated 60,000 or more on the market.

The landmark case that brought PFOA into the public spotlight began in 1999, when a farmer in West Virginia suspected that water draining from a DuPont waste dump was sickening and killing his dairy cows. The unlined waste dump turned out to have 7,000 tons of PFOA residues from Teflon production in a DuPont factory in nearby Parkersburg, West Virginia. The farmer sued DuPont, and the court ordered the company to turn over all documentation related to PFOA.

FIGURE 8.1 It's estimated that 20 percent of the PFOA in American blood comes from anti-grease coatings in pizza and popcorn containers.
CREATISTA/Shutterstock.com

Surprisingly, the records included confidential medical and health studies, some dating back half a century, documenting a variety of illnesses associated with PFOA. Major manufacturers, such as DuPont and 3M, had documented associations of PFOA exposure with birth defects and cancer in lab animals in the 1960s. They also knew that factory workers had high concentrations of PFOA in their blood, and that workers, as well as residents near industrial facilities and waste disposal sites, suffered from a variety of cancers, as well as kidney, liver, and skin ailments.

As in many environmental health issues, irresponsible waste disposal made up much of the problem. DuPont dumped PFOA and other waste materials into public sewers, waterways, and open waste dumps. These expedient methods avoided the much higher costs of incinerating waste or sending it to secure, contained landfills.

It quickly became apparent that DuPont had long recognized the adverse health effects of PFOA. They also knew that the chemicals were widespread in the environment and were present in local water supplies at unsafe concentrations. The company had been searching for an alternative to PFOA, and had announced in 1993 that they had a compound that appeared to be less toxic, and that stayed in the body for a much shorter time. Nevertheless, the company decided against shifting away from PFOA, which was worth an estimated $1 billion a year in profits.

In 2017 DuPont agreed to settle lawsuits related to the Parkersburg case for $671 million. And in 2018, 3M settled with the state of Minnesota for $850 million for dumping PFOA in the St. Paul suburbs. 3M claims that low levels of PFCs don't cause adverse health effects. Although PFCs were in the groundwater for decades, 3M couldn't find any measurable health effects. Nevertheless, DuPont, 3M, and six other major fluorocarbon chemical companies in the United States promised to eliminate PFOA and related compounds. But the substitutes they have come up with so far are short-chain PFCs that don't remain in the body as long as longer-chain varieties, but seem to be just as toxic.

The European Union also has ordered a phase-out of PFCs, but companies in Asia and other regions continue to manufacture and release these compounds. The 3M Company, which had been a major supplier of PFOA, ceased production in 2000, and DuPont stopped production and use in 2013. PFOA continues to turn up in groundwater in industrial areas, however, as more public health offices look for it.

This case study demonstrates a number of themes in environmental health. We have been tremendously successful in producing an amazing array of synthetic chemicals with highly useful properties. But we sometimes unwittingly set in motion a tsunami of unintended consequences when we release those materials into the environment. Monitoring new compounds and understanding their effects is extremely difficult. Regulating them is a challenge even when we have an active EPA and laws designed to protect public safety. On the other hand, relatively high exposure is sometimes needed to show up as a health risk. So how dangerous is low-level but widespread exposure to the thousands of new materials we've released? This is a central question of environmental toxicology.

8.1 ENVIRONMENTAL HEALTH

- *The global disease burden is changing. Infectious diseases are decreasing while chronic conditions, such as heart disease, stroke, and diabetes, are increasing.*
- *Emergent diseases are appearing with increasing frequency as we travel more, move into new habitats, and share our germs with others around the world.*
- *Conservation medicine combines ecology with health care.*

What is health? The World Health Organization (WHO) defines **health** as a state of complete physical, mental, and social well-being, not merely the absence of disease or infirmity. By that definition, we all are ill to some extent. Likewise, we all can improve our health to live happier, longer, more productive, and more satisfying lives if we think about what we do.

What is disease? A **disease** is an abnormal change in the body's condition that impairs important physical or psychological functions. Diet and nutrition, infectious agents, toxic substances, genetics, trauma, and stress all play roles in **morbidity** (illness) and **mortality** (death). **Environmental health** focuses on external factors that cause disease, including elements of the natural, social, cultural, and technological worlds in which we live. Figure 8.2 shows some major environmental disease agents, as well as the media through which we encounter them. Ever since the publication of Rachel Carson's *Silent Spring* in 1962, the discharge, movement, fate, and effects of synthetic chemicals have been a special focus of environmental health. Later in this chapter, we'll study these topics in detail. First, however, let's look at some of the major causes of illness worldwide.

The global disease burden is changing

Although there are many worries about pollution, environmental toxins, emerging diseases, and other health threats, we should take a moment to consider the remarkable progress we've made over the past hundred years in controlling many terrible diseases. Epidemics that once killed millions of people have been reduced or eliminated. Smallpox was completely wiped out in 1977. Polio has been eliminated everywhere in the world except for a few places in Afghanistan, Pakistan, Laos, and Madagascar. Guinea worms, river blindness, and yaws appear to be on their way to elimination. Epidemics of typhoid fever, cholera, and yellow fever that regularly decimated populations a century ago are now rarely encountered in developed countries. AIDS, which once was an immediate death sentence, has become a highly treatable disease. The average HIV-positive person in the United States now lives 24 years after diagnosis if treated faithfully with modern medicines.

One of the best indicators of health gains is a worldwide increase in child survival. Fifty years ago in South Asia, for

FIGURE 8.2 Major sources of environmental health risks.

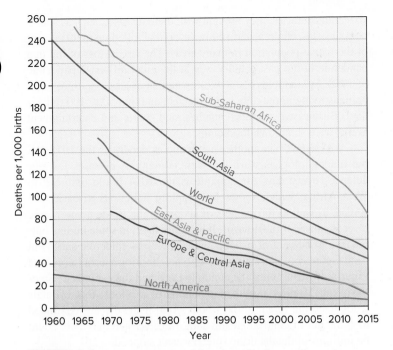

FIGURE 8.3 Child mortality has fallen dramatically over the past 50 years and is expected to continue to decline. East Asia has reached a level matching that of Europe and Central Asia, while sub-Saharan Africa has lagged behind—mainly because of wars, poverty, and poor health. Death rates for children under 5 years old are shown.

Source: Data from the World Health Organization, 2012

example, nearly one child in four died before their fifth birthday. Currently, the global five-year survival rate is over 95 percent (fig. 8.3). In another couple of decades, if current trends continue, the world under-5 mortality rate should approach that in the United States of less than 1 percent. Remarkably, East Asia and the Pacific region already have nearly as low a child mortality rate as the richer countries in Europe or North America, despite having a far lower per capita income. Much of this progress is due simply to better nutrition, improved sanitation, and clean water, which keep children healthy. According to the WHO, half of all gains in child survival are brought about by increases in women's education. Only sub-Saharan Africa has lagged in this health revolution, for a number of reasons that include political instability and poverty.

As chapter 7 shows, improved child survival has important demographic implications, including lowered birth rates and longer life expectancies. The fact that average global life expectancies have more than doubled over the past century from 30 to 64 years means many more productive years of life for most of us, but it also means that many more people are reaching old age.

The decrease in many infectious illnesses, along with an aging population, is producing a shift in global disease burden. Many people now live for years with chronic illness or disability. According to the WHO, chronic diseases now account for nearly three-fourths of the 56.5 million total deaths worldwide each year and about half of the global disease burden. To account for the costs of chronic illness, health agencies calculate disease burden for a population in terms of **disability-adjusted life years**

(DALYs). This statistic is the sum of overall years lost to disability or illness plus years lost to early deaths in a population. For example, a teenager permanently paralyzed by a traffic accident will have many years of suffering, health care costs, and lost potential. Similarly, many years of expected life are lost when a child dies of neonatal tetanus. A senior citizen who has a stroke has lost far fewer years to disease or disability.

Major causes of death have changed dramatically in recent decades. Chronic conditions, such as cardiovascular disease, cancer, and strokes were once common only in wealthy countries. These conditions are now among the leading causes of death in lower and middle income countries. Although the traditional killers in developing countries—infections, maternal and perinatal (birth) complications, and nutritional deficiencies—still take a terrible toll in the lowest income areas, diseases such as depression, diabetes, heart attacks, and stroke that once occurred mainly in rich countries are rapidly becoming increasingly common everywhere.

Look at the changes in the relative ranking of major causes of disease burden in table 8.1 Notice how many once-common diseases have been reduced in many cases, while chronic disorders related to lifestyle, age, and affluence are becoming more common. Much of this change in disease burden is occurring in the poorer parts of the world, where people are rapidly adopting the habits and diet of the richer countries. The largest source of unintentional injuries are traffic accidents, while the main sources of intentional injuries are self-harm and interpersonal violence.

Taking disability as well as death into account in our assessment of disease burden reveals the increasing role of mental health

TABLE 8.1 Leading Causes of Global Disease Burden

RANK	1990	RANK	2015
1	Respiratory infections	1	Heart disease
2	Diarrhea	2	Respiratory infections
3	Neonatal conditions	3	Stroke
4	Heart disease	4	Preterm birth[2]
5	Stroke	5	Diarrhea
6	COPD[1]	6	Road injury
7	Malaria	7	COPD
8	Tuberculosis	8	Diabetes
9	Malnutrition	9	Birth trauma
10	Neonatal encephalopathy	10	Congenital anomalies
11	Back and neck pain	11	HIV/AIDS
12	Unintentional Injuries	12	Tuberculosis
13	Congenital anomalies	13	Depression
14	Iron anemia	14	Iron anaemia
15	Depression	15	Back and neck pain

[1]Chronic obstructive pulmonary disease (COPD).

[2]Complications associated with preterm birth.

Source: Data from World Health Organization, 2018.

as a worldwide problem. WHO projections suggest that psychiatric and neurological conditions could increase their share of the global burden from the current 10 percent to 15 percent of the total load by 2020. Again, this isn't just a problem of the developed world. Depression is expected to be the second largest cause of all years lived with disability worldwide, as well as the cause of 1.4 percent of all deaths. For women in both developing and developed regions, depression is the leading cause of disease burden, while suicide, which often is the result of untreated depression, is the fourth largest cause of female deaths.

Pollution contributes to many of the health problems in table 8.1 and has become a leading cause of death and disability in many developing countries. The World Health Organization estimates that smog and toxic agents in outdoor air contribute to at least 3.7 million deaths per year from cardiovascular problems and cancer, particularly in urban areas. China and India account for most of those deaths.

Another 4.2 million people die from particulates exposure in indoor air from smoky stoves and cooking fires each year. About 1 million die from contaminants in soil and water, and nearly a million die from diseases linked to poor sanitation. At least 1.1 billion people smoke today, and this number is expected to increase by at least 50 percent, especially in developing countries. If current patterns persist, about 500 million people alive today will eventually be killed by tobacco. Dr. Gro Harlem Brundtland, former director-general of the WHO, observed that reducing tobacco use and air pollution could save billions of lives.

Infectious and emergent diseases still kill millions of people

Although the ills of modern life are becoming more important almost everywhere in the world, communicable diseases still kill millions of people, especially children (fig. 8.4). Humans are afflicted by a wide variety of pathogens (disease-causing organisms), including viruses, bacteria, protozoans (single-celled animals), parasitic worms, and flukes. Pandemics (worldwide epidemics) have changed the course of history. In the mid-fourteenth century, the "Black Death" (bubonic plague) swept out of Asia and may have killed half the people in Europe. When European explorers and colonists reached the Americas in the late fifteenth and early

sixteenth centuries, they brought with them diseases, such as smallpox, measles, cholera, and yellow fever, that killed up to 90 percent of the native population in many areas. The largest loss of life in a pandemic in the past century was in the great influenza pandemic of 1918. Epidemiologists now estimate that at least one-third of all humans living at the time were infected, and that 50 to 100 million died. Businesses, schools, churches, and sports and entertainment events were shut down for months.

We haven't had another pandemic as deadly as the 1918 flu, but epidemiologists warn that new contagious diseases test our defenses every day. An example is the 2014 Ebola outbreak in West Africa, which shows how rapidly infectious diseases can spread, as well as the dangers of lax oversight. Ebola hemorrhagic fever is caused by a filovirus. It is highly fatal; in some human outbreaks, 90 percent of patients died. The 2014 epidemic started in rural Guinea but quickly spread into neighboring Sierra Leone and Liberia. It wasn't until the contagion reached the capitals of Freetown and Monrovia, with populations of millions, that officials recognized the threat to the whole world if the disease continued to spiral out of control. Health authorities warned that there could be millions of deaths if the spread wasn't stopped.

It's estimated that Ebola infected 20,000 people in West Africa, and that at least 8,000 died (fig. 8.5). Both infected patients

Benchmark Data	
Among the ideas and values in this chapter, the following are a few worth remembering.	
1 in 6	Chance of dying from heart disease
1 in 188,364	Chance of dying in airplanes
1 in 5	Number of children in poorest countries who die before age 5
1 in 100	Number of children in richest countries who die before age 5
50%	Percentage of antibiotics prescribed for inappropriate conditions
20%	Estimate of PFOA in American blood from pizza and popcorn containers
2x	Increase in skin cancer risk from using tanning bed

FIGURE 8.4 (a) A group of influenza viruses magnified about 300,000 times. (b) Pathogenic bacteria magnified about 50,000 times. (c) *Giardia*, a parasitic intestinal protozoan, magnified about 10,000 times.

(a) CDC/SCIENCE PHOTO LIBRARY/Getty Images; (b) Image Source/Getty Images; (c) Source: Dr. Stan Erlandsen/Centers for Disease Control and Prevention.

FIGURE 8.5 Ebola workers need complete protection from bodily contact, which is difficult in a hot, humid climate. Protective gear also alienates locals, who tend to be suspicious of strangers and authority.
Kenzo Tribouillard/AFP/Getty Images

and health workers carried the disease to Europe and the United States. A massive investment from wealthy countries, coupled with improved local infrastructure brought the outbreak mostly under control by 2015, although sporadic cases continue to occur, and health officials worry about additional outbreaks in areas with poor health care infrastructure. In 2019, another Ebola outbreak occurred in the Democratic Republic of Congo. Second only to the 2014 epidemic in the number of deaths, the Congo situation was complicated by militants who fired at health care workers and prevented them from reaching patients. We have become more aware that increasing human penetration into once-remote areas is an emerging factor in many outbreaks.

A more widespread virus outbreak came to world attention in 2015. The Zika virus, which was first discovered in 1947 in Uganda, was long thought to cause only a mild fever and rash that dissipate in a few days. Most people who contract the disease never know they had it. The virus, which is transmitted by several mosquito species of the genus *Aedes,* moved slowly across tropical Africa and Asia. In 2013, however, the disease jumped to French Polynesia and infected two-thirds of the population in just seven months. It seemed to be both more virulent and more dangerous, causing 73 cases of Guillain–Barré syndrome, a life threatening paralysis. In 2015, the virus reached Brazil, and a new, frightening symptom was reported. Mothers who had Zika fever while pregnant were having babies with microcephaly, abnormally small heads and damaged brains. The virus had acquired the ability to cross the blood-brain barrier and infect the central nervous system.

The frequency of microcephaly when mothers have the virus during their first trimester is thought to be between 1 and 13 percent. In 2016 the World Health Organization (WHO) declared Zika fever a Public Health Emergency of International Concern, a rarely used designation. By then, the disease was present in 70 countries, including most of the tropical areas of the Americas and the Caribbean. Hundreds of U.S. citizens who traveled to tropical areas returned with Zika fever. Mosquito vectors for the virus already existed across much of the southern United States,

increasing the likelihood that the disease would spread rapidly across a broad swath of the country. There's evidence that the disease can be spread sexually between humans as well.

The most common carrier for the virus is the mosquito *Aedes egypti,* which also carries related viruses that cause dengue and chikungunya fever. These mosquitoes, which can breed in water containers as small as a bottle cap, are highly adapted to cities, especially where slum conditions exist and where there's much litter and standing water. Since most mosquitoes travel only a few hundred meters in their lifetime, living in these conditions is an important risk factor. If you do travel to places where *Aedes* are prevalent, the best protection is to cover up and use repellent. Experiments are underway to genetically modify mosquitoes so they can't transmit the disease or to cause virus-carrying species to simply die out. There are always uncertainties, though, about the unintended consequences of tampering with nature on a global scale.

The U.S. Centers for Disease Control and Prevention warn women who are pregnant, or might become pregnant, to avoid travel to countries where Zika is prevalent. These countries include areas of Asia as well as Latin American and Africa. According to the WHO, 2 billion people are currently at risk worldwide and 5 million babies are born annually in parts of the Americas with the Zika virus. Since the peak of the crisis passed in 2016, public concern has shifted away from the epidemic. Health officials now worry that inattention could lead to new outbreaks of the disease.

Emergent diseases are those not previously known or that have been absent for at least 20 years. Recent outbreaks of the Ebola and Zika virus are good examples. There have been hundreds of outbreaks of other emergent diseases over the past two decades (fig. 8.6), including measles, which had largely been extirpated from the United States but reemerged in California in 2014 among children whose parents had refused vaccination. Other examples include a new drug-resistant form of tuberculosis, now spreading in Russia; dengue fever, which is spreading through Southeast Asia and the Caribbean; malaria, which is reappearing in places where it had been nearly eradicated; and a new human lymphotropic virus (HTLV), which is thought to have jumped from monkeys into people in Cameroon who handled or ate bushmeat. These HTLV strains are now thought to infect 25 million people.

Most people now are aware of the threats from malaria, tuberculosis, and HIV, but fewer know about a silent health crisis spreading across Africa. Hepatitis is an infectious liver disease caused by a virus spread by exposure to infected blood or body fluids. The initial infection rarely causes death, but untreated chronic infections often lead to cirrhosis and liver cancer. Hepatitis now causes about 1.5 million deaths per year, surpassing HIV or malaria. About two-thirds of those deaths result from hepatitis B (HBV) infections. Most HBV infections occur around the time of birth when the infant immune system can't fight off the virus. The vaccine against the virus is highly effective, but in poor countries many people don't understand the disease, or can't afford the vaccine even when it's available.

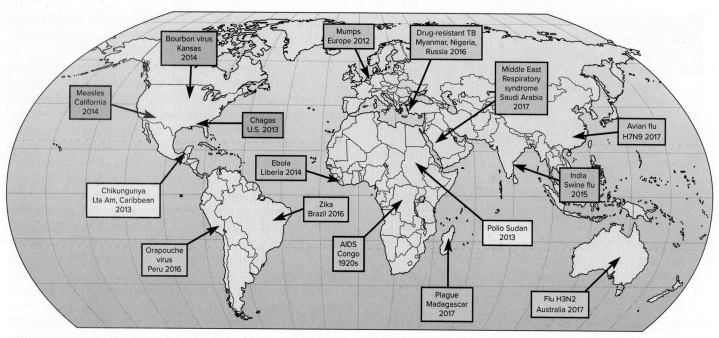

FIGURE 8.6 Some recent outbreaks of highly lethal infectious diseases. Why are supercontagious organisms emerging in so many different places?
Source: Data from U.S. Centers for Disease Control and Prevention.

Currently at least 350 million people are infected with HBV, and an additional 100 million new infections occur every year, mainly in sub-Saharan Africa and in East Asia. While HIV is more lethal than hepatitis, more than 7 times as many people have chronic HBV as have HIV. Both diseases can be treated with antiretroviral drugs, but there has been so much publicity about the threat of AIDS that patients with HBV are less likely to be treated. It's estimated that less than 10 percent of babies in Africa receive the HBV vaccine at birth. But several countries in sub-Saharan Africa are expanding their screen and inoculation efforts. Chronic HBV infections often are symptom free for decades, and many people who have the disease are unaware of it so screening is vital.

A frightening new disease is paralyzing children in the United States. The disease, called acute flaccid myelitis (AFM), causes limb weakness and paralysis resemblng the symptoms of polio. Many of those who develop the illness never recover and some have died. Some evidence suggests that an enterovirus called EV-D68 is causing the illness, but researchers have been able to find the pathogen in the spinal fluid of only a few sick children. The U.S. Centers for Disease Control and Prevention (CDC) have confirmed 484 cases of AFM in the United States since 2014. Blood samples from some people with AFM contain the virus. But many people who never developed AFM symptoms also have the virus in their blood. So far, the source of this mysterious and tragic disease remains elusive.

Conservation medicine combines ecology and health care

Humans aren't the only ones to suffer from new and devastating diseases. Domestic animals and wildlife also experience sudden and widespread epidemics, which are sometimes called **ecological diseases.** We are coming to recognize that the delicate ecological balances that we value so highly—and disrupt so frequently—are important to our own health. **Conservation medicine** is an emerging discipline that attempts to understand how our environmental changes threaten our own health as well as that of the natural communities on which we depend for ecological services. Although it is still small, this new field is gaining recognition from mainstream funding sources such as the World Bank, the World Health Organization, and the U.S. National Institutes of Health.

Ebola hemorrhagic fever, for example, is one of the most virulent viruses ever known. In 2002 researchers found that 221 of the 235 western lowland gorillas they had been studying in the Congo disappeared in just a few months. Many chimpanzees also died. Although the study team could find only a few of the dead gorillas, 75 percent of those tested positive for Ebola. Altogether, researchers estimate that 5,000 gorillas died in this small area. Extrapolating to all of central Africa, it's possible that Ebola has killed one-third of all the gorillas in the world.

Still more lethal was a disease discovered in 2006, when people living near a cave west of Albany, New York, reported something peculiar: Little brown bats (*Myotis lucifugus*) were flying outside during daylight in the middle of the winter. Inspection of the cave by the Department of Conservation found numerous dead bats near the cave mouth. Most had white fuzz on their faces and wings, a condition now known as white-nose syndrome (WNS). Little brown bats are tiny creatures, about the size of your thumb. They depend on about 2 grams of stored fat to get them through the winter. Hibernation is essential to making their energy resources last. Being awakened just once can cost a bat a month's worth of fat.

The white fuzz has now been identified as filamentous fungus (*Pseudogymnoascus destructans*), which thrives in the cool, moist conditions where bats hibernate. We don't know where the fungus came from, but, true to its name, the pathogen has spread like wildfire, reaching 33 states and 7 Canadian provinces. Biologists estimate that at least 7 million bats already have died from this disease. It isn't known how the pathogen spreads. Perhaps it moves from animal to animal through physical contact. It's also possible that humans introduce fungal spores on their shoes and clothing when they go from one cave to another.

Biologists have stated that WNS is the worst epidemic ever to affect bat populations, with over 6 million killed. So far, 12 species of bat, five of them endangered species, are known to be susceptible to this plague. In infected colonies, mortality can be 100 percent. Some researchers fear that bats could be extinct in 20 years in the eastern United States. Losing these important species would have devastating ecological consequences.

An even more widespread epidemic is wiping out amphibians worldwide: a disease called chytridiomycosis, which causes dramatic losses or even extinctions of frogs and toads (fig. 8.7). The fungus *Batrachochytrium dendrobatidis* causes the disease. It was first recognized in 1993 in dead and dying frogs in Queensland, Australia, and now is spreading rapidly, perhaps because the fungus has become more virulent or amphibians are more susceptible due to environmental change. Most of the world's approximately 6,000 amphibian species appear to be susceptible to the disease, and at least 2,000 species have declined or even become extinct in their native habitats as part of this global epidemic. Temperatures above 28°C (82°F) kill the fungus, and treating frogs with warm water can cure the disease in some species. Topical application of the drug chloramphenicol also has successfully cured some frogs. And certain skin bacteria seem to confer immunity to fungal infections. In some places, refuges have been established in which frogs can be maintained under antiseptic conditions until a cure is found. It's hoped that survivors can eventually be reintroduced to their native habitat and species will be preserved.

FIGURE 8.7 Frogs and toads throughout the world are succumbing to a deadly disease called chytridiomycosis. Is this a newly virulent fungal disease, or are amphibians more susceptible because of other environmental stresses?
Rosalie Kreulen/Shutterstock

Climate change also facilitates expansion of parasites and diseases into new territories. Tropical diseases, such as malaria, cholera, yellow fever, and dengue fever, have been moving into areas from which they were formerly absent, as mosquitoes, rodents, and other vectors expand into new habitat. Climate-related expansion of diseases affects other species, as well. A disease called dermo is spreading northward through oyster populations along the Atlantic coast of North America. This disease is caused by a protozoan parasite (*Perkinsus marinus*) that was first recognized in the Gulf of Mexico about 70 years ago. In the 1950s the disease was found in Chesapeake Bay. Since then the parasite has been moving northward, probably assisted by higher sea temperatures caused by global warming. It is now found as far north as Maine. This disease doesn't appear to be harmful to humans, but it is devastating oyster populations.

Another persistent epidemic, known as colony collapse disorder, is destroying honey bee colonies across North America and parts of Europe. Because bees are essential crop pollinators, at least 70 crops worth at least $15 billion annually are at risk. Many factors have been suggested to explain the mysterious bee die-off, including pesticides, parasites, viruses, fungi, malnutrition, genetically modified crops, and long-distance shipping of bees to pollinate crops. Although many authors claim to have discovered the cause of this disorder, it remains a mystery.

Resistance to antibiotics and pesticides is increasing

Population biologists have long known that constant exposure to a toxic substance produces resistance to it. Constant use of antibiotics, for example, kills off all but the resistant strains of microbes that tolerate the antibiotic. These multiply and come to dominate the population (fig. 8.8a). For bacteria, this problem is especially challenging: Bacteria readily exchange genes, increasing the likelihood of producing resistant populations (fig. 8.8b). Widespread overuse of antibiotics, in people and especially in livestock, is producing a rapidly growing problem of antibiotic-resistant infections.

Antibiotic resistance is expanding because we raise most of our cattle, hogs, and poultry in densely packed barns and feedlots. Confined animals are dosed constantly with antibiotics to keep them disease-free, and to make them gain weight faster. At least 80 percent of all antibiotics used in the United States are fed to livestock. Monitoring in and around livestock operations consistently finds multidrug-resistant pathogens in soil, water, and air. These multiple-resistant pathogens are increasingly found in people, a major public health worry. In some countries, such as Denmark, livestock have been raised for decades with only emergency uses of antibiotics. U.S. policymakers, however, have been slow to intervene in these practices (see chapter 9).

In humans, at least half of the 100 million antibiotics prescribed every year in the United States are unnecessary or are the wrong ones. In part, that's because at least one-third of Americans incorrectly believe antibiotics are an effective treatment for viral infections. They demand treatments their doctors know will be

(a) Mutation and selection create drug-resistant strains

(b) Conjugation transfers drug resistance from one strain to another

FIGURE 8.8 How microbes acquire antibiotic resistance. (a) Random mutations make a few cells resistant. When challenged by antibiotics, only those cells survive to give rise to a resistant colony. (b) Sexual reproduction (conjugation) or plasmid transfer moves genes from one strain or species to another.

ineffective, and doctors often find it easier to write a prescription than to argue with a patient. By some estimates, more than 20 percent of antibiotics prescribed for adults are for colds, upper respiratory tract infections, and bronchitis, three conditions for which antibiotics have little or no benefit. Furthermore, many people who start antibiotic treatment fail to carry it out for the time prescribed. Failing to complete an antibiotic treatment just aids the development of resistant strains of a disease.

Besides breeding resistant microbes, the increased use of antibiotics in the first 6 months of a child's life has been linked to a higher incidence of asthma, eczema, and allergic hypersensitivity. Parents should be told that antibiotics can't cure their children's colds and flus, and that symptoms should be treated with home remedies, including rest and fluids.

Hospitals are key centers of antibiotic resistance, but practices have begun to change. They are tracking prescriptions more closely, and they are examining closely how they clean and disinfect rooms, equipment, and the hands of doctors and nurses. Among the most lethal antibiotic-resistant bacteria are CRE (carbapenem-resistant enterobacteriaceae), which have been identified in health care facilities in 42 states. While CRE cause only 1 to 4 percent of all hospital-acquired infections, the CDC has dubbed them the "nightmare bacteria" because they are fatal in half of all cases.

Another serious infection is methicillin-resistant *Staphylococcus aureus* (MRSA). One U.S. study found that 60 percent of the estimated 80,000 MRSA infections in 2011 were related to outpatient hospital procedures, and another 22 percent occurred in the general community.

Clostridium difficile (*C. diff.*) is a bacterium that causes colon inflammation. Diarrhea and fever that can lead to other lethal complications are among the most common symptoms. In 2017 this species was estimated to cause about 500,000 infections in the United States and almost 30,000 deaths. Those most at risk were older adults. *C. diff.* infections are extremely difficult to eradicate.

Data from the CDC show that although hospital-acquired infections are now declining, about 1 in 25 patients in U.S. hospitals will get some type of infection. This amounts to more than 700,000 illnesses per year. About 75,000 of those patients—most often the elderly—will die. These infections cost U.S. hospitals around $30 billion annually. The average drug-resistant infection costs up to $37,000 to treat.

Dire as these hospital infections are in rich countries, the situation is much worse in poorer places. India may be the epicenter for antibiotic overuse and multidrug-resistant infections. Bacteria spread easily because at least half the population defecate outdoors, and much of the sewage generated by those who do use toilets is dumped, untreated, into rivers. As a result, Indians have among the highest infection rate in the world, and they collectively take more antibiotics, which are sold over the counter, than any other people. In 2016 at least 60,000 Indian babies died of multidrug-resistant infections. India's terrible sanitation and uncontrolled antibiotic use, coupled with substandard housing and a complete lack of monitoring of the problem, have created a tsunami of antibiotic resistance that is spreading throughout the world, say epidemiologists. Researchers have recently found "super bugs" carrying a genetic code first identified in India around the world, including to France, Japan, and the United States.

What would better health cost?

The heaviest burden of illness is borne by the poorest people, who can afford neither a healthy environment nor adequate health care. Women in sub-Saharan Africa, for example, suffer six times the disease burden per 1,000 persons as women in most European countries. The WHO estimates that 90 percent of all disease burden occurs in developing countries, where less than one-tenth of all health care dollars are spent. The group Médecins Sans Frontiéres (MSF, or Doctors Without Borders) calls this the 10/90 gap. While wealthy nations spend billions to treat cosmetic problems in humans or the ailments of their pets, billions of people are sick or dying from treatable infections and parasitic diseases to which little attention is paid. To counter this trend, benefactors such as the Bill and Melinda Gates Foundation have pledged $200 million for medical aid to developing countries to help fight AIDS, TB, and malaria.

Dr. Jeffrey Sachs of the Columbia University Earth Institute says that disease is as much a cause as a consequence of poverty and political unrest, yet the world's richest countries now spend just $1 per person per year on global health. He predicts that raising our commitment to about $25 billion annually (about 0.1 percent of the annual GDP of the 20 richest countries) not only would save about 8 million lives each year, but would boost the world economy by billions of dollars.

Section Review

1. What is PFOA and how might you be exposed to it?

2. What are emergent diseases? Give a few examples, and describe their cause and effects.

3. How does antibiotic resistance arise? How do hospitals and feed-lots contribute to this phenomenon?

8.2 TOXICOLOGY

- *Many toxic substances are harmful in concentrations as low as billionths, or even trillionths, of a gram.*
- *Allergens, carcinogens, and toxic substances occur in many common products.*
- *Hormone disruptors can interfere with growth, development, and physiology at very low doses.*

Toxicology is the study of **toxic substances** (poisons) and their effects, particularly on living organisms. Because many substances are known to be poisonous (whether plant, animal, or microbial), toxicology is a broad field, drawing from biochemistry, histology, pharmacology, pathology, and many other disciplines. Toxic agents damage or kill living organisms because they react with cellular components to disrupt metabolic functions. Because of this reactivity, many are harmful even in extremely diluted concentrations. In some cases, billionths, or even trillionths, of a gram can cause irreversible damage. Toxic substances produced naturally are generally known as toxins. Those made by human activities are called toxicants.

All toxic substances are hazardous, but not all hazardous materials are toxic. Some substances, for example, are dangerous because they're flammable, explosive, acidic, caustic, irritants, or sensitizers. Many of these materials must be handled carefully in large doses or high concentrations, but can be rendered relatively innocuous by dilution, neutralization, or other physical treatments. They don't react with cellular components in ways that make them poisonous at low concentrations.

Environmental toxicology, or ecotoxicology, focuses on the effects of toxic substances on living organisms, and on the way these substances interact and are transformed as they move through populations and ecosystems. In aquatic systems, special attention is devoted to the ways pollutants act at the interface of sediment and water, or of water and organisms, or at the water/air interface. In terrestrial environments, the emphasis tends to be on the effects of metals on the soil fauna community and population characteristics.

Table 8.2 presents a list of the top 20 toxic and hazardous substances considered the highest risk by the U.S. Environmental Protection Agency. Compiled from the 275 substances regulated by the Comprehensive Environmental Response, Compensation, and Liability Act (CERCLA), commonly known as the Superfund Act, these materials are listed in order of assessed importance in terms of human and environmental health. In 2019, there were 1,317 Superfund sites in the United States. More than 11 million people live within 1 mile (1.6 km) of these sites.

TABLE 8.2	Top 20 Toxic and Hazardous Substances
MATERIAL	**MAJOR SOURCES**
1. Arsenic	Treated lumber
2. Lead	Paint, gasoline
3. Mercury	Coal combustion
4. Vinyl chloride	Plastics, industrial uses
5. Polychlorinated biphenyls (PCBs)	Electric insulation
6. Benzene	Gasoline, industrial use
7. Cadmium	Batteries
8. Benzo(a)pyrene	Waste incineration
9. Polycyclic aromatic hydrocarbons (PAHs)	Combustion
10. Benzo(b)fluoranthene	Fuels
11. Chloroform	Water purification, industry
12. DDT	Pesticide use
13. Aroclor 1254	Plastics
14. Aroclor 1260	Plastics
15. Trichloroethylene	Solvents
15. Dibenz (a, h)anthracene	Incineration
17. Dieldrin	Pesticides
18. Chromium, hexavalent	Paints, coatings, welding, anticorrosion agents
19. Chlordane	Pesticides
20. Hexachlorobutadiene	Pesticides

Source: Data from U.S Environmental Protection Agency.

How do toxic substances affect us?

Allergens are substances that activate the immune system. Some allergens act directly as **antigens;** that is, they are recognized as foreign by white blood cells and stimulate the production of specific antibodies (proteins that recognize and bind to foreign cells or chemicals). Other allergens act indirectly by binding to and changing the chemistry of foreign materials so they become antigenic and cause an immune response.

Formaldehyde is a good example of a widely used chemical that is a powerful sensitizer of the immune system. It is directly allergenic and can also trigger reactions to other substances. Widely used in plastics, wood products, insulation, glue, and fabrics, formaldehyde concentrations in indoor air can be thousands of times higher than in normal outdoor air. Some people suffer from what is called **sick building syndrome:** headaches, allergies, chronic fatigue, and other symptoms caused by poorly vented indoor air contaminated by mold spores, carbon monoxide, nitrogen oxides, formaldehyde, and other toxic substances released from carpets, insulation, plastics, building materials, and other sources (fig. 8.9). The Environmental Protection Agency estimates that poor indoor

Attic — Asbestos (vermiculite insulation)

Formaldehyde (furniture, carpeting, particle board)
Fungicides (carpeting)
Nitrogen oxide (fireplace)

Living Room

Tetrachloroethylene (dry cleaning)

Bedroom

Tobacco smoke (cigarettes, pipes)
Lead (candle wicks, paint)

Mothballs

Oven cleaner
Nitrogen oxides (unvented stove)

Bathroom

Mold

Kitchen

Toilet bowl cleaner

Chloroform (chlorine-treated water in shower)

Disinfectants, pesticides
Cleansers, ammonia

Basement

Paints, solvents, paint stripper, glue
Pesticides (garden products)

Herbicides

Carbon monoxide

Garage

Mold, bacteria   Rust remover   Paint, glue
Radon (cracks in floor, foundation)
Nitrogen oxide } (furnace, water heater)
Carbon monoxide

Gasoline from auto, lawnmower

FIGURE 8.9 Some sources of toxic and hazardous substances in a typical home.

air quality may cost the United States $60 billion a year in absenteeism and reduced productivity.

Immune system depressants are pollutants that suppress the immune system rather than activate it. Little is known about how this occurs or which chemicals are responsible. Immune system failure is thought to have played a role, however, in the widespread deaths of seals in the North Atlantic and of dolphins in the Mediterranean. These dead animals generally contain high levels of pesticide residues, polychlorinated biphenyls (PCBs), and other contaminants that are suspected of damaging the immune system and making it susceptible to a variety of opportunistic infections.

Endocrine disrupters are chemicals that disrupt hormone functions. Hormones are chemicals that living organisms use to regulate the development and function of tissues and organs (fig. 8.10). You undoubtedly have heard about sex hormones and their powerful effects on how we look, behave, and develop, but these are only one example of the many regulatory hormones that rule our lives. Insulin is a hormone that controls our uptake of sugars from food, and adrenaline is a hormone that elevates our breathing and circulation in moments of stress. The hormone thyroxine helps regulate many functions, including heart rate, temperature, and growth. Chemicals that disrupt any of these normal hormone functions can severely damage an organism's health or reproduction.

As the opening case study for this chapter shows, we now know that some of the most insidious effects of persistent chemicals such as perfluorocarbons (PFCs) are that they interfere with normal growth, development, and physiology of a variety of animals—including humans—at very low doses. For some materials, picogram concentrations (trillionths of a gram per liter) may be enough to cause developmental abnormalities in sensitive organisms. Although the official EPA limit for PFOA is 0.4 ppb, Philippe Grandjean of the Harvard School of Public Health suggests that a "safe" level should be only 0.001 ppb or 1 part per trillion. In 2015, more

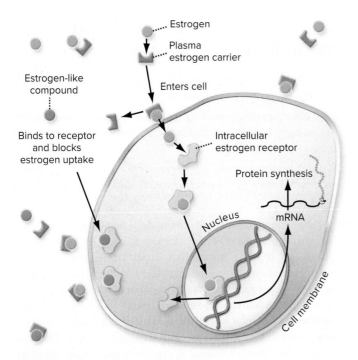

FIGURE 8.10 Steroid hormone action. Plasma hormone carriers deliver regulatory molecules to the cell surface, where they cross the cell membrane. Intracellular carriers deliver hormones to the nucleus, where they bind to and regulate the expression of DNA. Estrogen-like compounds bind to receptors and either block the uptake of endogenous hormones or act as a substitute hormone to disrupt gene expression.

than 200 scientists from 38 countries signed the Madrid Statement, which calls for the regulation of all PFCs due to their tendency to disrupt endocrine hormone functions.

Because these chemicals often cause sexual dysfunction (reproductive health problems in females or the feminization of males, for example), these chemicals are sometimes called environmental estrogens or androgens. They are just as likely, however, to disrupt other important regulatory functions as they are to obstruct sex hormones. Some researchers suggest that endocrine disrupters may be contributing to our obesity crisis.

Neurotoxins are a special class of metabolic poisons that specifically attack nerve cells (neurons). The nervous system is so important in regulating body activities that disruption of its activities is especially fast-acting and devastating. Different types of neurotoxins act in different ways. Heavy metals such as lead and mercury kill nerve cells and cause permanent neurological damage. Anesthetics (ether, chloroform, halothane, etc.) and chlorinated hydrocarbons (DDT, Dieldrin, Aldrin) disrupt nerve cell membranes necessary for nerve action. Organophosphates (Malathion, Parathion) and carbamates (carbaryl, zeneb, maneb) inhibit acetylcholinesterase, an enzyme that regulates signal transmission between nerve cells and the tissues or organs they innervate (for example, muscle). Most neurotoxins are both extremely toxic and fast-acting.

What Can You Do?

Tips for Staying Healthy

- Eat a balanced diet with plenty of fresh fruits, vegetables, legumes, and whole grains. Wash fruits and vegetables carefully; they may well have come from a country where pesticide and sanitation laws are lax.
- Use unsaturated oils such as olive or canola oil rather than hydrogenated or semisolid fats such as margarine.
- Cook meats and other foods at temperatures high enough to kill pathogens; clean utensils and cutting surfaces; store food properly.
- Wash your hands frequently. You transfer more germs from hand to mouth than by any other means of transmission.
- When you have a cold or flu, don't demand antibiotics from your doctor—they aren't effective against viruses.
- If you're taking antibiotics, continue for the entire time prescribed—quitting as soon as you feel well is an ideal way to select for antibiotic-resistant germs.
- Practice safe sex.
- Don't smoke, and avoid smoky places.
- If you drink, do so in moderation. Never drive when your reflexes or judgment are impaired.
- Exercise regularly: walk, swim, jog, dance, garden. Do something you enjoy that burns calories and maintains flexibility.
- Get enough sleep. Practice meditation, prayer, or some other form of stress reduction.
- Make a list of friends and family who make you feel more alive and happy. Spend time with one of them at least once a week.

Mutagens are agents that damage or alter genetic material (DNA) in cells. Chemicals and radiation can both act as mutagens. Damage to DNA can lead to birth defects if it occurs during embryonic or fetal growth. Later in life, genetic damage may trigger neoplastic (tumor) growth. When damage occurs in reproductive cells, the results can be passed on to future generations. Cells have repair mechanisms to detect and restore damaged genetic material, but some changes may be hidden, and the repair process itself can be flawed. It is generally accepted that there is no "safe" threshold for exposure to mutagens. Any exposure has some possibility of causing damage.

Teratogens cause birth defects. These are chemicals or other factors that cause abnormalities during embryonic growth and development. Some compounds that are not otherwise harmful can cause tragic problems in these sensitive stages of life. Perhaps the most prevalent teratogen in the world is alcohol. Drinking during pregnancy can lead to **fetal alcohol syndrome**—a cluster of symptoms that includes craniofacial abnormalities, developmental delays, behavioral problems, and mental defects that last throughout a child's life. Even one alcoholic drink a day during pregnancy has been associated with decreased birth weight.

Similarly, by some estimates, 300,000 to 600,000 children born every year in the United States are exposed in the womb to unsafe levels of mercury. The effects are subtle, but include reduced intelligence, attention deficit, and behavioral problems. The total cost of these effects is estimated to be $8.7 billion per year.

Carcinogens are substances that cause **cancer,** invasive, out-of-control cell growth that results in malignant tumors. Cancer rates rose in most industrialized countries during the twentieth century, and cancer is now the second leading cause of death in the United States, killing 600,000 people in 2018. According to the American Cancer Society, 1 in 2 males and 1 in 3 females in the United States will have some form of cancer in their lifetime. Some authors blame this cancer increase on toxic synthetic chemicals in our environment and diet. Others attribute it mainly to lifestyle (smoking, sunbathing, alcohol) or simply living longer. The U.S. EPA estimates that 200 million U.S. residents live in areas where the combined lifetime cancer risk from environmental carcinogens exceeds 1 in 100,000, or ten times the risk normally considered acceptable.

How does diet influence health?

Diet also has an important effect on health. For instance, there is a strong correlation between cardiovascular disease and the amount of salt and animal fat in one's diet. There's controversy about the role of high-fructose corn syrup, refined sugar, red meat, and other components of our food.

Fruits, vegetables, whole grains, complex carbohydrates, and dietary fiber (plant cell walls) often have beneficial health effects. Certain dietary components seem to have anticancer effects—these components include pectins; vitamins A, C, and E; substances produced in cruciferous vegetables (cabbage, broccoli, cauliflower, Brussels sprouts); and selenium, which we get from plants. Many of these substances act on the epigenetic system, which we'll discuss later in this chapter.

Eating too much food is a significant dietary health factor in developed countries and among the well-to-do everywhere. Seventy percent of all U.S. adults are now considered overweight, and the worldwide total of obese or overweight people is estimated to be over 1 billion. Every year in the United States, 300,000 deaths are linked to obesity (chapter 9).

The U.S. Centers for Disease Control and Prevention in Atlanta warn that one in three U.S. children will become diabetic unless many more people start eating less and exercising more. The odds are worse for Black and Hispanic children: Nearly half of them are likely to develop the disease. And among the Pima tribe of Arizona, nearly 80 percent of all adults are diabetic. More information about food and its health effects is available in chapter 9.

Section Review

1. What are endocrine disrupters and why are they dangerous?
2. What are teratogens and mutagens?
3. How does diet influence health?

8.3 THE MOVEMENT, DISTRIBUTION, AND FATE OF TOXIC SUBSTANCES

- *Solubility and mobility determine where and when chemicals move through the environment and in our bodies.*
- *Persistent materials, such as heavy metals and some organic compounds, concentrate in food webs and reach toxic concentrations in top predators.*
- *Synergistic interactions can increase the effects of combined toxic substances far beyond what any single substance might cause alone.*

There are many sources of toxic and hazardous chemicals in the environment, and many factors related to each chemical itself—its route or method of exposure, its persistence in the environment, and the characteristics of the target organism (table 8.3)—determine the dangerousness of the chemical. We can think of both individuals and ecosystems as sets of interacting compartments between which chemicals move, based on molecular size, solubility, stability, and reactivity (fig. 8.11). The dose (amount), route of entry, timing of exposure, and sensitivity of the organism all play important roles in determining toxicity. In this section, we will consider some of these characteristics and how they affect environmental health.

Table 8.3 Factors in Environmental Toxicity
Factors Related to the Toxic Agent
1. Chemical composition and reactivity
2. Physical characteristics (such as solubility, state)
3. Presence of impurities or contaminants
4. Stability and storage characteristics of toxic agent
5. Availability of vehicle (such as solvent) to carry agent
6. Movement of agent through environment and into cells
Factors Related to Exposure
1. Dose (concentration and volume of exposure)
2. Route, rate, and site of exposure
3. Duration and frequency of exposure
4. Time of exposure (time of day, season, year)
Factors Related to Organism
1. Resistance to uptake, storage, or cell permeability of agent
2. Ability to metabolize, inactivate, sequester, or eliminate agent
3. Tendency to activate or alter nontoxic substances so they become toxic
4. Concurrent infections or physical or chemical stress
5. Species and genetic characteristics of organism
6. Nutritional status of subject
7. Age, sex, body weight, immunological status, and maturity

Source: U.S. Department of Health and Human Services, 1995

Air
Photolysis
Oxidation
Precipitation

Source
Industry
Agriculture
Domestic, etc.

Biota
Metabolism
Storage
Excretion

Soil and Sediment
Photolysis and
metabolism
Evaporation

Water
Hydrolysis
Oxidation
Microbial degradation
Evaporation
Sedimentation

FIGURE 8.11 Living organisms are exposed to chemical compounds in water, soils, and the atmosphere. Commercial products move freely in the environment, increasing the likelihood of exposure.

Compounds dissolve either in water or in fat

Solubility is one of the most important characteristics in determining how, where, and when a toxic material will move through the environment or through the body. Chemicals can be divided into two major groups: those that dissolve more readily in water and those that dissolve more readily in oils or fats. Water-soluble compounds move rapidly and widely through the environment because water is everywhere around us. Water-soluble substances also tend to have ready access to most cells in the body, because aqueous solutions bathe all our cells. Molecules that are oil- or fat-soluble (usually these are organic, carbon-based molecules) generally need a carrier to move through the environment and into, and within, the body. Once inside the body, however, oil-soluble compounds penetrate readily into tissues and cells, because the membranes that enclose cells are themselves made of similar oil-soluble chemicals. Once they get inside cells, oil-soluble materials are likely to be accumulated and stored in lipid deposits, where they may be protected from metabolic breakdown and persist for many years.

Water
(drinking)

Cosmetics,
household accidents

(incidental and
accidental exposure)

Medicinal drugs
(oral, intravenous,
intramuscular)

Food
(eating)

Air
(breathing)

Occupational exposure
(oral, dermal, inhalation)

FIGURE 8.12 Routes of exposure to toxic and hazardous environmental factors.

How do these dangerous substances enter our bodies? There are many routes, although some cause damage more frequently than others (fig. 8.12). Airborne toxic particles generally cause more ill health than any other exposure source. We breathe far more air every day than the volume of food we eat or water we drink. Furthermore, the cellular lining of our lungs is designed to exchange gases very efficiently, but this also means foreign substances move easily from our lungs into our bloodstream. Epidemiologists estimate that millions of people die each year from diseases caused or exacerbated by air pollution (see chapter 16).

Food, water, and skin contact also can expose us to a wide variety of toxic substances. The greatest exposures to many of these toxicants are found in industrial settings (including industrial agriculture), where workers may encounter doses thousands of times higher than would be found anywhere else. In the opening case study for this chapter, workers in factories that synthesize perfluorocarbon products had much higher exposure to these chemicals and a greater incidence of cancers and other diseases than the general public. The European Agency for Safety and Health at Work warns that 32 million people (20 percent of all employees) in the European Union are exposed to unacceptable levels of carcinogens and other toxic substances in their workplace.

Condition of the organism and timing of exposure also have strong influences on toxicity. Healthy adults, for example, may be relatively insensitive to doses that would be very dangerous for young children or for someone already weakened by disease.

Similarly, exposure to a toxic substance may be very dangerous at certain stages of developmental or metabolic cycles, but may be innocuous at other times. A single dose of the notorious teratogen thalidomide, for example, taken in the third week of pregnancy (a time when many women aren't aware they're pregnant) can cause severe abnormalities in fetal limb development. A complication in measuring toxicity is that great differences in sensitivity exist between species. Thalidomide was tested on a number of laboratory animals without any deleterious effects. Unfortunately, however, it is a powerful teratogen in humans.

Bioaccumulation and biomagnification increase concentrations of chemicals

Cells selectively absorb and store (accumulate) a great variety of molecules. This process of storing substances in living tissues is known as **bioaccumulation.** Bioaccumulation allows cells to selectively take up and store nutrients and essential minerals. At the same time, though, cells can absorb and store harmful substances. Toxicants that are dilute in the environment, such as pesticides or metals, can reach dangerous levels inside cells and tissues through bioaccumulation.

This effect is magnified through food webs. **Biomagnification** (elevated concentrations of harmful chemicals in the cells of organisms at higher trophic levels) occurs when a predator consumes prey containing pesticides, heavy metals, or other substances. All the toxic substances selectively stored by the cells of the prey are then accumulated in the predator's tissues. One of the first dramatic demonstrations of biomagnification involved the pesticide DDT in aquatic ecosystems (fig. 8.13). Phytoplankton and bacteria take up DDT from water or sediments. Their predators—zooplankton and invertebrates—collect and retain the DDT from prey organisms, building up higher concentrations as they feed, and so on up the food chain. The cells and tissues of fish and fish-eating birds can contain DDT concentrations hundreds or thousands of times greater than concentrations in the tissues of bacteria and phytoplankton (see horizontal axis, fig. 8.13).

In the 1960s, bioaccumulation of DDT nearly wiped out a number of fish-eating bird species in North America. Inexpensive and poisonous to nearly all insect pests, DDT was sprayed on fields from airplanes, on beaches and wetlands, and used widely for household pest control starting in the late 1940s. Within 15 years it was found that extremely high concentrations of DDT in the tissues of peregrine falcons, brown pelicans, bald eagles, and other birds were interfering with the birds' ability to produce eggs and healthy chicks. Populations plummeted and disappeared from many areas. New controls on the use of DDT were largely responsible for rescuing these species from extinction.

Persistence makes some materials a greater threat

Some chemical compounds are very unstable and degrade (decompose to less harmful compounds) rapidly under most environmental conditions. When substances degrade rapidly, their concentrations

FIGURE 8.13 Biomagnification in a Lake Michigan food chain. The DDT tissue concentration in gulls, a tertiary consumer, was about 240 times that in the small insects sharing the same environment.

and toxicity decline quickly after they are released in the environment. Most modern herbicides and pesticides, for instance, are designed to break down in a few days and lose their toxicity. Other substances are more persistent and last for years or even centuries in the environment without breaking down or losing toxicity. Metals, such as lead or mercury, PVC plastics, chlorinated hydrocarbon pesticides, and asbestos are useful because they do not degrade rapidly. This stability makes the products valuable, but it is also a problem because these materials persist in the environment and have unexpected effects far from the sites of their original use.

Lead, for example, was one of the first persistent pollutants recognized by toxicologists. Lead interferes with the nervous system both in adults and children, causing decreased mental ability, kidney damage, anemia, miscarriages, and fertility losses. Lead was widely used in house paint, plumbing pipes, ceramic glazes, bullets, pesticides, and in tetraethyl lead, which was added to gasoline to improve combustion. Lead in gasoline was especially dangerous, because it was designed to be burned and released to the atmosphere, where it is widely distributed and easily taken into lung tissue. Developmental impacts on young children were especially disastrous, but they were also widespread and not concentrated near obvious sources. It took many years for epidemiologists to prove the dangers of leaded gasoline and other lead products. Manufacturers of lead-containing products disputed evidence of harm.

The epidemiologist Phillip Landrigan showed that even low lead levels in children's blood correlated with reduced IQ. Lead was banned from gasoline in 1975 and from paint in 1978. This was one of the most successful public health measures ever. By 2005, there was an 88 percent drop in American children's blood levels but the CDC estimates that 535,000 American children, mostly poor minorities living in large cities still have dangerous levels of lead in their bodies. In addition to links to cancer, genital deformities, reproductive problems, and diminished IQ, lead exposure can cause behavioral problems such as ADHD, aggression, and impulsivity. Recently, some epidemiologists have suggested that the decrease in violent crime over the past half-century tracks very closely with the decline of lead in the environment.

In 2016, there was a great concern about children's exposure to lead in Flint, Michigan. To save a few million dollars, the city had switched its source of drinking water from Lake Huron to the Flint River, a longtime dumping ground for industry that once operated in the city. The corrosive river water leached lead out of old plumbing (see figure 1.28, p. 30). But Flint's children are hardly the only ones facing this danger. In Flint, 4.9 percent of the children tested have elevated lead levels. In New York State, it's 6.7 percent. In Pennsylvania, it's 8.5 percent. And in parts of Detroit, it's 20 percent. Baltimore, Maryland, Sebring, Ohio, and even Washington, DC have similar problems.

POPs are an especially serious problem

Modern industry has given us many materials that make daily life more comfortable and convenient (fig. 8.14). But some of the 80,000 organic (hydrocarbon-based) chemicals in commercial production can cause long-lasting health problems. Many of these compounds persist a long time in the environment, which can be both a benefit (they do their job for a long time) and a drawback (they also are very hard to eliminate.) We call these **persistent organic pollutants (POPs).** Because of their effectiveness, they have been used abundantly and globally for decades. POPs are found now in the environment and in the tissues and blood of humans and animals everywhere from the tropics to the Arctic. They often accumulate in food webs and reach toxic concentrations in long-living top predators such as humans, sharks, raptors, swordfish, and bears. And they can act, even in low concentrations, as carcinogens, mutagens, neurotoxins, teratogens, and estrogen disrupters. In recent years, global treaties have begun to phase out some of these chemicals, but many remain in widespread use. The POPs of greatest concern include the following:

- **Chlorpyrifos** is an organophosphate insecticide used very widely to control insect pests in corn, soybeans, nut trees, and many fruits and vegetables. It is also used for mosquito and termite control, roach and ant traps, and golf courses, green houses, and wood products. It was introduced in 1965 by the Dow Chemical Company, and the EPA estimates that about 9,500 metric tons (21 million pounds) of this compound were used annually in the United States between 1987 and 1998. In 2007, chlorpyrifos which was the most widely used insecticide in the United States, was banned for most residential use, although commercial use has persisted. The EPA has estimated

FIGURE 8.14 Modern industry has given us many materials that make daily life more comfortable and convenient. But we're discovering that some of the 80,000 synthetic chemicals we use can cause long-lasting health problems.
William P. Cunningham

that 100,000 deaths per year could be traced to chlorpyrifos, which interferes with acetylcholine neurotransmitters.

In 2017 an EPA scientific panel recommended that chlorpyrifos be banned for all uses in the United States because of its high toxicity. But EPA administrator Scott Pruitt rejected the recommendations of his own scientists and sqashed efforts to regulate this chemical.

- **Dicamba** is a herbicide in the same family of chlorinated hydrocarbons as 2,4-D and is a plant hormone that causes broadleaf weeds to outgrow their nutrient supplies. In 2016 Monsanto introduced genetically modified crop varieties that tolerate dicamba, allowing it to be used liberally on fields without damaging yields. By the next year, 10 million ha (25 million acres) of soy and cotton had been planted in these Monsanto GMO crops and treated with dicamba (also sold by Monsanto). Because this herbicide is effective before weeds sprout, as well as after, much of the spraying was done in late spring or early summer when temperatures and humidity were high, and a significant portion of the herbicide evaporated and drifted into neighboring fields. Neighboring farmers who hadn't paid for the expensive GMO varieties complained that this drift damaged their crops. There are also worries about possible birth defects and reproductive risks from dicamba, especially given the widespread use and volatilization. The agricultural community is bitterly divided about the use of this compound, and there has been violence between neighbors about this dispute.

- **Glyphosate** is an extremely effective herbicide used to kill weeds and grasses marketed by Monsanto under the trade name Roundup®. The Environmental Working Group (EWG) reports that glyphosate is the most widely and heavily used herbicide in the world. The EWG estimates that 8.6 billion kg (18.9 billion pounds) have been used globally since its introduction in 1974.

Enough glyphosate was sprayed worldwide in 2014 to apply 0.5 kg per ha (0.75 pound/acre) over all the cropland in the world. The chemical is absorbed through the leaves and works by interfering with the synthesis of essential amino acids. An increasing number of crops have been genetically engineered to be tolerant of glyphosate. Monsanto's Roundup-Ready soybean was the first of these; others include corn, canola, alfalfa, sugar beets, cotton, and wheat. In 2017, 95 percent of soy, 89 percent of corn, and 89 percent of cotton produced in the United States were genetically modified to be herbicide-tolerant (see fig. 9.23).

In 2015 the European Agency for Research on Cancer listed glyphosate as "probably carcinogenic in humans," on the basis of epidemiological and lab studies. The U.S. EPA, on the other hand, considers glyphosate to be noncarcinogenic and relatively nontoxic. According to the EPA, there is no credible evidence for endocrine disruption by glyphosate. The agency stated that under the worst-case scenario of eating a lifetime of food derived entirely from glyphosate-sprayed fields with residues at maximum permitted levels, no adverse effects would be expected. The State of California disagreed, however, and in 2015 listed the chemical as "known to the State of California to cause cancer."

- **Bisphenol A (BPA),** a key ingredient of both polycarbonate plastics and epoxy resins, is one of the world's most widely used chemical compounds. In 2015, global production was about 5.4 million metric tons. BPA is used in items ranging from baby bottles, automotive accessories, eyeglasses, water pipes, the linings of cans and bottles, and dental sealants. It's estimated that 95 percent of American adults have measurable amounts of BPA in their bodies. The most likely source of contamination is from food and beverage containers. BPA can leach out, especially when plastic is heated, washed with harsh detergents, scratched, or exposed to acidic compounds. BPA has been linked to myriad health effects including mammary and prostate cancer, genital defects in males, early onset of puberty in females, obesity, and even behavior problems such as attention-deficit hyperactivity disorder. The U.S. Food and Drug administration has banned BPA in baby bottles and packaging for infant formula, but use in other food and water containers is still allowed.

Synergistic interactions can increase toxicity

Some materials interfere with each other, reducing their effects or breaking them down: For instance, vitamins E and A seem to reduce your body's response to some carcinogens. Other combinations of compounds make each other's impacts worse. This is called a **synergistic effect,** an interaction in which one substance exacerbates the effects of another. For example, occupational exposure to asbestos increases lung cancer rates 20-fold; smoking also increases lung cancer rates by the same amount. Asbestos workers who also smoke, however, have a 400-fold increase in cancer rates. How many other toxic chemicals are we exposed to that are below threshold limits individually but combine to give toxic results?

A fundamental concept in toxicology is that every material can be poisonous under some conditions, but most chemicals have some safe level or threshold below which their effects are undetectable or insignificant. Each of us consumes lethal doses of many chemicals over the course of a lifetime. One hundred cups of strong coffee, for instance, contain a lethal dose of caffeine. Similarly, 100 aspirin tablets, or 10 kilograms (22 lb) of spinach or rhubarb, or a liter of alcohol would be deadly if consumed all at once. Taken in small doses, however, most toxicants can be broken down or excreted before they do much harm. Furthermore, damage they cause can sometimes be repaired, and the repair can make you stronger and more resilient in the future.

Our bodies degrade and excrete toxic substances

Most organisms have enzymes that break down waste products and environmental poisons to reduce their toxicity. In mammals, most of these enzymes are located in the liver, the primary site of detoxification of both natural wastes and introduced poisons. Sometimes these reactions can produce more harmful compounds, however. Benzopyrene, for example, is not toxic in its original form, but our bodies break it down into substances that can cause cancer.

We also reduce the effects of waste products and environmental toxins by eliminating them from our body through excretion. Volatile molecules, such as carbon dioxide, hydrogen cyanide, and ketones, are excreted via breathing. Some excess salts and other substances are excreted in sweat. Primarily, however, excretion is a function of the kidneys, which can eliminate significant amounts of soluble materials through urine formation. Accumulation of toxins in the urine can damage this vital system, however, and the kidneys and bladder often are subjected to harmful levels of toxic compounds. In the same way, the stomach, intestine, and colon often suffer damage from materials concentrated in the digestive system and may be afflicted by diseases and tumors.

Our bodies also have mechanisms to repair damage. Our natural defenses include the ability of cells to repair damaged proteins and DNA. Similarly, tissues and organs that are exposed regularly to physical wear and tear or to toxic or hazardous materials often have mechanisms for damage repair. You have probably seen how your skin can repair damage rapidly. In the same way you continuously replace and repair tissues in your liver, lungs, the linings of your stomach and intestines, and other tissues. Each time tissues are damaged and reproduced, however, there is a chance that some cells will lose normal growth controls and run amok, creating a tumor. Thus, any agent that irritates tissues, such as smoking (which damages lungs) or alcohol (which damages the liver), is likely to be carcinogenic. And tissues with high cell-replacement rates are among the most likely to develop cancers.

Section Review

1. What is biomagnification, and why is it important in environmental health?
2. How do the physical and chemical characteristics of materials affect their movement, persistence, distribution, and fate in the environment?
3. Give three examples of persistent organic pollutants and describe their main effects.

8.4 TOXICITY AND RISK ASSESSMENT

- *The dose makes the poison. Almost everything is toxic at some level.*
- *LD50 and dose-response curves indicate how dangerous a substance is.*
- *Our risk assessment is often selective and nonrational.*

Almost 500 years ago, the Swiss scientist Paracelsus said "the dose makes the poison." He meant by this that almost everything is toxic at some level. This remains the most basic principle of toxicology. Table salt, for example, is necessary at low levels, but if you ate a kilogram of salt all at once, it would make you very sick. A similar amount injected into your bloodstream would kill you. Thus, the delivery method and rate, as well as the dose, matter in determining toxicity.

Substances vary dramatically in their degree of toxicity. Some are so poisonous that a single drop on your skin can kill you. Others require massive amounts injected directly into the blood to be lethal. Measuring and comparing the toxicity of different materials is difficult, because species differ in sensitivity to any substance, and individuals within a species respond differently to a given exposure. In this section, we will look at the ways we measure degrees of toxicity.

We usually test toxic effects on lab animals

Determining toxicity is extremely difficult. The most commonly used and widely accepted toxicity test is to expose a population of laboratory animals to measured doses of a specific substance under controlled conditions. This procedure is expensive, time-consuming, and often painful and debilitating to the animals being tested. It commonly takes hundreds—or even thousands—of animals, several years of hard work, and hundreds of thousands of dollars to thoroughly test toxicity. More humane tests using computer simulations of model reactions, cell cultures, and other substitutes for whole living animals are being developed. However, conventional large-scale animal testing is the method in which scientists have the most confidence and on which most public policies about pollution and environmental or occupational health hazards are based.

In addition to humanitarian concerns, a challenge for toxicologists and policymakers is that members of a population vary in their response to a toxic substance (fig. 8.15). Some individuals are very sensitive and respond at very low doses. Some respond only at very high doses. Most fall somewhere in the middle, forming a bell-shaped curve. The question for regulators and politicians is whether we should set pollution limits that will protect everyone, including the most sensitive people, or only aim to protect the average person. It might cost billions of extra dollars to protect a very small number of individuals at the extreme end of the curve. Is that a good use of resources?

Cumulative responses are usually depicted in a dose-response curve (fig. 8.16). This graph is like the histogram bell curve of figure 8.16, but it shows the percentage of the population on the

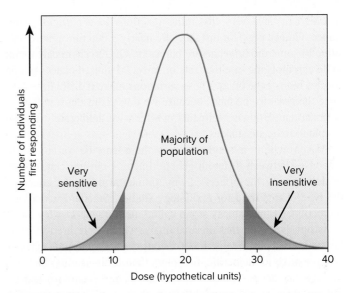

FIGURE 8.15 Probable variations in sensitivity in a population. Some members of a population may be very sensitive to a given substance, while others are much less sensitive. The majority of the population falls somewhere between the two extremes.

FIGURE 8.16 A dose-response curve shows the cumulative population response to increasing doses of a toxic substance. The LD50 is the dose that is lethal to half the population.

vertical axis, rather than the number of individuals. The percentage of organisms responding to low, medium, or high doses thus add up to 100 percent on the graph. A convenient way to describe the toxicity of a chemical is to determine the dose to which 50 percent of the test population is sensitive. In the case of a lethal dose (LD), this is called the **LD50.** In figure 8.16, for example, a dose of 3.5 units is lethal for 50 percent of the population.

Dose-response curves are often used to compare toxicity. One substance might be lethal to half the population at 3.5 units, while another might have an LD50 of only 1.5. This would mean the second substance is more than twice as dangerous as the first.

Unrelated species can react very differently to the same compound, not only because body sizes vary but also because of differences in physiology and metabolism. Even closely related species can have very dissimilar reactions to a particular substance. Guinea pigs, for instance, are nearly 5,000 times more sensitive to some dioxins than are hamsters. Of 226 chemicals found to be carcinogenic in either rats or mice, 95 caused cancer in one species but not the other. These variations make it difficult to estimate the risks for humans, because we don't consider it ethical to perform dangerous experiments in which we deliberately expose people to toxic substances.

Even within a single species there can be variations in responses between different genetic lines. A current controversy in determining the effects of environmental estrogens concerns the type of rats used for toxicology studies. Standard toxicology protocols call for a sturdy strain called the Sprague-Dawley rat. It turns out, however, that these animals, which were bred to grow fast and breed prolifically in lab conditions, are thousands of times less sensitive to endocrine disrupters than ordinary rats. Industry reports that declare certain chemicals to be harmless based on Sprague-Dawley rats are highly suspect.

There can be a wide range of toxicity

It is useful to group materials according to their relative toxicity. A moderately toxic substance takes about one gram per kilogram of body weight (about two ounces for an average human) to make a lethal dose. Very toxic materials take about one-tenth that amount, while extremely toxic substances take one-hundredth as much (only a few drops) to kill most people. Supertoxic chemicals are extremely potent; for some, a few micrograms (millionths of a gram—an amount invisible to the naked eye) make a lethal dose. These materials are not all synthetic. One of the most toxic chemicals known, for instance, is ricin, a protein found in castor bean seeds. It is so toxic that 0.3 billionths of a gram given intravenously will generally kill a mouse. If aspirin were this toxic, a single tablet, divided evenly, could kill 1 million people.

Many carcinogens, mutagens, and teratogens are dangerous at levels far below their direct toxic effect because they cause abnormal cell growth, which results in a kind of biological amplification. A single cell, perhaps altered by a single molecular event, can multiply into millions of tumor cells or an entire organism. Just as there are different levels of direct toxicity, however, there are different degrees of carcinogenicity, mutagenicity, and teratogenicity. Methanesulfonic acid, for instance, is highly carcinogenic, while the sweetener saccharin is a suspected carcinogen whose effects may be vanishingly small.

Acute and chronic doses and effects differ

Some toxic effects are sudden and severe, causing immediate impairment or death. We call these **acute effects.** An organism may be able to recover from acute effects, however. Often, if the individual experiencing an acute reaction survives this immediate crisis, the effects are reversible. **Chronic effects,** on the other hand, are long-lasting or even permanent. A chronic effect can result from a single dose of a very toxic substance, or it can be the result of a continuous or repeated sublethal exposure.

Exposure can also be acute or chronic. For example, frequent consumption of alcohol over many years (chronic exposure) can gradually degrade your liver, eventually causing long-lasting damage or even death. Consuming an extreme amount of alcohol all at once (acute exposure) can cause immediate and severe impairment or death. Often, it's difficult to identify or assess health risks of chronic exposures because other factors, such as aging or normal diseases, act simultaneously with the factor under study.

Figure 8.17 shows three possible results from low doses of a toxin. Curve (a) shows a baseline level of response in the population, even at a zero dose of the toxin. This suggests that some other factor in the environment also causes this response. Curve (b) shows a straight-line relationship from the highest doses to zero exposure. Many carcinogens and mutagens show this kind of response. Any exposure to such agents, no matter how small, carries some risks. Curve (c) shows a threshold for the response where some minimal dose is necessary before any effect can be observed. This generally suggests the presence of some defense mechanism that either prevents the toxin from reaching its target in an active form or repairs the damage that it causes. Low levels of exposure to the toxin in question may have no deleterious effects, and it might not be necessary to try to keep exposures to zero (What Do You Think?, p. 171).

Detectable levels aren't always dangerous

You may have seen or heard dire warnings about toxic materials detected in samples of air, water, or food. A typical headline announced recently that 23 pesticides were found in 16 food samples. What does that mean? Is any amount of dangerous materials unacceptable? Not necessarily. We have noted that the dose makes the poison. Toxicity depends on what is there, how much is there, how it is delivered, and other factors.

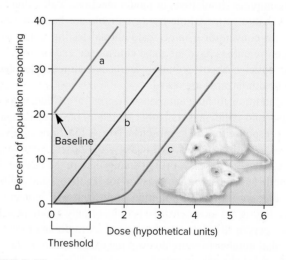

FIGURE 8.17 Three possible dose-response curves at low doses. (a) Some individuals respond, even at zero dose, indicating that some other factor must be involved. (b) Response is linear down to the lowest possible dose. (c) Threshold must be passed before any response is seen.

What's Acceptable Risk?

In late 2018, the EPA released a proposed rule titled "Strengthening Transparency in Regulatory Science Rule." Naturally, transparency is good in policies and rules, and this rule claimed to enhance rigor in scientific decision making. But hidden in the rule was language *weakening* regulation of toxic substances. Buried in a thicket of vague wording, the rule directs policymakers to assume that a little pollution is actually beneficial for your health.

In some cases, this makes sense: Eating a kilogram of table salt would kill you, but you need a little to stay healthy. Does the same principle apply to substances known to cause cancer even in minute doses? The proposed rule was cheered by industry, which expects to save huge amounts of money under looser pollution prevention rules. But what does this mean for the rest of us?

Public health scientists were alarmed by the implications of this rule. For decades we have followed the better-safe-than-sorry principle that any exposure to high-risk substances, such as mutagens and carcinogens, is risky to public health. In other words, we assume a linear dose-response rate all the way to zero exposure (like curves *a* or *b* in figure 8.17). At extremely low doses, that effect may be very small, say one in a million chance of getting cancer. But in a very large sample, say the U.S. population of over 300 million, there could be a significant response: In a sample of 300 million people exposed, you could expect 300 cases of cancer.

Some toxicologists suggest that many toxic substances have nonlinear dose-response relationships, like curve *c* in figure 8.17. It's hard to know, though, because extremely low doses are hard to evaluate. Effects of large doses are usually obvious and easy to document. To measure effects of extremely low doses, you might need millions of experimental animals and years of observations. This is impossibly expensive. So we're left with uncertainty about how dangerous many chemicals and physical factors may be at very low exposure. The 1958 Delaney Clause to the U.S. Food and Drug Act took a precautionary approach. It forbade the addition of any amount of a known carcinogen to foods or drugs, based on an assumption that any exposure to known carcinogens is an unacceptable risk. This standard was relaxed in 1996, however, by a "no reasonable harm" requirement, defined as a less than one cancer for every million people exposed over a lifetime. This change was supported by a report from the National Academy of Sciences, which concluded that synthetic chemicals in our diet represent an acceptable cancer risk. The question many consumer groups ask is "acceptable to whom?"

A similar debate has plagued the nuclear energy industry. Some evidence suggests that very low amounts of radiation may protect against certain cancers. This is perplexing, because ionizing radiation has long been recognized as a human carcinogen. It's thought, however, that very low radiation exposure may stimulate DNA repair along with enzymes that destroy free radicals (atoms with unpaired, reactive electrons in their outer shells). Activating these repair mechanisms might defend us from other, unrelated hazards.

These nonlinear effects, in which low exposure is beneficial but high exposure is harmful, are called hormesis. But not everyone is convinced that hormesis is relevant in toxic substance standards. Perhaps what looks like hormesis is just confounding influences that aren't recognized, especially when you're evaluating a very large population of humans over a very long time. The genetics and physiology of members of such a large group, along with what their actual exposure is and what other circumstances they may be exposed to, may be impossible to adequately assess.

The question remains, then, are we wasting money and missing opportunities by trying to eliminate every possible toxic exposure? Is cost to industry a sufficient reason to abandon a precautionary approach that demands conclusive evidence of no harm before approving risks in our environment? Or if risks are acceptable, what evidence makes that so? How would you feel if you were the one person in a million who developed cancer because of a known carcinogen in your diet, air, or water? If your senator or representative voted for weakened rules, what would you say to them?

Toxic pollutants may seem to be more widespread now than in the past, and this is surely a valid perception for many substances. The daily reports we hear of new materials found in new places, however, are also due in part to our more sensitive measuring techniques. Twenty years ago, parts per million was generally the limit of detection for most chemicals. Anything below that amount was often reported as zero or absent rather than more accurately stated as undetected. A decade ago, new machines and techniques were developed to measure parts per billion. Suddenly chemicals were found where none had been suspected. Now we can detect parts per trillion or even parts per quadrillion in some cases. Increasingly sophisticated measuring capabilities may lead us to believe that toxic materials have become more prevalent. In fact, our environment may be no more dangerous; we are just better at finding trace amounts.

Some symptoms can be erroneous

Sometimes outbreaks of apparent health problems have a psychological rather than physical basis. When your neighbors talk convincingly about being sick, you may begin to feel ill, too. Some physicians regard the so-called "wind turbine syndrome," for example, as a psychogenic condition with no physical basis. People who are opposed to wind turbines often claim the noise and shadow flicker caused by moving turbine blades make them sick, but there is no medical evidence of cause and effect. More than 155 different symptoms ranging from sleep loss, headaches, dizziness, nausea, muscle pain, and tinnitus to depression and suicidal thoughts have been blamed on wind turbines. Many stresses, including having large machines you don't like near your property may result in such symptoms, but there's little evidence that the motion of the blades by themselves can cause so many illnesses.

Risk perception isn't always rational

Risk is the possibility of suffering harm or loss. **Risk assessment** is the scientific process of estimating the threat that particular hazards pose to human health. This process includes risk identification, dose response assessment, exposure appraisal, and risk characterization. In hazard identification, scientists evaluate all available information to estimate the likelihood that a chemical will cause a certain effect in humans. The best evidence comes from human studies, such as physician case reports. Animal studies are also used to assess health risks. Risk assessment for identified toxicity hazards (for example, lead) includes collection and analysis of site data, development of exposure and risk calculations, and preparation of human health and ecological impact reports. Exposure assessment is the estimation or determination of the magnitude, frequency, duration, and route of exposure to a possible toxicant. Toxicity assessment weighs all available evidence and estimates the potential for adverse health effects to occur.

A number of factors influence how we perceive relative risks associated with different situations, such as the following:

- People with social, political, or economic interests—including environmentalists—tend to downplay certain risks and emphasize others that suit their own agendas. We also tend to tolerate risks that we choose—such as driving, smoking, or overeating—while objecting to risks we cannot control, such as exposure to wind turbines or pesticides.

- Most people have difficulty understanding and believing probabilities. We feel that there must be patterns and connections in events, even though statistical theory says otherwise. If the coin turned up heads last time, we feel certain it will turn up tails the next. In the same way, it is difficult to understand the meaning of a 1-in-10,000 risk of being poisoned by a chemical.

- Our personal experiences are often misleading. When we have not personally experienced a bad outcome, we feel it is more rare and unlikely to occur than it actually may be. Furthermore, we tend to rationalize the things we want to do and exaggerate risks we dislike (fig. 8.17).

- We have an exaggerated view of our own abilities to control our fate. We generally consider ourselves above-average drivers, safer than most when using appliances or power tools, and less likely than others to suffer medical problems, such as heart attacks. People often feel they can avoid hazards because they are wiser or luckier than others.

- News media give us a biased perspective on the frequency of certain kinds of health hazards, overreporting some accidents or diseases while downplaying or underreporting others. Sensational, gory, or especially frightful causes of death, like murders, plane crashes, fires, or terrible accidents, occupy a

disproportionate amount of attention in the public media. Heart diseases, cancer, and stroke kill nearly 15 times as many people in the United States as do accidents and 75 times as many people as do homicides, but the emphasis placed by the media on accidents and homicides is nearly inversely proportional to their relative frequency compared to either cardiovascular disease or cancer. This gives us an inaccurate picture of the real risks to which we are exposed.

- We tend to have an irrational fear or distrust of certain technologies or activities that leads us to overestimate their dangers. Nuclear power, for instance, is viewed as very risky, while coal-burning power plants seem to be familiar and relatively benign; in fact, coal mining, shipping, and combustion cause an estimated 10,000 deaths each year in the United States. An old, familiar technology seems safer and more acceptable than does a new, unknown one (see Data Analysis at the end of this chapter).

FIGURE 8.18 How dangerous is trick skating? Many parents regard it as extremely risky, while many students—especially males—believe the risks are acceptable. Perhaps the more important question is whether the benefits outweigh the risks.
Drpixel/Shutterstock

Risk acceptance depends on many factors

How much is it worth to minimize and avoid exposure to certain risks? Most people will tolerate a higher probability of occurrence of an event if the harm caused by that event is low. Conversely, harm of greater severity is acceptable only at low levels of frequency. A 1-in-10,000 chance of being killed might be of more concern to you than a 1-in-100 chance of being injured. For most people, a 1-in-100,000 chance of dying from some event or some factor is a threshold for changing what we do. That is, if the chance of death is less than 1 in 100,000, we are not likely to be worried enough to change our ways. If the risk is greater, we will probably do something about it. The EPA generally assumes that a risk of 1 in 1 million is acceptable for most environmental hazards. Is that the level you would choose? Why or why not?

For activities that we enjoy or find profitable, we are often willing to accept far greater risks than this general threshold. Conversely, for risks that benefit someone else, we demand far higher protection. For instance, your chance of dying in a motor vehicle accident in any given year is about 1 in 5,000, but that doesn't deter many people from riding in automobiles. Your lifetime chance of dying from lung cancer if you smoke one pack of cigarettes per day is about 1 in 4. By comparison, the risk from drinking water with the EPA limit of trichloroethylene is about 1 in 10 million. Strangely, many people demand water with zero levels of trichloroethylene, while continuing to smoke cigarettes.

Table 8.4	Lifetime Odds of Death for Selected Causes, United States	
Cause of Death	**Odds (1 in x)**
Heart disease	6
Cancer	7
Chronic Lower Respiratory Disease (COPD)	27
Suicide	88
Opioid overdose	96
Automobile accident	103
Fall	114
Gun assault	285
Pedestrian incident	556
Motorcyclist	858
Drowning	1,117
Fire or smoke	1,474
Choking on food	2,696
Bicyclist	4,047
Accidental gun discharge	8,527
Sunstroke	8,912
Electrocution, radiation, extreme temperatures, and pressure	15,638
Sharp objects	28,000
Storm	31,394
Hornet, wasp, and bee stings	46,562
Dog attack	115,111
Airplane accident	188,364

Source: National Safety Council 2017

Similarly, more than 1 million Americans are diagnosed with skin cancer each year. Some of these cancers are lethal, and most are disfiguring, yet only one-third of teenagers routinely use sunscreen. Tanning beds more than double your chances of cancer, especially if you're young, but about 10 percent of all teenagers admit regularly using these devices.

Table 8.4 lists lifetime odds of dying from a few leading diseases and accidents. These are statistical averages, of course, and there clearly are differences in where one lives or how one behaves that affect the danger level of these activities. Although the average lifetime chance of dying in an automobile accident is 1 in 103, there clearly are things you can do—like wearing a seat belt, following safety rules, and avoiding risky situations—that improve your odds. Still, it is interesting how we readily accept some risks while shunning others.

Our perception of relative risks is strongly affected by whether risks are known or unknown, whether we feel in control of the outcome, and how dreadful the results are. Risks that are unknown or unpredictable and results that are particularly gruesome or disgusting seem far worse than those that are familiar and socially acceptable.

Studies of public risk perception show that most people react more to emotion than to statistics. We go to great lengths to avoid some dangers while gladly accepting others. Factors that are involuntary, unfamiliar, undetectable to those exposed, catastrophic, or that have delayed effects or are a threat to future generations are especially feared, whereas those that are voluntary, familiar, detectable, or immediate cause less anxiety. Even though the actual number of deaths from automobile accidents, smoking, or alcohol, for instance, are thousands of times greater than those from pesticides, nuclear energy, or genetic engineering, the latter preoccupy us far more than the former.

Section Review

1. Why is there a threshold in response to some toxic substances?
2. What is an LD50?
3. What is the difference between acute and chronic doses and effects?
4. What is the difference between acute and chronic toxicity?

8.5 ESTABLISHING HEALTH POLICY

- *It's difficult to evaluate multiple risks and benefits simultaneously.*
- *In setting health standards, we need to consider combined exposures, different sensitivities, and differences between chronic and acute exposures.*

Risk management combines principles of environmental health and toxicology together with regulatory decisions based on socioeconomic, technical, and political considerations (fig. 8.19). Establishing policies to protect the public is difficult, because we face many sources of harm simultaneously, often without being aware of them. It is difficult to separate the effects of all these different hazards and evaluate their risks accurately. In spite of often vague and contradictory data, public policymakers still must make decisions.

A long-running, contentious debate surrounds the Endocrine Disrupter Screening Program. In 1996, Congress ordered the EPA to start testing 87,000 chemicals for their ability to disrupt endocrine hormone functions that regulate almost every aspect of reproduction, growth, development, and the functioning of our bodies. Twenty years later, out of a subset of 52 compounds, a smaller group of 18 had been identified for further testing, but no clear determinations of endocrine disruptors had been made. Some toxicologists argue that the exposure times may be too short, or the test animals may have been exposed to other chemicals besides the ones being studied. As mentioned earlier in this chapter, an albino rat breed called the Sprague-Dawley is stipulated in these tests. These rats were originally bred to be resistant to arsenic trioxide pesticides, and they may be unnaturally resistant to endocrine disrupters as well. Meanwhile, the chemical industry disputes the need for any endocrine testing at all. As you can see, establishing public policy isn't simple.

In setting standards for environmental toxins, we need to consider (1) the combined effects of exposure to many different sources of damage, (2) different sensitivities of members of the

Risk Assessment   **Risk Management**

FIGURE 8.19 Risk assessment organizes and analyzes data to determine relative risk. Risk management sets priorities and evaluates relevant factors to make regulatory decisions.

Table 8.5 Relative Risks to Human Welfare
Relatively High-Risk Problems
Habitat alteration and destruction
Species extinction and loss of biological diversity
Stratospheric ozone depletion
Global climate change
Relatively Medium-Risk Problems
Herbicides/pesticides
Toxics and pollutants in surface waters
Acid deposition
Airborne toxics
Relatively Low-Risk Problems
Oil spills
Groundwater pollution
Radionuclides
Thermal pollution

Source: Environmental Protection Agency

population, and (3) the effects of chronic as well as acute exposures. Because absolute proof is so elusive, as in the endocrine disrupter study, some European countries have adopted a "precautionary principle" that requires manufacturers to show that substances are safe, rather than for regulators to prove they are dangerous. This is a "better safe than sorry" approach. Critics of the precautionary principle argue that it's impossible to prove a substance is always safe, and it is costly to try to prevent all possible harm from all possible substances (fig. 8.20).

On the other hand, proponents of the precautionary principle argue that modern industry exposes us to countless new chemicals, and we need to stop releasing them into the environment without

FIGURE 8.20 "Do you want to stop reading those ingredients while we're trying to eat?"

Richard Guindon. Reprinted with permission.

understanding their effects. Each challenge to our cells by toxic substances is a stress on our bodies. Although each individual stress may not be life-threatening, the cumulative effects of all the environmental stresses, both natural and human-caused, to which we are exposed may seriously shorten or restrict our lives. Furthermore, some individuals in any population are more susceptible to those stresses than others. Should we set pollution standards so that no one is adversely affected, even the most sensitive individuals, or should the acceptable level of risk be based on the average member of the population?

Finally, policy decisions about hazardous and toxic materials also need to be based on information about how such materials affect the plants, animals, and other organisms that define and maintain our environment. In some cases, pollution can harm or destroy whole ecosystems with devastating effects on the life-supporting cycles on which we depend. In other cases, only the most sensitive species are threatened. Table 8.4 shows the EPA's assessment of relative risks to human welfare. This ranking reflects a concern that our exclusive focus on reducing pollution to protect human health has neglected risks to natural ecological systems. While there have been many benefits from a case-by-case approach in which we evaluate the health risks of individual chemicals, we have often missed broader ecological problems that may be of greater ultimate importance.

Section Review

1. Why has it been so difficult to regulate endocrine disrupters?
2. What is the "precautionary principle"?

Connecting the Dots

We have made marvelous progress in reducing some of the worst diseases that have long plagued humans. Infectious diseases such as guinea worms, and polio are nearly eradicated worldwide; bubonic plague, typhoid fever, cholera, yellow fever, tuberculosis, mumps, and other highly communicable diseases are rare in developed countries. Childhood mortality has decreased 90 percent globally, and life expectancy almost everywhere is twice what it was a century ago.

Now we worry about different kinds of conditions and exposures. The case of perfluorocarbon (PFC) compounds illustrate the insidious and slippery nature of many substances we are now exposed to regularly. The level of risk is uncertain and relatively small, but it is statistically demonstrated, and exposure is hard to track. Responses to small exposures are difficult to evaluate, and thousands of newly introduced compounds are present in products we use every day. They often make our lives more convenient, but we don't know the environmental and health costs of these ubiquitous substances.

Chronic conditions, such as cardiovascular disease, cancer, depression, dementia, diabetes, as well as traffic accidents and other risks that once were confined to richer countries have now become leading health problems nearly everywhere. Part of this change is that we no longer die at an early age of infectious disease, so we live long enough to develop the infirmities of old age. Another factor is that affluent lifestyles, lack of exercise, and unhealthy diets aggravate these chronic conditions.

Determining acceptable levels of environmental health risk is difficult. Toxicologists have methods to quantify effects, such as dose-response curves and LD50 levels, but low-dose effects are hard to identify. Sometimes we evaluate according to the type of impact: mutagens, teratogens, and carcinogens are taken seriously because they have especially tragic outcomes. Sometimes we are slow to identify lower-level but widespread impacts, as with endocrine disrupters. Almost always we evaluate risk according to whether it is familiar and chosen, as compared to unfamiliar and imposed by others. We engage in many risky practices because of this nonrational approach to risk evaluation.

Regulation of substances that affect public health is a complicated political process. Understanding the terms and principles involved, and familiarity with the substances in question, will help you engage in these debates. In addition, you can take many steps to protect your own health. A healthy diet, regular exercise, reduced sun exposure, moderate drinking, safe driving, and safe sex are among the most important.

Reviewing Key Terms

Can you define the following terms in environmental science?

acute effects 8.4
allergens 8.2
antigens 8.2
bioaccumulation 8.3
biomagnification 8.3
cancer 8.2
carcinogens 8.2
chronic effects 8.4
conservation medicine 8.1

disability-adjusted life
    years (DALYs) 8.1
disease 8.1
ecological diseases 8.1
emergent diseases 8.1
endocrine disrupters 8.2
environmental health 8.1
fetal alcohol syndrome 8.2
health 8.1

LD50 8.4
morbidity 8.1
mortality 8.1
mutagens 8.2
neurotoxins 8.2
persistent organic pollutants
    (POPs) 8.3
risk 8.4
risk assessment 8.4

sick building
    syndrome 8.2
synergistic effect 8.3
teratogens 8.2
toxic substances 8.2

Critical Thinking and Discussion Questions

1. What consequences (positive or negative) do you think might result from defining health as a state of complete physical, mental, and social well-being? Who might favor or oppose such a definition?

2. Do rich countries bear any responsibilities if the developing world adopts unhealthy lifestyles or diets? What could (or should) we do about it?

3. Why do we spend more money on heart or cancer research than on childhood illnesses?

4. What are the premises in the discussion of assessing risk? Could conflicting conclusions be drawn from the facts presented in this section? What is your perception of risk from your environment?

5. Should pollution levels be set to protect the average person in the population or the most sensitive? Why not have zero exposure to all hazards?

6. What level of risk is acceptable to you? Are there some things for which you would accept more risk than others?

7. Some of the mechanisms that help repair wounds or fight off infection can result in cancer when we're old. Why hasn't evolution eliminated these processes? Hint: Do conditions of postreproductive age affect natural selection?

8. What changes could you make in your lifestyle to lessen your risks from the diseases and health risks in table 8.1? What would have the greatest impact on your future well-being?

9. Why has it been so difficult to regulate endocrine disrupters? How should we act in the face of uncertainty?

10. Why should or shouldn't we adopt a "precautionary principle"? What different world views are represented in this debate?

Data Analysis

How Do We Evaluate Risk and Fear?

A central question in environmental health is how we perceive different risks around us. When we evaluate environmental hazards, how do we assess known factors, uncertain risks, and the unfamiliarity of new factors we encounter? Which considerations weigh most heavily in our decisions and our actions as we try to avoid environmental risks?

The graph shown next offers one set of answers to this question, using the aggregate responses of many people to risk. Go to Connect to find further discussion and a set of questions regarding risk, uncertainty, and fear.

Public perception of risk, depending on the familiarity, apparent potential for harm, and personal control over the risk.

Source: Data from Slovic, Paul, 1987. "Perception of Risk," Science, 236 (4799): 286–290.

9

Food and Hunger

▲ Farmers in Burkina Faso have used traditional methods and creativity to improve food production.
Lissa Harrison

Learning Outcomes

After studying this chapter, you should be able to:

9.1 Describe patterns of world hunger and nutritional requirements.

9.2 Identify key food sources, including protein-rich foods.

9.3 Explain new crops and genetic engineering.

9.4 Discuss how policy can affect food resources.

"It ain't the things we know that cause all the trouble; it's the things we think we know that ain't so."

– Will Rogers

Food Security in the Sahel

What causes hunger, and how do we resolve it? This ancient question has been especially urgent in Burkina Faso, one of the world's poorest and most resource-poor countries. Burkina Faso is situated in the Sahel, a vast transitional zone where central African forests meet the forbidding Sahara desert. Rainfall is seasonal, and farms must survive long dry months between the rains. The drier northern reaches of the Sahel can support only pastoral herders, who migrate seasonally with their cattle and goats. But the southern areas of the Sahel are moist enough for seasonal farming, despite sandy and easily eroded soils.

This small, landlocked country has few resources, but in the past 50 years, its population has quadrupled from 5 million to 20 million, about 80 percent of them living by subsistence agriculture. Drought and overuse have badly degraded available soils. Poverty remains stubbornly widespread.

Development and aid agencies have tried to resolve the problem of food supply, but they have seen little lasting success. Debates about solving hunger in Africa have frequently focused on ways that wealthy nations can share advanced technology; better seeds, machinery, or irrigation are often key strategies. But these systems often are ill-suited to resources in poor countries.

In a stunning contradiction of most development narratives, the farmers of Burkina Faso have initiated a homegrown revolution. Since the 1980s, farmers, villagers, and community organizations have been slowly turning the tide of land degradation and hunger using simple, traditionally based water conservation techniques. Low barriers built of stone or sticks, contoured to hillsides, slow runoff and also capture tree seeds, which collect and grow in the moist soil (fig. 9.1). Small, traditional planting pits, known as *zai,* are filled with manure or crop residue, to capture moisture and provide nutrients for seeds when the rains come. Half-moon-shaped berms, a half-circle hollow about 4 m wide with a raised edge, also capture runoff, allowing it to infiltrate into soil. Berms accumulate organic matter and nutrients, providing a rich environment for seeds.

Farming communities in Burkina Faso have reclaimed some 300,000 hectares of degraded land, improving productivity and

producing reliable sources of food to eat and to sell. Government agencies and nongovernmental organizations help educate new practitioners to spread the practices.

Part of the reasons these systems work is that people have a diverse diet. While most Americans subsist largely on wheat and on corn-fed meats, produced on vast fields with ever-larger machinery, people in Burkina Faso eat a wide variety of beans and peas (legumes that add nutrients to the soil), sorghum, millet (both of which grow in warm and dry conditions), and agroforestry products, including fruits and oils from shea, nere, cashew, baobab, moringa, and many other trees. Drought-tolerant trees have the advantage that they don't require seasonal cultivation, so they are a durable food source over time, and their fallen leaves help enrich soils. Maize and rice are also important crops, but the broad diversity of food sources can provide resiliency in dry years.

All these factors contribute to improving food security. But a new challenge has arisen in recent years. Growing violence from extremist groups from the north have driven farmers from their villages, causing new waves of hunger. Many observers attribute violence in part to climate change, which has expanded the deserts and destroyed the livelihoods of pastoralists in the region. Unreliable rains have also made farming harder where violence has not yet spread.

As you will read in this chapter, hunger often involves environmental degradation, climate change, or population growth. But famine is usually triggered by violent conflicts or political events that disrupt food supplies. Farmers are innovative, and they can often find strategic ways to produce food and increase food security, when given a chance. Clearly food security, including both enough food and good quality food for a healthy life, is a many-sided puzzle. This puzzle also has multiple solutions to explore. We examine some of these ideas in this chapter.

To read more, see C. Nyamekye, et al. 2018. Soil and Water Conservation in Burkina Faso, West Africa. *Sustainability 10, 3182; doi:10.3390/su10093182.*

FIGURE 9.1 This low bund (barrier) is a simple, low-cost strategy to capture moisture and improve food security in Burkina Faso.
Gary Tappan/USGS

Benchmark Data

Among the ideas and values in this chapter, the following are a few worth remembering.

2,200 kcal	Average human dietary energy requirements, according to the World Health Organization
>3,500 kcal	Average available per person in North America
40–60 g	Daily protein requirements
60 g	Average protein consumption in least-developed countries
110 g	Average in North America
70%; 5%	Percentage overweight; underweight, United States
20%; 10%	Percentage overweight; underweight, China
8 kg	Amount of feed to produce 1 kg of beef

9.1 WORLD FOOD AND NUTRITION

- *Food production has grown faster than the population, but chronic hunger remains widespread.*
- *Famines are political as well as environmental.*
- *Overeating is a growing world problem.*
- *A balanced diet is essential for health.*

The ways we produce food matter to everyone who eats. They also matter for water quality, land use, economies, and livelihoods. In this chapter, we examine concerns about hunger and its effects, as well as some major foods we eat. We focus here mainly on larger factors and processes that dominate our food systems. We address sustainable systems of food production in chapter 10.

Global hunger has long been a serious and widespread concern. Generations of policy analysts have predicted runaway population growth and famine. But world food supplies have more than kept up with increasing human numbers over the past two centuries. While population growth slowed to an average 1.08 percent per year during that time, world food production increased an average of 2.2 percent per year. Food availability has improved because of increasing use of irrigation and fertilizers, improved crop varieties, and better distribution systems. Globally, we consume an average of nearly 3,000 kcal per day, well above the 2,200 kcal considered necessary for a healthy and productive life (fig. 9.2a). Residents of industrialized countries consume over 3,500 kcal every day, on average. For most of the world, daily protein consumption is also growing, often beyond the 40–60 g per day needed for a healthy and productive life (fig. 9.2b).

The UN Food and Agriculture Organization (FAO) expects world food supplies to continue to increase faster than population growth. In some countries, such as the United States, the problem has long been what to do with surplus food. High production leads to low prices, which make farm profits chronically low. Farmers in these countries are paid billions of dollars per year to keep marginal land out of production, and crops are diverted to other uses such as biofuels to shore up prices. There have been many studies of how we could expand farmlands, and how much forest and water resources we should sacrifice. But the main question may not be whether we can produce enough food. More important concerns may be how can we improve access to food, how do we provide the right kinds of food, and what are the environmental costs of food production systems?

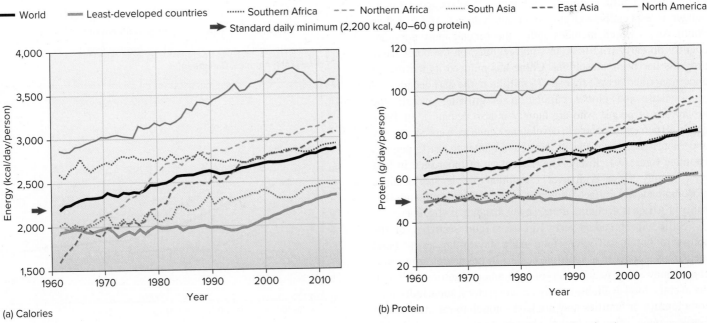

(a) Calories

(b) Protein

FIGURE 9.2 Changes in dietary energy (kcal) and protein consumption in selected regions. North America and other developed regions consume more calories and protein than are needed.

Data from Food and Agriculture Organization (FAO) 2018

FIGURE 9.3 Hunger rates worldwide. The most severe and chronic hunger is in developing regions, especially sub-Saharan Africa.
Source: Food and Agriculture Organization of the United Nations

Millions of people are still chronically hungry

Despite bountiful production, some 800 million people in the world today are considered **chronically undernourished,** getting less than the minimum 2,200 kcal per day. This number represents a tragic and persistent problem. It also represents a decline from around 1 billion chronically hungry 25 years ago. Chronic hunger is especially widespread in sub-Saharan Africa and in South Asia—which includes India, the second most populous country on earth (fig. 9.3). In some regions, on the other hand, progress has been substantial. China has reduced its number of undernourished people by over 75 million in the past decade.

Because the global population is growing overall, the *proportion* of people who are hungry is also declining. In 1960, nearly 60 percent of people in developing countries were chronically undernourished. Today that proportion has fallen to just 15 percent. Examine figure 9.4 closely. The percentage malnourished in East Asia, mainly China with over 1.4 billion people, has declined most dramatically. In fact, all regions have declined, although sub-Saharan Africa is declining more slowly than others (fig. 9.4).

Still, widespread poverty threatens **food security,** or the ability to obtain sufficient food on a day-to-day basis. Food security is a concern at multiple scales. In the poorest countries, hunger may affect nearly everyone. In other countries, although the average food availability may be satisfactory, some individual communities or families may not have enough to eat. And within families, males often get both the largest share and the most nutritious food, while women and children—who need food most—all too often get the poorest diet. At least 6 million children under

5 years old die every year (one every 5 seconds) from hunger and malnutrition. Providing a healthy diet might eliminate as much as 60 percent of all premature deaths worldwide.

People who are hungry get stuck in a cycle of poverty. Economist Robert Fogel has estimated that in 1790 about

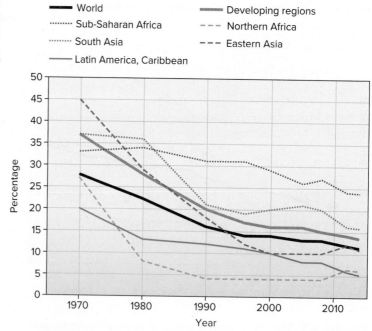

FIGURE 9.4 Changes in rates of malnourishment by region.
Data from UN Food and Agriculture Organization, 2015

20 percent of the population of England and France was effectively excluded from the labor force because they were too weak and hungry to work. Improved nutrition, he calculates, accounted for about half of all European economic growth during the nineteenth century. Because many developing countries are as poor now (in relative terms) as Britain and France were in 1790, his analysis suggests that reducing hunger could yield more than $120 billion in economic growth, produced by longer, healthier, more productive lives for several hundred million people.

Recognizing the role of women in food production is an important step toward food security for all. Throughout the developing world, women do 50 to 70 percent of all farm work but control only a tiny fraction of the land and rarely have access to capital or developmental aid. In Nigeria, for example, home gardens occupy only 2 percent of all cropland but provide half the food families eat. Making land, credit, education, and access to markets available to women could contribute greatly to family nutrition.

Famines usually have political and social causes

Although chronic hunger and malnutrition have declined steadily since the 1960s (see fig. 9.4), that decline has recently slowed or even reversed in some areas. Why is this? Drought, floods, and other natural disasters can cause widespread hunger, but most societies with a stable and well-operating government find ways to avoid widespread starvation. Generally it is violent conflict that leads to critical food shortages. Conflict disrupts both farming and food distribution, and the result can often be **famine,** large-scale food shortages accompanied by widespread starvation and even death. Starving people may be forced to eat seed grain and livestock in a desperate attempt to survive. Famines often involve mass migrations to towns, or to refugee camps as people seek safety and food (fig. 9.5).

The UN has recently reported famine conditions in a number of countries, including Yemen, Burkina Faso, northeastern Nigeria, South Sudan, and Ethiopia. All of these are experiencing, or have recently had, conflicts that have driven farmers from the land and devastated economic and social structures. Political corruption can also cause famine, as self-serving leaders restrict access to food or farmland for political gain. Nobel Prize–winning economist Amartya K. Sen has shown that armed conflict and political oppression almost always are at the root of famine. No democratic country with a relatively free press, he says, has ever had a major famine.

The aid policies of rich countries often don't help as much as we hope. Despite our best intentions, aid often serves as a way to get rid of surplus corn or grain rather than to stabilize local food production in recipient countries. Even emergency food aid has ambiguous effects. Herding people into feeding camps can badly destabilize communities, and crowding and lack of sanitation in the camps exposes people to diseases. There are no jobs in the refugee camps, so people can't support themselves. Corruption and violence can occur at food-dispensing centers, where aid recipients are vulnerable to theft, extortion, and sexual violence from aid providers. While emergency aid is often needed, the UN

FIGURE 9.5 Children wait for a ration of porridge at a feeding station, a common sight where violence has dismantled food production systems, economies, and families. These children are in East Timor, which suffered decades of ethnic conflict.
Antonio Dasiparu/EPA/Shutterstock

reports that the best strategy is usually to invest in building local capacity to govern and manage local food systems.

Recently, analysts have noted that climate change also contributes to widespread hunger, as multiple years of drought ultimately drives farmers from the land. Many migrate to cities, some migrate even farther: Desperate climate migrants from across Africa have beset European countries in recent years. Uprooted and desperate populations, of course, also contribute to violent conflict, as they are caught up by extremist groups, gangs, or warring factions.

Overeating is a growing world problem

Although hunger persists, world food supplies are increasing. This is good news, but the downside is increasing overweight and obese populations. In the United States, and increasingly in developing countries, highly processed foods rich in sugars and fats have become a large part of daily diets. Just over 70 percent of adult Americans are overweight, up from 50 percent in 1990. About one-third of us are seriously overweight, or **obese.** Obesity is quantified in terms of the body mass index (BMI), calculated as weight/height^2. For example, a person weighing 100 kg and standing 2 m tall (220 lb and 6 ft 6 inches) would have a body mass of (100 kg/4m^2) or 25 kg/m^2. Health officials consider a BMI greater than 25 kg/m^2 overweight; over 30 kg/m^2 is considered obese.

Globally, more than twice as many adults are overweight (nearly 2 billion) than underweight (870 million), according to a the World Health Organization. This number represents 39 percent of the world's adult population. About 13 percent of adults are obese. This trend is no longer limited to richer countries. Obesity is spreading around the world as Western diets and lifestyles are increasingly adopted in the developing world (fig. 9.6).

Being overweight substantially increases the risk of hypertension, diabetes, heart attacks, stroke, gallbladder disease, osteo-arthritis, respiratory problems, and some cancers. In the United States, about 400,000 people die from illnesses related to obesity every year. This number is approaching the number of deaths related to smoking (435,000 annually). Weight-related illnesses and disabilities are now a serious strain on health care systems and health care budgets worldwide.

Growing rates of obesity result partly from increased consumption of oily and sugary foods and soft drinks, and partly from lifestyles that involve less walking, less physical work, and more leisure than previous generations had. Changing these factors can be hard. Just walking to work regularly can be enough to keep weight down, but many of our daily routines are built around sitting still, at a desk or in a car. Many of our social activities, and our traditional holiday meals, focus on rich foods with gravies and sauces, or sweets. We are probably biologically adapted to prefer these energy-rich foods, which were rare and valuable for our ancestors. Today, it can take special effort to cut back on them.

An important cause is the economic necessity for food producers to increase profits. When we already have plenty to eat, and when food prices are low, food processors struggle constantly to ensure continuous growth in production and profits. Manufacturers can achieve better profits with "value added" products: Instead of selling plain oatmeal at 50 cents a pound, a manufacturer might convert oats into flavored, sweetened, instant microwavable oatmeal for $2.50 a pound. Better yet, processing the oats into sweetened, toasted oat flakes might bring $5 a pound. Increases in sugar and fat content, as well as constant exposure to advertising, encourages us to consume more than we might really need.

Food insecurity and poverty are also key contributors to overweight and obesity. Processed foods, high in sugar and fat, are often cheaper to acquire and to store than fresh and healthier foods. Food insecurity can contribute to overeating when food is available. And a lack of time for cooking, and limited access to healthy food choices along with ready availability of fast-food snacks and calorie-laden soft drinks, also lead to serious dietary imbalances.

We need the right kinds of food

Michael Pollan, who writes about food issues, says that plain, simple food is what our bodies are adapted to. He argues that products made of manufactured foodlike substances that your grandmother wouldn't recognize probably are not good for you. Pollan sums up the answer to health and obesity problems this way: "Eat food. Not too much. Mostly plants."

Generally, eating a good variety of foods provides the range of nutrients you need. In general, it's best to have whole grains and vegetables, with only sparing servings of meat, dairy, fats, and sweets. Based on observations of the health effects of Mediterranean diets, as well as a long-term study of 140,000 U.S. health professionals, Dr. Walter Willett and Dr. Meir Stampfer of Harvard University have recommended a dietary pyramid that minimizes red meat and starchy food such as white rice, white bread, potatoes, and pasta (fig. 9.7). Nuts, legumes (beans, peas, and lentils), fruits, vegetables, and whole grain foods form the basis of this diet. The base of this Harvard pyramid is regular, moderate exercise.

Food-insecure people often can't afford the protein, fruits, and vegetables that would ensure a balanced diet. Starchy foods like maize (corn), polished rice, and manioc (tapioca) form the bulk of the diet for poor populations, especially in developing countries. Even if they get enough calories, they may lack sufficient protein, vitamins, and trace minerals. **Malnourishment** is a term for nutritional imbalance caused by a lack of specific dietary components or an inability to absorb or utilize essential nutrients.

The FAO estimates that perhaps 3 billion people (nearly half the world population) suffer from vitamin, mineral, or protein deficiencies. Effects can include devastating illnesses and deaths, as well as slowed mental and physical development. These problems bring an incalculable loss of human potential.

Anemia (low hemoglobin levels in the blood, usually caused by dietary iron deficiency) is the most common nutritional problem in the world. More than 2 billion people suffer from iron deficiencies, especially women and children. The problem is most

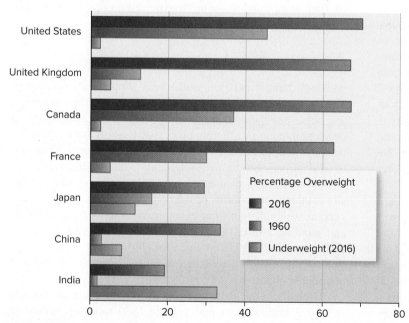

FIGURE 9.6 While over 800 million people are chronically undernourished, more than twice as many are at risk from eating too much. (Latest available data are shown.)
Source: Data from World Health Organization 2018

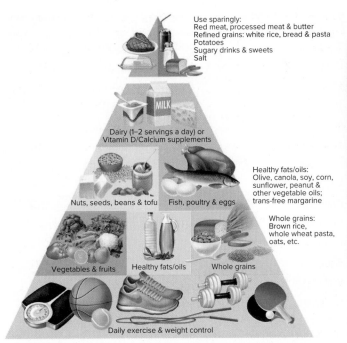

Use sparingly:
Red meat, processed meat & butter
Refined grains: white rice, bread & pasta
Potatoes
Sugary drinks & sweets
Salt

Dairy (1–2 servings a day) or
Vitamin D/Calcium supplements

Healthy fats/oils:
Olive, canola, soy, corn,
sunflower, peanut &
other vegetable oils;
trans-free margarine

Nuts, seeds, beans & tofu   Fish, poultry & eggs

Whole grains:
Brown rice,
whole wheat pasta,
oats, etc.

Vegetables & fruits   Healthy fats/oils   Whole grains

Daily exercise & weight control

FIGURE 9.7 The Harvard food pyramid emphasizes fruits, vegetables, and whole grains as the basis of a healthy diet. Red meat, white rice, pasta, and potatoes should be used sparingly.
Data from Willett and Stampfer, 2002

(a) Kwashiorkor

(b) Goiter

FIGURE 9.8 Dietary deficiencies can cause serious illness. (*a*) Kwashiorkor, resulting from a lack of protein, can cause a distended belly and thin limbs, and a weak immune system. (*b*) Goiter, a swelling of the thyroid gland, results from an iodine deficiency.
(a) Burger/Phanie/age fotostock; (b) Lester V. Bergman/Getty Images

severe in India, where 80 percent of all pregnant women may be anemic. Anemia increases the risk of maternal deaths from hemorrhage in childbirth and affects childhood development. Red meat, eggs, legumes, and green vegetables all are good sources of dietary iron.

Protein deficiency can cause conditions such as kwashiorkor and marasmus. **Kwashiorkor** is a West African word meaning "displaced child." (A young child is displaced—and deprived of nutritious breast milk—when a new baby is born.) This condition most often occurs in young children who subsist mainly on cheap starchy foods. Children with kwashiorkor often have puffy, discolored skin and a bloated belly (fig. 9.8*a*). **Marasmus** (from the Greek "to waste away") is caused by shortages of both calories and protein. A child suffering from severe marasmus is generally thin and shriveled. Children with these deficiencies have low resistance to infections and may suffer lifelong impacts on mental and physical development.

Iodine is essential for regulating metabolism and brain development. Chronic iodine deficiency causes goiter (a swollen thyroid gland, fig. 9.8*b*), stunted growth, and mental impairment. The FAO estimates that 740 million people—mainly in South and Southeast Asia—suffer from iodine deficiency and that 177 million children have stunted growth and development. Adding a few pennies' worth of iodine to table salt has nearly eliminated this problem in developed countries.

Vitamin A deficiencies affect 100–140 million children at any given time. At least 350,000 go blind every year from the effects of this vitamin shortage. Folic acid, found in dark green, leafy vegetables, is essential for early fetal development. Ensuring access to leafy greens can be one of the cheapest ways of providing essential vitamins.

Global factors can cause price spikes

If we have surplus production, why are high food prices still a problem for struggling families? Even nonindustrialized farming economies such as India have seen long-term price declines, yet impoverished populations still suffer acutely with shorter-term increases in prices for cooking oil, wheat, or other staples. Price changes are inconvenient for food-secure populations. But for impoverished families in developing regions, where more than half of the household income may be used to buy food, high prices can be devastating. Why do food prices rise despite global abundance?

Although crises like droughts and storms often trigger spikes in food prices, a growing factor is that food prices are often driven by global commodity prices. Increasingly, food is traded as an investment commodity, with bankers and investors buying and selling shares in multinational food companies. Traders in Chicago, London, and Tokyo purchase volumes of grain, sugar, coffee, or other commodities simply to make a profit on the trade. This trade can easily inflate prices paid by poor villagers in developing areas. Often, global trade involves speculation and trading in futures: I might promise to pay you $4 a bushel for next summer's corn crop, even though the planting season hasn't even started yet, just so I can reserve the crop and settle the price now. But if there's drought in the spring and the year's production looks poor, someone who really needs the corn might pay me $5 a bushel for the same crop that's not yet in the ground. I just made a 25 percent profit on a future corn crop, and my shareholders are delighted. Consumers somewhere else will cover the higher costs. Trading in commodities and futures, then, can drive food prices, even though the exchanges are far removed from the actual food that a farmer plants and a consumer eats. And expected future shortages can drive up prices today.

To complicate matters further, food prices are driven by non-food demands for crops. In 2007–2008, United States corn prices jumped from around $2 a bushel to over $5 a bushel when the U.S. Congress began to subsidize corn-based ethanol fuel and to require that ethanol be sold at gas stations nationwide. In that year, future speculation for ethanol drove up corn prices, and wheat and other grains followed in the excitement. Because of the ethanol boom, many small bakers and pasta makers couldn't afford wheat and were driven out of business, and U.S. consumers were pinched as food prices rose throughout the grocery store. Federal policies requiring the use of corn to produce ethanol continue to inflate corn demand, and prices, in many markets.

The same process occurred in 2008–2010 after the European Union passed new rules requiring biofuel use, with the idea that these fuels would be sustainable and climate neutral. Europe's biofuels are produced largely from palm oil, a tropical oily fruit grown mainly in Malaysia and Indonesia. European biofuel rules produced a boom in global palm oil demand. Unfortunately, palm oil is also a cooking staple for poor families across Asia, for whom a doubling of oil prices can be devastating. In developing countries across the globe, riots broke out over rising cooking oil prices, which were driven by well-meaning European legislation for Malaysian biofuel. The palm oil boom is also driving accelerated deforestation and wetland drainage across Malaysia, Indonesia, Ecuador, Colombia, and other palm-oil-producing regions, leading to further social and environmental repercussions (chapter 12).

Section Review

1. How many people in the world are chronically undernourished? What does chronically undernourished mean?

2. List at least five African countries with high rates of hunger (fig. 9.3; use a world map to help identify countries).

3. What are some of the health risks of overeating? What percentage of adults are overweight in the United States?

9.2 KEY FOOD SOURCES

- *Rice, wheat, and a few other crops provide most food.*
- *Meat and fish give excellent protein but consume resources.*
- *Antibiotic overuse is a serious concern in meat production.*

Table 9.1 Key Global Food Sources*

Crop	1965	1990	2016
Sugar cane	531	1,053	1,891
Maize	227	483	1,060
Rice	254	519	741
Wheat	264	592	749
Milk	358	524	798
Potatoes	271	267	377
Vegetables	66	140	290
Cassava	86	152	277
Soybeans	32	108	335
Meat	72	158	297
Barley	93	144	177
Sweet potatoes	108	123	105
Dry beans, pulses	31	41	44

*Production in million metric tons
Data from UN FAO, 2015

The ways we produce food, and the types of food we eat, can have important environmental impacts. Sustainable food production happens in many ways, in many parts of the world, and we examine some of these in chapter 10. It is also important to understand the dominant systems that produce the majority of foods that we eat. Advances in food production have revolutionized our lives. The key task is to understand the environmental impacts of these systems, so that we can seek ways to minimize them while still enjoying healthy and satisfying diets.

Of the thousands of edible plants and animals in the world, only about a dozen types of seeds and grains, three root crops, 20 or so common fruits and vegetables, six mammals, and two domestic fowl make up most of the food that humans eat (table 9.1). The three crops on which humanity depends for the majority of its nutrients and calories are wheat, rice, and maize (called corn in the United States). Together, over 2.5 billion metric tons of these three grains are grown each year. Wheat and rice are especially important as the staple foods for most of the 6 billion people in the developing countries of the world. These two grass species supply around 60 percent of the calories consumed directly by humans. As table 9.1 shows, production of all major crops has increased in the past 50 years.

Dominant crops often depend on local climates. Potatoes, barley, oats, and rye grow well in cool climates, and these are staples in mountainous regions and high latitudes (northern Europe, north Asia). Cassava, sweet potatoes, and other roots and tubers grow well in warm, wet areas and are staples in Amazonia, Africa, Melanesia, and the South Pacific. Sorghum and millet are drought-resistant, and they are staples in the dry regions of Africa.

Fruits, vegetables, and vegetable oils are usually the most important sources of vitamins, minerals, dietary fiber, and complex carbohydrates. In the United States, however, grains make

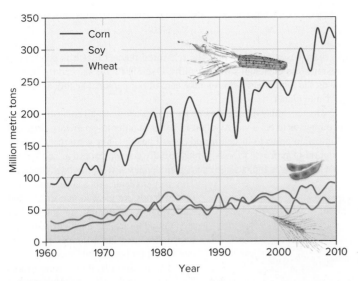

FIGURE 9.9 United States production of the three dominant crops, corn, soybeans, and wheat.
Data from USDA and UN FAO, 2015

FIGURE 9.10 Meat and dairy consumption have quadrupled in the past 40 years, and China represents about 40 percent of that increased demand.
William P. Cunningham

up a far larger part of agricultural production. Corn is by far the most abundant crop, followed by soybeans and wheat (fig. 9.9). Of these three, only wheat is primarily consumed directly by humans. Corn and soy are processed into products such as fuel, livestock feed, or high-fructose corn syrup (see section 9.4).

Rising meat production has costs and benefits

Dramatic increases in corn and soy production have led to rising meat consumption worldwide. In developing countries, meat consumption has risen from just 10 kg per person per year in the 1960s to over 26 kg today (fig. 9.10). In the same interval, meat consumption in the United States has risen from 90 kg to 136 kg per person per year. Meat is a concentrated, high-value source of protein, iron, fats, and other nutrients that give us the energy to lead productive lives. Dairy products are also a key protein source: Globally we consume more than twice as much dairy as meat. But health experts increasingly say we are eating more of both than is good for us.

Meat is a good indicator of wealth, because it is expensive to produce, in terms of the resources needed to grow an animal (fig. 9.11). As discussed in chapter 3, herbivores use most of the energy they consume for moving and growing; only a portion of energy consumed is stored for consumption by carnivores. A beef steer consumes over 8 kg of grain to produce just 1 kg of beef. Pigs are more efficient. Just 3 kg of pig feed are needed to produce 1 kg of pork. Chickens and herbivorous fish (such as catfish) are still more efficient.

Globally, over one-third of cereals are used as livestock feed each year. We could feed about eight times as many people by eating that cereal directly rather than converting it to meat. What differences do you suppose it would make if we did so (What Do You Think?, p. 186)?

A number of technological and breeding innovations have made this increased production possible. One of the most important is the **confined animal feeding operation (CAFO),** where animals are housed and fed—mainly on soy and corn—for rapid growth (fig. 9.12). These operations dominate livestock raising in the United States, Europe, and increasingly in China. Animals are housed in giant enclosures, with up to 10,000 hogs or a million chickens in an enormous barn complex, or 100,000 cattle in a feedlot. These systems require specially prepared mixes of corn, soy, and animal protein that maximize animals' growth rate. New breeds of livestock have been developed that produce meat rapidly, rather than simply getting fat. The turnaround time is getting shorter, too. A U.S. chicken producer can turn baby chicks into chicken nuggets after just eight weeks of growth. Steers reach full size by just 18 months of age.

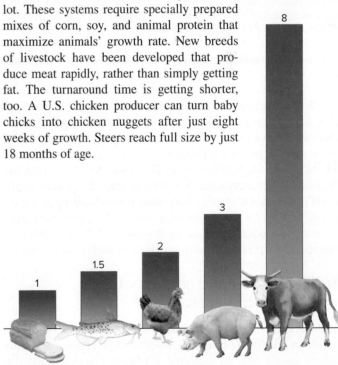

FIGURE 9.11 Number of kilograms of grain needed to produce 1 kg of bread or 1 kg live weight gain.

Diet for a Small Planet?

A central concern among global food analysts is this: How can we double food production in the next 50 years, without destroying our environment? As an informed reader, you are thinking, but wait, why? Will our population double in 50 years? No, but increasingly wealthy people across the globe have the right to expect the same meat- and dairy-rich diet enjoyed in wealthy countries today. That is only fair. Thus the question facing food analysts is, how do we achieve this growth?

A generation ago, though, many analysts framed the same question quite differently. Back in the 1970s, when many Americans were discovering environmentalism and when ecologists were sounding the alarm about unprecedented population growth and impending global food crises, the writer and activist Francis Moore Lappé wrote a seminal book, *Diet for a Small Planet*. Where Betty Crocker had taught generations to make better cakes and hot dishes (and, to be fair, to eat a healthy diet), Lappé encouraged eaters and cooks to think globally and act locally.

Fundamentally, thinking globally meant eating low on the food chain. Because producing meat and dairy products requires a disproportionate share of water, energy, land, and other resources, Lappé encouraged readers to rediscover more efficient foods, like lentils, beans, root vegetables, and greens. She provided dietary guidance, data on nutrition content, and of course recipes to help people lower their environmental impact while still enjoying good food (fig. 1). She explained how to use plant-based proteins, and she reassured readers that it is possible to consume less meat and still be healthy. Overall, Lappé showed that eating low on the food chain helps us take care of our personal health, as well as our finite planet.

FIGURE 1 Many traditional diets are low on the food chain, like this tabbouleh, in part because meat was an expensive and occasional food.
AS Food studio/Shutterstock

Doctors today make the same arguments about health and eating. They encourage us to eat a varied, plant-rich diet in order to get dietary fiber and vitamins, to protect our hearts and overall health. They warn that overconsumption of processed foods, which usually contain excessive and hidden amounts of sugar and fats, are causing epidemics of weight-related illnesses.

This is a challenge, of course, since processed foods are easy to get, quick, and satisfying, even though we know they aren't good for us. But health experts, like Lappé, are clear that a simple, plant-rich diet is best for health and fitness.

The question is, do we want to follow our doctors' advice? Many people find great pleasure in a meat-rich diet. Hamburgers and steaks, rare a few generations ago, have become central to our gastronomical identity. We know that exporting this rich diet to developing countries also means exporting a suite of weight-related illnesses: Heart disease and type II diabetes, once almost unheard of in developing countries, are now leading causes of illness and death. But if people are wealthy enough to eat meat, is it fair to say they should not?

If you were a global food analyst, trying to balance human wants and planetary limits, what would you say? Would you advise people to seek a diet for a small planet, and if so, how? What might you say to encourage more of us to eat within our planet's limits?

Because modern meat production is based on energy-intensive farming practices (see chapter 10), meat is also an energy-intensive product. It takes about 16 times as much fossil fuel energy to produce a kilogram of beef as it takes to produce a kilogram of vegetables or rice. The UN Food and Agriculture Organization estimates that livestock produce 20 percent of the world's greenhouse gases, more than is produced by transportation. In fact, by some estimates Americans could cut energy consumption more if we gave up just one-fifth of our meat consumption than if all of us were to drive a hybrid-electric Prius.

Seafood is our only commercial wild-caught protein source

Seafood is the main animal protein source for about 1.5 billion people in developing countries. Most of those people eat mainly locally caught fish. In wealthier countries, consumers eat fish from everywhere in the world, caught by huge, industrial-scale fishing ships. In the 1950s, the fishing industry developed freezer technology that allowed oceangoing factory ships to harvest the world's oceans and bring the catch home in huge volumes of frozen fish. Annual catches of ocean fish rose by about 4 percent annually between 1950 and 1988. Then catches began to fall, as fish populations declined and disappeared. Since the 1980s, all major marine fisheries have declined dramatically. An estimated 90 percent of global fisheries are now overfished or depleted, and most have become commercially unsustainable. Ever wider hunts for more, often smaller species keep fisheries going. An international team of marine biologists warns that if current trends continue, all the world's major fisheries will be exhausted by 2050.

The UN FAO, the main global source of fishery data, estimates that we harvest around 95 million metric tons of fish every year, slighty less than that produced now by fish farms (fig. 9.13). Fishery data are notoriously uncertain, however, as they are self-reported and difficult to monitor. A detailed study of multiple data sources by Daniel Pauly, of the University of British Columbia, and his colleagues, found that FAO data may be as low as

50 percent of actual catches. Illegal, unreported, and small-scale fishing are thought to make up the difference. Because the trade is global and difficult to track, much of the seafood you consume may be caught illegally or unsustainably.

Fish are the only wild-caught meat source still sold commercially on a large scale. Because nobody owns fish populations in the open ocean, fishers compete to deplete the population before competitors do. Even in exclusive national waters, illegal fishing is common. Rising numbers of boats, with increasingly efficient technology, exploit the dwindling resource. Ships as big as ocean liners travel thousands of kilometers and drag nets large enough to scoop up a dozen jumbo jets, sweeping a large patch of ocean clean of fish in a few hours. Most countries subsidize their fishing fleets to preserve jobs and to ensure access to fisheries. The FAO estimates that for the 4 million boats now harvesting wild fish, operating costs exceed sales by $50 billion per year.

The best hope for controlling these conditions is better monitoring, including satellite imagery and increased international regulation, as well as increasing establishment of no-fishing zones, which protect populations and allow a species to reproduce. Several Pacific island nations, such as Palau, have set up no-fishing zones in recent years, and both President Obama and President George W. Bush established vast marine sanctuaries in the Pacific (see chapter 6). The Trump administration eliminated many protections, however, despite early success in supporting fisheries, so their survival and effectiveness remain uncertain.

Most commercial fishing operates on an industrial scale

Commercial fishing uses many techniques, depending on the target species. The most common of these are trawling, seining, and long-lining. **Trawling** involves dragging a large net, often more than 50 m wide and 10 m tall. Bottom trawlers drag nets along the seafloor to catch shrimp or bottom-dwelling fish, such as halibut or sole (fig 9.14). This method is efficient, but the heavy nets destroy sensitive habitats; often this can include spawning grounds necessary for fish reproduction. Some critics say trawling is like harvesting forest mushrooms with a bulldozer; it's cheap, quick, and effective, but destroys the resource for future use. Trawling also produces tremendous amounts of "bycatch," unwanted marine life, which is usually thrown back dead or dying. In the North Pacific shrimp fishery, for example, as much as 98 percent of the catch from trawlers can be discarded as unwanted bycatch, producing over four million tons of waste seafood per year. Increasingly, however, bycatch may be kept to use as feed in fish farms (see below). There are reports that the Chinese shrimp trawl fleet discards very little bycatch, but rather uses it for aquaculture or human food. More damaging, however, are dredges, which use a heavy iron frame, often with rakes attached, to dredge up shellfish and other bottom-dwelling animals, which are then caught in a net attached to the dredge.

FIGURE 9.12 Concentrated feeding operations fatten animals quickly and efficiently, but they create enormous amounts of waste and expose livestock and human communities to unhealthy conditions.
Source: Photo by Tim McCabe, U.S. Department of Agriculture Natural Resources Conservation Service

Purse seining uses a huge net to surround a school of fish. The bottom of the net is then pulled closed, something like a drawstring purse. Floats prevent fish from escaping out the top. This method is used to catch schooling fish, such as tuna, sardines, herring, and mackerel. It also is used to catch species that gather

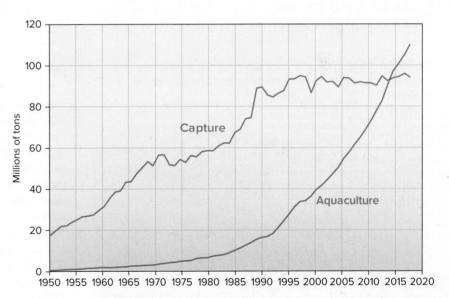

FIGURE 9.13 Estimates of the amount of fish produced globally by fishing and aquaculture. Despite continuously increasing fishing efforts, catches have not increased since the 1990s. Global fisheries kept production up by pursuing ever smaller, more remote species. Aquaculture in China, which accounts for 90 percent of farmed fish, has become an important fish source.
Data source UN FAO FishStatJ 2018

FIGURE 9.14 Trawlers drag a weighted net, often along the bottom, to capture everything in their path. Purse seines encircle a school of fish, then draw the "purse" closed.

to spawn in large groups, such as squid. Tuna seiners often set their nets on pods of dolphins, which frequently indicate schools of tuna. The seiners are supposed to allow the dolphins to escape before bringing in the nets, but they often don't do so, and dolphin mortality can be high. Turtles, sharks, sea lions, and other species may also be swept up in giant purse seines, which can be a mile in circumference.

Longlining uses a central fishing line that can be up to 50 miles (80 km) long. This line is strung with shorter lines with multiple baited hooks every couple of meters. A single longline might have 25,000 hooks. Longlines can be set near the surface to catch open-ocean fish, such as tuna or swordfish, or anchored to the bottom to catch bottom-dwellers, such as cod or halibut.

FIGURE 9.15 Fish farms produce about half our seafood, but most rely on wild-caught feed, and pollution and disease are common concerns.
WaterFame/Alamy *Stock Photo*

Shallow-set lines often catch dolphins, sharks, seabirds, or turtles that are attracted to the bait. By sinking lines deeper or using circular hooks, fishers can reduce unintended catch.

Aquaculture produces over half our seafood

Aquaculture (growing aquatic species in net pens or tanks) can involve marine species, such as salmon or tuna, or freshwater fish in ponds. Many freshwater fish are herbivores, such as carp, so they can be fed on grain or plant material. Most ocean fish, however, are carnivorous and require wild-caught fish for feed. About one-third of all wild-caught fish is used as food for fish in these operations. Because farmed carnivorous species such as salmon, sea bass, and tuna consume so much wild-caught fish, they also threaten wild fish populations and the seabirds and organisms that depend on them.

In salt-water (marine) aquaculture, net pens are anchored in near-shore areas, where open water circulates through, removing feces and uneaten food particles. A drawback of these systems is that they introduce diseases, with many fish crowded in a small area, as well as pollution and antibiotics into surrounding ecosystems (fig. 9.15). Farmed shrimp are usually produced in ponds built on former mangrove forests and wetlands. These environments provide essential nurseries for marine ecosystems. Studies have shown that the costs of lost ecosystem services can greatly outweigh the value of shrimp produced.

Freshwater aquaculture can allow better control of pollutants, if ponds are land-based, and isolated from other water bodies. Freshwater species are often plant-eating species, such as carp, catfish, or tilapia, which can have lower environmental impacts than carnivorous species.

About 60 percent of global aquaculture involves freshwater fish ponds in China, many of them producing fish for overseas markets. China has ancient traditions of fish farming, using sustainable, circular agricultural systems that produced carp, ducks, and rice together. Modern industrial-scale fish farms, however, are likely to rely on imported soy protein and antibiotics, and fish waste is an important water pollutant that affects the health of local communities.

Antibiotics are overused in intensive production

In any system that confines many animals in a small space, infections and diseases are likely. Fish farms, beef feedlots, large dairy herds, and pig and poultry operations all rely on antibiotics to keep their stock alive and healthy. In confined situations, bacteria thrive in the manure in the feedlots, in liquid wastes in manure storage lagoons (holding tanks) around hog farms, and in airborne dust around feedlots and manure tanks, which usually are uncontained and adjacent to barns (fig. 9.16). Antibiotics also make livestock put on weight faster.

For both reasons, antibiotics are often mixed with daily feed, keeping animals on a constant, low dose. Over 11 million kg of antibiotics are added to animal feed annually in the United States, about eight times as much as is used in human therapy. Nearly 90 percent of U.S. hogs receive antibiotics in their feed.

Constant exposure to antibiotics produces strains of bacteria that are resistant to these drugs, however. This is a simple case of evolution: Most of a population is killed off by an antibiotic, but a few resistant individuals are unaffected, and these strains grow and thrive. Antibiotic resistance is increasingly common around livestock operations, as well as in hospitals. This process is gradually rendering our standard antibiotics useless for human health care. The Centers for Disease Control and Prevention reports that 23,000 Americans die every year from resistant infections (chapter 8). The next time you are prescribed an antibiotic by your doctor, you might ask whether she or he worries about antibiotic resistance, and think about how you would feel if your prescription were ineffectual against your illness.

Because antibiotic resistance is such an important public health issue, several European countries have phased out the use of antibiotics for livestock, except in emergencies. Better monitoring of animal health is a necessary part of this phase-out. Denmark, one of the world's largest pork exporters, outlawed regular feeding of antibiotics in 2000. Since the ban, Denmark's ham production has grown, not declined, and the price is about the same per pound as American ham. The Netherlands has also banned routine antibiotic uses without reducing production. Observing this, U.S. pork (and chicken) producers have said they will try to curtail their use of antibiotics. However, producers in China, Brazil, India, and Russia are expected to double antibiotic use in the next 15 years.

Food systems are vulnerable to climate change

Most of the world's major crops have narrow temperature ranges for optimal growth. This narrow range means that a warming climate will reduce yields in many major crops. Crops suffer from heat in two major ways. First, physiologically, they produce seeds and plant biomass best within temperature ranges of a few degrees (fig. 9.17). Second, plants regulate temperature by transpiring (releasing) moisture more in hot conditions. This moisture loss helps cool plants,

FIGURE 9.16 This state-of-the-art lagoon was built to store manure from a hog farm. Odors, airborne dispersal of ammonia and bacteria, and overflows after storms are risks of open lagoons, but more thorough waste treatment is more expensive.
Source: Photo by Jeff Vanuga, U.S. Department of Agriculture Natural Resources Conservation Service

but it also increases water demand. Increased irrigation is therefore likely to become important with a warming climate.

Can we breed our way out of climate stress? Crop breeders have produced amazing varieties of food plants, including varieties that tolerate drought or slightly higher salinity in irrigation water. Usually these varieties don't have the high yields of plants

Vegetative Response Curve ———
Optimum Range ▬

Reproductive Response Curve ———
Optimum Range ▬

FIGURE 9.17 Temperature responses for vegetative growth and seed production for corn and soybeans.
USDA Agricultural Research Service, https://www.fs.fed.us/rm/pubs_other/rmrs_2008_backlund_p003.pdf

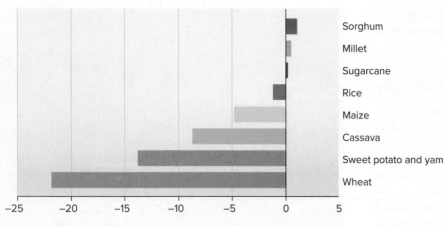

FIGURE 9.18 Expected changes in crop yields in Africa by 2050 with climate change, as compared to expected yields if the climate were not changing. Higher temperatures reduce plant growth in many crops, as well as increasing water demands.
Source: UN Environment Programme GEO5

FIGURE 9.19 Tropical winged beans (top) bear fruit year rounds, resist diseases, and both seeds and leaves can be eaten, as can fruits and leaves of baobab trees (bottom). These are among the thousands of traditional food sources that can diversify diets.
(a): William P. Cunningham (b): Jialiang Gao/Getty Images

optimized to grow under ideal moisture and nutrient conditions (see section 9.4). Breeding for higher heat tolerance is something that most plants have not been able to achieve, however, so rising temperatures remain a problem for major crops like corn and wheat.

Vulnerability is greatest in low-income regions, where small-scale farms produce most basic foods and family income. In this context, people are especially vulnerable to loss of crop productivity with rising temperatures, to increased costs of irrigation, or to increasing frequency of drought (fig. 9.18). As we noted in the opening case study, climate warming and unreliable rainfall have helped drive farmers from the land and contributed to political instability in sub-Saharan Africa. The same process can be seen in many developing regions of the world.

But wealthy regions are at risk because they often depend on just a few major crops, such as corn and soybeans. When crop varieties are optimized for high yields, not for drought tolerance, climate change seriously threatens production and income.

A key strategy could be more diversity in our diets. Many traditional diets, like that in Burkina Faso, include a wide variety of seasonal plant and animal products, and this diversity can contribute to food security as well as to healthy eating. Although most of the world's food now comes from only 16 species, at least 3,000 species of plants have been used for food at one time or another. For example, the winged bean is not widely used, but it could provide an excellent food source (fig. 9.19). This perennial plant grows well in hot climates, and the entire plant is edible (pods, mature seeds, shoots, flowers, leaves, and tuberous roots), it is resistant to diseases, and it enriches the soil. Knowing how to use diverse fruits, roots, beans, greens, and other foods—unfamiliar to most U.S. eaters—may help us adjust to a warming climate as well as to a smaller planet.

Section Review

1. Why does meat use more inputs than plant foods?
2. List several benefits and costs of fish and shrimp farming.
3. Why are antibiotics used in livestock production?

9.3 THE GREEN REVOLUTION AND GENETIC ENGINEERING

- *The green revolution increased production through breeding, fertilizer, and irrigation.*
- *Genetically modified crops have genes transferred from other organisms.*

The FAO predicts that most of the growth in agricultural production will come from higher yields and new crop varieties. How will we accomplish these changes? Since the 1950s, our main improvements in farm production have come from breeding hybrid varieties of a few well-known, widely used species. Yield increases often have been spectacular. A century ago, corn in the United States yielded an average of about 25 bushels per acre. Today, because of high-yield varieties, more fertilizer, and other factors, average yields are around 160 bushels per acre, and peak yields can be over 350 bushels per acre. It is these kinds of increases that have allowed us to send food aid overseas and to develop value-added markets for major crops, especially livestock and biofuels.

Starting in the 1950s and 1960s, agricultural research stations began to breed tropical wheat and rice varieties that would provide food for growing populations in developing countries. Crossbreeding plants with desired traits created new, highly productive hybrids known as "miracle" varieties. The first of these was a dwarf, high-yielding wheat developed by Norman Borlaug, who received a Nobel Peace Prize for his work, at a research center in Mexico (fig. 9.20). At about the same time, the International Rice Research Institute in the Philippines developed dwarf rice strains with three or four times the production of varieties in use at the time.

Green revolution crops are high responders

The dramatic increases obtained as these new varieties spread around the world has been called the **green revolution.** The success of these methods is one of the main reasons world food supplies have more than kept pace with the growing human population

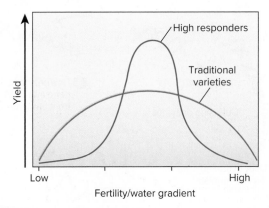

FIGURE 9.21 Green revolution miracle crops are really high responders, meaning that they have excellent yields under optimum conditions. For poor farmers who can't afford the fertilizer and water needed by high responders, traditional varieties may produce better yields.

over the past few decades. Miracle varieties were spread around the world as U.S. and European aid programs helped developing countries adopt new methods and seeds. The green revolution replaced traditional crop varieties and growing methods throughout the developed world, and nearly half of all farmers in the developing world were using green revolution seeds, fertilizers, and pesticides by the 1990s.

Most green revolution breeds really are "high responders." They yield more than other varieties if and only if they have steady inputs of fertilizer, water, and pest control (fig. 9.21). Without irrigation and fertilizer, these high responders may yield less than traditional varieties. New methods and inputs are also expensive. Poor farmers who can't afford hybrid seeds, fertilizer, machinery, fuel, and irrigation are put at a disadvantage, compared to wealthier farmers who can afford these inputs. Thus, the green revolution is credited with feeding the world, but it is also often accused of driving poorer farmers off their land, as rising land values and falling commodity prices squeeze them from both sides.

Recent studies by the FAO and the UN Environment Programme, aimed at finding less expensive methods for low-income farmers, have found that diversified, low-input farming can provide superior food security in poor regions. Soil can be enriched with nitrogen-fixing plants and manure. Rotating crops and interplanting crops reduces pest dispersal. Some data suggest that if farmers are taught better soil management, they can meet or exceed conventional production by improving fertility without costly inputs (see section 10.4).

Genetic engineering moves DNA among species

Genetic engineering involves removing genetic material from one organism and splicing it into the chromosomes of another (fig. 9.22). This technology introduces entirely new traits, at a much faster rate compared to cross-breeding methods. It is now possible to build entirely new genes by borrowing bits of DNA from completely unrelated species, or even synthesizing artificial DNA sequences to create desired characteristics in **genetically modified organisms (GMOs).**

FIGURE 9.20 Semi-dwarf wheat (*right*), bred by Norman Borlaug, has shorter, stiffer stems and is less likely to lodge (fall over) when wet than its conventional cousin (*left*). This "miracle" wheat responds better to water and fertilizer, and has played a vital role in feeding a growing human population.
William P. Cunningham

FIGURE 9.22 One method of gene transfer, using an infectious, tumor-forming bacterium such as agrobacterium. Genes with desired characteristics are cut out of donor DNA and spliced into bacterial DNA using special enzymes. The bacteria then infect plant cells and carry altered DNA into cells' nuclei. The cells multiply, forming a tumor, or callus, which can grow into a mature plant.

Genetically modified (GM) crops might provide dramatic benefits. Researchers are working to develop crops that resist drought, frost, or disease, or that tolerate salty, waterlogged, or nutrient-poor soils. These innovations could allow degraded or marginal farmland to become productive, and they could help reduce hunger in developing countries. Rice researchers have spent decades working on varieties such as "Golden Rice," enriched with vitamin A by inserting genes from daffodils, soil bacteria, or corn. This yellow-colored rice is proposed as a nutrition supplement for very poor populations in developing countries. Vandana Shiva, an Indian food activist, has argued that this rice will be more expensive and less nutritious than vegetables, greens, and other traditional foods. Proponents of Golden Rice believe that it could provide essential nutrients to children in developing areas.

Efforts are under way to develop crops such as bananas and potatoes containing oral vaccines that can be grown in developing countries where refrigeration and sterile needles are unavailable. Plants have been engineered to make industrial oils and plastics. Animals, too, are being genetically modified to grow faster, gain weight on less food, and produce pharmaceuticals such as insulin in their milk. It may soon be possible to create animals with human cell–recognition factors that could serve as organ donors.

Most GMOs have been engineered for pest resistance or herbicide tolerance

Humanitarian goals are admirable, but by far the dominant gene transfers involve either pest resistance or herbicide tolerance. The dominant pest resistance gene comes from a natural type of soil bacteria, *Bacillus thuringiensis* (Bt). The Bt gene produces toxins lethal to Lepidoptera (butterfly family) and Coleoptera (beetle family). The genes for some of these toxins have been transferred into crops such as maize (to protect against European cut worms), potatoes (to fight potato beetles), and cotton (for protection against boll weevils). This allows farmers to reduce insecticide spraying. Arizona cotton farmers, for example, report reducing their use of chemical insecticides by 75 percent on Bt cotton fields.

Constant exposure to Bt toxin has led to the emergence of Bt tolerance in insect pests. Tolerance develops when a pesticide kills off most of a population, leaving just a few individuals whose genetic traits allow them to survive the pesticide. These individuals then prosper and multiply, unaffected by the pesticide. Declining effectiveness of this natural pesticide—one of the few available to organic growers—is a significant worry for organic food producers, who often rely on natural Bt applications to combat insects. To protect the Bt trait, farmers are often required to plant a part of every field in non-Bt crops that will act as a refuge for nonresistant pests. The hope is that these "refuges" will maintain a nontolerant population of pests.

There also is a concern about the effects on nontarget species. In laboratory tests, about half of a group of monarch butterfly caterpillars died after feeding on plants dusted with pollen from Bt corn. Under field conditions, however, where many variables are present, it has been difficult to prove harm to butterflies.

The other major transgenic crops are engineered to tolerate herbicides. These crops are unaffected when fields are sprayed to kill weeds. These crops make up about three-quarters of all genetically engineered acreage. The two main products in this category are Monsanto's "Roundup Ready" crops—so-called because they tolerate Monsanto's best-selling herbicide, Roundup (glyphosate)—and AgrEvo's "Liberty Link" crops, which resist that company's Liberty (glufosinate) herbicide. Because crops with these genes can grow in spite of high herbicide doses, farmers can spray fields heavily to exterminate weeds. This practice helps conservation tillage and leaving more crop residue on fields

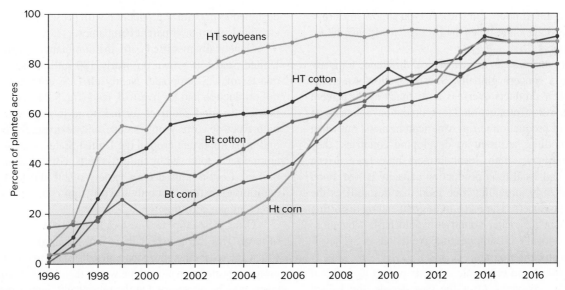

FIGURE 9.23 Growth of genetically engineered corn, cotton, and soybeans in the United States. HT (herbicide tolerant) varieties tolerate herbicides such as glyphosate (Monsanto's Roundup®); Bt crops contain insect-killing proteins from the bacterium *Bacillus thuringiensis*.
Source: USDA Economic Research Service, 2019

to protect topsoil from erosion, both good ideas, but it may also mean an increase in herbicide use.

Like antibiotics, heavily-used herbicides also produce herbicide-tolerant pests (see chapter 10). In 2014, the U.S. Environmental Protection Agency permitted the use of Enlist Duo, a combination of glyphosate and 2,4-D, two widely used herbicides developed by Dow AgroSciences. The new combined herbicide was designed in combination with new GM varieties of corn and soybeans, designed to tolerate both chemicals. As with Roundup Ready crops, the gene modification is designed to support expanded use of the herbicide, as previous weed-control chemicals became impotent.

Safety of GMOs is widely debated

GM crops have been introduced to the world's farmers even more rapidly than green revolution crops were in the 1960s and 1970s. A decade after their introduction in 1996, GM varieties were planted on 400 million hectares (1 billion acres) of farmland. Three years later, that number had doubled to 800 million ha (2 billion acres). This represents just over half of the world's 1.5 billion ha of cultivated land. The United States accounted for 56 percent of that acreage, followed by Argentina with 19 percent. In 2005, China approved GM rice for commercial production. This was the first GM cereal grain approved for direct human consumption and could move China into the forefront of GM crop production. The first GM animals developed for human consumption are GM Atlantic salmon, which grow much faster than normal because they contain growth hormone genes from an oceanic pout.

Federal monitoring agencies have decided that GM crops are "substantially equivalent" to non-GM varieties. Consumer groups question this judgment, arguing that we don't fully understand the potential effects of transferred genetic material. Meanwhile, a new gene modification method known as CRISPR-Cas9 may

be sidestepping debates about gene transfer. CRISPR allows crop developers to modify genes within a cell's own DNA. The environmental and ethical implications are broad and uncertain, although there are great hopes for medical applications (see chapter 3). We may soon learn what this means for crop and livestock production.

Ultimately, GM foods may be safe for consumers but risky for producers. Farmers producing GM crops must commit to more and more costly herbicides, pesticides, and seeds to stay in business. Although many consumers are wary of GM foods, you have almost certainly eaten them. Over 70 percent of U.S. corn has Bt traits or herbicide tolerance, or both. Nearly 95 percent of soybeans and 80 percent of cotton are modified for herbicide tolerance (fig. 9.23). It has been estimated that over 60 percent of all processed food in America contains GM ingredients.

Section Review

1. What is the "green revolution," and why is it important?
2. What are genetically modified organisms?
3. Why are most U.S. corn and soybeans GM varieties?

9.4 FOOD PRODUCTION POLICIES

- *Government support promotes farm production and innovation.*
- *Food production is an important economic question.*

Much of the increase in food production over the past 50 years has been fueled by government support for agricultural education, research, and development projects that support irrigation systems, transportation networks, crop insurance, and direct subsidies. The World Bank estimates that rich countries pay their own farmers $350 billion per year, or nearly six times as much as all developmental aid to poor countries. A typical cow in Europe enjoys annual subsidies three times the average yearly income for most African farmers.

Agricultural subsidies can make a critical difference for farmers, but they are a concern globally. Subsidies allow American farmers to sell their products overseas at as much as 20 percent below the actual cost of production. Government payments cover losses and costs of production to the American farmers, but when the cheap commodities are sent to developing countries, they undercut local farmers. Often, local farmers have been driven out of business, and local food production capacity is lost. Food aid functions similarly, providing free food that destabilizes the market for local producers. The FAO argues that ending distorting financial support in the richer countries would have a far more positive impact on local food supplies and livelihoods in the developing world than any aid program.

Powerful political and economic interests protect agricultural assistance in many countries. Over the past decade, the United States, for example, has spent $143 billion in farm support. This aid is distributed unevenly. According to the Environmental Working Group, 72 percent of all aid goes to the top 10 percent of recipients. One giant rice-farming operation in Arkansas, for example, received $38 million over a five-year period. Aid also is concentrated geographically. Just 5 percent (22) of the nation's 435 congressional districts collect more than 50 percent of all agricultural payments.

Most of this aid has been in direct payments for each bushel of targeted commodities, mainly corn, wheat, soybeans, rice, and cotton, as well as special subsidies for milk, sugar, and peanuts. Proponents insist that crop supports preserve family farms; critics claim that the biggest recipients are corporations that don't need the aid. In 2012, Congress voted to replace direct payments with insurance subsidies, by which the public pays insurance premiums on crops. Insurance reimburses farmers when crops fail because of weather or other risks, remain targeted at dominant commodity crops and are most beneficial for large producers.

An additional effect of these market interventions is to encourage the oil- and sugar-rich diets that lead to the spreading obesity epidemic. Subsidies help ensure that these processed foods are cheaper and more readily available than fresh fruits, vegetables, and whole grains. Many food policy analysts argue that we should support more vegetables and nutrient-rich foods and fewer commodity crops. Public attention to farm policy could help move us toward such policies.

Is genetic engineering about food production?

Much of the policy debate around which foods to subsidize involve the urgency of supporting high-yielding varieties that will ensure global food security. Usually this has meant supporting genetically modified crop varieties. Some development experts, however, question whether this is the most efficient strategy for reducing hunger. Most GM seed companies are owned by producers of agricultural chemicals: Monsanto, Dow, Syngenta, Bayer, and DuPont—and the world's largest pesticide manufacturers are also the world's largest seed producers. Their seeds are specially designed to maximize production with prescribed combinations of chemicals from their parent companies.

Genetically modified corn, the dominant crop produced in the United States, is grown on nearly 100 million acres, or nearly a quarter of U.S. cropland. Nearly all U.S. corn is used for ethanol production and for livestock feed (fig. 9.24). About 10 percent is used for food additives for industrial food production, such as high-fructose corn syrup, glucose, and dextrose. Soybeans the second most important crop in the United States, are used mainly for animal feed, followed by industrial oils, edible oils and meal, and biofuel. Cotton is used for a variety of oil and feed products, in addition to textiles, but only the oil from cotton seeds is used for direct human consumption.

Using these crops for livestock feed has greatly reduced the price of meat. This is true especially in industrialized nations, but also in China and other emerging economies. If we are to continue to expand protein consumption around the world, then GM soy, corn, and other crops may be necessary. On the other hand, increasing proportions of the world consume more protein and calories than may be healthy (see fig. 9.2). If the concern is to supply 9 billion people with a sufficient and healthy diet while sustaining our land and water systems, then producing meat is less efficient than producing food crops.

Like green revolution varieties, genetically modified seeds and pesticides are expensive for farmers in the developing world. Some food activists claim the GM revolution mainly benefits the corporations that sell seeds and pesticides. Specialized seeds and pesticides are expensive for farmers, who nonetheless must buy them in order to compete with other farmers in the market. When inputs are expensive, wealthy farmers have important advantages over poor farmers, especially poor farmers in developing countries, where financial supports are tenuous. Experience in India has shown that the costs of production with GM seeds drives farmers into chronic debt. Waves of farm losses and farmer suicides have swept across India as a result of these debt burdens.

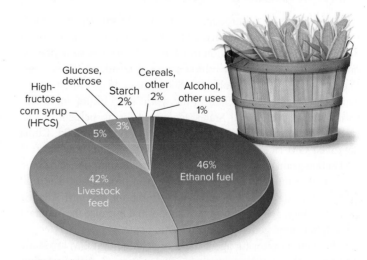

FIGURE 9.24 Little of the U.S. corn crop, our dominant agricultural product, is used directly for food. (Data are rounded to the nearest whole number.)
USDA ERS

Quantitative Reasoning

Figure 9.23 shows the adoption of genetically modified crops. What was the rate of use in soybeans when you were 5 years old? In corn? How much change has there been since then?

Farm policies can also protect the land

Every year, millions of tons of topsoil and agricultural chemicals wash from U.S. farm fields into rivers, lakes, and, eventually, the ocean. Farmers know that erosion both impoverishes their land and pollutes water, but they're caught in a bind. For every $1 the U.S. government pays farmers to conserve soil and manage nutrients, it pays $7 to support row-crop commodities that cause erosion and destroy soil quality through intensive cultivation and chemical use. The USDA estimates that if federal subsidies didn't promote these commodities, farmers would shift 2.5 million ha (6 million acres) of row crops into pasture, hay, and other crops that minimize erosion and build soil quality.

The United States tries to reduce soil erosion and overproduction of crops with the Conservation Reserve Program (CRP), which pays farmers to keep roughly 12 million ha (30 million acres) of highly erodible land out of production. The USDA reports

that CRP lands prevent the annual loss of 450 million tons of soil every year, protect 270,000 km (170,000 miles) of streams, and store 48 million tons of carbon per year. Keeping land enrolled in this soil conservation program is vulnerable to political and economic shifts, however, and the amount of land enrolled changes every year.

Land enrolled in the CRP has been declining, as farmers find it more profitable to plant corn for ethanol, and as farm policy commits less money to supporting land retirement. Between 2007, when new ethanol policies were introduced, and 2012, over a quarter of CRP lands were withdrawn from the program; in addition, tens of thousands of acres of uncropped grasslands were converted to corn and soy production. Many agronomists, though, say we should have more CRP land, not less. They argue that the United States could gradually shift payments from production subsidies and crop insurance to conservation programs that would truly support family farms while protecting the environment. We will discuss these and other aspects of farming and soil conservation in chapter 10.

Section Review

1. How is money spent supporting agriculture?
2. What are the primary uses of corn in the United States?
3. How can farm policy protect the land?

Connecting the Dots

World food supplies have increased dramatically over the past half century. Despite the fact that the human population has nearly tripled in that time, food production has increased even faster, and we now grow more than enough food for everyone. Because of uneven distribution of food resources, however, there are still some 800 million people who don't have enough to eat on a daily basis, and hunger-related diseases remain widespread. Severe famines continue to occur, although most result more from political and social causes (or a combination of political and environmental conditions) than from environmental causes alone.

While hunger persists in many areas, over a billion people consume more food than is healthy on a daily basis. Epidemics of weight-related illnesses are spreading to developing countries as they adopt the diets and lifestyles of wealthier nations. Obesity is challenging to respond to: It arises both from wealth and poverty, from food choices, and from the structural systems, such as farm subsidies and food marketing, that shape the ways we eat.

Most of the world's food comes from a few major crops, including grains, vegetables, wheat, rice, corn, and potatoes. In the United States, just three crops—corn, soybeans, and wheat—are principal farm commodities. Corn and soybeans are mainly used for livestock feed and fuel, rather than food. Increasing use of these

crops in confined feeding operations has dramatically increased meat production. For this and other reasons, global consumption of protein-rich meat and dairy products has climbed in the past 40 years. Protein gives us the energy to work and study, but raising animals takes a great deal of energy and food, so meat production can be environmentally expensive. Reliance on a few major crops and livestock species is also a risky strategy in a changing climate.

Most increases in food production in recent generations result from "green revolution" varieties of grains, which grow rapidly in response to fertilizer use and irrigation. More recent innovations have focused on genetically modified varieties. Some of these are being developed for improved characteristics, such as vitamin production or tolerance of salty soils. The majority of genetically modified crops are designed to tolerate herbicides, in order to improve competition with weeds.

As farmers have shown in Burkina Faso, and many other regions, resilient food systems are possible. Ingenuity and local solutions are often the key to success in developing areas. Protection from violence is also necessary. We can produce enough food for all. How we damage or sustain our environment while doing so is the subject of chapter 10.

Reviewing Key Terms

Can you define the following terms in environmental science?

anemia 9.1

aquaculture 9.2

chronically undernourished 9.1

confined animal feeding operation (CAFO) 9.2

famine 9.1

food security 9.1

genetically modified organisms (GMOs) 9.3

green revolution 9.3

kwashiorkor 9.1

longlining 9.2

malnourishment 9.1

marasmus 9.1

obese 9.1

purse seining 9.2

trawling 9.2

Critical Thinking and Discussion Questions

1. Do people around you worry about hunger? Do you think they should? Why or why not? What factors influence the degree to which people worry about hunger in the world?

2. Global issues such as hunger and food production often seem far too large to think about solving, but it may be that many strategies can help us address chronic hunger. Consider your own skills and interests. Think of at least one skill that could be applied (if you had the time and resources) to helping reduce hunger in your community or elsewhere.

3. Suppose you are a farmer who wants to start a confined animal feeding operation. What conditions make this a good strategy for you, and what factors would you consider in weighing its costs and benefits? What would you say to neighbors who wish to impose restrictions on how you run the operation?

4. Debate the claim that famines are caused more by human actions (or inactions) than by environmental forces. What kinds of evidence would be needed to resolve this debate?

5. Outline arguments you would make to your family and friends for why they should eat a mostly vegetarian diet, along the lines of *Diet for a Small Planet*. What reasons would be most compelling? What are some reasons why it is, or is not, fair to influence someone else's food practices?

6. Given what you know about GMO crops, identify some of the costs and benefits associated with them. Which of the costs and benefits do you find most important? Why?

7. Corn is by far the dominant crop in the United States. In what ways is this a good thing for Americans? How is it a problem? Who are the main beneficiaries of this system?

Data Analysis

Exploring Global Food Data

The UN Food and Agriculture Organization (FAO) is one of the most important sources of global food data. It lets you explore changing food production, and population growth in Burkina Faso and other countries.

Go to the website (http://www.fao.org/faostat/en/#compare). In the Compare Data section, you can specify Groups = Production, Domains = Crops, Country = Burkina Faso, Element = Area harvested, and Item = Maize and Cassava. Then click Compare Data. Scroll down the page to see the graph you have just produced. This is an excellent site to understand real-world changes, which affect peoples' lives around the world. What other countries and crops would be interesting? Try looking at Brazil's soy production (discussed in chapter 10). There is a wealth of data here, free for you to explore.

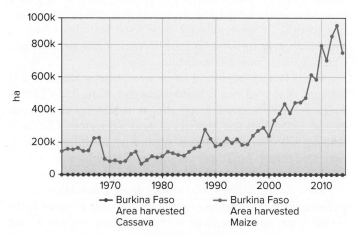

FIGURE 1 The UN FAO website lets you graph and examine changing production of many crops. Here data are shown for Burkina Faso.

Source: UN Food and Agriculture Organization, FAOSTAT http://www.fao.org/faostat/en/#compare

10

Farming: Conventional and Sustainable Practices

▲Soybean harvest in the Brazilian state of Mato Grosso, whose name means "dense forest."
Paulo Fridman/Corbis/Getty Images

Learning Outcomes

After studying this chapter, you should be able to:

10.1 Describe the components of soils.

10.2 Explain the ways we use, abuse, and conserve soils.

10.3 Discuss our principal pesticides and their environmental effects.

10.4 Describe several methods of organic and sustainable agriculture.

"We abuse the land because we regard it as a commodity belonging to us. When we see land as a community to which we belong, we may begin to use it with love and respect."

– Aldo Leopold

Farming the Cerrado

Brazil's presidential election of 2018 was watched closely by environmentalists around the world. This election put the industrialist and climate skeptic Jair Bolsonaro in charge of a country that contains, in the Amazon rainforest, one of the world's most important centers of biodiversity, carbon storage, and climate regulation. The agriculture industry, which has cleared much of the country's forests and grasslands, was thrilled to have a close friend as president. And Brazil's "king of soybeans," the world's largest soybean producer Blairo Maggi, who had also been Minister of Agriculture, was a key power behind the throne.

Soybeans and other major crops strongly shape Brazilian politics. Inexpensive land and expanding global markets for livestock feed and biofuels have made this one of the world's fastest-growing agricultural areas. The center of action has been the Cerrado, a rolling expanse of grasslands and woodlands, with at least 130,000 plant and animal species (fig. 10.1).

The region was long understood to be poor farmland. The iron-rich, red soils are too acidic to hold plant nutrients, and the hot, humid climate harbors innumerable pests and weeds. The Cerrado has long been cattle country, supporting grazers, not field crops. Then agronomists developed new varieties of soybeans specially adapted to tropical soils and climates. When soils are treated with powdered lime (calcium carbonate) and phosphorus, these new varieties can quadruple yields of soybeans, maize, cotton, and other crops. New breeds are genetically modified to tolerate herbicides, so growers can spray weeds freely without harming crops. Since 1975, Brazil's soybean crop has grown from near zero to over 36 million ha (89 million acres). Over half the Cerrado has now been converted to vast, level crop fields.

People don't eat these soybeans. Most soy is feed for pigs, chicken, and cattle. Industrial oils—for lubricants, plastics, and other materials—and biodiesel also use much of the soy produced here.

Rising income in China has driven much of Brazil's soy expansion. With more money to spend, the Chinese can afford to feed soy to livestock, especially pigs, chickens, and fish. (Still, Chinese eat only a fraction as much meat as Americans do.) Europe also represents a rapidly growing market for feed. Since 2007, Brazil has been the world's top soy exporter, shipping over 50 million metric tons per year, about 10 percent more than the United States. With two crops per year, cheap land, cheap labor, favorable tax rates, and high yields, Brazilian farmers can produce soybeans for half the cost in North America. Global soy production has nearly doubled in only about 15 years, with South America responsible for most of that growth. In addition to soy, Brazil now leads the world in exporting beef, maize, oranges, and coffee.

This rising production has global environmental effects. Some of the world's most biologically rich areas are retreating steadily as croplands expand. Rural farmworkers, displaced by large farm operations, have expanded into frontier forest areas, creating an "arc of destruction" as they clear still more forest (fig. 10.1). Deforestation also contributes to climate change, as the forest stores vast amounts of carbon. Drought appears to follow land conversion, as agriculture expands into the Amazon rainforest. Cleared land dries rapidly and stores little moisture, and satellite images show increasing amounts of dead and drought-stressed trees even deep in the rainforest. The worry is that dying forests will shift from a carbon sink to a net carbon source. A drying climate threatens agriculture, as much of the region's rainfall originates as evaporation and transpiration from the forest.

Conflicts over land have grown, as industrial farms invade indigenous areas or the lands of indigenous cultures communities and overrun small family farms. Conflicts have sometimes been violent: The Landless Workers Movement claims that assassinations and clashes over land rights killed 1,237 rural workers in Brazil between 1985 and 2000. Uncounted landless farmers and displaced families have been forced into squatter camps or migrated to Brazil's growing cities.

Expanding agriculture has far-reaching costs and benefits. What factors make this expansion possible? What sustainable approaches are available to help negotiate environmental and social priorities? In this chapter, we explore these questions and consider the nature and the future of agricultural production.

FIGURE 10.1 Brazil's Cerrado, 2 million ha of grassland and open woodland, is the site of the world's fastest expansion of soybeans. The continent's highest rate of forest loss, in the "arc of destruction," is a consequence, as displaced cattle ranchers and smallholders encroach on the Amazon rainforest.

Benchmark Data

Among the ideas and values in this chapter, the following are a few worth remembering.

1 mm	Amount of soil accumulated, per year, in ideal conditions
11%	Portion of global land area in cultivation
3 GT	Carbon capture possible on farm soils through "carbon farming" practices
1:4	Global pesticide use, 1965:2015 (million tons per year)
5	Years until resistant pests to appear, for most pesticides

10.1 WHAT IS SOIL?

- *Soils are composed of mineral grains (sand, silt, and clay), organic matter, organisms, water, and air.*
- *These affect soil fertility, erodibility, and other factors.*
- *Soils vary in texture, color, and horizon development.*

Agriculture has dramatically changed our environment, altering patterns of vegetation, soils, and water resources worldwide. The story of Brazil's Cerrado involves the conversion of millions of hectares of tropical savanna and rainforest to crop fields and pasture. This is one recent example of agricultural land conversion, but humans have been converting land to agriculture for thousands of years. Some of these agricultural landscapes are ecologically sustainable and have lasted for centuries or millennia. Others have depleted soil and water resources in just a few decades. What are the differences between farming practices that are sustainable and those that are unsustainable? What aspects of our current farming practices degrade the resources we depend on, and in what ways can farming help to restore and rebuild environmental quality?

As you have read in the opening case study, farm expansion has changed the landscape, the environment, and the economy of central Brazil. These changes are driven by financial investments and technological innovations from North American and European corporations. They are supported by rapidly expanding markets in Asia and Europe. But another essential factor has been the development of new ways to modify the region's nutrient-poor, acidic tropical soils. We will begin this chapter by exploring what soils are made of and how they differ from one place to another.

Soils are complex ecosystems

Is soil a renewable resource, or is it a finite resource that we are depleting? It's both. Over time, soil is renewable, because it develops gradually through weathering of bedrock and through the accumulation of organic matter, such as decayed leaves and plant roots. But these processes are extremely slow. Building a few millimeters of soil can take anything from a few years (in a healthy grassland) to a few thousand years (in a desert or tundra).

Under the best circumstances, topsoil accumulates at about 1 mm per year. This soil growth involves careful management that prevents erosion and allows organic material to accumulate in the soil. But most farming techniques deplete soil. Plowing exposes bare soil to erosion by wind or water, and annual harvests remove organic material such as leaves and roots. Severe erosion can carry away 25 mm or more of soil per year, far more than the 1 mm that can accumulate under the best of conditions.

Soil is a marvelous, complex substance. It is a combination of weathered rocks, plant debris, living fungi, and bacteria, an entire ecosystem that is hidden to most of us. In general, soil has six components:

1. *Sand and gravel* (mineral particles from bedrock, either in place or moved from elsewhere, as in windblown sand)
2. *Silts and clays* (extremely small mineral particles; clays are sticky and hold water because of their flat surfaces and ionic charges)
3. *Organic matter* (decaying plant matter that stores nutrients and gives soils a black or brown color)
4. *Soil fauna and flora* (living organisms, including soil bacteria, worms, fungi, roots of plants, and insects, that recycle organic compounds and nutrients)
5. *Water* (moisture from rainfall or groundwater, essential for soil fauna and plants)
6. *Air* (tiny pockets of air help soil bacteria and other organisms survive)

Variations in these components produce almost infinite variety in the world's soils. Abundant clays make soil sticky and wet. Abundant organic material and sand make the soil soft and easy to dig. Sandy soils drain quickly, often depriving plants of moisture. Silt particles are larger than clays and smaller than sand (fig. 10.2), so they aren't sticky and soggy, and they don't drain too quickly. Thus, silty soils are ideal for growing crops, but they are also light and blow away easily when exposed to wind. Soils with abundant soil fauna quickly decay dead leaves and roots, making nutrients available for new plant growth. Compacted soils have few air spaces, making soil fauna and plants grow poorly.

Soil texture—the amount of sand, silt, and clay in the soil—is one of the most important characteristics of soils. Texture helps determine whether rainfall drains away quickly or ponds up and drowns plants. Loam soils are usually considered best for farming because they have a mixture of clay, silt, and sand (fig. 10.3).

You can see some of the differences among soils just by looking at them. Reddish soils, including most tropical soils, often are colored by iron-rich, rust-colored clays, most of which (unless they are volcanic soils) store few nutrients for plants. Deep black soils, on the other hand, are rich in organic material, which is moist, rich in nutrients, and keeps soil soft and workable. Most Brazilian tropical soils are deeply weathered red clays. With frequent rainfall and year-round warm weather, organic material decays quickly and is taken up by living plants or washed away with rainfall. Red, deeply weathered, clay soils result. These reddish clays hold few nutrients and little moisture for growing fields of soybeans (fig. 10.4*b*).

FIGURE 10.2 Relative size of sand, silt, and clay particles (magnified about 100 times).

In contrast, the rich, black soils of the Corn Belt of the central United States have abundant organic matter and a good mix of sand, silt, and clay. These soils tend to hold enough moisture for crops without becoming waterlogged, and they tend to be rich in nutrients. Acidic tropical Brazilian soils can be improved by adding lime (calcium carbonate, as in limestone), which improves the soil's ability to retain nutrients applied in fertilizer. Liming vast areas was not economical until recently, but expanding markets for livestock feed in Asia and Europe now make it cost-effective for Brazilian farmers to apply lime to their fields. This is one of the innovations that has allowed recent expansion of Brazilian soy production.

Healthy soil fauna can determine soil fertility

Soil bacteria, algae, and fungi decompose and recycle leaf litter and other organic debris. These microorganisms thus make nutrients available to plants, and also help give soils structure and loose texture (fig. 10.5). The abundance of these organisms can be astonishing. One gram of soil can contain hundreds of soil bacteria and 20 m of tiny strands of fungal material. A cubic meter of soil can contain more than 10 kg of bacteria and fungal biomass. Tiny worms and nematodes process organic matter and create air spaces as they burrow through soil. Slightly larger insects, mites, spiders, and earthworms further loosen and aerate the soil. The sweet aroma of freshly turned soil is caused by actinomycetes, bacteria that grow in fungus-like strands and give us the antibiotics streptomycin and tetracycline. These organisms mostly stay near the surface, often within the top few centimeters. The roots of plants can reach deeper, however, allowing moisture, nutrients, and organic acids to help break down rocks or other parent material and form new soil.

Many plant species grow best with the help of particular species of soil fungi in relationships called **mycorrhizal symbiosis.**

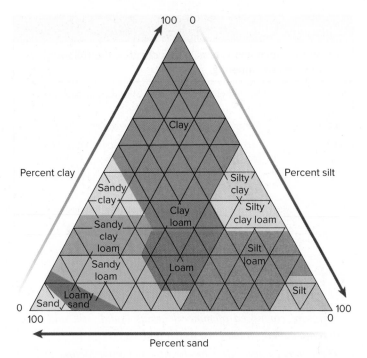

FIGURE 10.3 A soil's texture depends on its proportions of sand, clay, and silt particles. Read the graph by following lines across, up, or down from the axes. For example, "loam" has about 50–75 percent sand, 8–30 percent clay, and 18–50 percent silt. Loamy soils have the best texture for most crops, with enough sand to be loose and workable, yet enough silt and clay to retain water and nutrients.

FIGURE 10.4 A temperate grassland soil (a) has a thick, black organic layer. Tropical rainforest soils (b) have little organic matter and are composed mostly of nutrient-poor, deeply weathered iron-rich clays. Each of these profiles is about 1 m deep.

Source: (a) USDA Natural Resources Conservation Service; (b) Hari Eswaran, USDA Natural Resources Conservation Service

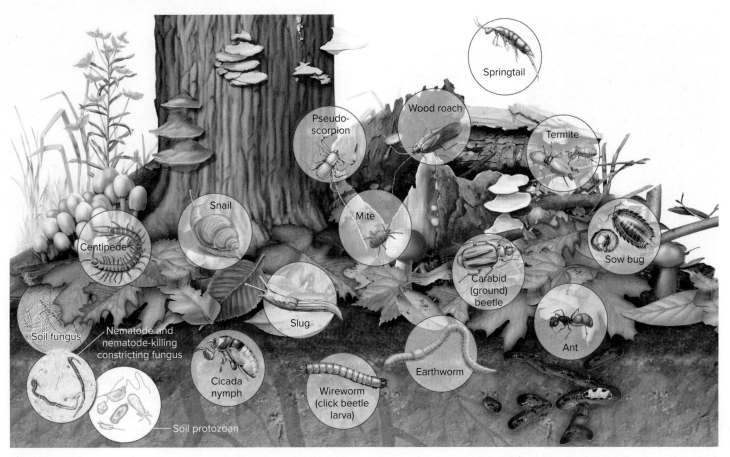

FIGURE 10.5 Soil ecosystems include countless organisms that consume and decompose organic material, aerate soil, and distribute nutrients through the soil.

In this relationship, the mycorrhizal fungus (a fungus growing on and around plant roots) provides water and nutrients to the plant, while the plant provides organic compounds to the fungus. Plants growing with their fungal partners often grow better than those growing alone.

The health of the soil ecosystem depends on a variety of factors, such as slope, rainfall, and frequency of disturbance. Too much rain washes away nutrients and organic matter, but soil fauna cannot survive with too little rain. In extreme cold, soil fauna recycle nutrients extremely slowly; in extreme heat, they may work so fast that leaf litter on the forest floor is taken up by plants in just weeks or months—so that the soil retains little organic matter. Frequent disturbance prevents the development of a healthy soil ecosystem, as does steep topography that allows rain to wash away soils. In the United States, the best farming soils tend to occur where the climate is not too wet or dry, on glacial silt deposits, such as those in the upper Midwest, and on silt- and clay-rich flood deposits, like those along the Mississippi River.

Most soil fauna occur in the uppermost layers of a soil, where they consume leaf litter. This layer is known as the "O" (organic) horizon. Just below the O horizon is a layer of mixed organic and mineral soil material, the "A" horizon (fig. 10.6). These layers are known as **topsoil,** or surface soil.

The B horizon, or **subsoil,** tends to be richer in clays than the A; the B horizon is below most organic activity. The B layer accumulates clays that seep downward from the A horizon with rainwater that percolates through the soil. If you dig a hole, you may be able to tell where the B horizon begins, because there the soil tends to become slightly more sticky and cohesive. If you squeeze a handful of B-horizon soil, it should hold its shape better than a handful of A-horizon soil.

Sometimes an E (eluviated, or washed-out) layer lies between the A and B horizons. An E layer is loose and light-colored because most of its clays and organic material have been washed down to the B horizon. The C horizon, below the subsoil, is mainly decomposed rock fragments. Parent materials underlie the C layer. Parent material is the sand, windblown silt, bedrock, or other mineral material on which the soil is built. About 70 percent of the parent material in the United States was transported to its present site by glaciers, wind, and water, and is not related to the bedrock formations below it.

Your food comes mostly from the A horizon

Ideal farming soils have a thick, organic-rich A horizon. The soils that support the Corn Belt farm states of the Midwest have a rich, black A horizon that can be more than 2 meters thick (although a century of farming has washed much of this soil away and down the Mississippi River). The A horizon in most soils is less than half a meter thick. Desert soils, with slow rates of organic activity, might have almost no O or A horizons (fig. 10.7).

FIGURE 10.7 In many areas, soil or climate constraints limit agricultural production. These hungry goats in Sudan feed on a solitary *Acacia* shrub. *Eye Ubiquitous/Newscom*

Organic layer:
O Decomposing leaves and organics

Topsoil:
A Organic matter, organisms, roots, inorganic minerals

Eluviated (washed-out) layer:
E Clays and humic acids leached to lower layers

Subsoil:
B Enriched in clays and other material leached from upper layers

FIGURE 10.6 Soil profile showing possible soil horizons. The actual number, composition, and thickness of these layers varies in different soil types. *Soil & Land Resources Division, University of Idaho/USDA*

Because topsoil is so important to our survival, we identify soils largely in terms of the thickness and composition of their upper layers. Soils vary endlessly in depth, color, and composition, but for simplicity we can describe a few general groups. The U.S. Department of Agriculture classifies the soils into 11 soil orders (table 10.1). In the Farm Belt of the United States, the dominant soils are Mollisols (*mollic,* "soft"; *sol,* "soil"). These soils have a thick, organic-rich A horizon that developed from the deep, dense roots of prairie grasses that covered the region until about 150 years ago (see fig. 10.4a). Another group that is important for farming is Alfisols. Alfisols have a slightly thinner A horizon than Mollisols do, and slightly less organic matter. Alfisols develop in deciduous forests where leaf litter is abundant. In contrast, the Aridisols (*arid,* "dry") of the desert Southwest have little organic matter, and they often contain accumulations of mineral salts. Mollisols and Alfisols dominate most of the farming regions of the United States.

Section Review

1. What are six major components of soil?
2. How are soil organisms important for good soils?
3. Why are Brazilian soils often red?
4. How fast (in mm per year) are soils understood to build under ideal conditions?

10.2 How Do We Use, Abuse, and Conserve Soils?

- *Arable land is limited, and expansion has environmental costs.*
- *Crops need many inputs, including water, fertilizer, and fuel.*
- *Wind and water erosion threatens soil fertility and abundance.*
- *Farming techniques, windbreaks, and other strategies can control erosion.*

Much of the world's land is too steep, soggy, salty, cold, or dry for farming. Only about 11 percent of the earth's land area (1.5 billion ha out of 13.4 billion ha of land area) is currently in crop production. In theory, up to four times as much land could potentially be converted to cropland, although that would mean destroying wetlands and forest lands that stabilize our climate and water resources.

Table 10.1	Soil orders defined by the USDA
Soil Order	**Characteristics**
Mollisol	organic-rich, black grassland soil
Alfisol	organic-rich, deciduous forest soil
Spodosol	leached by acidic litter, as from pine needles
Histosol	peaty, wet, largely organic matter
Aridisol	desert soils, may have salt deposits
Ultisol	old, stable, weathered soil
Oxisol	tropical, deeply weathered, red soil
Vertisol	clay-rich, sticky soil
Andisol	volcanic material; mineral-rich
Inceptisol	weak horizon development, as with recently deposited material
Entisol	little/no horizon development, as with sand or eroded material

Despite those costs, conversion to farmland is continuing in many areas, as global trade in agricultural commodities, such as corn, soy, and palm oil, becomes increasingly lucrative. Brazil's expansion of soy farming into the Cerrado (opening case study) is one of the most dramatic cases of land conversion, but ancient forests and grasslands are also turning into farmland in many parts of the developing world. The ecological costs of these land conversions are hard to calculate. Farmers can easily count the cash income from the farm products they sell, but it is never easy to calculate the value of biodiversity, clean water, and other ecological services of a forest or grassland, compared to the value of crops.

Arable land is unevenly distributed

The best agricultural lands occur where the climate is moderate—not too cold or too dry—and where thick, fertile soils are found. Take a look at the global map in the back of your book. What regions do you think of as the best agricultural areas? The poorest? Much of the United States, Europe, and Canada are fortunate to have temperate climates, abundant water, and high soil fertility. These produce good crop yields that contribute to high standards of living. Other parts of the world, although rich in land area, lack suitable soil, level land, or climates to sustain good agricultural productivity.

Most croplands occur in mid-latitudes, especially in areas where weather systems bring reliable moisture (fig. 10.8). Most interior continental areas have climates too cold, hot, or dry for much cropland. Note that areas of abundant cropland also correspond to regions of major world civilizations.

In developed countries, 95 percent of recent agricultural growth in the past century has come from a combination of improved crop varieties and increased use of fertilizers, pesticides, and irrigation. Conversion of new land to crop fields has contributed relatively little to increased production. In fact, less land is being cultivated now than 100 years ago in North America or 600 years ago in Europe. Productivity per unit of land has increased, and some marginal land has been retired. Careful management is important for preserving the remaining farmland.

Soil losses threaten farm productivity

Agriculture both causes soil degradation and suffers from it. Soil loss by erosion is the primary concern (fig. 10.9). Every year, erosion makes about 3 million ha of cropland unusable worldwide. Another 4 million ha are converted to nonagricultural uses, such as urban land, highways, factories, or reservoirs, according to the International Soil Reference and Information Center (ISRIC). In the United States alone, we've lost about 140 million ha of farmland in the past 30 years to urbanization, soil degradation, and other factors (fig. 10.10).

Land degradation is usually slow and hard to observe directly. Usually, fertility declines gradually, as soil washes and blows away, salts accumulate, and organic matter is lost. About 20 percent of vegetated land in Africa and Asia is degraded enough to

FIGURE 10.8 The world's agricultural lands are unevenly distributed, occurring mainly where moisture is available and temperatures are moderate. Areas with >10 percent arable land are shown.

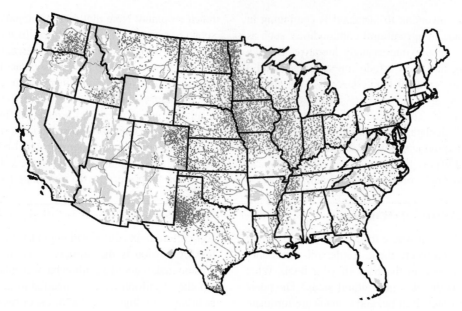

FIGURE 10.9 More than 43.7 million ha (108 million acres) in the United States are subjected to excess erosion by wind (red) or water (blue) each year. Each dot represents 200,000 tons of average annual soil loss.
Source: USDA Natural Resource Conservation Service

reduce productivity; 25 percent of lands in Central America and Mexico are degraded. Wind and water erosion are the primary causes of degradation. Additional causes of degradation include chemical deterioration (mainly salt accumulation from salt-laden irrigation water) and physical deterioration, such as compaction by heavy machinery or waterlogging.

As a consequence of soil loss, as well as population growth, the amount of arable land per person worldwide has shrunk from about 0.38 ha in 1970 to 0.21 ha in 2010. Consider that a hectare is an area 100 m × 100 m, or roughly the size of two football fields. On average, about five people are supported by that land area. By 2050, the arable land per person will decline to 0.15 ha. In the United States, farmland has fallen from 0.7 to 0.45 ha per person in the past 30 years, according to USDA data. To feed a growing population on declining land area, we are likely to need improvements in production and distribution, reduced production of non-food crops (chapter 9), and improved soil management.

Wind and water cause widespread erosion

When water washes away a thin layer of soil, we call it **sheet erosion.** When little rivulets of running water gather together and cut small channels in the soil, the process is called **rill erosion** (fig. 10.11*a*). When rills enlarge to form bigger channels or ravines that are too large to be removed by normal tillage operations, we call the process **gully erosion** (fig. 10.11*b*). Streambank erosion refers to the washing away of soil from the banks of established streams, creeks, or rivers, often as a result of removing streamside trees and brush or letting livestock trample streambanks.

Sheet and rill erosion are responsible for most soil loss on agricultural land. These types of erosion can be subtle, even invisible, but they can transport large amounts of soil. A farm field can lose 20 metric tons of soil per hectare during winter and spring runoff in rills so small they are erased by the first spring cultivation. That represents a loss of only a few millimeters of soil over the whole surface of the field, hardly an obvious change. But the maximum rate at which soils can rebuild, under the best conditions, is generally considered to be about 1 mm per year (largely through addition of organic matter) or about 11 metric tons per hectare (5 U.S. tons/acre). If soils rebuild slowly and erode rapidly, eventually they will be gone.

Wind can equal or exceed water in erosive force, especially in a dry climate. When plant cover and surface litter are removed from the land, wind lifts loose soil particles and sweeps them away (fig. 10.11*c*). In extreme conditions, windblown dunes encroach on useful land and cover roads and buildings. Over the past 30 years, China has lost 93,000 km² (about the size of Indiana) to **desertification,** or conversion of productive land to desert. Advancing dunes from the Gobi Desert are now only 160 km (100 mi) from Beijing.

Some of the highest erosion rates in the world occur in the United States and Canada. The U.S. Department of Agriculture reports that 69 million hectares (170 million acres) of U.S. farmland and range are eroding at rates that are reducing long-term productivity. Farming practices that expose soil, and intensive grazing that degrades plant cover are largely responsible for this situation. Row crops, such as corn and soybeans, leave soil exposed for much of the growing season (fig. 10.12). Deep plowing and heavy herbicide applications create fields that are weed-free but extremely vulnerable to erosion.

(a)

(b)

(c)

FIGURE 10.10 Disastrous erosion during the Dust Bowl years (a) led to national erosion control efforts that have reduced, but not eliminated, soil loss (b). Nationally, wind and water erosion have declined but continue to degrade farmland (c).

Source: (a) U.S. Department of Agriculture Natural Resources Conservation Service; (b) Photo courtesy of USDA Natural Resources Conservation Service

Since the Dust Bowl of the 1920s and 1930s, the USDA has been encouraging farmers to plow along contours rather than up and down hills, to avoid planting steep slopes, to leave windbreaks, and to protect grass-lined waterways in fields. These efforts have greatly reduced erosion, but as farm machinery gets larger and as prices rise for corn and soybeans, soil conservation measures are being abandoned in many areas. Farm policies, especially price supports for ethanol, encourage farmers to plow through grass-lined waterways, to disregard contours, and to pull out windbreaks and fencerows to accommodate the large machines and to get every last square meter into production. Pressed by economic conditions, many farmers have also abandoned traditional practices such as rotating crops or letting fields rest and go fallow from time to time.

(a) Sheet and rill erosion

(c) Wind erosion

(b) Gullying

FIGURE 10.11 Land degradation affects more than 1 billion ha yearly, or about two-thirds of all global cropland. Water erosion (a) and gullying (b) account for about half that total. Wind erosion affects a nearly equal area (c).

Source: (a) Tim McCabe, USDA Natural Resources Conservation Services; (b) Photo by Jeff Vanuga, U.S. Department of Agriculture Natural Resources Conservation Service; (c) Lynn Betts, USDA Natural Resources Conservation Service

FIGURE 10.12 Annual row crops leave soil bare and exposed to erosion by wind and water.
William P. Cunningham

Continuous monoculture cropping can increase soil loss tenfold over other farming practices. A soil study in Iowa showed that a three-year rotation of corn, wheat, and clover lost an average of about 6 metric tons per hectare. By comparison, continuous wheat production on the same land caused soil losses of more than 20 tons per hectare. Continuous corn cropping resulted in 40 tons per hectare, seven times as much soil loss as the rotation with wheat and clover.

Desertification affects arid-land soils

In dry regions, farms are especially vulnerable to land degradation and to unreliable rainfall. Both of these can lead to further dessication of soils. About one-third of the earth's surface and the livelihoods of at least one billion people are threatened by desertification (conversion of productive lands to desert), which contributes to food insecurity, famine, and poverty. Former UN secretary general Kofi Annan called this a "creeping catastrophe" that creates millions of environmental refugees every year. Forced by economic circumstances to overcultivate and overgraze their land, poor people often are both the agents and the victims of desertification.

Rangelands and pastures, which generally are too dry for cultivation, are highly susceptible to desertification. According to the UN, 80 percent of the world's grasslands are suffering from overgrazing and soil degradation, and three-quarters of that area has undergone some degree of desertification. The world's 3 billion domestic grazing animals provide livelihood and food for many people, but they can have severe environmental effects.

Areas of particular concern are the dry regions of sub-Saharan Africa and northern and western China. Arid lands, where rains are sporadic and infrequent and the economy is based mainly on crop and livestock raising, make up about two-thirds of the African continent. Nearly 400 million people live around the edges of these deserts. Rapid population growth and poverty create unsustainable pressures on the fragile soils of these areas. Stripping trees and land cover for fodder and firewood exposes the soil to erosion and triggers climate changes that spread desertification, which now affects nearly three-quarters of the arable land in Africa. The fringes of the two great African deserts, the Sahara and the Kalahari, are particularly vulnerable. Much of northern China, similarly, has little rainfall, a growing population, and increasing land degradation from overgrazing. Finding ways to reduce pressure and rebuild soils is one of the important tasks in stabilizing food security for these regions.

Irrigation is needed but can be inefficient

Agriculture accounts for the largest single share of global water use. About two-thirds of all fresh water withdrawn from rivers, lakes, and groundwater supplies is used for irrigation (chapter 17). Roughly 15 percent of all cropland is irrigated worldwide. Some regions are water rich and can readily afford to irrigate farmland; other regions are water poor and risk depleting or contaminating precious water resources through irrigation (fig. 10.13).

The efficiency of irrigation water use is low in most countries, although practices are improving in many areas as demands for water resources increase. Inefficiency starts with water delivery: In some places, evaporation and seepage from unlined canals lose 80 percent of water withdrawn for irrigation before it even reaches the field. Water is also lost to evaporation when farmers irrigate by flooding fields with water. Flood irrigation is cheap in terms of technology, and it can remove salts from desert soils, but much of the water simply evaporates in the hot summer sun. Farmers often overirrigate because water prices are low or because they lack the technology to deliver just the amount needed. Many farmers in the United States and Canada install drains under fields as well as irrigation systems, so they can both add and remove water.

Excessive water often results in **waterlogging.** Waterlogged soil is saturated with water, and plant roots die from lack of oxygen. **Salinization,** in which salts accumulate in the soil, occurs mainly when soils in dry climates are irrigated with salt-laden water. As the water evaporates, it leaves behind a salty crust on the soil surface that can be lethal to plants. Flushing with excess water can wash away this salt accumulation, but the result is even more saline water for downstream users. Extreme salinity is an important problem—for example, downstream of the fruit and vegetable growers of California's Central Valley and Imperial Valley.

Plants need nutrients, but not too much

In addition to water, sunshine, and carbon dioxide, plants need small amounts of nitrogen, potassium, phosphorus, calcium, magnesium, and other nutrients. Shortages of nitrogen, potassium,

FIGURE 10.13 Downward-facing sprinklers on this center-pivot irrigation system deliver water more efficiently than upward facing sprinklers.
cecilialim/123RF

and phosphorus, in particular, can limit plant growth. Adding these elements in fertilizer stimulates growth and increases crop yields.

Much of the doubling in worldwide crop production in recent decades has involved increased use of inorganic fertilizer, especially ammonia (NH_3). This form of nitrogen fertilizer, developed at the end of World War II by explosives manufacturers, is produced from natural gas (CH_4) and atmospheric nitrogen (N_2). Natural gas provides both the heat and the hydrogen atoms needed to produce ammonia. Since it began in 1947, production has risen to about 150 million metric tons per year. Applications of fertilizer have risen from 20 kg per hectare in 1950 to nearly 100 kg per hectare, on average, worldwide.

Fertilizer doesn't need to derive from this energy-intensive process. Manure from livestock and good soil management can produce yields as high as those from commercial fertilizers. Nitrogen-fixing crops such as beans or lentils, with nitrogen-fixing bacteria living in their roots, are especially valuable for making nitrogen available as a plant nutrient (chapter 9). Interplanting or rotating beans with nitrogen-hungry corn and wheat are traditional ways of increasing nitrogen availability.

In some regions, there is still considerable potential for increasing the world food supply by increasing fertilizer. Southern Africa, for instance, uses an average of only 19 kg of fertilizer per hectare, much less than the world average. It has been estimated that the developing world could triple its crop production by raising fertilizer use to the world average.

In many countries, though, farmers apply more fertilizer than plants can take up. They do this because they lack information about crop needs, because they fear underfertilizing, because fertilizers are relatively inexpensive, and because in most countries there is nobody whose job it is to prevent overuse. In the United States, for example, the EPA has responsibility for reducing water pollution and has worked to discourage overuse of fertilizer. North American farmers use about 60 kg of nitrogen fertilizer per hectare, whereas European farmers use about 130 kg, and Chinese farmers use nearly 200 kg. Yields with these different application rates are not significantly different, however.

Excess nutrients badly impair water quality. In the American Corn Belt, groundwater is often unsafe because of high nitrate levels. Young children are especially sensitive to the presence of nitrates. Using nitrate-contaminated water to mix infant formula can cause "blue baby syndrome," which can be fatal for newborns. The main environmental consequence of excess fertilizers is eutrification (excessive growth of algae, followed by decay and oxygen depletion). Rivers carry farmland soil and fertilizer into marine ecosystems around the world. Scores of major "dead zones," bays and seas depleted of oxygen and life, now occur around the world (see chapter 3).

The biggest dead zone in the United States is that in the Gulf of Mexico. Each summer, spring rains and snowmelt wash soil and nutrients down the Mississippi River, from Iowa, Illinois, and other farm belt states. Once these nutrients reach the Gulf, eutrophication depletes most marine life in areas as large as 57,000 km^2. Commercial and sport fisheries are severely depleted (chapter 3).

Conventional farming uses abundant fossil fuels

Farming in industrialized countries is energy-intensive. Reliance on fossil fuels began in the 1920s with the adoption of tractors, and energy use increased sharply after World War II when nitrogen fertilizer made from natural gas became available. Reliance on diesel and gasoline to run tractors, combines, and other machinery has continued to grow in recent decades. Agricultural analyst David Pimentel of Cornell University has calculated the many energy inputs, from fertilizer and pesticides to transportation and irrigation. His estimate amounts to an equivalent of 800 liters of oil (5 barrels of oil) per hectare of corn produced in the United States. A third of this energy is used in producing nitrogen fertilizer. Inputs for machinery and fuel make up another third; herbicides, irrigation, and other fertilizers make up the rest.

After crops leave the farm, additional energy is used in food processing, distribution, storage, and cooking. It has been estimated that the average food item in the American diet travels 2,400 km between the farm that grew it and the person who consumes it. The energy required for this complex processing and distribution system may be five times as much as that used directly in farming.

Contours and ground cover reduce runoff

Water runs downhill. The faster it runs, the more soil it carries off the fields. A bare field with a 5 percent slope loses eight times as much soil to erosion as a field with a 1 percent slope. **Contour plowing**—plowing across the hill rather than up and down—is one of the main strategies for controlling soil loss and water runoff. Contour plowing is often combined with strip farming, or planting different kinds of crops in alternating strips along the land contours (fig. 10.14a). The ridges created by cultivation also trap water and allow it to seep into the soil.

Terracing involves shaping the land to create level shelves of earth to hold water and soil. The edges of the terrace are planted with soil-anchoring plant species. This is an expensive procedure, requiring either much hand labor or expensive machinery, but it makes it possible to farm very steep hillsides. Rice terraces in Asia create beautiful landscapes as well as highly productive and sustainable agroecosystems (fig. 10.14b).

Annual row crops such as corn or beans generally cause the highest erosion rates because they leave soil bare for much of the year (table 10.2). On many steep lands or loose, highly erodible soils, the best way to keep soil in place is to plant perennial cover, such as grasses, rather than annual plants, which must be harvested every year. Establishing forests, orchards, grassland, or crops such as tea or coffee can minimize the need for regular cultivation.

Erosion control measures protect, or even build, soils

Careful management can build soils by improving organic matter, which improves a soil's ability to hold nutrients and moisture. Soil improvement has been practiced around the world for centuries. Many rice producing regions in Asia have been farmed continuously for a thousand years and remain fertile. Ancient civilizations in Amazon rainforests even created rich black soils in the intensely warm and wet climates (Exploring Science, p. 209). Mixed polyculture practices can also improve soils. Native Americans and pioneer farmers planted beans, which fix nitrogen, and pumpkins, which shade out weeds, along with corn plants. Traditional swidden cultivators in Africa and South America often plant as many as 20 different crops together, leaving little bare soil and providing balanced foods, while maintaining nutrient content.

Today, some farmers plant **cover crops** after harvest to hold and protect the soil, and to fix or store nitrogen at the same time. These cover crops, often rye, alfalfa, or clover, can be plowed under at planting time to provide "green manure." Keeping soil covered prevents oxidation and erosion, allowing soil carbon to accumulate. Cover crops that grow deep or complex roots are especially good at enriching soil carbon. Some agronomists have explored special varieties bred to maximize root growth for soil enrichment. Grazing management can also dramatically improve

(a)

(b)

FIGURE 10.14 Contour plowing (a) and terracing, as in these Balinese rice paddies (b), are both strategies to control erosion on farmed hillsides.

Source: (a) Photo by Lynn Betts, U.S. Department of Agriculture Natural Resources Conservation Service; (b) William P. Cunningham

soils: While overgrazing quickly kills grass and causes erosion, carefully rotating animals to different parts of a field can fertilize grass (with manure), stimulate plant regrowth, and keep livestock from overgrazing and killing forage plants.

Conservation tillage, a general term for practices that reduce plowing, has become an increasingly conventional way to conserve soil, and these practices also capture organic matter. Farmers can protect soil by leaving crop debris or standing stalks, which reduce wind and water erosion. Roots in the soil also aid water infiltration, improving moisture content and drought tolerance. In "no-till" planting, seeds are drilled into the ground directly through mulch and ground cover. This allows a cover crop to be interseeded with a subsequent crop (fig. 10.15).

Table 10.2 Soil Cover and Soil Erosion		
Cropping System	**Average Annual Soil Loss (Tons/Hectare)**	**Percent Rainfall Runoff**
Bare soil (no crop)	41.0	30
Continuous corn	19.7	29
Continuous wheat	10.1	23
Rotation: corn, wheat, clover	2.7	14
Continuous bluegrass	0.3	12

Source: Based on 14 years' data from Missouri Experiment Station, Columbia, Missouri

Ancient *Terra Preta* Shows How to Build Soils

Although it's ecologically rich, the Amazon rainforest is largely unsuitable for agriculture because of its red, acidic, nutrient-poor soils. But in many parts of the Amazon there are patches of dark, moist, nutrient-rich soils. These patches have long puzzled scientists. Locally known as *terra preta de Indio*, or "dark earth of the Indians," these patches of soil aren't associated with any particular environmental conditions or vegetation. Instead, the presence of bone fragments and pottery pieces hint that they may have a human origin.

Remote sensing surveys show that these dark earth patches, while usually rather small individually, collectively occupy somewhere between 1 and 10 percent of the Amazon. At the upper estimate, this would be about twice the size of Britain. Archaeologists now believe that these fertile soils once supported an extensive civilization of farms, fields, and even large cities in the Amazon basin for 1,000 years or more. After Europeans arrived in the sixteenth century, diseases decimated the indigenous population and cities were abandoned, but in many places the *terra preta* remains highly fertile 500 years later.

It's now believed the dark soils were created by native people who deliberately worked charcoal, human and animal manure, food waste, and plant debris into their gardens and fields. In some areas, these black soils, laced with bits of pottery, reach two meters (6 feet) in depth. Much of the dark color seems to come from charcoal that

Soils enriched by charcoal centuries ago *(left)* still remain darker and more fertile than the usual weathered, red Amazonian soils *(right)*.
Courtesy Dr. Bruno Glaser, University Bayreuth, Deutschland

has been added to the soil. Charcoal also improves the retention of nutrients, water, and other organic matter. Contrary to what scientists expected, the charcoal also seems to be beneficial for the soil-building activities of microorganisms, fungi, and other soil organisms. In short, what seems like a fairly simple practice of soil husbandry has turned extremely poor soils into highly productive gardens. Crops such as bananas, papaya, and mango are as much as three times more productive in *terra preta* than on nearby fields. And although most Amazonian soils need to be fallow for eight to ten years to rebuild nutrients after being farmed, these dark soils can recover after only six months or so.

Native people probably produced charcoal by burning biomass in low-temperature fires, in which fuel is allowed to smolder slowly in an oxygen-poor environment. Modern charcoal makers do this in an enclosed kiln. Some soil scientists are now advocating the use of charcoal, which they call "biochar," to help promote growth. But it turns out that charcoal can have another important benefit. When organic material is burned in an open fire or simply allowed to decompose in the open air, the carbon it contains is converted to CO_2 that contributes to global warming. Charcoal that is turned into the soil, on the other hand, can sequester carbon in the soil for centuries. Some of the Amazonian *terra preta* has five to ten times as much carbon as nearby soils. There's now an international movement to encourage biochar production and use, both to increase food production and to store carbon.

The use of charcoal as a soil amendment wasn't limited to the Amazon. Other places in South America, Africa, and Asia also have had similar soil management traditions, although soil scientists have only recently come to appreciate the benefits of this practice. At recent UN conventions on world food supplies, desertification, and global climate change, there have been discussions of global programs to make and distribute charcoal as a way to combat a whole series of environmental problems. It seems that the rediscovery of ancient methods may improve our soil management today.

(header photo): David L. Hansen

Conservation tillage often depends heavily on pesticides to control insects and weeds, especially if the same crops are grown year after year. Increased use of agricultural chemicals is a matter of great concern. Heavy use of pesticides is not, however, a necessary condition of reduced tilling and soil conservation. Crop rotation, trap crops, natural repellents, and biological controls are some of the many integrated pest management strategies that can combat pests and diseases without increased use of pesticides.

Carbon farming could be a key climate action

Carbon, in the form of organic matter, is a fundamental component of soils. Can carbon storage be increased to mitigate greenhouse gas emissions and climate change? Increasingly, studies show that practices that build up soil organic content, sometimes called carbon farming, are a necessary part of counteracting our greenhouse gas emissions. These practices are a double win, since increased

FIGURE 10.15 No-till planting involves drilling seeds through debris from last year's crops. Here, soybeans grow through corn mulch. Debris keeps weeds down, reduces wind and water erosion, and keeps moisture in the soil.
William P. Cunningham

organic matter makes soils richer, better able to hold nutrients and moisture for crops. Soil carbon storage has been proposed as an important aspect of **negative emissions technologies,** techniques to reduce greenhouse gas emissions or remove greenhouse gases from the atmosphere (chapter 15).

A number of well-understood strategies can enhance soil organic carbon storage (table 10.3). A 2018 study by Stephen Pacala and others found that these methods could capture 3 Gt of carbon per year globally, if implemented widely. This is nearly a third of the 10 Gt per year of carbon capture that would be needed to get us on track toward Paris Climate Agreement goals

TABLE 10.3 Examples of conventional practices that can increase soil organic carbon

Management Practice	Reduced C losses	Increased C inputs
Residue retention		X
Cover crops		X
No-tillage and other conservation tillage	X	X
Manure and compost addition		X
Convert to perennial grasses, legumes	X	X
Agroforestry	X	X
Rewetting organic (peat, wetland) soils	X	
Improved grazing land management	X	X

Source: Pacala et al. 2018 Negative Emissions Technologies and Reliable Sequestration: A Research Agenda. National Academy of Sciences. DOI 10.17226/25259

by mid-century. And the cost would be less than \$50/ton of CO_2 captured, less than one–tenth the cost of capturing CO_2 from the air or from power plant emissions.

Why don't all farmers practice carbon farming? Partly because in the complex system of growing and selling crops, most efforts focus on maximizing yields and profits. Without strong profits, farmers may be unable to repay loans for seeds, fertilizer, land, and other costs. Sometimes high crop values encourage farmers to disregard soil conservation measures. Perhaps as important, conventional practices don't emphasize carbon capture, and experimenting with new practices is risky. If we were to pay farmers for the ecosystem service of climate regulation (through carbon capture), a great many more might be eager to participate.

Section Review

1. What are three kinds of erosion?
2. How can excessive fertilizer and irrigation be a problem?
3. What are some strategies for reducing soil erosion?

10.3 Pests and Pesticides

- *Modern agriculture relies on herbicides, insecticides, and other chemical pest controls.*
- *Persistence in the environment is one reason DDT was so widely used and was so damaging.*
- *Pesticide use is growing steadily, with neonicotinoids and organophosphates among the most common.*

Every ecosystem has producers and consumers. In an agricultural system, we try to simplify the ecosystem to just one type of producer (the crop plant) and one type of consumer (humans). Other consumers that might compete with us, such as crop-eating insects or fungi, we consider "pests" that need to be controlled. Farmers and the agriculture industry spend a great deal of time and money on controlling crop pests, especially weeds, which compete with crops for space and sunlight, and insects that eat crop plants.

Pest control is an ancient practice. People in every culture have known that salt, smoke, and insect-repelling plants can keep away bothersome organisms and preserve food. The Sumerians controlled insects and mites with sulfur 5,000 years ago. Greeks and Romans used oil sprays, ash and sulfur ointments, lime, and other materials to protect themselves, their livestock, and their crops.

They also employed cover crops to reduce weeds. The Chinese developed plant-derived insecticides and introduced predatory ants in orchards to control caterpillars 1,200 years ago.

Planting practices also reduce pests and weeds. Crop rotation, for example, helps keep pest populations from getting established and spreading. Mixed cropping helps reduce concentrated outbreaks of pests. Planting cover crops helps keep weeds down (as well as enriching soil). Burning fields after harvest also reduces pests and weeds—although this has become a major source of air pollution in many areas. In addition, cultivation (hoeing and weeding) keep weeds down. Labor-intensive cultivation becomes much

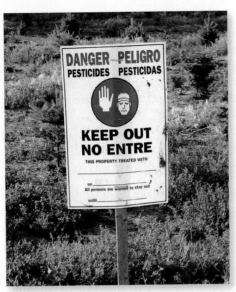

FIGURE 10.16 Broad-spectrum toxins can eliminate pests quickly and efficiently, but what are the long-term costs to us and to our environment?
William P. Cunningham

FIGURE 10.17 Before we realized the toxicity of DDT, it was sprayed freely on people to control insects, as shown here at Jones Beach, New York, in 1948.
Bettmann/Getty Images

harder as the scale of production increases from the family garden to the large-scale farm, however. So while there are important exceptions (What Do You Think? p. 212), most modern industrial farming relies heavily on pesticides.

Pesticide is a general term for a chemical that kills pests. Some pest-control compounds kill a wide range of living things and are called **biocides** (fig. 10.16). Chemicals such as ethylene dibromide, which are used to protect stored grain or to sterilize soils before planting strawberries, are biocides. In addition, there are chemicals aimed at particular groups of pests. **Herbicides** are chemicals that kill plants; **insecticides** kill insects; and **fungicides** kill fungi.

Synthetic (artificially made) chemical pesticides have been one of the main innovations of modern agricultural production. Our use of pesticides has increased dramatically in recent years, although pesticides receive relatively little public attention in most areas. Our current food system relies heavily on synthetic chemicals to control pests. These compounds have brought many benefits, but they also bring environmental health problems. Pesticide compounds vary a great deal in their characteristics and impacts, so it is useful to be familiar with some of the main types of pesticides and how they differ.

Modern pesticides provide benefits but also create health risks

The era of synthetic organic pesticides began in 1939 when Swiss chemist Paul Müller discovered the powerful insecticidal properties of dichloro-diphenyl-trichloroethane (DDT). Inexpensive, stable, easily applied, and highly effective, this compound seemed ideal for crop protection and disease prevention. DDT is remarkably lethal to a wide variety of insects but relatively nontoxic to mammals. Mass production of DDT started during World War II, when allied armies used it to protect troops from insect-borne diseases.

In less than a decade, manufacture of the compound soared from a few kilograms to thousands of metric tons per year. It was sprayed on crops and houses, dusted on people and livestock, and used to combat insects nearly everywhere (fig. 10.17).

By the 1960s, however, evidence began to accumulate that indiscriminate use of DDT and other long-lasting industrial toxins was having unexpected effects on wildlife. Peregrine falcons, bald eagles, brown pelicans, and other birds at the top of the food chain were disappearing from former territories in eastern North America. Studies revealed that eggshells were thinning in these species as DDT and its breakdown products were concentrated through food chains until they reached endocrine hormone-disrupting levels in top predators (see fig. 8.13). In 1962, biologist Rachel Carson published *Silent Spring,* warning that persistent organic pollutants, such as DDT, pose a threat to wildlife and perhaps to humans. DDT was banned for most uses in developed countries in the late 1960s, but it continues to be used in developing countries and remains the most prevalent contaminant on food imported to the United States.

Since the 1940s, many new synthetic pesticides have been invented. Many of them, like DDT, have proven to have unintended consequences on nontarget species. Assessing the relative costs and benefits of using these compounds continues to be a contentious topic, especially when unexpected complications arise, such as increasing pest resistance or damage to beneficial insects.

Information on pesticide use is often poorly reported, but the U.S. EPA estimates that world usage of conventional pesticides amounts to over 4 million metric tons per year of active ingredients (fig. 10.18). In addition "inert" ingredients are added to pesticides as carriers, stabilizers, and emulsifiers.

Roughly 80 percent of all conventional pesticides applied in the United States are used in agriculture or food storage and shipping. Some 90 million ha of crops in the United States—including 96 percent of all corn and about 95 percent of soybeans—are treated with herbicides. In addition, 25 million ha of agricultural fields and 7 million ha of parks, lawns, golf courses, and other lands are treated with insecticides and fungicides. Cotton probably has the highest rate of insecticide application of any crop.

It's important to note that most of us participate in pesticide use. Household uses in homes and gardens account for the fastest-rising sector, about 14 percent of total use, according to the most recent available EPA estimates. Three-quarters of all American homes use some type of pesticide, amounting to 20 million applications per year. Often, people use much larger quantities of chemicals in their homes, yards, or gardens than farmers would use to eradicate the same pests in their fields. While many people worry about exposure to pesticides in our food, the risk of exposure in the household may be greater.

Global use of pesticides is also hard to evaluate, but according to the UN Food and Agriculture Organization, global imports of pesticides have risen about 60-fold since data collection began in 1962 (fig. 10.18).

Organophosphates and chlorinated hydrocarbons are dominant pesticides

Pesticides vary in their chemical structure and composition. Most are organic compounds (carbon-based, such as DDT), with active components including phosphate, chlorine, or other compounds. Some are toxic metals (such as arsenic), and halogen fumigants

What Do You Think?

Shade-Grown Coffee and Cocoa

Are your purchases of coffee and chocolate contributing to the protection or destruction of tropical forests? Coffee and cocoa are examples of food products grown exclusively in developing countries but consumed almost entirely in wealthy nations (vanilla and bananas are some other examples). Coffee grows in cool, mountain areas of the tropics, while cocoa is native to the warm, moist lowlands. Both are small trees of the forest understory, adapted to low light levels.

Until a few decades ago, most of the world's coffee and cocoa were grown under a canopy of large forest trees. Recently, however, new varieties of both crops have been developed that can be grown in full sun. Yields for sun-grown crops are higher because more coffee or cocoa trees can be crowded into these fields, and they capture more solar energy than in a shaded plantation.

There are costs, however, in this new technology. Sun-grown trees die earlier from the stress and diseases common in these fields. Furthermore, ornithologists have found that the number of bird species can be cut in half in full-sun plantations, and the number of individual birds may be reduced by 90 percent. Shade-grown coffee and cocoa generally require fewer pesticides (or sometimes none) because the birds and insects residing in the forest canopy eat many of the pests. Shade-grown plantations also need less chemical fertilizer because many of the plants in these complex forests add nutrients to the soil. In addition, shade-grown crops rarely need to be irrigated because heavy leaf fall protects the soil, while forest cover reduces evaporation.

Currently, about 40 percent of the world's coffee and cocoa plantations have been converted to full-sun varieties and another 25 percent are in the process of converting. Traditional techniques for coffee and cocoa

Cocoa pods grow directly on the trunk and large branches of cocoa trees.
William P. Cunningham

production are worth preserving. Thirteen of the world's 25 biodiversity hot spots occur in coffee or cocoa regions. If all 20 million ha (49 million acres) of coffee and cocoa plantations in these areas are converted to monocultures, an incalculable number of species will be lost.

The Brazilian state of Bahia is a good example of both the ecological importance of these crops and how they might help preserve forest species. At one time, Brazil produced much of the world's cocoa, but in the early 1900s the crop was introduced into West Africa. Now Côte d'Ivoire alone grows more than 40 percent of the world total, and the value of Brazil's harvest has dropped by 90 percent. Côte d'Ivoire is aided in this competition by a labor system that reportedly includes widespread child slavery. Even adult workers in Côte d'Ivoire get only about $165 per year compared to a minimum wage of $850 per year in Brazil. As African cocoa production ratchets up, Brazilian landowners are converting their plantations to pastures or other crops.

The area of Bahia where cocoa was once king is part of Brazil's Atlantic forest, one of the most threatened forest biomes in the world. Only 8 percent of this forest remains undisturbed. Although cocoa plantations don't represent the full diversity of intact forests, they protect a surprisingly large sample of what once was there. And shade-grown cocoa can provide an economic rationale for preserving that biodiversity. Brazilian cocoa will probably never compete with that from other areas for lowest cost. There is room in the market, however, for specialty products. If consumers were willing to pay a small premium for organic, fair-trade, shade-grown chocolate and coffee, this provides an incentive to preserve biodiversity. As a consumer, what is it worth to you to help preserve biodiversity and environmentally responsible producers?

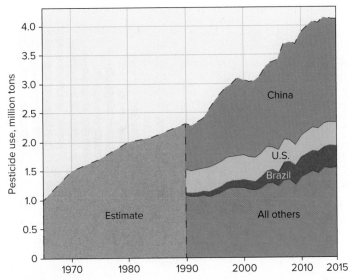

FIGURE 10.18 Global pesticide use. Standarized records exist only since 1990.
Source: UN FAO Pesticide database, 2018 (since 1990); Tilman et al. 2002 (1965–1990).

FIGURE 10.19 Usage of the top five pesticides in the United States.
Source: US EPA 2017

(gases such as bromine). Finally, there are natural organic compounds and biological agents.

Organophosphates, *organic* (carbon-based) compounds containing biologically reactive *phosphate*, are among the most heavily used synthetic pesticides. Glyphosate, the single most heavily used herbicide in the United States, is also known by the trade name Roundup. Glyphosate is applied to 90 percent of U.S. soybeans, as well as to other crops. "Roundup-ready" soybeans and corn—varieties genetically modified to tolerate glyphosate while other plants in the field are destroyed—are the most commonly planted genetically modified crops (chapter 9), and these tolerant varieties are one of the factors that make expanding soy production cost-effective in Brazil (opening case study). These "Roundup-ready" varieties have helped glyphosate surpass atrazine (an herbicide used most abundantly on corn) as the most-used herbicide (fig. 10.19).

Other organophosphates attack the nervous systems of animals, including humans. Parathion, malathion, dichlorvos, and other organophosphates were developed as an outgrowth of nerve gas research during World War II. These compounds can be extremely lethal. Because they break down quickly, usually in just a few days, they are less persistent in the environment than other pesticides. These compounds are very dangerous for workers, however, who are often sent into fields too soon after they have been sprayed (fig. 10.20).

Chlorinated hydrocarbons, also called organochlorines, are persistent and highly toxic to sensitive organisms. Atrazine was the most heavily used herbicide in the United States until the recent increase in glyphosate use (see fig. 10.19). Atrazine is applied to 96 percent of the corn crop in the United States to control weeds in cornfields (fig. 10.21). This widespread use has resulted in concerns about contamination of water supplies.

One study of Midwestern Corn Belt states found atrazine in 30 percent of community wells and 60 percent of private wells sampled. This is a worry because atrazine has been linked to sexual abnormalities and population crashes in frogs. Some studies find that atrazine is associated with higher cancer incidence and lower sperm counts in humans. Because of its persistence and uncertain health effects, atrazine was banned in Europe in 2003.

FIGURE 10.20 The United Farm Workers of America claims that 300,000 farmworkers in the United States suffer from pesticide-related illnesses each year. The World Health Organization estimates that globally 25 million people suffer from pesticide poisoning and 20,000 die each year from improper use or storage of pesticides.
David Litman/Shutterstock

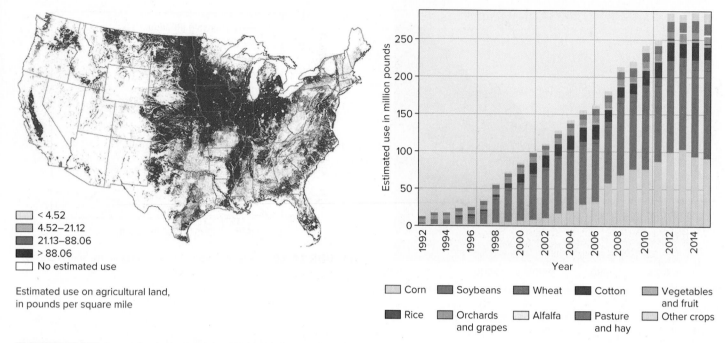

< 4.52	
4.52–21.12	
21.13–88.06	
> 88.06	
No estimated use	

Estimated use on agricultural land,
in pounds per square mile

Corn   Soybeans   Wheat   Cotton   Vegetables and fruit

Rice   Orchards and grapes   Alfalfa   Pasture and hay   Other crops

FIGURE 10.21 Distribution and growth of glyphosate (Roundup) applications. Crops genetically modified to tolerate this herbicide were introduced in the early 1990s.
Source: National Water Quality Assessment (NAWQA) project.

Among the hundreds of other organochlorines are DDT, chlordane, aldrin, dieldrin, toxaphene, and paradichlorobenzene (mothballs). Toxaphene is extremely toxic for fish and can kill goldfish at five parts per billion (5 µg/liter). This group also includes the herbicide 2,4-D, a widely used lawn chemical that selectively suppresses broad-leaf flowering plants, such as dandelions.

Chlorinated hydrocarbons can persist in the soil for decades, and they are stored in fatty tissues of organisms, so they become concentrated through food chains. DDT, which was inexpensive and widely used in the 1950s, has been banned in most developed countries, but it is still produced in the United States and is used in many developing countries.

Neonicotinoid pesticides, developed in the 1990s, are now the most widely used group of insecticides globally. Used to control insects, these compounds are powerful enough that minute concentrations can transfer to pollen and fruits. Insects pollinating these plants, notably honey bees, seem to be easily impaired by these compounds. Global disappearances of honey bee colonies, known as "colony collapse disorder," are widely blamed on neonicotinoid pesticides. Birds and aquatic organisms also appear to be affected by minute amounts of neonicotinoid pesticides.

Studies by the American Bird Conservancy have found that the EPA has done relatively little rigorous testing on these compounds, despite their dominance in the insecticide market. Because they are new, highly toxic, slow to break down, little understood, and little regulated, these pesticides worry fruit and vegetable farmers who rely on insects for pollination (fig. 10.22). Because of concerns over pollinators, the European Union imposed a trial ban on neonicotinoid pesticides in 2013. The same week, the USDA published position papers saying that diseases, not pesticides, were the primary cause of colony collapse disorder.

Fumigants are gases composed of small molecules, such as carbon tetrachloride, ethylene dibromide, and methyl bromide. These gases can be injected into soils or used to fumigate enclosed spaces. Fumigants are used to control fungus in strawberry fields and other low-growing crops, as well as to prevent decay or rodent and insect infestations in stored grain. Because these compounds are extremely dangerous for workers who apply them, many have been restricted or banned altogether in some areas.

Inorganic pesticides include compounds of toxic elements such as arsenic, sulfur, copper, and mercury. These broad-spectrum poisons are generally highly toxic and indestructible, remaining in the environment forever. They generally act as nerve toxins. Historically, arsenic powder was a primary pesticide applied to apples and other orchard crops, and traces remain in soil and groundwater in many agricultural areas.

Natural organic pesticides, or "botanicals," generally are extracted from plants. Some important examples are nicotine and nicotinoid alkaloids from tobacco, and pyrethrum, a complex of chemicals extracted from the daisy-like *Chrysanthemum cinerariaefolium* (fig. 10.23). These compounds also include turpentine,

Quantitative Reasoning

How has pesticide use changed since 1965 (fig. 10.18)? Calculate the percentage change and absolute change. What is the change in number of pesticide-resistant pest varieties from 1960 to 2010 (fig. 10.25)?

FIGURE 10.22 Honeybees, essential for much of our food production, have been disappearing, and many once-abundant bee species are now extremely rare. Neonicotinoid pesticides are thought to be largely responsible.
Daniel Prudek/Shutterstock

phenols, and other aromatic oils from conifers. All are toxic to insects, and many prevent wood decay.

Microbial agents and **biological controls** are living organisms or toxins derived from them that are used in place of pesticides. The natural soil bacterium *Bacillus thuringiensis* is one of the chief pest-control agents allowed in organic farming. This bacterium kills caterpillars and beetles by producing a toxin that ruptures the digestive tract lining when eaten. Parasitic wasps such as the tiny *Trichogramma* genus attack moth caterpillars and eggs, while lacewings and ladybugs are predators that control aphids.

Pesticides have profound environmental effects

Although we depend on pesticides for most of our food production, and for other purposes such as biofuel production, widespread use of these compounds brings a number of environmental and health risks. The most common risk is exposure of nontarget organisms. Many pesticides are sprayed broadly and destroy beneficial insects—those that pollinate crops or prey on pests—as well as pest species (fig. 10.24). The loss of insect diversity has been a growing problem in agricultural regions: At least a third of the crops we eat rely on pollinators, such as bees and other invertebrates, to reproduce.

Colony collapse disorder in honeybees has received particular attention in recent years. Many crops, including squash, tomatoes, peppers, apples, and other fruit, rely on bees for pollination. The California almond crop, for example, is worth more than $5 billion annually and is entirely dependent on bees for pollination. Honeybee colonies have been dying worldwide, sometimes with half or more disappearing in an area in a single year. While there are many possible explanations, including disease and other stresses, neonicotinoid insecticides are one of the chief suspects.

Pest resurgence, or the rebound of resistant populations, is another important problem in overuse of pesticides. This process occurs when a few resistant individuals survive pesticide treatments and those resistant individuals propagate a new pesticide-resistant population. The Worldwatch Institute reports that at least

FIGURE 10.23 Chrysanthemum flowers are a source of pyrethrum, a natural insecticide.
William P. Cunningham

FIGURE 10.24 This machine sprays insecticide on orchard trees—and everything nearby. Up to 90 percent of pesticides applied in this fashion never reach target organisms.
Balefire9/iStock/Getty Images Plus

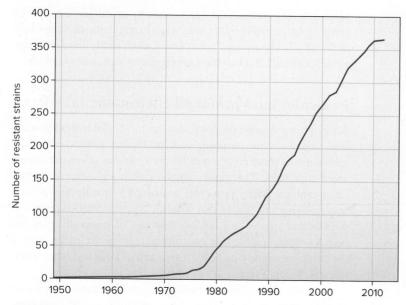

FIGURE 10.25 Pesticide-resistant varieties have increased, resulting in a "pesticide treadmill," in which farmers must use increasingly expensive combinations of chemicals to suppress weeds, disease, and pests.

Source: Data from US EPA: Weed Society of America http://www.weedscience.com

1,000 insect pest species and another 550 or so weeds and plant pathogens worldwide have developed chemical resistance. Of the 25 most serious insect pests in California, three-quarters or more are resistant to one or more insecticides. Cornell University entomologist David Pimentel reports that a larger percentage of crops are lost now to insects, diseases, and weeds than in 1944, despite the continuing increase in the use of pest controls (fig. 10.25).

As resistant pests evolve, there is an ever-increasing need for newer, better pesticides—this is called the **pesticide treadmill.** Glyphosate (Roundup), the dominant herbicide used in the United States and one of the primary herbicides in Brazil, Australia, and elsewhere, is no longer effective against a variety of superweeds. For most pesticides, it takes about 5 years before resistant weeds or insects begin to appear. More and more, farmers are advised to mix tanks of various pesticides—metachlor, Flexstar, Gramoxone, diuron, and other combinations—to keep down increasingly aggressive weeds such as pigweed and rye grass. At the same time, ever-larger amounts of glyphosate are needed to combat resistant weeds. In 2010, the U.S. Supreme Court reversed a ban on genetically modified glyphosate-tolerant alfalfa, a decision that crop scientists expect will increase pesticide-tolerant weeds on the 22 million acres of alfalfa grown in the United States. Increasing reliance on glyphosate and other herbicides is sure to increase environmental exposure, with uncertain effects on human health and ecosystems.

POPs accumulate in remote places

Many pesticides break down into less-harmful components several days or weeks after application. Certain compounds, such as DDT and other chlorinated hydrocarbons, are both effective and dangerous because they don't break down easily. **Persistent organic pollutants (POPs)** is a collective term for these chemicals, which are stable (or persistent), easily absorbed into fatty tissues (because they are fat-soluble organic compounds), and highly toxic (see chapter 8).

Because they persist for years, even decades in some cases, and move freely through air, water, and soil, they often show up far from the point of original application. Some of these compounds have been discovered far from any possible source and long after they most likely were used. Because they are easily absorbed by living cells, many chlorinated hydrocarbons are biomagnified (see fig. 8.13) and stored in the bodies of predators at the top of the food web—such as porpoises, whales, polar bears, trout, eagles, ospreys, and humans. In a study of human pesticide uptake and storage, Canadian researchers found that the level of chlorinated hydrocarbons in the breast milk of Inuit mothers living in remote arctic villages was five times that of women from Canada's industrial region some 2,500 km (1,600 mi) to the south. Inuit people have the highest levels of these persistent pollutants of any human population except those contaminated by industrial accidents.

These compounds accumulate in polar regions by what has been called the "grasshopper effect," in which contaminants evaporate from water and soil in warm areas and then condense and precipitate in colder regions. In a series of long-distance grasshopper-like jumps, they eventually collect in polar regions, where they accumulate in top predators. Polar bears, for instance, have been shown to have concentrations of certain chlorinated compounds 3 billion times greater than in the seawater around them. In Canada's St. Lawrence estuary, beluga (white whales), which suffer from a wide range of infectious diseases and tumors thought to be related to environmental toxins, have such high levels of chlorinated hydrocarbons that their carcasses must be treated as toxic waste.

Because POPs are so long-lasting and so dangerous, 127 countries agreed in 2001 to a global ban on the worst of them, including aldrin, chlordane, dieldrin, DDT, endrin, hexachlorobenzene, neptachlor, mirex, toxaphene, polychlorinated biphenyls (PCBs), dioxins, and furans. Most of this "dirty dozen" had been banned or severely restricted in developed countries for years. However, their production has continued, especially in developing countries where regulations are lax. Ironically, these pesticides may return to the United States on bananas and other imported crops. According to the 2001 Stockholm Convention on these substances, eight of the dirty dozen were banned immediately; PCBs, dioxins, and furans are being phased out; and use of DDT, still allowed for limited uses such as controlling malaria, must be publicly registered in order to permit monitoring. This treaty has been hailed as a triumph for environmental health and international cooperation. Unfortunately, other compounds—perhaps just as toxic—have been introduced to replace POPs.

FIGURE 10.26 Handling pesticides requires protective clothing and an effective respirator. Pesticide applicators in many countries can't afford safety equipment or don't fully understand risks.
Source: Photo by Tim McCabe, USDA Natural Resources Conservation Service

Pesticides often impair human health

Pesticide effects on human health can be divided into two categories: (1) acute effects, including poisoning and illnesses caused by relatively high doses and accidental exposures, and (2) chronic effects suspected to include cancer, birth defects, immunological problems, endometriosis, neurological problems, Parkinson's disease, and other chronic degenerative diseases.

The World Health Organization (WHO) estimates that 25 million people suffer pesticide poisoning and at least 20,000 die each year. At least two-thirds of this illness and death results from occupational exposures in developing countries where people use pesticides without proper warnings or protective clothing (fig. 10.26).

A tragic example of occupational pesticide exposure is found among workers in the Latin American flower industry. Fueled by the year-round demand in North America for fresh vegetables, fruits, and flowers, a booming export trade has developed in countries such as Guatemala, Colombia, Chile, and Ecuador. To meet demands in North American markets for perfect flowers, table grapes, and other produce, growers use high levels of pesticides, often spraying daily with fungicides, insecticides, and herbicides. Working in warm, poorly ventilated greenhouses with little protective clothing, the workers—70 to 80 percent of whom are women—find it hard to avoid pesticide contact. Almost two-thirds of nearly 9,000 workers surveyed in Colombia experienced blurred vision, nausea, headaches, conjunctivitis, rashes, and asthma. Although harder to document, serious chronic effects such as stillbirths, miscarriages, and neurological problems were also reported.

Rank	Food
1.	Strawberries
2.	Bell peppers
3.	Spinach
4.	Cherries (U.S.)
5.	Peaches
6.	Cantaloupe (Mexican)
7.	Celery
8.	Apples
9.	Apricots
10.	Green beans
11.	Grapes (Chilean)
12.	Cucumbers

Table 10.4 The 12 Most Contaminated Fruits and Vegetables

Source: Environmental Working Group

Pesticide use can expose consumers to agricultural chemicals. In studies of a wide range of foods collected by the USDA, the State of California, and the Consumers Union between 1994 and 2000, 73 percent of conventionally grown food had residue from at least one pesticide and were six times as likely as organic foods to contain multiple pesticide residues. Only 23 percent of the organic samples of the same groups had any residues. Using these data, the Environmental Working Group has assembled a list of the fruits and vegetables most commonly contaminated with pesticides (table 10.4).

Section Review

1. What is the difference between biocides, herbicides, insecticides, and fungicides?
2. Why was DDT considered a "magic bullet"? Why was it listed among the "dirty dozen" persistent organic pollutants (POPs)?
3. Name several of our dominant pesticides.

10.4 ORGANIC AND SUSTAINABLE AGRICULTURE

- *Proponents of conventional and sustainable farming disagree on strategies for feeding the world.*
- *"Organic" and other food labels can have multiple meanings.*
- *Sustainable strategies include IPM, crop rotation, intercropping, soil management, and other strategies.*

Industrial-scale farming produces most food in the United States and other wealthy regions, but it is important to remember that in most of the world, most food continues to be produced by smaller farms. According to the UN Development Programme, in 2015

FIGURE 10.27 The USDA finds more pesticides in commercial strawberries than in any other fruit. Organically grown berries like these are safer to eat.

William P. Cunningham

500 million family farms produced 80 percent of the food consumed worldwide. Most of these cannot afford to use many pesticides, so they are largely organic. These farms continue to feed the world despite dramatic expansion of corporate-scale producers, as described in the opening case study of this chapter.

Traditional methods require substantial labor inputs, however, in planting, weeding, and harvesting diverse crops. Where rural farming populations are declining, as they are in many urbanizing or industrialized countries, many farmers are seeking a compromise: They want to grow crops organically, protecting soil, water quality, and biodiversity, while still producing enough income to live a comfortable life. Is this compromise possible?

Global-scale trade and production like that in the Brazilian Cerrado is unlikely to work with sustainable methods, but at more moderate scales farmers are increasingly developing, or rediscovering, a multitude of strategies, to produce sustainably and to minimize use of pesticides (fig. 10.27). Planting nitrogen-fixing cover crops to avoid fertilizers, using crop rotation to minimize pesticides, strategic water management, mixed cropping, and use of perennial or tree crops are a few of these methods.

In general, soils stay healthier with these strategies than with chemical-intensive conventional farming. A long-running study in Switzerland has found that average yields on organic plots were 20 percent less than on adjacent fields farmed by conventional methods. But costs were so much lower that farmers' net income was higher with organic crops. Energy use was 56 percent less per unit of yield in organic farming than for conventional approaches. Beneficial root fungi were 40 percent higher, earthworms were three times as abundant, and spiders and other pest-eating predators were doubled in the organic plots. The organic farmers and their families reported better health and greater satisfaction than did their neighbors who used conventional farming methods.

Can sustainable practices feed the world's growing population?

The potential of sustainable farming is hotly debated. High-responding varieties and pesticides have been the foundation of commercial agriculture for over 50 years. The green revolution, and more recently the introduction of genetically modified crops, have produced staggering increases in the volume of harvests worldwide. Proponents of these input-intensive methods argue that sustainable farming is too inefficient and labor intensive to feed large populations.

Proponents of organic production argue that conventional farming has less to do with feeding people than with supporting industrial production, since after half a century of growth in commercial farming, most of the world's population continues to be fed by the small, traditional farms. Commercial crops, such as corn, soybeans, cotton, and sorghum, are used for livestock feed, which feed people indirectly but inefficiently, for biofuels, or for industrial uses (see chapter 9). Sustainable producers say their methods are more productive over time, because conventional practices degrade soils, water, ecosystems, and farm communities. The rapid spread of industrial farming, according to many food activists, serves mainly the multinational agrochemical corporations, which must constantly expand their global markets.

In 2011, the UN Commission on Human Rights and the UN Food and Agriculture Organization (FAO) both weighed in on the matter. Their studies indicate that if the aim is to provide food for impoverished regions, then developing countries should promote innovations in low-cost, low-input, sustainable soil-building and water-preserving methods. Data indicate that costs of irrigation, pesticides, fuel, and newly developed seed varieties have risen at least three times as fast as farm income in poor areas. These costs lead to high debt and widespread farm failures in poor regions. Debt drives many farmers off their land and into the already crowded cities of the developing world. In contrast, the FAO reports that areas that have invested in conservation-based farming innovations have increased yields at a rate similar to that of green revolution or genetically modified crops, while offering more sustainable food security in low-income regions.

Less than 1 percent of all American farmland is devoted to organic farming, but the market for organic products may stimulate more conversion to this approach in the future. Organic food is much more popular in Europe than in North America. Liechtenstein is probably the leader among industrialized nations with 18 percent of its modest land area in certified organic agriculture. Sweden is second with 11 percent of its land in organic production. Much of the developing world, where people can afford few fertilizers or pesticides, is effectively organic.

What does "organic" mean?

In general, organic food is grown without artificial pesticides and with only natural fertilizers, such as manure. Legal definitions of the term, though, are more exact and often more controversial. According to USDA rules, products labeled "100 percent organic" must be produced without hormones, antibiotics, pesticides, synthetic fertilizers, or genetic modification. "Organic" means that at least 95 percent of the ingredients must be organic. "Made with organic ingredients" means it must contain at least 70 percent organic contents. Products containing less than 70 percent organic ingredients can list them individually. Organic animals must be raised on organic feed, given access to the outdoors, given no steroidal growth hormones, and treated with antibiotics only to fight diseases.

Walmart has become the top seller of organic products in the United States, a step that has done much to move organic products into the mainstream. However, much of the organic food, cotton, and other products we buy from Walmart comes from overseas. Often, producers in China, India, Mexico, Thailand, and elsewhere have weak oversight compared to that in the United States. More than 2,000 farms in China and India are certified "organic," but how can we be sure what that means? With the market for organic food generating $11 billion per year, it's likely that some farmers and marketers try to pass off foods grown with pesticides as more valuable organic produce. Industrial-scale organic agriculture can also be hard on soils: It often depends on frequent cultivation for weed control, and the constant mechanical disturbance can destroy soil texture and soil microbial communities.

Many who endorse the concept of organic food are disappointed that legal definitions in the United States allow for partial organics and for unsustainable production methods. The term *organic* is also hard to evaluate clearly when you can buy organic fresh produce from half-way around the world. How do you incorporate the impacts of air transport and other inputs? Or what does it mean to buy highly processed snack foods labeled "organic"? Some farmers have declined to pay for organic certification because they regard the term as too broad to be meaningful. Often, farmers describe their operations instead as "sustainable" or "natural." Consumers can seek out local foods, both to support growers in their area and because it can be easier to find out about organic practices from local producers (fig. 10.28). Supporting local producers and farmers' markets also benefits the local community and economy.

Strategic management can reduce pests

Organic farming and sustainable farming use a multitude of practices to control pests. In many cases, improved management programs can cut pesticide use by 50 to 90 percent without reducing crop production or creating new diseases. Some of these techniques are relatively simple and save money while maintaining disease control and yielding crops with just as high quality and quantity as we get with current methods (see What Can You Do?). In this section, we will examine crop management, biological controls, and integrated pest management systems that could substitute for current pest-control methods.

Crop rotation involves growing a different crop in a field each year in a two- to six-year cycle. Most pests are specific to one crop, so rotation keeps pest populations from increasing from year to year: For example, a three-year soybean/corn/hay rotation is

FIGURE 10.28 Your local farmers' market is a good source of locally grown and organic produce.
William P. Cunningham

effective and economical protection against white-fringed weevils. Mechanical cultivation keeps weeds down, but it also increases erosion. Flooding fields before planting or burning crop residues and replanting with a cover crop can suppress both weeds and insect pests. Habitat diversification, such as restoring windbreaks,

What Can You Do?

Controlling Pests

Based on the principles of integrated pest management, the U.S. EPA releases helpful guides to pest control. Among their recommendations:

1. *Identify pests, and decide how much pest control is necessary.* Does your lawn really need to be totally weed-free? Could you tolerate some blemished fruits and vegetables? Could you replace sensitive plants with ones less sensitive to pests?

2. *Eliminate pest sources.* Remove from your house or yard any food, water, and habitat that encourages pest growth. Eliminate hiding places or other habitats. Rotate crops in your garden.

3. *Develop a weed-resistant yard.* Pay attention to your soil's pH, nutrients, texture, and organic content. Grow grass or cover varieties suited to your climate. Set realistic goals for weed control.

4. *Use biological controls.* Encourage beneficial insect predators such as birds, bats that eat insects, ladybugs, spiders, centipedes, dragonflies, wasps, and ants.

5. *Use simple manual methods.* Cultivate your garden and handpick weeds and pests from your garden. Set traps to control rats, mice, and some insects. Mulch to reduce weed growth.

6. *Use chemical pesticides carefully.* If you decide that the best solution is chemical, choose the right pesticide product, read safety warnings and handling instructions, buy the amount you need, store the product safely, and dispose of any excess properly.

Source: *Citizen's Guide to Pest Control and Pesticide Safety*: EPA 730-K-95-001 hedgerows, and ground cover on watercourses, provides habitat for insect predators, such as birds, and also reduces erosion. Adjusting the timing of planting or cultivation can help avoid pest outbreaks. Switching from vast monoculture fields to mixed polyculture (many crops grown together) makes it more difficult for pests to multiply beyond control.

Useful organisms can help us control pests

Biological controls such as predators (wasps, ladybugs, praying mantises; fig. 10.29) or pathogens (viruses, bacteria, fungi) can control many pests more cheaply and safely than broad-spectrum, synthetic chemicals. *Bacillus thuringiensis* or Bt, for example, is a naturally occurring bacterium that kills the larvae of lepidopteran (butterfly and moth) species but is generally harmless to mammals. A number of important insect pests such as tomato hornworm, corn rootworm, cabbage loopers, and others can be controlled by spraying the Bt bacterium on crops. Larger species are effective as well. Ducks, chickens, and geese, among other species, are used to rid fields of both insect pests and weeds. These biological organisms are self-reproducing and consume a wide variety of invertebrates. A few mantises or ladybugs released in your garden in the spring will keep producing offspring and protect your fruits and vegetables against a multitude of pests for the whole growing season.

Herbivorous insects have been used to control weeds. For example, the prickly pear cactus was introduced to Australia about 150 years ago as an ornamental plant. This hardy cactus escaped from gardens and found an ideal home in the dry soils of the outback. It quickly established huge, dense stands that dominated 25 million ha (more than 60 million acres) of grazing land. A natural predator from South America, the cactoblastis moth, was introduced into Australia in 1935 to combat the prickly pear. Within a few years, cactoblastis larvae had eaten so much prickly pear that the cactus has become rare and is no longer economically significant.

Some plants make natural pesticides and insect repellents. The neem tree (*Azadirachta indica*) is native to India but is now grown in many tropical countries (fig. 10.30). The leaves, bark, roots, and flowers all contain compounds that repel insects and can be used to combat a number of crop pests and diseases. Another approach is to use hormones that upset development or use sex attractants to bait traps containing toxic pesticides. To protect human health, many municipalities control mosquitoes with these techniques rather than aerial spraying of insecticides. Briquettes saturated with insect juvenile hormone are scattered in wetlands where mosquitoes breed. The presence of even minute amounts of this hormone prevent larvae from ever turning into biting adults (fig. 10.31).

FIGURE 10.30 Neem trees provide oil that is used for medicinal purposes, and the leaves, seeds, and bark provide a natural insecticide.
Dinodia Photos/Alamy Stock Photo

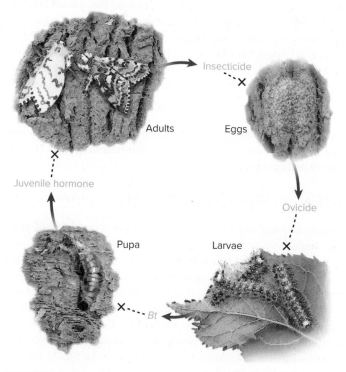

Insecticide

Adults

Eggs

Juvenile hormone

Ovicide

Pupa

Larvae

Bt

FIGURE 10.31 Different strategies can be used to control pests at various stages of their life cycles. *Bacillus thuringiensis* (Bt) kills caterpillars when they eat leaves with these bacteria on the surface. Releasing juvenile hormone in the environment prevents maturation of pupae. Predators attack at all stages.

FIGURE 10.29 The praying mantis looks ferocious and is an effective predator against garden pests, but it is harmless to humans. They can even make interesting and useful pets.
Millard H. Sharp/Science Source

Genetics and bioengineering can also help in our war against pests. Traditional farmers have long known to save seeds of disease-resistant crop plants or to breed livestock that tolerate pests well. Modern genetic methods have enhanced this process, especially by transferring Bt bacterial genes to corn, soy, and other crops. Heavy reliance on the Bt gene may dilute its effectiveness, however (chapter 9).

IPM uses a combination of techniques

Integrated pest management (IPM) is a flexible, ecologically based strategy that is applied at specific times and aimed at specific crops and pests. It often uses mechanical cultivation and techniques such as vacuuming bugs off crops as an alternative to chemical application (fig. 10.32). IPM doesn't give up chemical pest controls entirely but instead tries to use minimal amounts, only as a last resort, and avoids broad-spectrum, ecologically disruptive products. IPM relies on preventive practices that encourage growth and diversity of beneficial organisms and enhance plant defenses and vigor. Successful IPM requires careful monitoring of pest populations to determine economic thresholds, the point at which potential economic damage justifies pest-control expenditures, and the precise time, type, and method of pesticide application.

Trap crops, small areas planted a week or two earlier than the main crop, are also useful. These plots mature before the rest of the field and attract pests away from other plants. The trap crop then is sprayed heavily with pesticides so that no pests are likely to escape. The trap crop is then destroyed, and the rest of the field should be mostly free of both pests and pesticides.

IPM programs are used on a variety of crops. Massachusetts apple growers who use IPM have cut pesticide use by 43 percent in the past ten years while maintaining per-acre yields of marketable fruit equal to that of farmers who use conventional techniques. Some of the most dramatic IPM success stories come from the developing world. In Brazil, pesticide use on soybeans has been reduced up to 90 percent with IPM, while in Costa Rica, use of IPM on banana plantations has eliminated pesticides altogether in one region. In Africa, mealybugs were destroying up to 60 percent of the cassava crop (the staple food for 200 million people) before IPM was introduced in 1982. A tiny wasp that destroys mealybug eggs was discovered and now controls this pest in over 65 million ha (160 million acres) in 13 countries.

In Indonesia, rice farmers offer a successful IPM model for staple crops. There, brown planthoppers had developed resistance to virtually every insecticide and threatened the country's hard-won self-sufficiency in rice. Researchers found that farmers were spraying their fields habitually—sometimes up to three times a week—regardless of whether fields were infested. In 1986, President Suharto banned 56 of 57 pesticides previously used in Indonesia and declared a crash program to educate farmers about IPM and the dangers of pesticide use. By allowing natural predators to combat pests and spraying only when absolutely necessary with chemicals specific for planthoppers, Indonesian farmers using IPM raised yields and cut pesticide costs by 75 percent. Only two

FIGURE 10.32 This machine, nicknamed the "salad vac," vacuums bugs off crops as an alternative to treating them with toxic chemicals. *Tanimura and Antle, Inc.*

years after its initiation, the program was declared a success. It has been extended throughout the whole country. Because nearly half the people in the world depend on rice as their staple crop, this example could have important implications elsewhere (fig. 10.33).

Although IPM can be a good alternative to chemical pesticides, it also presents environmental risks in the form of exotic organisms. Wildlife biologist George Boettner of the University of Massachusetts reported that biological controls of gypsy moths, which attack fruit trees and ornamental plants, have also decimated populations of native North American moths. *Compsilura* flies, introduced in 1905 to control the gypsy moths, have a voracious appetite for other moth caterpillars as well. One of the largest North American moths, the Cecropia moth (*Hyalophora cecropia*), with a 15-cm wingspan, was once ubiquitous in the eastern United States but is now rare in regions where *Compsilura* flies were released.

Low-input agriculture aids farmers and their land

In contrast to the trend toward industrialization and dependence on chemical fertilizers, pesticides, antibiotics, and artificial growth factors common in conventional agriculture, some farmers are going back to a more natural, agroecological farming style. Finding that they can't—or don't want to—compete with factory farms, these producers are making money and staying in farming by returning to small-scale, low-input agriculture. Brent and Regina Beidler, for example, run one of nearly 200 small, organic dairies in Vermont (fig. 10.34). With just 40 cows, the Biedlers move the cattle every 12 hours, after each milking, to a different paddock. This rotation, with intensive grazing in small paddocks, method mimics the movement of herds of natural grazers. Traveling together in tight herds, the cows are forced to eat unpalatable plants as well as the sweet grasses, so that pastures stay healthy and diverse. With time to regenerate between visits, vegetation in the paddocks can remain healthy and diverse, and soil compaction is minimized.

Like other organic dairies, the Beidlers use no chemical fertilizers or pesticides on pastures. Also like other organic farmers, the Beidlers chose to go organic because it made sense for

Alternative Pest-Control Strategies

Pest management using chemicals

Pest management using biological control

Number of times insecticide used in rice season

4.5 applications

0.5 applications

Cost to farmers per hectare

7.5 rupiah

2.5 rupiah

Cost to government per hectare

27.5 rupiah

2.5 rupiah

Rice yield per hectare

6 metric tons

7.5 metric tons

FIGURE 10.33 Indonesia has one of the world's most successful integrated pest management (IPM) programs. Switching from toxic chemicals to natural pest predators has saved money while also increasing rice production.
Source: Tolba et al., World Environment, *1972–1992, p. 307; Chapman & Hall, 1992 United Nations Environment Programme*

FIGURE 10.34 Brent Beidler, of Randolph Center, VT, keeps his cows and his pasture healthy with intensive rotational grazing, without chemical pesticides or fertilizers.
Melanie Stetson Freeman/The Christian Science Monitor/Getty Images

sustainability of the land, as well as for the health of the cows and the quality of the milk produced. In an organic operation, antibiotics are used only to fight diseases. Cows are kept healthy on a diet of mainly grass and hay, with less grain and antibiotics than in a larger commercial dairy. Studies at Iowa State University have shown that animals on pasture grass rather than grain reduces nitrogen runoff by two-thirds while cutting erosion by more than half.

Similarly, the Franzens, who raise livestock on their organic farm near Alta Vista, Iowa, allow their pigs to roam in lush pastures where they can supplement their diet of corn and soybeans with grasses and legumes. Housing for these happy hogs is in spacious, open-ended hoop structures. As fresh layers of straw are added to the bedding, layers of manure beneath are composted, breaking down into odorless organic fertilizer.

Low-input farms such as these typically don't turn out the quantity of meat or milk that their intensive-agriculture neighbors do, but their production costs are lower and they get higher prices for their crops, so the all-important net gain is often higher. The Franzens, for example, calculate that they pay 30 percent less for animal feed, 70 percent less for veterinary bills, and half as much for buildings and equipment than neighboring confinement operations.

Consumers' choices play an important role

Preserving small-scale, family farms also helps preserve rural culture. As Marty Strange of the Center for Rural Affairs in Nebraska asks, "Which is better for the enrollment in rural schools, the membership of rural churches, and the fellowship of rural communities—two farms milking 1,000 cows each or twenty farms milking 100 cows each?" Family farms help keep rural towns alive by purchasing machinery at the local implement dealer, gasoline at the neighborhood filling station, and groceries at the community grocery store. Buying local can also help urban communities (What Do You Think? p. 223).

These are the arguments that lead many people to shift at least part of their diets to local foods. Eating locally can help sustain local businesses and farm communities. Most profits from

Organic Farming in the City

Farming is remote for most of us, with some 85 percent of Americans living in cities. We eat foods grown far away, processed in anonymous factories, delivered by national grocery store chains. But a growing movement has been reclaiming food production for city-dwellers. Urban farming, urban gardening, and community gardens are just a few of the names and strategies people in cities are using to bring some of their food closer to home.

One of the leading examples is Growing Power, an organization formed in Milwaukee, Wisconsin, by the former basketball player Will Allen, who received a MacArthur "genius" award for his work. Like many older industrial cities, Milwaukee has seen declining population, housing values, incomes, and economic opportunity for decades. Low-income or unemployed minority groups make up an increasing proportion of the city. Young people have few jobs, few training resources, and little food security.

Difficult conditions can make fertile ground for a movement promoting self-sustaining food production. Will Allen brought together unemployed teenagers and other community members, starting on just a 2-acre (0.8 hectare) plot of farmland. Allen's organization teaches kids and their parents to improve the soil with compost and mulch; to grow vegetables, tilapia, chickens, and other foods; and to manage a business and sell food. Growing Power serves kids by teaching them skills and providing internships and paid employment. The organization serves the community by providing a positive focus that brings people together. Who doesn't like fresh food grown by friends? The organization also serves the city by providing wholesome food resources that support the health and food security of low-income neighborhoods.

One of the first steps that members of Growing Power took was to form a land trust, an organization that could take long-term control of the land they work. This stability allows them to invest in the soil and in greenhouses, in projects, and plans. Another step they have taken is to provide workshops that spread their philosophy and techniques nationwide. Growing Power gives people access to fresh food, teaches kids about nurturing the land, and most important, it invests in the next generation of citizens of Milwaukee and other cities.

Urban farming and gardening movements are growing rapidly and have rich potential. One study found that East Lansing, Michigan, could produce 75 percent of its own vegetables on 4,800 acres (2,000 ha) of unbuilt land. Motivations for urban farming include, but are not limited to, issues of food security, community stability, youth employment, improving environmental quality for kids, and fun. Creative and enthusiastic projects abound, from Brooklyn to Detroit to Portland, Oregon, and many places between. Do you think community gardens or urban farming would be useful in areas where you live? Can you support these efforts in your area?

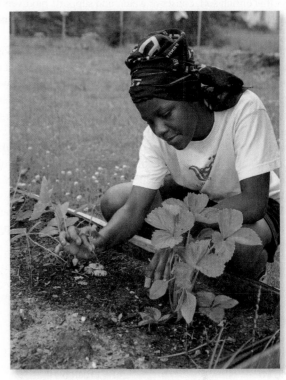

Urban farming helps young people and their communities grow stronger. The Capuchin Soup Kitchen/Earthworks Urban Farm in Detroit, Michigan, grows and supplies fresh food to the community.
Jim West/imageBROKER/Shutterstock

conventional foods, in contrast, go to a tiny number of giant food corporations: The top three or four corporations in each commodity group typically control 60 to 80 percent of the U.S. market. Where conventional foods were shipped an average of 2,400 km (1,500 mi) to markets, the average food item at a farmers' market traveled only 72 km (45 miles). Food from local, small-scale farms also often involves less energy for fertilizer, fuel for shipping, and plastic food packaging.

Many co-ops carry food that is locally grown and processed. An even better way to know where your food comes from and how it's produced is to join a community-supported agriculture (CSA) farm. In return for an annual contribution to a local CSA farm, you'll receive a weekly share of whatever the farm produces. CSA farms generally practice organic or low-input agriculture, and many of them invite members to visit and learn how their food is grown. Much of America's most fertile land is around major cities, and CSAs and farmers' markets are one way to help preserve these landscapes around metropolitan areas.

Section Review

1. List several arguments for and against sustainable farming.
2. Why is it controversial that Walmart is the largest seller of organic products?
3. What is IPM, and how is it used in pest control?

Connecting the Dots

Agriculture is necessary for life as we know it, but it is also responsible for many of the world's most devastating environmental changes. The transformation of Brazil's Cerrado, its Atlantic forests, and increasingly the Amazon rainforests all involve expansion of industrial-scale crop production. This production increases global food supplies, especially meat. Agriculture also holds exciting opportunities for innovation, and many farmers cultivate their land in ways that enrich soil, conserve water, and supports biodiversity as well as community. These producers use diverse strategies, including pest control with crop rotation, biological controls, cover crops, and integrated pest management, involving monitoring pest populations and using small, targeted applications of pesticides. What is the best policy for managing agricultural lands and feeding the world? These are difficult questions that involve many value judgements.

The first major input required for agriculture is soil—a complex and living ecosystem. Soils vary dramatically in character. Temperate grassland soils, like those in the American Midwest, have some of the best rich, black organic content. Tropical soils like those in Brazil are generally nutrient-poor and acidic, although archeologists have discovered ancient black cultivated soils. Inputs of lime and fertilizer, along with specialized soybean varieties, have been among the methods that make farming the Cerrado possible.

In many areas soil management is done unsustainably, in ways that allow severe rates of erosion. Most of the world's farmland soils are degraded or eroded. But soil can also be protected and enhanced, through cover crops, terracing, crop rotation, and other practices. Water is a second major input. Irrigation, like soil, often uses water extravagantly, especially when society sets water prices very low.

Pesticides, an important part of production on modern farms, are increasing dramatically in use. They bring both social benefits and environmental costs. In particular, pesticides harm nontarget organisms, such as bees, butterflies, and other pollinators. Our most abundantly used agricultural chemicals are organophosphates, including glyphosate, and organochlorines, such as atrazine. Glyphosate and atrazine are applied to nearly all soy and corn produced in the United States. Global consumption of these and similar agricultural chemicals continues to grow (although household use is the fastest-growing sector of pesticide applications). Production on the scale of that seen in Brazil, for example depends on glyphosate, among other inputs.

In short, agriculture has globally important value and impacts. As a consumer of food, you can consider how you would like to participate in and support these different farming systems.

Reviewing Key Terms

Can you define the following terms in environmental science?

biocide 10.3

biological controls 10.3

chlorinated hydrocarbons 10.3

contour plowing 10.2

cover crops 10.2

desertification 10.2

fumigants 10.3

fungicide 10.3

gully erosion 10.2

herbicide 10.3

inorganic pesticides 10.3

insecticide 10.3

integrated pest management (IPM) 10.4

microbial agents 10.3

mycorrhizal symbiosis 10.1

natural organic pesticides 10.3

negative emissions technologies 10.2

neonicotinoid 10.3

organophosphates 10.3

persistent organic pollutants (POPs) 10.3

pest resurgence 10.3

pesticide 10.3

pesticide treadmill 10.3

rill erosion 10.2

salinization 10.2

sheet erosion 10.2

subsoil 10.1

terracing 10.2

topsoil 10.1

waterlogging 10.2

Critical Thinking and Discussion Questions

1. As you consider the expansion of soybean farming and grazing in Brazil, what are the costs and what are the benefits of these changes? How would you weigh these costs and benefits for Brazilians? If you were a U.S. ambassador to Brazil, how would you advise Brazilians on their farming and grazing policies, and what factors would shape your advice?

2. The discoverer of DDT, Paul Müller, received a Nobel Prize for his work. Would you have given him this prize?

3. Are there steps you could take to minimize your exposure to pesticides, either in things you buy or in your household? What would influence your decision to use household pesticides or not to use them?

4. What criteria should be used to determine whether farmers should use ecologically sound techniques? How would your response differ if you were a farmer, a farmer's neighbor, someone downstream of a farm, or someone far from farming regions?

5. Should we try to increase food production on existing farmland, or should we sacrifice other lands to increase farming areas? Why?

6. Some rice paddies in Southeast Asia have been cultivated continuously for a thousand years or more without losing fertility. Could we, and should we, adapt these techniques to our own country? Why or why not?

7. *Terra preta* soils were a conundrum for soil scientists for decades. What expectations about tropical soils did these black soils violate? What would it take to make similar investments in soils today?

Data Analysis

Graphing Changes in Pesticide Use

The National Agricultural Statistics Service (NASS) keeps records of pesticide use in the United States, and you can access those records by going to http://www.pestmanagement.info/nass/ to see the kinds of data, and the date range, maintained by the U.S. Dapartment of Agriculture. This data source gives insights into changes in agricultural chemical use.

Go to Connect to graph and examine pesticide data in Excel, and to test your understanding of pesticide use.

Soybeans-All % of Acres Applied

Soybeans-All Application (Active Ingredients, 1,000 lb)

Linear (Soybeans-All % of Acres Applied)

Linear (Soybeans-All Application (Active Ingredients, 1,000 lb))

Plot trends in pesticide use using data provided in the Excel file on Connect.

11

Biodiversity
Preserving Species

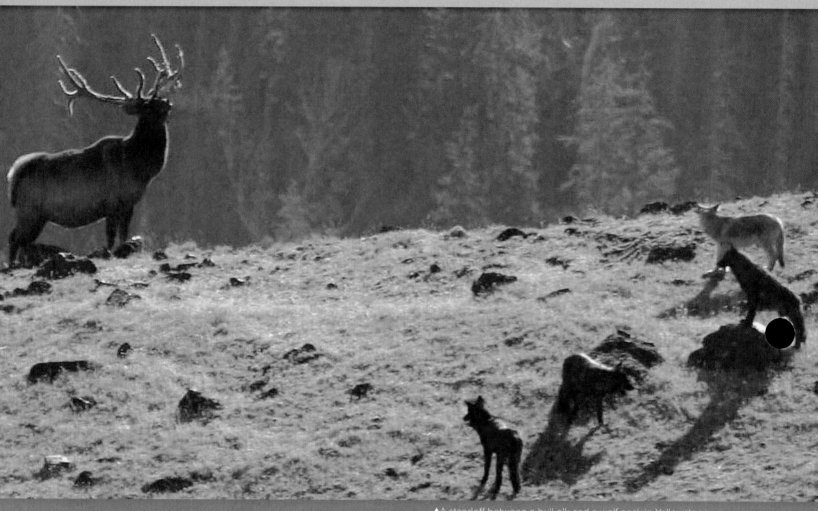

Learning Outcomes

▲A standoff between a bull elk and a wolf pack in Yellowstone National Park. Reintroducing a top predator stabilizes prey populations and revitalizes the ecosystem.
Source: NPS photo by Doug Smith

After studying this chapter, you should be able to:

11.1 Discuss biodiversity and the species concept, as well as some of the ways we benefit from biodiversity.

11.2 Characterize the threats to biodiversity.

11.3 Evaluate endangered species management.

11.4 Assess captive breeding and species survival plans.

"The first rule of intelligent tinkering is to save all the pieces."

– Aldo Leopold

How Wolves Can Change Rivers

How much does a top predator alter an ecosystem? The reintroduction of wolves in Yellowstone National Park nearly 25 years ago provides an interesting insight into this question. Yellowstone, the world's first national park, has long been renowned for its wildlife and natural wonders. More recently it has become a natural laboratory for studying the impacts of predator restoration.

Wolves were once abundant throughout the United States. In the nineteenth century it was thought that perhaps 100,000 wolves roamed the western forests and prairies. But as settlers moved west, they shot, trapped, and poisoned the predators they regarded as threats to themselves and their livestock. By the beginning of the twentieth century, wolves had been extirpated from almost all of their former range in the contiguous 48 states. In the absence of top predators, populations of elk, deer, moose, and other grazers began to expand rapidly. By the 1930s, Yellowstone National Park managers noticed that growing elk populations were over browsing willow and aspen. River valleys were becoming barren pastures, and populations of smaller mammals, such as beaver and fox, were dwindling.

Wildlife ecologists argued that the ecosystem needed wolves to restore a natural balance to the ecosystem. After decades of controversy, 31 wolves were captured in the Canadian Rockies and relocated to Yellowstone National Park in 1995 and 1996 (fig. 11.1). Wolves expanded quickly in their new home. With abundant elk for prey, as well as occasional moose and bison, the wolves tripled their number in just three years. There are now between 100 and 150 animals in 11 packs, meeting the recovery goals.

Following the wolf reintroductions, something remarkable happened. Elk numbers have fallen rapidly, as wildlife biologists expected. Where there were once more than 25,000 elk in the park, the number is now below 5,000 animals, but more importantly the behavior of the elk has changed dramatically. Where the elk had once spent most of their time grazing in the river bottoms, they now hang out on the ridges where they can watch for the wolves and move quickly exit if threatened.

Once-barren valley floors have regrown forests of aspen, willows, and cottonwoods. Many species of birds have returned to the park. The wolves have also killed coyotes, allowing populations of mice, voles, rabbits, weasels, and badgers to rebound.

With abundant vegetation along streams, beaver have returned, and the dams they build create ponds that support otters, muskrats, fish, turtles, amphibians, and ducks. Hawks feed on the small mammals, while ravens, eagles, and bears feast on the carrion left behind by the wolf packs. An ecological term for this is a trophic cascade: where the return of a top predator affects creatures in all the trophic levels below it. Another emerging concept is the "ecology of fear," in which elk avoid places where they might be ambushed by hunting wolves.

Perhaps most surprising is that the streams themselves are changing in response to the presence of wolves. Before wolf reintroduction, elk-grazed stream banks were bare and eroding. Sediment was filling streams and making them shallow, slow, warm, and silty. As regenerating vegetation has stabilized stream banks, channels have become deeper and clearer. Fast-running, cleaner streams now provide better habitat for endangered cutthroat trout. The wolves have changed the physical structure of the river.

The success of wolf reintroduction in Yellowstone has triggered talk of returning top predators to many places where they were extirpated. This remains controversial as some still regard wolves and other predators as dangerous and menacing something to be overcome and eliminated rather than encouraged. But this case study shows how a complete and diverse biological community contributes to the health and stability of an entire ecosystem. In this chapter we examine this and other aspects of biodiversity, what it means, and why it matters.

FIGURE 11.1 Biologists carry an anesthetized wolf into a pen prior to reintroduction.
Source: NPS photo by Barry O'Neill

11.1 BIODIVERSITY AND THE SPECIES CONCEPT

- *Biodiversity includes genetic, species, and ecological variety.*
- *A few "hot spots," mostly in the tropics, are a high conservation priority because they have both extremely high biodiversity and a high risk of disruption by humans.*
- *We benefit from other organisms in many ways.*

From the driest desert to the dripping rainforests, from the highest mountain peaks to the deepest ocean trenches, life on earth occurs in a marvelous spectrum of sizes, colors, shapes, life cycles, and interrelationships. Think for a moment how remarkable, varied, abundant, and important are the other living creatures with whom we share this planet (fig. 11.2). How will our lives be impoverished if this biological diversity diminishes?

What is biodiversity?

Previous chapters of this book have described some of the fascinating varieties of organisms and complex ecological relationships that give the biosphere its unique, productive characteristics. Three kinds of **biodiversity** are essential to preserve these ecological systems: (1) *genetic diversity* is a measure of the variety of different versions of the same genes within individual species; (2) *species diversity* describes the number of different kinds of organisms within individual communities or ecosystems; and (3) *ecological diversity* assesses the richness and complexity of a biological community, including the number of niches, trophic levels, and ecological processes that capture energy, sustain food webs, and recycle materials within this system.

FIGURE 11.2 This coral reef has both high abundance of some species and high diversity of different genera. What will be lost if this biologically rich community is destroyed?
Georgette Douwma/Getty Images

Within species diversity, we can distinguish between *species richness* (the total number of species in a community) and *species evenness* (the relative abundance of individuals within each species). Often, species evenness is considered more informative. Species richness could be high, for example, in a lawn that was 99 percent of one grass species and just a few individuals of nine other species. The total species count would be 10, but most people would not consider the lawn very diverse. A meadow containing even representation of 10 species would appear more diverse, and it would function differently, perhaps producing a greater variety of food sources or habitat for birds and butterflies.

Species are defined in different ways

As you can see, the species concept is fundamental in defining biodiversity, but what exactly do we mean by "species"? When Carl Linnaeus, the great Swedish taxonomist, created our system of scientific nomenclature in 1735, species classification was based entirely on the physical appearance of adult organisms. In recent years, taxonomists have introduced other characteristics as a means of differentiating species. As noted in chapter 3, species have often been defined in terms of the ability to breed and produce fertile offspring in nature. This *biological species concept* is based on the general idea of *reproductive isolation*—that is, organisms may be unable to breed because of physical characteristics, location, habitat, or even differing courtship behaviors. This definition has some serious weaknesses, especially among plants and protists, many of which either reproduce asexually or regularly make fertile hybrids.

Another definition favored by many evolutionary biologists and paleontologists is the *phylogenetic species concept*, which emphasizes the branching descent from a common ancestor, based on physical traits, regardless of whether organisms can interbreed successfully.

A third definition, favored by some conservation biologists, is the *evolutionary species concept*, which defines a species as a lineage of organisms that share a common ancestor and interbreed. The advantage of this definition is that it recognizes that there can be several "evolutionarily significant" populations within a genetically related group of organisms. Unfortunately, we rarely

Benchmark Data	
Among the ideas and values in this chapter, the following are a few worth remembering.	
5,560	Number of mammal species
22%	Mammal species endangered
100,000	Estimated number of insect species
$156 billion	Amount spent per year in the United States on bird and wildlife viewing and photography.
80%	Area of remaining sage grouse habitat opened for oil and gas drilling by Trump administration.

have enough information about a population to judge what its evolutionary importance or fate may be. Ecologists Paul Ehrlich and Gretchen Daily calculated that, on average, there are about 220 evolutionarily significant populations per species. This calculation could mean that there are up to 10 billion different populations in total. Deciding which ones to protect becomes an even more daunting prospect.

Molecular techniques are rewriting taxonomy

Increasingly, DNA sequencing and other molecular techniques are giving us insights into taxonomic and evolutionary relationships. Every individual has a unique hereditary complement called the *genome*. The genome is made up of the millions or billions of nucleotides in its DNA (see chapter 3). The specific sequence of these nucleotides spells out the structure of all the proteins that make up the cellular composition and machinery of every organism. As you know from modern court cases and paternity suits, we can use that DNA sequence to identify individuals with a high degree of certainty. Now this very precise technology is being applied to identify species in nature.

Because only a small amount of tissue is needed for DNA analysis, species classification—or even the identity of individual animals—can be based on samples such as feathers, fur, or feces when it's impossible to capture living creatures. For example, DNA analysis has shown that whale meat for sale in Japanese markets was from protected species. In North America, biologists can do genetic studies of elusive lynx and bears just by collecting samples of hair from scratching pads placed in the forest. This saves the expense and trauma of capturing the animals. Similarly, biologists identified a new tiger subspecies (*Tigris panthera jacksoni*) in Southeast Asia based on blood, skin, and fur samples from zoo and museum specimens, although no members of this subspecies are known to still exist in the wild (fig. 11.3).

This new technology can help resolve taxonomic uncertainties in conservation. In some cases, an apparently widespread and low-risk species may, in reality, comprise a complex of distinct species, some rare or endangered. Such is the case for a unique New Zealand reptile, the tuatara. Genetic marker studies revealed two distinct species, one of which needed additional protection. Similar studies have shown that the northern spotted owl (*Strix occidentalis caurina*) is a genetically distinct subspecies from its close relatives, the California spotted owl (*S. occidentalis occidentalis*) and the Mexican spotted owl (*S. occidentalis lucida*), and therefore deserves continued protection.

On the other hand, in some cases genetic analysis shows that a protected population is closely related to another much more abundant one. For example, the colonial pocket gopher from Georgia is genetically identical to the common pocket gopher and may not deserve endangered status.

In some cases, molecular taxonomy is causing a revision of the basic phylogenetic ideas of how we think evolution proceeded. Studies of corals and other cnidarians (jellyfish and sea anemones), for example, show that they share more genes with primates than do worms and insects. This evidence suggests a branching of the family

FIGURE 11.3 DNA analysis revealed a new tiger subspecies (*T. panthera jacksoni*) in Malaysia. This technology has become essential in conservation biology.
Royalty-Free/Corbis

tree very early in evolution rather than a single sequence from lower to higher animals.

How many species are there?

At the end of the great exploration era of the nineteenth century, some scientists confidently declared that every important kind of living thing on earth would soon be found and named. Most of those explorations focused on charismatic species such as birds and mammals. Recent studies of less conspicuous organisms, such as insects and fungi, suggest that millions of new species and varieties remain to be studied scientifically.

The 1.7 million species presently known (table 11.1) represent only a small fraction of the total number that exist. A firm number is hard to determine, since species classifications change over time, and many groups are poorly known. Taxonomists disagree on how narrowly distinctions between species should be defined. Inconspicuous species such as bacteria, algae, protists, and archaea are especially difficult to detect and identify. Taxonomists estimate, though, that there are around 8 to 11 million species, including

Quantitative Reasoning

Compare the estimates of known and threatened species in table 11.1. Are some groups overrepresented? Are we simply more interested in some organisms, or are we really a greater threat to some species?

Table 11.1 Estimates of Known and Threatened Species by Taxonomic Group

	Known	Threatened	Percentage
Mammals	5,560	1,220	22
Birds	11,121	1,492	13
Reptiles	10,450	1,367	18
Amphibians	7,635	2,157	32
Fish	33,500	2,494	14
Insects	1,000,000	1,578	19
Molluscs	85,000	2,231	26
Crustaceans	47,000	733	23
Other animals	173,250	577	28
Mosses	16,236	162	58
Ferns and allies	12,000	252	42
Gymnosperms	1,052	400	40
Flowering plants	268,000	13,536	43
Fungi, lichens, protists	65,434	110	69

Source: Known species from A. D. Chapman, 2009; threatened species (number and percentage) from IUCN Red List 2019.

animals, plants, bacteria, archaea, and other groups. The number of *known* species is probably less than 20 percent of the species that exist. Of marine species, taxonomists estimate we have described less than 10 percent.

Of known animal species, about 72 percent are invertebrates (insects, sponges, clams, worms, and other species without backbones). About 16 percent are plants. Fungi, lichens, and protists, and others add up to around 9 percent, and animals with backbones are a little over 3 percent.

The numbers of endangered species shown in table 11.1 are those officially listed by the International Union for Conservation of Nature and Natural Resources (IUCN). The number of species understood to be threatened or vulnerable to extinction has risen as biologists have had time to assess more species. Still, the actual number of threatened species is likely to be much greater than we yet know.

Hot spots have exceptionally high biodiversity

Of all the world's currently identified species, only 10 to 15 percent live in North America and Europe. The greatest concentration of different organisms tends to be in the tropics, especially in tropical rainforests and coral reefs. Norman Myers, Russell Mittermeier, and others have identified **biodiversity hot spots** that contain at least 1,500 *endemics* (species that occur nowhere else) and have lost at least 70 percent of their habitat owing to, for example, deforestation or invasive species. Using plants and

land-based vertebrates as indicators, they have proposed 34 hot spots that are a high priority for conservation because they have both high biodiversity and a high risk of disruption by human activities (fig. 11.4). Although these hot spots occupy only 1.4 percent of the world's land area, they house three-quarters of the world's most threatened mammals, birds, and amphibians. The hot spots also account for about half of all known higher plant species and 42 percent of all terrestrial vertebrate species. The hottest of these hot spots tend to be tropical islands, such as Madagascar, Indonesia, and the Philippines, where geographic isolation has resulted in large numbers of unique plants and animals. Special climatic conditions, such as those found in South Africa, California, and the Mediterranean Basin, also produce highly distinctive flora and fauna.

Some areas with high biodiversity—such as Amazonia, New Guinea, and the Congo basin—aren't included in this hot spot map because most of their land area is relatively undisturbed. Other groups prefer different criteria for identifying important conservation areas. Aquatic biologists, for example, point out that coral reefs, estuaries, and marine shoals host some of the most diverse wildlife communities in the world, and warn that freshwater species are more highly endangered than terrestrial ones. Other scientists worry that the hot spot approach neglects many rare species and major groups that live in less biologically rich areas (cold spots). Nearly half of all terrestrial vertebrates, after all, aren't represented in Myers's hot spots. Focusing on a few hot spots also doesn't recognize the importance of certain species and ecosystems to human beings. Wetlands, for instance, may contain just a few, common plant species, but they perform valuable ecological services, such as filtering water, regulating floods, and serving as nurseries for fish. Some conservationists argue that we should concentrate on saving important biological communities or landscapes rather than rare species.

Anthropologists point out that many of the regions with high biodiversity are also home to high cultural diversity as well. It isn't a precise correlation; some countries, like Madagascar, New Zealand, and Cuba, with a high percentage of endemic species, have only a few cultural groups. Often, however, the varied habitat and high biological productivity of places like Indonesia, New Guinea, and the Philippines that allow extensive species specialization also have fostered great cultural variety. By preserving some of the 7,200 recognized language groups in the world—more than half of which are projected to disappear in this century—we might also protect some of the natural settings in which those cultures evolved.

Quantitative Reasoning

Compare the "hot spot" map in figure 11.4 with the biomes map in figure 5.4. Which of the "hot spots" has the largest number of endemic species? Which has the least? Can you detect any patterns when you compare these two maps?

FIGURE 11.4 Biodiversity "hot spots," identified by Conservation International, tend to be in tropical or Mediterranean climates and on islands, coastlines, or mountains where many habitats exist and physical barriers encourage speciation. Numbers indicate endemic species.
Source: Data from Conservation International

We benefit from biodiversity in many ways

We benefit from other organisms in many ways, and seemingly obscure and insignificant organisms can play irreplaceable roles in ecological systems, or they may be a source of genes or drugs that someday may be indispensable.

Wild plant species could make important contributions to human food supplies, either as new crops or as a source of genetic material to provide disease resistance or other desirable traits to current domestic crops. Ecologist Norman Myers estimates that as many as 80,000 edible wild plant species could be utilized by humans. Villagers in Indonesia, for instance, are thought to use some 4,000 native plant and animal species for food, medicine, and other valuable products. Few of these species have been explored for possible domestication or more widespread cultivation. A 1975 study by the National Academy of Science (U.S.) found that Indonesia has 250 edible fruits, only 43 of which have been cultivated widely (fig. 11.5).

More than half of all modern medicines are either derived from or modeled on natural compounds from wild species (table 11.2). The United Nations Development Programme estimates the value of pharmaceutical products derived from developing world plants, animals, and microbes to be more than $30 billion per year. Indigenous communities that have protected and nurtured the biodiversity on which these products are based are rarely acknowledged—much less compensated—for the resources extracted from them. Many consider this expropriation "biopiracy," or theft of living things, and there are calls for royalties to be paid for folk knowledge and natural assets.

Consider the success story of vinblastine and vincristine. These anticancer alkaloids are derived from the Madagascar periwinkle (*Catharanthus roseus*) (fig. 11.6). They inhibit the growth

FIGURE 11.5 Mangosteens have been called the world's best-tasting fruit, but they are practically unknown beyond the tropical countries where they grow naturally. There may be thousands of other traditional crops and wild food resources that could be equally valuable but are threatened by extinction.
William P. Cunningham

Table 11.2	Some Natural Medicinal Products	
Product	**Source**	**Use**
Penicillin	Fungus	Antibiotic
Bacitracin	Bacterium	Antibiotic
Tetracycline	Bacterium	Antibiotic
Erythromycin	Bacterium	Antibiotic
Digitalis	Foxglove	Heart stimulant
Quinine	Chincona bark	Malaria treatment
Diosgenin	Mexican yam	Birth-control drug
Cortisone	Mexican yam	Anti-inflammation treatment
Cytarabine	Sponge	Leukemia cure
Vinblastine, vincristine	Periwinkle plant	Anticancer drugs
Reserpine	Rauwolfia	Hypertension drug
Bee venom	Bee	Arthritis relief
Allantoin	Blowfly larva	Wound healer
Morphine	Poppy	Analgesic

of cancer cells and are very effective in treating certain kinds of cancer. Before these drugs were introduced, childhood leukemias were invariably fatal. Now the remission rate for some childhood leukemias is 99 percent. Hodgkin's disease was 98 percent fatal a few years ago, but is now only 40 percent fatal, thanks to these compounds. The total value of the periwinkle crop is roughly $150 million to $300 million per year, although Madagascar gets little of those profits.

FIGURE 11.6 The rosy periwinkle from Madagascar provides anticancer drugs that now make childhood leukemias and Hodgkin's disease highly remissible.
William P. Cunningham

Pharmaceutical companies are actively prospecting for useful products in many tropical countries. The UN Convention on Biodiversity calls for a more equitable sharing of the gains from exploiting nature between rich and poor nations. Bioprospectors who discover useful genes or biomolecules in native species will be required to share profits with the countries where those species originate. This is not only a question of fairness; it also provides an incentive to poor nations to protect their natural heritage.

Biodiversity provides ecological services and aesthetic and cultural benefits

Human life depends on ecological services provided by other organisms. Bacteria and soil microbes form soil, dispose of waste, purify water, and recycle nutrients. Trees moderate our climate, and plants in healthy ecosystems capture water, build soil, and turn sunlight into nutrients. The total value of these ecological services is at least $173 trillion per year (see chapter 23) or about twice total world GNP.

Ecologists have argued that biodiversity contributes to ecosystem stability, although proof of this is elusive. It seems intuitively obvious that having more kinds of organisms would make a community better able to withstand or recover from disturbance, but few empirical studies show an unequivocal relationship. A landmark long-term study of diversity in prairie plots by David Tilman and his associates found that more diverse plots were more resilient, or more able to recover after a disturbance or drought. Other experiments, however, have found less clear correlation between diversity and stability.

Because we don't fully understand the complex interrelationships among organisms, we often are surprised at the effects of removing seemingly insignificant members of biological communities. For instance, wild insects provide a valuable but often unrecognized service in suppressing pests and disease-carrying organisms. It is estimated that 95 percent of the potential pests and disease-carrying organisms in the world are controlled by other species, which keep us safe by preying on or competing with potential pests. Unsuccessful efforts to control pests with synthetic chemicals (chapter 10) have shown that biodiversity provides essential pest-control services.

Biodiversity also contributes to human well-being by providing opportunities for fishing, wildlife watching, hunting, bird-watching, and other forms of recreation. These activities keep us healthy by providing invigorating physical exercise. Contact with nature can be psychologically and emotionally restorative. In some cultures, nature carries spiritual connotations, and a particular species or landscape may be inextricably linked to a sense of identity and meaning. Many moral philosophies and religious traditions hold that we have an ethical responsibility to care for creation and to save "all the pieces" as far as we are able.

Nature appreciation is economically important. The U.S. Fish and Wildlife Service estimates that Americans spend 156 billion annually on nature-based activities (fig. 11.7). This

FIGURE 11.7 The USFWS estimates that nearly one-third of all Americans participate in bird and wildlife watching and photography, spending more than $156 billion in these activities.
William P. Cunningham

is about as much as is spent every year on alcohol. Roughly one-third of all Americans enjoy wildlife, including those who watch, feed, or photograph wildlife. Ecotourism (tourism focused on experiencing wild places or viewing wildlife) can be a good form of sustainable economic development and environmental education.

For many people, the value of wildlife goes beyond the opportunity to hunt, photograph, or simply view a particular species. They argue that **existence value,** based on simply knowing that a species exists, is reason enough to protect and preserve it. We contribute to programs to save bald eagles, redwood trees, whooping cranes, whales, and a host of other rare and endangered organisms because we like to know they still exist somewhere, even if we may never have an opportunity to see them.

Section Review

1. Which group in table 11.1 has the most known species? Which has the highest proportion of endangered species?
2. What are the criteria for defining a biodiversity hot spot?
3. Describe four benefits we obtain from biodiversity.

11.2 WHAT THREATENS BIODIVERSITY?

- *Extinction rates are far higher now than in the past.*
- *Habitat destruction, climate change, invasive species, pollution, human population growth, and overharvesting are the greatest threats to biodiversity.*
- *Islands and specialized habitats are particularly susceptible to invasive species.*

Extinction, the elimination of a species, is a normal process of the natural world. Species die out and are replaced by others, often their own descendants, as part of evolutionary change. Over time, the average rate of extinction appears to be about one species lost every decade. In this century, however, human impacts on populations and ecosystems have accelerated that rate, causing hundreds or perhaps even thousands of species, subspecies, and varieties to become extinct every year. If present trends continue, we may destroy untold numbers of plants, animals, and microbes in the next few decades. In this section, we will look at some ways we threaten biodiversity.

Mass extinctions appear in the fossil record

Studies of the fossil record suggest that more than 99 percent of all species that ever existed are now extinct. Most of those species were gone long before humans came on the scene. Species arise through processes of mutation and natural selection, and they disappear the same way (chapter 4). Often, new forms replace their own parents. The tiny *Hypohippus,* for instance, has been replaced by the much larger modern horse, but most of its genes probably still survive in its distant offspring.

Periodically, mass extinctions have wiped out vast numbers of species and even whole families (fig. 11.8). Of the five major mass extinctions, the best known is most recent, at the end of the Cretaceous period 63 million years ago, which wiped out the dinosaurs. An asteroid impact is thought to have created enormous heat on impact extinguishing plant life and the animals that depended on it. Volcanic emissions and runaway climate warming may have contributed to the loss of 76 percent of genera.

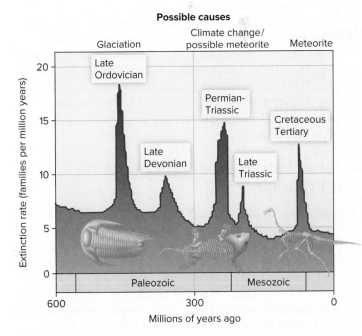

FIGURE 11.8 Major mass extinctions through history. We may be in a sixth mass extinction now, caused by human activities.

The extinction event at the end of the Permian period, 250 million years ago, eliminated 95 percent of genera in the oceans and nearly half of all land plant and animal families in just 10,000 years. Geologists believe that high concentrations of atmospheric CO_2, followed by extreme warming and ocean acidification, resulted from massive volcanic activity in Siberia. Climate change probably contributed to most mass extinctions (see fig. 14.6).

Are we entering a sixth extinction?

The rate at which species are disappearing appears to have increased dramatically over the last 150 years. It appears that from about 1600 to 1850, human activities caused extermination of two or three species per decade. By some estimates, we are now losing species at hundreds or even thousands of times natural rates. Many biologists now say we are entering a sixth mass extinction, caused this time by both climate change and habitat loss. In addition to converting landscapes to agriculture and other uses, we are changing the climate much faster in previous extinctions (Exploring Science, p. 239).

Accurate predictions of biodiversity losses are difficult when many species haven't yet been identified. Most predictions of human-caused extinction are based on assumptions about habitat area and species abundance: For example, biologist E. O. Wilson calculates that if you cut down 90 percent of a forest, you'll eliminate at least 50 percent of the species originally present. Estimates, assumptions, and observations vary, however. More than 90 percent of Costa Rica's dry seasonal forest, for instance, has been converted to pasture land, yet entomologist Dan Janzen reports that no more than 10 percent of the original flora and fauna appear to have been permanently lost. Wilson and others respond that remnants of the native species may be hanging on temporarily, but that in the long run they're doomed without sufficient habitat.

Still, it's clear that habitat is being destroyed in many places, and that numerous species are less abundant than they once were. E. O. Wilson has summarized human threats to biodiversity with the acronym **HIPPO,** which refers to Habitat destruction, Invasive species, Pollution, Population growth (of humans), and Overharvesting. The acronym HIPPCO is also used, which includes C for climate change. Note, though, that climate change contributes to all other HIPPO factors.

Habitat destruction is the principal HIPPO factor

Habitat loss is perhaps the most common of the HIPPO factors, and agricultural land conversion and forest clearing cause most extensive habitat loss. Habitat loss also results from other types of resource extraction and land clearing. Oil and gas drilling is a leading cost of deforestation (and pollution) from Ecuador to Nigeria to the western United States. Dam building drowns river valleys, and mining eliminates critical habitat. In the oceans, bottom-scraping fishing trawlers destroy spawning grounds and ecosystems.

One of the most long-running and contentious debates around habitat loss has been the controversy of the northern spotted owl.

Before European settlement, most of the Pacific Northwest from the Cascade Mountains to the coast was covered by a dense, structurally complex forest ideal for spotted owls. These birds nest deep in the forest, high in the ancient, mossy cedars and firs, subsisting on voles and other rodents, many of which spend their whole lives in and on the massive trees.

The ancient forest has also been a rich source of timber, and clear-cut logging has fragmented and eliminated much of the region's old-growth forest (fig. 11.9). Forest interior habitat has become rare, and edge habitat has expanded. This favors some species that occupy edge habitat, but spotted owls, and other old-growth species, have declined alarmingly. Since the 1970s, conservation advocates and logging interests have debated how to prevent the owl's extinction. At the same time, mechanization has eliminated most logging jobs in the region, and jobs have dwindled with the disappearance of the ancient forests. Protection of remnant old growth fragments angers struggling loggers, who blame the collapse of timber jobs on the spotted owl.

Expansive habitat is also necessary to preserve larger, wide-ranging predators like tigers or wolves, and to maintain complete ecological communities. Even for smaller species, gaps like those in fig. 11.9 create barriers to movement. If populations become small or isolated, they become vulnerable to events like bad weather or epidemics of disease.

One of the visible victims of habitat loss has been the monarch butterfly, a familiar, charismatic, and once ubiquitous species. As recently as the 1990s, around 1 billion of these beautiful black and orange butterflies spent the summer in the United States before migrating to Mexico, southern Florida, or southern California. Agricultural intensification, rising pesticide use, and logging in winter forest refuges have eliminated 90 percent of that enormous population. Widespread efforts are under way to restore this charismatic species (chapter 13).

FIGURE 11.9 Habitat degradation is a leading cause of biodiversity loss. Forest fragmentation, such as these clear-cut patches, destroys old-growth characteristics on which species, such as the northern spotted owl, depend.
William P. Cunningham

Invasive species displace resident species

Invasive species, the second HIPPO factor, are species that are especially able to colonize new territory. Humans have probably always transported organisms into new habitats, but the rate of movement has increased sharply in recent years with the huge increase in speed and volume of our travel by land, water, and air. Sometimes we deliberately transplant species we find attractive or useful. Some species hitch a ride in ship ballast water, in the wood of packing crates, inside suitcases or shipping containers, in the soil of potted plants, even on people's shoes. Sometimes we don't move species physically, but merely change habitat in ways that allow them to expand into new territories.

Over the past 300 years, an estimated 50,000 non-native species have become established in the United States. Many of these migrants don't thrive, and we don't notice them. By some estimates, only about 10 percent of introduced species become major problems. Those that do thrive, however, crowd out native species, or alter habitats, and may undermine ecosystem function. We call organisms that readily invade new territory **invasive species.** Non-native invasives often are especially successful because they are free of the predators, diseases, or resource limitations that controlled their populations in their native habitat (fig. 11.10).

Many ecologists regard invasive species as second only to habitat disruption in terms of threats to rare and endangered species. They also cause property damage. Invasive vines can kill trees and bury houses; invasive water weeds clog boats and reduce water quality and lake-side property values. Invasive weeds threaten crops and farm income. It's estimated that invasives cost the United States at least $138 billion annually.

Non-native invasive species are particularly destructive on islands where indigenous species have evolved without defenses against predators or competitors. For example, the New Guinea brown tree snake (*Boiga irregularis*) was introduced onto the Pacific Island of Guam sometime during or shortly after World War II (fig. 11.11*a*). This large, venomous snake quickly spread throughout the island, eating all manner of birds, lizards, small mammals, and bats. The snake exterminated 12 native bird species. Biodiversity on Guam has declined significantly, as have ecological functions: The decimated bird and bat populations include important pollinators in Guam's forests.

Some of our most destructive invasive species are feral pets or livestock. In the United States, domestic and feral and domestic house cats are estimated to kill at least 2.4 billion birds and 12.3 billion small mammals every year. Feral pigs are a major ecosystem threat in many U.S. forests. Rooting up plants, destroying

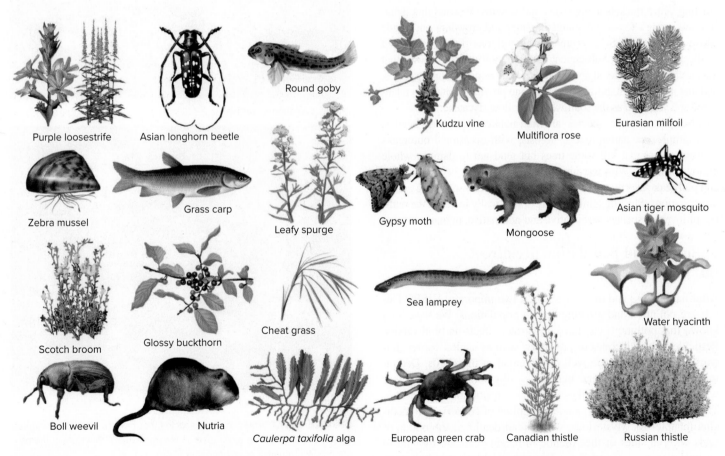

FIGURE 11.10 A few of the approximately 50,000 invasive species in North America. Do you recognize any that occur where you live? What others can you think of?

vegetation, and hard to capture, pigs are also dangerous to people who encounter them in the wild.

Disease organisms and their vectors can also be considered destructive invasive species. Consider the example of avian malaria. The protozoan parasites that cause this disease and the mosquitoes that spread them were accidentally introduced to the Hawaiian Islands sometime in the early 1900s. The disease has killed about one-half of the islands' native birds, many of which were already rare and endangered.

Sometimes native species can suddenly become invasive when environmental conditions change. The crown-of-thorns starfish, for example, is widely distributed in tropical oceans (fig. 11.11b). The large starfish preys on hard corals, digesting the living polyps and leaving a dead calcareous skeleton. When starfish densities are low (1–30/ha), coral regeneration can usually balance predation. But population explosions up to 1,000 starfish per hectare can destroy large areas of coral reef. We don't fully understand why these outbreaks occur, but overharvesting of starfish predators or algae blooms may play a role.

Often, we introduce species with good intentions. Tamarisk, or salt cedar (*Tamarix* sp), is a large, evergreen flowering shrub native to Eurasia and Africa (fig. 11.11c). Because it can grow in dry conditions and saline soil, it was considered useful in desert reclamation programs. The plants spread quickly by wind-borne or waterborne seeds and became widely established in the Southwestern United States, creating dense thickets along rivers and streams. Worried that tamarisk was using excessive amounts of water and crowding out fragile native species, the U.S. government spent millions of dollars on failed eradication programs. Subsequent research, however, has shown that the tamarisk uses water at about the same rate as native shrub species, and that it can help stabilize riverbanks and provide habitat for endangered species, such as the southwestern willow flycatcher.

Not all invasive species are non-native. Pine bark beetles, for example, are native to U.S. forests, with occasional outbreaks in warm years that kill some trees but tend not to damage whole regions of forest. With warming climates, however, beetle numbers and outbreak severity are increasing. They are also spreading across larger regions. Shifting climates may dramatically increase the number of invasive species, both native and non-native, in many areas.

Pollution and population are direct human impacts

After habitat loss and invasive species, two important HIPPO factors are pollution and growing human populations. We have long known that pollutants can have disastrous effects on biodiversity. Pesticide-linked declines of predators, such as eagles, osprey, falcons, and pelicans, were well documented in the 1970s. Declining populations of marine mammals, alligators, fish, and other wildlife alert us to the connection between pollution and health. This connection has led to a new discipline of conservation medicine (chapter 8). Mysterious, widespread deaths of thousands of seals on both sides of the Atlantic in recent years are thought to be linked to an accumulation of persistent chlorinated hydrocarbons, such as DDT, PCBs, and dioxins, in fat, causing weakened

(a) Brown tree snake

(b) Crown-of-thorns starfish

(c) Tamarisk or salt cedar

FIGURE 11.11 Three invasive species that have reduced or eliminated indigenous species: (a) brown tree snake from Guam, (b) crown-of-thorns starfish, (c) tamarisk or salt cedar.
(a) John Mitchell/Science Source; (b) Jeff Rotman/Getty Images; (c) William P. Cunningham

immune systems that make animals vulnerable to infections. Similarly, the mortality of Pacific sea lions, beluga whales in the St. Lawrence estuary, and striped dolphins in the Mediterranean are thought to be caused by accumulation of toxic pollutants.

Lead and mercury poisoning are another major cause of mortality for many species of wildlife. Bottom-feeding waterfowl, such as ducks, swans, and cranes, ingest spent shotgun pellets that fall into lakes and marshes (fig. 11.12). They store the pellets, instead of stones, in their gizzards, and the lead slowly accumulates in their blood and other tissues. The U.S. Fish and Wildlife Service estimates that 3,000 metric tons of lead shot are deposited annually in wetlands and that between 2 and 3 million waterfowl die each year from lead poisoning. There are several non-toxic alternatives for lead shot including steel, tungsteniron, and bismuth-tin alloys.

Mercury levels are elevated principally by burning coal, the fuel for many industrial processes and about 40 percent of global electricity production. Mining and waste incineration can be another source of this neurotoxin. In 1990, about 220 tons of mercury per year were emitted by these sources in the United States. Since the EPA started regulating smokestack emissions, that number has fallen to about 50 tons per year. Mercury enters the food web when aquatic bacteria convert it to methylmercury (CH_3Hg). This compound accumulates and concentrates as it passes from one trophic level to another. Birds and mammals that eat fish may accumulate toxic levels of this stable metal. Mercury poisoning is cumulative. It damages the brain and other internal organs. At high concentrations, harmful effects include reduced reproduction, slower growth and development, abnormal behavior, and death. For fish-eating species, such as loons, mercury poisoning is a serious threat.

Human population growth is a threat to biodiversity because everyone has some consumption needs—although some people consume more than others. As our numbers grow, we will need to harvest more timber, catch more fish, plow more land for agriculture, dig up more fossil fuels and minerals, build more houses, and use more water. All of these demands impact wild species. If our population grows to 9 to 11 billion as current projections predict, our impacts will certainly increase. The human population growth curve is leveling off (chapter 7), but it remains unclear whether we can reduce global inequality and provide a tolerable life for all humans while also preserving healthy natural ecosystems and a high level of biodiversity (fig. 11.13).

Climate change transforms ecosystems

Temperature and moisture conditions strongly define the kinds of organisms that occur in an area (chapter 5), so as regional climates become warmer or drier, many species decline, unable to tolerate new conditions or new competitors. Dry forest species, for example, may be replaced by savanna or grassland species, and grasslands may transition to arid desert conditions. Cold, high mountain habitats, already limited in extent may disappear entirely. Often secondary factors such as pests contribute to these shifts. Longer, hotter summers cause drought stress in conifer forests, for example, but infestations of pine beetles and spruce beetles, which occur naturally but thrive best in warm conditions, also kill millions of trees. Severe and extensive fires can result, potentially

FIGURE 11.12 A bald eagle's stomach contents include lead shot, which was consumed along with its prey. Fishing weights and shot remain a major cause of lead poisoning in aquatic and fish-eating birds.
Elliott Jacobson, University of Florida

FIGURE 11.13 Polar bears are threatened by rapidly shrinking sea ice, on which they depend for hunting. By mid-century, the Arctic Ocean could be largely free of summer ice, and polar bears may be limited to a small area around northern Greenland and northeastern Canada.
Vadim Balakin/Moment/Getty Images

transforming an ecosystem. In effect, climate change contributes to other HIPPO factors, especially habitat loss and invasive species.

Species that disperse easily to new habitat, and those that tolerate a wide range of environments, may simply shift their territories as we enter new climate regimes. Species with narrow habitat or climate requirements, or those that cannot easily disperse, may disappear as their current environment becomes inhospitable. High-mountain plants and animals, for example, adapted to alpine conditions, can only move upward so far as conditions warm. Invasive species are usually generalists, able to tolerate a wide range of conditions, and many of these are expanding their ranges as temperatures warm. In temperate regions, cold winters have normally prevented the expansion of warm-climate insects, ticks, and other invertebrates, as well as many invasive plants. Warming allows these species to expand to new ranges and to compete with resident species.

Important changes can be easy to overlook. Insects and other invertebrates, for example, are near the base of the food web in terrestrial ecosystems, but recent studies have found that insect populations are declining sharply (Exploring Science, p. 239). Shifts in species ranges or timing of migration or flowering can also disrupt ecosystems in ways that are not easy to predict. Biologists have documented hundreds of species whose migration patterns or behaviors have been altered by changes in temperature, the timing of spring seasons, rainfall, or other climate factors.

Warming especially threatens ice-dependent species such as polar bears, walrus, and penguins. Polar bears are highly adapted to hunt seals on sea ice of the Arctic Ocean (fig. 11.13). But rapidly disappearing and thinning sea ice makes it hard for polar bears to hunt and to feed their young. In 2008, the United States declared polar bears endangered because of their rapidly declining population. The situation has become worse since then. In recent years, summer sea ice covered only a quarter of the Arctic Ocean, down from half in the 1970s. As a consequence, bears have to swim farther to reach the ice. In some places, bears can no longer reach the summer ice because it is too far from their denning areas on shore. If these trends continue, polar bears may go extinct throughout much of their present range by mid-century.

An important effect of rising CO_2 concentrations is ocean acidification, which occurs as oceans absorb CO_2. Acidic conditions dissolve calcium carbonate, the material that makes up the shells of marine organisms. Marine plankton, snails, corals, and other organisms form shells poorly. Entire marine food webs, and unknown numbers of species, are at risk. Reefs are expected to stop growing and to begin to disintegrate around the globe when atmospheric CO_2 reaches 560 parts per million—double preindustrial levels—which will probably occur by the end of this century.

Overharvesting results when there is a market for wild species

Overharvesting, the last of the HIPPO factors, is responsible for the depletion or extinction of many species. A classic example is the extermination of the American passenger pigeon (*Ectopistes migratorius*), probably the world's most abundant bird 200 years ago. Although it inhabited only eastern North America, its population is

FIGURE 11.14 A museum display of passenger pigeons (*Ectopistes migratorius*). There may once have been 3 to 5 billion of these birds in the United States, but they were hunted to extinction in the nineteenth century. *Mary Ann Cunningham*

estimated to have been 3 to 5 billion birds (fig. 11.14), possibly one-quarter of all birds in North America. In 1830, John James Audubon saw a single flock of birds estimated to be ten miles wide and hundreds of miles long, and thought to contain perhaps a billion birds. In spite of this vast abundance, market hunters caused the entire population to crash in only about 20 years, between 1870 and 1890. They were able to eliminate so many birds so fast because of a ready abundance of firearms after the Civil War, as well as growing urban markets for wild bird meat. Rapid clearing of once-vast eastern forests for farmland also accelerated the decline. The last known wild passenger pigeon was shot in 1900, and the last existing passenger pigeon, a female named Martha, died in 1914 in the Cincinnati Zoo.

At about the same time that passenger pigeons were being extirpated, the American bison, or buffalo (*Bison bison*) was being hunted nearly to extinction on the Great Plains. In 1850, some 60 million bison roamed the western plains. Many were killed merely for sport or for their hides or tongues, leaving millions of carcasses to rot. Some of the bison's destruction was carried out by the U.S. Army so that native peoples who depended on bison for food, clothing, and shelter would be bereft of this resource and could then be forced onto reservations. By 1900, there were only about 150 wild bison left and another 250 in captivity.

EXPLORING SCIENCE

Where Are All the Insects?

If you have a chance, ask your parents or grandparents if they remember driving down a country road at twilight in midsummer, and having the windshield so covered with bugs that they needed to use windshield wipers to see. This once-common experience has become rare in most places. But few people noticed the decline. Who notices an absence of bugs? Then a 2017 publication stunned ecologists by showing that, over 27 summers, the biomass of flying insects in a set of 63 German nature reserves had decreased by 76 percent. Midsummer peak populations were down over 80 percent. "Defaunation," a widespread decline of animal abundance (like the more familiar deforestation), is a term ecologists use for this phenomenon. Few had used the term for invertebrate losses before the 2017 paper.

The insect study was shocking because insects fill critical ecological roles. They are a food source for higher trophic levels, such as birds, mammals, and amphibians. They cycle nutrients in ecosystems. They pollinate plants, making seed production and reproduction possible. Eighty percent of wild plants may depend on insect pollinators, and 60 percent of birds rely on insects as a food source. Three quarters of human foods rely on insect pollinators.

The German study design didn't specify a single cause of this surprising decline, but pesticides, habitat loss, fragmentation, deteriorating habitat quality, and climate change were proposed as possible factors. The areas sampled included brushy fields, grasslands, dunes, woodlands, and hedgerows, relatively small islands of natural habitat embedded in an ocean of cultivated farmland. Effects of changing farming practices on adjacent fields, especially pesticide spraying, may have spilled over into these natural areas.

Subsequent studies have confirmed that this change has been going on all around us,

(header photo): David L. Hansen

but no one knew, in part because there have been few long-term monitoring studies of insect abundance.

Even remote areas are affected. A 2018 study in Puerto Rico's pristine Luquillo Rainforest has found as little as one-sixtieth

FIGURE 1 Eighty percent of wild plants depend on insect pollinators, and 60 percent of birds rely on insects as a food source.
zefei/123RF

as much biomass of insects, beetles, spiders, and other arthropods compared to an assessment 35 years earlier. Insect-eating frogs and birds declined 50 percent or more. Few non-insectivorous species had reduced populations. The ruddy quail dove, which eats fruits and seeds, had no population change, but the endemic Puerto Rican tody, a small bird that eats only insects, declined by 90 percent.

The Puerto Rican study occurred in the Luquillo Long-term Ecological Monitoring site, a protected and relatively untouched tropical forest. Little farming occurs nearby, so pesticides were unlikely to explain the declines in this case. Climate change appeared responsible. Tropical arthropods, adapted to stable temperatures, have little temperature resilience, and they rapidly decline in fitness and

abundance with rising temperatures. Average high temperature in the forest increased 2°C (about 4°F) between 1973 and 2013, enough to explain the catastrophic decline in these species.

Long-term studies of ecosystem change require regular, repeated sampling, using constant methods so that values can be compared from year to year. This regularity is not easy to maintain, so long-term monitoring studies are relatively uncommon, especially for inconspicuous groups like insects. Shorter studies are easier to fund and carry out, but they are less likely to capture gradual systemic changes, such as long-term shifts in the structure of communities, climate impacts, or long-term effects of land use change and pesticides.

Long-term studies are often necessary for discerning gradual processes of ecosystem change, including the trophic cascades discussed in the opening case study for this chapter. The loss and then reintroduction of a top predator can profoundly alter lower trophic levels and even the structure and function of stream systems. The loss of insects is clearly creating a similar cascade of ecosystem effects, and we don't know how those changes will affect us.

To read more:

Hallmann CA, Sorg M, Jongejans E, Siepel H, Hofland N, Schwan H, et al. (2017) More than 75 percent decline over 27 years in total flying insect biomass in protected areas. PLoS ONE 12 (10): e0185809. https://doi.org/10.1371/journal.pone.0185809

Lister, BC, and A. Garcia. 2018. Climate-driven declines in arthropod abundance restructure a rainforest food web. PNAS 115 (44): E10397-E10406 https://doi.org/10.1073/pnas.1722477115

Don't Buy Endangered Species Products

You probably are not shopping for a fur coat from an endangered tiger, but there might be other ways you are supporting unsustainable harvest and trade in wildlife species. To be a sustainable consumer, you need to learn about the sources of what you buy. Many plant and animal products are farm-raised, not taken from wild populations. But some commercial products are harvested in unsustainable ways. Here are a few products about which you should inquire before you buy:

Seafood includes many top predators that grow slowly and reproduce only when many years old. Despite efforts to manage many fisheries, the following have been severely, sometimes tragically, depleted:

- Top predators: swordfish, marlin, shark, bluefin tuna, albacore ("white") tuna.
- Groundfish and deepwater fish: orange roughy, Atlantic cod, haddock, pollack (source of most fish sticks, artificial crab, generic fish products), yellowtail flounder, monkfish.
- Other species, especially shrimp, yellowfin tuna, and wild sea scallops, are often harvested with methods that destroy other species or habitats.
- Farm-raised species such as shrimp and salmon can be contaminated with PCBs, pesticides, and antibiotics used in their rearing. In addition, aquaculture operations often destroy coastal habitat, pollute surface waters, and deplete wild fish stocks to stock ponds and provide fish meal.

Pets and plants are often collected from wild populations, frequently depleting wild populations:

- Aquarium fish are usually collected in the wild, often using destructive practices that damage reef habitat and other fish.
- Reptiles: snakes and turtles, especially rare varieties, are often collected in the wild.
- Plants: orchids and cacti are the best-known, but not the only, group collected in the wild.

Herbal products such as wild ginseng and wild echinacea (purple coneflower) may be poached from the wild. Ask about the sources of these products.

Do buy some of these sustainably harvested products:

- Shade-grown (or organic) coffee, nuts, and other sustainably harvested forest products.
- Pets from the Humane Society, which works to protect stray animals.
- Organic cotton, linen, and other fabrics.
- Fish products that have relatively little environmental impact or have fairly stable populations: farm-raised catfish or tilapia, wild-caught salmon, mackerel, Pacific pollack, dolphinfish (mahimahi), squids, crabs, crayfish, and mussels.
- Wild freshwater fish like bass, sunfish, pike, catfish, and carp, which are usually better managed than most ocean fish.

More recently, overharvesting has reduced many of the world's fish populations to levels that may barely sustain themselves. A huge increase in the size and efficiency of fishing fleets in recent years is responsible. Of the world's 17 principal fishing zones, 13 are now reported to be commercially exhausted or in steep decline. At least three-quarters of all commercial oceanic species are overharvested. Fisheries biologists estimate that only 10 percent of the top predators, such as swordfish, marlin, tuna, and shark, remain in the Atlantic Ocean. Removal of these top predators from the ocean could have catastrophic ecological effects. You can avoid adding to this overharvest by eating only abundant, sustainably harvested varieties (What Can You Do?, above). Facebook initiated a campaign to ban longline fishing that threatens sea birds, turtles, and marine mammals. You could start or join similar efforts to raise awareness and change buying habits.

Overharvesting is often illegal and involves endangered species

Despite international bans on trade in products from endangered species, the smuggling of furs, hides, horns, live specimens, and folk medicines amounts to millions of dollars each year.

Developing countries in Asia, Africa, and Latin America with the richest biodiversity in the world are the main sources of wild animals and animal products. Europe, North America, and some of the wealthy Asian countries are the principal buyers. Japan, Taiwan, and Hong Kong buy three-quarters of all cat and snake skins, for instance; European countries buy a similar percentage of live birds. The United States imports 99 percent of all live cacti and 75 percent of all orchids sold each year, many of them rare and collected in the wild.

The profits to be made in wildlife smuggling are enormous. Tiger or leopard fur coats can bring (U.S.) $100,000 in Asia or Europe. And elephants are being slaughtered by the thousands every year for the illicit ivory trade (fig. 11.15). It's estimated that Africa has lost 90 percent of its original elephant population in just the past 50 years. Most of this trade is driven by rising wealth in Asia, where the desire for ivory carvings is intense. The enormous prices paid for black market ivory finance terrorists, despots, and criminal syndicates around the world. Similarly, rhino horn, valued in Asia because it supposedly enhances virility and allegedly cures a variety of diseases, has risen in value to as much as $300,000 for a single horn. As a consequence, the rhino population has plummeted from approximately 1 million in 1900 to only 30,000 today. Recently, breeding programs and better anti-poaching efforts have helped rhino populations stabilize in some African parks.

Plants also are threatened by overharvesting. Wild ginseng has been nearly eliminated in many areas because of the Asian demand for the roots, which are used as an aphrodisiac and folk medicine. In the American Southwest and Mexico, cactus "rustlers" steal

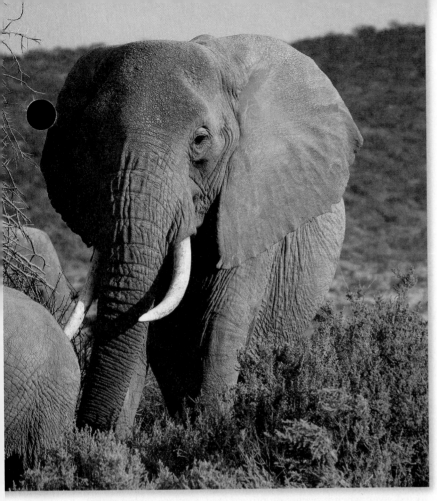

FIGURE 11.15 Poachers kill thousands of elephants every year for their ivory. It's estimated that Africa has lost 90 percent of its original elephant population in just the past 50 years.
Tom Finkle

cacti by the ton to sell for yard ornaments. With prices ranging as high as $1,000 for rare specimens, it's not surprising that many cacti are now endangered.

Island ecosystems are especially vulnerable to invasive species

New Zealand illustrates the damage that can be done by invasive species in island ecosystems. Having evolved for thousands of years without predators, New Zealand's flora and fauna are particularly susceptible to the introduction of alien organisms. Originally home to more than 3,000 endemic species, including flightless birds such as the kiwi and giant moas, New Zealand has lost at least 40 percent of its native flora and fauna since humans first landed there 1,000 years ago. More than 25,000 plant species have been introduced to New Zealand, and at least 200 of these have become pests that can create major ecological and economic problems. Many animal introductions (both intentional and accidental) also have become major threats to native species. Cats, rats, mice, deer, dogs, goats, pigs, and cattle accompanying human settlers consume native vegetation and eat or displace native wildlife.

One of the most notorious invasive species is the Australian brush-tailed possum, *Trichosurus vulpecula*. This small, furry marsupial was introduced to New Zealand in 1837 to establish a fur trade. In Australia, where their population is held in check by dingoes, fires, diseases, and inhospitable vegetation, possums are rare and endangered. Freed from these constraints in New Zealand, however, possum populations exploded. Now at least 35 million possums chomp their way through about 7 million tons of vegetation per year in their new home. They destroy habitat needed by indigenous New Zealand species, and also eat eggs, nestlings, and even adult birds of species that lack instincts to avoid predators.

Several dozen of New Zealand's offshore islands have been declared nature sanctuaries. Efforts are being made to eliminate invasive pests and to restore endangered species and native ecosystems.

Section Review

1. How do current rates of extinction compare to historic rates?
2. Explain HIPPO and its meaning for conserving biodiversity.
3. Why are islands especially sensitive to invasive species?

11.3 ENDANGERED SPECIES MANAGEMENT

- *The Endangered Species Act is one of the most important environmental laws ever passed in the United States.*
- *Protecting keystone, indicator, umbrella, or flagship species can be important beyond their mere survival.*
- *Endangered species protection is controversial.*

Over the years, we have gradually become aware of the harm we have done—and continue to do—to wildlife and biological resources. Slowly we are adopting national legislation and international treaties to protect these irreplaceable assets. Parks, wildlife refuges, nature preserves, zoos, and restoration programs have been established to protect nature and rebuild depleted populations. There has been encouraging progress in this area, but much remains to be done. Most people favor pollution control or protection of favored species such as whales or gorillas, but surveys show that few understand what biological diversity is or why it is important.

Hunting and fishing laws have been effective

In 1874, a bill was introduced in the U.S. Congress to protect the American bison, whose numbers were already falling to dangerously low levels. This initiative failed, however, because most legislators believed that all wildlife—and nature in general—was so abundant and prolific that it could never be depleted by human activity.

By the 1890s, most states had enacted some hunting and fishing restrictions. These laws protected game species, such as deer, ducks, and fish, that were useful to humans. They did not conserve wildlife for its own sake. The wildlife regulations and refuges established since that time have been remarkably successful for many species. In 1900, there were an estimated half million white-tailed deer in the United States; now there are at least 30 million—more in many places than the environment can support. Wild turkeys and wood ducks were nearly all gone 50 years ago. Restoring habitat, planting food crops,

FIGURE 11.16 At the beginning of the twentieth century, so many snowy egrets and other shore birds were killed for their plumes that the species seemed doomed to extinction. The Migratory Bird Act of 1918 protected these beautiful birds, and populations have rebounded.
Corinne Lamontagne/Getty Images

The Endangered Species Act is a powerful tool for biodiversity protection

Where earlier regulations had been focused almost exclusively on "game" animals, the Endangered Species Act (ESA) seeks to identify all endangered species and populations and to save as much biodiversity as possible, regardless of its usefulness to humans. As defined by the ESA, **endangered species** are those considered to be in imminent danger of extinction, whereas **threatened species** are those that are likely to become endangered—at least locally—in the foreseeable future. Gray wolves, brown (or grizzly) bears, sea otters, and a number of native orchids and other rare plants are a few of the species considered to be locally threatened, even though they remain abundant in other parts of their former range. **Vulnerable species** are naturally rare or have been locally depleted by human activities to a level that puts them at risk. Many of these are candidates for future listing. For vertebrates, protected categories include species, subspecies, and local races or ecotypes.

The ESA regulates a wide range of activities involving endangered species, including "taking" (harassing, harming, pursuing, hunting, shooting, trapping, killing, capturing, or collecting) either accidentally or on purpose; importing into or exporting out of the United States; possessing, selling, transporting, or shipping; and selling or offering for sale any endangered species. Prohibitions apply to live organisms, body parts, and products made from endangered species. In addition, critical habitat is legally protected, even when it occurs on private land. Critical habitat is that essential for a species survival. Protection of habitat is key to the success of the ESA. It has also made the act controversial, because habitat protection can compete with logging, mining, ranching, and especially oil and gas interests.

Currently, the ESA has 2,342 species on the endangered and threatened species lists, 1,660 of them in the United States. Worldwide, the International Union for Conservation of Nature and Natural Resources (IUCN) lists 28,659 endangered and threatened species. These numbers are rising as environmental degradation proceeds in biologically diverse regions.

Listing of endangered species favors those species that we find interesting or useful. In the United States, invertebrates make up about 75 percent of known species but only 9 percent of those listed for protection. The IUCN Red List describes 21.7 percent of mammals as threatened or endangered; only 0.01 percent of insects are listed as threatened. This is inequitable in two ways. First, there are probably far more endangered insect species than this, even among those we have identified. Furthermore, it's extremely rare to find a new mammal species, whereas the million known insect species may represent only a small fraction of the total insect species on earth.

Listing of new species in the United States has been very slow, sometimes taking decades from the first petition to final determination. Limited funding, political pressures, listing moratoria, and changing administrative policies have created long delays. Even after biologists have documented the distribution and population trends of a species, the U.S. Fish and Wildlife Service can take

transplanting breeding stock, building shelters or houses, protecting these birds during breeding season, and other conservation measures have made these beautiful and interesting birds common once again.

At the beginning of the twentieth century, the beautiful plumes from snowy egrets and other shorebirds were highly valued for women's hats (fig. 11.16). Market hunters shot so many birds that the species, survival was doubtful. Florida's Pelican Island was established in 1903 as the first wildlife refuge in America to protect birds. The Migratory Bird Act of 1918 extended this protection. Now, egret populations have rebounded.

In 2006, a survey of global fisheries projected that all fish and seafood species would collapse by mid-century. However, in 2013 both Norway and Russia reported that strict enforcement of fishing quotas for North Atlantic cod have resulted in rebounding populations. Similarly, in the northern Pacific, carefully regulated stocks of pollack and salmon have remained strong. This progress has led many scientists to hope that additional regulations may save marine species from extinction. Still, much of the ocean remains a free-for-all, and many developing countries lack the funds and technology to protect their territorial waters.

years to evaluate and decide. Administrators have often resisted any new listings, preferring to keep lands open to development.

Listing of inconspicuous species is especially difficult. The public is unlikely to champion an endangered clam or lizard as strongly as it does wolves, eagles, or condors. Still inconspicuous species are often more important than we recognize at first. Often they play important ecological roles. Protecting them usually preserves habitat and a host of unlisted species.

Recovery plans rebuild populations of endangered species

Once a species is officially listed as endangered, the Fish and Wildlife Service is required to prepare a recovery plan detailing how populations will be rebuilt to sustainable levels. It usually takes years to reach agreement on specific recovery plans. Reaching agreement is difficult for many reasons. Local politicians often resist setting aside land for conservation; recovery is costly; resource industries such as oil, logging, or mining industries lobby hard against conservation plans. The total cost of recovery plans for all currently listed species is estimated to be nearly $5 billion. (For comparison, this is about the cost of two modern Stealth bomber jets.)

The United States currently spends about $150 million per year on endangered species protection and recovery. About half that amount is spent on a dozen charismatic species, like the California condor, the Florida panther, and the grizzly bear, each of which receives around $13 million per year. By contrast, the 137 endangered invertebrates and 532 endangered plants get less than $5 million per year altogether. Our funding priorities often are based more on emotion and politics than on biology. But some species merit extra attention because they have special ecological value:

- **Keystone species** are those with major effects on ecological functions and whose elimination would affect many other members of the biological community; examples are gray wolf (*Canus lupus*) and bison (*Bison bison*).

- **Indicator species** are those tied to specific biotic communities or successional stages or a set of environmental conditions. They can be reliably found under certain conditions but not others; an example is brook trout (*Salvelinus fontinalis*).

- **Umbrella species** require large blocks of relatively undisturbed habitat to maintain viable populations. Saving this habitat also benefits other species. Examples of umbrella species are the northern spotted owl (*Strix occidentalis caurina*), the grizzly bear (*Ursus arctos*).

- **Flagship species** are especially interesting or attractive organisms to which people react emotionally. These species can motivate the public to preserve biodiversity and contribute to conservation; examples are the giant panda (*Ailuropoda melanoleuca*) and killer whales (*Orcinus orca*).

Some recovery plans have been remarkably successful. The American alligator was listed as endangered in 1967 because hunting (for meat, skins, and sport) and habitat destruction had reduced populations to precarious levels. Protection has been so effective that the species is now plentiful throughout its entire southern range. Florida alone estimates that it has at least 1 million alligators.

Sometimes restoring a single species can bring benefits to an entire ecosystem, especially when that species plays a keystone role in the community. Alligators, for example, dig out swimming holes, or wallows, that become dry-season refuges for fish and other aquatic species. American bison are being used in prairie restoration projects to reestablish the health and diversity of grassland ecosystems (see chapter 13).

As we discussed in the opening case study for this chapter, the gray wolf (*Canis lupis*) originally inhabited most of North America (fig. 11.17). By the 1950s, trapping and poisoning eliminated wolves from most of their territory. Only a few hundred wolves survived in northeastern Minnesota and on Isle Royale in Michigan. With the passage of the ESA in 1966, wolves were classified as endangered in the lower 48 states except in Minnesota, where they were listed as threatened. Under protected status, wolf populations rebounded, and in 2012 the species was declared recovered and delisted in the western Great Lakes and the Northern Rockies. Successful reintroduction and protection of this top predator, as we have noted, have had multiple beneficial ecological effects. Several states have reinstituted hunting seasons for wolves, but the Fish and Wildlife Service continues to monitor populations.

Other successful recovery programs include bald eagles, peregrine falcons, and whooping cranes. Forty years ago, due mainly

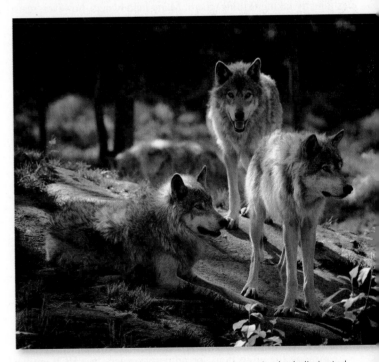

FIGURE 11.17 By the 1950s, poisoning and trapping had eliminated the gray wolf from nearly all the lower 48 United States. Protection and recovery programs have helped reestablish populations from Yellowstone National Park to the western Great Lakes states. A small population of wolves have had a profound effect on the 16,000 km^2 park.
gatordawg/123RF

to DDT poisoning, only 417 nesting pairs of bald eagles (*Haliaeetus leucocephalis*) remained in the contiguous United States. By 2007, the population had rebounded to more than 9,800 nesting pairs, and the birds were removed from the endangered species list. This doesn't mean that eagles are unprotected. Killing, selling, or otherwise harming eagles, their nests, or their eggs is still prohibited. In addition to eagles and falcons, several dozen other species, including mammals, fish, reptiles, birds, plants, and even one insect (the Tinian monarch), have been removed or downgraded from the endangered species list.

For some species, removing threats is extraordinarily challenging. The Columbia River salmon and steelhead, for example, are endangered by dams that block their migration to the sea. Opening floodgates to allow young fish to run downriver and adults to return to spawning grounds would have high economic costs to barge traffic, farmers, and electric rate payers who have come to depend on abundant water and cheap electricity. On the other hand, commercial and sport fishing for salmon was once worth $1 billion per year and employed about 60,000 people directly or indirectly. There can be economic benefits, as well as costs, in restoring species and their ecosystems.

Private land is vital for species protection

Eighty percent of the habitat for more than half of all listed species in the United States is on private land. The Supreme Court has ruled that destroying habitat is as harmful to endangered species as directly taking (killing) them. Many landowners, however, resist restrictions on how they use their own property to protect wildlife. This is especially true when the land has potential for economic development. If property is worth millions of dollars as the site of a housing development or shopping center, most owners don't want to be told they have to leave it undisturbed to protect some rare organism. Landowners may be tempted to "shoot, shovel, and shut up," if they discover endangered species on their property. Many feel they should be compensated for lost value caused by ESA regulations.

To reduce these tensions, the Fish and Wildlife Service negotiates agreements called **habitat conservation plans (HCPs)** with private landowners. Under these plans, landowners are allowed to harvest resources or build on part of their land as long as the species benefits overall. In return for improving habitat in some areas, funding conservation research, removing predators and competitors, or taking other steps that benefit the endangered species, developers are allowed to destroy habitat or even "take" endangered organisms.

Scientists and environmentalists often are critical of HCPs, claiming these plans often are based more on politics than biology, and that the potential benefits are frequently overstated. Defenders argue that by making the ESA more landowner-friendly, HCPs benefit wildlife in the long run.

Among the protections for property owners in a HCP are the so-called Safe Harbor and No-Surprises policies. Under the Safe Harbor clause, any increase in an animal's population resulting from a property owner's voluntary good stewardship would not increase their responsibility or affect future land-use decisions. As long as the property owner complies with the terms of the agreement, he or she can make any use of the property. The No-Surprises provision says that the property owner won't be faced with new requirements or regulations after entering into an HCP. Many ecologists worry that this protection allows too little room to accommodate unexpected risks to species survival, when we know ecosystems are dynamic and changeable.

Endangered species protection is controversial

Despite its strong popular support, the ESA has been contested from the beginning. The first major controversy involved the Tellico Dam, which threatened a tiny fish, the snail darter, on the Little Tennessee River. The Tennessee Valley Authority, a powerful federal agency, argued that a small fish was not significant enough to halt a multibillion dollar dam. In 1978 the Supreme Court ruled that it was the intent of Congress, as stated in the ESA, to prevent further extinctions, and that the law did protect the snail darter. The victory was short-lived, though, as Congress passed an exemption to the law the following year, allowing the completion of the dam.

Opponents of the law have often focused on the economic cost of species protection. They argue that it is foolish to spend millions of dollars to protect species that have no value to us. Supporters of species protection, meanwhile, point out that countless ecosystem benefits are lost when habitat is lost. As with the Columbia River salmon, the economic value of these resources can be huge.

Economic accounting of species protection finally became federal policy in 2019, when Trump administration's Secretary of the Interior and oil company lobbyist David Bernhardt determined that financial cost can be accounted for in decisions on species protection. The purpose of the rule change was to free up land for oil, gas, ranching, logging, and other development. This rule makes listing and species recovery more difficult. It could also demonstrate the many economic values of protections, as has occurred in economic accounting for the Clean Air Act and climate protection efforts.

The new rules also reduced the definition of critical habitat to include only areas where rare species currently occur, and they made it difficult to consider long-term impacts of climate change on species survival.

An important controversy in species management involves the sage grouse. When Lewis and Clark crossed the western prairies in 1804, they encountered flocks of thousands of these large, chicken-like birds (fig. 11.18). The total population must have been in the millions in the area between the Missouri River and the Sierra Nevada mountains. But ranching, mining, and other human activities reduced the population to fewer than 500,000 of this umbrella species by 2013. After many decades of wrangling, government agencies, ranchers, and landowners reached an agreement on how to manage the 9 million acres (about half of its historic range) of grouse habitat in 11 western states.

In addition to being essential for grouse, this area supports at least 297 species of birds, 87 species of mammals, 63 species of fish, and countless species of reptiles and insects that call the sagebrush steppe their home. The USFWS agreed to not put the

FIGURE 11.18 The sage grouse serves as an umbrella species because hundreds of other species benefit by its need for large blocks of undisturbed sagebrush steppe. The Trump administration has decided to abandon established cooperative conservation agreements to open 80 percent of its remaining grouse habitat to oil and gas drilling.
Tom Reichner/Shutterstock

sage grouse on the Endangered Species List, which would have restricted development within their habitat while ranchers and local officials agreed to create habitat conservation plans. But energy developers weren't happy.

After the 2016 election, oil and gas companies found they had a friend in the White House. The Trump administration overturned all Obama-era protections and agreements. Protection was eliminated from 80 percent of the grouse habitat when 7.2 million acres were opened to oil and gas extraction. The concept of cooperative conservation based on common goals was abandoned in favor of larger profits for the fossil fuel industry. This is also a loss in the battle against climate change.

Gap analysis promotes regional planning

An alternative strategy for species protection is to identify and preserve whole ecosystems that support maximum biological diversity, rather than battle for the rarest species one at a time. By focusing on a few high-profile populations already reduced to only a few individuals, we risk spending most of our conservation funds on species that may be genetically doomed no matter what we do. Furthermore, by concentrating on individual species we spend millions of dollars to breed plants or animals in captivity that have no natural habitat where they can be released. While flagship species such as mountain gorillas or Indian tigers are reproducing well in zoos and wild animal parks, the ecosystems they formerly inhabited have largely disappeared.

A leader of this new form of conservation is J. Michael Scott, who was project leader of the California condor recovery program in the mid-1980s and had previously spent ten years working on

endangered species in Hawaii. In making maps of endangered species, Scott discovered that even Hawaii, where more than 50 percent of the land is federally owned, has many vegetation types completely outside of natural preserves (fig. 11.19). The gaps between protected areas may contain more endangered species than are preserved within them.

This observation has led to an approach called **gap analysis** in which conservationists and wildlife managers look for unprotected landscapes that are rich in species. Computers and geographic information systems (GIS) make it possible to store, manage, retrieve, and analyze vast amounts of data and create detailed, high-resolution maps relatively easily. This broad-scale, holistic approach seems likely to save more species than a piecemeal approach.

Conservation biologist R. E. Grumbine suggests four management principles for protecting biodiversity in a large-scale, long-range approach:

1. Protect enough habitat for viable populations of all native species in a given region.

2. Manage at regional scales large enough to accommodate natural disturbances (fire, wind, climate change, and so on).

What Can You Do?

You Can Help Preserve Biodiversity

If you live in an urban area, as most Americans do, you may not think you have much influence on wildlife. But there are important ways you can help conserve biodiversity.

- Protect or restore native habitat. Environmental organizations or nature preserves near where you live may have volunteer opportunities to remove invasive species, gather native seeds, or replant native vegetation.
- Plant local, native species in your garden. Exotic nursery plants often escape and threaten native ecosystems.
- Don't transport firewood from one region to another. It may carry diseases and insects.
- Follow legislation and management plans for natural areas you value. Lobby or write letters supporting funding and biodiversity-friendly policies.
- Help control invasive species. Never release non-native animals (fish, leaches, turtles, etc.) or vegetation into waterways or sewers. If you boat, wash your boat and trailer when moving from one lake or river to another.
- Don't discard fishing bait worms in the woods. We think of earthworms as beneficial for soils, but many northern deciduous biomes evolved without them. Worms discarded by anglers are now causing severe habitat destruction in many places.
- Keep your cat indoors. House cats are major predators of woodland birds and other native animals. House cats in the United States kill billions of birds, small mammals, and lizards every year.

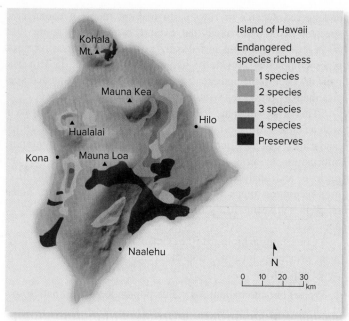

FIGURE 11.19 An example of the biodiversity maps produced by J. Michael Scott and the U.S. Fish and Wildlife Service. Notice that few of the areas of endangered species richness are protected in preserves, which were selected more for scenery or recreation than for biology.

3. Plan over a period of centuries so that species and ecosystems may continue to evolve.

4. Allow for human use and occupancy at levels that do not result in significant ecological degradation.

International treaties improve protection

The 1975 Convention on International Trade in Endangered Species (CITES) was a significant step toward worldwide protection of endangered flora and fauna. It regulated trade in living specimens and products derived from listed species, but it has not been foolproof. Smugglers remove species from countries where they are threatened or endangered, and falsify documents to make it appear that they have come from areas where the species are still common. Investigations and enforcement are especially difficult in developing countries, where wildlife is disappearing most rapidly. Still, eliminating markets for endangered wildlife is an effective way to stop poaching. Appendix I of CITES lists 700 species threatened with extinction by international trade.

Shark survival provides a good example of the need for international cooperation in endangered species protection. Every year, about 100 million sharks and rays are killed, primarily to provide fins to be made into shark fin soup, considered a delicacy in some Asian countries. Depletion of these wide-ranging key predators has risks for the health of entire ocean ecosystems. Marine scientists estimate that global shark populations have declined by as much as 70 to 80 percent, and that about one-third of all shark species are currently threatened with extinction.

Several Pacific Island nations, including Palau, French Polynesia, American Samoa, and the Cook Islands, have banned shark fishing in their territorial waters, but much shark fishing takes place far offshore and is difficult to control. We need joint action to monitor both fisheries and markets to protect these important species.

Section Review

1. Define endangered species and threatened species. Give an example of each.
2. Why are keystone, indicator, umbrella, and flagship species important?
3. What is gap analysis, and how is it related to ecosystem management and the design of nature preserves?

11.4 CAPTIVE BREEDING AND SPECIES SURVIVAL PLANS

- *Zoos and captive breeding are important in protecting biodiversity.*
- *There are limits to how many species we can preserve in captivity.*
- *We need to protect rare species in the wild.*

Breeding programs in zoos and botanical gardens are one way to attempt to save severely threatened species. Institutions like the Missouri Botanical Garden and the Bronx Zoo's Wildlife Conservation Society sponsor conservation and research programs. Botanical gardens, such as Kew Gardens in England, and research stations, such as the International Rice Institute in the Philippines, are repositories for rare and endangered plant species, some of which have ceased to exist in the wild. Valuable genetic traits are preserved in these collections, and in some cases, plants with unique cultural or ecological significance may be reintroduced into native habitats after being cultivated for decades or even centuries in these gardens and seed banks.

Zoos can help preserve wildlife

Until fairly recently, zoos depended primarily on wild-caught animals for most of their collections. This was a serious drain on wild populations, because up to 80 percent of the animals caught died from the trauma of capture and shipping. With better understanding of reproductive biology and better breeding facilities, most mammals in North American zoos now are produced by captive breeding programs.

Some zoos now participate in breeding programs, which can reintroduce endangered species to the wild. The California condor is one of the best-known cases of successful captive breeding. In 1986, only nine of these birds existed in their native habitat. Fearing the loss of these last condors, biologists captured them and brought them to the San Diego and Los Angeles zoos, which had begun breeding programs in the 1970s. By 2019, the population had reached nearly 450 birds, including 270 flying free in the wild.

Hawaii's endemic nene (*Nesochen sandvicensis*) also has been successfully bred in captivity and reintroduced into the wild. When Captain Cook arrived in the Hawaiian Islands in 1778, there were probably 25,000 of these land-dwelling geese. By the 1950s, however, habitat destruction and invasive predators had reduced the population to fewer than 30 birds. Today, there are about 800 wild

FIGURE 11.20 Nearly extirpated in the 1950s, the land-dwelling nene of Hawaii has been successfully restored by captive breeding programs. From fewer than 30 birds half a century ago, the wild population has grown to more than 800 birds.
William P. Cunningham

nene, and at least 1,000 in captivity, but there are worries about inbreeding that could reduce the fitness of the species (fig. 11.20).

One of the most successful captive breeding programs for mammals is that of the white rhino (*Ceratotherium simum*) in southern Africa. Although they once ranged widely across Africa, these huge animals were considered extinct in the south until a remnant herd was found in Natal, South Africa, in 1895. Today, there are an estimated 17,500 southern white rhinos in Africa, mainly in national parks and private game ranches. In 2015, a U.S. hunter paid $350,000 to shoot a male that wildlife managers said was too old to reproduce and was a threat to younger males. The money was used to support parks and further breeding programs. Animal rights activists condemned this "hunt" as legal execution.

Breeding programs have limitations, however. Some species, such as bats, whales, and many reptiles rarely reproduce in captivity, and zoo stock for these animals still comes mainly from the wild. We will never be able to protect the complete spectrum of biological variety in zoos. According to one estimate, if all the space in U.S. zoos were used for captive breeding, only about 100 species of large mammals could be maintained on a long-term basis.

These limitations lead to what is sometimes called the "Noah question": How many species can or should we save? How much are we willing to invest to protect the slimy, smelly, crawly things? Would you favor preserving disease organisms, parasites, and vermin, or should we use our limited resources to protect only beautiful, interesting, or seemingly useful organisms?

Even given adequate area and habitat conditions to perpetuate a given species, continued inbreeding of a small population in captivity can lead to the same kinds of fertility and infant survival problems described earlier for wild populations. To reduce genetic problems, zoos often exchange animals or ship individuals long distances to be bred. It sometimes turns out, however, that zoos far distant from each other unknowingly obtained their animals from the same source. Computer databases operated by the International Species Information System, located at the Minnesota Zoo, now keep track of the genealogy of many species. For some species, this system can tell the complete reproductive history of every animal of that species in every zoo in the world. Comprehensive species survival plans based on this genealogy help match breeding pairs and project resource needs.

The ultimate problem with captive breeding, however, is that natural habitat may disappear while we are busy conserving the species itself.

We need to save rare species in the wild

As long as umbrella species—large species such as rhinos and apes—persist in their native habitat, many other species survive as well. Renowned zoologist George Schaller says that ultimately "zoos need to get out of their own walls and put more effort into saving the animals in the wild." In many cases, the best thing we can do is to protect habitat. But sometimes individual animals need our help. For example, International Animal Rescue in Indonesia operates a rehabilitation center in Sumatra for orphaned and captive orangutans. Every year, thousands of orangutans are rescued

FIGURE 11.21 Rescued baby orangutans are transported by wheelbarrow to the Forest School in Sumatra where they will learn to climb, play, and find food in their natural habitat. Eventually, when they're able to fend for themselves, they'll be released back into the wild.
NurPhoto/Getty Images

from the pet trade. At the rehabilitation center babies are transported by wheelbarrow to the Forest School where they can learn to climb, play, and fend for themselves before being finally released back to their natural habitat (fig. 11.21). This is much better than taking them to live in cages for the rest of their lives.

Section Review

1. Give an example of a species whose population was rebuilt in zoos.
2. Why aren't zoos a good place to preserve whales, bats, and some reptiles?
3. Describe two examples of successful captive breeding programs.

Connecting the Dots

The story of wolf reintroductions in Yellowstone National Park illustrates some of the many ways in which biodiversity benefits whole ecosystems. A complete ecosystem is more likely to have stability, and to have resilience when change occurs. Presence of a top predator maintains healthy prey populations as well as plants, smaller species, and even physical systems like rivers.

Biodiversity is important for humans, too. We depend on animals, plants, and healthy ecosystems for food, fiber, medicines, clean water, and many other products and services. Yet nearly one-third of native species in the United States are at risk of disappearing. The acronym HIPPO (or HIPPCO) provides a summary of the major classes of threats to biodiversity. Often, these threats are multiple and overlapping: For example, habitat loss, invasive species introduction, and human population growth often proceed together.

Wolf reintroductions were done in accordance with the Endangered Species Act, because wolf numbers had become dangerously low in many U.S. ecosystems. This case shows, though, that the ESA benefits whole ecosystems when it protects one species. This law has been one of the most powerful tools we have for environmental protection. Increasingly, it tries to support cooperative solutions between landowners and wildlife managers. Because of its effectiveness, the act itself is endangered: Opponents have succeeded in limiting its scope and have sought to eliminate it altogether. Still, the ESA remains a cornerstone of our most basic environmental protections. It has given new hope for the survival of numerous species that were on the brink of extinction—less than 1 percent of species listed under the act have gone extinct since 1973, whereas 10 percent of candidate species still waiting to be listed are now gone.

Species protection is especially difficult when critical habitat has been degraded or destroyed. Reducing this problem requires both protecting and restoring critical habitats. These are topics we'll explore in the next two chapters.

Reviewing Key Terms

Can you define the following terms in environmental science?

biodiversity hot spots 11.1	flagship species 11.3	HIPPO 11.2	overharvesting 11.2
biodiversity 11.1	gap analysis 11.3	indicator species 11.3	threatened species 11.3
endangered species 11.3	habitat conservation plans	invasive species 11.2	umbrella species 11.3
existence value 11.1	(HCP) 11.3	keystone species 11.3	vulnerable species 11.3
extinction 11.2			

Critical Thinking and Discussion Questions

1. Many ecologists would like to move away from protecting individual endangered species to concentrate on protecting whole communities or ecosystems. Others fear that the public will respond to and support only glamorous "flagship" species such as gorillas, tigers, or otters. If you were designing conservation strategy, where would you put your emphasis?

2. Put yourself in the place of a fishing industry worker. If you continue to catch many species, they will quickly become economically extinct, if not completely exterminated. On the other hand, there are few jobs in your village, and welfare will barely keep you alive. What would you do?

3. Only a few hundred grizzly bears remain in the contiguous United States, but populations are healthy in Canada and Alaska. Should we spend millions of dollars for grizzly recovery and management programs in Yellowstone National Park and adjacent wilderness areas?

4. How could people have believed a century ago that nature is so vast and fertile that human actions could never have a lasting impact on wildlife populations? Are there similar examples of denial or misjudgment occurring now?

5. In the past, mass extinctions have allowed for new growth, including the evolution of our own species. Should we assume that another mass extinction would be a bad thing? Could it possibly be beneficial to us? To the world?

6. Some captive breeding programs in zoos are so successful that they often produce surplus animals that cannot be released into the wild because no native habitat remains. Plans to euthanize surplus animals raise storms of protests from animal lovers. What would you do if you were in charge of the zoo?

7. Compare the estimated numbers of known and threatened species in table 11.1. Are some groups overrepresented? Are we simply more interested in some organisms, or are we really a greater threat to some species?

8. Domestic and feral house cats are estimated to kill billions of birds and small mammals in the United States annually. In 2005, a bill was introduced in the Wisconsin legislature to declare an open hunting season year-round on cats that roam out of their owners' yards. Would you support such a measure? Why or why not? What other measures (if any) would you propose to control feline predation?

9. How many species can we afford to save? How much should we spend to protect, slimy, smelly, crawly things?

10. Is it okay to kill one species (say, barred owls) to protect another species (say, spotted owls)?

Data Analysis

Confidence Limits in the Breeding Bird Survey

A central principle of science is the recognition that all knowledge involves uncertainty. No study can observe every possible event in the universe, so there is always missing information. Scientists try to define the limits of their uncertainty, in order to allow a realistic assessment of their results. A corollary of this principle is that the more data we have, the less uncertainty we have. More data increase our confidence that our observations represent the range of possible observations.

One of the most detailed records of wildlife population trends in North America is the Breeding Bird Survey (BBS). Every June, volunteers survey more than 4,000 established 25-mile routes. The accumulated data from thousands of routes, over more than 40 years, indicate population trends, telling which populations are increasing, decreasing, or expanding into new territory. To examine a sample of BBS data, go to Connect, where you can explore the data and explain the importance of uncertainty in data.

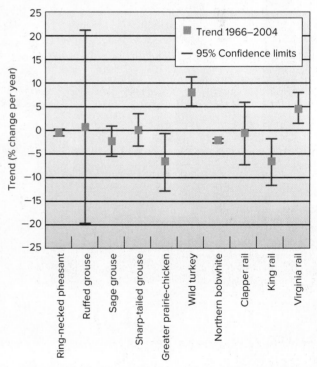

Eighty percent of wild plants depend on insect pollinators, and 60 percent of birds rely on insects as a food source

12

Biodiversity
Preserving Landscapes

▲Drought, longer fire seasons, and bark beetle infestations have combined to make forest fires bigger and more destructive across the American West in recent years.
Flickr/Getty Images

Learning Outcomes

After studying this chapter, you should be able to:

12.1 Discuss the types and uses of world forests.

12.2 Describe the location and state of grazing lands around the world.

12.3 Summarize the types and locations of nature preserves.

"If we destroy the land, God may forgive us, but our children will not."

– Togiak Elder

CASE STUDY

Ecosystems in Transition

Recent years have seen the most extensive—and most expensive—forest fire seasons ever in the American West. The fire season has lengthened by nearly 3 months, and fires have become larger and more intense. They are also more expensive, rising from around $250 million in 1985 to nearly $2.9 billion in 2018. What has caused the changes? A combination of factors, including longer, hotter summers and drought, play a role. On top of these is the explosive growth of billions of tiny beetles, the size of a grain of rice, that have been blamed for damaging or destroying nearly 17 million ha (42 million acres) of western forests. All of these factors are a natural part of western ecosystems. The question is, as all of them accelerate with climate change, what will become of the region's vast forests?

Fires are common in the region, periodically burning patches of spruce, fir, and pine forests in hot, dry summers and allowing young trees to emerge in aging stands. After a fire the forest in a particular area may grow for decades or centuries before burning again. Fire doesn't necessarily kill trees—prolonged or repeated drought is the more important cause of tree mortality. Many trees survive and grow, with bark gradually covering over scars from multiple fires.

Also common are bark beetles, mainly species of the genus *Dendroctonus* and the genus *Ips,* which burrow below the bark of conifers to consume the living inner bark, or phloem, and deposit their eggs. Bark beetles are especially attracted to damaged or drought-stressed trees: These cannot defend themselves, as a healthy tree can, by overwhelming the insects with oozing sap. Normally, long and intensely cold winters kill off many of the beetle larvae, so widespread destruction is infrequent.

In recent decades longer and hotter summers have allowed beetles to thrive. Larvae develop more rapidly in hot weather, and several generations may now occur in a summer. More beetles survive the shorter, milder winters. Outbreaks are now far more extensive and more devastating than western states have seen in the past, and the pests are moving north. Ever-larger areas of forest have become standing dry tinder, waiting for the next lightning strike for ignition (fig. 12.1).

Forests can recover from periodic disturbance, but increasingly ecologists are warning that a growing frequency and severity of beetle outbreaks and of fires, both driven by climate warming, are likely to lead to a state transition. What we have known as a forested western landscape is likely to transition to something new in many areas (fig. 12.2). Perhaps former forests will become patchy savannahs, shrublands, or even grasslands. It depends on how much a given area warms and how patterns of rain, snow, and below-freezing temperatures change.

To an ecologist, a transition from forest to some other type of ecosystem isn't surprising. Temperature and precipitation determine the general type of biome that occurs in an area, often influenced by related factors such as soil type and insect pests. We have usually considered biome regions stable, but shifting patterns of temperature and precipitation should cause a different type of ecosystem to develop in a place. So as climate regions change in coming decades, we can

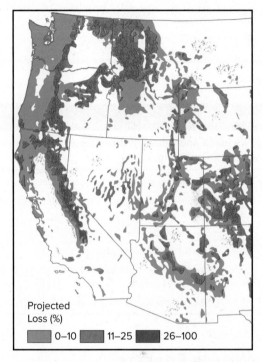

FIGURE 12.1 Projected forest losses from bark beetles and other insect pests by 2027.
Source: Data from US Forest Service

FIGURE 12.2 Dead trees killed by bark beetles are interspersed with green, still-living trees.
Wave Royalty Free/Science Source

expect that the map of biomes and ecological regions will shift. Where will shifts occur, and how dramatic will they be? That remains to be seen.

To residents of the West, a state transition may mean serious disruption. Much of the region is defined by its forested mountain landscapes. Economies depend on historically reliable forest resources, from timber to the scenery that supports tourism. Ecosystem services, especially provision of drinking water and irrigation from forested watersheds, support cities and farms throughout the region. A shift from abundant forests to another biome type could have dramatic impacts on the lives and livelihoods of people in the region.

Protecting biodiversity and preserving landscapes often are both interconnected and controversial. In this chapter, we'll look at a number of examples of how we can protect biodiversity and preserve landscapes.

Benchmark Data

Among the ideas and values in this chapter, the following are a few worth remembering.

29%	Percentage of land surface in forest and woodlands
18%	Percentage of anthropogenic climate impacts from forest clearing
17%	Area in preserves
11%	Percentage of land surface in cropland
11:47	Revenue: cost of managing Western grazing lands, in millions of dollars
3:58	Ratio of global protected areas, 1950:2017, in km^2

12.1 WORLD FORESTS

- *Forests provide habitat for biodiversity, valuable resources for humans, and essential ecological services.*

- *Tropical forests are particularly species rich and are threatened by human activities.*

- *Temperate forests also are at risk.*

Forests and woodlands occupy some 4 billion ha (roughly 15 million mi^2), or about 29 percent of the world's land surface (fig. 12.3). Grasslands (pastures and rangelands) cover almost as much area. Together these ecosystems supply many essential resources, such as lumber, paper pulp, and grazing lands for livestock. They also provide vital ecological services, including regulating climate, controlling water runoff, providing wildlife habitat, and purifying air and water. Forests and grasslands also have scenic, cultural, and historic values that deserve protection. These biomes are also among the most heavily disturbed (chapter 5), because they are places in which people prefer to live and work.

How do we maintain these ecosystems, with constant and growing demands for use? And how do we plan in the context of a warming climate? This perennial question has been the focus of many of our most contentious environmental policy debates.

This chapter examines the ways we use and abuse these biological communities, as well as some of the ways we can protect them and conserve their resources. We discuss forests first, followed by grasslands, and then strategies for conservation, restoration, and preservation. Chapter 13 focuses on restoration of damaged or degraded ecosystems.

Boreal and tropical forests are most abundant

Forests are widely distributed, but the largest remaining areas are in the tropical humid regions, especially Brazil, western Africa, and Southeast Asia, and in the boreal (northern) forests of Russia, Canada, and the United States (fig. 12.3 and fig. 12.4). There are differing definitions of what a "forest" is. The UN Food and Agriculture Organization (FAO) defines **forest** as any area where trees cover more than 10 percent of the land. Open wooded areas, with tree cover between about 10 and 20 percent of the ground, are often considered **savannas.** In **closed-canopy forests,** tree crowns overlap to cover most of the ground.

The largest tropical forests are in South America, which has about 22 percent of the world's forest area and by far the most extensive area of undisturbed tropical rainforest. Africa and Southeast Asia also have large areas of tropical forest that are highly important biologically, but both continents are suffering from rapid deforestation. North America and Eurasia have vast areas of relatively unaltered boreal forest. Although many of these forests are harvested regularly, both continents have a net increase in forest area and biomass because of replanting and natural regeneration.

Among the forests of greatest ecological importance are the ancient forests, such as the tropical rainforests described in the opening case study for this chapter, that are home to much of the world's biodiversity, ecological services, and indigenous human cultures. Sometimes called frontier, old-growth, or virgin forests, these are areas large enough and free enough from human modification that native species can live out a natural life cycle, and ecological relationships play out in a relatively normal fashion. A similar term used by the FAO is **primary forests,** which are "composed primarily of native species in which there are no clearly visible indications of human activity and ecological processes are not significantly disturbed."

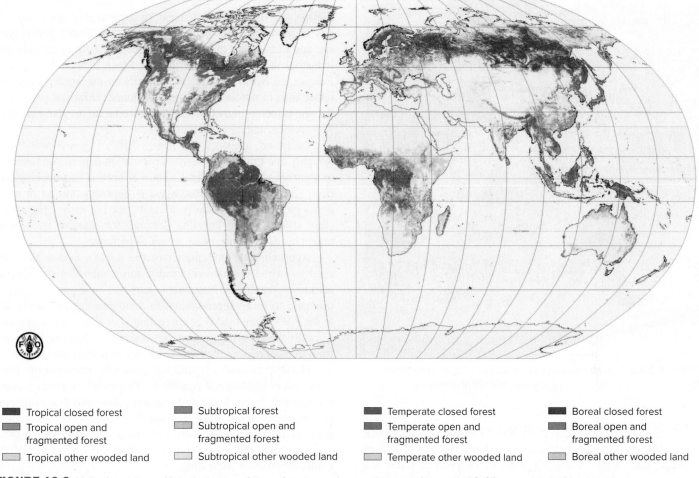

Tropical closed forest
Tropical open and fragmented forest
Tropical other wooded land

Subtropical forest
Subtropical open and fragmented forest
Subtropical other wooded land

Temperate closed forest
Temperate open and fragmented forest
Temperate other wooded land

Boreal closed forest
Boreal open and fragmented forest
Boreal other wooded land

FIGURE 12.3 Major forest types. Note that some of these forests are dense; others may have only 10–20 percent actual tree cover.
Source: UN Food and Agriculture Organization, 2002

This doesn't mean that all trees in a primary forest need be enormous or thousands of years old (fig. 12.5). In some biomes, such as lodgepole pine forests of the Rocky Mountains or Canada's boreal forests, most trees live only a century or so before being killed by disease or some natural disturbance. The successional processes (chapter 4) as trees die and are replaced create structural complexity and a diversity of sizes and ages important for

specialists, such as the northern spotted owl. Nor does it mean that humans have never been present. Where human occupation entails relatively little impact, a forest may be inhabited for millennia while still retaining its primary characteristics. Even forests that have been logged or converted to cropland often can revert to natural conditions if left alone long enough.

FIGURE 12.5 A tropical rainforest in Queensland, Australia. Primary, or old-growth, forests, such as this, aren't necessarily composed entirely of huge, old trees. Instead, they have trees of many sizes and species that contribute to complex ecological cycles and relationships.
Digital Vision/PunchStock

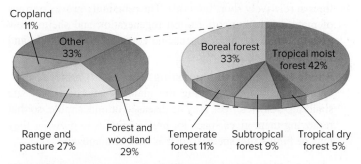

Cropland 11%
Other 33%
Range and pasture 27%
Forest and woodland 29%

Boreal forest 33%
Tropical moist forest 42%
Temperate forest 11%
Subtropical forest 9%
Tropical dry forest 5%

FIGURE 12.4 World land use and forest types. The "other" category includes tundra, desert, wetlands, and urban areas.
Source: UN Food and Agriculture Organization (FAO)

FIGURE 12.6 Firewood accounts for almost half of all wood harvested worldwide and is the main energy source for one-third of all humans.
William P. Cunningham

Even though forests still cover about half the area they once did worldwide, only one-quarter of those forests retain old-growth features. The largest remaining areas of old-growth forest are in Russia, Canada, Brazil, Indonesia, and Papua New Guinea. Together, these five countries account for more than three-quarters of all relatively undisturbed forests in the world. In general, remoteness rather than laws protects those forests. Although official data describe only about one-fifth of Russian old-growth forest as threatened, rapid deforestation—both legal and illegal—especially in the Russian Far East, probably puts a much greater area at risk.

Forests provide valuable products

Wood is generally considered the most important forest product. There is hardly any industry that does not use wood or wood products somewhere in its manufacturing and marketing processes. Think about the amount of printed material, cardboard boxes, and other paper products that each of us in developed countries handles, stores, and disposes of in a single day. Total annual world wood consumption is about 4 billion m^3. This is more than steel and plastic consumption combined. International trade in wood and wood products amounts to more than $100 billion each year. Developed countries consume about 80 percent of wood products, but they produce less than 50 percent of industrial wood. Less-developed countries, mainly in the tropics, produce more than half of all industrial wood but use only 20 percent.

Paper pulp, the fastest-growing type of forest product, accounts for nearly a fifth of all wood consumption. Most of the world's paper is used in the wealthier countries of North America, Europe, and Asia. Global demand for paper is also increasing rapidly as more countries become wealthy. The United States, Russia, and Canada are the largest producers of both paper pulp and industrial wood (lumber and panels). Much industrial logging in Europe and North America now occurs on managed plantations, rather than in untouched old-growth forest. However, paper production is increasingly blamed for deforestation in Southeast Asia, West Africa, and other regions.

Fuelwood accounts for nearly half of global wood use. Roughly one-third of the world's population depends on firewood or charcoal as their principal source of heating and cooking fuel (fig. 12.6). Demand for fuelwood is depleting forests in some developing areas, especially around growing cities. Improving stove efficiency and lowering demands for charcoal are seen as urgent needs to reduce these pressures on forests.

Approximately one-quarter of the world's forests are managed for wood production. Ideally, forest management involves scientific planning for sustainable harvests, with particular attention paid to forest regeneration. In temperate regions, according to the FAO, more land is being replanted or allowed to regenerate naturally than is being permanently deforested. Increased sourcing from tropical regions explains much of this forest regeneration.

Harvest methods depend on products, management practices, and environmental conditions. Pulpwood for paper can use all trees in a forest, while fine tropical hardwoods like mahogany may be extracted one at a time. Sometimes long-term management goals or requirements involve replanting after logging; in other cases foresters rely on natural re-seeding. These are some of the main categories of harvesting:

- **Clearcutting** (fig. 12.7a) entails removal of all trees in a stand. It can be a valuable tool for managing some species, such as lodgepole pine, that don't grow well in the shade of other trees. And it can be a rapid and economical technique, but it generally destroys wildlife habitat and exposes soil to erosion and invasive species.

- **Shelterwood** or seed tree systems (fig. 12.7b) are designed to remove a majority of mature trees but leave a protective overstory standing to shelter new growth. In some systems, the sheltering trees are selected to provide a seed source for regeneration.

- **Group selection** and corridor harvesting (fig. 12.7c) call for the harvest of mature trees or the thinning of intermediate size trees at relatively short intervals. The remaining groves or rows of mature trees support natural regeneration and shelter new growth, while the open spaces and corridors provide light and room for new growth.

- **Single tree selective harvesting** (fig. 12.7d) removes individual mature trees, leaving the majority of trees on a site standing. Selected trees may be diseased or undesirable, so that removing them gives space for preferred trees to grow, or they may be high-value species, such as teak or mahogany.

Reforestation for wood production, however, is not the same as maintaining a natural forest. Usually, production is most efficient in large plantations of a single species, in intensive cropping

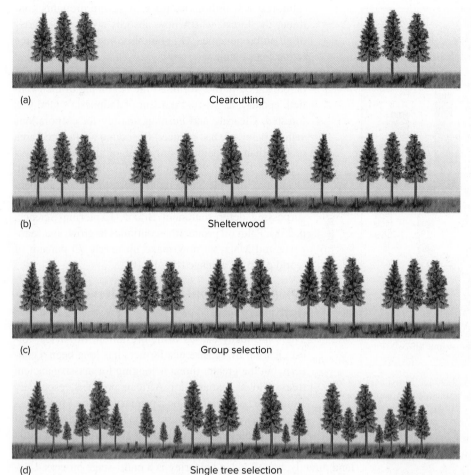

(a) Clearcutting

(b) Shelterwood

(c) Group selection

(d) Single tree selection

FIGURE 12.7 Different harvest methods are appropriate for specific landscapes, species combinations, management objectives, and climate.

planted billions of trees to control erosion and slow the spread of deserts. Korea and Japan also have had very successful forest restoration programs. After being almost totally denuded during World War II, both countries are now about 70 percent forested. However, these countries have accomplished gains largely by contributing to deforestation in other countries and regions.

Tropical forests are especially threatened

Tropical forests are among the richest and most diverse terrestrial systems. This is especially true of the moist forests (rainforests) of the Amazon and Congo River basins and of Southeast Asia. Although these forests now occupy less than 10 percent of the earth's land surface, they are thought to contain more than two-thirds of all higher plant biomass and at least half of all the plant, animal, and microbial species in the world.

A century ago, the world's tropical primary forest covered an area larger than the entire United States. At least half that forest has already been cleared or degraded. Every day, logging and burning destroy about 30,000 acres (11,000 ha) of tropical forest, while farming, grazing, or conversion to monoculture plantations degrades

called **monoculture agroforestry.** Cultivating a single species can maximize growth and simplify harvesting (fig. 12.8). However, a monoculture forest plantation generally supports little biodiversity and provides few ecological services, such as soil erosion control or clean water production. The United States Forest Service has stated that water resources are the most valuable product for U.S. forests, especially in arid Western states.

Some of the countries with the most successful reforestation programs are in Asia. China, for instance, cut down most of its forests 1,000 years ago and has suffered centuries of erosion and terrible floods as a consequence. Recently, however, timber cutting in the headwaters of major rivers has been outlawed, and a massive reforestation project has begun. Since 1990, China has

Quantitative Reasoning

Looking at figure 12.9, roughly what percentage of world forests was gained or lost during the period shown? If there are about 4 billion ha of forest in the world, how much area was lost per year between 2005 and 2010? How could Europe be shown to have so much forest area?

FIGURE 12.8 Monoculture forestry, such as this Wisconsin tree farm, produces valuable timber and pulpwood but has little biodiversity.
William P. Cunningham

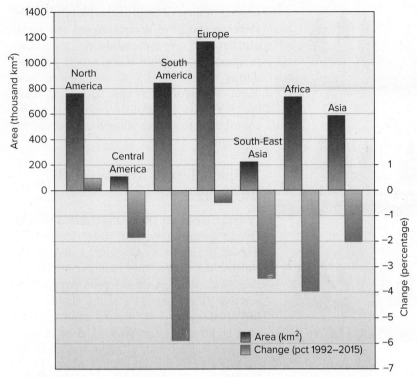

FIGURE 12.9 Forest area and percentage change 1992–2015. Forest clearing for agriculture accounts for most recent changes. (Note that statistics on forest change can depend on varying definitions of "forest.")

a roughly equal area. This amounts to an area about the size of Austria each year.

If current rates of destruction continue, no primary forest will be left in many countries outside of parks and nature reserves by the end of this century. The FAO reports that forest area is declining in most regions of the world, with the most dramatic changes in South America and Africa (fig. 12.9). A large percentage of these regions' vast forests has been eliminated in a few decades. Only North America appears to be making slight gains. Reforestation programs and low rates of corruption in forest management both help account for this. It is also important to note that North Americans buy many of their forest products—and foods grown on once-forested lands—from other parts of the world where protections are lax.

Not only is tropical deforestation a tragic loss of biological diversity, it also means lost livelihoods for the millions of people who depend on forests for part or all of their sustenance. In addition, carbon emissions from deforestation and land use change amounts to as much as 20 percent of all anthropogenic carbon emissions, or more than the emissions from all forms of transportation combined. Halting forest destruction and soil degradation would help significantly to avoid global climate change (see chapter 15). In a positive feedback, climate change also threatens forests. During severe droughts in 2005 and 2010, the Amazon

rainforest lost billions of trees. If severe droughts continue, the forest, which now absorbs about one-quarter of all anthropogenic CO_2, could become a carbon source rather than a sink.

Indonesia is now thought to have the highest rate of deforestation in the world. At the beginning of the twentieth century, at least 84 percent of Indonesia's land was forested. Clearing and burning, mainly for corporations selling palm oil, has reduced the country's forested area to less than 48 percent, only half of that primary forest. An estimated 80 percent of this clearing is illegal. Despite a 2014 pledge to reduce forest clearing, Indonesia continues to lose an estimated 840,000 ha per year, surpassing deforestation in Brazil (Exploring Science, p. 258). Palm oil production continues to grow, and Indonesia and Malaysia now make up nearly 75 percent of world palm oil acreage (fig. 12.10).

Local and global demand drive deforestation

There are many causes of deforestation. In Africa, forest clearing by subsistence farmers has long been a concern, but the greater threat is logging for massive ancient trees. Global demand for African products, especially from Asia, is thought to be a primary cause of forest loss (fig. 12.11). In Latin America, the largest single cause of deforestation is the expansion of soy farming and cattle ranching. Sometimes large-scale ranchers and soy producers clear land directly, but often land clearing is a multi-stage process. Loggers start the process by cutting roads into the forest (fig. 12.12) to harvest valuable hardwoods, such as mahogany or cocobolo. Logging roads allow subsistence farmers to move into the forest,

FIGURE 12.10 World palm oil production has become a leading driver of deforestation. The top four countries make up 85 percent of 21 million ha. *Data source: UN FAO 2018*

FIGURE 12.11 This massive table in a Chinese hotel was sliced from a single giant tropical tree. Trophy products like this help drive global demand for tropical timber.

Mary Ann Cunningham

which they clear for homesteads. Large-scale landowners, ranchers, and soy farmers then move in and remove most remaining forest (chapter 10).

Fires destroy about 350 million ha (1,350 mi²) of forest every year. Some fires are set intentionally to cover up illegal logging or land clearing. Others are started by natural causes. The greatest fire hazard in the world, according to the FAO, is in sub-Saharan Africa, which accounts for about half the global total. Uncontrolled fires tend to be worst in countries with corrupt or ineffective governments and high levels of poverty, civil unrest, and internal refugees. And as global warming brings drought and

insect infestations to many parts of the world, there's a worry that forest fires may increase catastrophically.

Indigenous groups often lead forest protection efforts

About 12.5 percent of all world forests are now in some form of protected status, but the effectiveness of that protection varies greatly. Costa Rica is attempting not only to rehabilitate the land (make an area useful to humans) but also to restore the ecosystems to naturally occurring associations. One of the best known of these projects is entomologist Dan Janzen's work in Guanacaste National Park. Like many dry tropical forests, the northwestern part of Costa Rica had been almost completely converted to ranchland. By controlling fires, however, Janzen and his coworkers are bringing back the forest. One of the keys to this success is involving local people in the project. Janzen also encourages grazing in the park. The original forest evolved, he reasons, together with ancient grazing animals that are now extinct. Ranching can be a force for forest destruction, but cattle can play a valuable role as seed dispersers (fig. 12.13).

Brazil has established vast reserves, but protection varies sharply with political winds. Many of Brazil's forest preservation struggles have been led by indigenous tribes and descendants of African slaves. These communities rely mainly on non-timber forest products, such as rubber tapping, nut collecting, or fishing. They have long fought the government in efforts to protect their land and livelihoods.

At the same time, however, Brazil is pressing ahead with building and paving a network of roads to connect the western Amazon with all-weather, high-speed roads to the Pacific. Critics warn that these projects will accelerate land invasions and will result in displacement of native people and wildlife throughout the forest.

1975

1989

2015

FIGURE 12.12 Forest destruction in Rondonia, Brazil, between 1975 and 2015. Construction of logging roads creates a feather-like pattern that opens forests to settlement by farmers.

Source: a–c: USGS Earth Resources Observation and Science (EROS) Center

Palm Oil and Endangered Species

Are your donuts, toothpaste, and shampoo killing critically endangered orangutans in Sumatra and Borneo? It seems remote, but they might be. Palm oil, a key ingredient in at least half of the packaged foods, cosmetics, and soaps in the supermarket, is almost entirely sourced from plantations that just 20 years ago were moist tropical forest. In Indonesia and Malaysia these forests were the habitat of orangutans, Sumatran tigers and rhinos, and other endangered species. As palm oil has become the world's most widely used vegetable oil, expanding palm oil plantations have become one of the greatest causes of tropical deforestation.

A 2017 study of orangutan populations in Borneo, an island shared by Malaysia and Indonesia, estimated that at least 100,000 of these rare and reclusive forest primates were killed in just 15 years, between 1999 and 2015 (fig. 1). This represents over half of the region's orangutans. By 2050 the population is expected to be only around 50,000, many of them in tiny, dispersed, and nonviable populations. The main reasons for this decline are the rapid conversion of primary forest to palm plantations, deforestation for wood products, and the increasing density of human populations as settlements expand to serve these industries. Habitat loss is a driving factor, but actual mortality in this study was attributed mainly to hunting in the forests around plantations, made possible by the expansion of the plantations and logging roads deep into the primary forest.

In Indonesian, *orang utan* means person of the forest. Orangutans are among our closest primate relatives, sharing at least 97 percent of our genes. Traditional cultures in Borneo may recognize this relationship, because taboos have discouraged hunting and eating them. These taboos seem to be diminishing, however, with the expansion of populations into once-forested regions.

Indonesia and Malaysia produce over 80 percent of the global palm oil supply. In 1960 the two countries together had about 100,000 ha (247,000 acres) of palm oil plantations. That number is now nearly 14 million hectares (34 million acres), according to the UN Food and Agriculture Organization. Expansion of palm plantations usually accompanies other deforestation practices. Often logging firms harvest the valuable hardwoods first; then logging debris is burned to clear the land for planting (and often to cover up illegal logging). Finally, a monoculture of palm trees is planted.

FIGURE 1 Orangutans are among the most critically endangered of all the great apes. Over the past 20 years, about 90 percent of their rainforest habitat in Borneo and Sumatra has been destroyed by logging and conversion to palm oil plantations.
G-miner/Getty Images

Thirsty oil palms need moist soil and a wet climate, so plantations are often established in lowland peat swamps. Peat is composed mainly of ancient, undecomposed plant material, so draining and burning of a hectare of peatland can release 15,000 tons of CO_2. More than 70 percent of the carbon released from Sumatran forests is from burning peat. Indonesia, which has the third largest area of rainforest in the world as well as the highest rate of deforestation, is now the world's third highest emitter of greenhouse gases. Smoke from burning peat often blankets Singapore, Malaysia, and surrounding regions.

At the 2014 UN Climate Summit in New York, 150 companies, including McDonald's, Nestlé, General Mills, Kraft, and Procter & Gamble, promised to stop using palm oil from recently cleared rainforest and to protect human rights in forest regions. Several logging companies, including the giant Asia Pulp and Paper, pledged to stop draining peat lands and to reduce deforestation by 50 percent by 2020.

Will these be effective promises or empty ones? It is difficult to trace oil origins or to monitor remote areas, but at least this movement sets a baseline for acceptable practices. In 2017 two of the world's largest palm oil traders, Wilmar International and Cargill, announced they would no longer do business with a Guatemalan company, Reforestadora de Palmas del Petén S.A. (REPSA), because of environmental and human rights abuses. REPSA was implicated in the murder of Rigoberto Lilma Choc, a 28-year-old Guatemalan schoolteacher who had protested when effluent from a REPSA palm oil operation poisoned the Pasión River, killing millions of fish. When a Guatemalan judge ordered REPSA to stop operations for 6 months, the ruling was quickly followed by the kidnappings of three human rights activists and by Choc's murder.

While the death of 100,000 orangutans has not had the impact of a human murder on global palm oil production and trade, growing awareness can help defend forests, along with forest-dwelling species and people. Throughout the world, monitoring and defending forests is key to protecting biodiversity, climate, water, and cultural diversity. In this chapter we look at the state of forest and grassland reserves, and at efforts to conserve them for future generations.

To read more, see Voigt et al., 2018, Global demand for natural resources eliminated more than 100,000 Bornean orangutans. *Current Biology* 28, 1–9. https://doi.org/10.1016/j.cub.2018.01.053

(header photo): David L. Hansen

FIGURE 12.13 Cattle ranching can increase pressure for forest destruction. However, in the proper setting, cattle also can assist forest regeneration by dispersing seeds.
William P. Cunningham

India has a long history of nonviolent, passive resistance movements—called *satyagrahas*—which often have been associated with forest preservation. Gandhi drew on this tradition in his protests of British colonial rule in the 1930s and 1940s. During the 1970s, commercial loggers began large-scale tree felling in the Garhwal region in the state of Uttar Pradesh in northern India. Landslides and floods resulted from stripping the forest cover from the hills. The firewood on which local people depended was destroyed, and the way of life of the traditional forest culture was threatened. In a remarkable display of courage and determination, the village women wrapped their arms around the trees to protect them, sparking the Chipko Andolan movement (literally, "movement to hug trees"). They prevented logging on 12,000 km² of sensitive watersheds in the Alakanada basin. Today, the Chipko Andolan movement has grown to more than 4,000 groups working to save India's forests.

Debt-for-nature swaps and REDD use finance for protection

Those of us in developed countries also can contribute toward saving tropical forests. Financing nature protection is often a problem in developing countries, where the need is greatest. One promising approach is called **debt-for-nature swaps.** Banks, governments, and lending institutions now hold nearly $1 trillion in loans to developing countries. There is little prospect of ever collecting much of this debt, and banks are often willing to sell bonds at a steep discount—perhaps as little as 10 percent of the debt's value. Conservation organizations buy debt obligations on the secondary market at a discount and then offer to cancel the debt if the debtor country agrees to protect or restore an area of biological importance.

There have been many such swaps. Conservation International, for instance, bought $650,000 of Bolivia's debt for $100,000—an 85 percent discount. In exchange for canceling this debt, Bolivia agreed to protect nearly 1 million ha (2.47 million acres) around the Beni Biosphere Reserve in the Andean foothills.

Another international finance approach is REDD, Reducing Emissions from Deforestation and Degradation. In this plan, wealthier countries pay developing countries to compensate them for protecting forests. In one of the first such agreements, Norway promised Indonesia up to $1 billion in aid, providing that Indonesia made acceptable progress toward sustainable forest management. Norway was motivated because Indonesia's deforestation impacts the global climate not just through loss of trees, but also through burning and decay of vast, ancient peatlands. Together, deforestation, land-use change, and the drying, decomposition, and burning of peatlands cause about 80 percent of Indonesia's current greenhouse gas emissions, which are currently the third highest in the world. Effectively Norway promised to pay for the ecosystem services of Indonesia's standing forest.

Political will has not been strong enough to defend the forests even under this agreement, however, and Indonesia's deforestation continues. Other smaller REDD programs have shown more success. The concept is not yet proven to give enduring protection, however.

Logging threatens temperate forests

In the United States and Canada, considerable forest protection controversy surrounds the ancient forests of the Pacific Northwest. Large areas of the temperate rainforest in this region have been set aside to protect endangered species. These forests have more standing biomass per square kilometer than any other ecosystem on earth (fig. 12.14). Because they're so wet, these forests rarely burn, and trees often live to be a thousand years old and many meters in diameter. A unique biological community has evolved in these dense, misty forests. Dozens of species of plants and animals spend their whole lives in the forest canopy, almost never descending to ground level.

In 1994, the U.S. government adopted the Northwest Forest Plan to regulate harvesting on about 9.9 million ha (24.5 million acres) of federal lands in Oregon, Washington, and northern California. This plan was an admirable example of using good science for natural resource planning. Teams of researchers identified specific areas of ancient forest essential for sustaining viable populations of endangered species, such as the northern spotted owl and the marbled murrelet (chapter 11). The plan prohibited most clear-cut logging, especially on steep hillsides and in riparian (streamside) areas where erosion threatens water quality and salmon spawning (fig. 12.15).

Still, logging has been allowed on the "matrix" lands surrounding these islands of ancient, old-growth forests. Conservationists lament the fact that fragmentation reduces the ecological value of the remaining forest, and they claim that many of the areas now lacking old-growth status could achieve the levels of structural complexity and age required for this classification if left uncut for a few more decades.

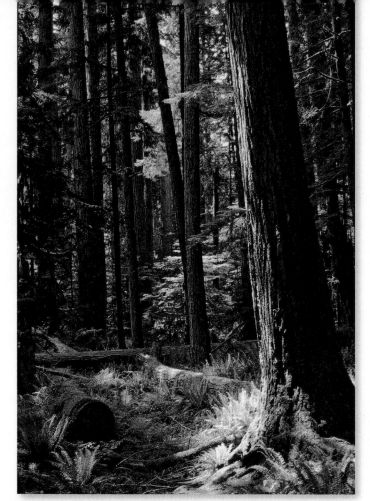

FIGURE 12.14 The huge old trees of the old-growth temperate rainforest accumulate more total biomass in standing vegetation per unit area than any other ecosystem on earth. They provide habitat to many rare and endangered species, but they are also coveted by loggers who can sell a single tree for thousands of dollars.

William P. Cunningham

One of the most controversial aspects of forest management in the United States in recent years has been road building in "de facto" wilderness areas. These areas are effectively wilderness—they have no roads and are not yet accessible to logging or mining—but they have no legal protection. Roads fragment forests, provide a route of entry for hunters and invasive species, often result in erosion, and destroy wilderness qualities. Shortly after passage of the Wilderness Act by the U.S. Congress in 1964, the Forest Service began a review of existing roadless (de facto wilderness) lands. Called the Roadless Area Review and Evaluation (RARE), this effort culminated in 1972 with the identification of 56 million acres (230,000 km²) suitable for wilderness protection. Some of these lands were subsequently included in state wilderness bills, but most remained vulnerable to logging, mining, and other extractive activities.

Roadless area protection gained traction in 2001, during the last days of the Clinton administration, when a national guideline called the **Roadless Rule** was established. This rule ended virtually all logging, road building, and development on virtually all the lands identified as deserving of protection in the 1972 RARE assessment. Despite repeated attempts by George W. Bush to either overturn the Roadless Rule or simply not defend it in lawsuits from timber companies, this rule continues to protect de facto wilderness in 38 states. The Obama administration maintained the rule, but actions of future administrations remain uncertain.

Global warming and fire are growing threats

As discussed in the opening case study, a still greater threat to temperate forests may be posed by climate change, insect infestations, and wildfires, all of which are interconnected. During the past few decades, the average temperature over much of North America has risen by more than 1°F (0.5°C). This may not sound like much, but it has caused the worst drought in 500 years. Hot, dry weather weakens trees and makes them more vulnerable to both insect attacks and fires. A research team led by ecologist Jerry Franklin has showed that tree mortality among a wide variety of species has increased dramatically across a wide area over the past few decades. Infestations by beetles have killed millions of hectares of conifers throughout western North America.

For 70 years, the U.S. Forest Service has had a policy of aggressive fire control. The aim has been to extinguish every fire on public land before 10 a.m. Smokey Bear was adopted as the forest mascot and warned us that "only you can prevent forest fires." Recent studies, however, of fire's ecological role suggest that our attempts to suppress all fires may have been misguided. Many biological communities are fire-adapted and require periodic burning for regeneration. Eliminating fire from these ecosystems has allowed woody debris to accumulate, greatly increasing chances for very big fires.

Forests that once were characterized by 50 to 100 mature, fire-resistant trees per hectare and an open understory now have a thick tangle of up to 2,000 small, spindly, mostly dead saplings in the same area. The U.S. Forest Service estimates that 33 million ha (73 million acres), or about 40 percent of all federal forestlands, are at risk of severe fires. To make matters worse, Americans increasingly

FIGURE 12.15 Clear-cuts, such as this one, threaten species dependent on old-growth forest and expose steep slopes to soil erosion.

Gary Braasch/Getty Images

live in remote areas where wildfires are highly likely. Because there haven't been fires in many of these places in living memory, many people assume there is no danger, but by some estimates 40 million U.S. residents now live in areas with high wildfire risk.

Much of the federal and state firefighting efforts are controlled, in effect, by these homeowners who build in fire-prone areas. A government audit found that 90 percent of the Forest Service firefighting outlays go to save private property. If people who build in forested areas would take more precautions to protect themselves, we could let fires play their normal ecological role in forests in many cases. For example, you shouldn't build a log cabin with a wood shake roof surrounded by dense forest at the end of a long, narrow, winding drive that a fire truck can't safely navigate. If you're going to have a home in the forest, you should use fireproof materials, such as a metal roof and rock or brick walls, and clear all trees and brush from at least 60 m (200 ft) around any buildings.

A recent prolonged drought in the western United States has heightened fire danger there, and in 2015 more than 68,000 wildfires burned 4 million ha (10 million acres) of forests and grasslands in the United States. Federal agencies spent about $2 billion to fight these fires, nearly ten times the average in the 1990s. The dilemma is how to undo years of fire suppression and fuel buildup. Fire ecologists favor small, prescribed burns to clean out debris. Loggers decry this approach as a waste of valuable timber, and local residents of fire-prone areas fear that prescribed fires will escape and threaten them. What do you think? What's the best way to restore forest health while also protecting property values and local jobs?

Ecosystem management seeks resilience

In the 1990s, the U.S. Forest Service began to shift its policies from a timber production focus to **ecosystem management,** which attempts to integrate sustainable ecological, economic, and social goals in a unified, systems approach. Some of the principles of this new philosophy include:

- Managing across whole landscapes, watersheds, or regions over ecological time scales.

- Considering human needs and promoting sustainable economic development and communities.

- Maintaining biological diversity and essential ecosystem processes.

- Utilizing cooperative institutional arrangements.

- Generating meaningful stakeholder and public involvement and facilitating collective decision making.

- Adapting management over time, based on conscious experimentation and routine monitoring.

Some critics argue that we don't understand ecosystems well enough to make practical decisions in forest management on this basis. They argue we should simply set aside large blocks of untrammeled nature to allow for chaotic, catastrophic, and unpredictable events. Others see this new approach as a threat to industry and customary ways of doing things. Still, elements of ecosystem management appear in the *National Report on Sustainable Forests* prepared by the U.S. Forest Service. Based on the Montreal

What Can You Do?

Lowering Your Forest Impacts

For most urban residents, forests—especially tropical forests—seem far away and disconnected from everyday life. There are things that each of us can do, however, to protect forests.

- Reuse and recycle paper. Minimize printing where possible.

- If you build, conserve wood. Use wafer board, particle board, laminated beams, or other composites, rather than plywood and timbers made from old-growth trees.

- Buy certified sustainably harvested wood. Look for the FSC (Forest Stewardship Council) logo.

- Don't buy products made from tropical hardwoods, such as ebony, mahogany, rosewood, or teak, unless the manufacturer can guarantee that the hardwoods were harvested from agroforestry plantations or sustainable-harvest programs.

- Don't patronize fast-food restaurants that purchase beef from cattle grazing on deforested rainforest land. Don't buy coffee, bananas, pineapples, or other cash crops if their production contributes to forest destruction.

- Do buy Brazil nuts, cashews, mushrooms, rattan furniture, and other nontimber forest products harvested sustainably by local people from intact forests. Contact the Forest Stewardship Council to learn about sustainable forestry at https://us.fsc.org/.

- If you hike or camp in forested areas, practice minimum-impact camping. Never transport firewood, which may carry pests that can destroy forests.

- Write to your congressional representatives and ask them to support forest protection and environmentally responsible government policies. Contact the U.S. Forest Service, and voice your support for recreational and nontimber forest values.

Table 12.1 Criteria for Sustainable Forestry
1. Conservation of biological diversity
2. Maintenance of productive capacity of forest ecosystems
3. Maintenance of forest ecosystem health and vitality
4. Maintenance of soil and water resources
5. Maintenance of forest contribution to global carbon cycles
6. Maintenance and enhancement of long-term socioeconomic benefits to meet the needs of legal, institutional, and economic framework for forest conservation and sustainable management

Source: U.S. Forest Service, 2002

Working Group criteria and indicators for forest health, this report suggests goals for sustainable forest management (table 12.1).

Section Review

1. What do we mean by *closed-canopy* forest and *primary* forest?

2. Which commodity is used most heavily in industrial economies: steel, plastic, or wood?

3. What is a *debt-for-nature swap*?

12.2 GRASSLANDS

- *Grasslands occupy more than one-quarter of the world's land surface and contribute crucially to human economy and biodiversity protection.*

- *Worldwide, the rate of grassland disturbance each year is three times that of tropical forests.*

- *Short-duration, rotational grazing can emulate the effects of wild herds and benefit grassland health.*

After forests, grasslands are among the biomes most heavily used by humans. Grasslands vary from humid to near-desert conditions and from open prairie to partially wooded savannas. Together, the various types of grasslands occupy about 27 percent of the world's land surface. Much of the U.S. Great Plains and the Prairie Provinces of Canada fall into this category (fig. 12.16). Globally, the 3.8 billion ha (12 million mi^2) of pastures and grazing lands make up about twice the area of all croplands. Livestock are also grazed on about 4 billion ha of other marginally useful land, such as forest, desert, tundra, marsh, and thorn scrub. More than 3 billion cattle, sheep, goats, camels, buffalo, and other domestic animals on these lands make a valuable contribution to human nutrition.

When grazing is not too severe, especially when grazers move from place to place, this use of grasslands can be sustainable: Historically, migratory herds of millions of buffalo were an essential part of the Great Plains ecosystem, rejuvenating grasslands as they traveled. The great migratory herds of wildebeest and zebras in Kenya and Tanzania today serve the same function. Unlike domestic grazers, they do not stay in place long enough to deplete biodiversity or expose erodible soils.

FIGURE 12.16 Grasslands are expansive, open environments that can support surprising biodiversity.
William P. Cunningham

Although they may appear uniform to the unfamiliar eye, native prairies support a rich variety of specialized plants and animals. According to the U.S. Department of Agriculture, more threatened plant species occur in grasslands than in any other major American biome. Nearly all of North America's moist grasslands, where rain is sufficient to support crops, has been converted to cropland or urban areas. Most remaining dry grasslands are managed as public range lands. Many of these areas have been grazed unsustainably for decades, resulting in lost vegetation and soil erosion.

Grazing can be sustainable or damaging

Historically, many societies have used grazing lands sustainably. **Pastoralists** (people who live by herding animals) have managed herds in many parts of the world, adjusting to variations in rainfall, seasonal plant conditions, and the nutritional quality of forage to keep livestock healthy and avoid overusing any particular area. Conscientious management can even improve the quality of the range, as grazing stimulates many grasses to grow rapidly, and short-term, intensive grazing creates openings that allow less common species to grow.

When grazing lands are abused by **overgrazing,** especially in arid areas, rain runs off quickly before it can soak into the soil to nourish plants or replenish groundwater. Springs and wells dry up. Seeds can't germinate in the dry, overheated soil. The barren ground reflects more of the sun's heat, changing weather patterns, driving away moisture-laden clouds, and leading to further desiccation. This process of converting once-fertile land to desert is called **desertification.**

The link between desertification and overgrazing has been recognized at least since Plato and Aristotle described land degradation in ancient Greece. More recently, range degradation has accelerated with the growth of both human and livestock populations. In many areas, as in sub-Saharan Africa, political instability also forces herders to overuse fragile lands. Overgrazing accelerates erosion, as well as reducing plant abundance and diversity (fig. 12.17). While natural grasslands often have deep, soft soils that absorb rainwater, the hard surface of bare, overgrazed pastures absorbs little rainfall. Instead, rain runs off the surface, carrying dirt particles with it. Erosion results, often leading to severe gullying that can eventually make pastures unusable. Nearly three-quarters of all rangelands in the world show signs of either degraded vegetation or soil erosion, and climate change is only making this situation worse. But degradation can be reversed with good management. If you were a livestock herder, are there arguments that would convince you to reduce your stocks?

Overgrazing threatens U.S. rangelands

The health of most public grazing lands, in the United States and many other countries, is not good. Political and economic pressures encourage managers to increase grazing allotments beyond the carrying capacity of the range. Lack of enforcement of existing regulations and limited funds for range improvement result in depletion of plant diversity. As livestock selectively consume

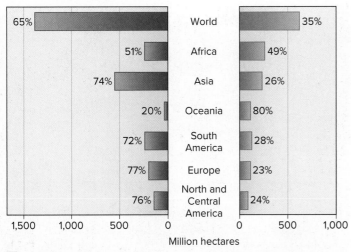

65%	World	35%
51%	Africa	49%
74%	Asia	26%
20%	Oceania	80%
72%	South America	28%
77%	Europe	23%
76%	North and Central America	24%

1,500　1,000　500　0　　0　500　1,000

Million hectares

■ Soil degradation due to overgrazing
■ Soil degradation due to other causes

FIGURE 12.17 Rangeland soil degradation due to overgrazing and other causes. Notice that in Europe, Asia, and the Americas, farming, logging, mining, urbanization, and so on are responsible for about three-quarters of all soil degradation. Grazing damage is higher in Africa and Oceania, where more grazing occurs and much of the range is desert or semiarid scrub.
Source: World Resources Institute, 2004

palatable plants, the range comes to be dominated by unpalatable or inedible species, such as sage, mesquite, cheatgrass, and cactus. Wildlife conservation groups regard cattle grazing as the most widespread form of ecosystem degradation and the greatest threat to endangered species in the southwestern United States. Some have called for a ban on cattle and sheep grazing on all public lands, arguing that those lands support only 2 percent of all livestock producers, and the soil erosion from overused lands degrades watersheds, stream valleys, recreational fisheries, and other valuable resources. The Natural Resources Defense Council reports that only 30 percent of public rangelands are in fair condition, and 55 percent are poor or very poor (fig. 12.18).

Like federal timber management policy, grazing fees charged for use of public lands often are far below market value and are an enormous hidden subsidy to western ranchers. Holders of grazing permits generally pay the government less than 25 percent of what it would cost to lease comparable private land. The 31,000 permits on federal range bring in only $11 million in grazing fees but cost $47 million per year for administration and maintenance. The $36 million difference amounts to a massive "cow welfare" system of which few people are aware.

On the other hand, ranchers defend their way of life as an important part of western culture and history. Although few cattle go directly to market from their ranches, public lands produce almost all the beef calves subsequently shipped to feedlots. And without a viable ranch economy, they claim, even more of the western landscape would be subdivided, to the detriment of both wildlife and environmental quality. Many conservation groups agree that well-managed ranches protect wildlife habitat.

FIGURE 12.18 Overgrazing depletes vegetation and reduces biodiversity. Exposed soils then wash away with rainfall or blow away in the wind. These cattle are grazing on dry public lands in Idaho managed by the Bureau of Land Management.
William H. Mullins/Science Source

What do you think? What are the best ways to protect traditional lifestyles and rural communities while also preserving natural resources?

Ranchers are experimenting with new methods

An experiment in cooperative range management is taking place in the Malpai desert on the U.S-Mexico border (fig. 12.19). In 1990, The Nature Conservancy bought the 1,300 km² (502 mi²) Gray Ranch. Neighboring ranchers worried the land would become a cattle-free wildlife refuge, further challenging an already shaky

FIGURE 12.19 In the Malpai borderlands, the Nature Conservancy is cooperating with neighboring ranchers to practice scientific range management on these arid lands.
William P. Cunningham

FIGURE 12.20 Intensive, rotational grazing encloses livestock in a small area for a short time (often only one day) within a movable electric fence to force them to eat vegetation evenly and fertilize the area heavily.
William P. Cunningham

FIGURE 12.21 Red deer (*Cervus elaphus*) are raised in New Zealand for antlers and venison.
William P. Cunningham

rural economy. But instead the Conservancy decided to turn the land into a model for scientific land use. The ranch is committed to cooperation with neighbors and to demonstrate landscape conservation.

The first area of cooperation involved fire. This desert biome needs periodic fire to suppress the growth of cactus, thorn scrub, and invasive trees, such as mesquite. But prescribed burns require coordination with neighbors. It took a lot of discussion to reach understanding and trust between people with very different backgrounds and philosophies, but eventually a Malpai Borderlands Group emerged and began to coordinate efforts.

During drought periods, western ranchers often face the choice of selling their livestock or running too many animals and degrading the land. The Nature Conservancy offered to serve as a grass bank. If the neighbors put their land into conservation easements, they could graze their cattle on the Conservancy ranch while their own lands recovered. The collaboration has worked splendidly. The health of the range has improved—not only on the Conservancy land but on neighboring ranches as well. Ranchers have cooperated in preserving endangered species. Scientific research projects have documented the role of fire in this ecosystem and its effects on wildlife. And the project now serves as a model for many other cooperative landscape conservation projects.

Rotational grazing can mimic natural regimes

Where livestock are free to roam a large pasture (free-range grazing), they generally eat the tender, best-tasting grasses first. Tough, unpalatable species are left to flourish, and gradually these rough species dominate the pasture. To reduce pasture degradation, some farmers and ranchers practice short-term intensive grazing, which mimics wild grazers. As South African range specialist Allan Savory observed, wild grazers such as African wildebeest or American bison normally form dense herds that graze briefly but intensively in one area, then move on to another. Similarly, herd owners can practice short-duration, **rotational grazing**—confining animals to a small area for a short time (often only a day or two) before shifting them to a new location (fig. 12.20). Livestock are forced to eat everything equally, to trample the ground thoroughly, and to fertilize heavily with manure before moving. Desirable forage species are able to compete with unpalatable weeds in this system. This approach doesn't work in all environments, but many ranchers find it a successful way to reduce damage to pastures.

Alan Savory has also argued that wild species, such as red deer, impala, wildebeest, or oryx (fig. 12.21) forage more efficiently, resist harsh climates, often are more pest- and disease-resistant, and fend off predators better than usual domestic livestock. Native species also may have different feeding preferences and needs for water and shelter than cows, goats, or sheep. The African Sahel, for instance, can provide only enough grass to raise about 20 to 30 kg (44 to 66 lb) of beef per hectare. Ranchers can produce three times as much meat with wild native species in the same area, because these animals browse on a wider variety of plant materials.

In the United States, ranchers find that elk and bison are hardier and require less supplemental feeding than cattle or sheep. Financial returns can be high because their lean meat can bring a better market price than beef or mutton. Media mogul Ted Turner has become both the biggest private landholder in the United States and the owner of more American bison than anyone other than the government.

Section Review

1. Are pastures and rangelands always damaged by grazing animals?
2. What are some results of overgrazing?
3. What is *rotational grazing,* and how does it mimic natural processes?

12.3 PARKS AND PRESERVES

- *Parks and preserves serve ecological, cultural, or recreational purposes.*
- *Different levels of protection are found in nature preserves.*
- *Native people can play important roles in nature protection.*

Although most forests and grasslands serve utilitarian purposes such as timber or livestock production, some natural areas have been set aside for ecological, cultural, or recreational values. Preserving land for non-extractive purposes is nothing new. Ancient Greeks and Druids, for example, protected sacred groves for religious purposes. Royal hunting grounds preserved wild areas in many countries. Although these areas were usually reserved for elite classes in society, they maintained biodiversity and natural landscapes in regions where most lands were heavily used.

The first public parks open to ordinary citizens may have been the tree-sheltered agoras in planned Greek cities. But the idea of providing natural space for recreation, or to preserve natural environments, has really developed in the past half century (fig. 12.22). According to the International Union for Conservation of Nature (IUCN), which maintains some of the best data on global protected areas, nearly 11.6 percent of the surface area of the earth is protected in some sort of park, preserve, or other natural area. This represents over 50 million ha (190,000 mi^2) of marine and terrestrial areas.

Levels of protection vary in preserves

Preserved lands vary greatly in how much protection they receive. The IUCN identifies five categories of protection (table 12.2). In ecological reserves and wilderness areas, little or no human impact is allowed, especially where particularly sensitive wildlife or natural features are present. In some wildlife sanctuaries, visitor numbers are minimized to avoid introducing invasive species or disrupting native species. The least restrictive categories are

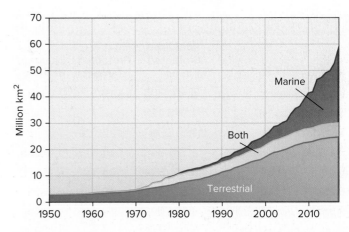

FIGURE 12.22 Growth of protected areas, 1950–2017. The total area has grown from 3 million to 58 million km^2 in this interval.
Source: Data from World Database on Protected Areas, 2018

designated for resource use and management. Most public lands in the United States, including most national forests and many areas leased for oil and gas or mining, are in this category.

Collectively, according to the World Commission on Protected Areas, Central America has 22.5 percent of its land area in some protected status (table 12.3). The Pacific region, at 1.9 percent in nature reserves, has both the lowest percentage and the second lowest total area. With land scarce on small islands, it's hard to find space to set aside for nature sanctuaries. Notice that the numbers reported in figure 12.22 and table 12.3 differ because they are based on different conservation categories.

Venezuela claims to have the highest proportion of its land area protected (70 percent) of any country in the world. About half this land is designated as preserves for indigenous people or for sustainable resource harvesting. With little formal management, there is minimal protection from poaching by hunters, loggers, and illegal gold hunters.

The United States, by contrast, has only about 15.8 percent of its land area in protected status, and less than one-third of

---

Quantitative Reasoning

Figure 12.22 represents shifts over time in global enthusiasm for protecting natural areas. How did total area change from 1950 to 1970? Roughly when did marine preserves increase sharply? Has the rate of terrestrial preservation increased or flattened recently?

Table 12.2 IUCN Categories of Protected Areas	
Category	**Allowed Human Impact or Intervention**
1. Ecological reserves and wilderness areas	Little or none
2. National parks	Low
3. Natural monuments and archaeological sites	Low to medium
4. Habitat and wildlife management areas	Medium
5. Cultural or scenic landscapes, recreation areas	Medium to high
6. Managed resource area	High

Source: Data from World Conservation Union, 1990

Table 12.3 Area in IUCN category 1–4 protection			
Region	Total Area Protected (km²)	Protected Percent	Number of Areas
North America	4,459,305	16.2	13,447
South America	1,955,420	19.3	1,456
North Eurasia	1,816,987	7.7	17,724
East Asia	1,764,648	14.0	3,265
Eastern, Southern Africa	1,696,304	14.1	4,060
Brazil	1,638,867	18.7	1,287
Australia/New Zealand	1,511,992	16.9	9,549
North Africa, Mid-East	1,320,411	9.8	1,325
Western and Central Africa	1,131,153	8.7	2,604
Southeast Asia	867,186	9.6	2,689
Europe	785,012	12.4	46,194
South Asia	320,635	6.5	1,216
Central America	158,193	22.5	781
Antarctic	70,323	0.5	122
Pacific	67,502	1.9	430
Caribbean	66,210	8.2	958
Total	**19,630,148**	**11.6**	**107,107**

The International Union for Conservation of Nature, 2010

that amount is in IUCN categories 1 or 2 (nature reserves, wilderness areas, national parks). The rest is in national forests or wildlife management zones that are designated for sustainable use. With hundreds of thousands of state and federal employees, billions of dollars in public funding, and a high level of public interest and visibility, U.S. public lands are generally relatively well managed.

Brazil, with more than one-quarter of all the world's tropical rainforest, is especially important in biodiversity protection. Currently, Brazil has the largest total area in protected status of any country. More than 1.6 million km² or 18.7 percent of the nation's land—mostly in the Amazon basin—is in some protected status. Some of the newest of these preserves are in the northern Brazilian state of Para, where nine new protected areas were declared along the border with Suriname and Guyana. These new areas, about half of which were planned as strictly protected nature preserves, link together several existing indigenous areas and nature preserves to create the largest tropical forest reserve in the world. More than 90 percent of the new 15 million ha (58,000 mi², or about the size of Illinois) Guyana Shield Corridor is in a pristine natural state. Conservation International president Russ Mittermeir says, "If any tropical rainforest on earth remains intact a century from now, it will be this portion of northern Amazonia."

These protections are vulnerable to shifts in the political winds, however. When Brazil elected the pro-industry president Jair Bolsonaro, in 2018, he reversed decades of protections and pledged to open up the Amazonian forest for further beef and soy production (see chapter 10). He promised to eradicate the Ministry of the Environment as well as all indigenous preserves.

Another area needing protection is the Pantanal, the world's largest wetland/savanna complex, which lies in southern Brazil and is richer in some biodiversity categories than the Amazon, is almost entirely privately owned. There are efforts to set aside some of this important wetland, but so far, little is in protected status.

Very large reserves also exist in Greenland (with a 972,000 km² national park that covers most of the northern part of the island) and Saudi Arabia (with a 640,000 km² wildlife management area in its Empty Quarter). These areas are relatively easy to set aside, however, being mostly ice covered (Greenland) or desert (Saudi Arabia). Canada's Quttinirpaaq National Park on Ellesmere Island is an example of a preserve with high wilderness values but little biodiversity. Only 800 km (500 miles) from the North Pole, this remote park gets fewer than 100 human visitors per year during its brief, three-week summer season (fig. 12.23). With little evidence of human occupation, it has abundant solitude and stark beauty, but very little wildlife and almost no vegetation. Brazil's Amazon has valuable timber, mineral, hydropower, and soil resources, as well as tremendous biodiversity. Slowing exploitation of these resources is very difficult.

Are protected lands representative of the world's ecosystems? Not really. Some biomes—often those that are not very useful to us—are well represented in nature preserves, while others are relatively underprotected. Figure 12.24 shows the percentage of each major biome in protected status. Not surprisingly, there's an inverse relationship between the percentage converted to human use (and where people live) and the percentage protected. Temperate grasslands and savannas (such as the American Midwest) and Mediterranean woodlands and scrub (as in Greece or southern France) are highly domesticated and therefore expensive to set

FIGURE 12.23 Canada's Quttinirpaaq National Park at the north end of Ellesmere Island offers plenty of solitude and pristine landscapes, but little biodiversity.
William P. Cunningham

Percent converted	Biome	Percent protected
45.8	Temperate grasslands and savannas	4.1
41.4	Mediterranean woodlands and scrub	10.2
48.5	Tropical/Subtropical seasonal forests	10.4
46.6	Temperate broadleaf and mixed forests	12.1
27.3	Tropical/Subtropical coniferous forests	8.7
32.2	Tropical/Subtropical moist forests	23.2
23.6	Tropical/Subtropical grasslands and savannas	15.9
26.6	Flooded grasslands and savannas	42.2
6.8	Deserts and scrublands	10.8
12.7	Montane grasslands and scrublands	27.9
12.6	Temperate conifer forests	15.2
2.4	Boreal forests/Taiga	8.5
0.4	Tundra	18.3

FIGURE 12.24 With few exceptions, the percent of each biome converted to human use is roughly inverse to the percent protected in parks and preserves. Rock and ice, lakes, and Antarctic ecoregions are excluded.
Source: World Database on Protected Areas, 2009

aside in large areas. Temperate and boreal conifer forests (as in Siberia or Canada's vast expanse of northern forest) are relatively uninhabited, and therefore easy to put into some protected category.

"Paper parks" are not really protected

Even parks and preserves designated with a high level of protection aren't always safe from exploitation or changes in political priorities. Serious problems threaten natural resources and environmental quality in many countries. In Greece, the Pindus National Park is threatened by plans to build a hydroelectric dam in the center of the park. Furthermore, excessive stock grazing and forestry exploitation in the peripheral zone are causing erosion and loss of wildlife habitat. In Colombia, dam building also threatens the Paramillo National Park. Ecuador's largest park, Yasuni National Park, which contains one of the world's most megadiverse regions of lowland Amazonian forest, has been opened to oil drilling, while miners and loggers in Peru have invaded portions of Huascaran National Park. In Palau, coral reefs identified as a potential biosphere reserve are damaged by dynamite fishing, while on some beaches in Indonesia, every egg laid by endangered sea turtles is taken by egg hunters. These are just a few of the many

problems faced by parks and preserves around the world. Many of the countries with the most important biomes lack funds, trained personnel, and experience to manage the areas under their control.

Even in rich countries, such as the United States, some of the "crown jewels" of the National Park System suffer from overuse and degradation. Yellowstone and Grand Canyon national parks, for example, have large budgets and are highly regulated, but are being "loved to death" because they are so popular. When the U.S. National Park Service was established in 1916, Stephen Mather, the first director, reasoned that he needed to make the parks comfortable and entertaining for tourists as a way of building public support. He created an extensive network of roads in the largest parks so that visitors could view famous sights from the windows of their automobiles, and he encouraged construction of grand lodges in which guests could stay in luxury.

Mather's plan was successful; the National Park System is cherished and supported by many American citizens. But sometimes entertainment seems to have trumped nature protection. Visitors were allowed—in some cases even encouraged—to feed wildlife. Bears lost their fear of humans and became dependent on an unhealthy diet of garbage and handouts (fig. 12.25). In Yellowstone and Grand Teton national parks, the elk herd was allowed to grow to 25,000 animals, or about twice the carrying capacity of the habitat. The excess population overgrazed the vegetation to the detriment of many smaller species and the biological

FIGURE 12.25 Wild animals have always been one of the main attractions in national parks. Many people lose all common sense when interacting with big, dangerous animals. This is not a petting zoo.
Emil Otto Hoppe/ullstein bild via Getty Images

FIGURE 12.26 Thousands of people wait for an eruption of Old Faithful geyser in Yellowstone National Park. Can you find the ranger who's giving a geology lecture?
William P. Cunningham

FIGURE 12.27 Off-road vehicles cause severe, long-lasting environmental damage when driven through wetlands.
Carl Lyttle/Getty Images

community in general. As we discussed earlier in this chapter, 70 years of fire suppression resulted in changes of forest composition and fuel buildup that made huge fires all but inevitable. In Yosemite, you can stay in a world-class hotel, buy a pizza, play video games, do laundry, play golf or tennis, and shop for curios, but you may find it difficult to experience the solitude or enjoy the natural beauty extolled by early conservationists as a prime reason for creating the park.

In many of the most famous parks, tremendous crowds and traffic congestion strain park resources (fig. 12.26). Some parks, such as Yosemite and Zion national parks, have banned private automobiles from the most congested areas. Visitors must park in remote lots and ride to popular sites in clean, quiet buses that run on electricity or natural gas. Other parks are considering limits on the number of visitors admitted each day. How would you feel about a lottery system that might allow you to visit some famous parks only once in your lifetime, but to have an uncrowded, peaceful experience on your one allowed visit? Or would you prefer to be able to visit whenever you wish even if it means fighting crowds and congestion?

Originally, the great wilderness parks of Canada and the United States were distant from development and isolated from most human impacts. This has changed in many cases. Forests are clear-cut right up to some park boundaries. Mine drainage contaminates streams and groundwater. At least 13 U.S. national monuments are open to oil and gas drilling, including Texas's Padre Island, the only U.S. breeding ground for endangered Kemps Ridley sea turtles. Even in the dry desert air of the Grand Canyon, where visibility was once up to 150 km, it's often too smoggy now to see across the canyon, due to air pollution from power plants just outside the park. Snowmobiles and off-road vehicles (ORVs) create pollution and noise and cause erosion while disrupting wildlife in many parks (fig. 12.27).

Chronically underfunded, the U.S. National Park System now has a maintenance backlog estimated to be at least $5 billion. Politicians from both major political parties vow during election campaigns to repair park facilities, but then find other uses for public funds once in office. Ironically, a recent study found that, on average, parks generate $4 in user fees for every $1 they receive in federal subsidies. In other words, they more than pay their own way and should have a healthy surplus if they were allowed to retain all the money they generate.

In recent years, the U.S. National Park System has begun to emphasize nature protection and environmental education over entertainment. This new agenda is being adopted by other countries as well. The IUCN has developed a **world conservation strategy** for protecting natural resources. This strategy has three objectives: (1) to maintain essential ecological processes and life-support systems (such as soil regeneration and protection, nutrient recycling, and water purification) on which human survival and development depend; (2) to preserve genetic diversity essential for breeding programs to improve cultivated plants and domestic animals; and (3) to ensure that any utilization of wild species and ecosystems is sustainable.

Marine ecosystems need greater protection

As ocean fish stocks become increasingly depleted globally (chapter 6), biologists are calling for protected areas where marine organisms are sheltered from destructive harvest methods. Research has shown that establishing no-fishing zones or marine reserves can replenish and maintain fish stocks in surrounding areas. In one study of 100 marine refuges around the world, the number of organisms inside no-fishing preserves was, on average, twice as high as in surrounding areas where fishing was allowed.

The biomass of organisms in preserves was three times as great and individual animals were, on average, 30 percent larger inside the refuge compared to outside. Closing reserves to fishing even for a few months can have beneficial results in restoring marine populations. The size necessary for a safe haven to protect flora and fauna depends on the species involved, but some marine biologists call on nations to protect at least 20 percent of their nearshore territory as marine refuges.

As discussed in chapter 11, coral reefs are among the most threatened marine ecosystems in the world. Remote sensing surveys show that globally, living coral covers only about 285,000 km^2 (110,000 mi^2), or less than half of previous estimates. More important, 90 percent of all reefs face threats from rising sea temperatures, destructive fishing methods, coral mining, sediment runoff, and other human disturbance. Healthy coral reefs are among the richest and most biologically productive communities in existence. But when damaged, these sensitive communities can take a century or more to recover from damage. Some researchers predict that if current trends continue, in 50 years there will be no viable coral reefs anywhere in the world.

What can be done to reverse reef destruction? Marine protected areas are a critical step, and many countries are starting to establish protected or no-fishing zones (figure 12.21). Australia, for example, has declared 2.3 million km^2 (890,000 mi^2) in six areas around its coastline to be marine reserves (fig. 12.28). And the Cook Islands and New Caledonia declared that 2.5 million km^2 (nearly 1 million mi^2) of their territorial waters would be managed sustainably for tourism and fishing. And an area twice as large will be managed to protect dwindling shark populations. Fourteen other small island Pacific nations, which collectively claim about 10 percent of the world's ocean area, have ambitious plans for sustainable management of their territorial waters. In 2016, President Obama expanded the Papahanaumokuakea National Monument around a chain of atolls northwest of Hawaii to 625,325 mi^2 (1.6 million km^2; see chapter 21). Commercial fishing companies complained, however, about being denied access to these rich fishing grounds. In 2017, despite abundant evidence that marine preserves increase fish populations in surrounding oceans, President Trump reduced the size of both of these national monuments and opened them to commercial fishing. This was part of his campaign to reduce protected areas throughout the United States (What Do You Think?, p. 270).

Recently, warming oceans have become a new threat that is even more dire, global, and possibly permanent than overfishing and pollution. In Australia's vast Great Barrier Reef, heat-related coral bleaching badly damaged at least one third of the reef system in 2018. Other reefs are suffering similar damage. This new threat makes marine preserves even more important, as safe havens for surviving reef systems.

Conservation and economic development can work together

Many of the most biologically rich communities in the world are in developing countries, especially in the tropics. These countries are the guardians of biological resources important to all of us.

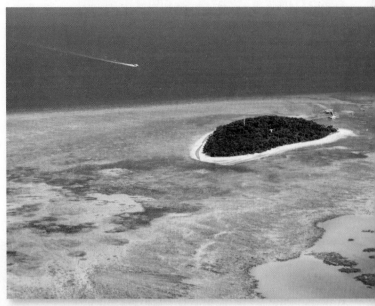

FIGURE 12.28 Australia's Great Barrier Reef is the world's largest coral reef complex. Stretching for nearly 2,000 km (1,200 mi) along Australia's northeast coast, this giant network is one of the biological wonders of the world. It is now threatened by warming oceans.
greenantphoto/Getty Images

Unfortunately, where political and economic systems fail to provide residents with land, jobs, food, and other necessities of life, people do whatever is necessary to meet their own needs. Immediate survival takes precedence over long-term environmental goals. Clearly, the struggle to save species and ecosystems can't be divorced from the broader struggle to meet human needs.

Residents of some developing countries are beginning to realize that their biological resources may be their most valuable assets, and that preserving those resources is vital for sustainable development. **Ecotourism** (tourism that is ecologically and socially sustainable) can be more beneficial in many places over the long term than extractive industries, such as logging and mining. The What Can You Do? box, shown later in this section, suggests some ways to ensure that your vacations are ecologically responsible.

Many preserves support traditional resource uses

The American ideal of wilderness parks untouched by humans is unrealistic in many parts of the world. Some natural areas are extremely vulnerable to human intrusions, but in many others, it is clear that indigenous people have been present for thousands of years and have a legitimate right to pursue traditional ways of life. Furthermore, many of the approximately 5,000 indigenous peoples that remain today possess ecological knowledge about their ancestral homelands that can be valuable in ecosystem management. According to author Alan Durning, "encoded in indigenous languages, customs, and practices may be as much understanding of nature as is stored in the libraries of modern science."

Monuments Under Attack

Land preservation has long been a focus of policy wars in the United States. A favorite ground for skirmishes has been the Antiquities Act of 1906, which authorizes the president to designate historic landmarks, structures, and "other objects of historic or scientific interest" as national monuments. President Theodore Roosevelt was the first to use this authority, establishing 17 monuments, many of which, such as Arizona's Grand Canyon, California's Lassen Peak, and Washington State's Olympic Mountains, have become flagship national parks.

The Antiquities Act frequently protects sites of archaeological and scientific value, as well as recreational resources. The original aim of this legislation was to protect prehistoric Indian ruins and antiquities on western federal lands that were being vandalized by visitors. That use has continued today with the recent designation of Bears Ears National Monument and the Grand Staircase-Escalante Monument in 2016. This area contains at least 100,000 archaeological sites associated with early Clovis, Fremont, and Anasazi cultures. The land was long occupied by native people, who consider it sacred. Protecting it was the culmination of years of campaigning by the neighboring Navajo, Hopi, Zuni, Ute, and Ute Mountain tribes.

The Antiquities Act has been used to protect a wide variety of natural wonders and resources. President Jimmy Carter set aside 56 million acres in Alaska in 17 national monuments, some of which were later promoted to national parks. George W. Bush established the world's first major marine preserve in Hawaii, protecting reproductive areas for fish and other marine life, as well as pristine environments for scientific research and ancient cultural sites. Barack Obama created or expanded 34 national monuments, bringing the total to 129 monuments encompassing 64,750 km^2 (25,000 mi^2) of land plus 3.1 million km^2 (1.2 million mi^2) of marine reserves.

Not everyone appreciates these monuments. Congress has tried more than 11 times over the past century to abolish national monuments, motivated by mining interests or by those who resent Washington's jurisdiction over western lands. These opponents found a new champion in President Trump. Among his first actions on taking office, Trump directed his secretary of the interior at the time, Ryan Zinke, to review the national monuments and marine reserves established by Presidents Clinton and Obama and to recommend which ones could be shrunk or abolished.

The first and most draconian reductions proclaimed by Trump were to Bears Ears and the Grand Staircase–Escalante national monuments in Utah. This rugged, rolling landscape, dissected by spectacular canyons, lies just west of the Glen Canyon Recreation Area in central Utah. It originally encompassed 1.9 million acres (0.77 million ha) of rarely visited wilderness. It's a hiker's paradise with a maze of narrow slot canyons hiding ephemeral streams and temporary pools in the red slickrock. But mining companies are eager to access mineral rights there. In the Grand Staircase-Escalante, the Kaiparowits Plateau is thought to contain 4 billion tons of coal. Bears Ears has uranium that could feed a now-closed mill just outside the monument. Trump and Zinke eliminated 85 percent of Bears Ears and half of the Grand Staircase-Escalante. In announcing his plan, Trump said, "We're returning this land to its rightful owners . . . the local residents." Critics responded that the rightful owners are all the citizens of the United States, because it is federal land.

Native American tribes have another perspective. Russell Begaye, president of the Navajo Nation, said, "If Trump wants to return the land to its rightful owners, he should give it back to indigenous people. This was our land for thousands of years before it was taken by the U.S. Government."

The last word hasn't been heard about these or other purged monuments. Both native tribes and environmental groups have vowed to sue to prevent the destruction of treasured public lands. This case represents one of the most high-profile cases in which environmental groups and Native Americans have joined forces to protect natural resources. It represents a welcome expansion of diversity in public discourse about our environment, as well as growing empowerment of native voices.

These policy debates may not affect most of us immediately, but ultimately this is a question of how we think about and manage public lands and resources. Who should make decisions about these lands? Voters? Industry advocates? If locals have a profit incentive in designating uses of public land, does that increase or decrease their right to determine use of the resources? As scientists and citizens, these are questions that concern all of us.

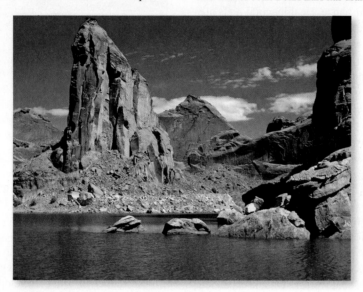

Bluffs and pinnacles of Navajo sandstone stand along the Escalante River near its junction with the Colorado River in Glenn Canyon.

William P. Cunningham

Land protection has sometimes displaced people, sometimes aggressively, and this is highly contentious (fig. 12.29). Historically, colonial governments often ignored the rights of tribal groups and evicted them from newly established parks. Sometimes park authorities have evicted local residents in the context of combatting poachers, often well-armed global traffickers in elephant ivory or tiger parts. Many parks struggle with steadily encroaching human populations, which gradually degrade resources. Kenya, for example, has sought to evict residents of the Mau Forest, whose water catchments supply some of the country's critical water resources. Decades of illicit village and farm expansion have dramatically reduced the forest, threatening nationally critical

What Can You Do?

Being a Responsible Ecotourist

1. *Pretrip preparation.* Learn about the history, geography, ecology, and culture of the area you will visit. Understand the do's and don'ts that will keep you from violating local customs and sensibilities.

2. *Environmental impact.* Stay on designated trails and camp in established sites if available. Take only photographs and memories, and leave only goodwill wherever you go.

3. *Resource impact.* Minimize your use of scarce fuels, food, and water resources. Do you know where your wastes and garbage go?

4. *Cultural impact.* Respect the privacy and dignity of those you meet, and try to understand how you would feel in their place. Don't take photos without asking first. Be considerate of religious and cultural sites and practices. Be as aware of cultural pollution as you are of environmental pollution.

5. *Wildlife impact.* Don't harass wildlife or disturb plant life. Modern cameras make it possible to get good photos from a respectful, safe distance. Don't buy ivory, tortoise shell, animal skins, feathers, or other products taken from endangered species.

6. *Environmental benefit.* Is your trip strictly for pleasure, or will it contribute to protecting the local environment? Can you combine ecotourism with work on cleanup campaigns or delivery of educational materials or equipment to local schools or nature clubs?

7. *Advocacy and education.* Get involved in letter writing, lobbying, or educational campaigns to help protect the lands and cultures you have visited. Give talks at schools or to local clubs after you get home, to inform your friends and neighbors about what you have learned.

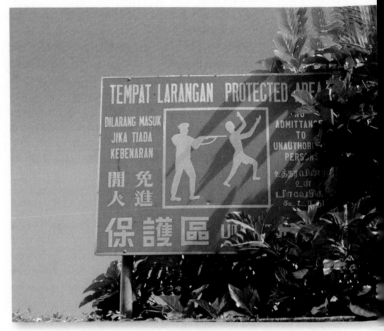

FIGURE 12.29 Some parks take Draconian measures to expel residents and prohibit trespassing. How can we reconcile the rights of local or indigenous people with the need to protect nature?
William P. Cunningham

resources. But some residents have now lived there for a generation or more, and they understandably resist eviction. Observers also point out that while poor settlers are evicted, large-scale deforestation and land clearing continue in other areas. Can rights of residents be protected while also protecting parks from expanding human impacts?

To encourage inclusion of these traditional societies, UNESCO (United Nations Educational, Scientific, and Cultural Organization) initiated in 1986 the **Man and Biosphere (MAB) program,** which encourages the designation of **biosphere reserves.** These reserves are protected areas designed to include low-impact or traditional human activities, rather than excluding them. In principle, reserves are divided into zones with different purposes. Critical ecosystem functions and endangered wildlife are protected in a central core region, where limited scientific study is the only human access allowed. Ecotourism and research facilities are located in a relatively pristine buffer zone around the core, while sustainable resource harvesting and permanent habitation are allowed in multiple-use peripheral regions (fig. 12.30).

A well-established example of a biosphere reserve is Mexico's 545,000 ha (2,100 mi^2) Sian Ka'an Reserve on the Tulum Coast of the Yucatán. The core area includes 528,000 ha (1.3 million acres)

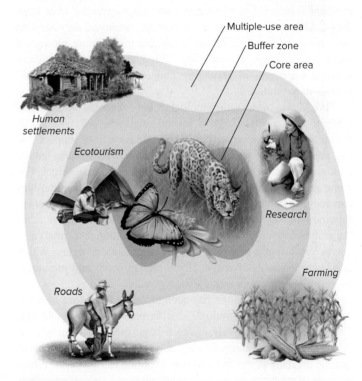

FIGURE 12.30 A model biosphere reserve. Traditional parks and wildlife refuges have well-defined boundaries to keep wildlife in and people out. Biosphere reserves, by contrast, recognize the need for people to have access to resources. Critical ecosystem is preserved in the core. Research and tourism are allowed in the buffer zone, while sustainable resource harvesting and permanent habitations are situated in the multiple-use area around the perimeter.

of coral reefs, bays, wetlands, and lowland tropical forest. More than 335 bird species have been observed within the reserve, along with endangered manatees, five types of jungle cats, spider and howler monkeys, and four species of increasingly rare sea turtles. Approximately 25,000 people (about the same number who live in the Great Bear Rainforest) reside in communities and the countryside around Sian Ka'an. In addition to tourism, the economic base of the area includes lobster fishing, small-scale farming, and coconut cultivation.

The Amigos de Sian Ka'an, a local community organization, played a central role in establishing the reserve and is working to protect natural resources while also improving living standards for local people. New intensive farming techniques and sustainable harvesting of forest products enable residents to make a living without harming their ecological base. Better lobster-harvesting techniques developed at the reserve have improved the catch without depleting native stocks. Local people now see the reserve as a benefit rather than an imposition from the outside. Similar success stories from many parts of the world show how we can support local people and recognize indigenous rights while still protecting important environmental features.

Species survival can depend on preserve size

Many natural parks and preserves are increasingly isolated, remnant fragments of ecosystems that once extended over large areas. As park ecosystems are shrinking, however, they are also becoming more and more important for maintaining biological diversity. Principles of landscape design and landscape structure become important in managing and restoring these shrinking islands of habitat.

For years, conservation biologists have disputed whether it is better to have a *single large or several small* reserves (the SLOSS debate). Ideally, a reserve should be large enough to support viable populations of endangered species, keep ecosystems intact, and isolate critical core areas from damaging external forces. For some species with small territories, several small, isolated refuges can support viable populations, and having several small reserves provides insurance against disease, habitat destruction, or other calamities that might wipe out a single population. But small preserves can't support species such as elephants or tigers, which need large amounts of space. Given human needs and pressures, however, big preserves aren't always possible.

One proposed solution has been to create **corridors** of natural habitat that can link multiple smaller habitat areas (fig. 12.31). Corridors could effectively create a large preserve from several small ones. Corridors could also allow populations to maintain genetic diversity or expand into new breeding territory. The effectiveness of corridors probably depends on how long and wide they are, and on how readily a species will use them.

Perhaps the most ambitious corridor project in the world today is the proposed Yellowstone to Yukon (Y2Y) corridor.

FIGURE 12.31 Corridors serve as routes of migration, linking isolated populations of plants and animals in scattered nature preserves. Although individual preserves may be too small to sustain viable populations, connecting them through river valleys and coastal corridors can facilitate interbreeding and provide an escape route if local conditions become unfavorable.

Linking more than two dozen existing parks, preserves, and wilderness areas, this corridor would stretch 3,200 km (2,000 mi) from the Wind River Range in Wyoming to northern Alaska. More than half this corridor is already forested. Some 31 different First Nations and American Indian tribes occupy parts of this land, and are being consulted in ecosystem management.

One of the reasons large preserves are considered better than small preserves is that they have more **core habitat**—areas deep in the interior of a habitat area that have better conditions for specialized species than do edges. **Edge effects** is a term generally used to describe habitat edges: For example, a forest edge is usually more open, bright, and windy than a forest interior, and temperatures and humidity are more varied. For a grassland, on the other hand, edges may be wooded with more shade and perhaps more predators than in the core of the grassland area. As human disturbance fragments an ecosystem, habitat is broken into increasingly isolated islands, with less core and more edge. Small, isolated fragments of habitat often support fewer species, especially fewer rare species, than do extensive, uninterrupted ecosystems. The size and isolation of a wildlife preserve, then, may be critical to the survival of rare species.

The study of these characteristics of habitat in the landscape is known as **landscape ecology,** a science that examines the

FIGURE 12.32 How small can a nature preserve be? In an ambitious research project, scientists in the Brazilian rainforest are carefully tracking wildlife in plots of various sizes, either connected to existing forests or surrounded by clear-cuts. As you might expect, the largest and most highly specialized species are the first to disappear.
William P. Cunningham

relationship between these spatial patterns and ecological processes, such as species movement or survival. Landscape ecologists explore the ways habitat size, shape, or arrangement influence the distribution—or survival—of species. Landscape ecologists also examine ecological functions of whole landscapes, including farm fields, hedgerows, and other cultivated features, rather than focusing only on pristine wilderness. By quantifying factors such as habitat extent and landscape complexity, landscape ecologists try to guide more effective design of nature preserves and parks.

A dramatic experiment in reserve size, shape, and isolation is being carried out in the Brazilian rainforest. In a project funded by the World Wildlife Fund and the Smithsonian Institution, loggers left 23 test sites when they clear-cut a forest. Test sites range from 1 ha (2.47 acres) to 10,000 ha. Clear-cuts surround some, and newly created pasture surrounds others (fig. 12.32); others remain connected to the surrounding forest. Selected species are regularly inventoried to monitor their survival after disturbance. As expected, some species disappear very quickly, especially from small areas. Sun-loving species flourish in the newly created forest edges, but deep-forest, shade-loving species disappear, particularly when the size or shape of a reserve reduces availability of core habitat. This experiment demonstrates the importance of maintaining core habitat in preserves.

Section Review

1. How do the size and design of nature preserves influence their effectiveness? What do landscape ecologists mean by core habitat and edge effects?

2. What is ecotourism, and why is it important?

3. What is a biosphere reserve, and how does it differ from a wilderness area or wildlife preserve?

Connecting the Dots

Forests and grasslands cover nearly 60 percent of global land area. The vast majority of humans live in these biomes, and we obtain many valuable materials from them. And yet these biomes also are the source of much of the world's biodiversity on which we depend for life-supporting ecological services. But new threats, such as climate change, invasive species, and the expansion of industrial oil palm plantations threaten forests. How we can live sustainably on our natural resources while also preserving enough nature so those resources can be replenished is one of the most important questions in environmental science.

There is some good news in our search for a balance between exploitation and preservation. Although deforestation and land degradation are continuing in some developing countries, many places are more thickly forested now than they were two centuries ago. Protection of Indonesia's rainforest and Australia's Great Barrier Reef shows that we can choose to protect some biodiverse areas in spite of forces that want to exploit them. Overall, nearly 12 percent of the earth's land area is now in some sort of protected status. Although the level of protection in these preserves varies, the rapid recent increase in number and area in protected status exceeds the goals of the United Nations Millennium Project.

We haven't settled the debate between focusing on individual endangered species versus setting aside representative samples of habitat, but pursuing both strategies seems to be working. Protecting charismatic umbrella organisms, such as orangutans in Indonesia or caribou in Alaska, can result in preservation of innumerable unseen species. At the same time, protecting whole landscapes for aesthetic or recreational purposes can also achieve the same end.

Reviewing Key Terms

Can you define the following terms in environmental science?

biosphere reserves 12.3
closed-canopy forest 12.1
core habitat 12.3
corridor 12.3
debt-for-nature-swap 12.1
desertification 12.2

ecosystem management 12.1
ecotourism 12.3
edge effects 12.3
forest 12.1
landscape ecology 12.3

Man and Biosphere (MAB) program 12.3
monoculture agroforestry 12.1
overgrazing 12.2
pastoralists 12.2

primary forest 12.1
Roadless Rule 12.1
rotational grazing 12.2
savannas 12.1
world conservation strategy 12.3

Critical Thinking and Discussion Questions

1. Paper and pulp are the fastest-growing sector of the wood products market, as emerging economies of China and India catch up with the growing consumption rates of North America, Europe, and Japan. What should be done to reduce paper use?

2. Conservationists argue that watershed protection and other ecological functions of forests are more economically valuable than timber. Timber companies argue that continued production supports stable jobs and local economies. If you were a judge attempting to decide which group is right, what evidence would you need on both sides? How would you gather this evidence?

3. Divide your class into a ranching group, a conservation group, and a suburban home-builders group, and debate the protection of working ranches versus building homes or establishing nature preserves. What is the best use of the land? What landscapes are most desirable? Why? How do you propose to maintain these landscapes?

4. Calculating forest area and forest losses is complicated by the difficulty of defining exactly what constitutes a forest. Outline a definition for what counts as forest in your area, in terms of size, density, height, or other characteristics. Compare your definition to those of your colleagues. Is it easy to agree? Would your definition change if you lived in a different region?

5. Why do you suppose dry tropical forest and tundra are well represented in protected areas, while grasslands and wetlands are protected relatively rarely? Consider social, cultural, geographic, and economic reasons in your answer.

6. Oil and gas companies want to drill in a number of parks, monuments, and wildlife refuges. Do you think this should be allowed? Why or why not? Under what conditions would drilling be allowable?

7. If you were superintendent of a major national park, how would you reconcile the demand for comfort and recreation with the need to protect nature? If no one comes to your park, you will probably lose public support. But if the landscape is trashed, what's the purpose of having a park?

Data Analysis

Detecting Edge Effects

Edge effects are a fundamental consideration in nature preserves. We usually expect to find dramatic edge effects in pristine habitat with many specialized species. But you may be able to find interior/edge differences on your own college campus, or in a park or other unbuilt area near you. Here are three testable questions you can examine using your own local patch of habitat: (1) Can an edge effect be detected or not? (2) Which species will indicate the difference between edge and interior conditions? (3) At what distance can you detect a difference between edge and interior conditions? Go to Connect to find a field exercise you can do to form and test a hypothesis regarding edge effects in your own area.

13

Restoration Ecology

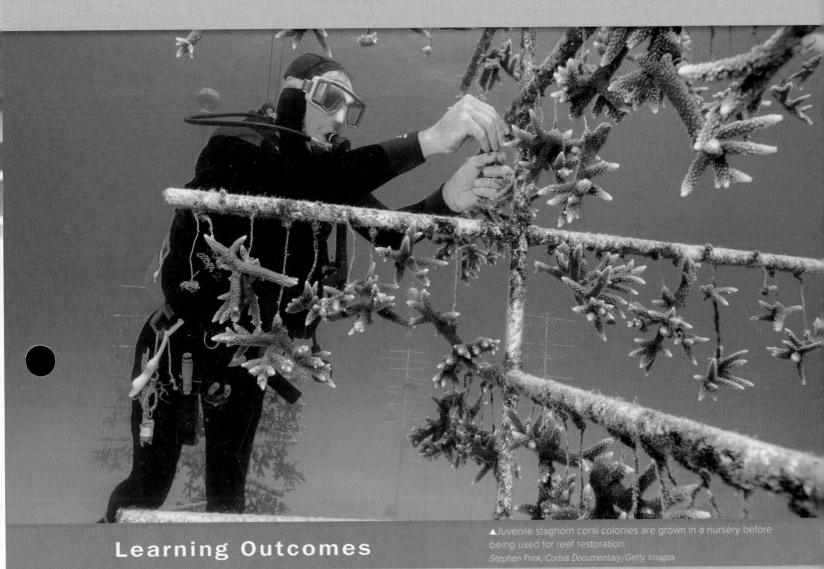

▲Juvenile staghorn coral colonies are grown in a nursery before being used for reef restoration.
Stephen Frink/Corbis Documentary/Getty Images

Learning Outcomes

After studying this chapter, you should be able to:

13.1 Illustrate ways we can help nature heal.
13.2 Explain how restoration can benefit society as well as nature.
13.3 Summarize plans to restore prairies.
13.4 Compare approaches to restoring wetlands and streams.

"When we heal the earth,
we heal ourselves."

– David Orr

Restoring Coral Reefs

Coral reefs are among the richest biological communities on Earth. They're the marine equivalent of tropical rainforests in diversity, productivity, and complexity. It's estimated that one quarter of all marine species spend some or all of their life cycle in the shelter of coral reefs. Globally, at least 17 percent of the protein we eat comes from species that occupy coral reef systems for at least part of their life cycle. In some coastal areas, that number can be as high as 70 percent. Reefs serve as nurseries and food sources for important commercial species, such as shrimp and tuna, and shelter ecologically important species, such as sharks. Reefs protect shorelines from storms, and are valuable recreation attractions for tourists.

But reefs are in serious trouble. According to recent surveys, we've already lost about 30 percent of coral worldwide, and another 60 percent of this valuable natural resource is threatened by climate change, destructive fishing methods, coral mining, sediment runoff, pollution, and other human-caused stressors. Some researchers warn that if current trends continue there won't be any viable coral reefs anywhere in the world by the end of this century.

Reefs are really colonies of tiny invertebrate animals embedded in calcium carbonate shells cemented together to create branches, digits, brackets, heads, and reefs. Individual animals are called polyps, which have minute fan-shaped tentacles to collect zooplankton and nutrients from the water. There can be thousands of polyps on a single coral branch. Nutrients are sparse in the clear, tropical waters where corals live, so reef-building corals form symbiotic relationships with microscopic algae, called zooxanthellae. Photosynthesis by the algae provide as much as 90 percent of the energy the corals need to grow and survive. Consequently, most corals need clear water and abundant sunlight.

One of the most visible and dangerous signs of reef damage occurs when water temperatures get too high. Under these conditions, the symbiotic algae produce toxic by-products that cause the host corals to expel them in a process called bleaching. This doesn't kill the corals immediately, but if they don't reacquire new symbionts, the coral will starve to death, leaving only stark, white carbonate skeletons. Entire reef ecosystems, starved of their primary producers, die off after a bleaching event. Reef bleaching events have become increasingly common around the world as global warming raises seawater temperatures. In 2016, the hottest year on record at that point, over 90 percent of Australia's Great Barrier Reef was affected by bleaching. In one-third of the areas surveyed, between 60 and 100 percent of corals were bleached.

Climate warming is a global risk to reefs, but scientists, volunteers, and community activists are working to protect and restore coral reef systems around the world. Many of these projects are aimed at reducing pollution and destructive human impacts. In Hawaii, large, barge-mounted vacuum cleaners hoover up invasive algae that are smothering reefs. In Palau, the government, together with international advisors, is training community organizations on how to protect priority marine and coastal areas. In Indonesia, conservation organizations are working with indigenous groups to stop destructive harvest techniques, such as cyanide and dynamite fishing. In the U.S. Virgin Islands, officials are working to reduce sediment, sewage, and pollution runoff from the land. And in Australia, divers are removing or killing crown of thorn urchins that destroy corals.

Some of the most exciting projects are studying ways to regrow—and even improve—corals. Some branched corals, such as staghorn and elkhorn, which are among the most threatened of all species, can grow and reproduce through fragmentation. If a branch breaks off and conditions are favorable, it can reattach to the rock substrate and begin to grow a new colony. Researchers are taking advantage of this feature by harvesting coral fragments and growing them in underwater nurseries until they're large enough to be relocated to suitable areas (fig. 13.1). Dozens of these nurseries are now in operation worldwide, and tens of thousands of baby corals have been transplanted to damaged or depleted reefs. Practitioners have found that it's best to create clustered colonies of different coral species so they can protect and support each other.

Some restoration projects are looking for corals with special characteristics to increase the success in restoration efforts. In Ofu lagoon in American Samoa, for example, corals have been found that can survive much warmer water than most corals can tolerate. If studies can unlock the secret of this unusual heat resistance, it could be valuable in restoration efforts. At this point, most coral reefs in the world have bleached, and many have recovered. What different environmental or biological conditions favor recovery? Similarly, an interesting example of natural selection has been

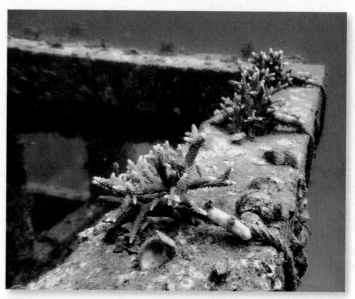

FIGURE 13.1 Fragments of staghorn and elkhom coral can be cultivated in nurseries and used to replenish damaged reef systems.
Placebo365/Getty Images

discovered in the Miami ship channel, where a rare colony of corals was discovered growing in highly polluted, turbid, unsanitary water. If studies can pinpoint why these specimens are so tough, it might help improve other colonies.

Other scientists aren't waiting for nature to produce resistant coral strains. In Australia and Hawaii, projects are working on "assisted evolution." By growing corals in conditions similar to those we expect to see in the future (warm temperatures, acidic water), they're attempting to select for advantageous genes. Breeding experiments are assisted by using species that reproduce continuously rather than only once a year, as most corals do. Interestingly, there's evidence that some coral adaptation to adverse conditions may be due to **epigenetic effects.** That is, environmental factors might alter gene expression (potentially affecting, for example, survival in warmer water) in ways that are passed to offspring. This response could provide added resiliency to coral reefs.

While we probably can't save all reefs, these efforts suggest that we may be able to protect and restore some of this wonderful resource. Restoration ecology is a new, exciting, and experimental field that applies ecological principles to healing natural systems like the coral reefs. Full restoration of an entire ecosystem is a staggering task, but even small steps can make a remarkable difference. And we're finding that nature can be more resilient and robust than we might imagine. In this chapter, we'll examine a number of other cases in which people are working to repair damage and rehabilitate, remediate, or restore ecosystems.

Benchmark Data

Among the ideas and values in this chapter, the following are a few worth remembering.

90%	Fraction of coral reefs worldwide that have been lost or threatened by human-caused damage
1903	President Teddy Roosevelt established Pelican Island as America's first wildlife refuge
2/3	Amount of Japan now forested after being almost completely denuded during WWII
50 billion	Number of trees planted in China to hold back the Gobi Desert
less than 2%	Area of original tall-grass prairie still remaining in the United States
90%	Decrease in monarch butterfly population in past 25 years
150	Number of wild bison in the United States in 1900
500,000	Number of bison in the United States today (mostly on bison ranches)

13.1 HELPING NATURE HEAL

- *A strict definition of restoration is to return a biological community as nearly as possible to a pristine, predisturbance condition.*
- *A broader definition of restoration is simply to develop a self-sustaining, useful ecosystem with as many of its original elements as possible.*
- *Restoration often is fraught with philosophical questions about what we want to restore, as well as pragmatic ones about how to do it.*

Humans have disturbed nature as long as we've existed. With the availability of industrial technology, however, our impacts have increased dramatically in both scope and severity. We've chopped down forests, plowed the prairies, slaughtered wildlife, filled in wetlands, and polluted air and water. Our greatly enhanced power also makes it possible, however, to repair some of the damage we've caused. The relatively new field of **ecological restoration** attempts to do this based on both good science and pragmatic approaches. Some see this as a new era in conservation history. As we discussed in chapter 1, the earliest phases in conservation and nature protection consisted of efforts to use resources sustainably and to protect special places from degradation.

Now, thousands of projects are under way to restore or rehabilitate nature (fig. 13.2). These range from individual efforts to plant native vegetation in urban yards or parks to huge efforts to restore millions of hectares of prairie or continent-size forests. Some restoration focuses on a particular species, such as the monarch butterfly, while others focus on whole ecosystems. Undoubtedly, the biggest reforestation project in history is the Chinese effort to create a "green wall" to hold back the encroaching Gobi and Taklamakan Deserts. More than a billion people are reported to have planted 50 billion trees in China over the past 30 years.

The success of these projects varies. In some cases, the land and biota are so degraded that restoration is impossible. In other situations, some sort of ecosystem can be re-created, but the available soil, water, nutrients, topography, or genetic diversity limit what can be done on a particular site. Figure 13.3 presents a schematic overview of restoration options. Note that restoration to an original pristine condition is rarely possible. Most often, the best choice is a compromise between ideal goals and pragmatic, achievable goals. Choosing these goals is one of the central questions in restoration ecology.

FIGURE 13.2 People in many places are working to restore native vegetation and to protect endangered species. These scouts are replanting the banks of a trout stream.
William P. Cunningham

Table 13.1   A Restoration Glossary
Some commonly used terms in restoration ecology:
• **Restoration:** (strict sense) to return a biological community to its predisturbance structure and function.
• **Restoration:** (broad sense) to reverse degradation and reestablish some aspects of an ecosystem that previously existed on a site.
• **Rehabilitation:** to rebuild a community to a useful, functioning state but not necessarily its original condition.
• **Intervention:** to apply techniques to discourage or reduce undesired organisms and favor or promote desired species.
• **Reallocation:** to use a site (and its resources) to create a new and different kind of biological community rather than the existing one.
• **Remediation:** to clean chemical contaminants from a polluted area using relatively mild or nondestructive methods.
• **Reclamation:** to use powerful chemical or physical methods to clean and repair severely degraded or even barren sites.
• **Re-creation:** to construct an entirely new ecosystem on a severely degraded site.
• **Mitigation:** to replace a degraded site with one of more or less equal ecological value somewhere else.

Restoration projects range from modest to ambitious

Management goals are described in a variety of ways. Table 13.1 summarizes some of the most common terms employed in ecological restoration. A strict definition of restoration would be to return a biological community as nearly as possible to a pristine, predisturbance condition. Often this isn't possible. A broader definition of restoration has a more pragmatic goal simply to develop a self-sustaining, useful ecosystem with as many of its original elements as possible. *Rehabilitation* may seek only to repair ecosystem functions. This may take the form of a community generally similar to the original one on a site, or it may aim for an entirely different community that can carry out the desired functions. Sometimes it's enough to leave nature alone to heal itself, but often we need to intervene in some way to remove or discourage unwanted organisms while also promoting the growth of more desirable species. *Reintroduction* generally implies transplanting organisms from some external source (often a nursery or hatchery where native species are grown under controlled conditions) to a site where they have been reduced or eliminated.

Remediation uses chemical, physical, or biological methods to remove pollution, generally with the intention of causing as little disruption as possible. *Reclamation* employs stronger, more extreme techniques to clean up severe pollution or create a newly functioning ecosystem on a seriously degraded or barren site. *Mitigation* implies compensation for destroying a site by purchasing or creating one of more or less equal ecological value somewhere else.

In some cases, invasion by exotic species coupled with physical changes caused by human actions bring about a state that restoration ecologists call a **novel ecosystem** (see upper right corner of fig. 13.3). In some places, this new hybrid biological community may be all we have to deal with, realistically. Weedy, invasive species may be the only ones able to survive under new conditions. Perhaps we will have to learn to appreciate those invasives if they're supplying ecological services on which we or remaining populations of native species depend. A drastic intervention to try to revert to an earlier state may only doom remaining endangered species. Take the example we discussed in chapter 11 of attempts to eliminate introduced Tamarisk (or salt cedar, *Tamarix* sp.) shrubs in the American Southwest. It turns out that Tamarisk plays a valuable role in stabilizing riverbanks and providing habitat for endangered species, such as the southwestern willow flycatcher.

Restore to what?

Restoration ecologists work in the real world, and although they may dream of returning a disturbed site to its untouched state, they have to deal with the constraints of a specific place, usually faced with multiple obstacles and limited budgets. Some restoration ecologists find it useful to express their goals in terms of ecosystem health or integrity, but others regard this as merely expressions of values or ideology rather than rigorous science. Often, the target

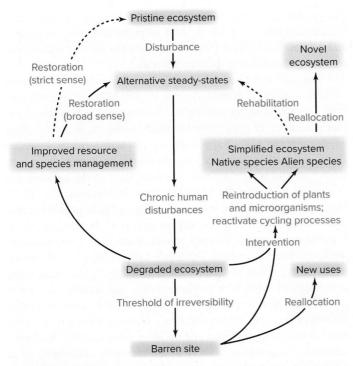

FIGURE 13.3 A model of ecosystem degradation and potential management options.
Source: Data from Walker and Moral 2003

of restoration is a matter of debate. In the Nature Conservancy's Hassayampa River Preserve near Phoenix, for instance, pollen grains preserved in sediments reveal that 1,000 years ago the area was a grassy marsh that was unique in the surrounding desert landscape. Some ecologists would try to rebuild a similar marsh. Corn pollen in the same sediments, however, show that 500 years ago Native Americans began farming the marsh. Which is more important, the natural or the early agricultural landscape?

It may be impossible to return to conditions of either 500 or 1,000 years ago because climate change may have made prehistoric communities incompatible with current conditions. In this case, should restoration ecologists attempt to restore areas to what they used to be, or create a community that will be more compatible with future conditions? Restoration is fraught with such philosophical questions, as well as pragmatic ones.

All restoration projects involve some common components

Different types of restoration involve separate challenges. Wetland restoration, for example, involves a variety of efforts at reestablishing hydrologic connections, plant and animal diversity, weed removal, and sometimes salinity control. Forest restoration, on the other hand, requires a different set of steps that may include controlled burning, selective cutting, weed removal, control of deer and other herbivores, and so on. Even so, there are at least five main components of restoration, and most efforts share a majority of these activities.

1. *Removing physical stressors.* The first step in most restoration efforts is to remove the cause of degradation or habitat loss. Physical stressors such as pollutants, vehicle traffic, or inadequate water supply may need to be corrected. In wetland restoration, water flow and storage usually must be corrected before other steps, such as replanting, can proceed. In forest restoration, clear-cut practices might be replaced with selective logging. In prairie restoration, cultivation might be ended so that land can be replanted with grassland plants, which then provide habitat for grassland butterflies, beetles, and birds.

2. *Controlling invasive species.* Often a few aggressive, "weedy" species suppress the growth of other plants or animals. These invasives may be considered biotic stressors. Removing invasive species can be extremely difficult, but without removal subsequent steps often fail. Some invasives can be controlled by introducing pests that eat only those species: For example, purple loosestrife (*Lythrum salicaria*) is a wetland invader that has succeeded in North America partly because it lacks predators. Loosestrife beetles (*Galerucella* sp.), introduced after careful testing to ensure that their introduction wouldn't lead to further invasions, have successfully set back loosestrife populations in many areas. Leafy spurge (*Euphorbia esula*), similarly, has invaded and blanketed vast areas of the Great Plains (fig. 13.4). Insects, including flea beetles (*Aphthona* sp.), have been a successful strategy to reduce the spread of spurge in many areas, allowing native grassland plants to compete again. Goats also are very effective in removing leafy spurge and other invasive weeds.

FIGURE 13.4 Invasive leafy spurge (*Euphorbia esula*) blankets a formerly diverse pasture. Introduced accidentally, and inedible for most herbivores, this plant costs hundreds of millions of dollars each year in lost grazing value and in weed control.
Mary Ann Cunningham

3. *Replanting.* Restoring a site or ecosystem usually involves some replanting of native plant species. Often, restorationists try to collect seeds from nearby sources (so that the plants will be genetically similar to the original plants of the area), and then they grow them in a greenhouse before transplanting to the restoration site. In some cases, restoration ecologists can encourage existing plants to grow by removing other plants that outcompete the target plants for space, nutrients, sunshine, or moisture.

4. *Captive breeding and reestablishing fauna.* In some cases, restoration involves reintroducing animals. Peregrine falcon restoration involved releasing captive-bred birds, which then managed to survive and nest on their own. In some situations, invertebrates, such as butterflies or beetles, may be released. Sometimes a top predator is reintroduced, or allowed to reinvade. Yellowstone National Park has had a 20-year experimental restoration of wolves. Evidence indicates that these predators are reducing excessive deer and elk populations, thus helping to restore vegetation, as well as reduce mid-level carnivores such as coyotes.

5. *Monitoring.* Without before-and-after monitoring, restoration ecologists cannot know if their efforts are working as hoped. Therefore, a central aspect of this science is planned, detailed, ongoing studies of key factors. Repeated counts of species diversity and abundance can tell whether biodiversity is improving. Repeated measures of water quality, salinity, temperature, or other factors can indicate whether suitable conditions have been established for target species.

Origins of restoration

As European settlers spread across North America in the eighteenth and nineteenth centuries, the woods, prairies, and wildlife populations seemed vast—much too large to be affected by anything humans could do. As we discussed in chapter 1, however, a few pioneers recognized that the rapid destruction of natural communities was unsustainable.

The most influential American forester was Gifford Pinchot. His first job after graduating from college was to manage the wealthy Vanderbilt family's Biltmore estate in North Carolina. Working on this private estate, Pinchot introduced a system of selective harvest and replanting of choice tree species that increased forest value while also producing a sustainable harvest. Pinchot went on to become the first head of the U.S. Forest Service, where he became the first in U.S. history to promote resource management, including the replanting of forests, based on principles of long-term use and scientific research.

Before Pinchot, companies had practiced "cut and run" logging—devastating a region, then moving on to another without any reclamation or replanting, leaving barren expanses of stumps where once had been primeval forests. Pinchot's concepts of science-based management were a vast improvement over these practices. Subsequent critics have complained that his policies have led to replanting of only commercially valuable trees, rather than ecologically important ones, and to a disregard for other ecological functions of a forest, such as protecting water resources and habitat. But Gifford Pinchot's ideas led to the first widespread restoration practices in American forests.

The person most often recognized as the pioneer of restoration ecology was Aldo Leopold. Born at the end of the nineteenth century, Leopold grew up hunting and fishing. He became one of the first generation of professional wildlife managers, in a time when wildlife habitat was disappearing rapidly across North America. Leopold believed that these changes should be turned around. In 1935, Leopold bought a small, worn-out farm on the banks of the Wisconsin River not far from his home in Madison (fig. 13.5). Originally intended to be merely a hunting camp, it quickly became a year-round retreat from the city, as well as a laboratory in which Leopold could test his theories about conservation, game management, and land restoration. The whole Leopold family participated in planting as many as 6,000 trees each spring. "I have read many definitions of what is a conservationist, and written not a few myself," Leopold wrote, "but I suspect that the best one is written not with a pen, but with an axe. It is a matter of what a man thinks about while chopping, or while deciding what to chop. A conservationist is one who is humbly aware that with each stroke he is writing his signature on the face of his land . . . A land ethic then, reflects the existence of an ecological conscience, and this in turn reflects a conviction of individual responsibility for the health of the land . . . Health is the capacity of the land for self-renewal. Conservation is our effort to understand and preserve this capacity."

FIGURE 13.5 Aldo Leopold's Sand County farm in central Wisconsin served as a refuge from the city and as a laboratory to test theories about land conservation, environmental ethics, and ecologically based land management.
William P. Cunningham

Many modern ecologists now regard goals of restoring the health, beauty, stability, or integrity of nature as unscientific, but they generally view Aldo Leopold as a visionary pioneer and an important figure in conservation history.

Sometimes we can simply let nature heal itself

As is the case in restoring coral reefs and other damaged ecosystems, the first step in conservation and ecological restoration is generally to stop whatever is causing damage. Sometimes this is all that's necessary. Nature has amazing regenerative power. If the damage hasn't passed a threshold of irreversibility, natural successional processes can often rebuild a diverse, stable, interconnected biological community, given enough time.

The first official wildlife refuge established in the United States was Pelican Island, a small sand spit in the Indian River estuary not far from present-day Cape Canaveral. The island was recognized by ornithologists in the mid-1800s as being especially rich in bird life. Pelicans, terns, egrets, and other wading birds nested in huge, noisy colonies. But other people discovered the abundant birds as well. Boatloads of tourists slaughtered birds just for fun, and professional hunters shot thousands of adults during the breeding season, when the birds' plumage was most beautiful, leaving fledglings to starve to death. In 1900, the American Ornithological Union, worried about the wanton destruction of colonies, raised private funds to hire wardens to protect the birds during the breeding season. And in 1903, President Theodore Roosevelt signed an executive order establishing Pelican Island as America's first National Bird Reservation. This set an important precedent that the government could set aside public land for conservation. Pelican Island also became the first of 51 wildlife refuges created by President Roosevelt that may have saved species, such as roseate spoonbills (fig. 13.6) and snowy egrets, from extinction.

Many of the forests and nature preserves described in chapter 12 suffered some degree of degradation before being granted protected status. Often, simply prohibiting logging, mining, or excessive burning is enough to allow nature to heal itself. Consider the forests of New England, for example. When the first Europeans arrived in America, New England was mostly densely forested. As settlers spread across the land, they felled the forest to create pastures and farm fields. In 1811, sheep were introduced in New England, and sheep farming expanded rapidly to provide wool to the mills in New Hampshire and Massachusetts. Just 30 years later, in 1840, Vermont had nearly 2 million sheep and 80 percent of its forests were gone. But competition soon ended this boom in wool.

In 1825, the Erie Canal opened access to western farmlands, which could raise both crops and sheep better than the cold Vermont hills. Eli Whitney invented the cotton gin, and cotton became a cheap and abundant (and more comfortable) alternative to woolen cloth. Within a few decades, most Vermont farmers had abandoned large-scale sheep farming, and fallow pastures reverted to forest. Today, 80 percent of the land in Vermont is once again forested, and less than 20 percent is farmland (fig. 13.7).

After a century of natural succession, much of this forest has reacquired many characteristics of old-growth forest: A mixture of native species of different sizes and ages gives the forest diversity and complexity; many of the original animal species—moose, bear, bobcats, pine martins—have become reestablished. There are reports of lynx, mountain lions, and even wolves migrating south from Canada. Now, Vermont law requires that before

FIGURE 13.6 In the nineteenth century, market hunters killed so many birds, such as this Roseate Spoonbill, for their plumes that Congress passed the Lacey Act protecting wildlife, and President Roosevelt created the nation's first Wildlife Refuge to protect them. Since then, populations have recovered.
Source: NPS photo

FIGURE 13.7 A mosaic of cropland, pasture, and sugar bush clothes the hills of Vermont. Two hundred years ago, this area was 80 percent cropland and only 20 percent forest. Today, that ratio is reversed as the forests have invaded abandoned fields. Many of these forests are reaching late successional stages and are reestablishing ecological associations characteristic of old-growth forests.
William P. Cunningham

woodlot owners log their land, they consult with a professional forester to develop a plan to sustain the biodiversity and quality of their forest.

Native species often need help to become reestablished

Sometimes rebuilding populations of native plants and animals is a simple process of restocking breeding individuals. In other cases, however, it's more difficult. Recovery of a unique indigenous seabird in Bermuda is an inspiring conservation story that gives us hope for other threatened and endangered species.

When Spanish explorers discovered Bermuda early in the fifteenth century, they were frightened by the eerie nocturnal screeching they thought came from ghosts or devils. In fact, the cries were those of extremely abundant ground-nesting seabirds now known as the Bermuda petrel, locally called the Bermuda cahow (*Pterodroma cahow*), endemic to the island archipelago (fig. 13.8*a*). There may originally have been half a million of these small, agile, gadfly petrels. Colonists soon found that the birds were easy to catch and good to eat. Those overlooked by humans were quickly devoured by the hogs, rats, and cats that accompanied settlers. Although Bermuda holds the distinction of having passed the first conservation laws in the New World, protecting native birds as early as 1616, the cahow was thought to be extinct by the mid-1600s.

In 1951, however, scientists found 18 nesting pairs of cahows on several small islands in Bermuda's main harbor. A protection and recovery program was begun immediately, including establishment of a sanctuary on 6-hectare (15-acre) Nonsuch Island, which has become an excellent example of ecological restoration.

Nonsuch was a near desert after centuries of abuse, neglect, and habitat destruction. All the native flora and fauna were gone, along with most of the soil in which the cahows once dug nesting burrows. This was a case of re-creating nature rather than merely protecting what was left. Sanctuary superintendent David Wingate, who devoted his entire professional career to this project, brought about a remarkable transformation of the island (fig. 13.8*b*).

The first step in restoration was to remove invasive species and reintroduce native vegetation. Wingate and many volunteers trapped and poisoned pigs and rats and other predators that threatened both wildlife and native vegetation. They uprooted millions of exotic plants and replanted native species, including mangroves and Bermuda cedars (*Juniperus bermudiana*). Initial progress was slow as trees struggled to get a foothold; once the forest knit itself into a dense thicket that deflected salt spray and ocean winds, however, the natural community began to reestablish itself. As was the case on New Zealand's islands (chapter 11), on Nonsuch Island native plants that hadn't been seen for decades began to reappear. Once the rats and pigs that ate seedlings were removed, and competition from weedy invasives was eliminated, native seeds that had lain dormant in the soil began to germinate again.

Still, there wasn't enough soil for cahows to dig the underground burrows they need for nesting. Wingate's crews built artificial cement burrows for the birds. Each pair of cahows lays only one egg per year, and only about half the young survive under ideal conditions. It takes 8 to 10 years for fledglings to mature, giving the species a low reproductive potential. They also compete poorly against the more common long-tailed tropic birds that steal nesting sites and destroy cahow eggs and fledglings. Wingate designed wooden baffles for the burrow entrances, with holes just large enough for cahows but too small for the larger tropic birds (fig. 13.8*c*). The round cement cap at the back of the burrow can be removed to monitor the nesting cahows.

It takes constant surveillance to eradicate exotic plant species that continue to invade the sanctuary. Rats, cats, and toxic toads

(a) A Bermuda cahow.

(b) David Wingate examines a cahow nest.

(c) Long-tall excluder at the mouth of the burrow.

FIGURE 13.8 (a) For more than three centuries, the Bermuda cahow, or Bermuda petrel, was thought to be extinct until a few birds were discovered nesting on small islets in the Bermuda harbor. David Wingate (b) devoted his entire career to restoring cahows and their habitat. (c) A key step in this project was to build artificial burrows. A long-tail excluder device (a board with a hole just the size of the cahow) keeps other birds out of the burrow. A round cement lump at the back end of the burrow can be removed to view the nesting cahows.
William P. Cunningham

FIGURE 13.9 This brackish pond on Nonsuch Island is a reconstructed wetland. Note the Bermuda cedars on the shore and the mangroves planted in the center of the pond by David Wingate.
William P. Cunningham

also swim from the mainland and must be removed regularly. Hurricane Fabian destroyed many nesting burrows on smaller islets in 2003. Fortunately, the restored native forest on Nonsuch Island withstood the winds and preserved the rebuilt soil. The total population is now over 115 breeding pairs, which produce around 60 chicks each year. Reestablishing a viable population of cahows (which are now Bermuda's national bird) has had the added benefit of rebuilding an entire biological community (fig. 13.9).

It's too early to know if the cahow population is large enough to be stable over the long term, but the progress to date is encouraging. Perhaps just as important as rebuilding this species is that Nonsuch has become a living museum of precolonial Bermuda that benefits many species besides its most famous resident. It is a heartening example of what can be done with vision, patience, and a great deal of hard work.

There are many other notable reintroduction programs. Once-rare peregrine falcons have been reestablished across North America. Where they once nested mostly on cliffs, they now have become urban birds and nest on chimneys, rooftops, window ledges, and bridges, where they hunt pigeons. California condors have been reintroduced and have begun to breed in the American West. Both of these species were locally extinct 30 years ago, but captive breeding programs produced enough birds to repopulate much of their former range. Similarly, Arabian Oryx have been successfully reestablished in the deserts of Saudi Arabia, and in Hawaii the endemic nene geese were raised in captivity and reintroduced to Volcano National Park, where they exist in a small but self-reproducing population (see fig. 11.20).

Section Review

1. Define restoration, rehabilitation, intervention, reallocation, remediation, reclamation, re-creation, and mitigation.
2. Describe five common components of restoration projects.
3. What were some of the steps needed to restore the Bermuda cahow?

13.2 RESTORATION IS GOOD FOR HUMAN ECONOMIES AND CULTURES

- *Most major restoration projects have often involved reforestation of cut-over or degraded forest lands.*
- *Fire is an important restoration tool.*
- *You can help restore and improve the quality of your local environment.*

Restoration has become a cornerstone of managing economic resources and a source of cultural pride in many places. Often restoration improves public amenities, as when retired surface mines are reclaimed, contoured, and replanted for new uses. Sometimes it also has commercial value, as when logging companies replant cut or degraded lands for future harvest (fig. 13.10).

These operations might be called silviculture, rather than restoration, because they produce an evenly-spaced stand of just one species. But they do follow Gifford Pinchot's philosophy of long-term resource conservation. Designed to produce wood efficiently, they have little resemblance to diverse, complex native forests (fig. 13.11). Still, these commercial forests supply ground cover,

FIGURE 13.10 Mechanical planters can plant thousands of trees per day.
Bates Littlehales/National Geographic

FIGURE 13.11 Monoculture forests, such as this tree plantation in Germany, often have far less biodiversity than natural forests.
Mary Ann Cunningham

FIGURE 13.12 Student volunteers plant native trees and shrubs to create an urban forest in the Mississippi River corridor within Minneapolis and St. Paul, Minnesota. This provides wildlife habitat, as well as beautifies the urban landscape.
William P. Cunningham

provide more habitat for wildlife than most agricultural lands do, and help moderate the climate.

Sometimes national policy is to protect local forests and to buy wood products from abroad. Japan, for example, was almost completely deforested at the end of World War II. In the decades since then, Japan has carried out a massive reforestation program. Now, more than two-thirds of the land is forest-covered. Tight restrictions on logging help preserve this forest, which has great cultural value for the Japanese people. Rather than cut their own forest, Japan buys wood from its neighbors. Is this fair to low-income, resource-dependent countries that now supply Japan's wood products? Perhaps not, but this practice is increasingly common in China, Europe, North America, and other regions wealthy enough to consume abundant paper, packaging, furniture, and other products.

Often, though, countries undertake forest restoration because their people depend on healthy forests. The small, war-torn country of Rwanda, for example, has undertaken an ambitious 25-year program to restore its degraded forest, soil, water, and wildlife. Poor forest management, damaging land-use practices, and war had caused the country's forests to shrink rapidly. Despite recent economic growth, 85 percent of the population still makes a living from subsistence farming on degraded lands. Among the new Forest Landscape Restoration Initiative priorities will be safeguarding the nation's rich wildlife, such as the critically endangered mountain gorilla, which is an important tourist attraction. For help in this project, Rwanda has turned to the International Union for Conservation of Nature (IUCN), the world's oldest and largest global environmental network. "This really is a good news story," said Stewart Maginnis, Director of Environment and Development of the IUCN. "For the first time, we're actually seeing a country recognize that part and parcel of its economic development trajectory has to be rooted in natural resources." If successful, this will be the first nationwide conservation effort in Africa, and one of very few in the world.

Tree planting can improve our quality of life

Planting trees within cities can be effective in improving air quality, providing shade, and making urban environments more pleasant. Figure 13.12 shows student volunteers planting native trees and bushes in a project called Greening the Great River. Over the past two decades, this nonprofit organization has mobilized more than 30,000 volunteers to plant nearly half a million native trees and shrubs on vacant land within the Mississippi River corridor as it winds through Minneapolis and St. Paul, Minnesota. This project helps beautify the cities, reduces global warming, and provides habitat for wildlife. Deer, fox, raccoons, coyotes, bobcats, and even an occasional wolf or cougar are now seen in the newly revegetated area.

A growing number of communities are recognizing the importance of urban trees. They provide shade and cool the city, sequester carbon, capture pollutants, and support biodiversity. The American Forestry Association has done an extensive remote sensing survey of the United States. They estimate that the national urban tree deficit now stands at more than 634 million trees. They suggest that everyone has a duty to plant at least one

What Can You Do?

Ecological Restoration in Your Own Neighborhood

Everyone can participate in restoring and improving the quality of their local environment.

1. *Pick up litter*
   - With your friends or fellow students, designate one day to carry a bag and pick up litter as you go. A small amount of collective effort will make your surroundings cleaner and more attractive, but also will draw attention to the local environment. Try establishing a competition to see who can pick up the most. You may be surprised how a simple act can influence others. Working together can also be fun!
   - Join a group in your community or on your campus that conducts cleanup projects. If you look around, you'll likely find groups doing stream cleanups, park cleanups, and other group projects. If you can't find such a group, you can volunteer to organize an event for a local park board, campus organization, fraternity, or other group.

2. *Remove invasive species*
   - Educate yourself on what exotic species are a problem in your area: In your environmental science class, assign an invasive species to each student and do short class presentations to educate each other on why these species are a problem and how they can be controlled.
   - Do your own invasive species removal: In your yard, your parents' yard, or on campus, you can do restoration by pulling weeds. (Make sure you know what you're pulling!) Find a local group that organizes invasive species removal projects. Park boards, wildlife refuges, nature preserves, and organizations such as the Nature Conservancy frequently have volunteer opportunities to help eradicate invasive species.

3. *Replant native species*
   - Once the invasives are eliminated, volunteers are needed to replant native species. Many parks and clubs use volunteers to gather seeds from existing native prairie or wetlands, and to help plant seeds or seedlings. Many of the parks and natural areas you enjoy have benefited from such volunteer labor in the past.
   - Create your own native prairie, wetland, or forest restoration project in your own yard, if you have access to available space. You can learn a great deal and have fun with such a project.

4. *Plant a tree or a garden*
   - Everyone ought to plant at least one tree in their life. It's both repaying a debt to nature and a gratifying experience to see a seedling you've planted grow to a mature tree. If you don't have a yard of your own, ask your neighbors and friends if they would like a tree planted.
   - Plant a garden, or join a community garden. Gardening is good for your mind, your body, and your environment. Flowering or fruiting plants provide habitat for butterflies and pollinating insects. Especially if you live in a city, small fragments of garden can provide critical patches of green space and habitat—even a windowsill garden or potted tomatoes are a good start. You might find your neighbors appreciate the effort, too!

tree every year to help restore our environment (What Can You Do?, above).

In 2007, the United Nations initiated a "billion tree campaign," encouraging people everywhere to plant trees in order to protect forest resources, water, climate stability, soil, and biodiversity. This campaign was inspired by the work of Nobel Peace Prize laureate Wangari Maathai. Dr. Maathai founded the Kenyan Green Belt Movement in 1976 as a way of controlling erosion, providing fodder and food, and empowering women. This network of more than 600 local women's groups from throughout Kenya has planted more than 30 million trees while mobilizing communities for self-determination, justice, equality, poverty reduction, and environmental conservation.

Fire is often an important restoration tool

Controlled or **prescribed burning** involves setting fires under carefully controlled conditions to remove brush and flammable material from a forest or grassland. In North America, indigenous people used fire to open up forests and to improve grazing lands long before European colonization. Land managers now recognize that prescribed burning can reduce the threat of catastrophic conflagrations and can benefit fire-adapted biological communities.

For example, oak savannas once covered a broad band at the border between the prairies of the American Great Plains and the eastern deciduous forest (see fig 13.16). Millions of hectares of the American Midwest had parklike savannas, with scattered "open-grown" trees, where dappled sunlight supported a variety of grasses and flowering plants (fig. 13.13). The most common tree species in this environment is bur oak (*Quercus macrocarpa*), but other species, including white and red oak, also occur in savannas. Due to reduced sunlight, however, some dominant prairie species, such as big and little bluestem grasses, most goldenrods, and asters, are generally missing from savannas.

Because people found them attractive places to live, most oak savannas were converted to agriculture or degraded by logging, overgrazing, and fire suppression. Wisconsin, for example, which probably had the greatest amount of savanna of any Midwestern state (at least 2 million ha) before European colonization, now has less than 200 ha (less than 0.01 percent of its original area) of high-quality savanna. Throughout North America, oak savanna rivals the tallgrass prairie as one of the rarest and most endangered plant communities.

It's difficult to restore, or even maintain, authentic oak savannas. Fire was historically important in controlling vegetation. Before settlement, periodic fires swept in from the prairies and removed shrubs and most trees. Mature oaks, however, have thick bark that allows

FIGURE 13.13 Oak savannas are parklike forests where the tree canopy covers 10 to 50 percent of the area, and the ground is carpeted with prairie grasses and flowers. This biome once covered a broad swath between the open prairies of the Great Plains and the dense deciduous forest of eastern North America.
William P. Cunningham

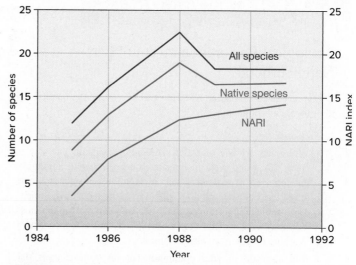

FIGURE 13.14 Biodiversity survey of the Somme Savanna restoration program. NARI, the Natural Area Rating Index, measures the frequency of native species associated with a high-quality community.
Source: Data from S. Packard and J. Balaban, 1994

them to survive low-intensity fires. Grazing by bison and elk may also have helped keep the savanna open. When settlers eliminated fire and grazing by native animals, savannas were invaded by a jumble of shrub and tree growth. Unfortunately, simply resuming occasional burning doesn't result in a high-quality savanna. If fires burn too frequently, they kill oak saplings—which are much more vulnerable than mature trees—and prevent forest regeneration. Excessive grazing can do the same thing. If fires are too infrequent, on the other hand, shrubs, dense tree cover, and dead wood accumulate so that when fire does occur, it will be so hot that it kills mature oaks as well as invading species. It's often necessary to clear most of the accumulated vegetation and fuel before starting fires if you want to maintain a savanna. Herbicide treatment may be necessary to prevent regrowth of invasive species until native vegetation can become established.

Challenges of restoration can be seen in the Somme Prairie Grove in Cook County, Illinois, one of the largest and oldest efforts to restore a native oak savanna, begun in the 1970s. A complex of wetland, prairie, and forest, much of the upland area was an oak savanna. Although the area was used as pasture for more than a century, most of it was never plowed. Limited grazing probably helped keep brush at bay and preserved the parklike character of the remaining forest. Spring burning was started in 1981 in an effort to preserve and restore the savanna. Burning alone, it was discovered, wasn't enough to eliminate aggressive invasive species, such as glossy buckthorn (*Rhamnus frangula*).

After several years of burning, the ground was mostly bare except for weedy species. In 1985, a more intensive management program was begun. Invasive trees were removed. Seeds of native species were collected from nearby areas and broadcast under the oaks. Weeds thought to be a threat to the restoration were reduced by pulling and scything; these included garlic mustard (*Alliaria officinalis*), burdock (*Arctium minus*), briers (*Rubus* sp.), and tall goldenrod (*Solidago altissima*). A research program monitored the results of this restoration effort. Four transects were established and repeatedly sampled for a wide variety of biota (trees, shrubs, herbs, cryptogams, invertebrates, birds, and small mammals) over a six-year period. Results from this survey are shown in fig. 13.14.

As you can see in this figure, the biodiversity of the forest increased significantly over the six years of sampling. Native species increased more than total biodiversity. NARI, the Natural Area Rating Index, is a measure of the relative abundance of native species characteristic of high-quality natural communities. Notice that this index shows the greatest increase of any of the measurements in the study. It also continues to increase after 1988, when native species began to replace invasive pioneer species. This suggests that the restoration efforts have been successful, resulting in an increasingly high-quality oak savanna.

Fire is now recognized as a key factor in preserving oak savannas and many other forest types. The Superior National Forest, which manages the Boundary Waters Canoe Area Wilderness in northern Minnesota, for example, has started an ambitious program of prescribed fires to maintain the complex mosaic of mixed conifers and hardwoods in the forest. Pioneering work in the 1960s by forest ecologist M. L. Heinselman showed that the mixture of ages and species in this forest was maintained primarily by burning. Any given area in the forest experiences a low-intensity ground fire about once per decade, on average, and an intense crown fire about once per century. As these fires burned randomly across the forest, they produced the complex forest we see today (fig. 13.15). But to maintain this forest, we need to allow natural fires to burn once again and to reintroduce prescribed fires where necessary to control fuel buildup.

Similarly, many national parks now recognize the necessity for fire to maintain forests. In Sequoia National Park, for example, rangers recognized that giant sequoias have survived for thousands

FIGURE 13.15 The conifer-hardwood mixture and complex matrix of ages and species in the Great Lakes Forest is dependent on regular but random fire. Maintaining this forest requires prescribed burning.
William P. Cunningham

of years because their thick bark is highly fire-resistant, and they shed lower branches so that fire can't get up into their crown. Decades of fire suppression, however, allowed a dense under-growth to crowd around the base of the sequoias. These smaller trees provide a "ladder" for fire to climb up into the sequoia and kill them. Fuel removal and periodic prescribed fires are now a regular management tool for protecting the giant trees.

Section Review

1. What is a prescribed fire?
2. What are oak savannas, and why are they difficult to restore?
3. Why are giant sequoias now threatened by fire?

13.3 RESTORING PRAIRIES

- *Fire is also crucial for prairie restoration and maintenance.*
- *Depopulation of the American Great Plains opens up the possibility of restoring a buffalo commons.*
- *Keystone animals, such as bison, help maintain prairies.*

Before European settlement, prairies covered most of the middle third of what is now the United States (fig. 13.16). The eastern edge of the Great Plains was covered by tallgrass prairies where big bluestem (*Andropogon gerardii*) reached heights of 2 m (6 ft). Their roots could extend more than 4 m into the soil and formed a dense, carbon-rich sod. This prairie has almost entirely disappeared, having been plowed and converted to corn and soybean fields. Less than 2 percent of the original 1 million km^2 (400,000 mi^2) of tallgrass prairie remains in its original condition. In Iowa, for example, which once was almost entirely covered by this biome, the largest sample of unplowed tallgrass prairie is only about 80 ha (200 acres), about the extent of many

Shortgrass prairie
Mixedgrass prairie
Tallgrass prairie
Savanna

FIGURE 13.16 The eastern edge of the Great Plains was covered by tallgrass prairie, where some grasses reached heights of 2 m (6 ft) and had roots more than 4 m long that formed a dense, carbon-rich sod. Less than 2 percent of the original 10 million km^2 (400,000 mi^2) of tallgrass prairie remains in its original condition. The middle of the Great Plains contained a mixed prairie with both bunch and sod-forming grasses. Few grasses in this region grew to heights of more than 1 m. The westernmost region of the Great Plains, in the rain shadow of the Rocky Mountains, had a shortgrass prairie where sparse bunch grasses rarely grew to more than 30 or 40 cm tall.

college campuses. The loss of this rich ecosystem to farming and development has had severe impacts on biodiversity even beyond this region (Exploring Science, p. 289).

The middle of the Great Plains contained a mixed prairie with both bunch and sod-forming grasses. The drier climate west of the 100th meridian meant that few grasses grew to heights of more than 1 m. The westernmost band of the Great Plains, in the rain shadow of the Rocky Mountains, received only 10 to 12 inches (25 to 30 cm) of rain per year. In this shortgrass region, the sparse bunch grasses rarely grew to more than 30 or 40 cm tall. In some areas, these drier prairies, less useful for crops and other purposes, still remain as pasture or public lands.

Fire is also crucial for prairie restoration

One of the earliest attempts to restore native prairie occurred at the University of Wisconsin. Starting in 1934, Aldo Leopold and others worked to re-create a tallgrass prairie on an abandoned farm field at the University Arboretum in Madison. Student volunteers and workers from the Civilian Conservation Corps gathered seed from remnant prairies along railroad rights-of-way and in pioneer cemeteries, and then hand-planted and cultivated them (fig. 13.17a). Prairie plants initially had difficulty getting established and competing against exotics, until it was recognized that fire is an essential part of this ecosystem. Fire not only kills many weedy species,

(a)

(b)

FIGURE 13.17 In 1934, workers from the Civilian Conservation Corps dug up old farm fields and planted native prairie seeds for the University of Wisconsin's Curtis Prairie (a). The restored prairie (b) has taught ecologists much about ecological succession and restoration.
(a) Courtesy of University Wisconsin-Madison Archives; (b) William P. Cunningham

but it also removes nutrients (especially nitrogen). This gives native species, which are adapted to low-nitrogen soils, an advantage. The Curtis Prairie is now an outstanding example of the tallgrass prairie community and a valuable research site (fig. 13.17b).

The Nature Conservancy (TNC) has established many preserves throughout the eastern Great Plains to protect fragments of tallgrass prairie. The biggest of these (and the largest remaining fragment of this once widespread biome) is in Oklahoma. Here TNC has purchased or obtained conservation easements on about 18,000 ha (45,000 acres) of land just northwest of Tulsa that was grazed but never plowed. In cooperation with the University of Tulsa, TNC has established the Tallgrass Prairie Ecological Research Station, which is carrying out a number of experimental ecological restoration projects on both public and private land. Approximately three dozen prescribed burns are conducted each year, totaling 15,000 to 20,000 acres (fig. 13.18). Since 1991, over 350 randomly selected prescribed burns have been conducted, totaling 210,000 acres. About one-third of each pasture is burned each year. About half of the burns are done in spring and the rest in late summer.

In addition, 300 bison were reintroduced to the preserve in the early 1990s. In 2016, the herd numbered about 2,700 animals on about 25,000 acres (10,900 ha). Patch burning and bison grazing create a habitat that supports the diverse group of plants and animals that make up the tallgrass prairie ecosystem. So far, more than three dozen research projects are active on the preserve, and 78 reports from these studies have been published in scientific journals.

The second largest example of this biome in the United States is the Tallgrass Prairie National Preserve in the Flint Hills region of Kansas. Because the land in this area is too rocky for agriculture, the land was grazed but never plowed. Most of this preserve was purchased originally by the Nature Conservancy, but it is now being managed by the National Park Service. Like its neighbor to the south, a part of this preserve—known as the Konza Prairie—is reserved for scientific research. A long-term ecological research (LTER) program, funded by the National Science Foundation and managed by Kansas State University, sponsors a wide variety of basic scientific research and applied ecological restoration experiments. Bison reintroduction and varied fire regimes are being studied at this prairie. The role of grasslands in carbon sequestration and responses of this biological community to climate change are of special interest.

Huge areas of shortgrass prairie are being preserved

Much of the middle, or mixed grass, section of the Great Plains has been converted to crop fields irrigated by water from the Ogallala Aquifer (chapter 17). As this fossil water is being used up, it may become impossible to continue farming over much of the area using current techniques. Many areas have declining populations (fig. 13.19). As farms and ranches fade away, the small towns that once supplied them also dry up. A great deal of worry and debate exist about the future of this region of the country. Some people argue that the best use for the land may be to restore it to the original prairie, as much as possible.

FIGURE 13.18 A burn-crew technician sets a back fire to control the prescribed fire that will restore a native prairie.
Tony Overman/Associated Press

EXPLORING SCIENCE

The Monarch Highway

Undoubtedly the most widely known and loved insect in the United States is the monarch butterfly (*Danaus plexippus*). As recently as 1996, an estimated 1 billion of these beautiful orange and black insects (fig. 1) migrated annually from breeding grounds across the United States and southern Canada to their winter refuges. Monarchs born west of the Rocky Mountains migrate to refuges in California, while some of those from the East Coast will fly south to Florida. But by far the largest of these migrations follows routes that converge on south Texas and down to the Sierra Madre mountains about 100 km northwest of Mexico City (fig. 2).

In Mexico, the butterflies overwinter in the cool, moist, high-altitude, oyamel fir forests. The next spring, those same animals start back north. The overwintering population may fly as far as Texas or Oklahoma before they lay their eggs and die. The second and third generations continue north to their summer habitat. The fourth generation born in the late fall will migrate back to the wintering grounds.

But in recent years this amazing migration has become alarmingly precarious. Monarch populations have declined by about 90 percent since 1996. One of the critical issues is loss of winter roosting habitat in Mexico to logging and to farmland expansion. Equally urgent is intensification and expansion of U.S. agriculture, and rapidly rising use of pesticides. A critical consequence of agricultural intensification is the disappearance of milkweed plants, the only food source for monarch caterpillars. Herbicide use on monoculture croplands has made milkweeds scarce along the migration route. And cultivation of every bit of land has eliminated many of the weedy corners and fallow fields where adult butterflies once found wildflowers on which they could feed.

Although there are still millions of monarchs, the U.S. Fish and Wildlife Service is considering the species for listing as endangered. When a species—especially one with as complex a life cycle as monarchs—loses so much of its population in so little time, it suggests a slide toward extinction. The monarch population overwintering in Mexico is now so small that a single snowstorm could wipe out the entire migration.

What can you do to help monarchs? First of all, you can plant a butterfly garden in your yard, if you have one, or on your campus, or in your community. Be sure to include some milkweed plants for the caterpillars as well as flowering plants for the adults. Or you

can participate in a prairie restoration project if there is one near where you live. You may even be able to take part in monarch

research. Tens of thousands of school children have participated in monarch tagging by placing tiny sticky id tags on their wings. This

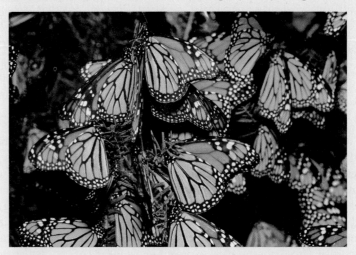

FIGURE 1 Monarch butterflies overwinter in central Mexico.
Jodi Jacobson/Getty Images

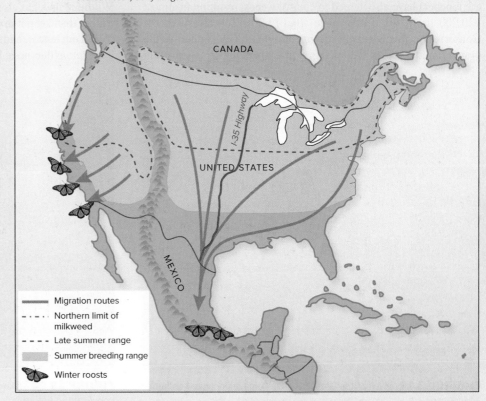

Legend:
— Migration routes
– · – · Northern limit of milkweed
– – – Late summer range
▨ Summer breeding range
🦋 Winter roosts

FIGURE 2 Most of the world's monarchs migrate across regions where development and agriculture have destroyed most habitat. The Interstate-35 highway could provide a corridor of nearly continuous habitat much of the way.
Source: https://www.monarchwatch.org/tagmig/index.htm

(header photo): David L. Hansen

is how we get most of our information about population sizes and migration routes.

In an ambitious, continental-scale restoration effort, a coalition of state and federal agencies, community organizations, departments of transportation, and universities have formed a Monarch Joint Venture, which is working together to rebrand U. S. Interstate 35 as the "Monarch Highway." The freeway runs 2,523 km from Duluth, Minnesota, to Laredo, Texas, right through the heart of the major monarch migration route. For much of that distance, the highway runs across what used to be tall-grass prairie, one of the most endangered biomes in North America. State departments of transportation are planting milkweed and other native flowers along parts of the route. They are adjusting mowing policies to allow wildflowers to bloom and support butterflies. States are writing habitat conservation plans and working with landowners to designate and protect butterfly habitat. They are also encouraging citizen volunteers, like yourself, to participate in supporting habitat restoration, both along the I-35 corridor and elsewhere.

You can learn about this and other butterfly restoration efforts by looking up the Monarch Highway online. Go to MonarchWatch.org for more information and resources.

There are two competing approaches to saving shortgrass prairie: with people or without them. Working for conservation that includes people, the Nature Conservancy (TNC) is cooperating with ranchers on joint conservation and restoration programs. On the 24,000 ha (60,000 acre) Matador ranch in Montana, which TNC bought in 2000, 13 neighboring ranchers graze cattle in exchange for specific conservation measures on their home range. The ranchers have agreed, for instance, to protect about 900 ha of prairie dog colonies and sage grouse leks (dancing grounds). All the ranchers have also agreed to control weeds, resulting in almost 120,000 ha of weed-free range (fig. 13.20). Together with the Conservancy, ranchers are experimenting with fire to improve the prairie, and they plan and manage a grass-banking arrangement in which ranchers access more Conservancy land in drought emergencies. In one case, TNC deeded one of its ranches to a young couple who promised to manage it sustainably and to protect some rare wetlands on the property. The Conservancy believes that keeping ranch families on the land is the best way to preserve both the social fabric and the biological resources of the Plains.

Near the Matador land, another group is pursuing a very different strategy for preserving the buffalo commons. The American Prairie Foundation (APF), which is closely linked to the World Wildlife Fund, has also bought about 24,000 ha of former ranchland. Rather than keep it in cattle production, however, this group intends to pull out fences, eliminate all the ranch buildings, and turn the land back into wilderness. Ultimately, the APF hopes to create a reserve

FIGURE 13.19 Much of the Great Plains is being depopulated as farms and ranches are abandoned and small towns disappear. This provides an opportunity to restore a buffalo commons much like that discovered two centuries ago by Lewis and Clark.
William P. Cunningham

FIGURE 13.20 Millions of hectares of shortgrass prairie are being converted to nature preserves. Some may be populated by bison and other wildlife rather than cattle, as envisioned by artist George Catlin in 1832.
William P. Cunningham

of at least 1.5 million ha in the Missouri Breaks region between the Charles M. Russell National Wildlife Refuge and the Fort Belknap Indian Reservation (fig. 13.21). The APF plans to reintroduce native wildlife, including elk, bison, wolves, and grizzly bears, to its lands. Neighboring ranchers don't mind having elk or bison nearby, but they object to reintroducing wolves and grizzlies, predators their parents and grandparents exterminated a century ago. Ranchers also bristle at the funding for the APF project. Many of the project's large donations come from Wall Street or California's Silicon Valley. Wealthy individuals, such as media mogul Ted Turner, are using their money to make striking changes in how western land is used. Often locals, who struggle just to stay on the land, resent outsiders coming in and competing to buy the range.

With donations of more than $12 million so far, and fund-raising goals of at least $100 million, the APF points out that it isn't forcing anyone from the land. The group is only buying from ranchers who want to sell. Locals worry, nonetheless, that the land will be restricted for hunting and other uses. The APF says it will allow tourism, bird-watching, and hunting on nearly all its land. Small towns also are anxious about how it will affect the local economy to take land out of production. The APF points out that without federal subsidies the average return on ranchland in the Missouri Breaks is less than $5 per acre. Tourism and hunting already bring in more income per acre than does raising cows.

Whether conservation should include human residents and economic activities has often been a deep divide among conservationists, who otherwise agree on the need to preserve biodiversity and ecosystems. Should people expect to maintain an economy in a preserve? How much should those economies be subsidized? Should the ecosystem be restored, predators and all? Or is it complete enough without predators? How would you decide these questions if you were in charge?

Bison help maintain prairies

American prairies coevolved with grazing animals. In particular, a keystone species for the Great Plains was the American buffalo (*Bison bison*). Perhaps 60 million of these huge, shaggy animals once roamed the plains from the Rocky Mountains to the edge of the eastern deciduous forest and from Manitoba to Texas. By 1900, there were probably fewer than 150 wild bison left in the United States, mostly in Yellowstone National Park. Wildlife protection and breeding programs have rebuilt the population to about 500,000 animals, but probably less than 4 percent of them are genetically pure.

Like fire, the bison's intensive grazing helped maintain native plant species. When put on open range, domestic cattle graze selectively on the species they prefer, giving noxious weeds a selective advantage. Bison, on the other hand, tend to move in dense herds, eating almost everything in their path and heavily fertilizing the area over which they pass (fig. 13.22). Their trampling and intense grazing disturb the ground and provide habitat for pioneer species, many of which disappear when bison are removed. Bison also create areas for primary succession by digging out wallows in which they take dust baths.

FIGURE 13.21 Both the Nature Conservancy and the American Prairie Foundation have bought large tracts of land in the Missouri Breaks between the Charles M. Russell National Wildlife Refuge and the Fort Belknap Indian Reservation. Ultimately, conservationists hope to protect and restore as much as 2 million ha of shortgrass prairie.

Having grazed an area heavily, bison will tend to move on, and if they have enough space to roam in, they won't come back for several years. This pattern of intensive, short-duration grazing and localized fertilization creates a mosaic of different successional stages that enhances biodiversity. This is one piece of evidence

FIGURE 13.22 Bison can be an important tool in prairie restoration. Their trampling and intense grazing disturb the ground and provide an opening for pioneer species. The buffalo chips they leave behind fertilize the soil and help the successional process. Where there were only about two dozen wild bison a century ago, there are now more than 400,000 on ranches and preserves across the Great Plains.
Mary Ann Cunningham.

supporting the idea of rotational grazing in sustainable livestock management. Bison increase plant productivity by increasing the availability of light and reducing water stress, both of which increase photosynthesis rates.

Grazing also affects the nutrient cycling in prairie ecosystems. Nitrogen and phosphorus are essential for plant productivity. By consuming plant biomass, bison return these nutrients to the soils in urine and buffalo chips. Bison are more efficient nutrient recyclers than the slow release from plant litter decay. Fire releases nitrogen by burning plant material. Bison, on the other hand, limit nitrogen loss by reducing the aboveground plant biomass and increasing the patchiness of the fire. These changes in nutrient cycling and availability in prairie ecosystems lead to increased plant productivity and species composition.

Many ranchers are coming to recognize the benefits of raising native animals. Where cows require shelter from harsh weather and lots of water to drink, bison are well adapted to the harsh conditions of the prairie. Bison meat is lean and flavorful. It brings a higher price than beef in the marketplace. It's estimated that there are now about 400,000 bison on ranches and farms in North America. This is probably less than 1 percent of the original population, but it's 10,000 times more than the tiny remnant left after the wanton slaughter in the nineteenth century.

Section Review

1. Why are fires essential for prairies?
2. Where did prairies exist in North America before European settlement?
3. Why are bison beneficial for prairies?

13.4 Restoring Wetlands and Streams

- *Wetlands and streams provide essential ecological services, but they've been abused and degraded by drainage and filling.*
- *Some major wetlands such as the Iraq marshes, Gulf of Mexico wetlands, and the American Everglades are being restored.*
- *Many streams need rebuilding.*

Wetlands, rivers, and streams provide important ecological services. They play irreplaceable roles in the hydrologic cycle. They also are often highly productive, and provide food and habitat for a wide variety of species. Although wetlands currently occupy less than 5 percent of the land in the United States, the Fish and Wildlife Service estimates that one-third of all endangered species spend at least part of their lives in wetlands. Storage of floodwaters in wetlands is worth an estimated $3 billion to $4 billion per year. Wetlands also improve water quality by acting as natural water purification systems, removing silt and absorbing nutrients and toxins.

Recognition of these services has come only recently, though. For many years, wetlands were considered disagreeable, dangerous, and useless. This attitude was reflected in public policies such as the U.S. Swamp Lands Act of 1850, which allowed individuals to buy swamps and marshes for as little as 10 cents per acre. Until recently, federal, state, and local governments encouraged wetland drainage and filling to create land for development. In an effort to boost crop production, the government paid farmers to ditch and tile millions of acres of wet meadows and potholes.

Coastal wetlands in Louisiana are especially important and especially vulnerable (fig. 13.23). The low-lying salt marshes around the Mississippi Delta region, built on millions of years of sediment from the Mississippi River, protect inland areas by absorbing hurricane storm surges. They support incredible biodiversity, as well as fisheries and human communities. Decades of development and river modification, though, have undermined these delta wetlands. Building levees and channelizing the Mississippi River has accelerated the movement of flood water and the sediment it carries out into the Gulf of Mexico. Soil that would have once built up the delta and nourished coastal wetlands is lost into the depths of the Gulf. Land subsidence and erosion are destroying about 75 km^2 (18,500 acres) per year in Louisiana alone. At the same time, cutting channels for oil and gas exploration and shortcuts for navigation through Gulf Coast marshes (fig. 13.24) has caused additional destruction of wetlands that once protected New Orleans from hurricanes.

Altogether, the draining and filling of wetlands has destroyed millions of hectares of wetlands in the United States. In Iowa, for example, 99 percent of the original wetlands have disappeared since European settlement. The 1972 Clean Water Act began protecting streams and wetlands by requiring discharge permits for dumping waste into surface waters. In 1977, federal courts interpreted this rule to prohibit both pollution and filling of wetlands (but not drainage). The 1985 Farm Bill went further with a "swamp buster" provision that blocked agricultural subsidies to farmers who drain, fill, or damage wetlands. Many states now have "no net loss" wetlands policies. Between 1998 and 2004, the United States

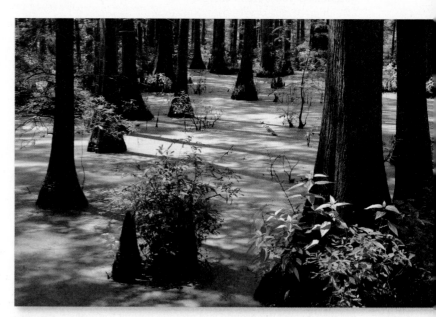

FIGURE 13.23 A Louisiana cypress swamp of the sort being threatened by coastal erosion and salt infiltration. Wetlands provide habitat for a wide variety of species, and play irreplaceable ecological roles.
William P. Cunningham

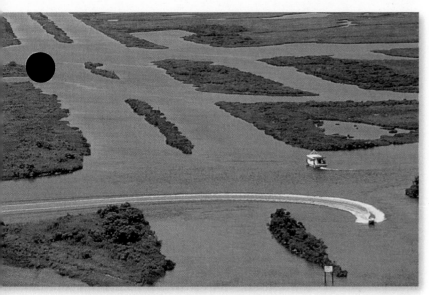

FIGURE 13.24 Louisiana has nearly 40 percent of the remaining coastal wetlands in the United States, but diversion of river sediments that once replenished these marshes and swamps, together with channels and boat wakes, are causing the Gulf shoreline to retreat about 4 m (13.8 ft) per year.
Chris Harris

FIGURE 13.25 Dams on the upper Mississippi River have created a series of large lakes. Currents created by wind, floods, and river traffic keep the soft sediment stirred up so that wetland plants can't take root. To restore these backwaters, the Army Corps is experimenting with wing dams (see the thin white lines between islands) that will allow the mud to solidify and turn into marshlands.
Source: U.S. Army Corps of Engineers

had a net gain of about 80,000 ha (197,000 acres) of wetlands. This total area concealed an imbalance, however. Continued losses of 210,000 ha of swamp and marsh wetlands were offset by a net gain of 290,000 ha of small ponds and shallow-water wetlands (which are easy to construct and good for duck production). And since 2004, another 700,000 ha have been added to the national total. This is a good start, but the United States needs much more wetland restoration to undo decades of damage. In this section, we'll look at some efforts to accomplish this goal.

Restoring river flow helps wetlands heal

As is the case with other biological communities, sometimes all that's needed is to stop the destructive forces. For wetlands, this often means simply to restore water and sediment supplies that have been diverted elsewhere.

To restore Gulf Coast wetlands, gaps are being cut in the levees below New Orleans, allowing fresh, muddy Mississippi River water to reenter the wetlands. Sediment deposited as the water slows is rebuilding eroded marshes, and fresh water encourages revegetation. In a few experimental restoration areas, such as the Caernarvon Diversion east of New Orleans, simply replenishing freshwater flow has reduced saltwater intrusion and already helped expand the area of healthy wetland.

One of the benefits of removing dams is replenishment of beaches. In many coastal areas of the United States, as well as elsewhere in the world, damming rivers has trapped sediment upstream that would once have been carried down to the sea to form deltas, beaches, and barrier islands. Coastal currents, wave action, and rising sea levels all are causing erosion that depletes beaches. Some communities pay millions of dollars annually to dredge sediment from the ocean, or to truck in sand from elsewhere. A more natural solution is to restore stream sediment flow.

In 2013, Washington State took down two century-old dams on the Elwha River, in order to restore salmon runs on the river. The first two years after the Elwha dams were removed, about 4 million cubic yards (nearly 10 million metric tons) of sediment was carried to the river mouth where it has formed about 70 acres (300,000 m^2) of new delta, estuary, and beach. This additional habitat has benefited crabs, clams, sea birds, fish, and other coastal creatures. It also provides new recreational opportunities for human neighbors.

Another simple physical solution to wetland degradation can be seen in the upper Mississippi River. When the U.S. Army Corps of Engineers built 26 locks and dams on the upper Mississippi to facilitate barge traffic in the 1950s, it created a series of large impoundments. Sediment from the surrounding farm fields began to fill these pools. This might have created a valuable network of wetlands, but waves and currents created by wind, floods, and river traffic kept the sediment constantly roiled up so that wetland plants couldn't take root. The result has been a series of wide, shallow, semisolid mud puddles that are too thick for fish but too liquid for vegetation. To remedy this situation, the Army Corps is experimenting with wing dams (see the thin white lines connecting islands in fig. 13.25) that separate the backwaters from the main river channel. Engineers hope these dams will allow the mud to solidify and turn into marshlands.

FIGURE 13.26 The Florida Everglades and the unique plants and wildlife that call this river of grass home are endangered by water pollution and diversion projects.
William P. Cunningham

FIGURE 13.28 The naturally meandering Kissimmee River (*right channel*) was straightened by the Army Corps of Engineers (*left*) for flood control 30 years ago. Now the Corps is attempting to reverse its actions and restore the Kissimmee and its associated wetlands to their original state.
Source: South Florida Water Management District

Replumbing the Everglades is one of the costliest restoration efforts ever

A huge, expensive restoration is now under way, theoretically, in the Florida Everglades (fig. 13.26). Famously described as a "river of grass," the Everglades covers nearly four million acres (1.6 million ha) across southern Florida. This vast marshland is created by a slow-moving sheet of water, which can be 80 km (50 mi) wide, but only a few centimeters deep. This river starts in springs near Orlando in the center of the state, and then moves through Lake Okeechobee and flows southward to the Gulf (fig. 13.27). Spreading over most of the southern tip of the peninsula, this landscape supports myriad fish, invertebrates, birds, alligators, and the rare Florida panther.

Farmers found the rich, black muckland of the Everglades could grow fantastic crops if it was drained. Ditching and diverting the water started more than a century ago. A series of floods that threatened the wealthy coastal cities also triggered a demand

FIGURE 13.27 Planned results of the Comprehensive Everglades Restoration Plan. The red dashed outline shows the national park boundary.

for more water management. Once-meandering rivers were straightened to shunt surplus water out to sea. Altogether, the Army Corps of Engineers built more than 1,600 km of canals, 1,000 km of levees, and 200 water-control structures to intercept normal water flow, drain farmlands, and divert floodwater (fig. 13.28). Ironically, many of the cities that demanded the water be diverted are now experiencing water shortages during the dry season. Water that might have been stored in the natural wetlands is no longer available. Nature, also, is suffering from water shortages. The Everglades National Park has lost 90 percent of its wading birds, and there are worries that the entire aquatic ecosystem may be collapsing.

After decades of debate and acrimony, the various stakeholders finally agreed in 2000 on a massive reengineering of the south Florida water system. This grand plan aimed to return some water to the Everglades, yet retain control to prevent flooding. More than 400 km of levees and canals would be removed. New reservoirs would store water currently lost to the ocean, and 500 million liters of water per day would be pumped into underground aquifers for later release. Rivers already are being dechannelized to restore natural meanders that store storm water and provide wildlife habitat (fig. 13.28). It's hoped that simply restoring some of the former flow to the Everglades will allow the biological community to recover, although whether it will be that easy remains to be seen. Altogether, this project was expected to cost at least $8 billion, making it one of the most expensive restoration projects ever undertaken.

Although announced with great fanfare, the restoration project was almost immediately derailed by an extraordinary lobbying blitz from the sugar industry—the largest polluter in the Everglades and one of the largest political donors in the state.

FIGURE 13.29 Volunteers plant dune grasses as part of Chesapeake Bay restoration. The large cylinder is a "bio log" made of biodegradable material that will help stabilize the shore.
Source: National Oceanic and Atmospheric Administration (NOAA)

FIGURE 13.30 Millions of prairie potholes once covered the Great Plains from Iowa and South Dakota, north to Canada's Prairie Provinces. More than half of these shallow ponds have been drained for agricultural crop production, but now many are being replaced.
Source: Photo by Don Poggensee, U.S. Department of Agriculture Natural Resources Conservation Service

Specifically, the industry refused to reduce their effluents or to pay for cleaning up pollution.

Over the past two decades, a number of small projects have been undertaken, but the results have been modest. Delays, corruption, lack of political will, and inflated costs have held back progress. Federal legislators and agencies have pushed for restoration to start, but state and local officials have delayed and impeded the project. Whether restoration will ever be successful remains to be seen. This case shows how large-scale restoration projects involving big financial interests often are complicated by politics.

Restoration of the Chesapeake Bay is another long, expensive, and contentious project. Volunteers are replanting 200,000 ha of coastline, salt marsh, and shallow water areas (fig. 13.29).

Wetland mitigation is challenging

Working in large, complex, highly political ecosystems that have been damaged by a large variety of human actions is especially challenging. Smaller ecosystems can be much easier to restore or replace. An encouraging example is the re-creation of **prairie potholes** in the Great Plains. Before European settlement, millions of these shallow ponds and grassy wetlands once spread across the prairies from Alberta, Saskatchewan, and Manitoba, to northeastern Kansas and western Iowa (fig. 13.30). They produced more than half of all North American migratory waterfowl and mediated flooding by storing enormous amounts of water. As agriculture expanded across the Plains, however, at least half of all prairie wetlands were drained and filled. In the 1900s, Canada's Prairie Provinces had about 10 million potholes; by 1964, an estimated 7 million were gone. The United States had similar losses.

Passage of the Migratory Bird Hunting Stamp Act (popularly known as the Duck Stamp Act) in 1934 marked a turning point in wetland conservation. Over the past 70 years, this program has collected $700 million that has been used to acquire

more than 5.2 million acres of habitat for the National Wildlife Refuge System. In 1934, when the Duck Stamp Act was passed, drought, predators, and habitat destruction had reduced migratory duck populations in the United States to about 26 million birds. John Phillips and Frederick Lincoln, in a 1930 report on the state of American waterfowl, wrote, "We believe it soon will be too late to save [wild-fowl] in numbers sufficient to be of any real importance for recreation in the future." By 2006, however, conservation efforts, a wetter climate, and habitat restoration had increased duck production to about 44 million birds. Duck stamps now provide about $25 million annually for wildlife conservation.

While it's relatively easy to dig new ponds, it's much more difficult to replace other wetland types. As we mentioned earlier in this chapter, between 1998 and 2004 the United States lost 210,000 ha of swamp and marshes. This was offset by repair or construction of 290,000 ha of small ponds and shallow-water wetlands, such as restored prairie potholes. Replacing a damaged wetland with a substitute is called **wetland mitigation.** Mitigation with a functionally similar wetland is now legally required, but wetlands are complex living communities that are not easy to create from scratch. Figure 13.31, for example, shows a replacement wetland created by a housing developer in Minnesota. In building a housing project, the developer destroyed about 10 ha of a complex, native wetland that contained rare native orchids and several scarce sedge species. To compensate for this loss, the developer simply dug a hole and waited for it to fill with rainwater. He wasn't required to replant wetland species. The law assumes that natural

FIGURE 13.31 In the past, developers weren't required to plant native vegetation in wetland mitigation. They were allowed to simply dig a hole and wait for it to fill with rainwater and invasive species. New rules now require a more ecological approach.
William P. Cunningham

succession will revegetate the disturbed area. In fact, it was soon revegetated, but entirely with exotic invasive species.

Wetland and stream restoration provide multiple benefits

Wetland restoration can have big economic benefits as well as environmental ones. The city of Arcata, California, for example, has used the natural water-filtering capacity of wetlands to avoid an expensive sewer plant upgrade. To comply with clean water requirements and keep excess nutrients out of the adjacent bay, Arcata transformed a 65-ha garbage dump into a series of ponds and marshes that serve as a simple, low-cost waste treatment facility. Arcata saved millions of dollars and improved its environment simultaneously. The marsh is a haven for wildlife and has become a prized recreation area for the city. Eventually, the purified water flows into Humbolt Bay, where marine life flourishes.

On a smaller scale, costs of stormwater management can be reduced with small wetlands and rain gardens. These can collect storm runoff and allow it to seep into the ground rather than run into rivers or lakes. And constructed marshes allow industrial cooling water to equilibrate before entering streams or other surface water bodies. For more on this topic, see chapter 18.

Rivers and streams suffer from pollution, pathogens, industrial contaminants, erosion, invasive species, and other factors. Only about half of U.S. streams fully support multiple uses, including drinking-water supply, fish and wildlife habitat, recreation, and agriculture, as well as flood prevention and erosion control. Sedimentation and excess nutrients were the most significant causes of degradation in the rest.

A common response to erosion and flooding in urban streams has been to turn them into cement channels that rush rainwater off into some larger body of water (fig. 13.32) or to bury them in underground culverts. The result is an artificial system with little resemblance to the living biological community that once made up the stream. More and more communities now recognize that natural streams can increase property values and improve the livability of the urban environment. Buried streams are being "daylighted," and channelized ditches are being turned back into living biological communities.

A variety of restoration techniques have been developed for streams. This field has become an important source of jobs for environmental science majors. A simple approach in which everyone can participate is to reduce sediment influx by planting ground cover on uplands and filling gullies with rocks or brush. For small streams, sometimes the quickest way to rebuild a channel is to use heavy earthmoving equipment to simply dig a new one. This can be very disruptive, however, stirring up sediment that can be harmful for fish and other aquatic organisms. Alternative, less intrusive stream improvement methods are available. Most of these methods involve placing barriers (weirs, vanes, dams, log barriers, brush bundles, root wads, or other obstructions) in streams to deflect current away from the banks or trap sediment. Often, these barriers will cause currents to scour out deep pools in the stream bottom that provide places for fish to hide and rest.

Other techniques also create fish habitat with **coarse woody debris.** Logs, root wads, brush bundles, and boulders can shelter fish. An expensive but effective way to create fish hiding places is the so-called "lunker" structure. This is a wood framework that rests on the stream bottom and is anchored securely to the shore. The top of the box is covered with rock, soil, and vegetation. Openings in the structure provide hiding places for fish (fig. 13.33).

FIGURE 13.32 Many former streams have been turned into concrete-lined ditches to control erosion and speed runoff. The result is an artificial system with little resemblance to the living biological community that once made up the stream.
Source: Federal Interagency Stream Corridor Restoration Handbook, NRCS, USDA

FIGURE 13.33 A lunker structure is a multilevel wooden framework that rests on the stream bottom and can be anchored to the shore. The top of the box is covered with rock, soil, and vegetation. Openings in the structure provide hiding places for fish.

Sometimes what's needed is to speed up the current rather than slow it down. Figure 13.34*a* shows a spring-fed Minnesota trout stream that was degraded by crop production in its uplands, and grazing that broke down the banks and filled the stream with sediment. The stream, which had been about 1 m wide and 1 m deep, spread out to be 5 m wide and only about 10 cm deep. The formerly cold, swiftly moving water was warmed by the sun as it passed through these shallows so that it became uninhabitable for trout. Furthermore, there was no place to hide from predators. Because

this was the last remaining trout stream in the Minneapolis/St. Paul metropolitan area, a decision was made to restore it. Two very different restoration approaches are shown in figures 13.34*b* and *c*. Trout Unlimited, an angler's group, offered to bring in a backhoe and completely rebuild the stream. A demonstration section they reconstructed is shown in figure 13.34*b*. They narrowed the channel with large stone blocks, which also provide a good surface for anglers to walk along the stream. There isn't much biological production in this stream model, but that isn't important to many anglers. They expected the Department of Natural Resources to stock the stream with hatchery-raised fish, which usually are caught before they have time to learn to forage for natural food.

Other groups involved in stream restoration objected to the artificial nature and lack of a native biological community in this design. Figure 13.34*c* shows an alternative approach that was ultimately adopted for most of the stream. This is the same section shown in figure 13.34*a*. Straw bales were placed in the deepest part of the stream. This narrowed the stream and increased the speed of the current, which then cut down into the soft sediment. Within three months, the stream had deepened about 50 cm. The shallow area between the straw bales and the original shore quickly filled with sediment and became a cattail marsh. The bales were anchored in place with green willow stakes, which rooted in the moist soil and sprouted to make saplings that arched over the stream and shaded it from the sun.

Seeds of native wetland vegetation were scattered on top of the straw bales. They sprouted, and after the first summer the stream was surrounded by a dense growth of vegetation that keeps the water cool, provides fish shelter, and supports a rich community of invertebrates that feed the native trout. An aquatic invertebrate survey found that while the stream was wide and shallow, 58 percent of the aquatic species were snails and copepods, whereas stone flies, caddis flies, and other preferred trout food made up only 20 percent of all invertebrates. After the straw bale restoration, snails and

(a)

(b)

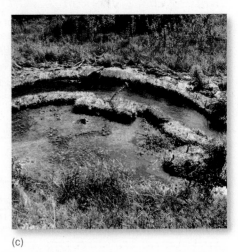

(c)

FIGURE 13.34 Different visions for restoring a trout stream. (a) Degraded by a century of agriculture and grazing, the stream had become too wide, shallow, and warm for native trout. (b) Trout Unlimited rebuilt a section of the stream to show their preferred option, which featured banks made of large stone blocks. (c) Other environmental organizations preferred a more organic approach. They used straw bales to narrow the stream and increase current flow. This washed away the soft sediment and re-created the deep, narrow, cool, deeply shaded channel that favored native trout.
William P. Cunningham

copepods made up only 15 percent of the invertebrate population, and stone flies and caddis flies had increased to 47 percent.

Stabilizing banks is an important step in stream restoration. Where banks have been undercut by erosion and stream action, they will continue to be unstable and cave into the stream. Ideally, the bank should be recontoured to a slope of no more than 45 degrees (fig. 13.35). Soil can then be held in place by rocks, planted vegetation, or other ground cover. It may be necessary to install erosion control fabric or mulch to hold soil until vegetation is established. If space isn't available for recontouring, steep banks may have to be supported by rock walls, riprap, or embedded tree trunks.

Beaver can be excellent wetland engineers. It's estimated that before the arrival of European immigrants, North America may have had as many as 200 million beaver ponds that inundated half a million km^2 of bottom land. The ponds provided habitat for plants, fish, amphibians, invertebrates, and other aquatic species. After the beaver moved on to new areas and the ponds silted in, they became productive wetlands, and meadows. Trappers de-beavered America in the eighteenth century, but these furry pond-builders have rebounded and are now becoming a landscaping force once again in many places.

Severely degraded or polluted sites can be repaired or reconstructed

In a relatively small area—say an old industrial site—it may be economical to simply excavate and replace contaminated soil. If the pollutants are organic, it may be possible to pass contaminated soil through an incinerator to eliminate toxins. After this treatment, the soil won't be worth much for growing vegetation, however.

For polluted surface or groundwater, bacteria can remove organic compounds such as oil and other contaminants. Naturally occurring bacteria in groundwater, when provided with oxygen and nutrients, can decontaminate many kinds of toxins. Experiments have shown that pumping air into aquifers can be more effective than pumping water out for treatment. For hostile environments or exotic, human-made chemicals that can't be metabolized by normal organisms, it's sometimes possible to genetically engineer new varieties of bacteria that can survive in extreme conditions and consume materials that would kill ordinary species.

Bioremediation is a growing strategy that uses living things, especially plants or bacteria, to selectively eliminate toxins from the soil. A number of plant species can selectively eliminate toxins from the soil, air, or water. Some types of mustard, for example, can extract lead, arsenic, zinc, and other metals from contaminated soil. Radioactive strontium and cesium have been removed from soil near the Chernobyl nuclear power plant using common sunflowers. And sunflowers and other plants can capture toxins, metals, and carcinogenic compounds from other industrial sites (fig. 13.36).

In some cases, bioremediation could have multiple benefits. A weedy species called field pennycress or stinkweed (*Thlaspi arvense*), for example, grows well on degraded, polluted soil, absorbing metals as it grows. Its seeds contain high concentrations of oils that make it competitive with canola as an oil seed crop. The oils aren't edible but can be easily converted into biodiesel fuel or can be used for bioplastic production. Furthermore, glucosinolates in the oil can be converted into a soil fumigant that could

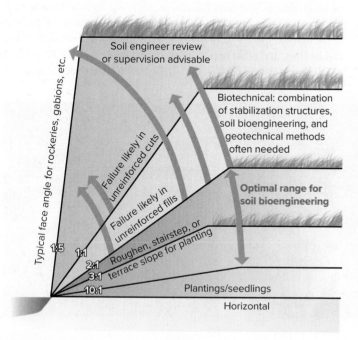

FIGURE 13.35 Steep streambanks need to be reinforced or recontoured to avoid erosion. Shallow slopes can be stabilized with vegetation or mulch. Steeper slopes need stabilization structures or reinforcement to hold the soil.

FIGURE 13.36 Sunflowers capture heavy metals and unburned hydrocarbons from a natural gas pumping station.
ProSha/Shutterstock.com

be an eco-friendly alternative to methyl bromide. Combining all these features, pennycress could be grown on contaminated inner-city lands, producing either fuel or chemical feedstocks while also decontaminating the soil.

Many cities are finding that decontaminating urban "brown fields" (abandoned, contaminated industrial sites) can turn unusable inner-city property into valuable assets. This is a good way to control urban sprawl and make use of existing infrastructure. Cleaning up hazardous and toxic wastes is now a big business in America, and probably will continue to be so for a long time in the future.

Reclamation implies using intense physical or chemical methods to clean and repair severely degraded or even totally barren sites. Historically, reclamation meant irrigation projects that brought wetlands and deserts (considered useless wastelands) into agricultural production. In the early part of this century, the Bureau of Reclamation and the Army Corps of Engineers dredged, diked, drained, and provided irrigation water to convert millions of acres of wild lands into farm fields. Many of those projects were highly destructive to natural ecosystems. Ironically, we are now using ecological restoration to restore some of these "reclaimed" lands to a more natural state.

Today, reclamation means the repairing of human-damaged land. The Surface Mining Control and Reclamation Act (SMCRA), for example, requires mine operators to restore the shape of the land to its original contour and revegetate it to minimize impacts on local surface water and groundwater. According to the U.S. Office of Surface Mining, more than 8,000 km^2 (3,000 mi^2) of former strip mines have been reclaimed and returned to beneficial uses, such as recreation areas, farming and rangeland, wetlands, wildlife refuges, and sites for facilities such as hospitals, shopping centers, schools, and office and industrial parks.

Ideally, if topsoil is set aside during surface mining, overburden and tailings (waste rock discarded during mining and ore enrichment) could be returned to the pit, smoothed out, and covered with good soil that will support healthy vegetation.

The largest mine pits will never be returned to their original contour. Figure 13.37 shows a view of the Berkeley mine pit in Butte, Montana. From the 1860s to the 1980s, this area was one of the world's richest sources of metals, including copper, silver, lead, zinc, manganese, and gold. In the early days, all mining was in deep shafts. In 1955, the Anaconda Mining Company switched to open-pit mining and dug the hole you see now. After mining ended in 1981, groundwater, previously controlled with pumps, began to fill the pit. It has now accumulated to make a lake 1.6 km wide and 300 m deep.

Like most similar mines, the Berkeley pit contained sulfide ores, which include sulfur as well as targeted metals. Exposed to air and water, the sulfur produces sulfuric acid in concentrations that make the pit deadly to most life. The water has a pH of 2.5

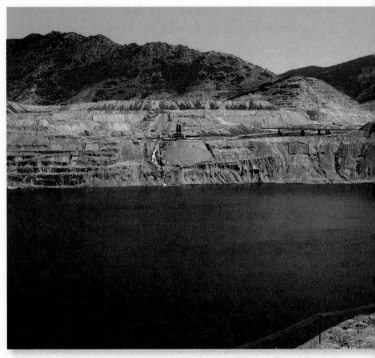

FIGURE 13.37 The Berkeley mine pit in Butte, Montana, may be the most toxic water body in the United States. Water entering the pit is now being treated, and eventually there may be an effort to pump water out of the pit and decontaminate it, but the pit itself will probably never be filled in.
William P. Cunningham

(about the same as vinegar) and is laden with heavy metals and toxic chemicals, such as arsenic, cadmium, zinc, and sulfuric acid. In 2016, a whole flock of several thousand migrating snow geese was found dead after landing by mistake in the pit. The pit is now a Superfund site. A water treatment plant built on the far shore of the pit (you can see a thin, white stream of treated water from the plant cascading into the lake) is now being expanded. It's expected to clean 10 million gallons of water per day, and to treat all the water in the pit before it reaches the critical elevation of 1,649 m above sea level, the height at which it will threaten groundwater used by the city of Butte. This reclamation project is difficult, expensive, ambitious, and like many others, critically important.

Section Review

1. Why is wetland restoration important?
2. What is wetland mitigation?
3. What is bioremediation?

Connecting the Dots

Humans have caused massive damage and degradation to a wide variety of biological communities, but there are many ways to repair this damage and to restore or rehabilitate nature. Efforts to rebuilt coral reefs illustrate some of the many strategies. Ideally, we might prefer to return a site to its pristine, predisturbance condition, but often that isn't possible. A more pragmatic goal may be simply to develop a useful, stable, self-sustaining ecosystem with as many of its original ecological elements as possible. Sometimes it's enough to leave nature alone to heal itself, but often we need to intervene in some way to remove or discourage unwanted organisms while also promoting the growth of more desirable species.

Restoration pioneer Aldo Leopold wrote, "A thing is right when it tends to preserve the integrity, stability, and beauty of the biotic community. It is wrong when it tends otherwise." Some modern ecologists object that ecosystems are highly dynamic and species appear and disappear stochastically and individually. Characteristics such as integrity, health, stability, beauty, and moral responsibility tend to be human interpretations rather than scientific facts. Still, we need to have goals for restoration that the public can understand and accept.

Many ecological restoration projects are now underway. Some are huge efforts, such as creating a Monarch Highway from Lake Superior to the Rio Grande Valley, rehabilitating the Florida Everglades, or returning a huge swath of prairie to its primeval state. Others are much more modest: building a rain garden to trap polluted storm runoff, planting trees on empty urban land, or turning a part of your yard into a native prairie or butterfly garden. Within this wide range, there are lots of opportunities for all of us to get involved.

Reviewing Key Terms

Can you define the following terms in environmental science?

bioremediation 13.4

coarse woody debris 13.4

ecological restoration 13.1

epigenetic effects 13.1

novel ecosystem 13.1

prairie potholes 13.4

prescribed burning 13.2

reclamation 13.4

wetland mitigation 13.4

Critical Thinking and Discussion Questions

1. Should we be trying to restore biological communities to what they were in the past, or modify them to be more compatible with anticipated future conditions? What future conditions would you consider most likely to be problematic?

2. How would you balance human preferences (aesthetics, utility, cost) with biological considerations (biodiversity, ecological authenticity, evolutionary potential)?

3. The Nature Conservancy's Hassayampa River Preserve near Phoenix illustrates a situation in which there may be more than one historic condition to which we may wish to restore a landscape. How would you reconcile these different values and goals?

4. Restoring savannas often requires the use of herbicides to remove invasive species. Some people regard this as dangerous and unnatural. How would you respond?

5. The Nature Conservancy believes it is essential to keep productive ranches on the land, both to sustain rural society and for effective protection of the range. The American Prairie Foundation (APF) is buying up ranches and converting them to wilderness where wild animals can roam freely. Which of these approaches would you favor?

6. Sometimes the quickest and easiest way to restore a stream is simply to reconstruct it with heavy equipment. You might even create something more interesting and useful (at least from a human perspective) than the original. Is it okay to replace real nature with something synthetic?

Data Analysis

Concept Maps

The figure below is a form of graphic representation we haven't used very often in this book. It's a concept map, or a two-dimensional representation of the relationship between key ideas. It could also be considered a decision flowchart because it is an organized presentation of different policy options. This kind of chart shows how we might think about a situation, and suggests affinities and associations that might not otherwise be obvious.

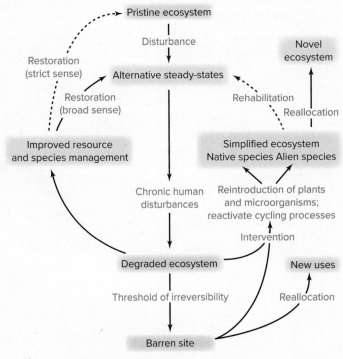

A model of ecosystem degradation and potential management options.
Source: Data from Walker and Moral, 2003

1. Using one of the examples presented in this chapter—or another familiar example—replace the descriptions in the colored boxes with brief descriptions of an actual ecosystem. Does this help you see the relationship between different states for the community you've chosen?

2. Now replace the terms associated with the arrows between boxes with actions that cause changes in your particular biological community, as well as restoration treatments that could accomplish the proposed restoration outcomes.

3. What do you suppose the authors meant by a "threshold of irreversibility"? If the system is irreversible, why are there arrows for reallocation or intervention?

4. The box for "simplified ecosystem" has two subcategories, "native species" and "alien species." What does this mean? What would be some examples for the ecological community you've chosen?

5. There are two arrows labeled "reallocation" in this diagram. One leads to "new uses," while the other leads to "novel ecosystem." What's the difference? Why are the two boxes separated in the diagram?

6. The arrow labeled "restoration (strict sense)" is a dotted line. What do you think the authors meant by this detail?

14

Geology and Earth Resources

▲ The Bristol Bay watershed supports the largest remaining sockeye salmon fishery on the planet, producing nearly half the world's wild sockeye harvest, but mining companies want to mine copper ore here.
Michael Melford/National Geographic

Learning Outcomes

After studying this chapter, you should be able to:

14.1 Summarize the processes that shape the earth and its resources.

14.2 Describe several earth resources of economic value.

14.3 Outline the environmental effects of resource extraction.

14.4 Identify several geological hazards, and explain how they occur.

"We live in a society exquisitely dependent on science and technology, in which hardly anyone knows anything about science and technology."

– Carl Sagan

CASE STUDY

Salmon or Copper?

Alaska has many natural wonders, but among the most spectacular is the migration of millions of salmon every year from the dark waters of Bristol Bay on the state's southwest coast up the Mulchatna and Nushagak Rivers into shallow headwater streams near Lake Clark and Katmai National Parks. It's one of the last, great wild salmon runs in the world (fig. 14.1).

The Bristol Bay watershed is a sprawling wilderness of winding streams and rivers, vast marshy wetlands, tundra, and forests of alder and spruce, filled with salmon, bears, caribou, moose, and wolves. This breathtaking place is one of the most productive marine ecosystems in the world. All five species of Pacific salmon—sockeye, Chinook, Coho, chum, and pink—spawn there and have sustained native salmon-based cultures for thousands of years.

The watershed supports the largest sockeye salmon fishery on the planet, producing nearly half the world's wild sockeye harvest. This fishery generates $1.5 billion a year and supports nearly 20,000 jobs throughout the United States. Over 4,000 Alaskans, including many native Yup'ik and Dena'ina, rely on fish as the mainstay of their diet.

But the Bristol Bay watershed contains other riches besides the salmon bonanza. Geologists estimate that the Pebble ore deposit in the uplands between the two largest river systems in the watershed may contain 26 million metric tons of copper, 1.5 million tons of molybdenum, 10,000 tons of silver, and 2,000 tons of gold, as well as palladium and rhenium. Mining engineers have claimed that the deposit contains over $300 billion worth of recoverable metals.

Since 2001, Northern Dynasty, a Canadian mining company, has sought to develop an open pit mine in the area. This pit could be deeper than the Grand Canyon, impact an area larger than Manhattan, and fill a major football stadium up to 3,900 times with mine waste. The liquid waste would be held behind earthen dams, some taller than a 50-story skyscraper. The EPA review of the project concluded that even without a collapse of those tailings ponds (as happened at similar mines in British Columbia in 2014 and Brazil in 2015 and again in 2019), the Pebble mine would destroy 150 km (94 mi) of salmon streams and 2,200 ha (5,400 acres) of wetlands, lakes, and ponds. The project would imperil not only the salmon fishery but the whole ecosystem.

The mine would also require building of a new deepwater port on Cook Inlet, a natural gas pipeline running across the ocean floor from the Kenai Peninsula to the mine site, and 100 km (60 mi) of private roads that would skirt the boundary of Lake Clark National Park and cross approximately 60 salmon streams. Routine release of silt, pollutants, and runoff into headwater streams, the EPA ruled, would irreparably harm the Bristol Bay fishery, wildlife, and people.

In 2014, the Obama administration, responding to concerns of a coalition of fishers, native Alaskans, environmental groups, and local businesses, denied mining permits for the Pebble mine, saying that it could have "significant and potentially catastrophic impacts" on the region. But in 2017, then-EPA administrator Scott Pruitt directed his staff to reverse previous protections and to expedite permitting for the mine.

What do you think about the trade-off of metals for salmon? We all use metals in our daily life. Every American uses about 6 kg of copper per year, on average, in various manufactured goods. Our demand for new metals is driven partly by the fact that we recycle very little. We recover only about one-third of commercial copper—and less than 9 percent of post-consumer copper—and we recycle far less of most of the other metals we consume. So we require a constant stream of new materials. At the same time, many of us enjoy fish, particularly salmon, too. Large mining companies, including Mitsubishi, Rio Tinto, and Anglo American, invested in Pebble years ago but pulled out, discouraged by local opposition and the dangers of operating in this cold, remote location.

Pebble Partnership claims that the mine will employ about 2,000 construction workers and provide about 850 permanent jobs over the 20-year life of the mine. Environmental groups point out that a healthy salmon fishery supports more than 20,000 jobs. The metals will provide billions in profits every year during the mine's

FIGURE 14.1 The Bristol Bay watershed lies in the headwaters of the Mulchatna and Nushagak Rivers near the borders of Lake Clark and Katmai national parks. It provides the spawning grounds for the world's last great sockeye salmon run, but it also contains valuable metal resources.

life, but little of those profits will go to Alaska or even the United States. Polling by mine opponents found that 58 percent of Alaskans overall, and 80 percent of Bristol Bay residents, oppose the project. Given our low rate of metal reuse, could we do better and preserve the salmon of Bristol Bay?

In this chapter, we'll look at the principles of geology, earth processes, and earth resources. We examine the ways our resource use affects our environment, and we also consider the ways in which we humans have become geologic agents.

14.1 EARTH PROCESSES AND MINERALS

- *Geological processes such as mineral formation, earthquakes, and mountain building result from tectonic plate movement.*
- *Oceanic crust is thinner, denser, and younger than continents.*
- *The rock cycle involves formation of three general rock types.*

All of us benefit from geological resources. Right now you are probably wearing several geological products: Plastics, including glasses and synthetic fabric, are made from oil; iron, copper, and aluminum mines produced your snaps and zippers and perhaps the chair you're sitting on; silver, gold, palladium, lithium, and other metals make up your jewelry or electronics. All of us also share responsibility for the environmental and social impacts of mining and drilling. Fortunately, there are many ways to reduce these costs, including recycling and using alternative materials.

Where do these resources come from? To understand this question, we must first examine the earth's structure and the processes that shape it.

Earth is a dynamic planet

Although we think of the ground under our feet as solid and stable, our planet is a dynamic and constantly changing structure. Titanic forces inside the earth cause continents to split, move, and crush into each other in slow, inexorable collisions. All this motion results from convection currents deep in the earth.

Benchmark Data	
Among the ideas and values in this chapter, the following are a few worth remembering.	
33%	Percentage of earth that is iron
30%	Percentage that is oxygen
1 cm/year	Approximate average rate of tectonic plate movement
250 MJ/kg	Energy needed to refine aluminum from ore
8 MJ/kg	Energy to recycle aluminum
14%	Percentage of U.S. household aluminum recycled

The earth is a layered sphere. The **core,** or interior, is composed of a dense, intensely hot mass of metal—mostly iron—thousands of kilometers in diameter (fig. 14.2). Solid in the center but more fluid in the outer core, this immense mass generates the magnetic field that envelops the earth.

FIGURE 14.2 Earth's cross-section. Slow convection in the mantle causes the thin, brittle crust to move.

Table 14.1	Eight Most Common Chemical Elements (Percent)		
Whole Earth		**Crust**	
Iron	33.3	Oxygen	45.2
Oxygen	29.8	Silicon	27.2
Silicon	15.6	Aluminum	8.2
Magnesium	13.9	Iron	5.8
Nickel	2.0	Calcium	5.1
Calcium	1.8	Magnesium	2.8
Aluminum	1.5	Sodium	2.3
Sodium	0.2	Potassium	1.7

Surrounding the molten outer core is a hot, pliable layer of rock called the **mantle.** The mantle is much less dense than the core, because it contains a higher concentration of lighter elements, such as oxygen, silicon, and magnesium.

The outermost layer of the earth is the lightweight, brittle **crust.** The crust below oceans is relatively thin (8–15 km), dense, and young (less than 200 million years old) because it is constantly recycled. This oceanic crust consists of rocks rich in iron, silicon, and magnesium, such as basalt. Continental crust is thicker (25–75 km), less dense, and older, some of it 3.8 billion years old. Rocks in the continental crust, such as granite or sandstone, generally are composed of lighter elements, such as oxygen, silicon, and aluminum, compared to the rest of the earth (table 14.1).

Tectonic processes move continents

The convection currents in the mantle provide the force to move the overlying crust, a mosaic of huge blocks called **tectonic plates** (fig. 14.3). These plates slide slowly across the earth's surface like ice sheets on water, in some places breaking up into smaller pieces, and in other places crashing ponderously into each other to create new, larger landmasses. Ocean basins form where continents crack and pull apart. The Atlantic Ocean, for example, is growing slowly as Europe and Africa move away from the Americas. **Magma** (molten rock) at these divergent boundaries forms new oceanic crust that piles up underwater in **mid-ocean ridges.** Creating the largest mountain range in the world, these ridges wind around the earth for 74,000 km. Although concealed from our view, this jagged range boasts higher peaks, deeper canyons, and sheerer cliffs than any continental mountains.

At the margins of converging plates, mountain ranges are pushed up, like those on the West Coast of North America and in Japan. The Himalayas are still rising as the Indian subcontinent grinds slowly into Asia. (These mountains are unusually tall because they occur at the convergence of two continental plates.) At many plate boundaries, two plates slide past each other. Southern California is slowly sliding north toward Alaska. In about 30 million years, Los Angeles will pass San Francisco. At these boundaries, earthquakes are common.

FIGURE 14.3 Map of tectonic plates. Plate boundaries are dynamic zones, characterized by earthquakes and volcanism and the formation of great rifts and mountain ranges. Arrows indicate direction of subduction where one plate is diving beneath another. These zones are sites of deep trenches in the ocean floor and high levels of seismic and volcanic activity.

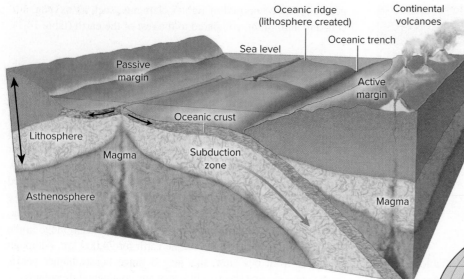

FIGURE 14.4 Tectonic plate movement. Where thin, oceanic plates diverge, upwelling magma forms mid-ocean ridges. A chain of volcanoes, like the Hawaiian Islands, may form as plates pass over a "hot spot." Where plates converge, melting can cause volcanoes, such as the Cascades.

Pangaea, 250 MYA

Laurasia and Gondwana, 210 MYA

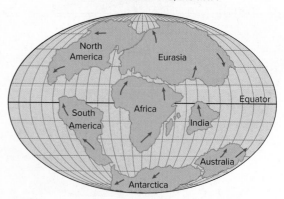

Most modern continents had formed by 65 MYA

FIGURE 14.5 Pangaea, an ancient supercontinent of 250 million years ago (MYA), combined all the world's continents in a single landmass. Continents have combined and separated repeatedly.

or biological processes, are generally not minerals, because they vary in chemical composition and lack a regular crystal structure. The ores from which metals are extracted are minerals. Purified metals such as iron, aluminum, or copper lack a crystal structure and thus are not minerals.

A **rock** is a solid, cohesive, aggregate of one or more minerals. Within the rock, individual mineral crystals are mixed together and held firmly in a solid mass (fig. 14.7). Mineral grains may be large

When an oceanic plate converges with a continental landmass, the continental plate usually rides up over the seafloor, while the oceanic plate is **subducted,** or pushed down into the mantle. As the subducted plate is forced downward, heat and pressure melt and recrystallize minerals. At great depth, melting material becomes fluid magma, which can rise back toward the surface (fig. 14.4). Deep ocean trenches mark these subduction zones. Volcanoes form where the magma erupts through vents and fissures in the overlying crust. Trenches and volcanic mountains ring the Pacific Ocean rim from Indonesia to Japan to Alaska and down the west coast of the Americas, forming a "ring of fire" where oceanic plates are being subducted under the continental plates. This ring is the source of more earthquakes and volcanic activity than any other region on the earth.

Over hundreds of millions of years, continents and oceans have continuously shifted, expanded, joined, and split apart (fig. 14.5). North America spent millions of years near the equator, with shallow seas covering much of what is dry land today. Antarctica and Australia once adjoined to Africa, also near the equator, and supported luxuriant forests. The redistribution of continents also affected the earth's climate. This may help explain some of the many mass extinctions that define different stages in earth history (fig. 14.6). In these events, high proportions of fossil species (and genera, families, and orders) disappear abruptly from one rock layer to another. Most famously, dinosaurs disappeared at the end of the Cretaceous period (see also section 11.2).

Rocks are composed of minerals

A **mineral** is a naturally occurring, inorganic, solid element or compound with a specific chemical composition and crystal structure. Organic materials such as coal, produced by living organisms

Era	Periods	Life on Earth
Cenozoic	Quaternary	First humans
	Tertiary	First important mammals
Mesozoic	Cretaceous	Extinction of dinosaurs
	Jurassic	
	Triassic	First dinosaurs
Paleozoic	Permian	First reptiles
	Pennsylvanian	
	Mississippian	
	Devonian	Fish become abundant
	Silurian	
	Ordovician	
	Cambrian	First abundant fossils
Precambrian		Earliest single-celled fossils
		(The Precambrian accounts for the vast majority of geologic time)
		Origin of the earth

Millions of years before present (ma): 3, 65, 245, 300, 400, 545, 3,500, 4,500

FIGURE 14.6 Periods and eras in geological time, and major life-forms that mark some periods.

or small, depending on how the rock was formed. Each rock type has a characteristic mixture of minerals, grain sizes, and ways in which the grains are mixed and held together. Granite, for example, is a mixture of quartz, feldspar, and mica crystals. Different kinds of rocks have distinct percentages of these or other minerals. Fine-grained crystalline rocks indicate a relatively rapid

FIGURE 14.7 Crystals of different minerals create beautifully colored patterns in a rock sample seen in a polarizing microscope.
Doug Sherman/Geofile

cooling environment near the surface. Large-grained rocks crystallized and solidified slowly, deep underground.

Rocks and minerals are recycled constantly

Like continents, rocks are endlessly reshaped and re-formed. They are crushed, folded, melted, and recrystallized. Heat, pressure, erosion, deposition, and other processes create new minerals and rocks. We call this cycle of creation, destruction, and metamorphosis the **rock cycle** (fig. 14.8).

There are three major rock classifications: igneous, metamorphic, and sedimentary. The most common rock-type in the earth's crust is solidified from magma, welling up from the earth's interior. These rocks are classed as **igneous rocks** (from *igni,* Latin for fire). Magma extruded to the surface from volcanic vents cools quickly to make basalt, rhyolite, andesite, and other fine-grained rocks. Magma that cools slowly in chambers deep below the surface or that intrudes into overlying layers makes granite, gabbro, or other coarse-grained crystalline rocks.

Extreme heat and pressure can transform mineral structures to create new forms called **metamorphic rock.** Metamorphism generally happens as buried layers of rocks are squeezed, folded, and heated by tectonic processes. Metamorphic rocks often show patterns of banding and folding, which make them desirable for

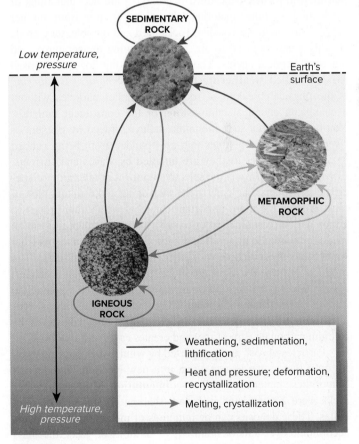

FIGURE 14.8 The rock cycle includes a variety of geological processes that can transform any rock.

FIGURE 14.9 Weathering slowly reduces an igneous rock to loose sediment. Here, exposure to moisture expands minerals in the rock, and frost may also force the rock apart.
Photo by David McGeary

FIGURE 14.10 Different colors of soft sedimentary rocks deposited in ancient seas during the Tertiary period 63 to 40 million years ago have been carved by erosion into the fluted spires and hoodoos of the Pink Cliffs of Bryce Canyon National Park.
Digital Vision/Getty Images

decorative building facades. Some common metamorphic rocks are marble (recrystallized from limestone), quartzite (from sandstone), and slate (from mudstone and shale). Metamorphic rocks produce a variety of economically important minerals such as talc, graphite, and gemstones.

Weathering breaks down rocks

Sedimentary rocks, the third major type, are accumulations of sand, mud, or other material deposited over time from another source. Although we consider rocks hard and durable, they erode steadily when exposed to air, water, changing temperatures, and reactive chemical agents. The gradual breaking down of exposed rocks is called **weathering** (fig. 14.9). *Mechanical weathering* is the physical breakup of rocks into smaller particles without a change in chemical composition of the constituent minerals. You have probably seen mountain valleys scraped by glaciers or river and shoreline pebbles that are rounded from being rubbed against one another as they are tumbled by waves and currents. *Chemical weathering* is the selective removal or alteration of specific components that leads to the weakening and disintegration of rock. Among the more important chemical weathering processes are oxidation (combination of oxygen with an element to form an oxide or hydroxide mineral) and hydrolysis (hydrogen atoms from water molecules combine with other chemicals to form acids). The products of these reactions are more susceptible to both mechanical weathering and dissolution in water. For instance, when acidic rainfall (formed naturally when CO_2 combines with rain) percolates through porous limestone layers in the ground, it dissolves the calcium carbonate (limestone) and creates caves and sink holes.

Particles of rock are transported by wind, water, ice, and gravity until they come to rest again in a new location. The deposition of these materials is called **sedimentation.** Most sedimentary rocks were accumulated in broad, shallow coastal seas or river deltas. These deposits can be hundreds of meters thick. As layers accumulate over tens or hundreds of millions of years, the mass compacts and hardens deeper layers, so that loose sand becomes sandstone, clay and mud become shale, or organic shell fragments

become limestone. Color and texture of rocks help identify environmental conditions in which they were deposited—near shore or in deep water, in arid and evaporating conditions, in shallow water, or exposed to the air. Some sedimentary rocks, such as sandstone and limestone, are relatively easily eroded and can be shaped into striking features (fig. 14.10).

Sedimentary "evaporite" minerals also form as salt layers accumulate in shallow, evaporating water bodies. The mineral halite, or rock salt, is deposited this way. Halite is the same as ordinary table salt (sodium chloride). Gypsum, the material used to make drywall in houses, is calcium sulfate, also deposited in shallow, evaporating water bodies.

Humans have become a major force in shaping landscapes. Geomorphologist Roger Hooke, of the University of Maine, estimates that activities such as housing excavations, road building, and mineral production move about 30 to 35 gigatons (Gt, or billion tons) per year worldwide. When combined with the 10 Gt each year that we add to river sediments through soil erosion, our earth-shaping impact may be greater than that of any other single agent except plate tectonics.

Section Review

1. Describe the layered structure of the earth.
2. Why do volcanoes and earthquakes mark the Pacific "ring of fire"?
3. Describe the processes that produce the three rock types in the rock cycle.

14.2 EARTH RESOURCES

- *Metals are valuable because they are strong and easily shaped.*
- *Metals derive from ores, which often include sulfur.*
- *Nonmetal resources are diverse and valuable.*
- *Conservation reduces energy consumption as well as pollution.*

The earth is unusually rich in mineral variety. Mineralogists have identified some 4,400 different mineral species—far more, we believe, than on any of our neighboring planets. What makes the difference? The processes of plate tectonics and the rock cycle on this planet have gradually concentrated uncommon elements and allowed them to crystallize into new minerals. But this accounts for only about one-third of our geological legacy. The biggest distinction is life. Most of our minerals are oxides, but there was little free oxygen in the atmosphere until it was released by photosynthetic organisms, thus triggering evolution of our great variety of minerals.

We use a tremendous variety of nonmetal minerals, from high-value gemstones, to industrial resources such as mica, talc, graphite, and sulfur (used in chemical industries), to everyday building materials such as sand, gravel, salts, and limestone. Of these, sand and gravel production, needed for road construction and fill, comprise by far the greatest volume and dollar value of all nonmetal mineral resources and a far greater volume than all metal ores. Sand and gravel are used mainly in brick and concrete construction and paving, as road fill, and for sandblasting. High-purity silica sand is our source of glass. These materials usually are retrieved from surface pit mines and quarries, where they were deposited by glaciers, winds, or ancient oceans.

Metals are especially valuable resources

Most attention to economic minerals is concerned with metal ores, which are valuable, versatile, and used in great quantities. The metals consumed in greatest quantity by world industry are iron (740 million metric tons annually), aluminum (40 million metric tons), manganese (22.4 million metric tons), copper and chromium (8 million metric tons each), and nickel (0.7 million metric tons). Most of these metals are consumed in the United States, China, Japan, and Europe. The largest sources are China, Australia, Russia, Canada, and the United States.

Metal-bearing ores are minerals with unusually high concentrations of metallic elements. Lead, for example, often comes from the mineral galena (PbS), and copper comes from sulfide ores, such as bornite (Cu_5FeS_4). Iron occurs in a variety of minerals, such as hematite (Fe_2O_3) and goethite ($Fe_2O_3s\ H_2O$).

Metals have been so important in human affairs that major epochs of human history are commonly known by their dominant materials and the technology involved in using them: The Stone Age was followed by the Bronze Age and the Iron Age. The mining, processing, and distribution of these materials have broad implications for both our culture and our environment (fig. 14.11). Metals are useful because they are light, strong, and malleable (easily shaped). Unlike stone, metals can hold a sharp edge. Most metals also conduct electricity well (table 14.2).

A new class of metals, known as rare earth metals, has become key to the production of small electronics, lightweight motors, as in wind turbines, and batteries for electric or hybrid cars. Because they are sparsely distributed, extracting them causes considerable environmental disturbance. Because they often occur, like other metals, in sulfur-rich ores, mining and processing these metals can produce wastewater rich in sulfuric acid. International competition for these metals has made them a new subject for trade tensions (Exploring Science, p. 310).

FIGURE 14.11 The availability of metals and the ways we extract and use them have profound effects on our society and environment. *Digital Vision/PunchStock*

Fossil fuels originated as peat and plankton

Coal, oil, and natural gas are another class of valuable and widely traded earth resources. These fossil fuels are deposits from living organisms. They are not minerals because they are not crystalline, and they are "organic" in that they are carbon-based compounds. Although we worry about supplies of these fossil fuels, they are globally distributed, and our estimates of

Table 14.2	Primary Uses of Some Major Metals Consumed in the United States
Metal	**Use**
Aluminum	Packaging foods and beverages (38%), transportation, electronics
Chromium	High-strength steel alloys
Copper	Building construction, electric and electronic industries
Iron	Heavy machinery, steel production
Lead	Leaded gasoline, car batteries, paints, ammunition
Manganese	High-strength, heat-resistant steel alloys
Nickel	Chemical industry, steel alloys
Platinum-group	Automobile catalytic converters, electronics, medical uses
Gold	Medical, aerospace, electronic uses; accumulation as monetary standard
Silver	Electronics, jewelry

EXPLORING SCIENCE

Rare Earth Minerals

What earth materials allow your cell phone to be so small? A group of elements known as "rare earth" metals are largely responsible. This group of metallic elements, including scandium, yttrium, and 15 lanthanides, such as neodymium, dysprosium, and gadolinium, have unusual properties. They conduct electricity efficiently. Small amounts of these metals can make motors 90 percent lighter and lights 80 percent more efficient. These make possible our small cell phones, high-efficiency lightbulbs, hybrid cars, superconductors, high-strength magnets, lightweight batteries, lasers, energy-conserving lamps, and a variety of medical devices.

Lightweight metals and batteries make it possible to build high-capacity wind turbines, whose motors must be extremely efficient and lightweight. The batteries in an electric or hybrid car are heavy, but vastly less so because they contain rare earth metals. A Toyota Prius uses about a kilogram of neodymium and dysprosium for its electric motor and as much as 15 kg of lanthanum for its battery pack.

Despite their name, these elements are not rare. They are widely distributed globally, although commercially viable concentrations are relatively uncommon. China, Russia, and the United States possess more than 68 percent of the world's known reserves of these materials.

China currently produces most of the world's rare earth metals (fig. 1). China dominates production partly because these metals are essential for manufacturing electronics, and partly because the Chinese government has been willing to overlook the environmental damage caused by extracting and processing

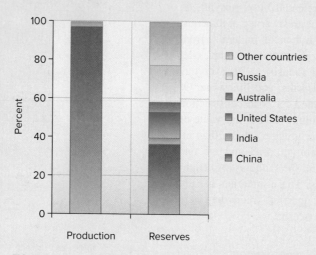

FIGURE 1 China controls a little more than one-third of known rare earth metals, but currently produces 97 percent of these important materials.

Source: USGS, 2010

Legend:
- Other countries
- Russia
- Australia
- United States
- India
- China

these metals. About half of all Chinese production of rare earth metals occurs in a single mine in Baotou in Inner Mongolia; much of the rest come from small, often unlicensed mines in southern China.

Unregulated mineral extraction can poison water resources and soils. Like gold, silver, and other precious metals, rare earth elements are often separated from ore by crushing ore-bearing rocks and washing the ore in strong acids. These acids extract metals from the ore, but when the metals are later separated from the acid slurry, tremendous amounts of toxic wastewater are produced. Often, metals are extracted by pumping acids straight into holes drilled in the ground. Acids dissolve metal from ores in place, and a toxic, metal-bearing slurry is then pumped to the surface for processing. Acidic

wastewater is frequently stored behind earthen dams, which can leak into surface and groundwater. Processing also releases the sulfur and radioactive uranium and thorium that frequently occur with rare earth elements.

These metals have become strategic resources globally. China has restricted exports of rare earth metals in recent years to protect its electronics industry. This move has made other nations worry about supplies for tactical needs (such as military guidance systems), as well as consumer products. Many firms are simply moving to China to ensure access to the resource. The General Motors division that deals with miniaturized magnet research, for example, shut down its U.S. office and moved its entire staff to China about a decade ago. The Danish wind turbine company Vestas also has moved much of its production to China.

Concerned about access, and wanting a part of the valuable market, other countries are developing their reserves. Australia, the United States, Russia, Myanmar, and India all have increased production. Impacts of expanding production on the environment, including water resources and soil contamination, will depend on how strong environmental policies are to enforce responsible mining practices in these different countries.

As with all metals, the more we can recycle, the better. Reusing electronics, or at least replacing them less frequently, also lowers your consumption of these resources. Recycling electronics can be difficult, but it is one way we can collectively help reduce the environmental impacts of rare earth element extraction.

(header photo): David L. Hansen

their abundance has risen recently, with new exploration technology (see chapter 19).

How did these resources form? All result from accumulation of organic material, mainly in shallow, warm swamps or seashore environments. Over hundreds of millions of years, there have been long periods, such as the Carboniferous period (about 360–300 million years ago), when these environments were widespread.

Coal derives from leaves and other plant material that accumulated in the bed of ancient swamps. During the Carboniferous era, when most coal was deposited, tree ferns and other large plants were abundant, as were giant dragonflies with wing spans up to 75 cm (30 in). (Dinosaurs were not yet around; they evolved about 50 million years later.) Modern swamps can accumulate tens of meters of undecayed plant material, or peat. Ancient swamps similarly accumulated peat to great depths but in vast expanses.

Coal beds are generally interleaved with layers of sandstone, shale, and limestone. These occurred as rising sea levels would flood coastal swamps, allowing clay deposits, then limestone, to form. Eventually the seas receded, and nearshore sands were deposited. Rising sea levels then created coastal wetlands, and gradually the cycle was repeated. In some places, up to a hundred of these interleaved cycles are evident from the Carboniferous period. Pressure, heat from the earth's interior, and time have converted these swampy deposits into rock-hard coal seams (fig. 14.12).

Oil, in contrast, derives from the remains of tiny plankton and algae (fig. 14.13). Also growing in generally warm, coastal environments, these microscopic organisms accumulated on the seafloor over millions of years. Deposits of dead algae and plankton were especially rich in delta environments, where rivers discharged nutrient-rich waters and sediment from onshore. Many of our richest oil deposits today occur in areas that have been deltas for hundreds of millions of years—the Mississippi Delta, the Niger Delta, and the McKenzie Delta off the coast of Alaska and northern Canada, for example.

Sand, mud, or clay intermixed with accumulating organic material so that oil compounds generally occur within shale or sandstone formations. Over millions of years, as with coal, pressure and heat and time have transformed these materials first into kerogen, a thick precursor to oil, then to simpler, more fluid molecules of oil. Less-developed oil formations, such as Alberta's tar sands, contain mostly thick, hard-to-extract kerogen (chapter 19). More developed oil bodies, as in Texas or Saudi Arabia, have more fluid forms of oil. Fluid molecules seep gradually upward through porous sandstone because they are less dense than the surrounding rocks. Folds or faults often trap accumulations of oil. The smallest, volatile molecules, mainly methane (CH_4, the main component of natural gas), collect in these folds above the oil. Methane can also collect above coal formations. Coalbed methane has been widely exploited in the American West.

Conserving resources saves energy and materials

Conservation offers great potential for extending our supplies of all these earth resources. Conservation can also help reduce the effects of mining and processing, including water use and contamination, air pollution, and energy use.

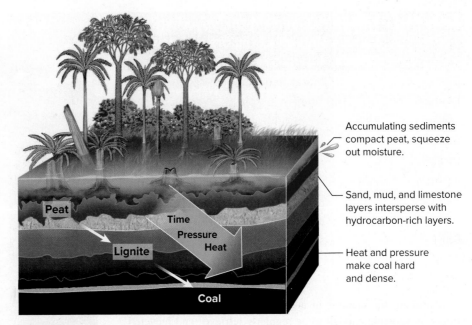

FIGURE 14.12 Coal was formed from vegetation in vast, ancient swamps. Pressure and heat gradually eliminated water and solidified deposits from soft peat to brownish lignite to rock-like coal.

Some waste products already are being exploited, especially for scarce or valuable metals. Aluminum, for instance, is extracted from bauxite ore by electrolysis, an expensive, energy-intensive process. Recycling waste aluminum such as beverage cans, on the other hand, consumes one-twentieth of the energy of extracting new aluminum. Today, nearly two-thirds of all aluminum beverage cans in the United States are recycled, up from only 15 percent 20 years ago. The high value of aluminum scrap ($650 a ton versus $60 for steel, $200 for plastic, $50 for glass, and $30 for paperboard) gives consumers plenty of incentive to deliver their cans for collection. Recycling is so rapid and effective that half of all the aluminum cans now on a grocer's shelf will be made into another can within two months. The energy cost of extracting other materials is shown in table 14.3.

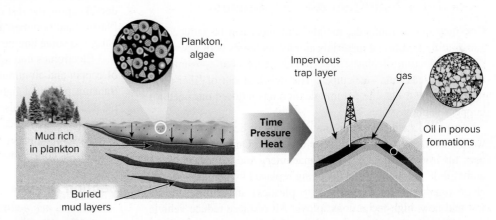

FIGURE 14.13 Oil and gas derive from buried deposits of plankton, algae, mud, and sand in ancient coastal seas. Over millions of years these low-density fluids can seep upward until they reach an impervious trap layer, which blocks and concentrates oil and gas.

Table 14.3	Energy Requirements in Producing Various Materials from Ore and Raw Source Materials	
	Energy Requirement (Mj/Kg)	
Product	New	From Scrap
Glass	25	25
Steel	50	26
Plastics	162	N/A
Aluminum	250	8
Titanium	400	N/A
Copper	60	7
Paper	24	15

Source: E. T. Hayes, Implications of Materials Processing, 1997

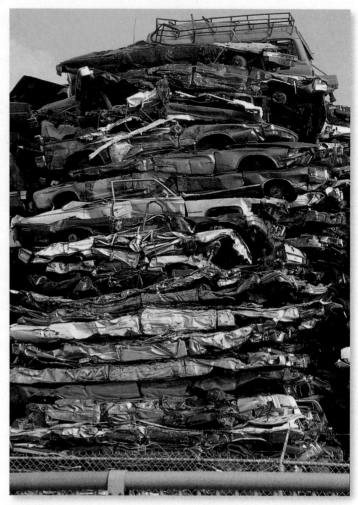

FIGURE 14.14 The richest metal source we have—our mountains of scrapped cars—offers a rich, inexpensive, and ecologically beneficial resource that can be "mined" for a number of metals.
Joseph Nettis/Science Source

Platinum, the catalyst in automobile catalytic exhaust converters, is valuable enough to be regularly retrieved and recycled from used cars (fig. 14.14). Other metals commonly recycled are gold, silver, copper, lead, iron, and steel. The latter four are readily available in a pure and massive form, including copper pipes, lead batteries, and steel and iron auto parts. Gold and silver are valuable enough to warrant recovery, even through more difficult means. See chapter 21 for further discussion of recycling.

Many American and European cities were built around giant steel mills that turned raw iron ore into steel. While these large mills still exist, recycling has become an important source of steel. Minimills, which remelt and reshape scrap iron and steel, are smaller and cheaper to operate than traditional integrated mills that perform every process from preparing raw ore to finishing iron and steel products. Minimills produce steel at $225 to $480 per metric ton, whereas steel from integrated mills averages $1,425 to $2,250 per metric ton. The energy cost is likewise lower in minimills: 5.3 million Btu/ton of steel compared to 16.08 million Btu/ton in integrated mill furnaces.

Resource substitution reduces demand

Consumption of minerals, metals, and fuels can be reduced by substituting traditional materials and technologies with new ones. The introduction of plastic pipe has decreased our consumption of copper, lead, and steel pipes. The development of fiber-optic technology and satellite communication has reduced our need for copper telephone wires.

Iron and steel have been the backbone of heavy industry, but we are now moving toward other materials. One of our primary uses for iron and steel has been machinery and vehicle parts. In automobile production, steel is being replaced by polymers (long-chain organic molecules similar to plastics), aluminum, ceramics, and new, high-technology alloys. All of these reduce vehicle weight and cost while increasing fuel efficiency. Some of the newer alloys that combine steel with titanium, vanadium, or other metals wear much better than traditional steel. Ceramic engine parts provide heat insulation around pistons, bearings, and cylinders, keeping the rest of the engine cool and operating efficiently. Plastics and polymers reinforced with glass fiber are used in body parts and some engine components.

Electronics and telephone technology, once major consumers of copper and aluminum, now use ultra-high-purity glass cables to transmit pulses of light, instead of metal wires carrying electron pulses. Computers also contain a fraction of the copper, lead, and other metals they once contained. Once again, this technology has been developed for efficiency, convenience, and lower cost, but it also affects consumption of our most basic metals.

Section Review

1. Why are rare earth metals important, and what are their environmental impacts?
2. What are the origins of coal and oil?
3. Identify several reasons for recycling metals.

14.3 ENVIRONMENTAL EFFECTS OF RESOURCE EXTRACTION

- *Mining produces dust, acids, and toxic metal pollutants.*
- *Metal is extracted by smelting or chemical extraction.*
- *The value of minerals and the cost of cleanup lead to conflict over these resources.*

The Pebble mine controversy (opening case study) derives from the many environmental consequences of mineral extraction. Each of us depends daily on geological resources mined from all around the world. We use scores of metals and minerals, many of which we've never even heard of, in our lights, computers, watches, fertilizers, and cars. The most obvious effect of mining is often the disturbance or removal of the land surface. Farther-reaching effects, though, include air and water pollution. The EPA lists more than 100 toxic air pollutants, from acetone to xylene, released from U.S. mines every year. Nearly 80,000 metric tons of particulate matter (dust) and 11,000 tons of sulfur dioxide are released from nonmetal mining alone. Pollution from chemical and sediment runoff is a major problem in many local watersheds.

Mining can affect water quality in several ways. Gold, copper, and other metals are often found in sulfide ores. When these minerals are exposed to air and water, they produce sulfuric acid, which is highly mobile and strongly acidic. Ores contain very low concentrations of gold, platinum, or other valuable metals, just 10 to 20 parts per billion, that may be economically extractable. Consequently, vast quantities of ore must be crushed and washed to extract metals. Toxic substances such as cyanide are often used to chemically separate metals from ore.

Different mining techniques pose different risks to water and air

There are many techniques for extracting geological materials. The most common methods are open-pit mining, strip mining, and underground mining. Less widespread but possibly more ancient is placer mining, in which nuggets of gold, diamonds, or coal are washed from stream sediments. Since the California gold rush of 1849, placer miners have used water cannons to blast away gravel hillsides. Mobilized sand and gravel clog streams and destroy fisheries, but this inexpensive method is still used in Alaska, Canada, and many other regions.

Underground mining is another ancient and much more dangerous method. Mine tunnels occasionally collapse. In underground coal mines, coal dust, as well as natural gas associated with coal beds, can cause explosions and uncontrollable fires. One mine fire in Centralia, Pennsylvania, has been burning since 1962; control efforts have cost at least $40 million, but the fire continues to expand. China, which depends on coal for much of its heating and electricity, has hundreds of smoldering mine fires; one has been burning for 400 years. Hundreds of coal mines smolder in the United States, China, Russia, India, South Africa, and Europe. The inaccessibility and size of these fires can make

them impossible to extinguish or control. According to a recent study from the International Institute for Aerospace Survey in the Netherlands, these fires consume up to 200 million tons of coal every year and emit as much carbon dioxide as all the cars in the United States.

Open-pit mines are used to extract massive beds of metal ores, coal, and other minerals. The size of modern open pits can be hard to comprehend. The Bingham Canyon copper mine, near Salt Lake City, Utah, is 800 m (2,640 ft) deep and nearly 4 km (2.5 mi) wide at the top. More than 5 billion tons of copper ore and waste material have been removed from the hole since 1906. A chief environmental challenge of open-pit mining is that groundwater accumulates in the pit. In metal mines, a toxic soup results. No one yet knows how to detoxify these lakes, which can endanger wildlife.

Surface or strip mines produce more than half the coal used in the United States (fig. 14.15). Because coal is often found in expansive, horizontal beds, the entire land surface can be stripped away to cheaply and quickly expose the coal. The overburden, or surface material, is placed back into the mine. Historically, this debris, or spoil, was dumped in long ridges called spoil banks, which eroded easily and allowed sulfur, mercury, and other contaminants to enter surface and groundwater. Lacking topsoil, spoil banks revegetate slowly, if at all.

The 1977 federal Surface Mining Control and Reclamation Act (SMCRA) has required better restoration of strip-mined lands, especially where mines replaced prime farmland. Since then, the record of strip-mine reclamation has improved substantially. Complete mine restoration is expensive, however, and compacted, acidic soil makes restoration difficult.

FIGURE 14.15 The scale of surface mining is enormous. Some giant mining machines stand as tall as a 20-story building and can scoop up thousands of cubic meters of rock per hour.
Gates Frontiers Fund Wyoming Collection within the Carol M. Highsmith Archive, Library of Congress, Prints & Photographs Division

Mountaintop removal is a still more controversial and damaging technique for mining coal. Used on the long mountain ridges of Appalachia, this process involves blasting off the top of a mountain to expose coal seams and shoveling the debris (the overburden) into stream valleys (fig. 14.16). Streamside habitats are destoyed, and water quality is degraded. The debris can be laden with selenium, arsenic, sulfur, and other toxic substances. At least 900 km (560 mi) of streams have been buried in West Virginia alone. In 2011, the EPA stopped one of the largest proposed mountaintop mines in U.S. history, the Spruce No. 1 in West Virginia, because the company's plan to dispose of mining waste in valley streambeds violated the Clean Water Act.

Ore processing emits acids and metals

Metals are extracted from ores by heating or with chemical solvents. Both processes release large quantities of toxic materials that can be even more environmentally hazardous than mining. Smelting—roasting ore to release metals—is a major source of air pollution. One of the most notorious cases of ecological devastation from smelting was a wasteland near Ducktown, Tennessee. In the mid-1800s, mining companies began excavating the rich copper deposits in the area. To extract copper from the ore, the company built huge, open-air wood fires, using timber from the surrounding forest. Dense clouds of sulfur dioxide released from sulfide ores poisoned the vegetation and acidified the soil over a 130 km^2 area. Rains washed the soil off the denuded land, creating a barren moonscape.

Sulfur emissions from Ducktown smelters were reduced in 1907 after Georgia sued Tennessee over air pollution. In the 1930s, the Tennessee Valley Authority (TVA) began treating the soil and replanting trees to cut down on erosion. Costly restoration efforts are still ongoing, a century after initial efforts to control sulfur emissions. While cases like this are unusual today in North America, they continue to occur elsewhere. The smelters of Norilsk, in Siberia, continue to make that city one of the world's most devastated places.

Heap-leach extraction is often used to dissolve and remove high-value metals, especially gold, from ore. A cyanide solution is sprayed on a large heap of crushed rock to dissolve gold. The solution is captured to recover the gold, but cyanide can leak into surface or groundwater. At the Beartrack Mine in east-central Idaho, for example, $220 million in gold was extracted with cyanide. When the mine closed in 2000, the company left 15 million metric tons of mine waste and huge ponds laced with cyanide. Meridian Gold, the mine owner, posted a $2 million bond to cover the costs of reclamation, but it may well cost 30 times that much to restore the area. Meanwhile, the acids, metals, and cyanide in the heap threaten the nearby Salmon River and its $2 billion annual tourism and fishing industry.

Ore washing and leaching consume a great deal of water. This is a concern in arid mining regions, such as the western United States. The USGS estimates that in Nevada, mining consumes about 230,000 m^3 (60 million gal) of fresh water per day. After use in ore processing, much of this water contains sulfuric acid, arsenic, and heavy metals, which can contaminate lakes and streams.

The Mineral Policy Center in Washington, D.C., estimates that mine drainage contaminates 19,000 km (12,000 mi) of rivers and streams in the United States. The EPA estimates that cleaning up impaired streams, along with 550,000 abandoned mines, in the United States could cost $70 billion. Worldwide, mine closing and rehabilitation costs are estimated in the trillions of dollars. Because of the volatile prices of metals and coal, many mining companies have gone bankrupt before restoring mine sites, leaving the public responsible for cleanup (fig. 14.17).

In 2015, a contractor working for the EPA was trying to drain toxic water from the abandoned Gold King mine outside the historic town of Silverton, Colorado. The mine, which closed in 1923 had been leaking into a nearby creek for decades and had killed almost all aquatic life there. The workers misjudged the volume and pressure of water in the mine. A blowout occurred, and at least 11 million liters (3 million gal) of highly toxic sludge gushed out into the Animas River, which flows into the Colorado River, a body that provides drinking water to 40 million people. The EPA apologized profusely for the error and tried to clean up the mess. Agency opponents howled that the EPA was deliberately trying to poison them; they never questioned why the mine was filled with toxic waste in the first place.

High-value minerals can support corruption

Political and human costs also result from mineral extraction and trade. Durable, highly valuable, and easily portable, gemstones and precious metals have long been a way to store and transport

FIGURE 14.16 Mountaintop removai mining is a highly destructive, and deeply controversial, method of extracting Appalachian coal.
David T. Stephenson/Shutterstock

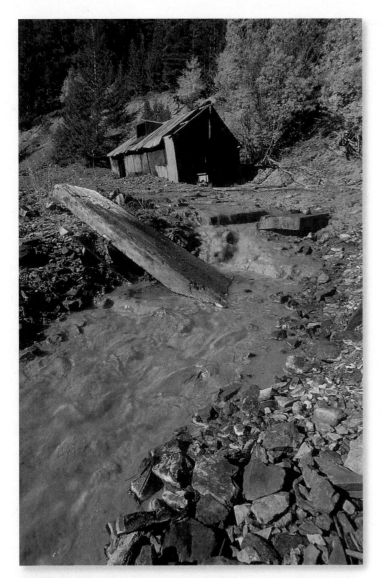

FIGURE 14.17 Thousands of "zombie" mines on public lands poison streams and groundwater with acid, metal-laced drainage.
Bryan F. Peterson

wealth. These properties have also led to illicit and criminal trade. Often, these valuable materials have bankrolled political dictators, criminal gangs, and terrorism around the world. In recent years, brutal civil wars in Africa have been financed—and often motivated—by gold, diamonds, tantalum ore (used in consumer electronics such as cell phones), and other high-priced commodities.

Although international agreements often ban trade that finances war and genocide, illegal trade continues. Much of the "blood diamond" trade ends up in the $100-billion-per-year global jewelry trade, two-thirds of which sells in the United States. Many people who treasure a diamond ring or a gold wedding band as a symbol of love and devotion are unaware that it may have been obtained through inhumane labor conditions and environmentally destructive mining and processing

methods. Civil rights organizations are campaigning to require better documentation of the origins of gems and precious metals to prevent their use as financing for crimes against humanity. Conflict over oil fields, as in Iraq and Kuwait, is another form of violence over mineral resources. These conflicts are one argument that is often made for reducing our dependence on oil. Even where outright war is absent, violence and human rights abuses are widespread in oil-producing regions, as in Nigeria and Ecuador.

In 2004, a group of Nobel Peace Prize laureates asked the World Bank to overhaul its policies on lending for resource extractive industries. "War, poverty, climate change, and ongoing violations of human rights—all of these scourges are all too often linked to the oil and mining industries," wrote Archbishop Desmond Tutu, winner of the 1984 Nobel Peace Prize for helping to eliminate apartheid in South Africa. In response, the World Bank appointed an Extractive Industries Review committee. In its final report, the committee recommended that some areas of exceptionally high biodiversity value should be "no-go" zones for extractive industries, and it recommended that the rights of those affected by extractive projects need better protection. Such recommendations remain hard to follow, however, when so much money is at stake.

Every year the organization PureEarth.org compiles a list of the world's worst pollution issues. You can see the list at worstpolluted.org. Among the top threats to human health are lead, mercury, chromium, and arsenic. Lead smelting and small-scale gold mining, which uses abundant mercury and has no environmental controls, are among the worst causes. These problems are especially disastrous in the developing world and in states of the former Soviet Union, where there is little money or political will to deal with environmental or health effects of pollution.

In the United States, where legal controls often are in place for environmental and public protection, mining also remains controversial. Often, the public bears the cost of cleanup, and must also provide the resources nearly free of charge. On the other hand, mining on public lands is a way of boosting the economy, as long as costs don't outweigh benefits (What Do You Think? reading, "Should We Revise Mining Laws?").

Section Review

1. Why is sulfuric acid common in mine runoff?
2. Describe smelting and some of its environmental effects.
3. Why might restoration be hard after mountaintop removal?

Quantitative Reasoning

If the United States has 550,000 abandoned mines, and the estimated cleanup cost is $70 billion, what is the average cleanup cost per mine? How much is that per person, if there are around 325 million Americans?

Should We Revise Mining Laws?

In 1872, the U.S. Congress passed the General Mining Law, intended to encourage prospectors to open up the public domain and promote commerce. This law, which has been in effect for more than a century, says, "All valuable mineral deposits in lands belonging to the United States are hereby declared to be free and open to exploration and purchase . . . by citizens of the United States." Claim holders can "patent" (buy) the land for $2.50 to $5 per acre (0.4 hectares), depending on the type of claim. Once the patent fee is paid, the owners can do anything they want with the land, just like any other private property. Although $2.50 per acre may have been a fair market value in 1872, many people regard it as ridiculously low today, amounting to a giveaway of public property.

In Nevada, for example, a Canadian mining company paid $9,000 for federal land that contains an estimated $20 billion worth of precious metals. Similarly, Colorado investors bought about 7,000 ha (17,000 acres) of rich oil-shale land in 1986 for $42,000 and sold it a month later for $37 million. You don't actually have to find any minerals to patent a claim. A Colorado company paid a total of $400 for 65 ha (160 acres) it claimed would be a gold mine. Almost 20 years later, no mining has been done, but the property—which just happens to border the Keystone Ski Area—is being subdivided for condos and vacation homes.

According to the Bureau of Land Management (BLM), some $4 billion in minerals are mined each year on U.S. public lands. Under the 1872 law, mining companies pay nothing to the government for the ores they take. Furthermore, they can deduct a depletion allowance from taxes on mineral profits. Former Senator Dale Bumpers of Arkansas, who called the 1872 mining law "a license to steal," has estimated that the government could derive $320 million per year by charging an 8 percent royalty on all minerals and probably could save an equal amount by requiring larger bonds to be posted to clean up after mining is finished.

On the other hand, mining companies argue they would be forced to close down if they had to pay royalties. Jobs would be lost, and the economies of western mining towns would collapse if mining becomes uneconomic. We provide subsidies and economic incentives to many industries to stimulate economic growth. Why not support mining for metals essential for our industrial economy? Mining is a risky and expensive business. Without subsidies, mines would close down and we would be completely dependent on unstable foreign supplies.

Mining critics respond that other resource-based industries have been forced to pay royalties on materials they extract from public lands. Coal, oil, and gas companies pay 12.5 percent royalties on fossil fuels obtained from public lands. Timber companies—although they don't pay the full costs of the trees they take—have to bid on logging sales and clean up when they are finished. Even gravel companies pay for digging up the public domain. Ironically, we charge for digging up gravel, but give gold away for free.

Numerous bills have been introduced in Congress to revise the mining law. Bills supported by public interest groups would require companies mining on federal lands to pay a higher royalty on their production. They also would eliminate the patenting process, impose stricter reclamation requirements, and give federal managers authority to deny inappropriate permits. In contrast, bills offered by western legislators and backed by mining supporters would leave most provisions of the 1872 bill in place.

A current controversy involves a subsidiary of Antofagasta, a Chilean company with a terrible environmental record that wants to dig a 300 m (1,000 ft) deep open pit copper–nickel mine next to a river flowing into the Boundary Canoe Area Wilderness in northern Minnesota. The million acre (400,000 ha) wilderness is the most heavily visited wilderness in the United States, and the only one devoted to canoeing on lakes and rivers currently clean enough to drink. The ore is less than 1 percent copper, so more than 99 tons of rock have to be dug up to get one ton of ore. The ore is copper sulfide. When crushed and exposed to air and water, the sulfide oxidizes to form sulfuric acid. Opponents argue that the acids, heavy metals (including mercury, arsenic, lead, and selenium) leached out of waste rock will poison water and kill fish, wildlife, and native vegetation in the wilderness. The company owns the mineral rights, but wants to pile up the 400 million tons of waste they'll produce on state and federal land. The mine would only be in operation for about 20 years, but wastewater would need to be treated forever. Is this a worthwhile trade off?

The Obama Administration rejected requests to store waste rock on federal lands, ruling that the threat to water quality in the Wilderness area is too great. The Trump administration, however, reversed this ruling and ordered approval of the leases in 2018. This action was taken late on Friday afternoon just before Christmas without any public notice or input. It would be inconvenient for Antofagasta, Trump argued, to have to wait for an environmental review. As it happens, the chairman of Antofagasta also is an acquaintance of the Trump family and rented a house to the president's daughter Ivanka Trump. This relationship has raised questions among critics of the sudden approval.

A study by the University of Minnesota estimated that tourism in Northeast Minnesota (including the BWCA) generates $1 billion per year and supports 20,000 jobs. The proposed copper mine would produce $150 million and support 427 jobs over the 20-year life of the mine.

This case, like so many before it, hinges on the balance between legal policies that favor mining, including low fees and lenient environmental standards, a persistent hunger for employment in rural areas, and growing concern about other, non-mineral values of a place.

What do you think we should do about our mining laws? Do they remain necessary and justifiable, or should they be revoked or revised? How would you write the law if you had the chance?

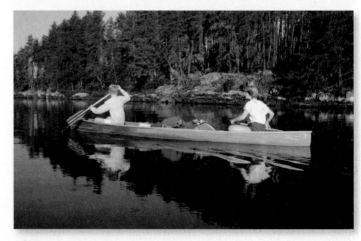

A Chilean company with a poor environmental record wants to dig an enormous open-pit copper–nickel mine immediately adjacent to Minnesota's Boundary Waters Canoe Area Wilderness. Does 20 years of a few dozen local jobs justify risking a million acres of wilderness?

William P. Cunningham

14.4 GEOLOGICAL HAZARDS

- *Earthquakes result from sudden shifts of tectonic plates.*
- *Earthquakes and floods cause the greatest damage and mortality of all geologic hazards.*
- *Erosion and landslides frequently cause property damage.*

Earthquakes, volcanic eruptions, floods, and landslides are among the catastrophic geological events that have shaped the world around us. Some of these have had global effects: Vast volcanic lava flows in Siberia 250 million years ago and in India 66 million years ago are thought to have caused, or contributed to, major mass extinctions that marked the end of the Permian and Cretaceous eras (see fig. 14.6). Volcanic emissions of sulfur dioxide and carbon dioxide may have devastated ecosystems and warmed the climate sharply during these and other major volcanic events. Meteor impacts (fig. 14.18) are also leading suspects in global mass extinctions. Most geological hazards are more regional, though. The San Andreas Fault (fig. 14.19), for example, has shaped California's landscapes. It has also caused devastation when major earthquakes have hit population centers.

Although floods are less surprising than earthquakes, they take more human lives than any other geological hazard. Earthquakes, landslides, and volcanic eruptions also cause tremendous losses of life and property in local areas.

Earthquakes usually occur on plate margins

Earthquakes are sudden movements in the earth's crust that occur along faults (planes of weakness) at transform boundaries where one plate slides past another. Only floods have caused more widely devastating events. A 2010 earthquake in Haiti killed perhaps a quarter million people; the impoverished island nation is still reeling from its effects. Still worse was a 1976 earthquake in Tangshan, China. Government officials reported 655,000 deaths (some geologists doubt the total was that high).

The San Andreas fault in California is one of the most notorious and highly visible faults in the world (fig. 14.19). It runs about 1,300 km from the Salton Sea in southern California to Point Reyes, just north of San Francisco, where it veers offshore to

FIGURE 14.19 The Elkhorn Escarpment shows where the San Andreas fault runs across the Carrizo Plain northwest of Los Angeles.
StockTrek/Getty Images

follow the coast north nearly to Oregon. The fault occurs where the Pacific plate is moving to the northwest, past the North American plate, which is moving south. A number of major quakes have originated along this fault, including those that struck San Francisco in 1906 and 1989.

Earthquakes vary dramatically in the amount of energy and movement released (table 14.4). When movement along faults occurs gradually and relatively smoothly, it is called creep or seismic slip and may be undetectable to the casual observer. When friction prevents rocks from slipping easily, stress builds up until it is finally released with a sudden jerk, which can cause the ground to shake like jelly.

Most earthquake deaths result from the collapse of weakly constructed buildings (fig. 14.20). Cities built on soft landfill or poorly consolidated soil, like much of San Francisco or Port-au-Prince, Haiti, usually suffer the greatest damage from earthquakes. Water-saturated soil can liquefy when shaken. Under these

(a) A meteor impact crater in Arizona

(b) Volcanic eruption

(c) Glacier in Alaska

FIGURE 14.18 Geological events such as meteor or asteroid impacts (a), massive volcanic eruptions (b), or climate change (c) are thought to trigger mass extinctions that mark major eras in the earth's history.
(a) StockTrek/Getty Images; (b) ammit/123RF; (c) McGraw-Hill Education

Table 14.4	Worldwide Frequency of Earthquakes	
Richter Scale Magnitude*	Description	Average Number per Year
2–2.9	Unnoticeable	300,000
3–3.9	Smallest	felt 49,000
4–4.9	Minor earthquake	6,200
5–5.9	Damaging earthquake	800
6–6.9	Destructive earthquake	120
7–7.9	Major earthquake	18
>8	Great earthquake	1 or 2

* For every unit increase in the Richter scale, ground displacement increases by a factor of 10, while energy release increases by a factor of 30. There is no upper limit to the scale, but the largest earthquake recorded was 9.5, in Chile in 1960.

Source: Gutenberg, B. In Earth, edited by F. Press and R. Siever. W. H. Freeman & Company, 1978.

conditions, buildings sometimes sink out of sight or fall like rows of dominoes.

Increasingly, builders in earthquake zones are designing buildings to withstand tremors. The primary methods used are heavily reinforced structures, strategically placed weak spots in the building that can absorb vibration from the rest of the building, and pads or floats beneath the building on which it can shift harmlessly with ground motion.

Human-induced earthquakes are becoming more common

We used to think that humans couldn't cause earthquakes, but surges in seismic activity associated with well drilling and disposal of waste water are changing our assumptions. Between 2000 and 2007, Oklahoma, for example, averaged about 20 measurable earthquakes per year. Starting about 2008, that number began shooting up, reaching more than 6,000 tremors in 2015 (fig. 14.21). This was more than all the other 48 continental states combined. Oklahoma has become the earthquake capital of the United States. Some towns that had never experienced a quake had as many as 50 in a single day in 2015. This swarm of quakes correlates with a rush of oil and gas well drilling, and in particular with a proliferation of waste water wells. Between 2008 and 2014, oil and gas production in Oklahoma more than doubled.

The problem seemed to be that Oklahoma had a wastewater disposal problem. Some oil wells there produce as much as 65 barrels of salty wastewater for every barrel of oil. The waste is much too toxic to dump into surface water, and it would cost more than the oil is worth to treat it, so most of it is pumped back down into deep geological formations. In north central Oklahoma alone, about 1,000 wells injected some 17 billion liters (850 million barrels) of wastewater into deep strata in 2014. The result, many seismologists believe, was a flood of earthquakes.

Arkansas, Kansas, and Colorado also have experienced dramatic upticks in seismic activity associated with disposal wells. Geologists now think injection water lubricates fault lines causing slippage that releases energy and results in quakes. Not all injection wells are equally guilty. In Arkansas, it was only necessary to plug four wells to prevent earthquakes. Oklahoma is considering limits on deep well disposal, but the economy may do that for them. In 2016, with oil at less than $30 per barrel, thousands of wells were capped, and earthquakes decreased by at least 20 percent. The question remains, though, whether earthquakes have become an inevitable corollary of gas extraction.

FIGURE 14.20 Earthquakes are most devastating where building methods can't withstand shaking. The 2010 earthquake in Haiti is thought to have killed 230,000 people. Collapsed buildings, food and water shortages, infectious diseases, and exposure to the elements all contributed to the death toll.

Source: U.S. Geological Survey photo by Walter D. Mooney

FIGURE 14.21 Earthquakes, historically uncommon in Oklahoma, have become frequent and often large around injection wells (blue) used for disposing of fracking wastewater. Earthquakes (orange) greater than magnitude 3.0 for 2016–17 are shown.

Source: Oklahoma Geological Survey

Tsunamis can be more damaging than the earthquakes that trigger them

Tsunamis are powerful waves triggered by earthquakes or landslides. The name is derived from the Japanese word for "harbor wave," because the waves often are noticed only when they approach shore. Tsunamis can be more damaging than the earthquakes that create them. The tsunami that struck the coast of Japan on March 11, 2011, for example, was triggered by a magnitude 9.0 underwater earthquake about 72 km (45 mi) out to sea. Although the earthquake shook buildings, it did relatively little damage because Japan has high construction standards. The waves that followed, however, were up to 38 m (124 ft) high. In some low-lying areas, the waves traveled up to 10 km (6 mi) inland, washing away buildings, boats, cars, and even whole trains. As the wall of water smashed through towns and villages, it carried a thick slurry of vehicles, building materials, mud, and debris that was much more destructive than the water alone. At least 25,000 people were listed as dead or missing, hundreds of thousands of buildings were damaged or destroyed (fig. 14.22).

The most long-lasting effect of that tsunami will be the damage it did to the Fukushima Daiichi nuclear power plant on Japan's northeast coast. The reactors shut down after the earthquake, as they were designed to do, but the tsunami destroyed the cooling systems (as well as the backup generators) needed to keep the reactors under control. Three of the six reactors at the Fukushima Daiichi complex were damaged by fires and hydrogen gas explosions after partial meltdown of the reactor cores, and a fourth was damaged when water levels dropped in the waste fuel storage pool, exposing fuel rods to the air.

Some 230,000 residents were evacuated from around the damaged reactors, and many will never move back. It's estimated that five million tons of debris (some of it radioactive) washed into the sea. Within a year, some of that waste reached North America. Japan estimated economic losses of as much as $300 billion, not including costs such as the cleanup of the nuclear reactors, several of which will probably have to be entombed in concrete forever.

Arguably, the highest tsunami risk in North America is the Pacific coast from British Columbia to northern California. A complex of subduction zones called the Cascadia fault stretches for 1,000 km (620 mi) along this coast. The last recorded major earthquake along this fault was in 1700. Records of an "orphan" tsunami in Japan match Native American memories of whole villages being swallowed by the sea, and physical evidence of up to 2 meters of subsidence along the coast. The fault has been quiet for 300 years, but similar subduction zones tend to average earthquakes about every 100 to 200 years. A mega earthquake and tsunami along the fault today could cause huge damage because most cities, including Seattle, Portland, and Vancouver, are poorly prepared. Some geologists predict a 10 to 15 percent chance of a magnitude 9 earthquake and a tsunami 30 meters or more high in the next 50 years. Millions of people could be killed or displaced. Volcanic activity along this section of the "ring of fire" also presents a substantial risk.

FIGURE 14.22 In 2011, a magnitude 9.0 earthquake just off the coast of Japan created a massive tsunami that destroyed homes, killed at least 25,000 people, and destroyed four nuclear power plants.
JIJI PRESS/AFP/Getty Images

Volcanoes eject gas and ash, as well as lava

Volcanoes and undersea magma vents produce much of the earth's crust. Over hundreds of millions of years, gaseous emissions from these sources formed the earth's earliest oceans and atmosphere. Many of the world's fertile soils are weathered volcanic materials. Volcanoes have also been an ever-present threat to human populations. One of the most famous historic volcanic eruptions was that of Mount Vesuvius in southern Italy, which buried the cities of Herculaneum and Pompeii in 79 C.E. The mountain had been giving signs of activity before it erupted, but many citizens chose to stay and take a chance on survival. On August 24, the mountain buried the two towns in ash. Thousands were killed by the dense, hot, toxic gases that accompanied the ash flowing down from the volcano.

Nuées ardentes (French for "glowing clouds") are deadly, denser-than-air mixtures of hot gases and ash like those that inundated Pompeii and Herculaneum. These pyroclastic clouds can have temperatures exceeding 1,000°C, and they move at more than 100 km/hour (60 mph). In November 2010, pyroclastic clouds rolled down the slopes of Mt. Merapi just outside Yogyakarta, Indonesia. At least 325 people were killed, and more than 300,000 were forced from their homes (fig. 14.23).

It is not just a volcano's dust that blocks sunlight. Sulfur emissions from volcanic eruptions combine with rain and atmospheric moisture to produce sulfuric acid (H_2SO_4). The resulting droplets of H_2SO_4 interfere with solar radiation and can significantly cool the world climate. In 1991, Mt. Pinatubo in the Philippines emitted 20 million tons of sulfur dioxide, which combined with water to form tiny droplets of sulfuric acid. This acid aerosol reached the stratosphere, where it circled the globe for two years. The high, thin

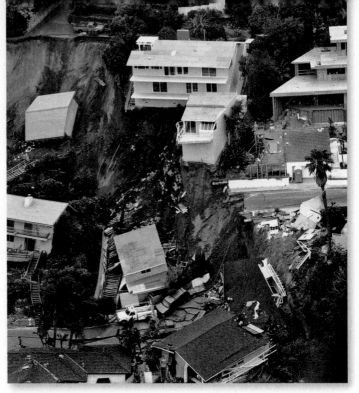

FIGURE 14.23 Volcanic ash covers damaged houses and dead vegetation after the November 2010 eruption of Mt. Merapi *(background)* in central Java. More than 300,000 residents were displaced and at least 325 were killed by this multiday eruption.
andersen_oystein/Getty Images

FIGURE 14.24 Landslides tend to occur on steep, unconsolidated hillsides, like this one in Laguna Beach, California.
NICK UT/AP Images

haze cooled the entire earth by 1°C and postponed global warming for several years. It also caused a 10 to 15 percent reduction in stratospheric ozone, allowing increased ultraviolet light to reach the earth's surface.

Landslides and mass wasting can bury villages

Gravity constantly pulls downward on every material everywhere on earth, causing a variety of phenomena collectively termed **mass wasting** or mass movement, in which geological materials are moved downslope from one place to another. The resulting movement is often slow and subtle, but some slope processes such as rockslides, avalanches, and land slumping can be swift, dangerous, and very obvious. *Landslide* is a general term for rapid downslope movement of soil or rock. In the United States alone, over $1 billion in property damage is done every year by landslides and related mass wasting.

Activities such as road construction, forest clearing, agricultural cultivation, and building houses on steep, unstable slopes increase both the frequency and the damage done by landslides. In some cases, people are unaware of the risks they face by locating on or under unstable hillsides. In other cases, they simply ignore clear and obvious danger. Southern California, where people build expensive houses on steep hills of relatively unconsolidated soil, is often the site of large economic losses from landslides. Especially after chaparral fires remove plant cover, winter rains can saturate and loosen the exposed soil. These conditions frequently cause slope failure (fig. 14.24). In narrow canyons, mudslides can carry away whole neighborhoods and bury downslope areas in debris flows.

Floods are the greatest geological hazard

Like earthquakes and volcanoes, floods are normal events that cause damage when people get in the way. As rivers carve and shape the landscape, they build broad floodplains and level expanses that are periodically inundated. Many cities have been built on these flat, fertile plains, which are both good for agriculture and convenient to the river. When floods occur infrequently, people develop a false sense of security. But eventually most floodplains do flood.

Among direct natural disasters, floods take the largest number of human lives and cause the most property damage. A flood on the Yangtze River in China in 1931 killed 3.7 million people, making it the most deadly natural disaster in recorded history. In another flood, on China's Yellow River in 1959, about 2 million people died mostly due to resultant famine and disease.

Flooding is a major hazard in low-lying delta regions, such as Bangladesh. In 2004, torrential monsoon rains displaced some 25 million people and flooded about one-fourth of the low-lying country (fig. 14.25). Similarly, in 2008 the cyclone Nargis hit the Irrawaddy Delta of Myanmar. The government made the situation worse by refusing international assistance. It took weeks for aid to reach some areas. An estimated 140,000 people died, either directly from the flood or from starvation and disease that followed. Some governments are learning from experience, though. When Bangladesh suffered hurricane-force storms and flooding in 2019, the government had learned from previous disasters. Over a million coastal villagers were evacuated to storm shelters, above the storm surge, and few casualties occurred.

Flooding associated with extreme precipitation is increasingly frequent, as a warming climate causes more evaporation, more convection in the atmosphere, and more extreme rainfall. Changes in land use also increase flooding. Covering the land with impervious surfaces, such as roads, parking lots, and building roofs, reduces water infiltration into the soil and speeds the rate of runoff into streams and lakes. Clearing forests for agriculture and destroying

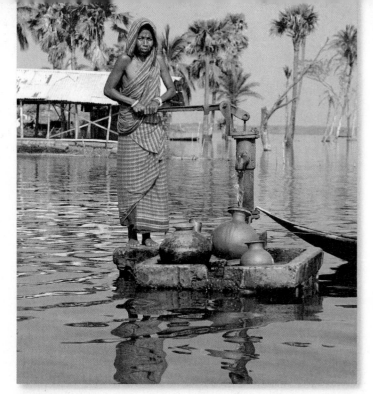

FIGURE 14.25 In 2004, unusually strong monsoon rains caused rivers to overflow their banks, and about one-quarter of Bangladesh was under water. About 25 million people were displaced.
Digital Vision/PunchStock

natural wetlands also increases both the volume and the rate of water discharge after a storm.

In an effort to control floods, many communities build levees and floodwalls to keep water within riverbanks, and river channels are dredged and deepened to move water along faster. Every flood-control structure simply transfers the problem downstream, however, because the water has to go somewhere.

Rather than spend money on levees and floodwalls, hydrologists increasingly argue that we should restore wetlands as natural "sponges" for floodwaters and protect forests or other undeveloped areas at headwaters of streams. They also argue that we could get out of the way, removing houses and other vulnerable development from natural floodplains. Undeveloped floodplains can provide wildlife habitat, parks, and recreation areas that are resilient to flood damage.

Flood insurance often complicates efforts to reduce development on floodplains. The U.S. federal government, in particular, subsidizes flood insurance, which helps, or forces, people to stay and rebuild in flood-prone areas. Even when people would prefer to move, they can only collect insurance payments if they rebuild in the same place as before. Sometimes homeowners rebuild repeatedly with insurance payments adding up to far more than the value of the home.

Some areas are acknowledging the growing severity and cost of flooding, and they are seeking alternatives to this cycle. Nashville, Tennessee, for example, has avoided millions of dollars in flood damages by buying flood-prone properties and converting them to recreational areas. Using state and local money, reimbursable from federal funds, the city has produced huge savings in money, trauma, and disruption of floods.

Beaches erode easily, especially in storms

Beach erosion occurs on all sandy shorelines because the motion of the waves is constantly redistributing sand and other sediments. One of the world's longest and most spectacular sand beaches runs down the Atlantic Coast of North America from New England to Florida and around the Gulf of Mexico. Much of this beach lies on some 350 long, narrow barrier islands that stand between the mainland and the open sea.

Early inhabitants recognized that exposed, sandy shores were hazardous places to live, and they settled on the bay side of barrier islands or as far upstream on coastal rivers as was practical. Modern residents, however, place a high value on living where they have an ocean view and ready access to the beach. Construction directly on beaches and barrier islands can cause irreparable damage to the whole ecosystem. Under normal circumstances, fragile vegetative cover holds the shifting sand in place. Damaging this vegetation and breaching dunes with roads can destabilize barrier islands. Storms then wash away beaches or even whole islands. Hurricane Katrina in 2005 caused $100 billion in property damage along the Gulf Coast of the United States, mostly from the storm washing over barrier islands and coastlines (fig. 14.26). In the future, intensified storms and rising sea levels caused by global warming will make barrier islands and low-lying coastal areas even riskier places to live.

Cities and individual property owners often spend millions of dollars to protect beaches from erosion and repair damage after storms. Sand is dredged from the ocean floor or hauled in by the truckload, only to wash away again in the next storm. Building

FIGURE 14.26 The aftermath of Hurricane Katrina on Dauphin Island, Alabama. Since 1970, this barrier island at the mouth of Mobile Bay has been overwashed at least five times. More than 20 million yd³ (15 million m³) of sand has been dredged or trucked in to restore the island. Some beach houses have been rebuilt, mostly at public expense, five times. Does it make sense to keep rebuilding in such an exposed place?
Source: USGS

artificial barriers, such as groins or jetties, can trap migrating sand and build beaches in one area, but they often starve downstream beaches and make erosion there even worse.

As is the case for inland floodplains, many government policies encourage people to build where they probably shouldn't. Subsidies for road building and bridges, support for water and sewer projects, tax exemptions for second homes, flood insurance, and disaster relief are all good for the real estate and construction businesses but invite people to build in risky places. In some areas, beach houses have been rebuilt, mostly at public expense, four or five times in the past 20 years. Does it make sense to keep rebuilding in these places?

The Coastal Barrier Resources Act of 1982 prohibited federal support, including flood insurance, for development on sensitive islands and beaches. In 1992, however, the U.S. Supreme Court ruled that ordinances forbidding floodplain development amount to an unconstitutional "taking," or confiscation, of private property. The debate over property rights and sound environmental policy is likely to continue.

Section Review

1. What are the main reasons for mortality with earthquakes?
2. Describe different types of volcanic emissions and their effects.
3. Is it a good idea to build on barrier islands? Why or why not?

Connecting the Dots

We need materials from the earth to sustain our modern lifestyle, but many of the methods we use to get those materials have severe environmental consequences. Some of these consequences can be seen in the controversies over copper mining in Alaska's Bristol Bay (case study) and on the edge of Minnesota's Boundary Waters wilderness. We can extend resources through recycling and by substituting old methods with new materials and more efficient ways of using them. We also can do much to repair the damage caused by resource extraction, although open-pit mines and mountains with their tops removed will never be returned to their original condition. Water contamination is one of the main environmental costs of mining. Pollutants

include acids, cyanide, mercury, heavy metals, and sediment. Air pollution from smelting also can cause widespread damage.

Geological hazards, including floods, earthquakes, volcanoes, landslides, and coastal erosion, can cause tremendous damage and loss of life. Usually, poor building practices or land use decisions contribute to these disasters. Many of these events also occur rarely, so the risks are easy to ignore until catastrophe strikes. Although the earth seems stable most of the time, floods, earthquakes, and other hazards remind us that the same forces that have modified the earth in the past are still at work, continuing to reshape the land we live on.

Reviewing Key Terms

Can you define the following terms in environmental science?

core 14.1

crust 14.1

earthquakes 14.4

igneous rocks 14.1

magma 14.1

mantle 14.1

mass wasting 14.4

metamorphic rock 14.1

mid-ocean ridges 14.1

mineral 14.1

rock 14.1

rock cycle 14.1

sedimentary rock 14.1

sedimentation 14.1

subduction 14.1

tectonic plates 14.1

tsunami 14.4

volcano 14.4

weathering 14.1

Critical Thinking and Discussion Questions

1. Look at the walls, floors, appliances, interior, and exterior of the building around you. How many earth materials were used in their construction?

2. Is your local bedrock igneous, metamorphic, or sedimentary? How do you know? If you don't know, who might be able to tell you?

3. Suppose you live in a small, mineral-rich country, and a large, foreign mining company proposes to mine and market your

minerals. How should revenue be divided between the company, your government, and citizens of the country? Who bears the greatest costs? How should displaced people be compensated? Who will make sure that compensation is fairly distributed?

4. Geological hazards affect a minority of the population that builds houses on unstable hillsides, in flood-prone areas, or on faults. What should society do to ensure safety for these people and their property?

5. A persistent question in this chapter is how to reconcile our responsibility as consumers with the damage from mineral extraction and processing. How responsible are you? What are some steps you could take to reduce this damage?

6. If gold jewelry is responsible for environmental and social devastation, should we stop wearing it? Should we worry about the economy in producing areas if we stop buying gold and diamonds? What further information would you need to answer these questions?

Data Analysis

Mapping Geological Hazards

Find an animated map of volcanoes in the animated National Atlas at www.nationalatlas.gov/dynamic.html. Where are most of the volcanoes in the lower 48 states? Roll over the red dots to see photos and elevations of the volcanoes.

You can explore recent earthquakes, where they were, and how big they were, by looking at the USGS earthquake information page at www.earthquake.usgs.gov/. Click on the map to investigate recent events, and go to Connect to find study questions for both volcanoes and earthquakes.

Probability of earthquake within 50 years

- 1.00
- 0.77
- 0.59
- 0.46
- 0.35
- 0.27
- 0.21
- 0.16
- 0.12
- 0.10
- 0.08
- 0.06
- 0.04
- 0.03
- 0.02
- 0.01

Go to Connect to explore earthquake and volcano occurrences.

▲ California's farming economy depends entirely on reliable water supplies from mountain snowpack. Californians are motivated to reduce climate change to protect their economy, environment, and way of life.
Robert Glusic/Exactostock/Superstock

Learning Outcomes

After studying this chapter, you should be able to:

15.1 Describe the general composition and structure of the atmosphere.

15.2 Explain processes that shape regional and seasonal weather patterns.

15.3 Outline some factors in natural climate variability.

15.4 Explain how we know that recent climate change is human-caused.

15.5 List some effects of climate change.

15.6 Identify some solutions being developed to slow climate change.

"I was born in 1992. You have been negotiating all my life. You cannot tell me you need more time."

– Christina Ora, youth delegate from the Solomon Islands addressing the plenary at COP15, 2009

CASE STUDY

Climate Action in California: No Longer Just Talking About the Weather

Mark Twain said that everybody talks about the weather, but no one ever does anything about it. With climate change, that is no longer true: We are doing a lot to change the weather everywhere, as we warm our climate.

But as awareness grows, we are also working to reduce our climate impacts, and one of the leaders in climate action is California. Californians know that their state is vulnerable to climate change. The agricultural economy depends almost entirely on water derived from winter snow in the Sierra Nevada and other mountains. Longer, drier summers have reduced water supplies and produced intense, expensive, and deadly fire seasons. Heat waves threaten low-income and urban populations. Rising sea levels threaten the state's coastal cities and its iconic beaches and coastlines.

California politics has a long tradition of leading environmental policy, including air quality, vehicle efficiency, solar and wind energy, and other measures that have improved public health and economies nationally. In 2006, the state Assembly passed Assembly Bill 32 (AB32), the Global Warming Solutions Act, which planned to reduce greenhouse gas (GHG) emissions below 1990 levels by 2020. After that, goals get more aggressive. The aim is 40 percent below 1990 levels by 2030 and 80 percent by 2050.

How is the state taking action? By implementing strategies already well established in energy technology and economics. These include both standards (rules) and economic incentives. They also identify and target the largest sectors first. Globally, electricity, heating, industry, and agriculture dominate greenhouse gas (GHG) emissions (fig. 15.1). Strategies to decarbonize these sectors are well known, with much of the work on these strategies done by scholars at California's universities (see Climate Action, section 15.6). The difficulty is finding the political will to act.

In California, the most important sectors for GHG emissions are transportation (40 percent) and industry (23 percent). So the first priorities are a 50 percent reduction in gasoline-powered vehicles and getting 50 percent of electricity from renewable sources. More renewable electricity will also be needed as we electrify industry and transportation, to get them off of fossil fuels. Weaning these sectors off of oil and gas is hard to imagine now, but a few generations ago it was hard to imagine the infrastructure that would be needed to replace horse-drawn wagons with gas-fueled vehicles. And electric vehicles are far more efficient and cheaper to operate, in part because internal combustion engines use over 80 percent of fuel to produce heat, not motion.

California's AB32 plan has lots of benefits beyond climate protection. Vehicle efficiency will save Californians $10 billion by 2030. Doubling of efficiency in existing buildings will save money for businesses and residents. Initiatives under AB32 are expected to produce tens of thousands of new jobs that can't be exported overseas. Already, the state's clean energy industry employs over 500,000 people and has brought in two thirds of all clean-energy investment dollars nationwide. A carbon trading market puts a price on carbon and provides economic incentives for innovation. New electric-powered city buses will make urban air healthier to breathe and reduce smog as well as health impacts. New houses are required to have solar panels by 2030, a controversial rule but one that is sure to accelerate discussions of how to maximize production and upgrade distribution grids.

State policies also outline plans for sequestering (capturing and storing) CO_2 through land use practices and forest management. This is critical because to avoid the worst effects of climate change, we need not only to drastically cut emissions but also to develop "negative emissions," or carbon capture systems (fig. 15.2).

At present about 30 percent of California's electrical generation is renewable or small hydropower. Another 20 percent is large hydropower (which has larger environmental impacts than small hydro) and nuclear, so over 50 percent comes from non-fossil sources. The state is not abandoning nuclear-powered electricity, a point that has been controversial. While many argue that nuclear power is expensive and unsafe, others say that because it is low-carbon (zero-carbon at the point of power generation) and large-scale, nuclear power is an important part of avoiding catastrophic climate change.

Most U.S. states and more than 500 cities and counties have taken steps to promote renewable energy and reduce greenhouse gas emissions. While federal policies often discourage climate progress, this case demonstrates the power of state and local action. A transition to renewable energy has countless co-benefits. It reduces pollution, protects infrastructure, and improves quality of life for many. It will also save trillions of

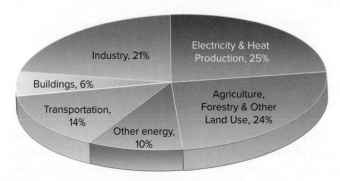

FIGURE 15.1 Global greenhouse gas emissions, by sector.

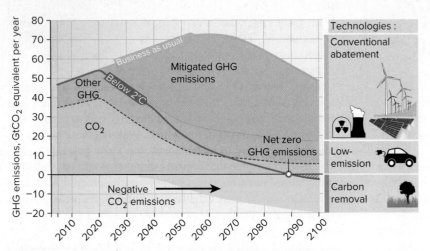

FIGURE 15.2 Reducing climate change to relatively "safe" 1.5°C will require not just reversing emissions (red line) but also using "negative emissions" technologies to capture CO₂ already emitted.
Source: UNEP 2017

dollars in military costs if we no longer need to fight wars to defend oil and gas fields abroad. We have a long way to go in reversing climate change, and the task is daunting, but it is also exciting. In this chapter we discuss some of the many ways to take action on climate change. We also examine climate processes and the impacts of climate change that make action so important.

To read more about strategies of the California Air Resources Board, see https://www.arb.ca.gov/cc/pillars/pillars.htm and California Climate Change, https://www.climatechange.ca.gov/

15.1 WHAT IS THE ATMOSPHERE?

- *The atmosphere is layered, with more massive molecules near the ground surface.*
- *The troposphere is heated by the warmed earth surface; convection cells result as warmed air rises.*
- *The composition of the atmosphere influences how much heat energy is stored.*
- *Atmospheric circulation redistributes heat and moisture.*

Of all the planets in our solar system, only the Earth has an atmosphere that makes life possible, as far as we know. The atmosphere retains solar heat, protects us from deadly radiation in space, and distributes the water that makes up most of your body. The atmosphere consists of gas molecules, held near the earth's surface by gravity and extending upward about 500 km. All the weather we see is in just the lowest 10–12 km of the atmosphere. **Weather** is a term for the short-lived and local patterns of temperature and moisture that result from this circulation. In contrast, **climate** refers to long-term patterns of temperature and precipitation. Understanding the difference between short-term variations and long-term patterns is important in understanding our climate.

The earth's earliest atmosphere probably consisted mainly of lightweight hydrogen and helium. Over billions of years, most of that hydrogen and helium diffused into space. Volcanic emissions added carbon, nitrogen, oxygen, sulfur, and other elements

Benchmark Data

Among the ideas and values in this chapter, the following are a few worth remembering.

5–18 km	Approximate depth of the atmosphere, at poles and equator
280 ppm	Pre-industrial (1750) concentration of CO₂
350 ppm	Concentration that limits climate warming to < 2 degrees C
415 ppm	Concentration in 2018
40 GT	Approximate amount of CO₂ released annually
5–20% of GDP	Cost to global economy of current climate change trajectory
1% of GDP	Cost to aggresively control CO₂ emissions

to the atmosphere. Virtually all the molecular oxygen (O₂) that we breathe was probably produced by photosynthesis in blue-green bacteria, algae, and green plants.

Clean, dry air is mostly nitrogen and oxygen (table 15.1). Water vapor concentrations vary from near zero to 4 percent, depending on air temperature and available moisture. Minute particles and liquid droplets—collectively called **aerosols**—also are suspended in the air. Atmospheric aerosols are important in capturing, distributing, or reflecting energy.

Table 15.1	Present Composition of the Lower Atmosphere*	
Gas	Symbol or Formula	Percent by Volume
Nitrogen	N_2	78.08
Oxygen	O_2	20.94
Argon	Ar	0.934
Carbon dioxide	CO_2	0.035
Neon	Ne	0.00182
Helium	He	0.00052
Methane	CH_4	0.00015
Krypton	Kr	0.00011
Hydrogen	H_2	0.00005
Nitrous oxide	N_2O	0.00005
Xenon	Xe	0.000009

*Average composition of dry, clean air.

The atmosphere has four distinct zones of contrasting temperature, which result from differences in absorption of solar energy (fig. 15.3). The layer of air immediately adjacent to the earth's surface is called the **troposphere** (*tropein* means to "turn" or "change" in Greek). Air in this layer is in constant motion, redistributing heat and moisture around the globe. Within the troposphere, air absorbs energy from the sun-warmed earth's surface, and from moisture evaporating from oceans. Warmed air circulates in great vertical and horizontal **convection currents,** which occur when warm, low-density air rises above a cooler, denser layer. (You can observe a similar process in a pot of simmering water on the stove: Water heated at the hot bottom of the pot rises up above the cooler layers at the top, creating convective circulation patterns.) Convection constantly redistributes heat and moisture around the globe. The depth of the troposphere ranges from about 18 km (11 mi) over the equator, where heating and convection are intense, to about 8 km (5 mi) over the poles, where air is cold and dense. The troposphere contains about 75 percent of the total mass of the atmosphere. Within the troposphere, temperatures drop rapidly with increasing distance from the earth, reaching about −60°C (−76°F). At this point, air is no longer warmer than its surroundings, and it ceases to rise. We call this boundary, where mixing ends, the tropopause.

The **stratosphere** extends from the tropopause up to about 50 km (31 mi). This layer is vastly more dilute than the troposphere, but it has similar composition—except that it has almost no water vapor and nearly 1,000 times more **ozone** (O_3). Ozone is a pollutant near the earth's surface, but in the stratosphere it serves a very important function: Ozone absorbs certain wavelengths of ultraviolet solar radiation, known as UV-B (290–330 nm; see fig. 3.11). This absorbed energy warms the upper stratosphere, so temperature increases with elevation. Stratospheric UV absorption also protects life on the earth's surface, because UV radiation damages living tissues, causing skin cancer, genetic mutations,

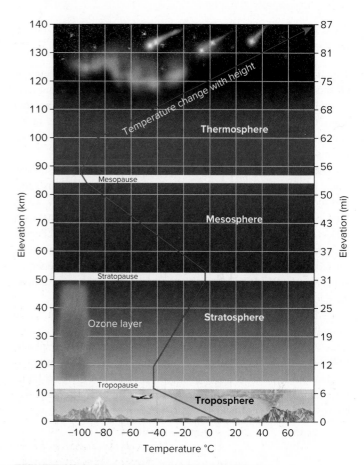

FIGURE 15.3 Layers of the atmosphere vary in temperature and composition. Our weather happens in the troposphere. Stratospheric ozone is important for blocking ultraviolet solar energy.

and crop failures. A number of air pollutants, including Freon, once used in refrigerators, and bromine compounds, used as pesticides, deplete stratospheric ozone, especially over Antarctica. This has allowed increased amounts of UV radiation to reach the earth's surface (see fig. 16.19).

Unlike the troposphere, the stratosphere is relatively calm, because warm layers lie above colder layers. There is so little mixing that when volcanic ash or human-caused contaminants reach the stratosphere, they can remain in suspension there for years.

Above the stratosphere, the temperature diminishes again in the mesosphere, or middle layer. The thermosphere (heated layer) begins at about 80 km. This is a region of highly ionized (electrically charged) gases, heated by a steady flow of high-energy solar and cosmic radiation. In the thermosphere, intense pulses of high-energy radiation cause electrically charged particles (ions) to glow. We call this phenomenon the *aurora borealis* and *aurora australis,* or northern and southern lights.

No sharp boundary marks the end of the atmosphere. The density of gas molecules decreases with distance from the earth until it becomes indistinguishable from the near-vacuum of interstellar space.

The land surface absorbs solar energy to warm our world

The sun supplies the earth with an enormous amount of energy. Of the solar energy that reaches the outer atmosphere, about one-quarter is reflected by clouds and atmospheric gases, and another quarter is absorbed by carbon dioxide, water vapor, ozone, methane, and a few other gases (fig. 15.4). This energy absorption warms the atmosphere.

About half of incoming energy reaches the earth's surface. Some of this energy is reflected by bright surfaces, such as snow and ice. The rest is absorbed by the earth's surface and by water. Surfaces that *reflect* energy have a high **albedo** (whiteness, or reflectivity). Most of these surfaces appear bright to us because they reflect light (that is, visible wavelengths of radiative energy). Surfaces that *absorb* energy have a low albedo and generally appear dark. Black soil, pavement and open water, for example, have low albedos (table 15.2).

Absorbed energy heats the absorbing surface (such as an asphalt parking lot that becomes hot in the summer sun), evaporates water, or provides the energy for photosynthesis in plants. Following the second law of thermodynamics, absorbed energy is gradually reemitted as lower-quality heat energy. A brick building, for example, absorbs energy in the form of light and reemits that energy in the form of heat.

Table 15.2 Albedo (Reflectivity) of Surfaces	
Surface	**Albedo (%)**
Fresh snow	80–85
Dense clouds	70–90
Water (low sun)	50–80
Sand	20–30
Water (sun overhead)	5
Forest	5–10
Black soil	3
Earth/atmosphere average	30

Water generally has low albedo and is extremely efficient at absorbing energy and storing it as heat. This is why increasing open water at the poles (see fig. 15.23) worries climatologists. For hundreds of thousands of years, the Arctic has been mostly white, reflecting most energy that reached the icy surface. Now the open water absorbs and stores that energy as heat, further accelerating ice melting and atmospheric warming. This is a good example of a **positive feedback loop,** in which melting leads to further melting, with probably dramatic consequences.

FIGURE 15.4 The global energy budget. The atmosphere is transparent to most incoming solar light energy, although about one-third is reflected or absorbed by the atmosphere. Of the energy that reaches the earth's surface, some is reflected, but most is absorbed by land and water. This energy is re-emitted as longer wave (terrestrial, or infrared) wavelengths. The atmosphere is much less transparent to these wavelengths, so it reflects or absorbs most terrestrial energy. The atmosphere is warmed by this retained energy. We call this selective transmission and retention of energy in the atmosphere the greenhouse effect.

Greenhouse gases capture energy selectively

Most solar energy is intense, high-energy visible or ultraviolet wavelengths (see fig. 3.11). These short wavelengths pass relatively easily through the atmosphere to the earth's surface. Some of this energy is reflected back into space. Most of the rest is absorbed by the earth's surface or used in evaporating water. As we have noted, surfaces that absorb this visible-wavelength energy gradually re-radiate it as lower-intensity, longer-wavelength infrared or heat energy. Certain gases in the atmosphere, especially carbon dioxide and water vapor, absorb much of this long-wavelength energy and re-release it in the lower atmosphere. This long-wave energy provides most of the heat in the lower atmosphere (see fig. 15.4). If these gases were not in the atmosphere, the earth's average surface temperature would be about −18°C (0°F), rather than the current average of 15°C (59°F). There would probably be no liquid water on earth, and life as we know it would not exist.

The retention of long-wave terrestrial energy in the atmosphere is called the **greenhouse effect** because the atmosphere, loosely comparable to the glass of a greenhouse, allows sunlight to travel through while trapping heat inside. The greenhouse effect is a natural atmospheric process that is necessary for life on earth. **Greenhouse gases** is a general term for trace gases that are especially effective at capturing the long-wavelength heat energy from the earth's surface. Water vapor (H_2O) is the most abundant greenhouse gas, but concentrations vary, and human inputs are modest, so it is not a concern like other GHGs. Carbon dioxide (CO_2), methane (CH_4), nitrous oxide (N_2O), and a variety of fluorine gases are among the most effective greenhouse gases.

Atmospheric circulation redistributes energy

Incoming solar energy is unevenly distributed around the globe. Near the equator, the sun shines straight overhead, and solar heating (and water evaporation) is intense. At mid-latitudes, around 30° to 60° north or south, the angle of the sun is lower, and heating is less intense. Near the poles, the sun shines low on the horizon much of the year, and little heat is available for warming the ground or evaporating water.

Solar energy absorbed by the ground eventually re-radiates as long-wave infrared energy or heat, and this energy warms the lowest layers of air. Warmed air expands, becomes less dense than the cooler air above it, and begins to rise. Eventually, this warm air cools and then sinks, forming a circulation pattern known as a convection cell.

Circulation is more vigorous near the equator than at higher latitudes because of the more intense heating from the sun. We call the equatorial convection cells (one north and one south of the equator) Hadley cells (fig. 15.5). We call the mid-latitude and polar cells Ferrell cells and polar cells, respectively. These have less intense heating and less vigorous circulation than the equatorial Hadley cells.

Where air rises in convection cell currents, air pressure at the surface is low. Where air is sinking, or subsiding, air pressure is high (fig. 15.6). On a weather map, you can see these high- and

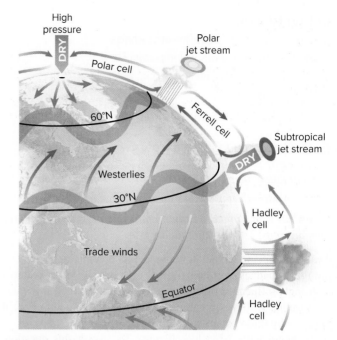

FIGURE 15.5 Convection cells circulate air, moisture, and heat around the globe. Jet streams develop where cells meet, and surface winds result from convection. Convection cells expand and shift seasonally.

low-pressure centers—spiraling currents of rising and sinking air—move across continents. In the Northern Hemisphere, these pressure centers generally travel from west to east.

Incoming solar energy is also used to evaporate water. Each gram of evaporating water absorbs 580 calories of energy as it transforms from liquid to gas. This stored energy is known as **latent heat.** As rising air cools with altitude water vapor condenses to liquid droplets (which you can see as clouds). As it condenses, each gram of water releases 580 calories of heat.

Moving water vapor carries latent heat energy from one place to another. Imagine the sun shining on the Gulf of Mexico in early spring. Warm sunshine and plenty of water allow continuous evaporation that converts an immense amount of solar energy into latent heat. Convection carries that warm humid air northward. Somewhere over the Midwest, this Gulf air is likely to encounter a cold, dry air mass moving south from Canada. The collision of these two air masses can cause the warm air to rise vigorously. Cooling, condensing water vapor releases more heat, accelerating the rising air. Rain, snow, thunderstorms, or even tornadoes may result.

Atmospheric circulation carries a great deal of heat and moisture from warm humid places to colder, drier places. The redistribution of heat and water around the globe is essential to life on earth.

Pressure differences are an important cause of wind. There is always someplace with high-pressure (sinking) air and someplace with low-pressure (rising) air. Air moves from high-pressure centers toward low-pressure areas, and we call this movement wind.

Section Review

1. Describe the troposphere and stratosphere.
2. What is albedo, and why is it important?
3. Explain the idea of a greenhouse gas, and list four of them.
4. What is latent heat?

15.2 REGIONAL PATTERNS OF WEATHER

- *Precipitation happens when moisture cools and condenses.*
- *The Coriolis effect describes the apparent curving of winds.*
- *Seasonal rain, frontal weather, and cyclonic storms are three important types of precipitations.*

"Weather" refers to temperature, wind, and precipitation (rain and snow) that changes over hours, days, or weeks. Weather patterns are strongly influenced by uneven solar heating on the earth's surface, together with the spinning of the earth. In this section, we'll examine why those patterns occur.

To begin with, why does it rain? To answer this question, remember two things: Water condenses as air cools, and air cools as it rises. Any time air is rising, clouds, rain, or snow might form. Cooling occurs because of changes in pressure with altitude: Air cools as it rises (as pressure decreases); air warms as it sinks (as pressure increases). Air rises in convection currents where solar heating is intense, such as over the equator. Air rises when two air masses collide and one must rise over the other. Air also rises when it encounters mountains. If the air is moist (if it has recently come from over an ocean or an evaporating forest region, for example), condensation and rainfall are likely as the air is lifted (fig. 15.6). Precipitation tends to be abundant in regions with intense solar heating, frequent colliding air masses, or mountains.

Where air is sinking, on the other hand, it tends to warm because of increasing pressure. As it warms, available moisture evaporates. Rainfall occurs relatively rarely in areas of high pressure. High pressure and clear, dry conditions occur where convection currents are sinking. High pressure also occurs where air sinks after flowing over mountains. Figure 15.6 shows sinking, dry air at about 30° north or south latitude. If you look at a world map, you will see a band of deserts at approximately these latitudes.

Another ingredient is usually necessary to initiate condensation of water vapor: condensation nuclei. Tiny particles of smoke, dust, sea salts, spores, and volcanic ash all act as condensation nuclei. These particles form a surface on which water molecules can begin to coalesce. Without them, even supercooled vapor can remain in gaseous form.

The Coriolis effect explains why winds seem to curve on a weather map

In the Northern Hemisphere, winds generally appear to bend clockwise (right), and in the Southern Hemisphere they appear to bend counterclockwise (left). Examples include the trade winds that brought Columbus to the Americas and the mid-latitude Westerlies

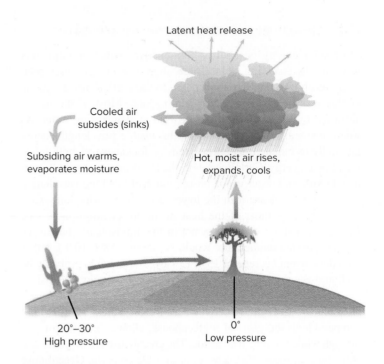

FIGURE 15.6 Convection currents distribute latent energy (heat in evaporated water) around the globe.

that bring hurricanes north from Florida to North Carolina (see fig. 15.5). Ocean currents similarly curve clockwise in the Northern Hemisphere (the Gulf Stream) and counterclockwise in the Southern Hemisphere (the Humboldt Current near Peru). This curving pattern results from the fact that the earth rotates in an eastward direction as the winds move above the surface.

The apparent curvature of the winds is known as the **Coriolis effect.** How does this effect work? Imagine you are looking down on the North Pole of the rotating earth. The earth is spinning like a merry-go-round in a playground, with the North Pole at its center and the equator around the edge. As it spins counterclockwise (eastward), the outer edge moves very fast (a full rotation, 39,800 km, every 24 hours for the real earth, or more than 1,600 km/hour). Near the center, though, there is very little eastward velocity, because the distance around a circle near the pole is relatively short.

Suppose you were standing at the center of an eastward spinning merry-go-round, looking at a person standing at the edge. You are looking straight at the other person, but she is moving much faster (in an eastward direction) than you are. If you threw a ball toward the person standing on the edge, she would have rotated eastward by the time the ball reached the edge. To you, it would appear that the ball had swerved to the right, even though you actually threw it in a straight line. If you stood at the outer edge of a spinning merry-go-round and threw toward the center, the ball would leave your hand with your fast-moving eastward velocity, and it would maintain this eastward momentum as it traveled toward the center. Again it would appear to you to bend to the right.

Winds move freely above the earth's surface much as the ball does. That is why on a weather map, we see weather patterns curving to the right in the Northern Hemisphere. Ocean currents also curve to the right in the North. In the Southern Hemisphere, winds and currents curve to the left. (It's a myth that bathtubs and sinks spiral in opposite directions in the Northern and Southern Hemispheres. Those movements are far too small to be affected by the spinning of the earth.)

On a global scale, this effect produces predictable weather patterns and currents. On a regional scale, the Coriolis effect produces "cyclonic" winds, or wind movements controlled by the earth's spin. Cyclonic winds spiral clockwise out of an area of high pressure (sinking air) in the Northern Hemisphere. They spiral counterclockwise into a low-pressure zone, because the air is moving upward: Imagine you were looking up from the ground into a rising column of air that is spinning to the right as you look up. On a weather map, looking down from the sky on that same rising air, you see it spinning counterclockwise. This is why hurricanes (extremely strong cyclonic storms) in the North spin counterclockwise. (Note, however, that they generally travel in a path that bends to the right.)

Jet streams deflect weather systems

Major weather systems and storms often follow the undulating path of **jet streams,** powerful, high-altitude winds that circle the earth high in the troposphere (fig. 15.7). What causes jet streams? Like all winds, they result from differences in atmospheric pressure. These pressure differences result from different rates of warming to north and south, or over land and sea.

To understand how jet streams form, first remember that the atmosphere is denser near the poles, where air is cold, while the atmosphere at lower latitudes is warm and buoyant (compare the Polar, Ferrell, and Hadley cells in fig. 15.5). Now suppose you were to ascend to 10 km above the surface in northern Canada. At that height you are near the tropopause in the Arctic, and air pressure is low. But farther south, still at 10 km elevation, you would be well below the tropopause, and air pressure is higher. At that elevation, then, atmospheric pressure is higher to the south than to the north of you.

Whenever there is a difference in pressure, air moves *from* high pressure *toward* the lower pressure area. In your position at 10 km up, air is drawn from south to north, toward lower pressure at high latitudes. As we know from examining the Coriolis effect, though, poleward moving air "bends" to the right in the northern hemisphere. Strong, high-altitude winds result, blowing generally eastward, more than 200 km per hour (fig. 15.7). Cells of high and low pressure, resulting from differences in heating on the earth's continents and oceans, push and pull the jet streams these north and south, making them sinuous and discontinuous, as shown in figure 15.7.

When there is a strong north-south temperature gradient, there is a strong north-south pressure gradient, and jet stream winds travel faster and straighter. In winter, when the poles are extremely cold, strong jet streams form a boundary between north and south, keeping the poles very cold and mid-latitudes milder. In summer, when long days of sun warm the Arctic, the pressure gradient weakens, and so do jet stream winds.

Jet stream behavior is changing, though, as climate change creates a warmer Arctic year-round. With smaller differences between arctic and mid-latitude temperatures, north-south

FIGURE 15.7 The polar jet stream, typically 5–10 miles up, consists of powerful winds. In this NASA visualization, fastest winds are in red, slower winds are blue. Note circular swirling patterns of rising (low pressure) and sinking (high pressure) cells in the curves of the jet stream.
Source: NASA

pressure gradients decline as well. A weak temperature/pressure gradient allows jet stream winds to become more irregular. Great loops reach north and south, drawing arctic air far south to freeze the orange groves of Florida, while unusually warm temperatures reach far into Alaska (fig. 15.8). Ironically, then, climate warming

Difference from average temperature (°F)

−27   0   27

FIGURE 15.8 Loopy jet streams (black) can draw the polar vortex (cold, high-pressure conditions) to southern states in winter, while Alaska is far warmer than normal. Temperatures in early January 2014 are shown here.
Source: NOAA Climate.gov

is occasionally producing extreme cold spells in the eastern United States, as well as increasingly snow-less winters.

Ocean currents redistribute heat

Ocean currents strongly influence climate conditions on land, depending upon whether the ocean water is warm or cold. Surface ocean currents result from wind pushing on the ocean surface, as well as from the Coriolis effect. As surface water moves, deep water wells up to replace it, creating deeper ocean currents. Differences in water density—depending on the temperature and saltiness of the water—also drive ocean circulation. Huge cycling currents called gyres carry water north and south, redistributing heat from low latitudes to high latitudes (fig. 15.9).

The Gulf Stream, one of the best-known currents, carries warm salty Caribbean water north past Canada to northern Europe. This current is immense, some 800 times the volume of the Amazon, the world's largest river. The heat transported from the Gulf Stream keeps Europe much warmer than it should be for its latitude. As the warm Gulf Stream passes Scandinavia and swirls around Iceland, the water cools and evaporates, becomes more dense, and sinks downward, creating a strong, deep, southward current. Oceanographer Wallace Broecker has called this the ocean conveyor system.

Ocean circulation patterns were long thought to be unchanging, but now oceanographers believe that currents can change over time. About 11,000 years ago, for example, as the earth was gradually warming at the end of the Pleistocene ice ages, a huge body of meltwater, called Lake Agassiz, collected along the south margin of the North American ice sheet. At its peak it

Quantitative Reasoning

Find London and Copenhagen on a globe or a map. Then, in Canada find Churchill, Manitoba, and Edmonton, Alberta, at roughly the same latitudes. Temperatures in London and Copenhagen rarely fall much below freezing, but Churchill and Edmonton have freezing weather seven to eight months each year. What factors described on these pages could explain the difference?

contained more water than all the current freshwater lakes in the world. Drainage of this lake to the east was blocked by ice covering what is now the Great Lakes. When that ice dam suddenly gave way, an estimated 163,000 km^3 of fresh water roared down the St. Lawrence Seaway—450 times the river's annual flow today. Flowing out into the North Atlantic, the fresh river water apparently interrupted the ocean conveyor circulation that carried warm water north. This event may have initiated another ice age (called the Younger Dryas, after a small tundra flower that became more common in colder conditions) that lasted for another 1,300 years.

Could this happen again? Freshwater inputs from melting glaciers and higher river flow around the Arctic are now flooding into the North Atlantic just where the Gulf Stream sinks and creates the deep south-flowing current. Already evidence shows that the deep return flow has weakened by about 30 percent. Even minor changes in the strength or path of the Gulf Stream might give northern Europe a climate more like that of Siberia—an ironic consequence of polar warming.

FIGURE 15.9 Ocean currents act as a global conveyor system, redistributing warm and cold water around the globe. These currents moderate our climate: For example, the Gulf Stream keeps northern Europe much warmer than northern Canada. Variations in ocean salinity and density, low (*blue*) to high (*yellow*), help drive ocean circulation.

Seasonal rain supports billions of people

Large parts of the world, especially near the tropics, receive seasonal rains that sustain both ecosystems and human communities. Seasonal rains give life, but when they fail to arrive, crop failures and famine can result (fig. 15.10). Seasonal rains can also cause disastrous flooding, as in the 2010 floods in Pakistan, which left 2 million homeless, or the 2003 floods in China, which forced 100 million people from their homes.

The most regular seasonal rains are known as **monsoons.** In India and Bangladesh, monsoon rains come when seasonal winds blow hot, humid air from the Indian Ocean (fig. 15.11). The hot land surface produces strong convection currents that lift this air, causing heavy rain across the subcontinent. When the rising air reaches the Himalayas, it rises even further, creating some of the heaviest rainfall in the world. During the five-month rainy season of 1970, one weather station in the foothills of the Himalayas recorded an astonishing 25 m (82 ft) of rain.

Tropical and subtropical regions around the world have seasonal rainy and dry seasons (see chapter 5). The main reason for this variable climate is that the region of most intense solar heating and evaporation shifts through the year. To understand this, first note that the earth's axis (from north to south poles) is not perpendicular to the sun's rays. Instead, the axis is at a slight angle (23.5 degrees: see fig. 15.15). Because of this, in December the most intense heating, evaporation, and rainfall are just south of the equator. In June, the sun is most intense just north of the equator.

As the earth orbits the sun, the tilt of its axis creates seasons with varying amounts of wind, rain, and heat or cold. Seasonal rains support seasonal tropical forests, and they fill some of the world's greatest rivers, including the Ganges and the Amazon. As the year shifts from summer to winter, solar heating weakens, the rainy season ends, and little rain may fall for months.

Frontal systems occur where warm and cold air meet

The boundary between two air masses of different temperature and density is called a front. When cooler air pushes away warmer air, we call the moving boundary a **cold front.** Cold, dense air of a cold front tends to hug the ground and push under the lighter, warmer air as it advances. As warm air is forced upward, it cools, and its cargo of water vapor condenses to water droplets or ice crystals. Air masses near the ground move slowly because of friction and turbulence near the ground surface, so upper layers of a moving air mass often move ahead of the lower layers (fig. 15.12). Notice that the region of cloud formation and precipitation is relatively narrow. Cold fronts can generate strong convective currents as they push warmer air rapidly upward. Violent winds and thunderstorms can result, with towering thunderheads. The weather after the cold front passes is usually clear, dry, and invigorating.

In a **warm front,** the advancing air mass is warmer than surrounding air. Because warm air is less dense than cool air, an

FIGURE 15.10 A shortage of monsoon rains brings drought, starvation, and death to both livestock and people in the Sahel desert margin of Africa. *Mamadou Toure Behan/AFP/Getty Images*

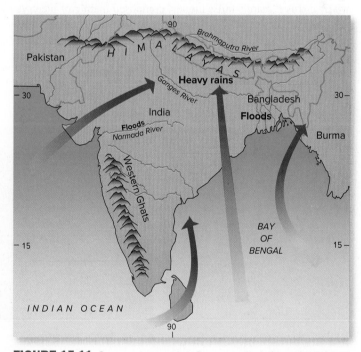

FIGURE 15.11 Summer monsoon air flows over the Indian subcontinent. Warming air rises over the plains of central India in the summer, creating a low-pressure cell that draws in warm, wet oceanic air. As this moist air rises over the Western Ghats or the Himalayas, it cools and heavy rains result.

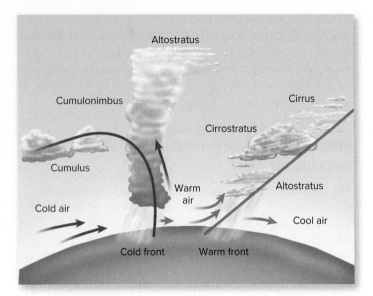

FIGURE 15.12 An advancing cold front has a steep profile, often producing vigorous storms as it forces warmer air upward. An advancing warm front, in contrast, rises over cooler air masses in a long, wedge-shaped zone. Warm fronts often produce light rain and cloudy days.

advancing warm front will slide up over cooler air masses, creating a long, wedge-shaped profile with a broad band of clouds and precipitation (fig. 15.12, *right*). Gradual lifting and cooling in a warm front lacks the violent updrafts and strong convection currents that accompany a cold front. A warm front can have many layers of clouds at different heights. The highest layers are often wispy cirrus ("mare's tail") clouds, composed mainly of ice crystals, which can extend 1,000 km (600 mi) ahead of the front we detect at the ground level. Often, we see these wispy cirrus clouds a day or two before the warm front arrives on the ground. A moist warm front can bring days of drizzle and cloudy skies.

Cyclonic storms can cause extensive damage

Vigorously rising air can produce especially strong storm systems. If water vapor is abundant, as it is over a warm ocean, rising and cooling air releases latent heat. This released heat intensifies circulation, producing swirling winds that capture still more moisture and latent heat energy. These storms swirl in a direction dictated by the Coriolis effect (spiraling upward in a counterclockwise direction in the Northern Hemisphere, clockwise in the South) and are called **cyclonic storms.**

Over the warm low-latitude Atlantic, these cyclonic storms are called **hurricanes,** storms that can be hundreds of kilometers across with winds up to 320 km/hr (200 mph). Powerful winds drive storm surges (high water) far inland (fig. 15.13*a*). Hurricane winds can tear apart buildings, but most death and damages result from storm surges. Flooding after Hurricane Katrina, which devastated coastal Louisiana and the U.S. Gulf Coast in 2005, was as much as 9 m (29 ft) deep in some areas. Funneled inland by canals built for shipping and oil drilling, the surge destroyed large parts of New Orleans and several other cities, many of which still have not recovered (fig. 15.13*b*). Winds and flooding in the 2017 hurricanes Harvey, Maria, and Irma together amounted to around $330 billion in damages and lost economic productivity (this amounts to about one-third the annual economic productivity of New York State).

Tornadoes, swirling funnel clouds that form over land, also are considered cyclonic storms. Though never as large or powerful as hurricanes, tornadoes can be just as destructive in the limited areas where they touch down (fig. 15.13*c*). Tornadoes occur frequently in the central United States, where Canadian cold fronts collide with warm, humid air moving north from the Gulf of Mexico.

Greater air temperature differences cause more powerful storms. This is why most tornadoes occur in the spring, when arctic cold fronts reach far south over the warming plains. As warm air rises rapidly over dense, cold air, intense vertical convection currents generate towering thunderheads with anvil-shaped

(a) Hurricane Katrina, 2005

(b) Gulf Shores, Alabama, 2005

(c) A tornado touches down

FIGURE 15.13 (a) Hurricane Katrina was hundreds of kilometers wide as it approached Louisiana in 2005. Note the hole, or eye, in the center of the storm. (b) Destruction caused by Hurricane Katrina in 2005. More than 230,000 km^2 (90,000 mi^2) of coastal areas were devastated by this massive storm, and many cities were almost completely demolished. (c) Tornadoes are much smaller than hurricanes, but can have stronger local winds.
(a) StockTrek/Getty Images; (b) Shutterstock; (c) 2010 Willoughby Owen/Getty Images

leading edges and domed tops up to 20,000 m (65,000 ft) high. Water vapor cools and condenses as it rises, releasing latent heat and accelerating updrafts within the supercell. Extremely intense, swirling winds result, capable of lifting houses from their foundations and destroying everything in their path.

Section Review

1. Why does it rain?
2. What is the Coriolis effect, and how might it cause trade winds?
3. Describe a monsoon, a cold front, and a warm front.
4. What is a cyclonic storm?

15.3 NATURAL CLIMATE VARIABILITY

- *Milankovitch cycles drive long-term climate changes.*
- *Ice cores contain CO₂ and oxygen isotopes used to reconstruct past temperatures and atmospheric composition.*
- *El Niño is one example of an ocean–atmosphere oscillation.*

Climatologists have studied many different sources of evidence to understand long-term climate changes. The main drivers in these long-term changes are known as orbital cycles, because they result from shifts in the earth's movement, or **Milankovitch cycles,** after Serbian scientist Milutin Milankovitch, who first described them in the 1920s. These cycles influence how close the earth is to the sun and control how much of the sun's energy we receive, which in turn affects the earth's climate. Long-term climate shifts can be seen in sedimentary rocks, as temperatures and rainfall shifted over millions of years (fig. 15.14).

These cycles are regular shifts in the earth's orbit and tilt (fig. 15.15), which change the distribution and intensity of

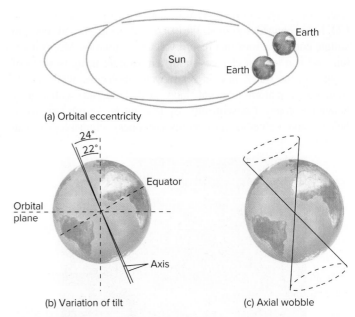

(a) Orbital eccentricity

(b) Variation of tilt

(c) Axial wobble

FIGURE 15.15 Milankovitch cycles, which may affect long-term climate conditions: (a) changes in the elliptical shape of the earth's orbit, (b) shifting tilt of the axis, and (c) wobble of the earth.

sunlight reaching the earth's surface. The earth's elliptical orbit stretches and shortens in a 100,000-year cycle, while the axis of rotation changes its angle of tilt in a 40,000-year cycle. Over a 26,000-year period, the axis wobbles like an out-of-balance spinning top. Bands of sedimentary rock laid in the oceans seem to match these Milankovitch cycles. Periodic cold spells associated with worldwide expansion of glaciers every 100,000 years or so also correspond to these cycles.

Ice cores tell us about climate history

Geologists and climatologists use many lines of evidence to reconstruct past climates. One is examining the widths of tree rings, which let them count back to find dry, drought-stressed years, or humid years of fast growth. Another is remains of plants, pollen, or microscopic organisms in lake beds and marine mud. Over really long time periods of millions of years, geologists use fossils and rock types to infer what the climate must have been like.

Ice cores are especially important because they provide detailed records reaching back hundreds of thousands of years. How? Every time it snows, small amounts of air are trapped in the snow layers. In Greenland, Antarctica, and other places where cold is persistent, yearly snows slowly accumulate over the centuries. New layers compress lower layers into ice, but still tiny air bubbles remain, even thousands of meters deep into glacial ice (fig. 15.13). Each bubble is a tiny sample of the atmosphere at the time the bubble was sealed off.

Climatologists have discovered that by drilling deep into an ice sheet, they can extract ice cores, from which they can collect and analyze air from these bubbles. Samples taken every few centimeters show how the atmosphere has changed over time. Ice core records have revolutionized our understanding of climate history

FIGURE 15.14 Banded rocks in these cliffs on the coast of France show a record of long-term climate cycles.
Muriel Hazan/Science Source

(fig. 15.16). We can now see how concentrations of atmospheric CO_2 have changed over time. We can detect ash layers and spikes in sulfate concentrations that record volcanic eruptions. Most important, we can look at changes in air temperature using isotopes of oxygen in the water. In cold years, water molecules with slightly lighter oxygen atoms evaporate more easily than water with slightly heavier isotopes. Consequently, by looking at the proportions of heavier and lighter oxygen atoms (isotopes), climatologists can

(a)

(b)

FIGURE 15.16 (a) Ice cores contain air bubbles that give samples of the atmosphere thousands of years ago, as in this sample of 45,000-year-old ice from Antarctica. (b) Dr. Mark Twickler, of the University of New Hampshire, holds a section of the 3,000-m Greenland ice sheet core, which records 250,000 years of climate history.

Source: (a) Ludovic Brucker/NASA's Goddard Space Flight Center; (b) Jim West/ agefotostock/Newscom

reconstruct temperatures over time and plot temperature changes against concentrations of CO_2 and other atmospheric gases.

The first very long record was from the Vostok ice core, which reached 3,100 m into the Antarctic ice and gives us a record of temperatures and atmospheric CO_2 over the past 420,000 years. A team of Russian scientists worked for 37 years at the Vostok site, about 1,000 km from the South Pole, to extract this ice core. A similar core has been drilled from the Greenland ice sheet. More recently, the European Project for Ice Coring in Antarctica (EPICA) has produced a record reaching back over 800,000 years (fig. 15.15). All these cores show that climate has varied dramatically over time but that there is a close correlation between atmospheric temperatures and CO_2 concentrations. From these ice cores, we know that CO_2 concentrations have varied between 180 to 300 ppm (parts per million) in the past 800,000 years. Therefore, we know that today's concentrations, above 415 ppm, are about one-third higher than the earth has seen in nearly a million years. We also know that present temperatures are as warm as any in the ice core record. Further warming in the coming decades is likely to exceed anything in the ice core records, as CO_2 gradually traps heat in the atmosphere.

The climate is warmer now than it has been since the development of agriculture and of civilization as we know it. We know from historical accounts that slight climate shifts can be seriously destabilizing. During the "little ice age" that began in the 1400s (possibly caused by volcanic ash in the atmosphere), cold summers caused crop failures and famine across much of Europe. The Baltic Sea froze, and fish migrations shifted. Scandinavian settlements in Greenland, founded during the warmer period around 1000 C.E. lost contact with Iceland and Europe as ice blocked shipping lanes, and disappeared. Meanwhile, in Asia, cool and wet conditions may have expanded the grasslands that allowed Genghis Khan and his Mongolian armies to invade China and Europe.

Evidence from ice cores drilled in the Greenland ice cap suggests that world climate can change abruptly. It appears that during the last major interglacial period, 135,000 to 115,000 years ago, temperatures flipped suddenly from warm to cold or vice versa over a period of decades rather than centuries.

El Niño is an ocean–atmosphere cycle

On a shorter time scale, there are oscillations, or periodic shifts, in ocean currents and atmospheric circulation over years or decades. One important example is known as El Niño/Southern Oscillation.

Peruvian fishermen were the first to name these irregular cycles, because the weakened upwelling currents and warming water led to the disappearance of the anchovy fishery. They named these events **El Niño** (Spanish for "the Christ child") because they were observed around Christmas time. Increased attention to these patterns has shown that sometimes, between El Niño events, coastal waters become extremely cool, and these extremes have come to be called **La Niña** (or "the little girl"). Together, this cycle is called the El Niño Southern Oscillation (ENSO).

The core of the ENSO system is a huge pool of warm surface water in the Pacific Ocean that sloshes slowly back and forth between

FIGURE 15.17 Atmospheric CO_2 concentrations (red line) correspond with temperatures (blue, derived from oxygen isotopes) in the Antarctic EPICA ice core. This 800,000-year record has no evidence of temperatures or CO_2 higher than that anticipated within the coming century.
Source: UN Environment Programme; Bouzel, J., et al. 2007. "EPICA Dome C Ice Core 800KYr Deuterium Data and Temperature Estimates."

Indonesia and South America. In most years, steady equatorial trade winds push ocean surface currents westward, holding this warm pool in the western Pacific, roughly between Tahiti and Indonesia (fig. 15.16). Air above this warm region of the ocean is heated and moist, so it rises, producing strong convection cells in the atmosphere. Towering thunderheads created by rising air bring torrential summer rains to the tropical rainforests of northern Australia and Southeast Asia. Surface winds are drawn in from across the Pacific to replace the rising air. In the eastern Pacific, dry subsiding air currents contribute to arid conditions from Chile to southern California.

Trade winds also help drive ocean currents westward, setting up currents in the Pacific. Upwelling currents along the coast of South America draw cold, nutrient-rich water up from deep in the ocean. These nutrients support the dense schools of anchovies and other fish that traditionally support many fishermen in Peru.

Every three to five years, for reasons we don't fully understand, these convection currents weaken. The pool of warm surface water surges back east across the Pacific, toward Peru and California. The shift in position of the low-pressure area (the rising convection currents) has repercussions in weather systems across North and South America and perhaps around the world.

How does the ENSO cycle affect you? During an El Niño year, the northern jet stream—which normally is over Canada—splits and is drawn south over the United States. This pulls moist air inland from the Pacific and Gulf of Mexico, bringing intense storms and heavy rains from California across the Midwestern states. La Niña years bring extreme hot, dry weather to these same areas. El Niño events have brought historic floods to the Mississippi River basin, but Oregon, Washington, and British Columbia tend to be warm and dry in El Niño years. Droughts in Australia and Indonesia during El Niño episodes can cause disastrous crop failures and forest fires.

ENSO-related droughts and floods are expected to intensify and become more irregular with global climate change, in part because the pool of warm water is warming and expanding. Extreme El Niños may occur more frequently than they have in the past. High sea surface temperatures spawn larger and more violent storms such as hurricanes. On the other hand, increased cloud cover would raise

(a) Normal

(b) El Niño

FIGURE 15.18 Normally, surface trade winds drive currents from South America toward Indonesia, and cold, deep water wells up near Peru. During El Niño years, winds and currents weaken, and warm, low-pressure conditions shift eastward, bringing storms to the Americas.

the albedo while upwelling convection currents generated by these storms could pump heat into the stratosphere. This might have an overall cooling effect—a negative feedback in the warming climate system.

Climatologists have observed many decade-scale oscillations. The Pacific Decadal Oscillation (PDO), for example, involves a vast pool of warm water that moves back and forth across the North Pacific every 30 years or so. From about 1977 to 1997, surface water temperatures in the middle and western parts of the North Pacific Ocean were cooler than average, while waters off the western United States were warmer. During this time, salmon runs in Alaska were bountiful, while those in Washington and Oregon were greatly diminished. In 1997, however, ocean surface temperatures along the coast of western North America turned significantly cooler, perhaps marking a return to conditions that prevailed between 1947 and 1977. Under this cooler regime, Alaskan salmon runs declined while those in Washington and Oregon improved somewhat. A similar North Atlantic Oscillation (NAO) occurs between Canada and Europe.

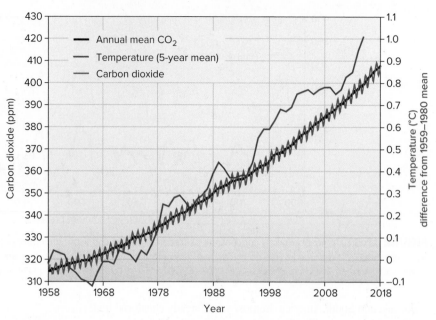

FIGURE 15.19 Measurements of atmospheric CO_2 taken at the top of Mauna Loa, Hawaii, show an increase of about 2.2 percent per year in recent years. Monthly mean CO_2 (*red*) and annual average CO_2 (*black*) track closely with the general trend in global temperatures (*blue*).

Source: Data from National Aeronautics and Space Administration and National Oceanic and Atmospheric Administration

Section Review

1. What are three Milankovitch cycles?
2. How might past climate be reconstructed from ice cores?
3. How do ocean and atmosphere interact in El Niño cycles?

15.4 ANTHROPOGENIC CLIMATE CHANGE

- *The Keeling curve documents changing CO_2 concentrations.*
- *Human activities alter greenhouse gas concentrations.*
- *There is clear evidence for a human role in current climate change.*

Many scientists regard anthropogenic (human-caused) global climate change to be the most important environmental issue of our times. The idea that humans might alter world climate is not new. In 1895, Svante Arrhenius, who subsequently received a Nobel Prize in Chemistry, predicted that CO_2 released by coal burning could cause global warming. At the time, this idea was mostly theoretical, though, and real impacts seemed unlikely.

The first data showing human impacts on atmospheric CO_2 came from an observatory on top of the Mauna Loa volcano in Hawaii. The observatory, established in 1957 as part of an International Geophysical Year, was built in a remote location to capture data on air chemistry in an undisturbed, pristine environment. Surprisingly, measurements showed CO_2 concentrations increasing about 0.5 percent per year. Levels have risen from 315 ppm in 1958 to over 415 ppm in 2019 (fig. 15.17). This graph, first produced by Charles David Keeling at the Mauna Loa observatory, is one of the first and most important pieces of evidence that demonstrates Svante Arrhenius's prediction.

Keeling's graph has some distinctive patterns. One is the annual variation in CO_2 concentrations: Every May, CO_2 levels drop slightly as plant growth on the northern continents, which are much larger than the southern continents, capture CO_2 in photosynthesis; during the northern winter, levels rise again as respiration releases CO_2. Another pattern is that CO_2 levels are rising at an accelerating rate, currently more than 2 ppm each year. We are on track to double the pre-industrial concentration of CO_2, which was 280 ppm, in about a century's time.

The IPCC assesses climate data for policymakers

The climate system is extraordinarily complex, so a great deal of effort has been invested in analyzing observations like those from Mauna Loa. Since 1988, the **Intergovernmental Panel on Climate Change (IPCC)** has brought together scientists and government representatives from around the world to review scientific evidence on the causes and likely effects of climate change. The fifth Assessment Report (known as AR5) was published in 2013–2014, representing six years of work by 2,500 scientists. The report assessed evidence regarding rates of change, possible causes, impacts, and available solutions to reduce harmful impacts of climate change.

Scientists rarely agree fully on anything, but one point that climate scientists do agree on is that human activities are raising global temperatures rapidly. The IPCC projects warming of about 1–6°C by 2100, depending on what policies we follow to curb climate change (fig. 15.20). The IPCC's "best estimate" for the most likely scenario is 2–4°C (3–8°F). To put that in perspective, the average global temperature change between now and the middle of the last glacial period is about 5°C. Subsequent reports, though, have shown that keeping change below 1.5°C would vastly reduce the social, environmental, and economic costs of climate change (see section 15.5).

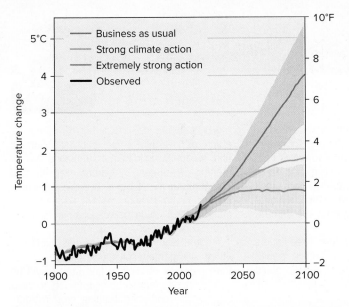

FIGURE 15.20 Global average temperature change, observed (black) and in climate models (colors). Note that the black line shows accelerating change.

Source: USGCRP 2018 Fourth National Climate Assessment

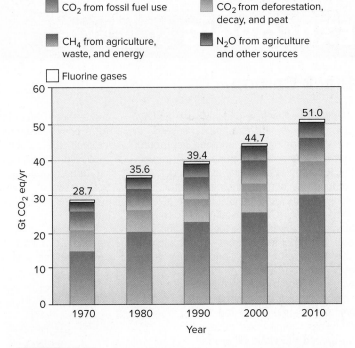

FIGURE 15.21 Contributions to climate change by different gases.
Source: IPCC

Major greenhouse gases include CO_2, CH_4, and N_2O

Earlier in this chapter, we described the greenhouse effect: Gases in our atmosphere prevent long-wavelength (terrestrial) energy leaving the earth's surface from escaping to space (see fig. 15.5). The energy retained in our atmosphere keeps our earth warm enough for life as we know it. The problem now is an enhanced greenhouse effect, with slowed heat loss and increased heat retention in the atmosphere.

Four types of greenhouse gases are primary drivers in climate change: carbon dioxide (CO_2), methane (CH_4), nitrous oxide (N_2O), and a group of gases containing fluorine, such as the CFCs (chlorofluorocarbons) often used in air conditioners (fig. 15.21). These are trace gases, making up less than 0.1 percent of the atmosphere, but multiplying even these low concentrations has increased energy capture in the atmosphere. Water vapor (H_2O) is also an important greenhouse gas, but human activity is not introducing water in the atmosphere as it is the other gases. In addition to these gases, a number of other factors are discussed below, including land use change, sea ice loss, and black carbon particulate emissions (Exploring Science, p. 340).

Carbon dioxide is the most important greenhouse gas because it is produced abundantly, it lasts decades to centuries in the atmosphere, and it is very effective at capturing long-wave energy. Emissions of CO_2 more than doubled in the 45 years from 1970 to 2015, from about 14 Gt/yr to more than 40 Gt/yr (see fig. 15.21). Carbon

dioxide contributes to about 80 percent of human-caused climate impacts (up from 60 percent a few decades ago). Burning fossil fuels is by far the greatest source of CO_2. Deforestation and other land-use changes are the second biggest factor. Deforestation releases carbon stored in standing trees, and organic material in the exposed soil oxidizes and decays, producing still more CO_2 and CH_4. Cement production is also an important contributor, with CO_2 released both from fuel used in baking limestone and from carbon stored in the limestone itself. Cement used in construction has recently helped push China into the lead for global CO_2 emissions (fig. 15.22).

Methane (CH_4) from agriculture and other sources is the second most important greenhouse gas, accounting for 14 percent of

FIGURE 15.22 Carbon emissions by region since 1850. The two largest emitters, China (24%) and the United States (21%), produce nearly half of all emissions.
Source: Data from Boden, T. A., G. Marland, and R. J. Andres. 2010. Carbon Dioxide Information Analysis Center, Oak Ridge National Laboratory, U.S. Department of Energy

Black Carbon

Greenhouse gases, especially such as CO_2, CH_4, and N_2O, are the main concerns in climate warming, but another contributor that gets less attention is black carbon. Black carbon is soot, tiny airborne particles produced by burning organic material. It is highly mobile, and being black, it absorbs solar energy. A major source is fossil fuel burning, in power plants, factories, and vehicles, but other key sources include biofuels (such as wood or dung), and burning of forests and crop residue. All these types of combustion produce smoke, soot, or ash that can linger in the atmosphere for days or weeks, sometimes traveling far around the globe before falling to the ground.

Human activities release an estimated 6 to 8 million metric tons of black carbon per year. Approximately 40 percent is from fossil fuel combustion, 40 percent is from open biomass burning, and 20 percent comes from burning biofuels. Some researchers report that about 60 percent of all black carbon comes from India and China. But they point out that per-capita emissions of black carbon from the United States and some European countries are comparable to those in South and East Asia.

Although there is far less particulate material in the atmosphere than CO_2, black carbon absorbs a million times more solar energy, weight for weight, than CO_2, making black carbon a critical driver of climate change. When it falls on snow and ice, black carbon darkens the surface, increasing absorption of sunlight. While this might seem like a small effect, it has profound impacts on the albedo (reflectance, or whiteness) of ice caps, glaciers,

(header photo): David L. Hansen

and sea ice. Darker ice melts faster, producing a positive freedback: As the snow and ice melt, the carbon on the surface is concentrated, absorbing more energy and causing further melting. Eventually, bare rock or open water is exposed, which absorbs even more sunlight and warms the climate further.

A recent study found that Greenland's ice and snow have been darkening significantly over the past 20 years. Albedo could

FIGURE 1 Black carbon in soot and smoke from fires, power plants, and diesel engines contributes to both health problems and global warming.
John Thoeming/McGraw-Hill Education

decrease 10 percent or more by the end of this century. The black carbon causing melting isn't all from recent combustion. Smoke, soot, and ash from ancient forest fires and volcanic activity are stored in glaciers and snowfields. As ice and snow melt, some of

this stored carbon is washed away, but as much as one-third remains on the surface and gradually builds up to enhance further melting. Research by James Hansen and coauthors at the National Oceanic and Atmospheric Administration (NOAA) suggests that black carbon could be responsible for more than 30 percent of Arctic warming.

What can we do about this problem? Since 40 percent of black carbon comes from biomass burning, a first step would be to reduce burning forests to clear land, and burning crop residue. Another 40 percent comes from fossil fuel combusion, so targeting brick kilns and coke ovens, inefficient stoves, and diesel vehicles everywhere can help reduce black carbon emissions. Black carbon releases also can be reduced by requiring cleaner fuel and new standards for engines and boilers. The U.S. EPA estimates that a switch to fuel-efficient automobiles and trucks (especially electric ones) could cut U.S. transportation emissions by 86 percent. Ships burning heavy bunker fuel are especially polluting, and more of them sailing through the Arctic as sea lanes open up will deposit heavy soot more directly on Arctic ice cover. Requiring these ships to burn cleaner fuel is imperative. Similarly, switching from coal-fired power plants to solar and wind energy would slash emissions. Because particulates remain in the atmosphere for a relatively short time, cutting emissions will immediately reduce the rate of warming, particularly the rapidly changing Arctic. As with many climate change issues, we know what to do. The question is, can we take action to reduce the problem?

our greenhouse output. Methane absorbs 23 times as much energy per gram as CO_2 does, and it is accumulating at a faster rate than CO_2. Methane is produced when plant matter decays in oxygen-free conditions, as in the bottom of a wetland. (Where oxygen is present decay produces mainly CO_2.) Methane is also released from natural gas wells. Rice paddies are a rich source of CH_4, as are ruminant animals, such as cattle. In a cow's stomach, which has little oxygen, digestion produces CH_4, which cows then burp into the atmosphere. A single cow can't produce much CH_4, but the global population of about 1 billion cattle produces enough methane to double the concentration naturally present in the atmosphere.

Nitrous oxide (N_2O), our third most important greenhouse gas, accounts for 8 percent of greenhouse gases. This gas is also released from agricultural fertilizers, plant decay, vehicle engines, denitrification of soils, and other sources. Even though we don't produce as much N_2O as we do other greenhouse gases, this is an important gas because it is especially effective at capturing heat.

A large family of gases containing fluorine (a reactive element similar to chlorine) is increasingly important because of a ballooning number of air conditioners and refrigerators worldwide. These gases are excellent long-lasting refrigerants, but when they escape to the atmosphere, they capture thousands of times more heat per molecule than CO_2 (table 15.3). Like N_2O and CH_4, these are emitted in relatively small amounts, but their ability to absorb specific energy wavelengths gives them a disproportionate effect.

One way to compare the importance of these various sources is to convert them all to equivalents of our most important greenhouse gas, CO_2. The units used on the Y-axis in fig. 15.19, gigatons of CO_2-equivalent per year (Gt CO_2 eq/yr), let us compare the effects of these sources. All four have increased, but fossil fuel burning rose the most between 1970 and 2010. This is a reason transportation and coal-burning power plants are two of the key sectors addressed in efforts to slow climate change.

TABLE 15.3 Global Warming Potential (GWP[1]) of Sample Gases, Compared to CO_2

Gas	Lifetime[2]	20-yr GWP	100-yr GWP
Carbon dioxide (CO_2)	100	1	1
Methane (CH_4)	12.4	86	34
Nitrous oxide (N_2O)	121	268	298
CFC-12 (CCl_2F_2)	100	10,800	10,900
Sulfur hexafluoride (SF_6)	3,200	16,300	23,900

[1] Radiative effects per unit mass, integrated over a 20-year or 100-year time span, relative to an equal mass of CO_2 emissions. CO_2 is set as 1 for comparison. *Source: Carbon Dioxide Information Analysis Center, 2011*

[2] Average residence time in atmosphere; actual range for CO_2 is decades to centuries.

The large orange bars in fig. 15.19 show that burning fossil fuels is also our most abundant source of greenhouse gases. Electricity production, transportation, heating, and industrial activities that depend on fossil fuels together produce 50 percent of our greenhouse gases. Deforestation and agriculture account for another 30 percent. The remaining 20 percent is produced by industry.

Melting ice accelerates change

As noted earlier in this chapter, the melting of polar ice is a concern because it will increase energy absorption (water has a low albedo, while ice reflects 80–90 percent of solar energy) and enhance warming globally. In the past 30 years, late-summer sea ice in the Arctic has declined to about one-third its historical extent (fig. 15.23). Long-lasting sea ice (more than 4 years old), which supports marine ecosystems and melts less easily than fresh annual ice, is almost entirely gone from the region.

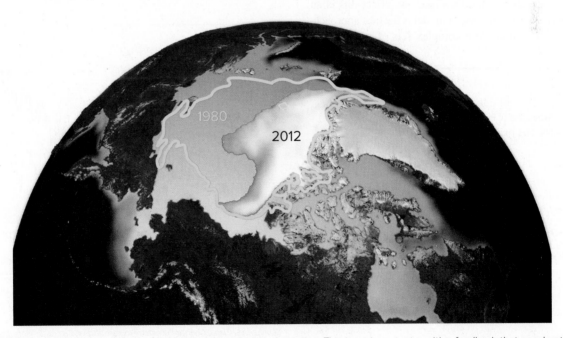

FIGURE 15.23 Declining sea ice reduces albedo, increasing energy absorption. This is an important positive feedback that accelerates warming.

Another important feedback is the CO_2 and CH_4 release from warming and drying peat and permafrost. Peat is soggy, semi-decayed plant matter accumulated over thousands of years across the vast expanses of tundra, especially in Canada and Siberia. As peat thaws and dries, it oxidizes and decays, releasing additional CO_2 and CH_4. The world's peatlands sequester (capture and hold) 0.37 Gt of carbon each year, more than all other vegetation types: Tropical forests capture abundant carbon but also release it through decay. But decaying peatlands are a major source of greenhouse gases, releasing as much as 6 percent of anthropogenic CO_2 emissions.

A general term for these positive feedbacks is **arctic amplification,** that is, more extreme changes at high latitudes, because of declining albedo and loss of permafrost. High latitudes also warm more because winter heat loss is slowed. As a consequence, regions near the poles have warmed much faster than the rest of the world. In Alaska, western Canada, and eastern Russia, average temperatures have increased as much as 4°C (7°F) over the past 50 years. These changes don't affect just the Arctic, though. As shown by the unstable jet stream discussed earlier, Arctic change alters climate systems and weather globally.

How do we know that recent change is caused by humans?

It is clear that climate warming corresponds closely to the expansion of industrialization, fossil fuel burning, agriculture, and other activities since the eighteenth century (fig. 15.24). But we also know that *correlation* is not proof of *cause*. How do we know that human activities are responsible?

The best way to demonstrate cause is to run an experiment with replicates and controls. But we have only one earth, and we cannot control the multiple factors involved—in effect this is an uncontrolled experiment. In this case, a model is the best approach. In a computer model, you can use complex sets of equations to calculate likely temperature patterns from known input factors. You can start with all the known natural fluctuations (such as solar variations and Milankovitch cycles). You also include variables for all the known human-caused inputs (CO_2, CH_4, aerosols, soot, and so on). Then, you run the model and see if it can recreate past changes in temperatures.

If it does a good job of replicating what happened in the past, then your model is a good description of how the system works—how the atmosphere responds to more CO_2, how oceans absorb heat, how reduced snow cover contributes feedback, and so on.

If you can create a model that represents the system quite well, then you can rerun the model, but this time you leave out the extra CO_2 and other factors we know that humans have contributed. If the model *without* human inputs is *inconsistent* with observed changes in temperature, and if the model *with* human inputs is *consistent* with observations, then you can be confident that the human inputs have made the difference. Models with human activities incorporated into them are the only way to explain recently observed increases in air temperatures (fig. 15.25).

We also know that current rates of change are much higher than anything we see in the geologic record. In data from ice cores (fig. 15.15), we can see that at the end of Pleistocene ice ages, temperatures rose several degrees over hundreds or thousands of years. The earth's current air temperature is rising ten times faster than during the transition from a glacial to interglacial period, and atmospheric CO_2 concentrations are increasing one thousand times faster.

Section Review

1. Approximately how much has atmospheric CO_2 changed since 1959?

2. List four important greenhouse gases, and identify major sources.

3. How do we know recent climate change is human-caused?

FIGURE 15.24 Changes in atmospheric concentration of the major greenhouse gases over time. Rapid increases began with the industrialization after 1800.
Source: USGS 2009

FIGURE 15.25 Comparison of observed changes (black line) to climate forcing from natural and human-caused driving factors. Natural drivers include volcanoes, solar activity, orbital cycles, and other factors. Human drivers include fuel burning, land use, aerosols, and other factors. Shading shows a 95 percent confidence interval.
Source: U.S. Global Change Research Program https://nca2018. globalchange.gov/chapter/2/, KM1

15.5 What Effects Are We Seeing?

- *Polar regions and continental interiors are warming most rapidly.*
- *Effects of climate change include loss of glaciers and intensified drought.*
- *Public confusion about climate change persists for various reasons.*

The American Geophysical Union, one of the nation's leading scientific organizations, says, "As best as can be determined, the world is now warmer than it has been at any point in the last two millennia, and, if current trends continue, by the end of the century it will likely be hotter than at any point in the last two million years" (fig. 15.26).

Fortunately, we have many options for addressing these changes, if we choose to use them. Mitigating climate change doesn't mean reverting to the Stone Age; it mostly means investing our resources in different kinds of energy. In this section, we'll examine some of the consequences of recent climate changes; we then explore some of the many steps we can take as individuals and as a society to work for a better future.

Warming affects crops, health, and ecosystems

Over the last century, the average global temperature climbed about 0.9°C (1.6°F); continued warming will be most extreme near the poles and in mid-continents (fig. 15.26). Sixteen of the warmest years in the past 150 have occurred since 1998. New records for hot years are observed with increasing frequency. Here are some effects that have been observed:

- Summer heat waves are increasing in severity and frequency. Extreme heat causes more deaths than all other weather-related hazards in some years, according to the U.S. Department of Homeland Security. Storm intensity and flooding are also increasing in some areas and drought in others.

2090–2099

| 0 | 0.5 | 1 | 1.5 | 2 | 2.5 | 3 | 3.5 | 4 | 4.5 | 5 | 5.5 | 6 | 6.5 | 7 | 7.5 |

Surface temperature changes, °C

FIGURE 15.26 Surface temperature is projected to rise sharply, even under conservative scenarios, with rapid adoption of new energy technology, reduction of fossil fuel use, and a declining population after 2050.
Source: IPCC AR4, WG1

FIGURE 15.27 Expanding drought is expected to increase water scarcity and displace far more climate refugees.
Bartosz Hadyniak/Getty Images

- Water resources are becoming scarce in many areas. Irregular rains, longer hot seasons, and increased evapotranspiration together decrease moisture available for ecosystems and human communities alike (fig. 15.27).

- Drought has expanded and deepened in many agricultural regions, including much of the United States (fig. 15.28). In North America, recent wet winters and hot, dry summers are consistent with climate models. In Africa, droughts have increased about 30 percent since 1970. Extreme droughts in the Amazon rainforest, associated with high temperatures, have killed billions of trees, releasing an estimated 8 billion metric tons of CO_2 (more than China produced in 2009). A 2°C temperature rise (the best-case scenario) will destroy 20–40 percent of the Amazon forest, turning it from a carbon sink to a carbon source.

- Arctic sea ice is half as thick and half as extensive as it was 30 years ago. By 2040, the Arctic Ocean could be entirely ice-free in the summer. This is bad news for ice-dependent species, such as polar bears and walruses. An aerial survey in 2005 found bears swimming across as much as 260 km (160 mi) of open water to reach pack ice where they could hunt seals. But sea ice provides critical habitat for most Arctic ecosystems, including marine algae, plankton, and fish that underpin the marine food web.

- Ice loss also impacts Antarctic marine ecosystems, and 90 percent of the glaciers on the Antarctic Peninsula are now retreating an average of 50 m per year. Recent studies found that the world's largest King Penguin colony declined by 88 percent in just the past 35 years (fig. 15.29).

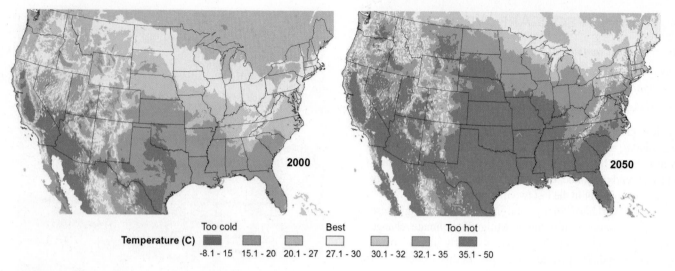

Temperature (C)

Too cold			Best			Too hot
-8.1 - 15	15.1 - 20	20.1 - 27	27.1 - 30	30.1 - 32	32.1 - 35	35.1 - 50

FIGURE 15.28 Areas with suitable summer temperatures for growth of major crops, such as corn, will shift northward. July maximum temperature (average for 2000 ± 10 years and 2050 ± 10 years) is shown.
Source: Data from CCAFS (Climate Change, Agriculture and Food Security), CMIPS and WorldClim climate models, RCP 8.5

- Glaciers are disappearing at an accelerating rate. Half of the world's small glaciers will disappear by 2100, according to a study of 120,000 such glaciers. This is creating create crises in many regions that depend almost entirely on glaciers for irrigation and drinking water, such as Bolivia, Chile, Nepal, and Central Asia.

- Sea level has risen worldwide approximately 15–20 cm (6–8 in.) in the past century, with accelerating sea-level rise in the past 20 years. Thermal expansion of seawater is the main cause, along with melting land-based glaciers. Melting of Greenland's massive ice cap is a matter of particular concern, because it holds enough water to raise sea level by about 7 m (about 23 ft) if it all melts. Melting glaciers and ice caps are contributing about 1 cm per decade to sea-level rise (fig. 15.30).

FIGURE 15.29 Marine and ice-dependent species, such as this colony of King penguins, are especially vulnerable to climate change.
moodboard/Glow Images

- Biologists report that many animals are breeding earlier or extending their ranges into new territory as the climate changes. Other species are disappearing altogether. Insect pests and diseases have also expanded their range as hard winters have retreated northward. Beetle infestations, expanding with warmer winters, are killing millions of hectares of pine and spruce across western North America, and increasing risk of fires.

- Coral reefs worldwide are "**bleaching**" (dying as they lose their photosynthetic algae), as water temperatures rise above 30°C (85°F). With reefs nearly everywhere threatened by pollution, overfishing, and other stressors, biologists worry that rapid climate change could be the final blow for many species in these complex, biologically rich ecosystems.

- The oceans have been both absorbing CO_2 and storing heat, thus postponing the impacts of our GHG emissions. Deep-diving sensors show that the oceans are absorbing 0.85 watts per m^2 more than is radiated back to space. This capacity to absorb heat is diminishing, however. Recent studies show that the oceans are warming 40 times faster than was believed just 5 years ago. This implies the atmosphere will soon start warming faster. Absorbed CO_2 is also acidifying the oceans. Because shells of mollusks and corals dissolve at low pH, ocean acidity is likely to alter marine communities.

Climate change costs far more than prevention

There are strong economic reasons to act on climate change. The U.S. Global Change Research Program reported in 2018 that U.S. costs of climate change are likely to reach $500 billion per year, in storm damage, lost agricultural productivity, and other impacts. A decade earlier, Sir Nicholas Stern, former chief economist of the World Bank, calculated that global, near-term costs of climate change will be at between 5 and 20 percent of the global GDP each

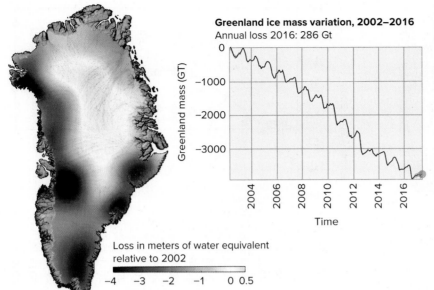

Greenland ice mass variation, 2002–2016
Annual loss 2016: 286 Gt

Loss in meters of water equivalent relative to 2002

−4 −3 −2 −1 0 0.5

FIGURE 15.30 Greenland is losing ice at an accelerating rate. This contributes to sea level rise and amplifies Arctic warming.
Source: NASA GRACE mission

year. Meanwhile, the IPCC and the Stern report both found that that stabilizing the world climate and avoiding the worst impacts of change would cost less than 1 percent of annual global GDP per year. That means that $1 invested now could save us $20 later in this century. Energy production, Stern suggests, will have to be at least 80 percent decarbonized by 2050 to stabilize our global climate.

Many residents of wealthier countries will likely have resources to control impacts on their lives, but low-income people, and residents of poorer countries, will have fewer options. The Stern report projected that drought and other disasters could drive at least 200 million people from their homes, creating widespread refugee crises.

The Stern review recommended four key elements for combating climate change: (1) *emissions trading* to promote cost-effective emissions reductions; (2) *technology sharing* that would double research investment in clean-energy technology and accelerate the spread of that technology to developing countries; (3) *reduction of deforestation,* which is a quick and highly cost-effective way to reduce emissions; and (4) *helping poorer countries* by honoring pledges for development assistance to adapt to climate change.

Rising sea levels will flood many cities

Flooding and storm surges associated with higher sea level are one of the most important economic risks of climate change. Many of the world's largest cities lie on vulnerable coastlines. Even the 75 cm (30 in.) sea-level rise expected by 2050 will flood much of south Florida, Bangladesh, Pakistan, and many other low-lying coastal areas (fig. 15.31). Fast-growing cities in developing regions, such as Jakarta, Kolkata, or Manila are likely to be largely submerged. Small island countries such as the Maldives,

the Bahamas, Kiribati, and the Marshall Islands could become uninhabitable if sea levels rise a meter or more. The South Pacific nation of Tuvalu has already announced that it is abandoning its island homeland. All 11,000 residents will move to New Zealand, a visible example of the world's growing number of climate refugees.

Insurance companies worry that the $2 trillion in insured property along U.S. coastlines is at increased risk from a combination of high seas and catastrophic storms. At least 87,000 homes in the United States within 150 m (500 ft) of a shoreline are in danger of coastal erosion or flooding in the next 50 years. These losses, together with damage to fishing stocks, agriculture, and water supplies, could rapidly multiply the financial costs of climate change.

Why do we still debate climate evidence?

Scientific studies have long been unanimous about the direction of climate trends. The United States is rare in still having public and political debates about climate evidence. Why is this? One reason may be that change is threatening, and many of us would rather ignore it or dispute it than acknowledge it. Another reason may be a lack of information. Another is that while scientists tend to look at trends in data, the public might be more impressed by one or two recent events, such as an especially snowy winter in their local area. On radio and TV, colorful opinions capture more attention than data and graphs. Finally, and perhaps most important, there is an established industry dedicated to disputing scientific evidence (fig. 15.32). Climate scientists offer the following responses to some of the following claims in the popular media:

Remaining land
Current shoreline

0   100 Km
0   50 Miles

FIGURE 15.31 Approximate change in land surface with the 1-m (3-ft) sea-level rise that the IPCC says is possible by the year 2100. Some analysts expect a 2-m (6-ft) rise if no action is taken.

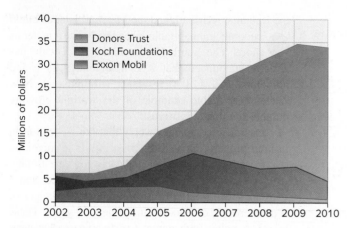

FIGURE 15.32 Funding given to groups promoting climate denial in popular media, blogs, and political lobbying. Sources include oil and gas industries and anonymous contributions to the Donor's Trust.
Source: Greenpeace

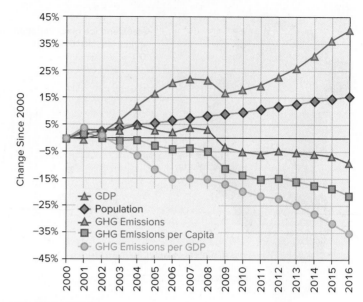

FIGURE 15.33 Decoupling of economy and emissions can be seen in the percentage change in California's GDP, population, and greenhouse gas emissions since 2000.
Source: California Air Resources Board

"Reducing climate change requires abandoning our current way of life." Reducing climate change requires that we use different energy. By replacing coal-powered electricity with wind, solar, natural gas, and improved efficiency, we can drastically cut our emissions but keep our computers, TVs, cars, and other conveniences. Reducing coal dependence will also reduce financial costs of pollution damage to health and vegetation.

"There is no alternative to current energy systems." Alternative energy and conservation could easily reduce our fossil fuel consumption to safe levels if we chose to invest in them. Already, new wind and even solar electricity are cheaper than new fossil-fuel power plants. Alternative energy and improved efficiency have also created new markets and jobs.

"Alternative energy requires subsidies." Fossil fuels receive far more subsidies and supports than alternatives do. Global direct subsidies for fossil fuels reached nearly $1,000 billion in 2015, compared to about $140 billion for renewables, according to the International Energy Agency.

"We can't have economic growth without climate impacts." As noted in the opening case study, California and many other states and cities have seen a **decoupling** of economic growth and energy use, showing that prosperity need not be tied to carbon emissions (fig. 15.33). Most northern Europeans produce less than half the CO_2 of North Americans, per capita. Yet they have higher standards of living (in terms of education, health care, life span, vacation time, financial security) than North Americans. Residents of San Francisco consume about one-sixth as much energy as residents of Kansas City, yet quality of life is not necessarily six times greater in Kansas City than in San Francisco.

"Natural changes such as solar variation can explain observed warming." Solar input fluctuates, but the changes are slight; these and other natural changes do not coincide with the direction of changes in temperatures or with the amount of warming (see fig. 15.17).

"We had cool temperatures and snowstorms last year, not heat and drought." Regional differences in temperature and precipitation trends are predicted by climate models. Most of the United States is expected to see wetter, warmer winters and drier, hotter summers (fig. 15.26).

"Climate scientists have gaps in their knowledge and have made errors in the past." The gaps and uncertainties in climate data are minute compared to the evident trends. There are many unknowns—details of precipitation change or interaction of long-term cycles such as El Niño—but the trends are unequivocal. Climatologist James Hansen has noted that while most people make occasional honest mistakes, fraud in data collection is almost unheard of. This is because transparency in the scientific process ensures public visibility of errors. There is much less public accountability in popular media, however, where climate scientists are regularly subjected to personal attacks from climate-change deniers.

Section Review

1. Why might drought be associated with climate warming?
2. List five to ten effects of changing climate.
3. List several arguments in public disputes over climate evidence.

15.6 CLIMATE ACTION

- *The Paris Agreement seeks to keep warming below 2°C.*
- *About seven 1-Gt "wedges" are needed to stabilize climate.*
- *Negative emissions technologies are needed.*

The good news is that we already know how to fix climate change. We are not moving fast enough, but we have countless available solutions, from energy efficiency to forest management to innovative agricultural management and urban planning. Almost every

Unburnable Carbon

Climatologists have shown that the earth will become largely uninhabitable if we continue to burn oil, gas, and coal at the rate we have done in recent decades. Major cities will be under water, and most farmland will be desert. Alternative energy sources are therefore a priority. But fossil fuels are abundant. We have enough natural gas and coal to last for centuries and enough oil to last 40–60 years even at current high-use rates. Because we can burn only a fraction of these fuels without making our planet uninhabitable, analysts often refer to these as **unburnable carbon.** Financial investments in these resources are often called **stranded assets,** because they have no monetary value if they cannot be sold.

Three important obstacles have blocked wider adoption of alternative energy and conservation in recent years. One is inertia: We already have trillions of dollars invested in existing cities, roads, and power infrastructure. Replacing all this infrastructure is not easy to budget, even though most of it is gradually and steadily replaced and upgraded.

A second factor is ready availability of abundant fossil fuels. As long as they are easy to use and comparatively cheap to produce, we will have difficulty weaning ourselves from them.

A third obstacle is that owners of this carbon strongly resist energy and climate policies that would turn their wealth into stranded assets. Understandably enough, they have invested millions of dollars in campaigns to block fossil fuel regulation, to support lawmakers who deny climate change, and to fund research that seeks alternative explanations for climate change, such as sunspots or random variability.

At the same time global subsidies for fossil fuel development, exploration, and marketing exceed $550 billion per year, according to the International Energy Agency (IEA). This is about 4.5 times the $120 billion for all renewable energy. The United States is the world's most generous supporter of fossil fuels, providing about one-fifth of global subsidies. The IEA and other groups have begun to question whether this ratio can be changed. If you were in charge of energy policy, what would you do? Would you compensate oil and gas companies for their unburnable carbon? Would you try to restrict corporate influence on climate policy, or not? If so, how?

---

sector of society can be part of the solution, and indeed, many actors at all scales have stepped up to innovate and lead the way (What Can You Do?, p. 350).

The main challenge is entrenched interests and embedded infrastructure. Fossil fuel industries, or conventional agricultural interests, have strong motivations to resist change. Infrastructure designed around the automobile or around cheap oil make change difficult even for many who are eager to participate.

Despite the challenges, it is important to recognize that many countries have made real progress. The 1997 **Kyoto Protocol,** the first major global accord on climate change, called for countries to voluntarily set targets for reducing emissions. Although the agreement failed to reduce overall emissions, many countries, mostly in Europe, have met or exceeded their targets of 5 or 10 percent below 1990 emissions.

At the same time, many countries have shown that economic growth is now increasingly decoupled from growth in carbon emissions, as in California (see fig. 15.33). Economies in Europe, Asia, the United States, and the world as a whole have all shown this decoupling.

The Paris Climate Agreement establishes new goals

The most important international agreement since the Kyoto Protocol was the 2015 Paris Climate Agreement. The 195 countries in attendance at the Paris meeting agreed on a number of major points, including these:

- Holding the global average temperature increase to well below 2°C (above pre-industrial levels) is necessary. Below 1.5°C should be the target.

- Zero carbon emissions, ultimately negative emissions, is a global goal, to avoid compounding climate effects already committed by past emissions.

- Each participating country establishes voluntary emission reductions goals, and those goals, and progress toward them, must be publicly visible.

- Reduction plans submitted thus far are not sufficient to keep warming below 2°C, so plans must be revised every five years.

- Climate finance is necessary: Advanced economies agreed to strive toward donations of $100 billion per year to a "green carbon fund" to support low-carbon development in emerging economies.

This agreement calls for all countries to invest a great deal of effort and resources in emissions reductions. Many people are skeptical that countries will follow through on pledges. But others point to ongoing innovations in renewable energy, in greenhouse gas regulation, and in policy strategies that can make goals achievable. And many developing regions are not waiting for the green carbon fund to start investing in clean energy. In Bangladesh and India, for example, both central banks and private funds are investing in "green bonds" that support renewable energy and that are producing hundreds of thousands of new jobs. Perhaps most important, agreement on the need for better policies is almost universally accepted now. And there are increasing numbers of local and regional examples. These are showing that efficiency and renewable energy can be good for financial stability, jobs, energy security, and public health, (see chapter 20 and What Do You Think?, above).

The Paris Accord is, like any policy, imperfect, but enthusiasm for it remains overwhelming. At 2018 UN meetings in Katowice,

Poland, countries reaffirmed and strengthened commitments, clarified definitions, and inched closer to needed changes. And while federal governments of the United States and Brazil have declared they would move away from the accord (after a required waiting period), a growing movement of U.S. states, cities, businesses, colleges, and universities are declaring "we are still in," and making their own plans to meet Paris goals.

Drawdown strategies abound

Resolving climate change is not rocket science. Solutions are known. But the problem has seemed too big and too complex to solve. To help us out of a quagmire of indecision, a Princeton ecologist and an engineer, Steven Pacala and Robert Scolow, proposed that big problems are always more tractable when broken down into smaller pieces. In a 2004 paper in the journal *Science,* they calculated that the world needs to cut 7 Gt of carbon emissions (or nearly 25.7 Gt of CO_2, because the molecule is heavier than C alone). That is a lot, but 7 1-Gt pieces is easier to imagine. They showed that currently available technologies—efficient vehicles, buildings, power plants, alternative fuels—could solve our problems, starting immediately (fig. 15.34).

Because small steps today multiply into big impacts decades from now, Pacala and Socolow called this approach **wedge analysis.** They further modeled three trajectories: a business as ususal scenario, which heads toward tripling CO_2 emissions by 2100, a stabilization scenario, and a dramatic reduction scenario. Even the third, they pointed out, is achievable with current technology.

In the 2017 book *Drawdown,* Paul Hawken and his colleagues offered a still wider range of solutions within easy reach. Solutions included many options that had little to do with energy

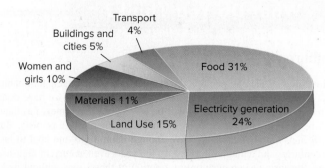

FIGURE 15.35 Categories of the 100 Project Drawdown solutions, by percentage contribution to needed climate impact reductions.
Data source: Project Drawdown (plausible scenario): Drawdown.org

technology and that had great co-benefits. This book's title, *Drawdown,* acknowledged that we need to draw emissions down to *negative* levels, because we are already exceeding safe concentrations of GHGs (see fig. 15.2). Their "plausible" estimate of strategies, which they calculated could be implemented with reasonable but vigorous efforts, amounted to reductions of 1,034 Gt CO_2-equivalent by 2050. Better yet, while these efforts would cost around U.S. $29 trillion (a large amount), they would save over U.S. $74 trillion in other costs.

The solutions benefit human health and welfare, in addition to protecting the climate. The top 12 strategies included refrigerant management (especially for air conditioning), wind turbines, reducing food waste, plant-rich diets, tropical forest conservation, educating girls, family planning, solar farms, silvopasture, rooftop solar, regenerative agriculture, and conservation of temperate forests (fig. 15.35). You can learn much more, and find detailed explanations and documentation of models, at Drawdown.org.

Policymakers are increasingly aware of these options, and of their importance. At the start of 2019, *The Washington Post* published a list of 11 policies to protect the planet, drawn from policymakers and experts from around the country. The piece started with the fundamental question, "the world has until 2030 to drastically cut our emissions. Where do we begin?" Table 15.4 shows the answers offered.

Carbon capture is needed

As we noted in the opening case study, climate action will require capturing CO_2 in addition to reducing emissions. **Carbon capture** and **negative emissions technologies** are two of the terms used for this. As with other approaches, many strategies exist, some very expensive and some cheap. Direct capture of CO_2 from the air costs around $600 per ton CO_2, but capturing CO_2 from a powerplant's smokestack might be less than $100 per ton. Conserving tropical forest costs much less. Capturing and selling methane that leaks from landfills can earn money.

In a 2018 study for the National Academy of Sciences, Stephen Pacala and colleagues compared the potential of different technologies. They concluded that four practices are likely to make the biggest contributions: bioenergy with carbon capture

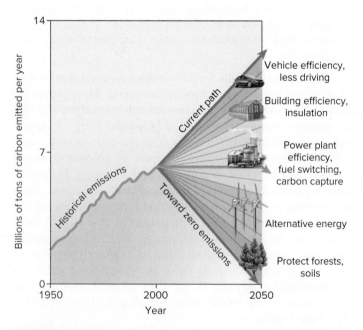

FIGURE 15.34 A wedge approach could use multiple strategies to reduce or stabilize carbon emissions rapidly and relatively cheaply.

TABLE 15.4 Eleven Policy Ideas to Protect the Planet[1]

Set local emissions goals
Be smart about your air conditioner
Encourage electric vehicles
Be smart about nuclear power
Make it easier to live without cars
Prevent wasted food—the right way
Incentivize carbon farming
Curb the effects of meat and dairy
Adopt a carbon tax
Open electric markets to competition
Pass a Green New Deal

[1]Source: *The Washington Post,* January 2, 2019

(that is, making biofuel from corn or soybeans, but capturing CO_2 in the biofuel factory), increasing carbon content in agricultural soils (planting cover crops and reduced plowing; see chapter 10), improved forest management, and reforestation (see chapter 12).

By 2050, the need for carbon capture is likely to be around 10 Gt per year, with higher needs after that. The Pacala study suggested that scaling up to this level is possible, but half of that is more likely (fig. 15.36).

Carbon capture and storage is already widely practiced at oil and gas facilities and some fossil-fuel power plants, although it is expensive. Oil and gas companies can pump CO_2 deep into the ground, usually into porous rocks or salt-water aquifers. Injecting

CO_2 into the ground near wells increases pressure on oil reservoirs and enhances oil recovery. In countries like Norway, a leading oil and gas producer, carbon taxes of $50 per ton further incentivize carbon storage deep in the ground.

Economic solutions make progress possible

Committing to a good idea is always easier when it also saves money. A central idea is to put a price on carbon emissions. Traditionally carbon emissions have been free to polluters, although society as a whole bears the cost. Putting a price on carbon emissions provides a direct incentive to find ways to reduce emissions.

Economists point out that eliminating **environmentally harmful subsidies** is necessary for reducing carbon emissions. These subsidies include benefits such as direct payments, tax breaks, cheap loans, and publicly funded research and development, and other practices that aid environmentally harmful industries. Global subsidies for fossil fuels were nearly $1 trillion per year in 2015, according to Oil Change International. But a 2018 report from the International Monetary Fund further calculated that indirect subsidies—monetary costs of air pollution, illness, and economic losses from climate change—raise the subsidy rate to around $5.3 trillion per year. In addition, U.S. military analysts point out that about half the $700 billion U.S. defense budget is used to protect security in the oil-rich Persian Gulf. Estimates for the cost of that protection range from about $0.28 to $0.70 per gallon of gasoline on top of what you pay at the pump.

The size of fossil fuel subsidies depends on which factors are included in accounting, but they clearly exceed the roughly $140 billion in annual subsidies to renewable energy. Most countries agreed in 2016 to gradually phase out their direct subsidies for fossil fuels. Many observers question how well countries are holding to promises, however.

Emissions trading markets are the most widely adopted way to put a price on carbon emissions. In this system, a government issues emissions permits to industries or other major emitters. If permit holders can find low-cost ways to cut emissions, they can sell their permits for a profit. Europe established the first emissions trading scheme in 2005, but it was quickly followed by California (2006), and the northeastern U.S. states (2007). By 2016, trade had grown to 140 billion tons worth $53 billion.

Emissions trading schemes are efficient because the cheapest fixes are taken care of first. The cost of implementing the plan is low, and industries can meet standards in any way they choose. Carbon trading is not perfect: Carbon prices are often too low to provide real incentives for some industries, and many question whether a "right to pollute" is the best strategy. However, these compacts are widely considered successful—and palatable—approaches to reducing emissions. Carbon credit auctions for the northeastern Regional Greenhouse Gas Initiative (RGGI) have generated more than $700 million in

FIGURE 15.36 Negative emissions technologies, and estimated maximum potential carbon removal globally, at $100/ton CO_2. (Direct air is technologically available but not at this cost.) Models suggest these measures could capture up to 9–11 Gt of CO_2 per year, globally.

source: Pacala, S., et al., 2018. "Negative Emissions Technologies and Reliable Sequestration: A Research Agenda." National Academies of Sciences.

What Can You Do?

Climate Action

The main thing we all need to do is reduce energy consumption and work for policies that support low-carbon energy. The most important steps are changing national energy policies. At the same time, millions of individual actions also have powerful positive impacts.

Use your phone. Contact your senators and representatives to tell them to protect your future. Vote for climate-conscious candidates.

Save gas. Drive less, use an efficient vehicle, and make plans to live where you can use public transit or bicycles. Vehicles may represent our biggest individual climate impact.

Buy efficient appliances. Avoiding catastrophic climate change is about more than just changing lightbulbs, but better bulbs and appliances are also important, and the impact multiplies as millions of us follow better practices.

Eat low on the food chain. Meat and dairy production are important sources of methane and CO_2.

Weatherize your home. Improve your comfort and reduce heat loss in winter or heat gain in summer. Turn down the thermostat and dress for the season.

Buy wind and solar energy. Many utilities allow you to purchase alternatives.

Practice energy policy at school and at work. Ask administrators what they are doing to reduce energy use, and help them write plans to do better.

Often they are used to offset other taxes, so that they don't raise overall government revenue; instead they redistribute costs to activities that are harmful.

Carbon taxes are an increasingly popular idea, and the 2018 Nobel Prize in economics was given to William Nordhaus of Yale for his work on the topic. To read more, see chapter 23.

Wind, water, and solar could meet all our needs

Energy technology is also advancing on many fronts. Some of the most thorough analysis of transitioning to renewables has been by Mark Jacobson and colleagues, who have calculated how U.S. states could meet all their energy needs with wind, water, and solar (WWS) technologies. They calculated that all U.S. energy—for transportation, heating, and industrial uses, as well as electricity—could be 80 percent wind, water, and solar (WWS) by 2030, and 100 percent WWS by 2050, with existing and feasible technologies. They also found savings of $10,000 per person per year in financial benefits from this transtion, including energy cost savings, increased employment, reduced health care costs, and avoided storm- and heat-related losses (see chapter 20).

A transition to renewables has shown dramatic success in countries around the globe. Among the notable cases is Germany, which has raised its target to 80 percent renewable electricity by 2020. Germany has been switching from coal to wind, solar, and gas and encouraging energy efficiency throughout society. Denmark, a world leader in wind power, now gets over 40 percent of its electricity from wind. Plans are to generate half of the nation's electricity from offshore wind farms by 2030. Atmospheric scientist Steve Schneider called this a "no regrets" policy—even if we didn't need to stabilize our climate, many of these steps save money, conserve resources, and have other environmental benefits.

Section Review

1. What are some points in the Paris Climate Agreement?
2. List seven climate "wedges" and 12 "drawdown" strategies.
3. What are some options for carbon capture?

---

revenue to support conservation and alternative energy initiatives in participating states.

Carbon taxes or carbon fees are also seen as an efficient means to use price to encourage efficiency. Carbon fees are a cost per ton of CO_2, usually charged at the point of purchase like other taxes. These taxes have helped reduce consumption and fund renewable energy in Sweden, Denmark, Canada, and elsewhere.

Connecting the Dots

Climate change may be the most far-reaching issue in environmental science today. Although the challenge is almost inconceivably large, solutions are possible if we choose to act as both individuals and as a society. Temperatures are now higher than they have been in thousands of years, and climate scientists say that if we don't reduce greenhouse gas emissions soon, drought, floods, and

conflict may be inevitable. The "stabilization wedge" proposal is a list of immediate and relatively modest steps that could be taken to accomplish needed reductions in greenhouse gases.

Understanding the climate system is essential to understanding the ways in which the changing composition of the atmosphere (more carbon dioxide, methane, and nitrous oxide, in particular) matters to us. Basic concepts to remember about the climate system include how the earth's surfaces absorb solar heat, how atmospheric convection transfers heat, and that different gases in the atmosphere absorb and store heat that is reemitted from the earth. Increasing heat storage in the lower atmosphere can cause increasingly vigorous convection, more extreme storms and droughts, melting ice caps, and rising sea levels. Changing patterns of monsoons, cyclonic storms, frontal weather, and other precipitation patterns could have extreme consequences for humans and ecosystems.

Just as evidence for climate change already surrounds us, so do climate solutions. The example of California's climate action is exemplary, but this is just one of countless states, cities, counties, businesses, schools, and other organizations working to protect the climate. While the challenges are daunting, the variety of creative solutions is also exciting and inspiring.

From the Paris Agreement to drawdown strategies to energy transitions, we have boundless opportunities to take action. The benefits for our communities, economies, health, and well-being hold tremendous promise if we choose to work together.

Reviewing Key Terms

Can you define the following terms in environmental science?

aerosols 15.1	decoupling 15.5	jet streams 15.2	stratosphere 15.1
albedo 15.1	El Niño 15.3	Kyoto Protocol 15.6	tornado 15.2
arctic amplification 15.4	emissions trading 15.6	La Niña 15.3	troposphere 15.1
bleaching 15.5	environmentally harmful	latent heat 15.1	unburnable carbon 15.6
carbon capture 15.6	subsidies 15.6	Milankovitch cycles 15.3	warm front 15.2
carbon taxes 15.6	greenhouse effect 15.1	monsoon 15.2	weather 15.1
climate 15.1	greenhouse gases 15.1	negative emissions	wedge analysis 15.6
cold front 15.2	hurricanes 15.2	technology 15.6	
convection currents 15.1	Intergovernmental Panel	ozone 15.1	
Coriolis effect 15.2	on Climate Change	positive feedback loop 15.1	
cyclonic storms 15.2	(IPCC) 15.4	stranded assets 15.6	

Critical Thinking and Discussion Questions

1. Weather patterns change constantly over time. From your own memory, what weather events can you recall? Can you find evidence in your own experience of climate change? What does your ability to recall climate changes tell you about the importance of data collection?

2. Many people don't believe that climate change is going on, even though climate scientists have amassed a great deal of data to demonstrate it. What factors do you think influence the degree to which a person believes or doesn't believe climatologists' reports?

3. How does the decades-long, global-scale nature of climate change make it hard for new policies to be enacted? What factors might be influential in people's perception of the severity of the problem?

4. What forces influence climate most in your region? In neighboring regions? Why?

5. Of the drawdown or wedge solutions noted in this chapter, which would you find most palatable? Least tolerable? Why? Can you think of any additional wedges that should be included?

6. Would you favor building more nuclear power plants to reduce CO_2 emissions? Why or why not?

Data Analysis

The U.S. National Climate Assessment

The Fourth National Climate Assessment, produced by the U.S. federal agencies and led by the U.S. Global Change Research Program, provides an overview of what we know about climate science, and in particular what we know about regional impacts of climate change. The online report can be found at nca2018.globalchange.gov. Go to the Key Messages, https://nca2018.globalchange.gov/chapter/2/#key-message-1, and read up to interactive Figure 2.1, reproduced here.

In the interactive figure, click on the legend to show and hide different drivers.

Answer these questions: 1) How much has temperature changed? When did the increase become visibly steeper? 2) Which of the natural drivers, if any, correspond to observed changes? 3) Which human drivers correspond best to observed changes? 4) Effects of land cover change involve changes in surface albedo. Refer to table 15.2. What land cover changes might explain this? 5) Aerosols are mainly pollutants. What will happen if we succeed in eliminating air pollution?

Source: *U.S. Global Change Research Program, https://nca2018.globalchange.gov/chapter/2/*

16

Air Pollution

Learning Outcomes

▲ Like many fast-growing cities, Beijing is struggling to control its air pollution.
Mark Ralston/AFP/Getty Images

After studying this chapter, you should be able to:

16.1 Identify natural and human-caused sources of air pollution.

16.2 Explain how atmospheric circulation affects air quality.

16.3 Compare the effects of air pollution.

16.4 Evaluate air pollution control efforts and progress.

"You shouldn't have to leave your neighborhood to live in a better one."

– Majora Carter, Activist and

Community Advocate

Beijing Looks for Answers to Air Pollution

As the capital of China, Beijing is famous for its history, culture, cuisine, and architecture. Increasingly, the city also has a global reputation for its air pollution. Visibility is often limited to a few hundred meters, and air pollution index values rank between "unhealthy" and "hazardous" over half of the time. In especially bad spells, schools have closed because of air pollution. Causes of pollution are well known: Coal-burning power plants are the primary source, followed by industry, vehicles, and home oil and coal burners. Epidemiologists—and the city's residents—have long been interested in understanding the effects of these conditions for the 25 million people in greater Beijing, and for hundreds of millions more in other Chinese cities. The Chinese government calculated that the economic cost of air pollution amounts to around 3 percent of China's GDP.

Understanding the overall impacts of this pollution requires a lot of data. The Chinese government has developed a network of automated sampling stations, 945 in 190 cities across the country as of 2016, with hundreds more not yet in the network. Each station evaluates ambient (surrounding) air every hour and automatically submits hourly reports on six of the most important air pollutants.

For human health impacts, the most important pollutant in most cities is PM2.5—particulate matter (tiny airborne particles) less than 2.5 micrometers (0.0025 mm) in diameter. These tiny particulates can penetrate deep into lungs, obstructing breathing and exacerbating heart disease, asthma, and other conditions. Also important is slightly larger but still breathable particulate matter 10 micrometers (0.01 mm) or less in diameter (PM10). Together, PM2.5 and PM10 are the main health concerns in most Chinese cities, and in most other world cities, increasing illness and contributing to deaths from stroke, heart disease, lung cancer, and obstructive pulmonary disease.

Among the other major considerations for both human health environmental quality are ozone (O_3), nitrogen dioxide (NO_2), sulfur dioxide (SO_2), and carbon monoxide (CO). Ozone, less stable than normal oxygen (O_2) molecules, is highly reactive. It irritates sensitive tissues such as eyes, throat, and lungs. Nitrogen dioxide and sulfur dioxide, also highly reactive, contribute to the formation of acid rain and damage plant tissues and buildings, as well as health. Carbon monoxide bonds to blood cells, restricting the flow of oxygen in the blood stream.

How many deaths result from these pollutants? A 2015 study by University of California, Berkeley scientists Robert Rohde and Richard Muller examined this question. The study drew on data from over 1500 air monitoring sites, including most of the 945 nationally networked sites and hundreds of additional sites in China and neighboring countries. Hourly data from these sites were collected for four straight months in 2014, and then combined with weather data to map the movement of airborne pollutants. Rohde and Muller could then model hourly pollution exposures by population for all of eastern China, where 97 percent of the population lives. After evaluating population size and exposure to the major pollutants, they used World Health Organization models to project the likely number of deaths per year.

For the whole of China, this study calculated that air pollution contributes to 1.6 million premature deaths per year, about 17 percent of all deaths. Numbers like these help strengthen incentives for Beijing and other cities to push harder for air pollution controls. Recent policy shifts have included plans to eliminate coal-burning power plants (replacing them with renewable energy and also with nuclear power) and establish policies promoting electric cars and energy efficiency. The central government also promises to improve the enforcement of rules, so as to ensure that polluters comply with established health standards.

Beijing is one of hundreds of cities in the rapidly industrializing world to suffer from these conditions. In China, India, Southeast Asia, and elsewhere, rapidly growing cities are powered by coal plants or by poorly regulated oil and gas burners. Coal is often used for household heating and for industrial furnaces. Oil and gas production are also major pollutant sources, and regions that produce or process oil and gas often have especially poor air quality.

FIGURE 16.1 Improved data monitoring makes it possible to evaluate air quality in Beijing and elsewhere. The Beijing-based organization AQICN.org created this map to deliver real-time air quality data for China and much of the rest of the world. Colors range from purple and red (unsafe air quality) to green (good air quality).

Source: AQICN.org, U.S. Embassy Beijing, China Ministry of Environmental Protection

You can see real-time air quality index reports for cities in China and others worldwide by looking up the Beijing-based non-governmental organization AQICN.org online (fig. 16.1). This group began in 2007 by reporting data from the U.S. embassy in Beijing but continues by reporting real-time air monitoring data provided by the Chinese Ministry of Environmental Protection and other agencies worldwide. Like the Rohde and Muller study, this group draws on increasingly available monitoring data. By delivering it in standard units, the site allows easy comparison of world cities. The same group provides world air quality index data at WAQI.info.

Further Reading:

Rohde, R. A., and Muller, R. A. 2015. *Air Pollution in China: Mapping of Concentrations and Sources,* PLoS ONE 10(8): e0135749. doi:10.1371/journal.pone.0135749.

Benchmark Data	
Among the ideas and values in this chapter, the following are a few worth remembering.	
2.5	Diameter of most harmful particulate pollutants, in μm
10	Diameter of course particulate pollutants, in μm
6	Number of criteria pollutants regulated by the EPA
$2 trillion	Economic benefits of the Clean Air Act, 1990–2020
30:1	Economic benefits:costs of enforcing the Clean Air Act

16.1 MAJOR POLLUTANTS IN OUR AIR

- *Air pollutants affect health and environmental quality.*
- *The EPA regulates six major criteria pollutants.*
- *Mercury and organic compounds are examples of hazardous air pollutants.*

Beijing, Delhi, Bangkok, and other rapidly growing cities face serious air pollution challenges today, but only a few decades ago, many American and European cities also endured similar conditions. Chronic bad air, and occasional severe smog events (see section 16.2) gradually led to the adoption of pollution controls. Legal enforcement has improved, and many students today don't realize how bad air quality once was in their home towns. Many American and European cities still have bad air: Major port cities, oil and gas extraction areas, and industrial cities are particularly bad (fig. 16.2). But in the past 40 years, air quality protections have increased in number and in effectiveness, greatly improving public health. In most developed economies, there are established, legally enforceable rules to protect the air we all breathe. Industry has also relocated to hungry regions of the developing world, where environmental and health protections are poorly enforced.

In this chapter, we examine what major pollutants are, what their sources are, and how we can implement policies to control them. Air pollution impairs human health, damages crops and ecosystems, and corrodes buildings and infrastructure. Greenhouse gas pollutants are altering our climate. **Aesthetic degradation,** such as odors and lost visibility are also important consequences of air pollution. These factors rarely threaten life or health directly, but they can strongly impact our quality of life. They also increase stress, which affects health.

Many natural factors degrade air quality degradation, including ash and gases from volcanoes and desert dust. Trees emit volatile organic compounds (terpenes and isoprenes), and decaying vegetation in swamps produces methane, as do termites and ruminant animals. Forest fires produce particulate matter, nitrogen oxides, and carbon monoxide. For the purposes of this discussion, however, these sources are treated as background levels. They are normally too diffuse to cause severe damage to living systems, and most of the contaminants discussed in this chapter are produced in greater abundance by human activities (fig. 16.3). When we refer to pollutants, we generally mean human-caused emissions.

In many Chinese cities, airborne dust, smoke, and soot often are ten times higher than levels considered safe for human health.

FIGURE 16.2 Poor air quality and photochemical smog are less common in U.S. cities than they were 40 years ago, although they persist in many major cities (like Los Angeles, shown here) and industrial or port areas.
Daniel Stein/Getty Images

FIGURE 16.3 Dust, also known as particulate matter, occurs naturally but is disturbed and mobilized by human activity. Fine particulates are the most common cause of air-related illnesses.
MarketPlace/Media Bakery

China's city dwellers are four to six times more likely than rural people to die of lung cancer. Poorly regulated industrial cities of India, Russia, Pakistan, and many other countries have similar hazards. While pollution contributes to about 1.6 million deaths each year, in China, it also contributes to at least 1.4 million deaths in India and some 7 million worldwide, according to the World Health Organization. In India, the primary risk has historically been particulate matter from household fires and other biomass burning, but rapid industrialization is expanding the range of pollutants.

The Clean Air Act designates standard limits

Air pollution is one of our most diffuse and hard to control environmental problems. Industry continues to challenge clean air rules in every country, and public attention is always needed to protect the safeguards we now rely on. However, epidemiologists and economists are increasingly making it clear that pollution control rapidly pays for itself in social costs of health care and infrastructure damage. Cost-benefit analysis of air regulation is discussed in section 16.4.

There have been countless efforts throughout history to control objectionable smoke, odors, and noise, but nearly all of these efforts were local. For example, the 1963 Clean Air Act, the first national air quality legislation in the United States, was careful to preserve states' rights to set and enforce air quality regulations. But polluting industries can easily move across state boundaries to avoid local regulations. It soon became obvious that piecemeal, local standards did not resolve the problem, because neither pollutants nor the markets for energy and industrial products are contained within state boundaries.

Amendments to the U.S. Clean Air Act in 1970 essentially rewrote the law. Congress designated new standards, to be applied evenly across the country, for six major pollutants: sulfur dioxide, nitrogen oxides, carbon monoxide, ozone (and its precursor volatile organic compounds), lead, and particulate matter. These standards were set according to health criteria and environmental quality. Transportation and power plants are the dominant sources of criteria pollutants (fig. 16.4). We'll examine each of these, and then we'll look at additional pollutants that are also monitored under the Clean Air Act.

Conventional pollutants are most abundant

National ambient air quality standards (NAAQS) identify maximum allowable limits for six **conventional** or **criteria pollutants** in **ambient air** (the air around us). The Clean Air Act addressed these first because they contributed the largest volume of air quality degradation and also are considered the most serious threats to human health and welfare. Primary standards (table 16.1) are intended to protect human health.

Note that for each of the six criteria pollutants in table 16.1, standards of acceptable exposure are specified as concentrations in the air you breathe (micrograms, µg, per cubic meter of air) averaged over time. For example, the annual average concentration of particulate matter in the air you breathe should not exceed $50\,\mu g/m^3$. This means that if you were to sample the air in your neighborhood repeatedly over a year, you want the average concentration to be less than $50\,\mu m/m^3$. Over a shorter period of 24 hours, most people can tolerate three times that concentration ($150\,\mu m/m^3$). Sulfur dioxide, nitrogen dioxide, and ozone have similar standards. But carbon monoxide (CO) and lead have much lower standards because they are much more hazardous even with small exposure. For these, exposure should not exceed $10\,\mu m/m^3$ averaged over 8 hours (for CO) or $1.57\,\mu m/m^3$ averaged over a quarter year (for lead).

While these primary standards are designed to protect human health, we also have secondary standards to protect crops, materials, climate, visibility, and personal comfort.

Table 16.1	National Ambient Air Quality Standards (NAAQS)	
Pollutant	**Primary (Health-Based) Averaging Time**	**Standards (Allowable Concentrations)**
PM[a]	Annual geometric mean[b]	$50\ \mu g/m^3$
	24 hours	$150\ \mu g/m^3$
SO$_2$	Annual arithmetic mean[c]	$80\ \mu g/m^3$ (0.03 ppm)
	24 hours	$120\ \mu g/m^3$ (0.14 ppm)
CO	8 hours	$10\ mg/m^3$ (9 ppm)
	1 hour	$40\ mg/m^3$ (35 ppm)
NO$_2$	Annual arithmetic mean	$80\ \mu g/m^3$ (0.05 ppm)
O$_3$	Daily max 8 hour avg.	$157\ \mu g/m^3$ (0.08 ppm)
Lead	Maximum quarterly avg.	$1.5\ \mu g/m^3$

[a] Total suspended particulate material, PM2.5 and PM10.

[b] The geometric mean is obtained by taking the *n*th root of the product of *n* numbers. This tends to reduce the impact of a few very large numbers in a set.

[c] An arithmetic mean is the average determined by dividing the sum of a group of data points by the number of points.

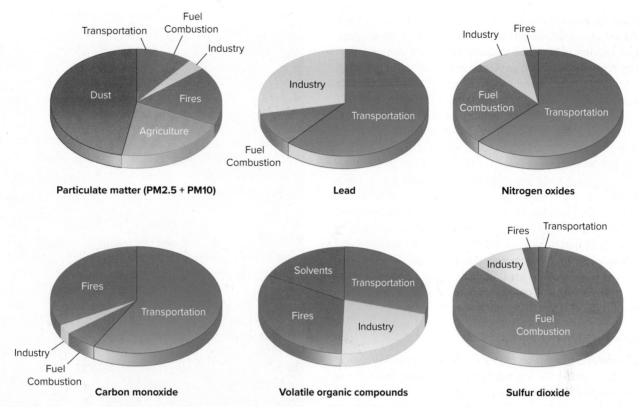

FIGURE 16.4 Major sources of primary "criteria" pollutants in the United States. Volatile organic compounds and nitrogen oxides are important precursors of ozone, one of the six criteria pollutants.
Source: EPA Air Emission Sources, 2016

In addition to the six conventional pollutants, the Clean Air Act regulates an array of **unconventional pollutants,** compounds that are produced in less volume than conventional pollutants but that are especially toxic or hazardous, such as asbestos, benzene, mercury, polychlorinated biphenyls (PCBs), and vinyl chloride. Most of these are uncommon in nature or have no natural sources.

We also distinguish pollutants according to how they are produced. **Primary pollutants** are those released directly from the source into the air in a harmful form (fig. 16.5). **Secondary pollutants** are converted to a hazardous form after they enter the air or are formed by chemical reactions as components of the air mix and interact. Solar radiation often provides the energy for these reactions. Photochemical oxidants and atmospheric acids formed by these mechanisms (see fig. 16.2) are among our most important pollutants in terms of health and ecosystem damage.

Unlike pollutants from a point source, such as a smokestack, **fugitive emissions** are those that do not go through a smokestack. By far the most massive example of this category is dust from soil erosion, strip mining, rock crushing, and building construction (and destruction). Fugitive industrial emissions are hard to monitor, but they are extremely important sources of air pollution. Leaks around valves and pipe joints, and evaporation of volatile compounds from oil-processing facilities, contribute as much as 90 percent of the hydrocarbons and volatile organic chemicals emitted from oil refineries and chemical plants.

Sulfur Dioxide (SO₂)

Natural sources of sulfur in the atmosphere include evaporation of sea spray, erosion of sulfate-containing dust from arid soils, fumes from volcanoes and hot springs, and biogenic emissions of hydrogen sulfide (H_2S) and organic sulfur-containing compounds.

FIGURE 16.5 Primary pollutants are released directly from a source into the air. Extremely dirty factories like this coal-powered steel plant are now rare in places where pollution control rules exist, and are enforced, to protect public health and the environment.
Denys Bilytskyi/Alamy Stock Photo

Total yearly emissions of sulfur from all sources amount to some 114 million metric tons. Worldwide, anthropogenic sources represent about two-thirds of all the airborne sulfur, but in most urban areas they contribute as much as 90 percent of the sulfur in the air. The predominant form of anthropogenic sulfur is **sulfur dioxide** (SO_2) from combustion of sulfur-containing fuel (coal and oil), purification of sour (sulfur-containing) natural gas or oil, and industrial processes, such as smelting of sulfide ores.

Sulfur dioxide is a colorless, corrosive gas that is directly damaging to both plants and animals (fig. 16.6). Once in the atmosphere, it can react with atmospheric oxygen and water vapor to form sulfuric acid (H_2SO_4), a major component of acid rain. Very small solid particles or liquid droplets can transport the acidic sulfate ion (SO_4^{-2}) long distances through the air or deep into the lungs, where it is very damaging. Sulfate particles and droplets reduce visibility, sometimes dramatically. Some of the smelliest and most noxious air pollutants are sulfur compounds, such as hydrogen sulfide from pig manure lagoons or mercaptans (organosulfur thiols) from paper mills (fig. 16.7).

Nitrogen Oxides (NO_x)

Nitrogen oxides are highly reactive gases formed when nitrogen in fuel or in air is heated (during combustion) to temperatures above 650°C (1,200°F) in the presence of oxygen. The initial product, nitric oxide (NO), oxidizes further in the atmosphere to nitrogen dioxide (NO_2), a reddish-brown gas that gives photochemical smog its distinctive color. In addition, nitrous oxide (N_2O) is an intermediate form that results from soil denitrification. Nitrous oxide absorbs ultraviolet light and is an important greenhouse gas. Because nitrogen readily changes from one of these forms to

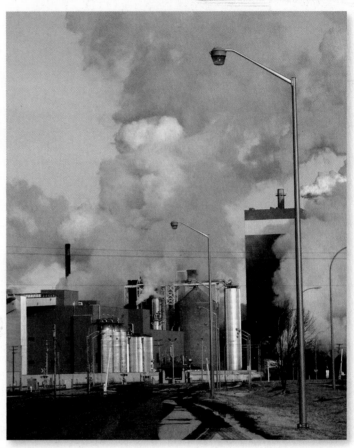

FIGURE 16.7 The most noxious pollutants from this paper mill are pungent organosulfur thiols and sulfides. Chlorine bleaching can also produce extremely dangerous organochlorines, such as dioxins.
William P. Cunningham

another by gaining or losing O atoms, the general term NO_x is used to describe the gases NO and NO_2. Nitrogen oxides react with water to make nitric acid (HNO_3), a major component of acid rain.

Anthropogenic sources account for 60 percent of the global emissions of about 230 million metric tons of reactive nitrogen compounds each year. About 95 percent of all human-caused NO_x in the United States is produced by fuel combustion in transportation and electric power generation. Because we continue to drive more miles every year, and to consume abundant electricity, we have had less success in controlling NO_x than other pollutants.

Excess nitrogen from agricultural fertilizer use and production is also an important contributor to airborne NO_x.

Carbon Monoxide (CO)

Carbon monoxide (CO) is a colorless, odorless, nonirritating, but highly toxic gas. CO is produced mainly by incomplete combustion of fuel (coal, oil, charcoal, or gas), as in furnaces, incinerators, engines, or fires, as well as in the decomposition of organic matter. CO blocks oxygen uptake in blood by binding irreversibly to hemoglobin (the protein that carries oxygen in our blood), making hemoglobin unable to hold oxygen and deliver it to cells. Human activities produce about half of the 1 billion metric tons

FIGURE 16.6 High concentrations of sulfur dioxide damage plants directly. This soybean plant was exposed to 2.1 mg/m³ sulfur dioxide for 24 hours. White patches show where chlorophyll has been destroyed.
William P. Cunningham

of CO released to the atmosphere each year. In the United States, two-thirds of the CO emissions are created by internal combustion engines in transportation. Land-clearing fires and cooking fires also are major sources. About 90 percent of the CO in the air is converted to CO_2 in photochemical reactions that produce ozone. Catalytic converters on vehicles are one of the important methods to reduce CO production by ensuring complete oxidation of carbon to carbon dioxide (CO_2).

Carbon dioxide is the predominant form of carbon in the air. Growing recognition of the health and environmental risks associated with climate change have led to recent regulations on CO_2, which are discussed below.

Ozone (O_3) and Photochemical Oxidants

Ozone (O_3) high in the stratosphere provides a valuable shield for the biosphere by absorbing incoming ultraviolet radiation. But at ground level, O_3 is a strong oxidizing reagent that damages vegetation, building materials (such as paint, rubber, and plastics), and sensitive tissues (such as eyes and lungs). Ozone has an acrid, biting odor that is a distinctive characteristic of photochemical smog. Ground-level O_3 is a product of photochemical reactions (reactions initiated by sunlight) between other pollutants, such as NO_x or volatile organic compounds. A general term for products of these reactions is **photochemical oxidants** (fig. 16.8). One of the most important of these reactions involves splitting nitrogen dioxide (NO_2) into nitrous oxide (NO) and oxygen (O). This single O atom is then available to combine with a molecule of O_2 to make ozone (O_3).

Hydrocarbons in the air contribute to the accumulation of ozone by combining with NO to form new compounds, leaving single O atoms free to form O_3 (see fig. 16.7). Many of the NO compounds are damaging photochemical oxidants. A general term for organic chemicals that evaporate easily or exist as gases in the air is **volatile organic compounds (VOCs).** Plants are the largest source of VOCs, releasing an estimated 350 million tons of isoprene (C_5H_8) and 450 million tons of terpenes ($C_{10}H_{15}$) each year, but normally concentrations are low.

The greater health threats from VOCs involve a large number of other synthetic organic chemicals, such as benzene, toluene, formaldehyde, vinyl chloride, phenols, chloroform, and trichloroethylene, which are released into the air by human activities. About 28 million tons of these compounds are emitted each year in the United States, mainly unburned or partially burned hydrocarbons from transportation, power plants, chemical plants, and petroleum refineries (fig. 16.9). These chemicals play an important role in the formation of photochemical oxidants.

Lead

Our most abundantly produced metal air pollutant, lead, is toxic to our nervous systems and other critical functions. Lead binds to enzymes and to components of our cells, such as brain cells, which

FIGURE 16.8 The heat of fuel combustion causes nitrogen oxides to form from atmospheric N_2 and O_2. Nitrogen dioxide (NO_2) interacts with water (H_2O) to form nitric acid (HNO_3), a component of acid rain. In addition, solar radiation can force NO_2 to release a free oxygen atom, which joins to atmospheric O_2, creating ozone (O_3). Fuel combustion also produces incompletely burned hydrocarbons (including volatile organic compounds, VOCs). Both O_3 and VOCs contribute to photochemical oxidants, in reactions activated by sunlight. The VOC shown here is benzene, a ring of six carbon atoms with a hydrogen atom attached to each carbon.
Hisham F. Ibrahim/Photodisc/Getty Images

FIGURE 16.9 Many of our most serious pollutants are hazardous organic compounds that escape as fugitive emissions from petroleum facilities, such as this one in Baton Rouge, Louisiana. Hazardous pollutants from chemical plants, petroleum refineries, and paper mills have declined nearly 70 percent since the EPA began regulating them.
Mary Ann Cunningham

then cannot function normally. Airborne lead is produced by a wide range of industrial and mining processes. The main sources are smelting of metal ores, mining, and burning of coal and municipal waste, in which lead is a trace element, and burning of gasoline to which lead has been added. From the 1920s to 1980s, leaded gasoline was the main source of lead in the United States, but leaded gas was phased out in the 1980s. Since 1986, when the ban was enforced, children's average blood lead levels have dropped 90 percent and average IQs have risen three points. Banning leaded gasoline in the United States was one of the most successful pollution-control measures in American history. Now, 50 nations have renounced leaded gasoline. The global economic benefit of this step is estimated to be more than $200 billion per year.

Worldwide atmospheric lead emissions amount to about 2 million metric tons per year, or two-thirds of all metallic air pollution. Globally, most of this lead is still from leaded gasoline, as well as metal ore smelting and coal burning.

Particulate Matter

Particulate matter includes solid particles or liquid droplets suspended in the air. Very fine solid or liquid particulates suspended in the atmosphere are **aerosols.** This includes dust, ash, soot, lint, smoke, pollen, spores, algal cells, and many other suspended materials. Particulates often are the most obvious form of air

pollution, because they reduce visibility and leave dirty deposits on windows, painted surfaces, and textiles.

While particulate matter is not as directly toxic as many pollutants, fine particulates can readily enter deep into lungs and damage lung tissues. These are the most common cause of air pollution-related health problems, in part because they are so abundant. Particulates smaller than 2.5 micrometers in diameter are particularly dangerous. Smoke, haze, power plant and factory emissions, car exhaust, and many other sources produce **PM2.5.** Very fine aerosol droplets of sulfuric acid, from coal smoke, are also considered **PM2.5** particulates.

Coarse inhalable particles larger than 2.5 micrometers but less than 10 micrometers in diameter are known as **PM10.** Heavier than PM2.5, they do not travel as far, and they are typically found near roads or other dust sources. The American Dust Bowl of the 1930s involved mainly this kind of particulates. At that time, farmland soils were often left bare, especially during severe drought, and billions of tons of topsoil blew away from farmlands. Soil conservation on farmlands is one strategy for reducing PM10; another strategy is better management of dust at construction sites.

Dust and aerosols are most noticeable in local concentrations, but they travel around the world. Satellite imagery from NASA shows particulates from dust storms, fires, and urban industrial centers circulating with weather systems far from their source (fig. 16.10). Dust

FIGURE 16.10 Global transport of aerosols and dust means that land changes and industrial activities are not isolated In one region. Colors show sources: dust (red), sea salt (blue), organic/black carbon (green), and sulfates (white). Visualization from NASA, https://svs.gsfc.nasa.gov/30637.
Source: Visualization from NASA's Goddard Space Flight Center/Global Modeling and Assimilation Office

from Africa's Sahara desert regularly crosses the Atlantic and raises particulate levels above federal health standards in Miami and San Juan, Puerto Rico. Amazon rainforests receive mineral nutrients carried in dust from Africa; more than half the 50 million tons of dust transported to South America each year has been traced to the bed of the former Lake Chad in Africa. In China, vast dust storms blow out of the Gobi desert every spring, choking Beijing and closing airports and schools in Japan and Korea. The dust plume follows the jet stream across the Pacific to Hawaii and then to the west coast of North America, where it sometimes makes up as much as half the particulate air pollution in Seattle, Washington. Some Asian dust storms have polluted the U.S. skies as far east as Georgia and Maine.

Mercury, from coal, is particularly dangerous

In addition to criteria pollutants or conventional pollutants, many other substances are regulated to protect public health and our environment. Standards for these pollutants continue to evolve, as do definitions of which pollutants require regulation. These changes reflect increases in certain pollutants, such as airborne mercury; the introduction of new pollutants, such as newly developed organic compounds; and increasing recognition of risks, as in the case of carbon dioxide.

Many toxic metals are released into the air by burning coal and oil, mining, smelting of metal ores, or manufacturing. Lead, mercury, cadmium, nickel, arsenic (a highly toxic metalloid), and others are released in the form of metal fumes or suspended particulates by fuel combustion, ore smelting, and burning garbage, coal, or other metal-laden materials. Among these, lead and mercury are the most abundantly produced toxic metals.

Mercury is a powerful neurotoxin that damages the brain and central nervous system at high doses. Minute amounts can cause nerve damage and developmental defects in children. About 75 percent of human exposure to mercury comes from eating fish. This is because aquatic bacteria are mainly responsible for converting airborne mercury (which falls or washes into water bodies) into methyl mercury, a form that is taken up and stored by living organisms. Methyl mercury accumulates in the tissues of fish, becoming especially concentrated in top predators. Large, long-lived, predatory fish contain especially high levels of mercury in their tissues, and these are most dangerous to eat. In a survey of freshwater fish from 260 lakes across the United States, the EPA found that every fish sampled contained some level of mercury.

Tuna fish alone is responsible for about 40 percent of all U.S. exposure to mercury (fig. 16.11). Swordfish, shrimp, and other seafood are also significant sources of mercury in our diet, and these species should be avoided. They also should be avoided because their populations are dwindling in many areas.

FIGURE 16.11 Airborne mercury accumulates in seafood, especially in top predators such as tuna.
Source: McGraw-Hill Education

Politics, Public Health, and the Minamata Convention

The dangers of mercury have been understood for decades, but action has been slow. In the United States, it wasn't until 1994 that the EPA declared mercury a hazardous pollutant, to be regulated under the Clean Air Act. Municipal and medical incinerators were required to reduce their mercury emissions by 90 percent within 5 years. Industrial and mining operations also agreed to cut emissions. However, the law did not address the 1,032 coal-burning power plants, which produced nearly half of total annual U.S. emissions.

In 2000, the EPA finally declared mercury from coal plants, like that from other sources, a public health risk. But the agency opted for a "cap and trade" market approach, expected to reduce mercury 70 percent in about 30 years, unlike the 90 percent reductions in 5 years required of other sources. Cap and trade approaches set limits (caps) and allow utilities to buy and sell unused pollution credits. This strategy uses a profit motive rather than rules, and it allows industries to find the cheapest reduction methods. It also allows continued emissions if credits are cheaper than emission controls. Public health advocates argue that this slow-going approach is inappropriate for a substance like mercury, which is toxic at very low levels, and which should be eliminated as quickly as possible.

Meanwhile, on the global stage, United Nations negotiators began working in earnest to control mercury by 2003, but negotiations stalled until 2009, with U.S. delegates insisting on voluntary emissions controls. Voluntary regulations are much like students writing their own tests and assigning their own grades. It works well for individual students, but it doesn't ensure that progress is being made.

In 2009, with a change in U.S. presidential administrations, progress started up again, and finally in 2013–14, delegates from 140 countries signed the Minamata Convention on Mercury. The agreement is named for the city of Minamata, Japan, which experienced tragic cases of severe mercury poisoning in the 1950s. Like many industries at the time, a chemical factory in Minamata regularly discharged mercury-laden waste into the bay. In Japan, where people consume a great deal of seafood, this meant that the local population was directly exposed to unusually concentrated levels of mercury. Babies whose mothers ate mercury-contaminated fish suffered profound neurological disabilities, including deafness, blindness, mental retardation, and cerebral palsy (fig. 1). In adults, mercury poisoning caused numbness, loss of muscle control, and dementia. The connection between "Minamata disease" and mercury was established in the 1950s, but waste dumping didn't end for another ten years.

The Minamata Convention establishes global rules for monitoring and reporting on mercury emissions. It calls for elimination of mercury in many uses where alternatives exist, such as household batteries and thermometers. It also requires control of mercury released in gold mining, where it is used to separate gold from ores and is often released into rivers with little or no regulation in developing areas. The convention also requires that countries monitor and limit pollution from coal-burning power plants.

The United States now falls under the international agreement, and standards will be monitored by the international community, which after all is affected by U.S. emissions. Subsequently, in 2016, the EPA issued rules requiring that most power plants capture most mercury from emissions, along with fine particulate matter.

The Minamata Convention would seem an easy agreement to reach because the dangers of mercury are well understood. But broad agreements to control diffuse and widespread pollutants are often difficult. Regulating air pollution often pits a diffuse public interest (improving general health levels or child development) against specific private interests (utilities that must pay millions of dollars per year to control pollutants).

If you were in charge, how would you set rules for controlling mercury emissions? Would you impose rules or allow for trading of mercury emission permits, or would you allow voluntary self-regulation? How would you negotiate the responsibility for controlling pollutants?

FIGURE 1 Minamata disease is a severe form of mercury poisoning, which causes nerve damage and other conditions. The international treaty to control mercury emissions was named after the city of Minamata, Japan.
Source: AP Photo

Global air circulation also deposits airborne mercury on land. Half or more of the mercury that falls on North America probably comes from Asian coal-burning power plants. Similarly, North American mercury travels to Europe. A 2009 report by the U.S. Geological Survey found that mercury levels in Pacific Ocean tuna have risen 30 percent in 20 years, with another 50 percent increase projected by 2050. Increased coal burning in China, which for years built new coal-burning power plants at the rate of one or two per week, is understood to be the main cause of growing mercury emissions in the Pacific.

The U.S. National Institutes of Health (NIH) estimates that one in 12 American women has more mercury in her blood than the 5.8 µg/l considered safe by the EPA. Between 300,000 and 600,000 of the 4 million children born each year in the United States are exposed in the womb to mercury levels that could cause diminished intelligence or developmental impairments. According to the NIH, elevated mercury levels cost the U.S. economy $8.7 billion each year in higher medical and educational costs and in lost workforce productivity. The EPA has been slow to regulate mercury, however (What Do You Think?, above).

Carbon dioxide, methane, and halogens are key greenhouse gases

Carbon dioxide (CO_2) is produced abundantly, mainly by living organisms and combustion. In general, CO_2 is nontoxic and innocuous, but atmospheric levels are steadily increasing due to human activities and are now causing global climate change, with serious implications for both human and natural communities (chapter 15).

Regulating CO_2 has been a subject of intense debate since the 1990s. On the one hand, policymakers have widely acknowledged that climate change is already causing disastrous effects. On the other hand, CO_2 is difficult to control because we produce it in so many ways, in manufacturing, food production, transportation, and heating our houses.

Methane (CH_4) is an important greenhouse gas that has over 30 times the global warming potential (GWP) per unit mass of CO_2. The GWP of nitrous oxide is nearly 300 times that of CO_2 (see table 15.3). Like CO_2, methane was not a primary consideration when the Clean Air Act was passed in 1970, but now these gases are dominant factors in climate change. Methane occurs naturally as organic matter, such as plant material, and decays in the absence of oxygen. Wetlands and rice paddies produce methane, as does anaerobic (oxygen-free) digestion by termites and ruminant animals such as cattle (fig. 16.12). Dramatic growth in livestock populations in recent decades and increasing rice production have sharply increased CH_4 concentrations in the atmosphere. Leaking natural gas wells and pipelines are also an important new source of atmospheric CH_4. Since the development of hydrofracturing (fracking) for natural gas and oil wells after about 2008, oil and gas producing regions around the world have introduced a great, but not well measured, amount of methane to the atmosphere (see chapter 15).

The federal government has been deeply divided over regulation of greenhouse gases. Many members of Congress have been intent on eliminating this and other pollution regulation, arguing that it is too costly for industry and the economy (see further discussion in section 16.4). Energy companies and their representatives, in particular, have lobbied to prevent legal limits on greenhouse gases. Meanwhile, public health and climate coalitions continue to push for controls.

The question of whether the EPA should regulate greenhouse gases was so contentious that it went to the Supreme Court in 2007. The Court ruled that it was the EPA's responsibility to limit these gases, on the grounds that greenhouse gases endanger public health and welfare within the meaning of the Clean Air Act. The Court, and subsequent EPA documents, noted that these risks include increased drought, more frequent and intense heat waves and wildfires, sea-level rise, and harm to water resources, agriculture, wildlife, and ecosystems. In addition to these risks, the U.S. military has cited climate change as a security threat. A coalition of generals and admirals signed a report from the Center for Naval Analyses stating that climate change "presents significant national security challenges" including violence

CO_2

N_2O

CH_4

FIGURE 16.12 Where do GHGs come from? Fossil fuel burning produces about 60 percent of GHG emissions, followed by deforestation (17 percent) and industrial and agricultural N_2O (14 percent). Fracking, rice paddies, belching livestock, and tropical dams produce CH_4 (9 percent of emissions).
Evgenii Bakhchev/Shutterstock; Larry Lee Photography/Corbis; David Wasserman/Brand X Pictures/PunchStock; Source: Brand X Pictures/ Punchstock; Charles Brutlag/Shutterstock; Hisham F. Ibrahim/Photodisc/Getty Images; ginosphotos/Getty Images; Jeffrey Schreier/Getty Images; G.K. & Vikki Hart/Photodisc/Getty Images; Barry Barker/ McGraw-Hill Education

resulting from scarcity of water, and migration due to sea-level rise and crop failure.

Since the Supreme Court ruling, the EPA is charged with regulating six greenhouse gases: carbon dioxide, methane, nitrous oxide, hydrofluorocarbons, perfluorocarbons, and sulfur hexafluoride. These are gases whose emissions have grown dramatically in recent decades.

Three of these six greenhouse gases contain halogens, a group of lightweight, highly reactive elements (fluorine, chlorine, bromine, and iodine). Because they are generally toxic in their elemental form, they are commonly used as fumigants and disinfectants, but they also have hundreds of uses in industrial and commercial products. Chlorofluorocarbons (CFCs) have been banned for most uses in industrialized countries, but about 600 million tons of these compounds are used annually worldwide in foams, and especially in air conditioners and refrigerators. They diffuse into the stratosphere, where they release chlorine and fluorine atoms that destroy ozone molecules that protect the earth from ultraviolet radiation (see section 16.2).

Halogen compounds are also powerful greenhouse gases: They trap more energy per molecule than does CO_2, and they persist in the atmosphere for decades to centuries. Perfluorocarbons will persist in the atmosphere for thousands of years. The global warming potential (per molecule, over time) of some CFCs is thousands of times greater than that of CO_2 (fig. 16.13).

Hazardous air pollutants (HAPs) can cause cancer and nerve damage

Although most air contaminants are regulated because of their potential adverse effects on human health or environmental quality, a special category of toxins is monitored by the U.S. EPA because they are particularly dangerous. Called **hazardous air pollutants (HAPs),** these chemicals include carcinogens, neurotoxins, mutagens, teratogens, endocrine system disrupters, and other highly toxic compounds (chapter 8). Most of these hazardous air pollutants are emitted from industrial sources, such as petrochemical (oil and gas) refineries, plastics and chemical manufacturers, and paper mills (see fig. 16.9).

EPA regulations have cut emissions of HAPs by nearly 70 percent. The most persistent compounds require special reporting and management because they remain in ecosystems for long periods of time and accumulate in animal and human tissues. The tendency to bioaccumulate makes many of these hazardous air pollutants especially dangerous. Most of these chemicals are either metal compounds, chlorinated hydrocarbons, or volatile organic compounds. Gasoline vapors, solvents, and components of plastics are all HAPs that you may encounter on a daily basis.

Only about 50 locations in the United States regularly measure concentrations of HAPs in ambient air. Often, the best source of information about these chemicals is the **Toxic Release Inventory (TRI)** collected by the EPA as part of the community right-to-know program. Established by Congress in 1986,

FIGURE 16.13 Global warming potential of several CFCs over 100 years, compared to CO_2. Methane (CH_4) and nitrous oxide (N_2O) have a 100-year GWP of about 30 and 300, respectively.
Source: Carbon Dioxide Information Analysis Center, 2011

the TRI requires 23,000 factories, refineries, hard rock mines, power plants, and chemical manufacturers to report on toxin releases (above certain minimum amounts) and waste management methods for 667 toxic chemicals. Although this total is less than 1 percent of all chemicals registered for use, and represents a limited range of sources, the TRI is widely considered the most comprehensive source of information about toxic pollution in the United States.

Most HAP releases are decreasing, but discharges of mercury and dioxins—both of which bioaccumulate and are toxic at extremely low levels—have increased in recent years. Dioxins are created mainly by burning plastics and medical waste containing chlorine. The EPA reports that 100 million Americans live in areas where the cancer rate from HAPs exceeds 10 in 1 million, or ten times the normally accepted standard for action. Benzene, formaldehyde, acetaldehyde, and 1,3 butadiene are responsible for most of this HAP cancer risk. To help residents track local air quality levels, the EPA has estimated the concentration of HAPs in localities across the continental United States (over 60,000 census tracts). You can access this information on the Environmental Defense Fund web page at http://scorecard.goodguide.com/env-releases/hap/us.tcl.

Indoor air can be worse than outdoor air

We have spent a considerable amount of effort and money to control the major outdoor air pollutants, but we have only recently begun to address indoor air pollutants. The EPA has found that concentrations of toxic air pollutants are often higher indoors than outdoors. Because, people generally spend more time inside than out, they are also exposed to higher doses of these pollutants.

In some cases, indoor air in homes has concentrations of chemicals that would be illegal outside or in the workplace. The EPA has found that concentrations of such compounds as chloroform, benzene, carbon tetrachloride, formaldehyde, and styrene can be 70 times higher in indoor air than in outdoor air, as plastics, carpets, paints, and other common materials off-gas these compounds. Finding less-toxic paints and fabrics can make indoor spaces both healthier and more pleasant.

In developing regions, smoke from wood and charcoal cooking fires is the greatest source of indoor air pollution (fig. 16.14). The World Health Organization (WHO) estimates that 2.5 billion people—over a third of the world's population—are adversely affected by pollution from this source. Women and small children spend long hours each day around open fires or unventilated stoves in enclosed spaces. Levels of carbon monoxide, particulates, aldehydes, and other toxic chemicals can be 100 times higher than would be legal for outdoor ambient concentrations in the United States. Designing and building cheap, efficient, non-polluting energy sources for the developing countries would not only save shrinking forests but would make a major impact on health as well.

FIGURE 16.14 Smoky cooking and heating fires may cause more ill health effects than any other source of indoor air pollution except tobacco smoking. Some 2.5 billion people, mainly women and children, spend hours each day in poorly ventilated kitchens and living spaces where carbon monoxide, particulates, and cancer-causing hydrocarbons often reach dangerous levels.
Source: China Tourism Press/Stone/Getty Images

Section Review

1. Define *primary air pollutants* and *secondary air pollutants*.
2. What are the six criteria pollutants in the original Clean Air Act? Why were they chosen?
3. List several additional hazardous air toxins that are regulated.

16.2 ATMOSPHERIC PROCESSES

- *Inversions trap still, contaminated air.*
- *Long-distance transport distributes pollutants.*
- *Reactive chlorine destroys stratospheric ozone.*

Topography, climate, and physical processes in the atmosphere play an important role in the transport, concentration, dispersal, and removal of many air pollutants. Cities concentrate dust and pollutants in urban "dust domes"; winds cause mixing between air layers. Precipitation and atmospheric layering also influence air quality. For example, a rain storm or a high-pressure system can transform air quality from one day to the next (fig. 16.15). We examine some of these factors in this section.

Temperature inversions trap pollutants

Normally, air near the ground is warmed in the daytime, as the ground absorbs solar energy and radiates it into the lower layers of air. As air near the ground warms, it expands, becomes less dense, and rises, causing turbulence or winds that circulate air. Circulation also removes pollutants generated near the ground in a city, so it is important to keeping a city livable. Sometimes, however, **temperature inversions** occur. At night or on cold days, air near the ground cools, or cold air may sink down into a valley from surrounding hills. A stable layer of cold air settles near the ground, while warmer air sits above. In these stable conditions, pollutants can accumulate to very high concentrations. Inversions might last from a few hours to a few days, until the weather warms or a warm front moves in.

The most stable inversion conditions are usually created by rapid nighttime cooling in a valley or basin where air movement is restricted. Los Angeles is a classic example, with conditions that create both temperature inversions and photochemical smog (fig. 16.16). The city is surrounded by mountains on three sides and the climate is dry, with abundant sunshine for photochemical oxidation and ozone production. Millions of automobiles and trucks create high pollution levels. Skies are generally clear at night, allowing heat to radiate from the ground. The ground and the lower layers of air cool quickly at night, while upper air layers remain relatively warm. During the night, cool, humid, onshore breezes also slide in under the contaminated air, which is trapped by a wall of mountains to the east and by the cap of warmer air above.

Morning sunlight is absorbed by the concentrated aerosols and gaseous chemicals caught near the ground by the inversion.

FIGURE 16.15 Air quality can change dramatically when a shift in the weather brings fresh air to a city. Here, Shanghai is shown before and after a cold front brought in clean air, pushing smog out of the city.
William P. Cunningham

This complex mixture quickly cooks up a toxic brew of hazardous compounds. As the ground warms later in the day, convection currents break up the temperature gradient and pollutants are carried back down to the surface, where more contaminants are added. Nitric oxide (NO) from automobile exhaust is oxidized to a brownish haze of nitrogen dioxide (NO₂, see

FIGURE 16.16 Atmospheric temperature inversions occur where ground-level air cools more quickly than air at upper levels. This temperature differential prevents mixing and traps pollutants close to the ground.

fig. 16.8). As nitrogen oxides are used up in reactions with unburned hydrocarbons, the ozone level begins to rise. By early afternoon an acrid brown haze fills the air, making eyes water and throats burn. In the 1970s, before pollution controls were enforced, afternoon concentrations of NO_x and ozone in the Los Angeles basin often would reach levels hazardous to health (see fig. 16.2). Usually, inversions are minor nuisances for healthy people. Sometimes, however, they can have dramatic effects (Exploring Science, p. 367).

Wind currents carry pollutants worldwide

Dust and contaminants can be carried great distances by the wind. Areas downwind from industrial complexes often suffer serious contamination, even if they have no pollution sources of their own (fig. 16.17). Pollution from the industrial belt between the Great Lakes and the Ohio River Valley, for example, regularly contaminates the Canadian Maritime Provinces, and sometimes can be traced as far as Ireland. As noted earlier, long-range transport is a major source of Asian mercury in North America.

Studies of air pollutants over southern Asia reveal a 3-km-thick toxic cloud of ash, acids, aerosols, dust, and photochemical reactants that regularly covers the entire Indian subcontinent and can last for much of the year. Produced by forest fires, the burning of agricultural wastes, and fossil fuels, the Asian smog layer cuts

EXPLORING SCIENCE

The Great London Smog and Pollution Monitoring

How do pollution control rules originate? Usually in a crisis too big to ignore. London, for example, was once legendary for its pea-soup fogs. In the days of Charles Dickens and Sherlock Holmes, darkened skies and blackened buildings, saturated with soot from hundreds of thousands of coal-burning fireplaces, were a fact of life. Londoners had been accustomed to filthy air since the Industrial Revolution, but over a period of four days in 1952 they experienced the worst air pollution disaster on record, which helped dramatically change approaches to air pollution.

On a cold afternoon in early December 1952, a dense blanket of coal smoke and fog settled on the city. Because it was a cold day, home coal burners and industrial furnaces began pumping out smoke at full force. During the afternoon, visibility plummeted and traffic came to a halt as drivers were blinded by the smoke and fog. Hundreds of cattle at a cattle market were the first to go. With lungs blackened by soot, they suffocated while standing in their pens. Concerts were canceled because of blackened air in the halls, and books in the British Museum were tainted with

soot. Visibility fell to one foot in some places by the third day of the inversion (fig. 1a).

Thousands died, especially the ill and elderly. Hospitals filled with victims of bronchitis, pneumonia, lung inflammations, and heart failure. Patients' lungs were clogged by smoke and microscopic soot particles, their lips turned blue, and many asphyxiated due to lack of oxygen.

A temperature inversion initiated the crisis. Normally, atmospheric circulation moves polluted air away from the city and out over the countryside. Stable inversions, with cold air resting near the ground, can be unpleasantly chilly and damp. In this case, the inversion trapped coal dust particulates and tiny droplets of sulfuric acid (from sulfur in coal) in the city.

After four days, a change in the weather brought fresh winds into London, and the inversion dissipated. At least 4,700 deaths were attributable to air pollution during and immediately after the inversion. More-recent studies have found that lingering ailments killed perhaps another 8,000 in the months that followed, bringing the total death toll to over 12,000.

London had been debating what to do about air pollution for centuries, since at

least 1300, when controls on smoke were attempted (and failed). Pollution was normal, too pervasive to change. The crisis of 1952 pushed the city to come up with new policies. Coal fireplaces were phased out, replaced with oil burners. In 1956, Parliament passed a Clean Air Act, and in 1968 the act was strengthened and expanded to address additional industrial emissions. The United States, watching closely, adopted a Clean Air Act in 1963, with major amendments in 1970.

These laws would not be possible without good quality air monitoring data. As with any environmental policy, regular and reliable collection of data is a first requirement for setting emissions limits. Atmospheric chemists can then compare observed sulfur dioxide levels, for example, to acceptable health limits (fig. 1b). Epidemiologists can tie rates of hospitalization and deaths to high levels of sulfur dioxide in the air. Monitoring data make it possible to justify and legally defend emissions limits. Legal enforcement of rules is the main reason the events of London in 1952 are now, for many places, fading into history.

(a)

(b)

FIGURE 1 Sulfur dioxide was a major cause of deaths during the London smog of December 1952 (a). During this event, sulfur dioxide levels reached 2 mg/m³ of air, well above the EPA standard limit of 0.08 mg/m³ (lower dashed line (b)).

a. *Source: Central Press/Hulton Archive/Getty Images*

(header photo): David L. Hansen

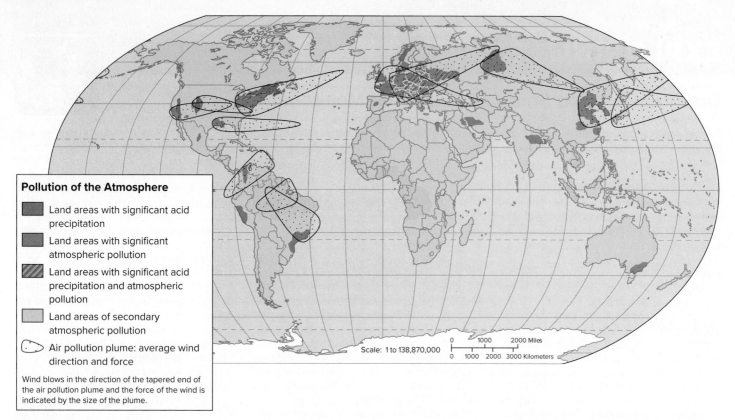

Pollution of the Atmosphere

- Land areas with significant acid precipitation
- Land areas with significant atmospheric pollution
- Land areas with significant acid precipitation and atmospheric pollution
- Land areas of secondary atmospheric pollution
- Air pollution plume: average wind direction and force

Wind blows in the direction of the tapered end of the air pollution plume and the force of the wind is indicated by the size of the plume.

Scale: 1 to 138,870,000

0   1000   2000 Miles
0 1000 2000 3000 Kilometers

FIGURE 16.17 Long-range transport carries air pollution from source regions thousands of kilometers away into formerly pristine areas. Secondary air pollutants can be formed by photochemical reactions far from primary emissions sources.

by up to 15 percent the amount of solar energy reaching the earth's surface beneath it. Meteorologists suggest that the cloud—80 percent of which is human-made—could disrupt monsoon weather patterns and may be disturbing rainfall and reducing rice harvests over much of South Asia. As UN Environment Programme executive director Klaus Töpfer said, "There are global implications because a pollution parcel like this, which stretches 3 km high, can travel half way round the globe in a week."

An increase in monitoring activity has revealed industrial contaminants in places usually considered among the cleanest in the world. Samoa, Greenland, Antarctica, and the North Pole all have heavy metals, pesticides, and radioactive elements in their air. A process called "grasshopper" transport, or atmosphere distillation, helps deliver contaminants to the poles. Volatile compounds evaporate from warm areas, travel through the atmosphere, and then condense and precipitate in cooler regions (fig. 16.18). Over several years, contaminants accumulate in the coldest places, generally at high latitudes where they bioaccumulate in food chains. Whales, polar bears, sharks, and other top carnivores in polar regions have been shown to have dangerously high levels of pesticides, metals, and other HAPs in their bodies. The Inuit people of Broughton Island, well above the Arctic Circle, have higher levels of polychlorinated biphenyls (PCBs) in their blood than any other known population except victims of industrial accidents. Far from any source of this industrial by-product, these people accumulate PCBs from the flesh of fish, caribou, and other animals they eat.

Chlorine destroys ozone in the stratosphere

Ozone near the ground is a pollutant because it irritates living tissues and contributes to smog. Far out in the stratosphere, however, ozone absorbs harmful ultraviolet (UV) solar energy, high-energy wavelengths that can damage plant and animal tissues, including the eyes and the skin. **Stratospheric ozone** provides a critical protective shield against UV radiation.

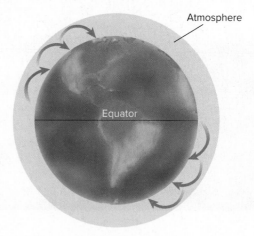

Atmosphere

Equator

FIGURE 16.18 Air pollutants evaporate from warmer areas and then condense and precipitate in cooler regions. Eventually, this "grasshopper" redistribution leads to accumulation in the Arctic and Antarctic.
Source: NASA

In 1985, the British Antarctic Atmospheric Survey announced a startling and disturbing discovery: Stratospheric ozone concentrations over the South Pole were dropping precipitously during September and October every year as the sun reappeared at the end of the long polar winter (fig. 16.19). This ozone depletion has been occurring at least since the 1960s but was not recognized because earlier researchers programmed their instruments to ignore changes in ozone levels that were presumed to be erroneous.

Chlorine-based aerosols, especially **chlorofluorocarbons (CFCs)** and other halon gases, are the principal agents of ozone depletion. Nontoxic, nonflammable, chemically inert, and cheaply produced, CFCs were extremely useful as industrial gases and in refrigerators, air conditioners, Styrofoam inflation, and aerosol spray cans for many years. From the 1930s until the 1980s, CFCs were used all over the world and widely dispersed through the atmosphere.

What we often call an ozone "hole" is really a vast area of reduced concentrations of ozone in the stratosphere. The thinning happens in the early Antarctic spring: Antarctica's exceptionally cold winter temperatures (–85 to –90°C) help break down ozone. During the long, dark winter months, strong winds circle the pole, isolating Antarctic air and allowing stratospheric temperatures to drop low enough to create ice crystals at high altitudes—something that rarely happens elsewhere in the world. Ozone and chlorine-containing molecules are absorbed on the surfaces of these ice particles. When the sun returns in the spring, it provides energy to liberate chlorine ions, which readily bond with ozone,

FIGURE 16.20 Each chlorine atom in the stratosphere can bond with many ozone (O_3) molecules, one after another, converting them to oxygen (O_2) molecules.

breaking it down to molecular oxygen (fig. 16.20). It is only during the Antarctic spring (September through December) that conditions are ideal for rapid ozone destruction. During that season, temperatures are still cold enough for high-altitude ice crystals, but the sun gradually becomes strong enough to drive photochemical reactions.

As the Antarctic summer arrives, temperatures moderate, the circumpolar vortex breaks down, and air from warmer latitudes mixes with Antarctic air, replenishing ozone concentrations in the ozone hole. Slight decreases worldwide result from this mixing, however. Ozone re-forms naturally, but not nearly as fast as it is destroyed. Because the chlorine atoms are not themselves consumed in reactions with ozone, they continue to destroy ozone for years. Eventually, they can precipitate out, but this process happens very slowly in the stable stratosphere.

About 10 percent of all stratospheric ozone worldwide has been destroyed in recent years, and levels over the Arctic have averaged 40 percent below normal. Ozone depletion has been observed over the North Pole as well, although it is not as concentrated as that in the south.

The Montreal Protocol was a resounding success

The discovery of stratospheric ozone losses brought about a remarkably quick international response. In 1987, an international meeting in Montreal, Canada, produced the **Montreal Protocol**, the first of several major international agreements on phasing out most use of CFCs by 2000. As evidence accumulated, showing that losses were larger and more widespread than previously thought, the deadline for the elimination of all CFCs (halons, carbon tetrachloride, and methyl chloroform) was moved up to 1996, and a $500 million fund was established to assist poorer countries in switching to non-CFC technologies. Fortunately, alternatives to CFCs for most uses already exist. The first substitutes are

FIGURE 16.19 The region of stratospheric ozone depletion grew steadily to an area of nearly 30 million km² in 2006 (shown here). This ozone "hole" has shown signs of decline since the Montreal Protocol went into effect.
Source: NASA

hydrochlorofluorocarbons (HCFCs), which release much less chlorine per molecule. These HCFCs are also being phased out, as newer halogen-free alternatives are developed.

The Montreal Protocol is often cited as the most effective international environmental agreement ever established. Global CFC production has been cut by more than 95 percent since 1988 (fig. 16.21). Some of that has been replaced by HCFCs, which release chlorine, but not as much as CFCs. The amount of chlorine entering the atmosphere already has begun to decrease.

The size of the ozone "hole" increased steadily from its discovery until the mid-1990s, when the Montreal Protocol began having an effect. Since then, it has varied from year to year, but it has gradually gotten smaller, and stratospheric O_3 levels should be back to normal by about 2049.

The Montreal Protocol had an added benefit in slowing climate change, because CFCs and other ozone-destroying gases are also powerful greenhouse gases (see fig. 16.12). Reductions in emissions of these gases under the Montreal Protocol amount to one-quarter of all greenhouse gas emissions worldwide. This reduction has had a greater impact on climate-changing gases than the Kyoto Protocol.

In 2016, another little-recognized but landmark step was made in the **Kigali Amendment** to the 1987 Montreal protocol. At a meeting in Kigali, Rwanda, the world's countries agreed to accelerate the phase-out of hydrofluorcarbons (HFCs, see fig. 16.13), and to replace them with climate-friendly alternatives. While the Montreal Protocol targets ozone protection, its climate impact is increasingly recognized. The Kigali negotiations were headed by African countries that face dire threats from climate change. This step alone could prevent 0.5 degrees of warming by 2100 and is probably the world's biggest single accomplishment toward both climate and ozone protection.

Section Review

1. What is an atmospheric temperature inversion, and how can it concentrate pollutants in a city?
2. What is the difference between ambient and stratospheric ozone? What is destroying stratospheric ozone?
3. What did the Montreal Protocol aim to accomplish?

16.3 EFFECTS OF AIR POLLUTION

- *Lung tissues are vulnerable to acids and particulates.*
- *Sulfur and nitrogen produce acidic deposition.*

Air pollution is a problem of widespread interest because it affects so many parts of our lives. The most obvious effects are on our health. Damage to infrastructure, vegetation, and aesthetic quality—especially visibility—are also important considerations.

The World Health Organization estimates that some 5 to 6 million people die prematurely every year from illnesses related to air pollution. Heart attacks, respiratory diseases, and lung cancer all are significantly higher in people who breathe dirty air, compared to matching groups in cleaner environments. Residents of the most polluted cities in the United States, for example, are 15 to 17 percent more likely to die of these illnesses than those in cities with the cleanest air. This can mean as much as a five- to ten-year decrease in life expectancy for those who live in the worst parts of Los Angeles or Baltimore, compared to a place with clean air. Of course, the likelihood of suffering ill health from air pollutants depends on the intensity and duration of exposure, as well as the age of the person and their prior health status.

In industrialized countries, one of the biggest health threats from air pollution is particulate matter less than 2.5 micrometers in diameter. These particulates have been linked to heart attacks, asthma, bronchitis, lung cancer, immune suppression, and abnormal fetal development, among other health problems (fig. 16.22). Fine particulates have many sources. Power plants, especially coal plants, are usually the largest source, except where clean air rules require filters and precipitators to capture particulates. The U.S. EPA estimates that at least 160 million Americans—more than half the population—live in areas with unhealthy concentrations of fine particulate matter. PM2.5 levels have decreased about 30 percent over the past 25 years, but health conditions will improve if we can make further reductions.

Diesel engines have long been a major source of both soot and SO_2 in the United States. Until 2006, diesel-powered engines were allowed to use fuel that was up to 3,400 parts per million sulfur. In 2006, the allowable limit was cut to 500 ppm sulfur, and by 2010, the EPA phased in rules requiring that trucks, buses, and cars use ultra-low-sulfur fuel, with sulfur concentrations of less than 15 ppm. This dramatic reduction in sulfur also allows the use of advanced emission control technology in engines, further reducing health risks and odors of diesel emissions. These standards now

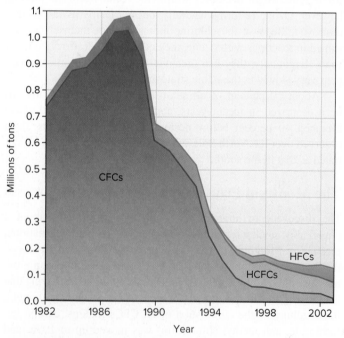

FIGURE 16.21 The Montreal Protocol has been remarkably successful in eliminating CFC production. The remaining HFC and HCFC use is primarily in developing countries, such as China and India.

FIGURE 16.22 Residents of many Chinese cities know to protect themselves. Soot and fine particulates from diesel engines, power plants, and other combustion sources have been linked to diseases including asthma and heart attacks.
Hung_Chung_Chih/Getty Images

also apply to off-road vehicles, such as tractors, bulldozers, locomotives, and barges, whose engines previously emitted more soot than all the nation's cars, trucks, and buses together.

How does pollution make us sick?

The most common route of exposure to air pollutants is by inhalation, but direct absorption through the skin or contamination of food and water also are important pathways. Because they are strong oxidizing agents, sulfates, SO_2, NO_x, and O_3 act as irritants that damage delicate tissues in the eyes and respiratory passages. Fine particulates, irritants in their own right, penetrate deep into the lungs and carry metals and other HAPs on their surfaces. Inflammatory responses set in motion by these irritants impair lung function and trigger cardiovascular problems as the heart tries to compensate for lack of oxygen by pumping faster and harder. If the irritation is really severe, so much fluid seeps into the lungs through damaged tissues that the victim actually drowns.

Carbon monoxide binds to hemoglobin and decreases the ability of red blood cells to carry oxygen. Asphyxiants such as this cause headaches, dizziness, and heart stress, and can be lethal if concentrations are high enough. Lead also binds to hemoglobin, reducing its oxygen-carrying capacity at high levels. At lower levels, lead causes long-term damage to critical neurons in the brain that results in mental and physical impairment.

Bronchitis and emphysema are two common, chronic health effects of air pollutants. Bronchitis is a persistent inflammation of bronchi and bronchioles (large and small airways in the lung) that causes mucus buildup, a painful cough, and involuntary muscle spasms that constrict airways. Severe bronchitis can lead to emphysema, an irreversible **chronic obstructive pulmonary disorder (COPD)** in which airways become permanently constricted and alveoli are damaged or even destroyed. COPD is one of the world's leading causes of death.

Sulfur and nitrogen emissions produce acid rain

Uncontrolled industrial fumes from furnaces, smelters, refineries, and chemical plants destroy vegetation and have created desolate, barren landscapes around mining and manufacturing centers. The copper–nickel smelter at Sudbury, Ontario, is a spectacular and notorious example of air pollution effects on vegetation and ecosystems. In 1886, the corporate ancestor of the International Nickel Company (INCO) began open-bed roasting of sulfide ores at Sudbury. Sulfur dioxide and sulfuric acid released by this process destroyed nearly all plant life within about 30 km of the smelter. Rains washed away the exposed soil, leaving a barren moonscape of blackened bedrock (fig. 16.23). Super-tall, 400-m smokestacks were installed in the 1950s, and sulfur scrubbers were added 20 years later. Emissions were reduced by 90 percent and the surrounding ecosystem is beginning to recover.

Most people in the United States became aware of problems associated with **acid deposition** (the deposition of wet acidic solutions or dry acidic particles from the air) in recent decades, but English scientist Robert Angus Smith coined the term *acid rain* in his studies of air chemistry in Manchester, England, in the 1850s. By the 1940s, it was known that pollutants, including atmospheric acids, could be transported long distances by wind currents. This was thought to be only an academic curiosity until it was shown that precipitation of these acids can have far-reaching ecological effects.

We describe acidity in terms of pH (see chapter 3). Values below pH 7 are acidic, while those above pH 7 are basic. Normal, unpolluted rain generally has a pH of about 5.6 due to carbonic acid created by CO_2 in air. Sulfur, chlorine, and other elements also form acidic compounds as they are released in sea spray, volcanic emissions, and biological decomposition. These sources can lower the pH

FIGURE 16.23 In 1975, acid precipitation from the copper–nickel smelters (tall stacks in background) had killed all the vegetation and charred the pink granite bedrock black for a large area around Sudbury, Ontario.
William P. Cunningham

of rain well below 5.6. Other factors, such as alkaline dust can raise it above 7. In industrialized areas, anthropogenic acids in the air usually far outweigh those from natural sources.

Acid deposition is one of our important success stories in pollution control. A few decades ago, it was a critical issue for many U.S. ecosystems and cities. Lawsuits led to regulations that reduced sulfur dioxide emissions, mainly from from coal-burning power plants and factories. As a consequence, acid deposition has declined dramatically (fig. 16.24). Residual issues persist in many areas, however, and rapidly industrializing regions of the developing world are now experiencing these effects on aquatic systems, forests, and urban infrastructure.

Acid deposition damages ecosystems and infrastructure

Lakes and streams can be especially sensitive to acid deposition, especially where vegetation or bedrock makes them naturally acidic to start with. This problem was first publicized in Scandinavia, which receives industrial and automobile emissions—principally H_2SO_4 and HNO_3—generated in northwestern Europe.

1986

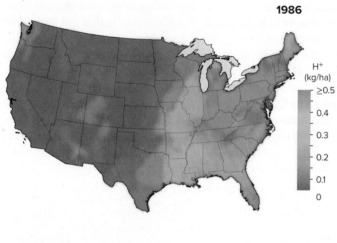

H⁺
(kg/ha)

- ≥0.5
- 0.4
- 0.3
- 0.2
- 0.1
- 0

2013

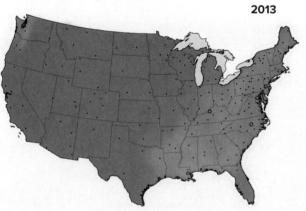

FIGURE 16.24 Acid precipitation over the United States in 1986 and 2013. Controls on sulfur dioxide emissions have been one of the greatest successes in pollution control.

Source: National Atmospheric Deposition Program/National Trends Network http://nadp.isws.illinois.edu

The thin, acidic soils and oligotrophic lakes and streams in the mountains of southern Norway and Sweden have been severely affected by this acid deposition. Some 18,000 lakes in Sweden are now so acidic that they will no longer support game fish or other sensitive aquatic organisms.

Generally, reproduction is the most sensitive stage in fish life cycles. The eggs and young of many species are killed when the pH drops to about 5.0. This level of acidification also can disrupt the food chain by killing aquatic plants, insects, and invertebrates on which fish depend for food. At pH levels below 5.0, adult fish die as well. Trout, salmon, and other game fish are usually the most sensitive.

Some 48,000 lakes in Ontario are currently endangered, and nearly all of Quebec's surface waters, including about 1 million lakes, are believed to be highly sensitive to acid deposition.

Impacts on forests became widely recognized in the early 1980s, following alarming reports of rapid forest declines in both Europe and North America. One of the earliest was a detailed ecosystem inventory on Camel's Hump Mountain in Vermont. A 1980 survey showed that seedling production, tree density, and the viability of spruce-fir forests at high elevations had declined about 50 percent in 15 years. A similar situation was found on Mount Mitchell in North Carolina, where almost all red spruce and Fraser fir above 2,000 m (6,000 ft) are in a severe decline. Nearly all the trees are losing needles and about half of them are dead (fig. 16.25). The stress of acid rain and fog, other air pollutants, and attacks by an invasive insect called the woody aldegid are killing the trees.

In the longest-running forest-ecosystem monitoring record in North America, researchers at the Hubbard Brook Experimental Forest in New Hampshire have shown that forest soils have become depleted of natural buffering reserves of basic cations such as calcium and magnesium through years of exposure to acid

FIGURE 16.25 A Fraser fir forest on Mount Mitchell, North Carolina, killed by acid rain, insect pests, and other stressors.
William P. Cunningham

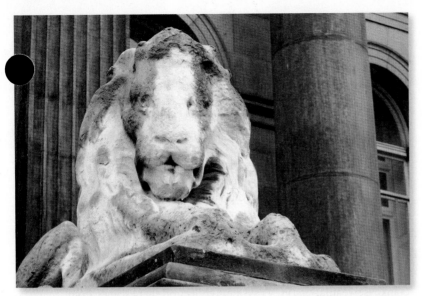

FIGURE 16.26 Atmospheric acids, especially sulfuric and nitric acids, have almost completely eaten away the face of this medieval statue. Each year, the total loss from air pollution damage to buildings and materials amounts to billions of dollars.
William P. Cunningham

rain. Replacement of these cations by hydrogen and aluminum ions seems to be one of the main causes of plant mortality.

Often it is infrastrucure damage, especially to limestone monuments, that is most obvious to us. In cities throughout the world, some of the oldest and most glorious buildings and works of art are being destroyed by air pollution. Smoke and soot coat buildings, paintings, and textiles. Limestone and marble are destroyed by atmospheric acids at an alarming rate. The Parthenon in Athens, the Taj Mahal in Agra, the Colosseum in Rome, frescoes and statues in Florence, medieval cathedrals in Europe (fig. 16.26), and the Lincoln Memorial and Washington Monument in Washington, D.C., are slowly dissolving and flaking away because of acidic fumes in the air. Medieval stained glass windows in Cologne's gothic cathedral are so porous from etching by atmospheric acids that pigments disappear and the glass literally crumbles away. Restoration costs for this one building alone are estimated at up to 3 billion euros.

Section Review

1. List several illnesses that are made worse by dirty air.
2. Explain the idea of "synergistic effects."
3. What is acid deposition? Identify two of the pollutants that cause it.

16.4 POLLUTION CONTROL

- *Pollutants can be captured after production or avoided.*
- *Clean air legislation has reduced CO and SO_4 but not NO_x.*
- *Data show that the CAA has saved money and has not diminished economic growth.*

"Dilution is the solution to pollution" was our main approach to air pollution control for most of history. Tall smokestacks were built to send emissions far from the source, where they became unidentifiable and largely untraceable. But dispersed and diluted pollutants are now the source of some of our most serious pollution problems. We are finding that there is no "away" to which we can throw our waste products. Although most of the discussion in this section focuses on industrial solutions, each of us can make important personal contributions to this effort (What Can You Do?, below).

Because most air pollution in the developed world is associated with transportation and energy production, the most effective strategies involve conservation and renewable energy. Reducing electricity consumption, insulating homes and offices, and developing better public transportation could all greatly reduce air pollution. Alternative energy sources, such as wind and solar power, produce energy with little or no pollution, and these and other technologies are becoming increasingly available (see chapter 20). In addition to conservation, pollution control technology can capture pollutants at the source.

Pollutants can be captured after combustion

Particulate removal involves filtering air emissions. Filters trap particulates in a mesh of cotton cloth, spun glass fibers, or asbestos-cellulose. Industrial air filters are generally giant bags 10 to 15 m long and 2 to 3 m wide. Effluent gas is blown through the bag, much like the bag on a vacuum cleaner. Every few days or weeks, the bags are opened to remove the dust cake. Electrostatic

What Can You Do?

Reducing Pollution and Saving Energy

- Reduce your electric power and fossil fuel demands: Aim to live where you can carpool, bike, walk, and use public transport. Buy energy-efficient appliances.
- Write to your representatives and ask them to support a transition to an energy-efficient, low-carbon economy.
- Avoid spray-can products. Household products expose you directly to hazardous pollutants.
- Buy clothes that can be washed rather than dry-cleaned: Dry cleaners are important sources of VOCs.
- Don't buy polluting two-cycle gasoline lawn mowers, boat motors, or other small engines.
- If green-pricing options are available in your area, buy renewable energy.
- If your home has a fireplace, install a high-efficiency, clean-burning, two-stage insert that conserves energy and reduces pollution up to 90 percent.
- Have your car tuned every 10,000 miles (16,000 km), and make sure that its anti-smog equipment is working properly. Turn off your engine when idling longer than one minute.

precipitators are the most common particulate controls in power plants. Ash particles pick up an electrostatic surface charge as they pass between large electrodes in the effluent stream (fig. 16.27). Charged particles then collect on an oppositely charged collecting plate. These precipitators consume a large amount of electricity, but maintenance is relatively simple, and collection efficiency can be as high as 99 percent. The ash collected by both of these techniques is a solid waste (often hazardous due to the heavy metals and other trace components of coal or other ash source) and must be buried in landfills or other solid-waste disposal sites.

Sulfur removal is important because sulfur oxides are among the most damaging of all air pollutants in terms of human health and ecosystem impacts. As with particulate matter, sulfur dioxide can be captured after combustion, or fuel (especially coal) can be cleaned before burning. Coal can be crushed, washed, and gasified to remove sulfur and metals before combustion. These processes cost money, but when sulfur is captured from fuels, it can also be sold as a useful industrial product. Elemental sulfur, sulfuric acid, and ammonium sulfate can all be produced using catalytic converters to oxidize or reduce sulfur.

Switching to low-sulfur fuels is the surest way to reduce sulfur emissions: from soft coal with a high sulfur content to low-sulfur coal, or from coal to oil or gas or renewable energy. We often use high-sulfur coal for political or economical reasons. In the United States, most high-sulfur coal is produced in Appalachia, a region with chronic poverty but substantial political influence. China, similarly, has long relied on sulfur-rich domestic coal. Both the United States and China have begun switching to lower-sulfur coal, and to cleaner oil or gas, but the change takes time. Sulfur emissions from transportation have been reduced by new rules requiring low-sulfur diesel fuel.

Nitrogen oxides (NO_x) can be reduced in both internal combustion engines and industrial boilers by as much as 50 percent by carefully controlling the flow of air and fuel. Staged burners, for example, control burning temperatures and oxygen flow to prevent formation of NO_x. The catalytic converter on your car uses platinum-palladium and rhodium metals to catalyze reactions that convert NO_x, unburned hydrocarbons, and carbon monoxide, to water, nitrogen gas, and carbon dioxide. Catalytic converters became required on new car engines in the United States in the 1980s, and EPA data show that new vehicles today emit 96 percent less carbon monoxide, 98 percent less hydrocarbons, and 90 percent less NO_x than new vehicles in the 1970s.

Hydrocarbon controls mainly involve complete combustion or controlling evaporation. Hydrocarbons and volatile organic compounds are produced by incomplete combustion of fuels or by solvent evaporation from chemical factories, paints, dry cleaning, plastic manufacturing, printing, and other industrial processes. Closed systems that prevent escape of fugitive gases can reduce many of these emissions. In automobiles, for instance, positive crankcase ventilation (PCV) systems collect oil that escapes from around the pistons and unburned fuel and channels them back to the engine for combustion. Controls on fugitive losses from industrial valves, pipes, and storage tanks can have a significant impact on air quality. Afterburners are often the best method for destroying volatile organic chemicals in industrial exhaust stacks.

Clean air legislation is controversial but effective

Since 1970, the Clean Air Act has been modified, updated, and amended many times. Amendments have involved acrimonious debate. As in the case of CO_2 restrictions, discussed earlier, victims of air pollution demand more protection, while industry and energy groups insist that controls are too expensive. Bills have sometimes languished in Congress for years because of disputes over burdens of responsibility, cost, and definitions of risk. Many states have argued that air pollution is a necessary consequence of economic growth—an argument not well supported by evidence (fig. 16.28). The EPA reports that simply by enforcing existing clean air legislation, the United States could prevent at least 6,000 deaths and 140,000 asthma attacks every year.

The most significant amendments were in the 1990 update, which addressed a variety of issues, including acid rain, urban air pollution, and toxic air emissions. These amendments also restricted ozone-depleting chemicals in accordance with the Montreal Protocol.

One of the most contested aspects of the act has been the "new source review," which was established in 1977. This provision was adopted because industry argued that it would be intolerably expensive to install new pollution-control equipment on old power plants and factories that were about to close down anyway. Congress agreed to "grandfather" existing equipment, or exempt it from new pollution limits, with the stipulation that when they were upgraded or replaced, more stringent rules would apply. The result was that owners have kept old facilities operating precisely because they were exempted from pollution control. In fact, corporations poured millions into aging power plants and factories, expanding their capacity, to avoid having to build new ones. Forty years later, many

FIGURE 16.27 An electrostatic precipitator can remove 99 percent of unburned particulates in the effluent (smoke) from power plants. Electrodes transfer a static electric charge to dust and smoke particles, which then adhere to collector plates. Particles are then shaken off of the plates and collected for reuse or disposal.

Clean flue gas

Separator plates

Particle-laden flue gas

Electrodes

Fly ash to the storage silo

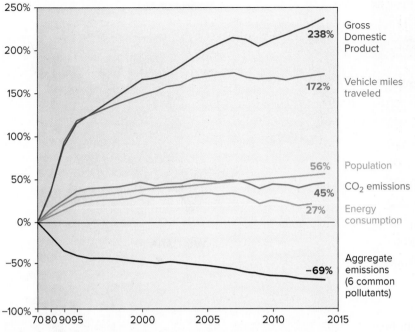

FIGURE 16.28 Decoupling of economic growth and emissions. Common assumptions that growth requires pollution has not been supported at least since the 1970s.
Source: EPA, 2016

FIGURE 16.29 Air pollution trends in the United States, 1970 to 2014. Since Congress passed the Clean Air Act , most criteria air pollutants decreased significantly, even while population has increased. Pollution protections also aid the economy by reducing illness and creating jobs in pollution prevention.
Source: EPA, 2016

of those grandfathered plants are still going strong and continue to be among the biggest contributors to smog and acid deposition.

In 2015, the Obama administration's EPA issued the Clean Power Plan, designed to accelerate reductions in CO_2, soot, and other pollutants. In addition, the plan would have saved an estimated $155 billion in health and pollution costs, and met 2030 U.S. climate protection targets. However, industry lawsuits blocked the plan's implementation, and the Trump EPA quickly rescinded it, and a newly conservative Supreme Court has blocked further challenges to the Trump EPA's actions.

Clean air protections help the economy and public health

Despite these disputes, the Clean Air Act has been extremely successful in saving money and lives. The EPA estimates that between 1970 and 2014, lead emissions fell 99 percent, SO_2 declined 84 percent, and CO shrank 72 percent (fig. 16.29). Filters, scrubbers, and electrostatic precipitators on power plants and other large stationary sources are responsible for most of the particulate and SO_2 reductions. Catalytic converters on cars are responsible for most of the CO and O_3 reductions. For 23 of the largest U.S. cities, air quality now reaches hazardous levels 93 percent less frequently than a decade ago. Forty of the 97 metropolitan areas that failed to meet clean air standards in the 1980s are now in compliance, many for the first time in a generation.

In a 2011 study of the economic costs and benefits of the 1990 Clean Air Act, the EPA found that the direct benefits of air quality protection by 2020 will be $2 trillion, while the direct costs of implementing those protections was about one-thirtieth of that, or $65 billion (fig. 16.30). The direct benefits were mainly in prevented costs of premature illness, death, and work losses (table 16.2). About half of the direct costs were improvements in cars and trucks, which now burn more cleanly and efficiently than they did in the past. This

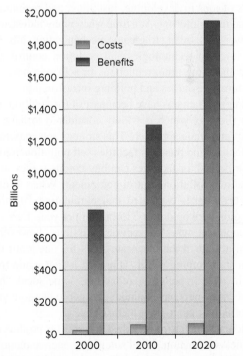

FIGURE 16.30 Direct costs and benefits of the Clean Air Act provisions by 2000, 2010, and 2020, in billions of 2006 U.S. dollars.
Source: EPA 2011, Clean Air Impacts Summary Report

Table 16.2 Reductions of Health Impairments Resulting from Ozone and Particulate Reductions Since 1990		
Health Effect Reductions (PM2.5 and Ozone Only)	Year 2010[1]	Year 2020[1]
Adult Mortality - particles	160,000	230,000
Infant Mortality - particles	230	280
Mortality - ozone	4,300	7,100
Chronic Bronchitis	54,000	75,000
Heart Disease	130,000	200,000
Asthma Exacerbation	1,700,000	2,400,000
Emergency Room Visits	86,000	120,000
School Loss Days	3,200,000	5,400,000
Lost Work Days	13,000,000	17,000,000

[1] Number of cases reduced

Source: EPA, 2011 Clean Air Impacts Summary Report

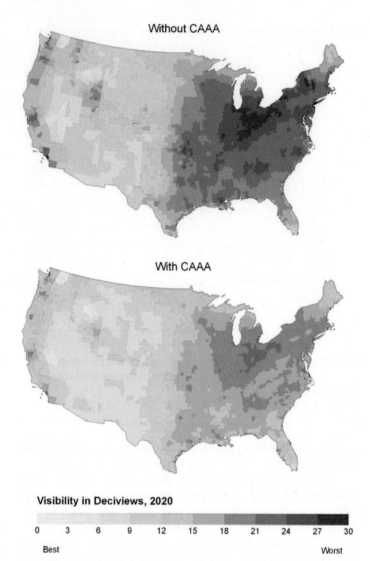

FIGURE 16.31 Projected visibility impairments, shown with dark colors, would be considerably worse in 2020 without the 1990 Clean Air Act amendments (CAAA, *top*) than they will be with the amendments (*bottom*). Units are deciviews, a measure of perceptible change in visibility.
Source: EPA 2011, Clean Air Impacts Summary Report

cost has been distributed to vehicle owners, who also benefit from lower expenditures on fuel. A quarter of costs involved cleaner furnaces and pollutant capture at electricity-generating power plants and other industrial facilities. The remaining costs involved pollution reductions at smaller businesses, municipal facilities, construction sites, and other sources. Overall, emission controls have not dampened economic productivity, despite widespread fears to the contrary. Emissions of criteria pollutants have declined in recent decades, whereas economic indicators have grown (see fig. 16.28).

In addition to these savings, the Clean Air Act has created thousands of jobs in developing, installing, and maintaining technology and in monitoring. At a time when many industries are providing fewer jobs, owing to greater mechanization, jobs have been growing in clean technologies and pollution control and monitoring. At the same time, reductions in acid rain have decreased losses to forest resources and building infrastructure.

Market mechanisms have been part of the solution, especially for sulfur dioxide, which is widely considered to have benefited from a cap-and-trade approach. This strategy sets maximum limits for each facility and then lets facilities sell pollution credits if they can cut emissions, or facilities can buy credits if they are cheaper than installing pollution-control equipment. When trading began in 1990, economists estimated that eliminating 10 million tons of sulfur dioxide would cost $15 billion per year. Left to find the most economical ways to reduce emissions, however, utilities have been able to reach clean air goals for one-tenth that price. A serious shortcoming of this approach is that while trading has resulted in overall pollution reduction, some local "hot spots" remain where owners have found it cheaper to pay someone else to reduce pollution than to do it themselves.

Particulate matter (mostly dust and soot) is produced by agriculture, fuel combustion, metal smelting, concrete manufacturing, and other activities. Industrial cities, such as Baltimore, Maryland, and Baton Rouge, Louisiana, also have continuing problems. Eighty-five other urban areas are still considered nonattainment

regions. In spite of these local failures, however, 80 percent of the United States now meets the National Ambient Air Quality Standards (fig. 16.31). This improvement in air quality is perhaps the greatest environmental success story in our history.

In developing areas, rapid growth can outpace pollution controls

The outlook is not so encouraging in many parts of the world. The major metropolitan areas of many developing countries are growing at explosive rates to incredible sizes (chapter 22), and environmental quality can be abysmal. Pollution has increased especially rapidly in the growing cities of Asia. Although Chinese cities brought global attention to the human cost of air pollution, those in South Asia have become even worse in recent years. Of the 20 cities with the worst levels of PM2.5 in 2018, only two were in China. The rest were in

India (15), Pakistan (2), or Bangladesh (1). A more complete list is available from the group IQAir and its website AirVisual.

In all these cities, leading causes of air pollution include coal-burning power plants and vehicle exhaust, along with industrial pollution. These sources have multiplied in recent years as demand for cars, trucks, and electricity has exploded, but rules controlling emissions have not kept pace.

Even when laws on the books are strong, enforcement is often weak. In China, despite legal restrictions, many of the 400,000 factories have no air pollution controls. Experts estimate that home coal burners and factories emit 10 million tons of soot and 15 million tons of sulfur dioxide annually. Shenyang, an industrial city in northern China, has reported particulate matter concentrations in winter of over 700 mg/m^3 (nine times U.S. maximum standards). Airborne particulates in Shenyang exceed WHO standards on 347 days per year.

Every year the organization Pure Earth (formerly the Blacksmith Institute) compiles a list of the world's worst-polluted places. Globally, smelters, mining operations, petrochemical industries—which release hazardous organic compounds to the air and water—and chemical manufacturing are frequently the worst sources of pollutants. Often, these are in impoverished and developing areas of Africa, Asia, or the Americas, where government intervention is weak and regulations are nonexistent or poorly enforced. Funds and political will are usually unavailable to deal with pollution, much of which is involved with materials going to wealthier countries or waste that is received from developed countries. You can learn more about these places at www.pureearth.org.

Norilsk, Russia (one site highlighted on Pure Earth's list of worst places), is a notorious example of toxic air pollution. Founded in 1935 as a prison labor camp, this Siberian city is considered one of the most polluted places on earth. Norilsk houses the world's largest nickel mine and heavy metals smelting complex, which discharge over 4 million tons of cadmium, copper, lead, nickel, arsenic, selenium, and zinc into the air every year. The snow turns black as quickly as it falls, the air tastes of sulfur, and the average life expectancy for factory workers is ten years below the Russian average (which already is the lowest of any industrialized country). Difficult pregnancies and premature births are far more common in Norilsk than elsewhere in Russia. Children living near the nickel plant are ill twice as much as Russia's average, and birth defects are reported to affect up to 10 percent of the population. Why do people stay in such a place? Many were attracted by high wages and hardship pay, and now that they're sick, they can't afford to move.

Air quality improves where controls are implemented

Despite global expansion of chemical industries and other sources of air pollution, there have been some spectacular successes in air pollution control. Sweden and West Germany (countries affected by forest losses due to acid precipitation) cut their sulfur emissions by two-thirds between 1970 and 1985. Austria and Switzerland

FIGURE 16.32 Cubatao, Brazil, was once considered one of the most polluted cities in the world. Better environmental regulations and enforcement, along with massive investments in pollution-control equipment, have improved air quality significantly.
William P. Cunningham

have gone even farther, regulating even motorcycle emissions. The Global Environmental Monitoring System (GEMS) reports declines in particulate levels in 26 of 37 cities worldwide. Sulfur dioxide and sulfate particles, which cause acid rain and respiratory disease, have declined in 20 of these cities.

Twenty years ago, Cubatao, Brazil, was described as the "Valley of Death," one of the most dangerously polluted places in the world. Every year, a steel plant, a huge oil refinery, and fertilizer and chemical factories churned out thousands of tons of air pollutants that were trapped between onshore winds and the uplifted plateau on which São Paulo sits (fig. 16.32). Trees died on the surrounding hills. Birth defects and respiratory diseases were alarmingly high. Since then, however, the citizens of Cubatao have made remarkable progress in cleaning up their environment. The end of military rule and the restoration of democracy allowed residents to publicize their complaints. The environment became an important political issue. The state of São Paulo invested about $100 million and the private sector spent twice as much to clean up most pollution sources in the valley. Particulate pollution was reduced 75 percent, ammonia emissions were reduced 97 percent, hydrocarbons that cause ozone and smog were cut 86 percent, and sulfur dioxide production fell 84 percent. Fish are returning to the rivers, and forests are regrowing on the mountains. It proves that progress is possible.

Section Review

1. How are sulfur and particulate matter removed from effluent?
2. What is the ratio of direct costs and benefits of the Clean Air Act? What costs are mainly saved?
3. Which conventional pollutants have decreased most and least?
4. What are some sources of air pollution in developing areas?

Connecting the Dots

As we can see in the opening case study, air pollution in China, as in many rapidly industrializing countries, is a severe and widely recognized problem. Citizens, officials, and industry recognize the costs. But without uniformly applied regulations, changing practices is not easy.

Air pollution causes millions of deaths every year in China, and millions more in other countries. Health impacts include respiratory diseases, birth defects, heart attacks, developmental disabilities in children, and cancer. Environmental impacts include the destruction of stratospheric ozone, the poisoning of forests and waters by acid rain, and corrosion of building materials.

Solutions to these problems are well known, and in countries where air quality standards are enforced, conditions have widely improved. Many people aren't aware of how much worse air quality was in the industrial centers of North America and Europe a century or two ago compared to today. Cities such as London, Pittsburgh, Chicago, Baltimore, and New York had air quality as bad as or worse than most cities in the developing world now. The progress in reducing air pollution in these cities gives us hope that residents can do so elsewhere as well.

Global agreements, as well as national clean air policies, have made important progress. The success of the Montreal Protocol in eliminating CFCs is a landmark in international cooperation on an environmental problem. The Kigali Amendment to this protocol is also path-breaking, as it provides protections for the climate as well as for stratospheric ozone. The Minamata Convention to control mercury is another example of important global cooperation.

Developing areas face severe challenges in air quality. Most of the worst air pollution in the world occurs in large cities of developing countries. As countries like China transition to developed states, public pressure to improve the environment—as seen in the online map in figure 16.1, becomes a stronger force for action. The world has a long way to go in protecting air quality, but the methods for doing so are known. As the Chinese philosopher Lao Tsu wrote, "A journey of a thousand miles must begin with a single step."

Reviewing Key Terms

Can you define the following terms in environmental science?

acid deposition 16.3
aerosols 16.1
aesthetic degradation 16.1
ambient air 16.1
carbon monoxide 16.1
chlorofluorocarbons (CFCs) 16.2
chronic obstructive pulmonary disorder (COPD) 16.3

conventional or criteria pollutants 16.1
fugitive emissions 16.1
hazardous air pollutants (HAPs) 16.1
Kigali Amendment 16.2
Montreal Protocol 16.2
nitrogen oxides (NO_x) 16.1
ozone 16.1

particulate matter (PM2.5, PM10) 16.1
photochemical oxidants 16.1
primary pollutants 16.1
secondary pollutants 16.1
stratospheric ozone 16.2
sulfur dioxide 16.1

temperature inversion 16.2
Toxic Release Inventory (TRI) 16.1
unconventional pollutants 16.1
volatile organic compounds (VOCs) 16.1

Critical Thinking and Discussion Questions

1. What might be done to improve indoor air quality? Should governments mandate such changes? What values or worldviews are represented by different sides of this debate?

2. Debate the following proposition: Our air pollution blows onto someone else; therefore, installing pollution controls will not bring any direct economic benefit to those of us who have to pay for them.

3. Utility managers once claimed that it would cost $1,000 per fish to control acid precipitation in the Adirondack lakes and that it would be cheaper to buy fish for anglers than to put scrubbers on power plants. Suppose that is true. Does it justify continuing pollution?

4. Developing nations claim that richer countries created global warming and stratospheric ozone depletion, and therefore they should bear the responsibility for fixing these problems. How would you respond?

5. If there are thresholds for pollution effects, is it reasonable or wise to depend on environmental processes to disperse, assimilate, or inactivate waste products?

6. How would you choose between government "command and control" regulations versus market-based trading programs for air pollution control? Are there situations where one approach would work better than the other?

Data Analysis

How Is the Air Quality in Your Town?

How does air quality in your area compare to that in other places? You can examine trends in major air pollutants—both national and local trends in your area—on the EPA's website. The EPA is the principal agency in charge of protecting air quality and informing the public about the air we breathe and how healthy it is.

Go to Connect to find a link to data and maps showing trends in SO_2 emissions since 1980. At the same site, you can see trends in NO_x, CO, lead, and other criteria pollutants. Examine national trends, and then look at your local area on the map on the same page to answer questions about trends in your area and to also compare your area to others.

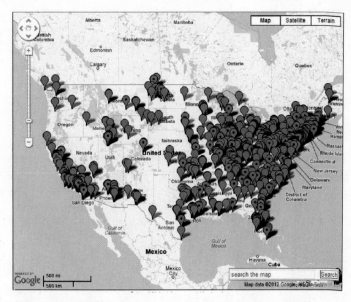

Examine pollutant trends in your area on the EPA website.

Water Use and Management

▲Lake Mead, the largest reservoir on the Colorado River, has fallen dramatically as the western climate has become warmer and drier. Both hydroelectric power and urban water resources are at risk.
David McNew/Getty Images

Learning Outcomes

After studying this chapter, you should be able to:

17.1 Discuss why water is a precious resource and why shortages occur.

17.2 Summarize water availability and use.

17.3 Explain causes of freshwater shortages.

17.4 Describe how we might get by with less water.

"I tell you gentlemen; you are piling up a heritage of conflict and litigation of water rights, for there is not sufficient water to supply the land."

– John Wesley Powell, 1893

When Will Lake Mead Go Dry?

The Colorado River is the lifeblood of the American Southwest. More than 30 million people and a $1.2 trillion regional economy in cities such as Los Angeles, Phoenix, Las Vegas, and Denver depend on its water (fig. 17.1). Hydroelectric dams, which created the two large reservoirs Lake Powell and Lake Mead, provide electricity to the region. But the reliability of this essential resource is in doubt. Demand for water has grown, in cities and farms, even as longer, hotter summers and declining snowpack have reduced supplies.

In 2008, Tim Barnett and David Pierce from the Scripps Institute in California published a provocative article projecting that in a decade or so, both Lake Mead and Lake Powell could fall so low that neither would be able to produce power or provide water for urban or agricultural use. These huge lakes constitute more than 85 percent of the water storage for the entire Colorado system. If they reach "dead pool" levels (at which water could not drain by gravity), it would be a catastrophe for a region that has come to depend on these artificial reservoirs. This warning was based on historical records and on climate models that project a 10 to 30 percent runoff reduction in the area by 2050.

The roots of this problem can be traced to the Colorado Compact of 1922, which allocated a portion of the river to each state in the watershed (outlined in fig. 17.1). Indulging in wishful thinking, the agreement dramatically overestimated the amount of water available to share. Negotiators estimated the annual river flow at 18 million acre-feet (22 billion m^3), about 20 percent more than the twentieth-century average.

The error didn't matter much at the time, because none of the states had the technology or need to withdraw its entire share. But cities have grown and agriculture has expanded since 1922, and the federal government has spent billions of dollars on massive water diversion projects to encourage this growth. The Colorado River Aqueduct delivers water to Los Angeles; the All-American Canal irrigates California's Imperial Valley; and the Central Arizona Project transports water over the mountains and across the desert to Phoenix and Tucson. By 1944, hardly any water was left in the river when it reached Mexico. So the United States agreed to allocate 1.5 million acre-feet to Mexico, although it might have been of dubious quality.

Now, climate change is expected to decrease western river flows drastically. The Southwest has seen drought conditions in most of the last 20 years. The maximum water level in Lake Mead (an elevation of 1,220 feet or 372 m) was last reached in 2000. Since then, the lake level has been dropping about 12 feet (3.6 m) per year, and it fell to 1,071 feet in July 2016, the lowest level since the lake was filled in the 1930s. If it drops another 20 feet, the Hoover Dam will be unable to produce electricity. Already it has been downgraded from 2 GW to 1.6 GW of production, and it runs mainly for peak demand hours. Lake Powell, upstream at the northern

Arizona border, has been lowered to help keep Lake Mead's water levels up. Barnett and Pierce estimated that without changes in current management plans, there was a 50 percent chance that minimum power pool levels in both Lakes Mead and Powell would be reached by 2021. Shifting management practices, including releasing water from upstream dams, has helped push back this crisis, but dam managers still predict that critical shortages could emerge in the next few years.

Already, we're at or beyond the sustainable limits of the river. Currently, Lake Powell holds less than half its maximum volume, and Lake Mead is less than 40 percent full. The shores of both lakes now display a wide "bathtub ring" of deposited minerals left by the receding water (see the chapter-opening photo). One suggestion has been to drain Lake Powell to ensure a water supply for Lake Mead. This solution would keep the lights on and faucets running in Los Angeles, Phoenix, Las Vegas, and other cities. But it is strenuously opposed by many of the 3 million recreational users of Lake Powell.

The American Southwest isn't alone in facing this problem. The United Nations warns that water supplies are likely to become one of the most pressing environmental issues of the twenty-first

FIGURE 17.1 The Colorado River flows 2,330 km (1,450 mi) through seven western states. Its water supports 30 million people and a $1.2 trillion regional economy, but drought, climate change, and rapid urban growth threaten the sustainability of this resource.

century. By 2025, two-thirds of all humans could be living in places where water resources are inadequate. In this chapter, we'll look at the sources of our fresh water, what we do with it, and how we might protect its quality and extend its usefulness.

For further reading, see:

Barnett, T. P., and D. W. Pierce. 2008. When will Lake Mead go dry? *Journal of Water Resources Research,* vol. 44, W03201.

Benchmark Data	
Among the ideas and values in this chapter, the following are a few worth remembering.	
2.4%	Percentage of all water on earth that is fresh
0.8%	Percentage of fresh water that is liquid surface water
13,000	Liters water needed to produce 1 kg of beef (average)
600 m³	Global water withdrawal, per person
75,000	Large dams in the United States

Table 17.1   Some Units of Water Measurement
One cubic kilometer (km^3) equals 1 billion cubic meters (m^3), 1 trillion liters, or 264 billion gallons.
One acre-foot is the amount of water required to cover an acre of ground 1 foot deep. This is equivalent to 325,851 gallons, or 1.2 million liters, or 1,234 m^3, about the amount consumed annually by a family of four in the United States.
One cubic foot per second of river flow equals 28.3 liters per second or 449 gallons per minute.

Note: See the table at the end of the book for conversion factors.

17.1 WATER RESOURCES

- *Less than 3 percent of all the water on the earth is fresh, and 87 percent of that water is locked up in ice and snow.*

- *The hydrologic cycle constantly purifies and redistributes fresh water.*

- *Some places get too much rain, while others get almost none at all.*

Water is a marvelous substance—flowing, rippling, swirling around obstacles in its path, seeping, dripping, trickling, constantly moving from sea to land and back again. Water can be clear, crystalline, icy green in a mountain stream, or black and opaque in a cypress swamp. Water bugs skitter across the surface of a quiet lake; a stream cascades down a stairstep ledge of rock; waves roll endlessly up a sand beach, crash in a welter of foam, and then recede. Rain falls in a gentle mist, refreshing plants and animals. A violent thunderstorm floods a meadow, washing away streambanks. Water is a most beautiful and precious resource.

Water is also a great source of conflict. Some 2 billion people now live in countries with insufficient fresh water. Some experts estimate this number could double in 25 years. To understand this resource, let's first examine where our water comes from, and why it is so unevenly distributed.

The hydrologic cycle constantly redistributes water

The water we use cycles endlessly through the environment. The total amount of water on our planet is immense—more than 1.4 billion km^3 (table 17.1). This water evaporates from moist surfaces, falls as rain or snow, passes through living organisms, and returns to the ocean in a process known as the **hydrologic cycle** (see fig. 3.19). Every year, about 500,000 km^3, or a layer 1.4 m thick, evaporates from the oceans. More than 90 percent of that moisture falls back on the ocean. The 47,000 km^3 carried onshore joins some 72,000 km^3 that evaporate from lakes, rivers, soil, and plants to become our annual, renewable freshwater supply. Plants play a major role in the hydrologic cycle, absorbing groundwater and pumping it into the atmosphere by transpiration (transport plus evaporation). In tropical forests, as much as 75 percent of annual precipitation derives from transpiration by plants.

Solar energy drives the hydrologic cycle by evaporating surface water, which becomes rain and snow. Because water and sunlight are unevenly distributed around the globe, water resources are very uneven. At Iquique in the Chilean desert, for instance, no rain has fallen in recorded history. At the other end of the scale, 27 m (1,041 in) of rain were recorded in a single year at Cherrapunji in India. Figure 17.2 shows broad patterns of precipitation around the world. Most of the world's rainiest regions are tropical, where heavy rainy seasons occur, or in coastal mountain regions. Deserts occur on every continent just outside the tropics (the Sahara, the Namib, the Gobi, the Sonoran, and many others). Rainfall is also slight at very high latitudes, another high-pressure region.

Water supplies are unevenly distributed

Rain falls unevenly over the planet (fig. 17.2). While some places get almost no precipitation, others receive heavy rain almost daily. Three principal factors control these global water deficits and

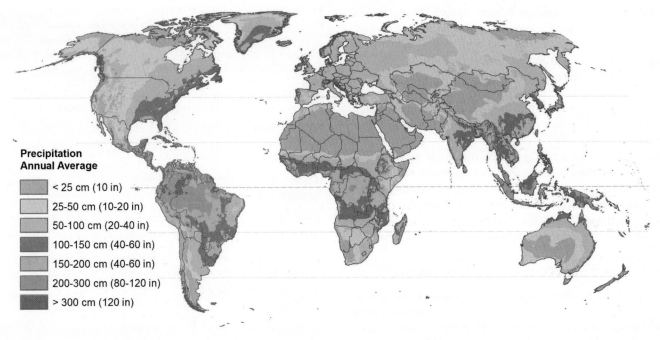

Precipitation
Annual Average

- < 25 cm (10 in)
- 25-50 cm (10-20 in)
- 50-100 cm (20-40 in)
- 100-150 cm (40-60 in)
- 150-200 cm (40-60 in)
- 200-300 cm (80-120 in)
- > 300 cm (120 in)

FIGURE 17.2 Average annual precipitation. Note wet areas that support tropical rainforests occur along the equator, while the major world deserts occur in zones of dry, descending air between 20° and 40° north and south.

surpluses. First, global atmospheric circulation creates regions of persistent high air pressure and low rainfall about 20° to 40° north and south of the equator (chapter 15). These same circulation patterns produce frequent rainfall near the equator and between about 40° and 60° north and south latitudes. Second, proximity to water sources influences precipitation. Where prevailing winds come over oceans or large lakes, they bring moisture to land. Areas far inland—in a windward direction—are usually relatively dry.

A third factor in water distribution is topography. Mountains act as both cloud formers and rain catchers. As air sweeps up the windward side of a mountain, air pressure decreases and air cools. As the air cools, it reaches the saturation point, and moisture condenses as either rain or snow. Thus, the windward side of a mountain range, as in the Pacific Northwest, is usually wet much of the year. Precipitation leaves the air drier than it was on its way up the mountain. As the air passes the mountaintop and descends the other side, air pressure rises, and the already-dry air warms, increasing its ability to hold moisture. Descending, warming air rarely produces any rain or snow. Places in the **rain shadow,** the dry, leeward side of a mountain range, receive little precipitation. A striking example of the rain shadow effect

is found on Mount Waialeale, on the island of Kauai, Hawaii (fig. 17.3). The windward side of the island receives nearly 12 m of rain per year, while the leeward side, just a few kilometers away, receives just 46 cm.

Usually, a combination of factors affects precipitation. In Cherrapunji, India, monsoon winds carrying moisture from the warm Indian Ocean crash against the high ridges of the

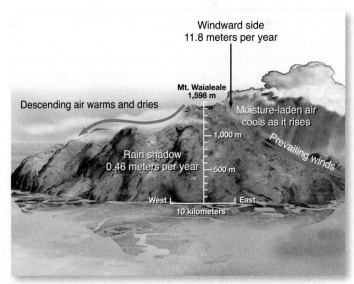

FIGURE 17.3 Rainfall on the east side of Mount Waialeale in Hawaii is more than 20 times as much as on the west side. Prevailing trade winds bring moisture-laden sea air onshore. The air cools as it rises up the flanks of the mountain, and the water it carries precipitates as rain, as much as 11.8 m (38 ft) per year!

Quantitative Reasoning

How do the 27 meters of rain that fall in a normal year in Cherrapunji, India, compare to the rainfall where you live? If you don't know your average annual precipitation, you can approximate it from the map in figure 17.2. How might your life change if you lived in a place with 27 meters of rain per year?

Himalayas to release a seasonal deluge. Iquique, Chile, lies in the rain shadow of the Andes and in a high-pressure desert zone. Prevailing winds are from the east, so even though Iquique lies near the ocean, it is far from the winds' moisture source—the Atlantic. In the American Southwest, Australia, and the Sahara, high-pressure atmospheric conditions tend to keep the air and land dry. The global map of precipitation represents a complex combination of these forces of atmospheric circulation, prevailing winds, and topography.

Human activity also explains some regions of water deficit. As noted earlier, plant transpiration recycles moisture and produces rain. When forests are cleared, falling rain quickly enters streams and returns to the ocean. In Greece, Lebanon, parts of Africa, the Caribbean, South Asia, and elsewhere, desert-like conditions have developed since the original forests were destroyed.

Oceans hold 97 percent of all water on earth

If explorers from another galaxy were to discover our planet, they'd probably call it Aqua rather than Earth, because water is its dominant characteristic. Oceans cover 71 percent of our planet's surface and contain more than 97 percent of all the *liquid* water in the world (fig. 17.4). Seawater is too salty for most human uses, but contains 90 percent of the world's living biomass.

Water distribution often is described in terms of interacting compartments in which water resides, sometimes briefly and sometimes for eons (table 17.2). The length of time water typically stays in a compartment is its **residence time.** On average, a water molecule stays in the ocean for about 3,000 years, for example, before it evaporates and starts through the hydrologic cycle again.

Residence time is short in the atmosphere: An individual water molecule resides in the atmosphere for about ten days, on average. While water vapor makes up only a small amount (4 percent maximum at normal temperatures) of the total volume of the air, movement of water through the atmosphere provides the mechanism for distributing fresh water over the landmasses and replenishing terrestrial reservoirs. The atmosphere is also one of the smallest of the major water reservoirs of the earth in terms of water volume, containing less than 0.001 percent of the total water supply.

Table 17.2   Earth's Water Compartments

Compartment	Volume (1,000 km³)	Percent of Total Water	Average Residence Time
Total	1,386,000	100	2,800 years
Oceans	1,338,000	96.5	3,000 to 30,000 years*
Ice and snow	24,364	1.76	1 to 100,000 years*
Saline groundwater	12,870	0.93	Days to thousands of years*
Fresh groundwater	10,530	0.76	Days to thousands of years*
Fresh lakes	91	0.007	1 to 500 years*
Saline lakes	85	0.006	1 to 1,000 years*
Soil moisture	16.5	0.001	2 weeks to 1 year*
Atmosphere	12.9	0.001	1 week
Marshes, wetlands	11.5	0.001	Months to years
Rivers, streams	2.12	0.0002	1 week to 1 month
Living organisms	1.12	0.0001	1 week

*Depending on depth and other factors.
Source: Data from UNEP, 2002

Glaciers, ice, and snow contain most surface fresh water

Of the 2.4 percent of all water that is fresh, 87 percent is stored in glaciers, ice caps, and snowfields (fig. 17.5). Most of this ice covers Antarctica, Greenland, and remote alpine ice fields of Alaska. At lower latitudes, though, mountain glaciers and snowfields supply water to some of the world's most populous regions, including northern India, Southeast Asia, and much of China. In the American Southwest and California, economies have been built on runoff from winter snowpack on the western slope of the Rocky Mountains, which provides 75 percent of the flow in the Colorado River, discussed in the opening case study.

Vast glacial fields in the Himalayas are the source of Asia's greatest rivers, including the Ganges, Yangtze, and Mekong.

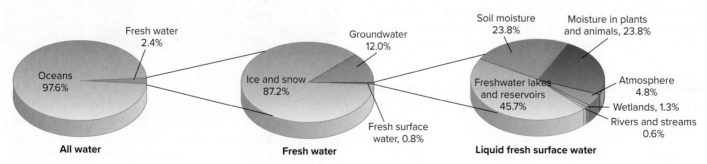

FIGURE 17.4 Less than 1 percent of fresh water, and less than 0.02 percent of all water, is the fresh, liquid surface water on which terrestrial life depends.
Source: U.S. Geological Survey

FIGURE 17.5 Glaciers and snowfields exist in high mountains or at high latitudes, but many are critical, and diminishing, reservoirs of fresh water.
Source: Photograph by Bruce F. Molnia, U.S. Geological Survey

FIGURE 17.6 Himalayan glacial runoff supports farms and cities in much of Asia, including this part of Nepal. Rapidly declining Himalayan ice threatens about one quarter of the world's population.
Erica Simone Leeds

About a quarter of the world's population depends on this region's glacier-fed rivers for drinking water and farming (fig. 17.6). Industry and electricity production across much of Asia also rely on glacial runoff and snowmelt. In South America, many regions depend entirely on glacier melt for survival. Much of Bolivia, Peru, and Chile have no other water source, and diminishing glaciers are an urgent problem.

If a glacier is in long-term equilibrium, it accumulates snow in winter, which compacts over time into ice, and it melts mainly in summer. In especially snowy winters, snow accumulates faster. In cold summers, less melting occurs. When warm seasons are longer and hotter, melting increases and the glacier loses mass.

As climate warming produces longer melt seasons, most of the world's glaciers are retreating—even Greenland and Antarctica are losing ice mass at accelerating rates (see fig. 15.30). Loss of glacial ice storage poses a growing threat around the world. Drought conditions already have reduced snowfall (and runoff) in the western United States. Even more worrisome are Himalayan glaciers. Two-thirds of these may be gone in 80 years, under current greenhouse gas emission trends.

Groundwater stores large resources

After glaciers, the next largest reservoir of fresh water is held in the ground as **groundwater.** Precipitation that does not evaporate back into the air or run off over the surface percolates through the soil and into fractures and spaces of permeable rocks in a process called **infiltration** (fig. 17.7). Upper soil layers that hold both air and water make up the **zone of aeration.** Moisture for plant growth comes primarily from these layers. Depending on rainfall amount, soil type, and surface topography, the zone of aeration may be very shallow or quite deep. Lower soil layers where all spaces are filled with water make up the **zone of saturation.** The top of this zone is the **water table.** The water table is not flat, but undulates according to the surface topography and subsurface structure. Water tables also rise and fall seasonally, depending on precipitation, infiltration, and withdrawal rates.

Porous layers of sand, gravel, or rock lying below the water table are called **aquifers.** Aquifers are always underlain by relatively impermeable layers of rock or clay that keep water from seeping out at the bottom (fig. 17.8).

Folding and tilting of the earth's crust by geological processes can create areas of high water pressure in confined aquifers (those

FIGURE 17.7 Precipitation that does not evaporate or run off over the surface percolates through the soil in a process called infiltration. The upper layers of soil hold droplets of moisture between air-filled spaces. Lower layers, where all spaces are filled with water, make up the zone of saturation, or groundwater.

trapped between two impervious, confining rock layers). When a pressurized aquifer intersects the surface, or if it is penetrated by a pipe or conduit, the result is an **artesian well** or spring, from which water gushes without being pumped.

Areas where water infiltrates into an aquifer are called **recharge zones.** The rate at which most aquifers are refilled is very slow, however, and groundwater presently is being removed faster than it can be replenished in many areas. Urbanization, road building, and other development often block recharge zones and prevent replenishment of important aquifers. Contamination of surface water in recharge zones and seepage of pollutants into abandoned wells have polluted aquifers in many places, making them unfit for most uses (chapter 18). Many cities protect aquifer recharge zones from pollution or development, both as a way to drain off rainwater and as a way to replenish the aquifer with pure water.

Some aquifers contain very large volumes of water. The groundwater within 1 km of the surface in the United States is more than 30 times the volume of all the freshwater lakes, rivers, and reservoirs on the surface. Although water can flow through limestone caverns in underground rivers, most movement in aquifers is a dispersed and almost imperceptible trickle through tiny fractures and spaces. Depending on geology, it can take anywhere from a few hours to several years for contaminants to move a few hundred meters through an aquifer.

Rivers, lakes, and wetlands cycle quickly

Precipitation that does not evaporate or infiltrate into the ground runs off over the surface, drawn back toward the sea by the force of gravity. Rivulets accumulate to form streams, and streams join to form rivers. Although the total amount of water contained at any one time in rivers and streams is small compared to that in the other water reservoirs of the world (table 17.2), these surface waters are vitally important to humans and most other organisms. Most rivers, if they were not constantly replenished by precipitation, meltwater from snow and ice, or seepage from groundwater, would begin to diminish in a few weeks.

We measure the size of a river in terms of its **discharge,** the amount of water that passes a fixed point in a given amount of time. This is usually expressed as liters or cubic feet of water per second. The 16 largest rivers in the world carry nearly half of all surface runoff on earth. The Amazon is by far the largest river in the world (table 17.3), carrying roughly ten times the volume of the Mississippi. Several Amazonian tributaries, such as the

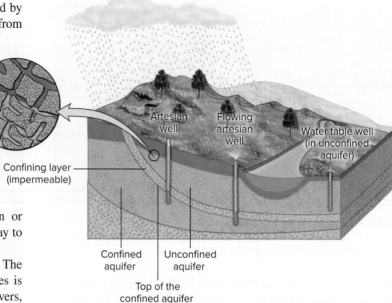

FIGURE 17.8 An aquifer is a porous or cracked layer of rock. Impervious rock layers (aquicludes) keep water within a confined aquifer. Pressure from uphill makes an artesian well flow freely. Pumping can create a cone of depression, which leaves shallower wells dry.

Madeira, Rio Negro, and Ucayali, would be among the world's top rivers in their own right.

Ponds are generally considered to be small temporary or permanent bodies of water shallow enough for rooted plants to grow over most of the bottom. Lakes are inland depressions that hold standing fresh water year-round. Maximum lake depths range from a few meters to over 1,600 m (1 mi) in Lake Baikal in Siberia. Surface areas vary in size from less than a hectare to large

Table 17.3	Major Rivers of the World	
River	**Countries in River Basin**	**Average Annual Discharge at (m³/sec)**
Amazon	Brazil, Peru	175,000
Orinoco	Venezuela, Colombia	45,300
Congo	Congo	39,200
Yangtze	Tibet, China	28,000
Bramaputra	Tibet, India, Bangladesh	19,000
Mississippi	United States	18,400
Mekong	China, Laos, Burma, Thailand, Cambodia, Vietnam	18,300
Parana	Paraguay, Argentina	18,000
Yenisey	Russia	17,200
Lena	Russia	16,000

1 m³ = 264 gallons
Source: World Resources Institute

Quantitative Reasoning

Figure 17.6 shows that only 2.4 percent of all water on the earth is fresh (nonsalty). How much of that water is liquid in rivers and lakes? Assuming this is the main source of drinking water for most animals, what percentage is this of the total world water supply?

inland seas, such as Lake Superior or the Caspian Sea, covering hundreds of thousands of square kilometers. Both ponds and lakes are relatively temporary features on the landscape because they eventually fill with silt or are emptied by cutting an outlet stream through the barrier that creates them.

While lakes contain nearly 100 times as much water as all rivers and streams combined, they are still a minor component of the total world water supply. Their water is much more accessible than groundwater or the water in glaciers, however, and they are important in many ways for humans and other organisms.

Wetlands play a vital and often unappreciated role in the hydrologic cycle. Their lush plant growth stabilizes soil and holds back surface runoff, allowing time for infiltration into aquifers and producing even, year-long stream flows. In the United States, about 20 percent of the 1 billion ha of land area was once wetland. In the past 200 years, more than one-half of those wetlands have been drained, filled, or degraded. Agricultural drainage accounts for the bulk of the losses.

When wetlands are disturbed, their natural water-absorbing capacity is reduced and surface waters run off quickly, resulting in floods and erosion during the rainy season, and creating dry, or nearly dry, streambeds the rest of the year. This has a disastrous effect on biological diversity and productivity, as well as on human affairs.

Section Review

1. What is the hydrologic cycle, and how does it redistribute water around the globe?
2. What percentage of water on the planet is fresh (nonsalty)?
3. What is an aquifer, and how is it recharged?

17.2 WATER AVAILABILITY AND USE

- *Many countries suffer water scarcity and stress.*
- *Water use is increasing.*
- *Agriculture is the greatest water user worldwide.*

Clean, fresh water is essential for nearly every human endeavor. Perhaps more than any other environmental factor, the availability of water determines the location and activities of humans on earth (fig. 17.9). **Renewable water supplies** are made up, in general, of surface runoff plus the infiltration into accessible freshwater aquifers. About two-thirds of the water carried in rivers and streams every year occurs in seasonal floods that are too large or violent to be stored or trapped effectively for human uses. Stable runoff is the dependable, renewable, year-round supply of surface water. Much of this occurs, however, in sparsely inhabited regions or where the lack of technology, finances, or other factors make it difficult to use it productively. Still, the readily accessible, renewable water supplies are very large, amounting to some 1,500 km^3 (about 400,000 gal) per person per year worldwide.

FIGURE 17.9 Water has always been the key to survival in arid regions. Who has access to this precious resource and who doesn't has long been a source of tension and conflict.
Mieszko9/Shutterstock

Many countries suffer water scarcity or water stress

The United Nations considers 1,000 m^3 (264,000 gal) of water per person per year to be the minimum necessary to meet basic human needs. **Water scarcity** occurs when the demand for water exceeds the available amount or when poor quality restricts its use. **Water stress** occurs when renewable water supplies are inadequate to satisfy essential human or ecosystem needs, bringing about increased competition among potential demands. Water stress is most likely to occur in poor countries where the per capita renewable water supply is low.

As you can see in figure 17.2, South America, West Central Africa, and South and Southeast Asia all have areas of very high rainfall. The highest per capita water supplies generally occur in countries with wet climates and low population densities. Iceland, for example, has about 600 million liters per person per year. In contrast, Bahrain, where temperatures are extremely high and rain almost never falls, has essentially no natural fresh water. Almost all of Bahrain's water comes from imports and desalinized seawater. Egypt, in spite of the fact that the Nile River flows through it, has only about 42,000 liters of water annually per capita, or nearly 15,000 times less than Iceland.

Ironically, Cherrapunji, India, described at the beginning of this chapter as the wettest place on earth, is now experiencing water shortages during the dry season. Growing population, increased per capita water use, deforestation of surrounding hillsides, higher industrial demand, and changing monsoon patterns caused by global warming all contribute to this problem.

There is no simple definition of drought. What would be an amazingly wet year in a desert might be considered a catastrophic drought elsewhere. In general, a drought is an extended period of consistently below-average precipitation that has a substantial impact on ecosystems, agriculture, and economies.

As the opening case study shows, much of the western United States has been exceptionally dry in recent years. Many places are experiencing water crises. Is this just a temporary cycle or a long-term climatic change? In 2011–2015, the United States experienced the most severe drought in at least 50 years. The U.S. Department of Agriculture reports that this drought reduced yields on 80 percent of agricultural land in the country.

Even so, we continue to increase production of high-value crops, such as avocados, almonds, and fruit in California, alfalfa for dairy feed throughout the West, and corn for biofuel across the Midwest. These crops rely on steadily increasing amounts of irrigation, even as water supplies become more precarious. Once these economies begin to establish and grow, however, changing course becomes difficult.

The West has always had droughts

For many, recent drought conditions in the United States recall the "dust bowl" days of the 1930s, when wind stripped topsoil from millions of hectares of land, and billowing dust clouds turned day into night. Thousands of families were forced to leave farms and migrate to other areas.

Droughts in the American West aren't a new phenomenon. In fact, the dry spells in recent years have been of relatively short duration compared to historic events. The dust bowl of the 1930s, for example, lasted only about six years. By contrast, the megadroughts that destroyed the Ancient Puebloan (or Anasazi) cultures in the twelfth and thirteenth centuries (fig. 17.10) lasted 25 to 50 years.

If the government had listened to Major John Wesley Powell, settlement patterns in the western United States would be very different from what we see there today. Powell, who led the first expedition down the Colorado River, went on to be the first head of the U.S. Geological Survey. In that capacity, he did a survey of the agricultural and settlement potential of the western desert. His conclusion, quoted at the beginning of this chapter, was that there isn't enough water there to support a large human population.

Powell recommended that the political organization of the West be based on watersheds, so that everyone in a given jurisdiction would be bound together by the available water. He thought that farms should be limited to local surface water supplies, and that cities should be small oasis settlements. Instead, we've built huge metropolitan areas, such as Los Angeles, Phoenix, Las Vegas, and Denver, in places where there is little or no natural water supply. Will those cities survive predicted shortages?

Water use is increasing

Water consumption is increasing rapidly as populations grow (fig. 17.11). The average amount of water withdrawn worldwide is slightly more than

FIGURE 17.10 The stunning cliff dwellings of Mesa Verde National Park were abandoned by the Ancient Puebloan people during a severe megadrought between 1275 and 1299. Will some of our modern cities face the same fate?
William P. Cunningham

600 m³ (150,000 gal) per person per year. This overall average hides great discrepancies in the proportion of annual runoff withdrawn in different areas. Some countries with a plentiful water supply withdraw a very small percentage of the water available to them. Canada, Brazil, and the Congo, for instance, withdraw less than 1 percent of their annual renewable supply.

By contrast, in desert countries such as Libya, Yemen, and Israel, groundwater and surface water withdrawal together amount to more than 100 percent of their renewable supply. They are

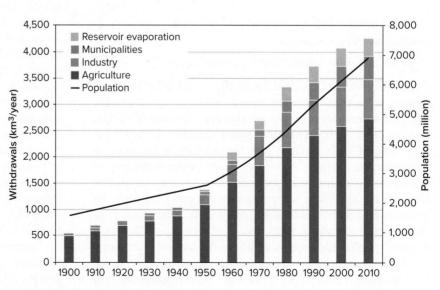

FIGURE 17.11 Global water use has increased steadily, as has the global population.
Source: http://www.fao.org/nr/water/aquastat/water_use/image/WithTimePopAxis_eng.pdf

essentially "mining" water—extracting groundwater faster than it is being replenished. Obviously, this isn't sustainable in the long run.

The total annual renewable water supply in the United States amounts to an average of about 9,000 m³ (nearly 2.4 million gal) per person per year. We now withdraw about one-fifth of that amount, or some 5,000 liters (1,300 gal) per person per day, including industrial and agricultural water. By comparison, the average water use in Haiti is less than 30 liters (8 gal) per person per day.

Unlike some resources, water can be used over and over if it is not too badly contaminated. Water **withdrawal** is the total amount of water taken from a water body. Much of this water could be returned to circulation in a reusable form. Water **consumption,** on the other hand, is loss of water due to evaporation, absorption, or contamination. Essentially, consumption is when the water gets moved into a different part of the water cycle.

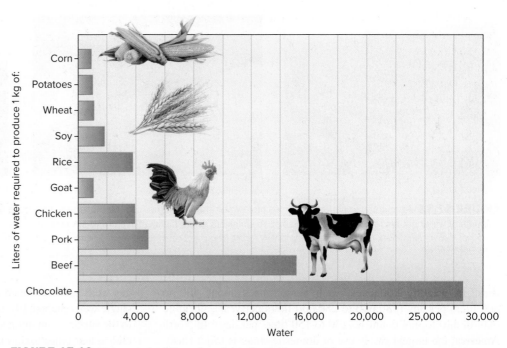

FIGURE 17.12 Water required to produce 1 kg of some important foods.

Agriculture dominates water use

Water consumption is especially high in evaporation and transpiration from crop plants. Worldwide, agriculture claims about 70 percent of total water withdrawal, and that use has increased sharply in recent decades (fig. 17.11). Agricultural use ranges from 93 percent of all water withdrawn in India to only 4 percent in Kuwait, which cannot afford to spend its limited water on crops.

Most of us rarely think about the water required to grow and prepare the food we eat (fig. 17.12). Raising a kilogram of beef in a concentrated feeding operation, for example, takes more than 15,000 liters of water, because cattle in these facilities are fed grain, which they process very inefficiently, that often was grown with water-intensive irrigation systems. The result is that the average hamburger takes 2,400 liters (630 gal) to produce. Raising goats requires at least 90 percent less water from all sources than cattle because they eat a much wider diet and more efficiently metabolize what they do eat.

Rice cultivation (which generally occurs in wet paddies) takes about three times as much water as raising potatoes or wheat. For some foods, the greatest water use isn't in growing the crop, but in preparing it for our consumption. Chocolate, for example, is highly processed before we eat it.

Irrigation can be very inefficient. Traditionally, the main method has been flood or furrow irrigation, in which water runs across the land surface (fig. 17.13a). Up to half of this water can be lost through evaporation. Much of the rest runs off before it is used

by plants. In arid lands, flood irrigation helps remove toxic salts from soil, but these salts contaminate streams, lakes, and wetlands downstream. Repeated flooding also waterlogs the soil, eventually reducing crop growth. Sprinkler systems can also be wasteful (fig. 17.13b). For decades, field sprinkler systems sprayed water high in the air, where much of it evaporated before reaching crops. In recent years, though, growing pressure on water resources has led to more-efficient sprinkler systems that hang low over crops to reduce evaporation.

Drip irrigation (fig. 17.13c) dramatically reduces irrigation water use. These systems release carefully regulated amounts of water just above plant roots, so that nearly all water is used by plants. Only about 1 percent of the world's croplands currently use these expensive systems, however.

Irrigation infrastructure, such as dams, canals, pumps, and reservoirs, is expensive. Irrigation is also the economic foundation of many regions. In the United States, for nearly a century, the federal government has taken responsibility for providing irrigation. The argument for doing so is that irrigated agriculture is a public good that cannot be provided by individual farmers. A consequence of this policy has frequently been heavily subsidized crops whose costs, in water and in dollars, far outweigh their value.

Industry and households withdraw less but often contaminate water

Worldwide, domestic water use accounts for only about 6 percent of water withdrawals. Because most of this water is returned to a river or lake after use, actual consumption is low, about 10 percent of the world total, on average. Where sewage treatment is unavailable, however, water can be badly degraded by urban

(a) Flood irrigation

(b) Rolling sprinklers

(c) Drip irrigation

FIGURE 17.13 Agricultural irrigation consumes more water than any other use. Methods vary from flood and furrow (a), which uses extravagant amounts of water but also flushes salts from soils, to sprinklers (b), to highly efficient drip systems (c).
Source: (a) Photo by Jeff Vanuga, U.S. Department of Agriculture Natural Resources Conservation Service; (b) cecilialim/123; (c) Photo by Lynn Betts, USDA Natural Resources Conservation Service

uses. In wealthy countries, each person uses about 500 to 800 liters per day (180,000 to 280,000 liters per year), far more than in developing countries (30 to 150 liters per day). In North America, the largest single use of domestic water is toilet flushing, despite improved toilet design standards. On average, each person in the United States uses about 50,000 liters (13,000 gal) of drinking-quality water annually to flush toilets. Showering and washing accounts for nearly a third of water use, followed by laundry. In western cities such as Palm Desert and Phoenix, landscape irrigation is also a major water user.

Urban and domestic water use have grown approximately in proportion with urban populations, about 60 percent between 1960 and 2020. Although individual water use seems slight on the scale of world water withdrawals, the cumulative effect of inefficient appliances, long showers, lavish lawn-watering, and other uses is enormous. California has established increasingly stringent standards for washing machines, toilets, and other appliances, in order to reduce urban water demands. Many other cities and states are following this lead to reduce domestic water use. Some people are replumbing houses so that gray water (from showers, dishwashing, etc.) can be reused for irrigating plants or flushing toilets rather than simply dumped directly in a sewer. We'll discuss more water conservation practices later in this chapter.

Industry accounts for 20 percent of global freshwater withdrawals. Industrial use rates range from 70 percent in industrialized parts of Europe to less than 5 percent in countries with little industry. Power production, including hydroelectric, nuclear, and thermoelectric power, makes up 50 to 70 percent of industrial uses, and industrial processes make up the remainder. As with domestic water, little of this water is made unavailable after use, but it is often degraded by defouling agents, chlorine, or heat when it is released to the environment.

The greatest industrial producer of degraded water is mining and energy production. Ores must be washed and treated with chemicals such as mercury and cyanide (chapter 14). As much as 80 percent of water used in mining and processing is released with only minimal treatment. Fracturing (or fracking) oil and natural gas wells takes as much as 23 million liters (6 million gal) for a single well (chapter 19). Up to 40 percent of this water comes back to the surface contaminated with petrochemicals, arsenic, radionuclides, and other toxic materials. Providing this water is difficult in dry regions, and disposal of contaminated water is a problem everywhere.

Section Review

1. Define water scarcity and water stress.
2. What is drip irrigation, and why is it beneficial?
3. What percentage of all water withdrawals are for agricultural uses?

17.3 FRESHWATER SHORTAGES

- *Billions of people lack access to clean water.*
- *Groundwater is being depleted in many places.*
- *Diversion projects redistribute water, but often with social and environmental costs.*

Clean drinking water and basic sanitation are necessary to prevent communicable diseases and to maintain a healthy life. For many of the world's poorest people, one of the greatest environmental threats to health is polluted water. In Mali, for example, 88 percent of the population lacks safe water; in Ethiopia, 94 percent do. Rural people often have less access to good water than do city dwellers. This often results in diarrhea and a variety of other diseases linked to contaminated water and lack of sanitation. Every year, about 1.6 million people, 90 percent of them children under 5 years old, die from these diseases.

Shortage of water resources compounds these problems. The United Nations estimates that about a third of the world's population lives in countries where water supplies don't meet essential needs (fig. 17.14). By 2025, two-thirds of the world's people may be living in countries that are water-stressed—defined by the United Nations as consuming more than 10 percent of their renewable freshwater resources. As noted earlier, the World Health

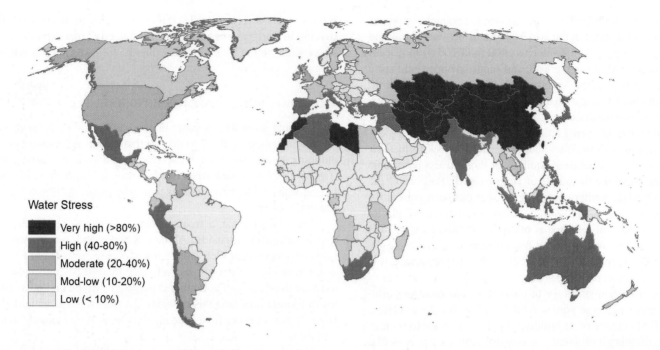

Water Stress

- ■ Very high (>80%)
- ◼ High (40-80%)
- ▨ Moderate (20-40%)
- ▧ Mod-low (10-20%)
- □ Low (< 10%)

FIGURE 17.14 Water stress can be measured as the ratio of water withdrawals to supply. Data are aggregated by country; even in low-stress regions, such as West Africa, unreliable rains and limited infrastructure can make water shortages locally severe.
Source: Data World Resources Institute

Organization considers an average of 1,000 m³ (264,000 gal) per person per year to be a necessary amount of water for modern domestic, industrial, and agricultural uses. Some 45 countries, most of them in Africa or the Middle East, cannot meet the minimum essential water needs of all their citizens.

Severe conditions of water stress and scarcity also occur in north Africa and in central and southern Asia, where rainfall is low and poor countries can't afford to build expensive infrastructure. Growing populations, industrial demands, and agricultural irrigation only compound these stresses.

More than two-thirds of the world's households have to fetch water from outside the home (fig. 17.15). This is heavy work, done mainly by women and children, often taking hours every day. Improved public water systems bring many benefits to these poor families, but availability doesn't always mean affordability. A typical poor family in Lima, Peru, for instance, uses only one-sixth as much water as a middle-class American family but pays three times as much for it. If they followed government recommendations to boil all water to prevent cholera, up to one-third of the poor family's income could be used just in acquiring and purifying water. These challenges are widespread in developing areas (chapter 18).

Groundwater is an essential but declining resource

Groundwater provides nearly 40 percent of the fresh water for agricultural and domestic use in the United States. Nearly half of all Americans and about 95 percent of the rural U.S. population depend on groundwater for drinking and other domestic purposes.

Overuse of these supplies causes several kinds of problems, including drying of wells and natural springs, disappearance of surface water features such as wetlands, and loss of aquifer storage capacity.

In many areas of the United States, groundwater is being withdrawn from aquifers faster than natural recharge can replace it.

FIGURE 17.15 Women and children often spend hours every day collecting water—which often is unsafe for drinking—from local water sources.
Layton Thompson

The Ogallala High Plains Regional Aquifer, for example, underlies eight agricultural states in the arid high plains between Texas and North Dakota (fig. 17.16). As deep as 400 m (1,200 ft) in its center, this porous bed of sand, gravel, and sandstone once held more water than all the freshwater lakes, streams, and rivers on earth. Steady pumping for irrigation and other uses has removed so much water that wells have dried up in many places, and farms, ranches, even whole towns are being abandoned.

Aquifers in shallow, fractured, or permeable rock formations can recharge rapidly, but deep aquifers, or those in less permeable formations, may take thousands of years to refill. Portions of the southern Ogallala aquifer contain water that has been present for thousands of years, since the wetter climate conditions of glacial periods. This ancient resource is often called "fossil water," and is not a renewable resource on a human time scale. Where aquifers are regularly recharged by rainfall on the ground surface, it is important to protect recharge areas.

Many cities and states try to prevent development or pollution in aquifer recharge zones. Arid Los Angeles, for example, spent its first century or so building drains and canals to remove rainwater, draining it efficiently as possible to the ocean. Now that water shortages are clearly the city's future, it is working to establish unpaved areas where rainwater can infiltrate groundwater.

Beijing, another chronically thirsty and fast-growing city, is striving to find ways to become a "sponge city" that captures and stores water in the ground, rather than rushing stormwater out of the city.

Groundwater overdrafts have long-term impacts

Drawing out groundwater faster than it can be replaced can cause long-term damage to the resource. At the scale of individual wells, pumping lowers the water table, forming a cone of depression (fig. 17.17). Often, new wells must be drilled deeper, to reach the falling water table, and this makes nearby older, shallower wells run dry.

On a larger scale, a frequent consequence of aquifer depletion is **saltwater intrusion.** Along coastlines and in areas where saltwater deposits are left from ancient oceans, overuse of freshwater reservoirs often allows saltwater to intrude into aquifers used for domestic and agricultural purposes. Miami and other cities in Florida now face this problem. As freshwater is withdrawn from Florida's porous limestone aquifers, seawater is moving in to replace it.

Withdrawing large amounts of groundwater causes porous formations to collapse, resulting in **subsidence,** or settling of the surface. The cause is loss of water pressure in the pore spaces of deep formations. Once these pore spaces have collapsed, they can never be refilled, and the resource is lost. The U.S. Geological Survey estimates that the San Joaquin Valley in California has sunk more than 10 m in the last 50 years because of excessive groundwater pumping (fig. 17.18). Since this photo was taken, continuing groundwater extraction has lowered the ground surface by several more meters in some parts of the valley.

Around the world, many cities are experiencing subsidence. Most were built on former lake beds, river deltas, or other unconsolidated sediments. Flooding is often a problem where coastal areas sink below sea level. Mexico City is one of the most notorious examples. Built on an old lake bed, it has probably been sinking since Aztec times. In recent years, however, rapid population

FIGURE 17.16 The Ogallala/High Plains regional aquifer supports a multimillion-dollar agricultural economy, but withdrawal far exceeds recharge. Some areas are down to less than 3 m of saturated thickness.

Water level change
1980–1996, in feet

- dewatered
- <−20
- −5 to −20
- −5 to 5
- 5 to 20
- > 20

0 50 100 Miles

FIGURE 17.17 Saltwater intrusion into a coastal aquifer as the result of groundwater depletion. Many coastal regions of the United States are losing freshwater sources due to saltwater intrusion.

FIGURE 17.18 A photo taken in 1977 showing subsidence caused by groundwater pumping in California's San Joaquin Valley. The 1925 sign indicates the ground level in that year.
Source: Richard Ireland, U.S. Geological Survey

growth and urbanization have caused groundwater overdrafts. Some areas of the city have sunk as much as 8.5 m (25.5 ft). The Shrine of Guadalupe, the cathedral, and many other historic monuments are sinking at odd and perilous angles.

Diversion projects redistribute water

Regional conditions within a country can be much worse than the countrywide averages shown in figure 17.14. For example, conditions in the American Southwest (described in the opening case study) are especially severe. Similarly, China is vulnerable to water shortages in dry northern regions, including the capital Beijing. Two-thirds of the country's population lives in the north, but two-thirds of its rainfall is in the south. China has undertaken vast water transfer projects, redirecting water from the Yangtze River region hundreds of km northward to Beijing. By most accounts, this project, known as the South to North Water Transfer Project, is the largest, and most expensive, infrastructure project ever undertaken.

Dams and canals are a foundation of civilization because they store and redistribute water for farms and cities. Many great civilizations, including the ancient empires of Sumeria, Egypt, and India, have been organized around large-scale water control systems. As modern dams and water diversion projects have grown in scale and number, though, their environmental costs have raised serious questions about efficiency, costs, and the loss of river ecosystems.

One of the most disastrous diversions in world history is that of the Aral Sea. Situated in arid Central Asia, on the border of Kazakhstan and Uzbekistan, the Aral Sea is a huge saline lake fed by rivers from distant mountains. Starting in the 1950s, the Soviet Union began diverting these rivers to water cotton and rice fields. Gradually, the Aral Sea has shrunk, leaving vast salt flats contaminated by toxic chemical runoff from the farm fields (fig. 17.19). The economic value of the cotton and rice probably never equaled the cost of lost fisheries and villages, and the health of local residents.

FIGURE 17.19 Starting in the 1950s, rivers feeding the Aral Sea were diverted to irrigate cotton and rice fields. The sea has lost more than 90 percent of its water. The "Small Aral" (upper right lobe) has separated from the main lake, and is now being refilled.
Source: U.S. Geological Survey

Measuring Invisible Water

Farms and cities around the world depend on groundwater. Powerful pumps extract water, sometimes from hundreds of meters below the ground, to irrigate rice fields in California's Central Valley and to flood alfalfa fields in the baking Imperial Valley, to water cotton fields in north Texas, and to fill urban water lines in Miami. In California and many other places, surface water is tightly monitored and regulated, but groundwater is unregulated. If you can get it out, it's yours. This is a classic tragedy of the commons, with predictable results: Groundwater is declining steadily. Some of it is irreplaceable. Much of it is unlikely to recover for centuries—especially as the climate warms and rain becomes less reliable.

Farmers and water districts know their groundwater is declining, but it's hard to know how much, or how fast. The only way to watch groundwater resources has been to monitor wells, to see when they started to run dry. It is expensive and difficult to collect records from tens of thousands of scattered well logs, and to assess regional changes from these records. But as the resource dwindles, it becomes essential to understand it better.

A new approach is available in the Gravity Recovery and Climate Experiment (GRACE) satellites. The system is a pair of satellites, which travel in a polar orbit (passing over the poles as the earth rotates below them) 500 km (310 mi) above the earth. The two satellites are separated by about 220 km (137 mi). Subtle differences in gravitational pull cause slight acceleration or deceleration in the satellites, which they record as slight changes in the distance between the pair. The GRACE satellites have been monitoring variations in gravity since 2002, and recent years of drought have turned the world's spotlight on the data they are producing.

What causes variation in gravity on the earth's surface? The main factor is geological formations. Dense rock types, such as those rich in iron and other metals, have more gravitational pull than less massive ones, such as limestone formations. But changes in ice and water storage, over months or years, are also detectable as gravitational changes. Greenland's mass has declined as its ice melts and runs off into the Arctic Ocean. Major lakes lose mass as water is

(header photo): David L. Hansen

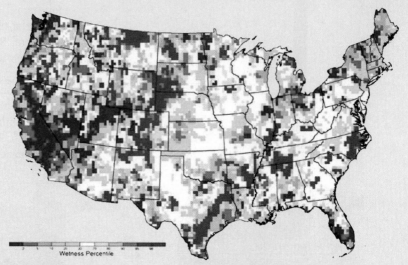

GRACE-Based Shallow Groundwater Drought Indicator

FIGURE 1 Groundwater and soil moisture, in much of the American West, was low in 2013, compared to the range of normal conditions. Dark brown (2nd percentile) indicates areas in which 98 percent of historical conditions are wetter.

Source: University of Nebraska GRACE Data Assimilation Project

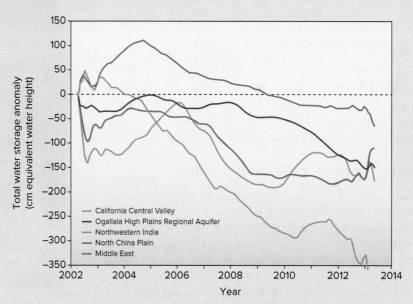

FIGURE 2 Withdrawals are reducing water storage in many of the world's most important aquifers, including those in politically tense regions. Groundwater change calculated from NASA's Gravity Recovery and Climate Experiment (GRACE) satellites.

Source: Adapted from J. S. Famiglietti, 2014. "The Global Groundwater Crisis," Nature Climate Change, 4: 945–948.

diverted or evaporates. Even changes in the density of ocean currents can be detected by the GRACE satellites. And now it is possible to monitor changes in groundwater storage in major aquifers from Los Angeles, to Beijing, to Delhi.

In the United States, GRACE data show that droughts extend well beyond California, covering the Colorado River Basin, and populous regions of the eastern and southern states (fig. 1). GRACE data also show that heavy rains in early 2016 did little to alter the deep drawdowns of previous years.

Global analysis shows that most of the world's major aquifers are becoming seriously overdrawn. Northeastern India is among the worst in drawdown (fig. 2). This information can be critical for policymakers who need to decide whether to subsidize irrigation or promote water conservation measures.

Satellites are providing more detailed and more global perspectives on our environment. You can learn more by finding NASA and GRACE websites online. You can also learn more by studying earth science, geography, physics, or any of the many fields that make use of satellite data.

For further reading: J. S. Famiglietti, 2014. "The global groundwater crisis," *Nature Climate Change* 4: 945–948.

Recently some river flow has been restored to the "Small Aral" or northern lobe of the once-great sea. Water levels have risen 8 m and native fish are being reintroduced. It's hoped that one day commercial fishing may be resumed. The fate of the larger, southern remnant is more uncertain. There may never be enough water to refill it, and if there were, contaminants left in the lake bed could make it unusable anyway.

Dams have diverse environmental and social impacts

More than half of the world's 227 largest rivers have been dammed or diverted. Half of the world's large dams are in China, and China continues to build and plan dams on its remaining rivers. Dams and diversion projects provide flood control, water storage, and electricity. But large dams also have multiple social and environmental impacts. Some analysts have argued that, in terms of economic losses and gains (not counting impacts on biodiversity and ecosystems), at least one-third of the world's large dams probably should never have been built.

In hot, dry climates, evaporative water losses are one of the most important impacts. Lake Mead, on the Colorado River, loses more than 1 billion m^3 of water to evaporation and seepage every year. Lake Powell loses nearly as much through evaporation and through seepage into the sandstone bedrock that lines the reservoir. Some studies estimate that 6 percent of the river's water is lost this way every year (fig. 17.20).

In tropical areas, where vegetation is abundant, decaying plant matter in reservoirs produces methane. Recent studies have calculated that methane from tropical dams has greater global warming potential than equivalent coal or oil power plants would have. Local governments and dam-building companies see a great deal of money to be made in these projects, however.

International Rivers, an environmental and human rights organization, reports that dam projects have forced more than 23 million people from their homes and land and that years later many are still suffering the impacts of dislocation. Many of the people being displaced are ethnic minorities. On India's Narmada River, for example, a proposed series of 30 dams have displaced about 1 million villagers and tribal people.

There is also increasing concern that big dams in seismically active areas can trigger earthquakes. In more than 70 cases worldwide, large dams have been linked with increased seismic activity. Geologists suggest that a devastating 7.9-magnitude earthquake in China's Sichuan province, which killed an estimated 90,000 people in 2008, may have resulted from filling the reservoir behind the nearby Zipingpu Dam on the Min River. If true, it would be the world's deadliest dam-induced earthquake.

Dams also eliminate migratory fish, such as salmon. Adult fish are blocked from migrating to upstream spawning areas.

FIGURE 17.20 Hoover Dam powers Las Vegas, Nevada. Lake Mead, behind the dam, loses about 1.3 billion m^3 per year to evaporation.
Source: Photo by Lynn Betts, USDA Natural Resources Conservation Service

And juvenile fish die if they go through hydroelectric turbines. The slack water in reservoirs behind dams is also a serious problem. Juvenile salmon evolved to ride the surge of spring runoff downstream to the ocean in two or three weeks. Reservoirs slow this journey to as much as three months, throwing off the time-sensitive physiological changes that allow the fish to survive in salt water when they reach the ocean. Reservoirs expose young salmon to predators, and warm water in reservoirs increases disease in both young and older fish. About half of the roughly 400 stocks of salmon, steelhead, and sea-run cutthroat trout in the United States are endangered, and more than 100 are already extinct due to dam construction.

Despite the many social and environmental concerns surrounding large dams, the world's governments remain eager to build them on the remaining available sites. In places where there are still undammed rivers and governments hungry for industrial growth, dams are planned on many major rivers, including the Mekong River and tributaries of the Amazon and the Congo. Chinese dam building contractors are world leaders in these efforts, having developed the industry first at home. The Yangtze River and its tribuaries alone have more than 50,000 dams. Now the Mekong River, running from the Tibetan Plateau through China, Laos, and Cambodia, is a major focus of dam development. A planned series of over 360 dams will convert this international river into a sequence of pools and reservoirs (fig. 17.21).

Damming the Mekong will provide industrial and urban power to undeveloped regions of western China and Laos. It will also eliminate fisheries, wildlife habitat, villages, and economies along the river. Cambodia's economy depends largely on seasonal flooding on the Mekong, which waters rice fields and sustains fisheries throughout the country. Cambodia has struggled to intercede in dam-building plans but with little success.

Dams have a limited lifespan

As water shortages on the Colorado River become more severe, there have been increasing calls to save Lake Mead and the Hoover Dam by eliminating Lake Powell, the reservoir behind the Glenn Canyon Dam (see opening case study). Already low water has cut power production in half at the Glenn Canyon Dam. This may be the most prominent of many U.S. dams whose purpose has been questioned. In 1998, the Army Corps of Engineers announced that it would no longer build large dams and diversion projects, in part because few available sites remain but also because political opposition is greater than it was a generation ago. Now we are starting to remove dams and restore rivers and habitat. Former interior secretary Bruce Babbitt said, "Of the 75,000 large dams in the United States, most were built a long time ago and are now obsolete, expensive, and unsafe. They were built with no consideration of the environmental costs. As operating licenses come up for renewal, removal and restoration to original stream flows will be one of the options." The tallest dam demolished in the United States, so far, is the 64 m (210 ft) Glines Canyon Dam on the Elwha River in Washington State. For the first time in a century, the Elwha flows freely to the sea, and, it's hoped, some of

FIGURE 17.21 A series of 364 dams is transforming the seasonally dynamic Mekong River and its tributaries into a sequence of flat pools.
Source: CGIAR Research Program on Water, Land and Ecosystems - Greater Mekong. https://wle-mekong.cgiar.org/maps/

the 400,000 salmon per year that once migrated up this river may be restored.

In addition to low water levels and declining power production, sediment is a common threat to dams. Rivers with high sediment loads can fill reservoirs quickly (fig. 17.22). In 1957, the Chinese government began building the Sanmenxia Dam on China's Huang He (Yellow River). From the beginning, engineers warned that the river carried so much sediment that the reservoir would have a very limited useful life. Dissent was crushed, however, and by 1960 the dam began filling the river valley and inundating fertile riparian fields that once had been part of China's traditional granaries.

Within two years, sediment accumulation behind the dam had become a serious problem. It blocked the confluence of the Wei and Yellow Rivers and backed up the Wei so much that it threatened to flood the historic city of Xi'an. By 1962, the reservoir was almost completely filled with sediment, and hydropower production dropped by 80 percent. The increased elevation of the riverbed raised the underground water table and

FIGURE 17.22 This dam is now useless because its reservoir has filled with silt and sediment.
Source: Photo by Tim McCabe, USDA Natural Resources Conservation Service

caused salinization of wells and farm fields. By 1991, the riverbed was 4.6 m above the surrounding landscape. The river is kept in check only by earthen dams that frequently fail and flood the surrounding countryside. By the time the project was complete, more than 400,000 people had been relocated, far more than planners expected.

Climate change threatens water supplies

The Intergovernmental Panel on Climate Change (IPCC) warns us that climate change threatens to exacerbate water shortages caused by population growth, urban sprawl, inefficiency, and pollution. The IPCC Fifth Assessment reported that higher evaporation rates caused by higher temperatures as well as reduced precipitation in some areas and lost snow cover in others, will sharply increase water scarcity and water stress in many areas. Models assessed by the IPCC generally show dry areas becoming drier and wet areas becoming wetter. Areas with declining precipitation include the southwestern United States, northern Africa, South Africa, central Asia, and southern Europe.

Rising temperatures will increase evaporation, as well as consumption of water by plants (transpiration). Evaporation and transpiration will reduce available water resources even in many areas where precipitation increases.

Drying vegetation also promotes wildfires. Australia, a notable example, has been experiencing extraordinary heat waves, with temperatures reaching 46.6 °C (116 °F) in early 2019. Massive wildland fires, water deficits, and dying livestock, crops, and wildlife, have become more common. The Australian government has acknowledged that recent droughts and devastating wildfires have resulted from climate change. Wealthy countries have a higher adaptive capacity—an ability to provide infrastructure and aid to their populations, but they cannot stop widespread wildfires or drought.

Declining water supplies lead to conflict and humanitarian crises. The Syrian civil war and refugee crisis followed years of the most extreme drought on record in the region. Agricultural failure resulted as irrigation water dried up, and farmers were forced to migrate to cities that could not provide them with jobs. The Syrian situation was made worse by a generation of subsidies for water-wasteful cotton production, which collapsed especially rapidly in a drying climate. But Syria is just one of many countries that subsidize extravagant agricultural water consumption. Many countries continue to do so despite an awareness of dwindling supplies—including parts of Central Asia, southern Europe, and also the United States, which raises cotton in the arid Plains of north Texas and Oklahoma and raises rice and other water-intensive crops in the intensely arid Central Valley of California.

Changing precipitation and surface water resources also affect groundwater recharge. Yemen, for example, depends entirely on groundwater, which is being depleted much more rapidly than it is replenished. Some wells in Sanaa are now 800 to 1,000 meters deep, and many are no longer usable because of the sinking water table. Millions of people may have to abandon Sanaa and other mountain cities for the coastal plain. The shrinking resource base played a role in the civil wars and the rise in international terrorism in Yemen in recent years.

Water is a growing cause of conflict

Defense analysts warn that declining water supplies could lead to wars between nations. *Fortune* magazine wrote, "Water will be to the 21st century what oil was to the 20th." For its 2009 World Water Day, the United Nations focused on transboundary water supplies. Nearly 40 percent of the world's population lives in river and lake basins shared by two or more countries. These 263 watersheds include the territory of 145 countries and cover nearly half the earth's land surface. Great reservoirs of fresh water also cross borders. There are more than 270 known transboundary aquifers.

Already we've seen skirmishes—if not outright warfare—over water access. An underlying factor in hostilities between Israel and its neighbors has been control of aquifers and withdrawals from the Jordan River. India, Pakistan, and Bangladesh also have confronted each other over water rights; and Turkey and Iraq threatened to send armies to protect access to water in the Tigris and Euphrates rivers.

Many analysts expect emerging conflict zones in some of the densely populated transboundary river basins of South and Southeast Asia (fig. 17.23). These regions rely on shrinking Himalayan glacial meltwater, at the same time as populations and water demands are growing rapidly. Most conflict so far has been restricted to diplomatic and legal channels, but violence could easily break out in some of these regions. In Africa, violence has been more prevalent. Water scarcity is an underlying cause of the conflict across sub-Saharan Africa, including ongoing genocide in the Darfur region of Sudan. When rain was plentiful, Arab pastoralists and African farmers coexisted peacefully. Drought upsets that truce. Many consider the hundreds of thousands who have fled to Chad to be climate refugees

FIGURE 17.23 Major transboundary rivers in water-stressed regions are already seeing conflicts over water rights. Many of these major rivers support densely populated regions (darker browns), and have declining water volume.

rules in place, farmers can shift in dry years from planting low-value crops, such as alfalfa, to using just part of their land to grow higher-value crops, thus keeping their income up while still using less water and fertilizer.

A description often used for such a plan is a "land bank." Other California water districts are using this same approach. Los Angeles, for example, is paying farmers to agree to fallow land in dry years. Farmers get enough income to cover their fixed costs—buildings, equipment, mortgages, and taxes—while still staying in business until better years come. The city can ensure a supply of water in bad years at a much lower cost than other alternatives.

Farmers in the Klamath Basin have also agreed to a similar approach for wetlands. They agreed to take turns flooding fields on a rotating basis so that waterfowl have a place to rest and feed. Some call this a "walking" wetlands program. No one loses his or her fields permanently, but there's a guarantee of more habitat for birds (fig. 17.24). Money to pay for both wetland mitigation and crop reductions will come from a $1 billion budget provided mainly for endangered species protection.

as well as war victims. How many more tragedies such as these might we see in the future as people struggle for declining water resources?

Section Review

1. How does agriculture affect water resources in the Aral Sea?
2. Describe some effects of dams and diversion projects.
3. How might climate change affect our water supplies?

17.4 WATER CONSERVATION

- *Conservation involves both policy and technology.*
- *Recycled water has many uses.*
- *Water prices and policies can encourage efficient water use.*

In many cases, we simply have to adapt to less water. A breakthrough agreement for the Klamath River in California, for example, shows how cooperation can be negotiated after decades of conflict. To keep enough water in the river to rebuild fish populations sufficient for sustainable tribal, recreational, and commercial fisheries, farmers have had to reduce their withdrawals. But the agreement gave farmers more reliable and predictable water allocations, as well as funding for conservation measures, in exchange for a 10 to 25 percent reduction in water consumption. In most years, farmers have to get by on less water than they had historically. In very dry years, they may have to pump groundwater or fallow—temporarily dry up—some cropland. With clearer

Desalination is expensive but needed

There have been many attempts to enhance local supplies and redistribute water. As we see in the opening case study for this chapter, building dams and enormous aqueducts has allowed us to establish farms and major cities in arid regions. Another widely attempted approach is cloud seeding—distributing condensation

FIGURE 17.24 A flock of snow geese rises from the Lower Klamath National Wildlife Refuge. Millions of migrating birds use these wetlands for feeding and resting.
Alan and Sandy Carey/Getty Images

How Does Desalination Work?

The ocean has a vast amount of water. Unfortunately, it's too salty for most human uses. As cities become more desperate for water, they are developing more technologies to remove salt from sea water. The process is energy-intensive and expensive. The most common form of desalination is distillation, which produces 85 percent of all desalinated water in the world. Water is boiled, and the steam is collected and condensed to make fresh water.

At a large scale, distillation is usually done in several stages that gradually extract freshwater from salt water. The world's largest desalination plant is the Jebel Ali Plant in the United Arab Emirates. It uses multistage flash distillation and is capable of producing 300 million cubic meters of water per year.

In a multistage distillation facility, water evaporates in a series of compartments containing heat exchangers and condensate collectors. Each compartment has a partial vacuum, which allows water to boil at a much lower temperature than 100°C (212°F). As water passes from one compartment to another, temperatures and pressures are adjusted to match the boiling point as salt concentration increases. The total evaporation in all the stages is about 15 percent of the input water.

(header photo): David L. Hansen

The main limitations of this design are the corrosion caused by the warm brine and the energy required to heat and pump water and to create a vacuum. In oil-rich countries, such as Saudi Arabia, where water costs more than pumping oil out of the ground, energy costs aren't important, but in other places it becomes a concern. However, coupling a desalination facility with a power plant can cut costs by as much as two-thirds. Waste heat from the power plant can preheat seawater, simultaneously cooling the power plant. Disposal of warm, salty brine can cause serious pollution, though, in coastal environments.

The other principal method of desalination is reverse osmosis. This filtration process removes large molecules from solutions by applying pressure to the solution on one side of a selectively permeable membrane. This is a process used by every cell in your body, as the selectively permeable cell membrane admits some molecules and excludes others. This process is one of the things that make life possible. The plasma membrane around each cell has tiny pores that allow small molecules, such as water, to pass through but that exclude large molecules, such as salt.

Ordinarily, osmosis causes small molecules to move from an area of high concentration (pure water) to areas of lower concentration (a salt solution, which has fewer water molecules per unit area because some space is occupied by the salt). You may have observed osmosis in a biology lab. If you put an amoeba in a high-salt solution, it shrivels (water is drawn out). If you put it in pure water, it bloats and explodes (the interior is saltier than the water).

Reverse osmosis drives this process backward by applying a pressure to the high-salt side of a semipermeable membrane to filter the water. Membranes, with pore sizes as small as 1 nanometer (0.000001 mm), are packed in tens of thousands of pressurized tubes. Reverse osmosis systems can range from industrial-size facilities capable of purifying hundreds of thousands of gallons per day, to a pen-size straw that you can carry in your pocket to sip water from a contaminated source.

Reverse osmosis systems produce much less freshwater globally than distillation, in part because they can be vulnerable to fouling of membranes. But because they are smaller and require less energy, they can be more mobile and easier to operate than a distillation plant. While both systems produce expensive water, they are increasingly important. Especially if you live in a dry region, you may well encounter them more in the future.

---

nuclei in humid air to help form raindrops. This approach has had mixed success, however, and there are concerns, about diverting rain from areas where it might otherwise have fallen, as well as about contamination from cloud-seeding materials.

A technology that is becoming important for increasing freshwater supplies is desalination of ocean water or brackish saline lakes and lagoons. Worldwide, over 18,000 desalination plants produce over 50 billion liters (23 billion gallons) of water a day (Exploring Science, above). Middle Eastern oil-rich states produce about 60 percent of desalinated water. Saudi Arabia is the largest single producer, at about one-third of the world total. The United States is second, at 20 percent of the world supply. Some American cities, such as Tampa, Florida, and San Diego, California, now depend on desalination for a significant part of their water resources. Although desalination is still three to four times more expensive than most other sources of fresh water, it provides a welcome water source in such places as Oman and Bahrain, where there is no other access to fresh water.

Domestic conservation has important impacts

We could probably save as much as half of the water we now use for domestic purposes without great sacrifice or serious changes in our lifestyles. Simple steps, such as taking shorter showers, stopping leaks, and washing cars, dishes, and clothes as efficiently as possible, can go a long way toward forestalling the water shortages that many authorities predict. Isn't it better to adapt to more conservative uses now, when we have a choice, than to be forced to do it by scarcity in the future?

The use of resource-conserving appliances, such as low-volume shower heads and efficient dishwashers and washing machines, can reduce water consumption greatly (What Can You Do?, p. 400). If you live in an arid region, you might consider whether you really need a lush green lawn that requires constant watering, feeding, and care. Planting native ground cover in a "natural lawn" or developing a rock garden or landscape in harmony with the surrounding ecosystem can be both ecologically sound and aesthetically pleasing (fig. 17.25). There are about

Saving Water and Preventing Pollution

Each of us can conserve much of the water we use and avoid water pollution in many simple ways.

- Take short showers; don't flush every time you use the toilet; wash your car infrequently.
- Don't let the faucet run while washing hands, dishes, food, or brushing your teeth. Draw a basin of water for washing and another for rinsing dishes. Don't run the dishwasher when half full.
- Dispose of used motor oil and household hazardous waste responsibly. Don't dump anything down a storm sewer that you wouldn't want to drink.
- Avoid using toxic or hazardous chemicals for simple cleaning or plumbing jobs. A plunger or plumber's snake will often unclog a drain just as well as caustic acids or lye.
- If you have a lawn, use water sparingly. Water your grass and garden at night, not in the middle of the day. Consider planting native plants, low-maintenance ground cover, a rock garden, or some other xeriphytic landscaping.
- Use water-conserving appliances: low-flow showerheads, low-flush toilets, and aerated faucets.
- Use recycled (gray) water for lawns, house plants, and car washing.
- Check your toilet for leaks. A leaky toilet can waste 50 gallons per day. Add a few drops of dark food coloring to the tank and wait 15 minutes. If the tank is leaking, the water in the bowl will have changed color.

FIGURE 17.25 By using native plants in a natural setting, residents of Phoenix save water and fit into the surrounding landscape.
William P. Cunningham

In many developing countries, as much as 70 percent of all the agricultural water used is lost to leaks in irrigation canals, application to areas where plants don't grow, runoff, and evaporation. Better farming techniques, such as leaving crop residue on fields and ground cover on drainage ways, intercropping, use of mulches, and low-volume irrigation, could reduce these water losses dramatically.

30 million ha (75 million acres) of cultivated lawns, golf courses, and parks in the United States. They receive more water, fertilizer, and pesticides per hectare than any other kind of land.

Recycling can reduce consumption

Significant amounts of water can be reclaimed and recycled. In California, water recovered from treated sewage constitutes the fastest-growing water supply, growing about 30 percent per year. Despite public squeamishness, purified sewage effluent is being used for everything from agricultural irrigation to flushing toilets (fig. 17.26). In a statewide first, San Diego is currently piping water from the local sewage plant directly into a drinking-water reservoir. Residents of Singapore and Queensland, Australia, also are now drinking purified sewage effluent. While most people find the notion distasteful, treated water can be made just as clean as any other source for drinking. Most modern wastewater treatment should, theoretically, make water safe enough to swim in. Higher-level treatment routinely makes water even cleaner than many standard drinking water sources.

FIGURE 17.26 Recycled water is being used in California and Arizona for everything from agriculture to landscaping to industry. While we are often squeamish about drinking recycled water, it is increasingly necessary—and anyway, all water is recycled one way or another.
William P. Cunningham

Prices and policies have often discouraged conservation

Through most of U.S. history, water policies have generally worked against conservation. In the well-watered eastern United States, water policy was based on riparian use rights—those who lived along a river bank had the right to use as much water as they liked as long as they didn't interfere with its quality or availability to neighbors downstream. It was assumed that the supply would always be endless and that water had no value until it was used. In the drier western regions where water is scarce, water law is based primarily on the Spanish system of prior appropriation rights, or "first in time are first in right." Even if the prior appropriators are downstream, they can legally block upstream users from taking or using water flowing over their property. But the appropriated water has to be put to "beneficial" use by being consumed. This creates a policy of "use it or lose it." Water left in a stream, even if essential for recreation, aesthetic enjoyment, or to sustain ecological communities, is not being appropriated or put to "beneficial" (that is, economic) use. Under this system, water rights can be bought and sold, but water owners frequently are reluctant to conserve water for fear of losing their rights.

In most federal "reclamation" projects, customers are charged only for the immediate costs of water delivery. The costs of building dams and distribution systems is subsidized, and the potential value of competing uses is routinely ignored. Farmers in California's Central Valley, for instance, for many years paid only about one-tenth of what it cost the government to supply water to them. This didn't encourage conservation. Subsidies created by underpriced water amounted to as much as $500,000 per farm per year in some areas.

Growing recognition that water is a precious and finite resource has changed policies and encouraged conservation across the United States. Despite a growing population, the United States is now saving some 144 million liters (38 million gal) per day, compared to 1980. With 100 million more people in the United States now than in 1980, we get by with 10 percent less water. New requirements for water-efficient fixtures and low-flush toilets in many cities help to conserve water on the home front. More efficient irrigation methods on farms also are a major reason for the downward trend.

Charging a higher proportion of real costs to users of public water projects has helped encourage conservation, and so have water marketing policies that allow prospective users to bid on water rights. Both the United States and Australia have had effective water pricing and allocation policies that encourage

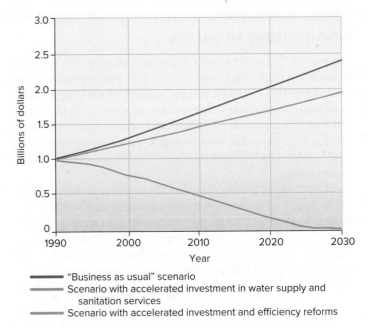

FIGURE 17.27 Three scenarios for government investments on clean water and sanitation services, 1990 to 2030.

Source: World Bank estimates based on the research paper by Dennis Anderson and William Cavendish, "Efficiency and Substitution in Pollution Abatement: Simulation Studies in Three Sectors."

the most socially beneficial uses and discourage wasteful uses of water. Market mechanisms for water allotment can be sensitive, however, in developing countries where farmers and low-income urban residents could be outbid for irreplaceable water supplies.

It will be important, as water markets develop, to be sure that environmental, recreational, and wildlife values are not sacrificed to the lure of high-bidding industrial and domestic uses. Given prices based on real costs of using water and reasonable investments in public water supplies, pollution control, and sanitation, the World Bank estimates that everyone in the world could have an adequate supply of clean water by the year 2030 (fig. 17.27).

Section Review

1. How have farmers in California's Klamath River Basin reduced their water use?

2. Describe some examples of domestic water conservation.

3. Give an example in which water policies and prices have encouraged conservation.

Connecting the Dots

Water is our most universally important resource. We use it to produce food and energy and to run cities and industries. Most fundamentally, we need clean water to drink. In the United States, Lake Mead demonstrates how our dependency on water has grown and how precarious this resource can become. Lake Mead, and the Colorado River that runs through it, provide irrigation, urban water, and the hydropower that lights Las Vegas and other cities in the region.

Like many of the world's rivers, the Colorado River originates in the nearby mountains, where summer rains and winter snows provide year-round (though seasonally variable) runoff. Water is unevenly distributed in space as well as in time in the Colorado's watershed. Moisture is abundant in the Rocky Mountains, where conditions are cool and mountains capture atmospheric moisture. The arid lands downstream see little rain, and cities and farms depend almost completely on runoff from elsewhere. Groundwater also provides critical resources, but overpumping has diminished this resource in many areas. As climate warming brings shorter winters and longer, hotter summers, the reliability of runoff, and the future of cities and farms that depend on it, are uncertain.

These conditions characterize many of the world's great rivers. Especially in the populous river basins of Asia, billions of people depend on rivers, groundwater, and seasonal rainfall for survival. Those relying on glacial melt are especially at risk.

We do have strategies for reducing these risks, however. Smarter policies about water pricing, better irrigation and farming practices, and low-flow household appliances are a few of these. Careful stewardship of the world's water resources is likely to become an increasing theme in years to come.

Reviewing Key Terms

Can you define the following terms in environmental science?

aquifers 17.1

artesian well 17.1

consumption 17.2

discharge 17.1

groundwater 17.1

hydrologic cycle 17.1

infiltration 17.1

rain shadow 17.1

recharge zone 17.1

renewable water supplies 17.2

residence time 17.1

saltwater intrusion 17.3

subsidence 17.3

water scarcity 17.2

water stress 17.2

water table 17.1

withdrawal 17.2

zone of aeration 17.1

zone of saturation 17.1

Critical Thinking and Discussion Questions

1. What changes might occur in the hydrologic cycle if our climate were to warm or cool significantly?

2. Farms dominate water consumption in most regions. You are a policymaker in charge of agricultural water conservation California or Arizona. What policies would you promote?

3. Are there ways you could use less water in your own personal life? What obstacles prevent you from taking these steps?

4. Should we use up underground water supplies now or save them for some future time?

5. How should we compare the values of free-flowing rivers and natural ecosystems with the benefits of flood control, water diversion projects, hydroelectric power, and dammed reservoirs?

6. Would it be feasible to change from flush toilets and using water as a medium for waste disposal to some other system? What might be the best way to accomplish this?

Data Analysis

Graphing Global Water Stress and Scarcity

According to the United Nations, water stress is when annual water supplies drop below 1,700 m^3 per person. Water scarcity is defined as annual water supplies below 1,000 m^3 per person. More than 2.8 billion people in 48 countries will face either water stress or scarcity conditions by 2025. Of these 48 countries, 40 are expected to be in West Asia or Africa. By 2050, far more people could be facing water shortages, depending on both population projections and scenarios for water supplies based on global warming and consumption patterns.

To explore some of the issues and questions about future water scarcity, go to Connect, and answer questions about this figure and others from this chapter.

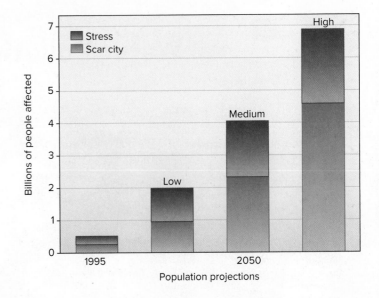

18

Water Pollution

▲ Daily bathing in the Ganges River confers ritual purity, but pollution makes this practice physically risky.
dbimages /Alamy Images

Learning Outcomes

After studying this chapter, you should be able to:

18.1 Identify major water contaminants and their sources.
18.2 Compare sources of pollutants.
18.3 Explain water pollution control.
18.4 Describe major water legislation.

"If there is magic on this planet, it is contained in water."

– Loren Eiseley

India's Holy River

Early each morning, residents of Varanasi, India, go to the river to purify themselves in the water of the holy Ganges. Mother Ganga, as the river is known, is considered a goddess on earth, and the river confers purity on bathers making their daily ablutions. Pilgrims travel to the Ganges from across India, bringing home bottles of the river's sacred water. After death, many hope to be cremated on the banks of the river, especially in the ancient and holy city of Varanasi, and then have Mother Ganga carry away their ashes.

The Ganges is also one of the world's most important rivers. Starting as an icy stream rushing from the mouth of the Gangotri Glacier high in the Himalayas, the river falls over 3,500 m (over 11,300 ft) to the plains at the foot of the mountains. Flowing through India's most populous region (fig. 18.1), the river provides water to more than 500 million people—nearly 40 percent of all Indians and 1 in 15 of the world's entire human population. The Ganges River Plain holds some of India's most fertile farmland, much of it irrigated by the river. The river also supplies water to industries and power plants, and its hydroelectric dams provide electricity for India's cities. Far downstream, 2,500 km from its glacial headwaters, the river deposits Himalayan sediment in the Ganges delta, home to rich mangrove forests and millions of farmers and fishers.

The river also carries away the waste of hundreds of polluting industrial sites—chemical factories, fabric dying industries, tanneries, mines, and other sources—as well as plastic and other trash, animal carcasses, agricultural runoff (including fertilizers and pesticides), and the ashes of cremated bodies. And it carries sewage, most of it untreated, from the half a billion people living in the watershed. Even while it is a holy river, the Ganges is one of the world's most severely polluted rivers. Sewage effluent accounts for some 80 percent of the river's pollution, and industrial effluent makes up the other 20 percent.

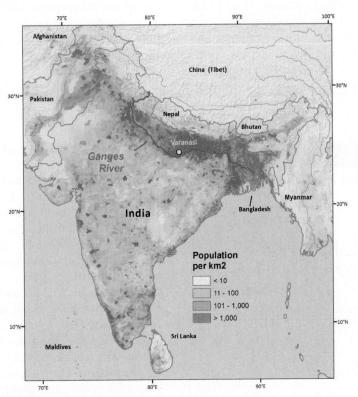

FIGURE 18.1 Population density in South Asia. The densely populated Ganges River Plain follows the southern edge of the Himalayas, which are the source of most of Asia's most important rivers.
Source: Center for International Earth Science Information Network (CIESIN) Columbia University. 2015. Gridded Population of the World, Version 4.

The effects of water pollution, and the strategies for controlling it, depend on what the pollutants are. Organic waste and nutrients support bacteria, which can cause illnesses ranging from diarrhea to cholera. Coliform bacteria, which grow in nutrient-rich water, occur at 100 times the legal standards in many areas of the Ganges. Decomposition of organic waste also depletes oxygen in the water, asphyxiating fish and other aquatic life and leading to "dead zones" with little oxygen and few living organisms. Industrial waste, especially chemical effluent or metals, such as chromium or arsenic, is toxic in low concentrations, especially with frequent exposure. These pollutants, and agricultural pesticides, cause cancer or damage nerves or organs.

Cleaning the Ganges is a stated goal of many Indians, including Prime Minister Narendra Modi. The river has extraordinary cultural and historical importance—the bathing steps in Varanasi are just the most famous of the countless cultural and religious places. While the Ganges is undeniably filthy, it also deserves the respectful treatment owed to a sacred river. In 2014, Mr. Modi's newly elected government initiated a plan to protect the river, its cultural heritage, and public health, starting with new efforts to monitor pollution sources, and then to control them in the dozens of major cities, hundreds of factories, and thousands of sewage outfalls that line the river and its tributaries.

This was far from the first such effort. Studies in 1981 led to the Ganga Action Plan, an effort to identify and address dominant pollution sources. Over the decades, though, ineffective enforcement has led to a lack of action at all levels of government and industry. Another cleanup campaign was announced in 2010 to treat municipal sewage waste and regulate industrial effluent. A year later, the World Bank gave $1 billion to support cleanup efforts. Prime Minister Modi's 2014 initiative included higher budgets, an army battalion designated as an Eco Task

Force, new legislative efforts, and better coordination among ministers. With international funding, the National Ganga River Basin Authority pledged to stop all untreated sewage and industrial inputs by 2020. Industrial cooperation has been slow thus far, however.

Like many rivers of the world's rivers, the Ganges also faces serious threats from climate change. With warming conditions, Himalayan glaciers have been retreating for decades. Many of the world's major agricultural areas and cities depend on Himalayan ice melt, and declining supplies threaten cities, farmlands, and

hydropower. Lower water levels also reduce a river's ability to absorb and dilute pollutants.

Clean water is one of the most important resources we have. Protecting clean water, or restoring it, is one of our most difficult public challenges. On the other hand, water is constantly recycled and replenished. If we can reduce pollution inputs, rivers become clean again. This has been observed in many countries that have managed to institute and enforce policies for protecting water quality. In this chapter, we examine the major types of water pollution, where they come from, and how to control them to protect clean water for everyone.

Benchmark Data	
Among the ideas and values in this chapter, the following are a few worth remembering.	
2 ppm	Oxygen concentration in water needed for most aquatic life
679	Global dead zones recorded 1950–2000
24	Global dead zones recorded 1700–1940
1	Rank of pathogens among U.S. water impairments
2	Rank of nutrients among U.S. water impairments

18.1 WATER POLLUTION

- *Waterborne pathogens kill hundreds of millions of people.*
- *Nutrient enrichment leads to eutrophication.*
- *Organic pollutants include drugs, pesticides, and other chemicals.*

The pollutants plaguing the Ganges reflect the impacts of growing, urbanizing, industrializing populations the world over. Rapidly growing cities, especially in low-income regions, cannot build sewage treatment facilities fast enough to keep up. Indian cities have the capacity to treat only one-third of the sewage produced, and more than two-thirds of the country's monitored river miles are severely polluted. India is an extreme case: It is one of the world's fastest-growing countries, and it is soon to be the most populous, with over 1.3 billion people, many of them living in deep poverty. But many developing countries share similar growing pains. How can these conditions be improved?

Most students today are too young to appreciate that water in most industrialized countries was once far more polluted and dangerous than it is now. Until the 1970s, it was common to dump toxic solvents, organic chemicals, mine waste, and raw sewage into U.S. surface waters, or onto the ground, where it seeps into

groundwater. We are still paying billions of dollars every year to clean up those contaminants of the past. These practices began to change in the United States when President Nixon signed the Clean Water Act in 1972. This law has been called the United States' most successful and popular environmental legislation. It established a goal that all the nation's waters should be "fishable and swimmable." While this goal is far from being achieved, the Clean Water Act remains popular because it protects public health (thus saving taxpayer dollars) and reduces environmental damage. Clean water is also valued for its aesthetic appeal. The view of a clean lake, river, or seashore makes people happy, and water provides for recreation, so many people feel their quality of life has improved as water quality has been restored. It is clear that strong policies with reliable enforcement can make a difference.

We still have a long way to go in improving water quality. In the United States and other wealthy countries, pollution from factory pipes has been vastly reduced in recent decades, but erosion from farm fields, construction sites, and streets has gotten worse in many areas. Airborne mercury, sulfur, and other substances continue to contaminate lakes and wetlands. Concentrated livestock production and agricultural runoff threaten underground water as well as surface water systems. Increasing industrialization in developing countries has led to widespread water pollution in impoverished regions with little environmental regulation. Much of this chapter outlines the U.S. experience with water quality, but the same types of pollutants, and similar challenges for pollution control, exist worldwide.

Water pollution is anything that degrades water quality

Any physical, biological, or chemical change in water quality that adversely affects living organisms or makes water unsuitable for desired uses might be considered **water pollution.** There are natural sources of water contamination, such as poison springs, oil seeps, and sediment from erosion, but in this chapter we focus primarily on human-caused factors that affect water quality.

FIGURE 18.2 Point sources of pollution, such as sewer outfalls, industrial effluent pipes, acid draining out of abandoned mines, are relatively easy to recognize and to monitor.
Vastram/Shutterstock

FIGURE 18.3 Non-point sources are the leading unresolved cause of water pollution. This scene looks peaceful and idyllic, but allowing cows free access to streams and ponds is a major cause of bank erosion and water contamination.
David Forman/Image Source

Pollution-control standards and regulations usually distinguish between point and nonpoint pollution sources. Factories, power plants, sewage treatment plants, underground coal mines, and oil wells are classified as **point sources** because they discharge pollution from specific locations, such as drain pipes, ditches, or sewer outfalls (fig. 18.2). These sources are discrete and identifiable, so they are relatively easy to monitor and regulate. It is generally possible to divert effluent from the waste streams of these sources and treat it before it enters the environment.

In contrast, **nonpoint sources** of water pollution are scattered or diffuse, having no specific location where they discharge into a particular body of water. Nonpoint sources include runoff from farm fields and livestock feedlots and pastures (fig. 18.3), golf courses, lawns and gardens, construction sites, logging operations, roads, streets, and parking lots. Whereas point sources may be fairly uniform and predictable throughout the year, nonpoint sources are often highly episodic. The first heavy rainfall after a dry period may flush high concentrations of gasoline, lead, oil, and rubber residues off city streets, for instance, while subsequent runoff may have lower levels of these pollutants. In some areas, spring snowmelt carries high levels of atmospheric acid deposition into streams and lakes. The irregular timing of these events, their multiple sources, and their scattered locations make them much more difficult to monitor, regulate, and treat than point sources.

A particularly diffuse type of nonpoint pollution is **atmospheric deposition** of contaminants carried by air currents and precipitated into watersheds or directly onto surface waters as rain, snow, or dry particles. The Great Lakes, for example, have been found to be accumulating industrial chemicals such as PCBs and dioxins, as well as agricultural toxins such as the insecticide toxaphene, that cannot be accounted for by local sources alone. The nearest sources for many of these chemicals are sometimes thousands of kilometers away (chapter 16).

Amounts of these pollutants can be quite large. It is estimated that there are 600,000 kg of the herbicide atrazine in the Great Lakes, most of which is thought to have been deposited from the atmosphere. Several studies have indicated health problems among people who regularly eat fish, which accumulate toxic compounds in the food chain, from the Great Lakes.

Ironically, lakes can be pollution sources as well as recipients. In the past 12 years, about 26,000 metric tons of PCBs "disappeared" from Lake Superior. Apparently these compounds evaporate from the lake surface and are carried by air currents to other areas where they are redeposited.

Although the types, sources, and effects of water pollutants are often interrelated, it is convenient to divide them into major categories for discussion (table 18.1). Examine this table closely as you consider important sources and effects of each type of pollutant in the sections that follow.

Infectious agents, or pathogens, cause diseases

The most serious water pollutants in terms of human health worldwide are pathogenic organisms (chapter 8). Among the most important waterborne diseases are typhoid, cholera, bacterial and amoebic dysentery, enteritis, polio, infectious hepatitis, and schistosomiasis. Malaria, yellow fever, and filariasis are transmitted by insects that have aquatic larvae. Altogether, at least 25 million deaths each year are blamed on these water-related diseases. Nearly two-thirds of the mortalities of children under 5 years old are associated with waterborne diseases.

The main source of these pathogens is untreated or improperly treated human wastes. Animal wastes from feedlots or fields near waterways and food-processing factories with inadequate waste treatment facilities also are sources of disease-causing organisms.

In wealthier countries, sewage treatment plants and other pollution-control techniques have reduced or eliminated most of the worst sources of pathogens in inland surface waters, and drinking water is generally disinfected by chlorination, so epidemics of waterborne diseases are rare in these countries. The United

Table 18.1 Major Categories of Water Pollutants

Category	Examples	Sources
A. Causes Health Problems		
1. Infectious agents	Bacteria, viruses, parasites	Human and animal excreta
2. Organic chemicals	Pesticides, plastics, detergents, oil, and gasoline	Industrial, household, and farm use
3. Inorganic chemicals	Acids, caustics, salts, metals	Industrial effluents, household cleansers, surface runoff
4. Radioactive materials production, natural sources	Uranium, thorium, cesium, iodine, radon	Mining and processing of ores, power plants, weapons
B. Causes Ecosystem Disruption		
1. Sediment	Soil, silt	Land erosion
2. Plant nutrients	Nitrates, phosphates, ammonium	Agricultural and urban fertilizers, sewage, manure
3. Oxygen-demanding wastes	Animal manure and plant residues	Sewage, agricultural runoff, paper mills, food processing
4. Thermal	Heat	Power plants, industrial cooling

Nations estimates that 90 percent of the people in developed countries have adequate (safe) sewage disposal, and 95 percent have clean drinking water.

The situation is quite different in poor countries, as noted in the opening case study for this chapter. The United Nations estimates that at least 2.5 billion people in these countries lack adequate sanitation, and 780 million lack access to clean drinking water. Conditions are especially bad in remote rural areas where sewage treatment is usually primitive or nonexistent and purified water is either unavailable or too expensive. The World Health Organization estimates that 80 percent of all sickness and disease in less-developed countries can be attributed to waterborne infectious agents and inadequate sanitation.

The World Bank calculates that if everyone had pure water and satisfactory sanitation, 200 million fewer episodes of diarrheal illness would occur each year, and 2 million childhood deaths would be avoided. Furthermore, 450 million people would be spared debilitating roundworm or fluke infections. Access to clean water is a major focus of the UN's Sustainable Development Goals (chapter 1).

Detecting specific pathogens in water is difficult, time-consuming, and costly; thus, water quality tests usually analyze water for the presence of **coliform bacteria,** any of the many types that live in the colon or intestines of humans and other animals. The most common of these is *Escherichia coli* (or *E. coli*). Many strains of bacteria live symbiotically, and beneficially, in mammals, but some, such as *Shigella, Salmonella,* or *Lysteria,* can cause fatal diseases. It is usually assumed that if any coliform bacteria are present in a water sample, infectious pathogens are present also.

How do we test for coliform bacteria? Water is filtered through a paper filter, which is then placed in a dish containing a nutrient medium that supports bacterial growth. After 24 hours in an incubator, living cells will have produced small colonies. If *any* colonies are found in drinking water samples, the U.S. Environmental Protection Agency considers the water unsafe and requiring disinfection. The EPA-recommended maximum coliform count for swimming water is 200 colonies per 100 ml of water, but some cities and states

allow higher levels. If the limit is exceeded, the contaminated pool or beach usually is closed to swimming (fig. 18.4).

Low oxygen levels indicate nutrient contamination

The amount of oxygen dissolved in water is a good indicator of water quality and of the kinds of life it will support. Water with an oxygen content above 6 parts per million (ppm) will support game fish and other desirable forms of aquatic life. Water with less than 2 ppm oxygen will support mainly worms, bacteria, fungi, and other detritus feeders and decomposers. Oxygen is added to water by diffusion from the air, especially when turbulence and mixing rates are high, and by photosynthesis of green plants, algae, and cyanobacteria. Oxygen is removed from water by respiration and chemical processes that consume oxygen.

Organic waste, such as sewage, paper pulp, or food waste, is rich in nutrients, especially nitrogen and phosphorus. These

FIGURE 18.4 Sewage waste is a leading cause of water contamination. Consequences can include high nutrient concentrations and the presence of fecal bacteria.

budgetstockphoto/Getty Images

nutrients stimulate the growth of oxygen-demanding decomposing bacteria. **Biochemical oxygen demand (BOD)** is thus a useful test for the presence of organic waste in water. Most BOD tests involve incubating a water sample for five days, and then comparing oxygen levels in the water before and after incubation. An alternative method, called the chemical oxygen demand (COD), uses a strong oxidizing agent to completely break down all organic matter in a water sample. This method is much faster than the BOD test, but it records inactive organic matter as well as bacteria, so it is less useful as an indicator of biological activity. A third method of assaying pollution levels is to measure **dissolved oxygen (DO) content** directly, using an oxygen electrode. The DO content of water depends on factors other than pollution (for example, temperature and aeration), so it indicates general conditions in the water body, rather than nutrients and bacteria in particular.

The effects of oxygen-demanding wastes on rivers depends to a great extent on the volume, flow, and temperature of the river water. Aeration occurs readily in turbulent, rapidly flowing rivers, which are therefore often able to recover quickly from oxygen-depleting processes. Downstream from a point source, such as a municipal sewage plant discharge, a characteristic decline and restoration of water quality can be detected either by measuring dissolved oxygen content or by observing the flora and fauna that live in successive sections of the river.

The oxygen decline downstream of a pollutant source is called the **oxygen sag** (fig. 18.5). Upstream from the pollution source, oxygen levels support normal populations of clean-water organisms. Immediately below the source of pollution, oxygen levels begin to fall as decomposers consume waste materials. Rough fish, such as carp, bullheads, and gar, are able to survive in this oxygen-poor environment where they eat both decomposer organisms and the waste itself. Farther downstream, the water may become so oxygen-depleted that almost nothing but the most resistant microorganisms and invertebrates can survive (below 2 ppm). We sometimes call this low-oxygen environment a "dead zone." Eventually most of the nutrients are used up, decomposer populations decline, and oxygen levels recover as oxygen diffuses into water from the air, or is released by aquatic plants. Depending on the volumes and flow rates of the effluent plume and the river receiving it, normal communities may not appear for several miles downstream.

Nutrient enrichment leads to cultural eutrophication

Water clarity (transparency) is affected by sediments, chemicals, and the abundance of plankton organisms, and is a useful measure of water quality and water pollution. Rivers and lakes that have clear water and low biological productivity are said to be **oligotrophic** (*oligo* = little + *trophic* = nutrition). By contrast, **eutrophic** (*eu* + *trophic* = truly nourished) waters are rich in organisms and organic materials. Eutrophication is an increase in nutrient levels and biological productivity. Some amount of eutrophication is a normal part of successional changes in most lakes. Tributary streams bring in sediments and nutrients that stimulate plant growth. Over time, ponds or lakes may fill in, eventually becoming marshes. The natural rate of eutrophication and succession depends on water chemistry and depth, volume of inflow, mineral content of the surrounding watershed, and the biota of the lake itself.

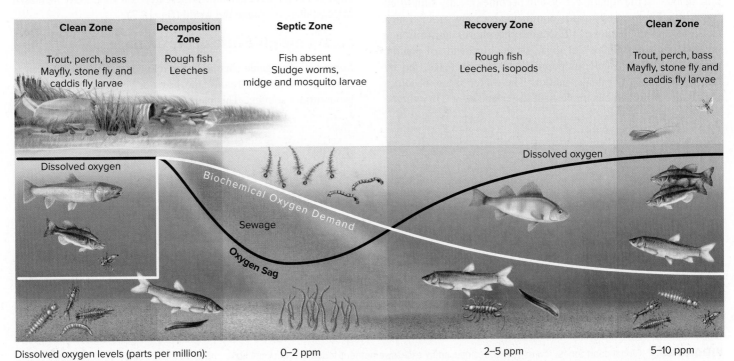

FIGURE 18.5 Oxygen sag downstream of an organic source. A great deal of time and distance may be required for the stream and its inhabitants to recover.

FIGURE 18.6 Eutrophic lake. Nutrients from agriculture and domestic sources have stimulated growth of algae and aquatic plants. This reduces water quality, alters species composition, and lowers the lake's recreational and aesthetic values.
William P. Cunningham

As with BOD, nutrient enrichment sewage, fertilizer run-off, even decomposing leaves in street gutters, can produce a human-caused increase in biological productivity called **cultural eutrophication.** Cultural eutrophication can also result from higher temperatures, more sunlight reaching the water surface, or a number of other changes.

Water bodies can vary from mildly eutrophic to extremely eutrophic, as nutrient enrichment increases. In strongly eutrophic instances, the aquatic ecosystem becomes greatly simplified (with little biodiversity) and is dominated by a few groups, such as algae and decomposers. Elevated phosphorus and nitrogen levels stimulate "blooms" of algae or thick growths of aquatic plants (fig. 18.6). Bacterial populations also increase, fed by larger amounts of organic matter. The water often becomes cloudy or turbid and has unpleasant tastes and odors. In extreme cases, plants and algae die and decomposers deplete oxygen in the water.

Eutrophication can cause toxic tides and "dead zones"

Eutrophication in marine waters often contributes to a **red tide,** a bloom of deadly aquatic microorganisms. Red tides occur naturally, starting with blooms of toxic algae that develop offshore. As they wash in toward nutrient-rich coastal waters, however, those nutrients enhance the growth of toxic algae, making it reach farther and last longer.

In recent years, Florida's tourism and fishing industries have been devastated by red tides that spread up the Gulf Coast, fed by agricultural and urban nutrients. Producing noxious odors and toxins that can kill fish, birds, and other organisms, events like these have become increasingly common in nearshore waters and partially enclosed bays or estuaries.

Urban, agricultural, and industrial development are chiefly responsible for eutrophication. During the tourist season, the coastal population of the Mediterranean, for example, swells to 200 million people. Eighty-five percent of the effluent from large cities goes untreated into the sea. Beach pollution, fish kills, and contaminated shellfish result. Enclosed bays or water bodies, such as the Gulf of Mexico, the Caspian Sea, the Baltic, and China's Bohai Bay, tend to be in especially critical condition. Extensive "dead zones" often form where rivers carry oxygen-depleting nutrients into estuaries and shallow seas. One of the largest hypoxic (oxygen-depleted) zones in the world occurs during summer months in the Gulf of Mexico at the mouth of the Mississippi River (see chapter 3). As human populations, cities, and agriculture have expanded, these hypoxic zones have increased in coastal areas around the world (fig. 18.7).

Heavy metals cause nerve damage

Some toxic inorganic chemicals are released from rocks by weathering, are carried by runoff into lakes or rivers, or percolate into groundwater aquifers. This pattern is part of natural mineral cycles

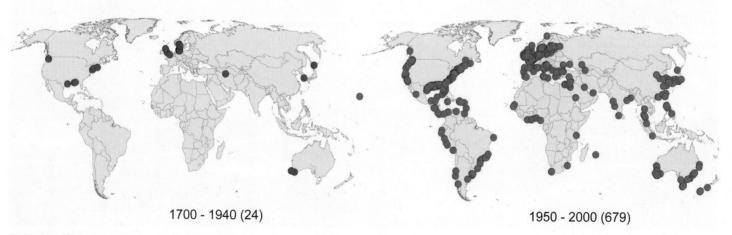

1700 - 1940 (24)   1950 - 2000 (679)

FIGURE 18.7 Global dead zones, low-oxygen zones caused by excessive nutrient loads and eutrophication, have multiplied since 1950. Once rare, they now occur in urban and industrial ports and estuaries worldwide.

(chapter 3). Humans often accelerate the transfer rates in these cycles thousands of times above natural background levels through the mining, processing, using, and discarding of minerals. These processes introduce a variety of heavy metals to waterways, such as mercury, lead, tin, and cadmium. Other toxic elements, such as selenium and arsenic, also have reached hazardous levels in some waters.

The primary metal contaminants of health concern include mercury, lead, cadmium, tin, and nickel, all highly toxic in minute concentrations. These are widely used in industrial processes, such as battery production or metal processing and plating, and they are released in coal combustion. Metals persist in the environment, so they accumulate in food webs and have stronger effects upon top predators—including humans.

Mercury, released from coal-burning power plants, is the most widespread toxic metal contamination problem in North America and many other industrialized regions. Consuming fish is our main source of human mercury exposure. As noted in chapter 16, a 2007 study tested more than 2,700 fish from 636 rivers and streams in 12 western states, and mercury was found in every one of them. More than half the fish contained mercury levels unsafe for women of childbearing age, and three-quarters exceed the safe limit for young children. Fifty states have issued advisories about eating freshwater or ocean fish, most often because of mercury contamination. Top marine predators, such as shark, swordfish, marlin, king mackerel, and blue-fin tuna, should be avoided completely.

Naturally occurring arsenic in groundwater is an important inorganic contaminant in some regions, as in West Bengal, India, and eastern Bangladesh (fig. 18.8). Arsenic occurs naturally in the sediments that make up the Ganges River delta, one of many mineral components eroded from the Himalayan source of the river

over millions of years. Rapid population growth, industrialization, and intensification of irrigated agriculture have depleted or polluted limited surface water supplies. In an effort to provide clean drinking water for local residents, thousands of tube wells were drilled during the 1960s. But by the 1980s, health workers became aware of widespread signs of chronic arsenic poisoning among Bengali villagers. Symptoms include watery and inflamed eyes, gastrointestinal cramps, gradual loss of strength, dry skin and skin tumors, anemia, confusion, and eventually death. Health workers estimate that the total number of potential victims in India and Bangladesh may exceed 100 million people. Fortunately, arsenic can be filtered from water supplies once it is discovered (Exploring Science, p. 423).

Acidic runoff can destroy aquatic ecosystems

Acids are released as by-products of industrial processes such as leather tanning, metal smelting and plating, petroleum distillation, and organic chemical synthesis. Coal mining is an especially important source of acid water pollution. Sulfur compounds in coal react with oxygen and water to make sulfuric acid. Thousands of kilometers of streams in the United States have been acidified by acid mine drainage, some so severely that they are essentially lifeless.

Coal and oil combustion also leads to formation of atmospheric sulfuric and nitric acids (chapter 16), which are deposited with precipitation (acidic rain, snow, fog, or dry deposition) in surface waters. Where soils are rich in such alkaline material as limestone, these atmospheric acids have little effect because they are neutralized. In high mountain areas or recently glaciated regions where bedrock is granitic and lakes are oligotrophic, however, there is little buffering capacity (ability to neutralize acids), and aquatic ecosystems can be severely disrupted. These effects were first recognized in the mountains of northern England and Scandinavia about 50 years ago.

Aquatic damage due to acid precipitation has been reported in about 200 lakes in the Adirondack Mountains of New York State and in several thousand lakes in eastern Quebec, Canada. Game fish, amphibians, and sensitive aquatic insects are generally the first to be killed by increased acid levels in the water. If acidification is severe enough, aquatic life is limited to a few resistant species of mosses and fungi. Increased acidity may result in leaching of toxic metals, especially aluminum, from soil and rocks, making water unfit for drinking or irrigation as well.

Organic pollutants include drugs, pesticides, and industrial products

Thousands of different natural and synthetic organic (carbon-based) chemicals are used in the chemical industry to make pesticides, plastics, pharmaceuticals, pigments, and other products that we use in everyday life. Many of these chemicals are highly toxic (chapter 8). Exposure to very low concentrations (perhaps even parts per quadrillion, in the case of dioxins) can cause birth defects, genetic disorders, and cancer. Some can persist in the environment

FIGURE 18.8 Bangladesh occupies the low-lying delta of the Ganges River. Arsenic, deposited in delta sediments over millions of years, often contaminates well water.

Data source: Center for International Earth Science Information Network (CIESIN) Columbia University. 2015. Gridded Population of the World, Version 4.

because they are resistant to degradation and toxic to organisms that ingest them. Contamination of surface waters and groundwater by these chemicals is a serious threat to human health.

The two most important sources of toxic organic chemicals in water are improper disposal of industrial and household wastes and runoff of pesticides from farm fields, lawns, golf courses, and other areas where they are used in large quantities. The U.S. EPA estimates that about 500,000 metric tons of pesticides are used in the United States each year. Because these substances are organic (carbon-based) and dissolve in fatty tissues, they tend to be stored in living organisms. As they pass through the food chain, they accumulate, often to levels that can cause nerve damage and reproductive failure. This phenomenon became notorious when widespread use of the pesticide DDT decimated populations of bald eagles, pelicans, and other predatory birds in the 1960s.

More recently, studies have found high concentrations of dioxins and other chlorinated hydrocarbons (hydrocarbon molecules that contain chlorine atoms in the fat of salmon, fish-eating birds, and humans. Many of these compounds disrupt endocrine hormone pathways that have widespread effects on growth and development.

Atrazine, the second most widely used herbicide in America, has been shown to disrupt normal development in frogs at concentrations as low as 0.1 ppb. This level is very common in farming regions. Could this be a problem for humans as well as frogs? Another of the most widely distributed water contaminants is triclosan, an antibiotic used in thousands of products including toothpaste, hand soap, face wash, body lotion, and cosmetics. Triclosan survives wastewater treatment, and chlorination, the final stage of drinking water treatment, converts triclosan to dioxin, which is both a carcinogen and an endocrine hormone disrupter. According to the EPA, at least 75 percent of all Americans have trace amounts of this compound in their bodies. Washing with ordinary soap is just as effective as using antibiotic compounds. We could easily reduce this ubiquitous pollutant.

Cosmetic and medical compounds occur in water almost everywhere now. How do they get there? In some cases, people simply dump unwanted food, medicines, and health supplements down the toilet or sink. More often, we consume more than our bodies can absorb, and we excrete the excess, which passes through sewage treatment facilities relatively unchanged. Numerous studies have found high levels of caffeine, birth-control hormones, antibiotics, recreational drugs, and other compounds downstream from major cities. This often results in developmental and behavioral changes in fish and other aquatic organisms.

In a nationwide study of pharmaceuticals and hormones in 130 streams, the U.S. Geological Service Scientists found detectable levels of 95 contaminants, including antibiotics, natural and synthetic hormones, detergents, plasticizers, insecticides, and fire retardants (fig. 18.9). One stream had 38 of the compounds tested. Drinking water standards exist for only 14 of the 95 substances. A similar study found the same substances in groundwater, which is much harder to clean than surface waters. What are the effects of these widely used chemicals on our environment or on people consuming the water? Nobody knows.

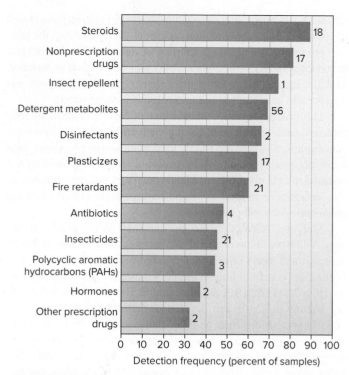

FIGURE 18.9 Detection frequency of organic contaminants in wastewater in a USGS survey. Numbers indicate maximum concentrations, in micrograms/liter. Dominant substances included caffeine (a steroid), drugs such as ibuprofen and acetaminophen, DEET (an insect repellent), and triclosan, from antibacterial soaps.
Source: USGS

Oil spills are common and often intentional

Oil spills are the dominant source of toxic organic compounds into water bodies. Although they are invisible to most of us, oil spills occur constantly around the world, as ships carrying oil leak or spill their cargo, or as ships leak engine oils and fuel oil into the ocean. Onshore, oil pipelines can rupture, spilling oil into surface waters. Likely ruptures in oil pipelines have been the chief concern in disputes over oil pipelines, such as the long-debated Keystone XL pipeline, planned to carry Canadian oil across sensitive Midwestern aquifer regions to Texas refineries.

Oil spills can be dramatic, large-scale disasters. Oil well blowouts, such as the 2010 explosion of the *Deepwater Horizon* in the Gulf of Mexico, devastate ecosystems and local fishing economies (fig. 18.10). According to the Smithsonian Institution, however, by far the greatest amount of oil leaked into the ocean every year comes from routine, intentional oil dumping. By their estimates, vehicle maintenance, urban runoff, bilge pumping, and other human activities release about 2.4 billion liters (644 million gal) of oil annually, most of which eventually ends up in the sea. This is ten times as much as from natural oil seeps and about 20 times as much as shipwrecks and well spills. However, as we move into more remote and more dangerous areas for drilling and shipping oil, accidents may become a larger and more destructive cause of ocean pollution (see chapter 19).

FIGURE 18.10 Oil spills kill thousands of birds, fish, and reptiles, and as we hunt for oil in increasingly remote and extreme places, accidents are likely to increase.
Source: NOAA and Georgia Department of Natural Resources

Sediment also degrades water quality

Rivers have always carried sediment to the oceans, but erosion rates in many areas have been greatly accelerated by human activities. Some rivers carry astounding loads of sediment. Erosion and runoff from croplands contribute about 25 billion metric tons of soil, sediment, and suspended solids to world surface waters each year. Forests, grazing lands, urban construction sites, and other sources of erosion and runoff add at least 50 billion additional tons. This sediment fills lakes and reservoirs, obstructs shipping channels, clogs hydroelectric turbines, and makes purification of drinking water more costly. Sediments smother gravel beds in which insects take refuge and fish lay their eggs. Sunlight is blocked so that

FIGURE 18.11 A plume of sediment and industrial waste flows from this drainage canal into Lake Erie.
Lawrence Lowry/Science Source

plants cannot carry out photosynthesis, and oxygen levels decline. Murky, cloudy water also is less attractive for swimming, boating, fishing, and other recreational uses (fig. 18.11).

In many river systems, though, sediment is also beneficial. Mud carried by rivers nourishes floodplain farm fields. Sediment deposited in the ocean at river mouths creates valuable deltas and islands. The Ganges River, for instance, builds up islands in the Bay of Bengal that are eagerly colonized by land-hungry people of Bangladesh. Louisiana's coastal wetlands require constant additions of sediment from the muddy Mississippi to counteract coastal erosion and subsidence. In recent decades, this sediment source has been lost, as levees direct river sediment out to the Gulf, in order to keep a deep shipping channel open. As a consequence, Louisiana's wetlands are disappearing at a disastrous rate.

Thermal pollution threatens sensitive organisms

Thermal pollution (heated water entering water bodies) results from a variety of sources, such as warm water discharged from power plants, or runoff from warm, paved urban land surfaces. Normally, water temperatures are much more stable than air temperatures, so aquatic organisms tend to be poorly adapted to rapid temperature changes. Raising or lowering the temperature of tropical oceans by even one degree can be lethal to some corals and other reef species. For many organisms, warming water is a problem because it reduces oxygen levels. Cold water holds more dissolved oxygen than warm water. As temperatures increase, species requiring high oxygen levels, such as trout, may not survive.

Industry and power plants often create thermal pollution: The cheapest way to remove heat from an industrial facility is to draw cool water from an ocean, river, lake, or aquifer, run it through a heat exchanger to extract excess heat, and then return the heated water back to the original source. A **thermal plume** of heated water is often discharged into rivers and lakes, where raised temperatures can disrupt natural ecosystems. Warm-water species proliferate, and cold-water species are driven away. Nearly half the water we withdraw is used for industrial cooling. Electric power plants, metal smelters, petroleum refineries, paper mills, food-processing factories, and chemical manufacturing plants all use and release large amounts of cooling water.

To minimize thermal pollution, power plants frequently are required to construct artificial cooling ponds or cooling towers in which heat is released into the atmosphere and water is cooled before being released into natural water bodies.

Some species find thermal pollution attractive. Warm water plumes from power plants often attract fish, birds, and marine mammals that find food and refuge there, especially in cold weather. This artificial environment can be a fatal trap, however. Organisms dependent on the warmth may die if they leave the plume or if the flow of warm water is interrupted by a plant shutdown. Endangered manatees in Florida, for example, are attracted to the abundant food and warm water in power plant thermal plumes and are enticed into spending the winter much farther north than they normally would. On several occasions, a midwinter power plant breakdown has exposed a dozen or more of these rare animals to a sudden thermal shock that they could not survive.

Section Review

1. What are point sources and nonpoint sources?
2. What is a dead zone? What causes it?
3. List eight major categories of water pollutants and give an example for each category.

18.2 WATER QUALITY TODAY

- *The Clean Water Act provides important water quality protection.*
- *Globally, over two billion people gained access to improved water supplies in the past 20 years.*
- *Problems remain in cleaning surface waters and providing modern sanitation to everyone.*

Surface-water pollution is often highly visible and is one of the most common threats to environmental quality. In more developed countries, reducing water pollution has been a high priority over the past few decades. Billions of dollars have been spent on control programs, and considerable progress has been made. But much remains to be done; poor water quality often remains a serious problem. In this section, we will look at progress as well as continuing obstacles in this important area.

The Clean Water Act protects our water

In wealthy countries with strong enforcement of environmental policies, there has been encouraging progress in protecting and restoring water quality in rivers and lakes over the past 40 years. In 1948, only about one-third of Americans were served by municipal sewage systems, and most of those systems discharged sewage without any treatment or with only primary treatment (the bigger lumps of waste are removed). Most people depended on cesspools and septic systems to dispose of domestic wastes.

In 1972, the U.S. Clean Water Act established a National Pollution Discharge Elimination System (NPDES), which requires an easily revoked permit for any industry, municipality, or other entity discharging wastes in surface waters. The permit requires public reporting of discharges, which gives regulators valuable data and evidence for enforcement. Because of these rules, point-source pollution has declined dramatically since 1972. Only about 10 percent of our water pollution now comes from industrial or municipal point sources.

Since the Clean Water Act was passed in 1972, the United States has spent more than $180 billion in public funds and perhaps ten times as much in private investments on water pollution control. Most of that effort has been aimed at point sources, especially to build or upgrade thousands of municipal sewage treatment plants. As a result, nearly everyone in urban areas is now served by municipal sewage systems and no major city discharges raw sewage into a river or lake except as overflow during heavy rainstorms.

These efforts have led to significant improvements in surface-water quality in many places. Fish and aquatic insects have returned to waters that formerly were depleted of life-giving oxygen. Swimming and other water-contact sports are again permitted

FIGURE 18.12 Not all rivers and lakes are "fishable or swimmable," but we've made substantial progress since the Clean Water Act was passed in 1972.
Tom Finkle

in rivers, lakes, and at ocean beaches that once were closed by health officials.

The Clean Water Act goal of making all U.S. surface waters "fishable and swimmable" hasn't been fully met, but by 1999 the EPA reported that 91.4 percent of all monitored river miles and 87.5 percent of all assessed lake acres were suitable for their designated uses (fig. 18.12). However, the designated goal for some rivers and lakes is merely to be "boatable." Water quality doesn't have to be very high to be able to put a boat in it. Even in "fishable" rivers and lakes there isn't a guarantee that you can catch anything other than rough fish like carp or bullheads, nor can you be sure that what you catch is safe to eat.

In 1998, a new regulatory approach to water quality assurance was instituted by the EPA, focusing on watershed-level monitoring and protection. States are required to identify waters not meeting water quality goals and to specify **total maximum daily loads (TMDLs)** for each pollutant and each listed water body. A TMDL is the amount of a particular pollutant that a water body can receive from both point and nonpoint sources. It considers seasonal variation and includes a margin of safety. Some 4,000 watersheds are monitored for water quality. The intention of TMDL reporting is to give the public more and better information about the health of their watersheds. In addition, states have greater flexibility as they identify impaired water bodies and set priorities, and new approaches can be used to achieve goals.

You can find information about your watershed on the EPA website (search online for *How's My Waterway?*). The EPA reports the kinds of contaminants that fail to meet TMDL standards for your local surface waters. High levels of bacteria or algae, for example, may indicate nutrient or sewage contamination; excess sediment loading often indicates erosion from construction sites or farmlands; metals or acids could indicate industrial pollution sources. Take a look at this source for your area. What kinds of contaminants and sources are primary concerns where you live?

Nonpoint sources are difficult to control

The greatest impediments to achieving national goals in water quality in both the United States and Canada are sediment, nutrients, and pathogens (fig. 18.13), especially from nonpoint discharges of pollutants. These sources are harder to identify and to reduce or treat than are specific point sources. About three-fourths of the water pollution in the United States comes from soil erosion, fallout of air pollutants, and surface runoff from urban areas, farm fields, and feedlots. In the United States, as much as 25 percent of the 50 million tons of fertilizer spread on farmland each year is carried away by runoff.

Lake Erie provides an example of how complex water quality protection can be. When urban industrial pollution was cleaned up in the 1970s, it was thought the lake's problems were solved. Bacterial counts and algal blooms decreased more than 90 percent. Water that once was murky brown cleared up.

Decades later, runoff from farms is now stimulating algal blooms once again. Dramatic summer air photos of green swirling algae on Lake Erie show the effects of agricultural runoff. It was much easier to stop highly concentrated "end of the pipe" effluents from specific industrial sources than low-level, nonpoint runoff from many individual farms.

Cattle in feedlots produce some 129,600,000 metric tons (144 million tons) of manure each year, and the runoff from these sites is rich in viruses, bacteria, nitrates, phosphates, and other contaminants. A single cow produces about 30 kg (66 lbs) of manure per day, or about as much as that produced by ten humans. Some feedlots have 100,000 animals with little provision for capturing or treating runoff water. Imagine drawing your drinking water downstream from such a facility. Pets also can be a problem. It is estimated that the wastes from about a half million dogs in New York City are disposed of primarily through storm sewers, and therefore do not go through sewage treatment.

Loading of both nitrates and phosphates in surface water has decreased from point sources but has increased about fourfold since 1972 from nonpoint sources. Fossil fuel combustion has become a major source of nitrates, sulfates, arsenic, cadmium, mercury, and other toxic pollutants that find their way into water. Carried to remote areas by atmospheric transport, these combustion products now are found nearly everywhere in the world. Toxic organic compounds, such as DDT, PCBs, and dioxins, also are transported long distances by wind currents.

Water pollution is especially serious in developing countries

Japan, Australia, and most of western Europe also have improved surface-water quality in recent years. Sewage treatment in the wealthier countries of Europe generally equals or surpasses that in the United States. Sweden, for instance, serves 98 percent of its population with at least secondary sewage treatment (compared with 70 percent in the United States), and the other 2 percent have primary treatment.

Poorer countries have much less to spend on sanitation (fig. 18.14). In China, for example, hundreds of millions of people live in areas without sufficient fresh water. Pollution makes much of the limited water unusable. An estimated 70 percent of China's surface water is unsafe for human consumption, and the water in half the country's major rivers is so contaminated that it's unsuited for any

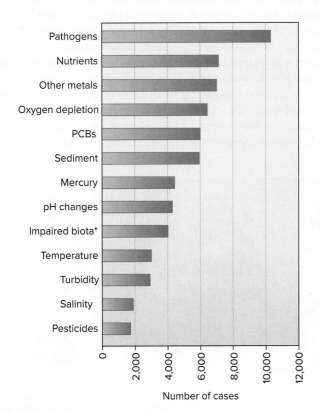

FIGURE 18.13 Leading causes of surface-water impairment in the United States. The * represents undetermined causes.
Source: U.S. EPA, 2015

FIGURE 18.14 Ditches in this Haitian slum serve as open sewers into which all manner of refuse and waste are dumped. The health risks of living under these conditions are severe.
Robert Nickelesberg/Getty Images

FIGURE 18.15 As high-rise apartment blocks move in to China's agricultural areas, development takes priority over environmental protection. Half of the water in China's major rivers is too polluted to be suitable for any human use.
Mary Ann Cunningham

use, even agriculture. Economic growth has been a much bigger priority than environmental quality (fig. 18.15).

According to the Chinese Environmental Protection Agency, the country's ten worst polluted cities are all in Shanxi Province. Factories there have been allowed to exceed pollution discharges with impunity. For example, 3 million tons of wastewater are produced every day in the province, with two-thirds of it discharged directly into local rivers without any treatment. Locals complain that the rivers, which once were clean and fresh, are now visibly polluted with industrial waste. Among the 26 rivers in the province, 80 percent are rated Grade V (unfit for any human use) or worse. More than half the wells in Shanxi are reported to have dangerously high arsenic levels. Many of the 180,000 reported public protests in China in 2012 involved complaints about air and water pollution. In 2013, more than 16,000 dead pigs were pulled from the Huangpu River just upstream from Shanghai's drinking water intake. They apparently were dumped by farmers to avoid fines for having diseased pigs. Because of issues like these, China's central government has become increasingly vigilant about water quality. Demands for a cleaner environment are growing stronger, and agencies are working harder to control pollution inputs in many areas.

As discussed in the opening case study for this chapter, water quality remains a severe problem in India. Two-thirds of India's surface water is so contaminated that even coming into contact with it is considered dangerous to human health. Yet millions of people drink and bathe in this water. Although the Ganges receives most attention, its tributaries are often much worse. Consider the Yamuna River, which flows through New Delhi and past the magnificent Taj Mahal in Agra. About 57 million people depend on the Yamuna for agricultural, domestic, and industrial use. Much of the runoff from these activities goes back into the river either untreated or only partially cleaned. Coliform bacterial counts can be millions of times greater than the level considered safe for drinking or bathing. Although the Indian government has spent more than $500 million in recent years to upgrade the sewage system, the Yamuna, and Ganges into which it flows, remain badly polluted.

Water treatment improves safety

Progress is possible, despite the challenges. China has spent at least $125 billion over the past five years to reduce water pollution and provide clean water to its cities and villages. Already there are indications of success. In 1990, the United Nations declared, as one of its Millennium Development Project targets, the goal of reducing by half the proportion of people without sustainable access to safe drinking water by 2015. That goal was met in 2010. Over those 20 years, more than 2 billion people gained access to "improved" water. Half of that progress occurred in India and China.

"Improved" water sources are not necessarily free of all pathogens and contaminants. Health agencies classify surface water and hand-dug, shallow wells as "unimproved" and especially likely to be polluted. "Improved" water supplies include tube (drilled and sealed) wells, water piped from springs and other clean sources, and treated water supplies. These sources may not be completely safe to drink, but they're usually better than lakes, rivers, or ponds. Improved water also doesn't necessarily mean it's available in the home. You may still have to carry it from a central spigot or well. Still, that's better than having to carry it from a source hours away. Having water piped into the home saves hours of labor and frees women and children for studies or income-producing work.

Overall, about 96 percent of all urban areas in the world enjoy improved water. The greatest remaining problems are in rural areas, especially in sub-Saharan Africa (fig. 18.16). Although we've made progress in supplying this vital resource, about 780 million

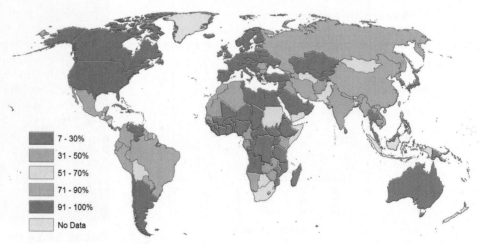

7 - 30%
31 - 50%
51 - 70%
71 - 90%
91 - 100%
No Data

FIGURE 18.16 Percentage of rural populations with access to improved drinking water. Hundreds of millions of rural residents, especially in sub-Saharan Africa, lack access to safe drinking water.
Source: UNESCO and WHO, 2012

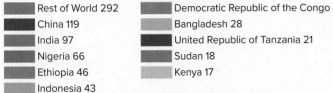

Millions without clean water

▇	Rest of World 292	▇	Democratic Republic of the Congo 36
▇	China 119	▇	Bangladesh 28
▇	India 97	▇	United Republic of Tanzania 21
▇	Nigeria 66	▇	Sudan 18
▇	Ethiopia 46	▇	Kenya 17
▇	Indonesia 43		

FIGURE 18.17 Just ten countries represent two-thirds of the 780 million people who still lack access to improved drinking water sources. *Sources: UNESCO and WHO, 2012*

people still lack access to improved water. Two-thirds of that number are in just ten countries (fig. 18.17).

Water treatment is not always adequate even in wealthier countries. Data collected by the EPA in 2008 show that about 30 million U.S. residents get water from community systems that don't meet all health-based drinking water standards. Most of these systems are small, under-funded systems. Every year epidemiologists estimate that around 1.5 million Americans fall ill from infections caused by fecal contamination. Preventive measures such as protecting water sources and aquifer recharge zones, providing basic treatment for all systems, installing modern technology and distribution networks, consolidating small systems, and strengthening the Clean Water Act and the Safe Drinking Water Act would cost far less. Unfortunately, in the present climate of budget cutting and antiregulation, these steps seem unlikely.

In 2016, water contamination put the city of Flint, Michigan, in the news. Two years earlier, as part of a cost-saving measure, the city had begun drawing water from the chlorine-contaminated Flint River, rather than its cleaner Lake Huron source. Because chlorine leads to corrosion of aging lead pipes, most cities are required to add anti-corrosion agents (mainly orthophosphate) to the water. Flint overlooked this step, and household taps began to deliver water contaminated with lead from city water lines. Two years later, as Flint residents reported widespread illness and rashes, and untold rates of impaired child development, the crisis erupted, rapidly spreading across the state and prompting new investigations of water quality in cities across the United States (see fig 1.28).

Is bottled water safer?

Every year, Americans buy over 40 billion bottles of water at a cost of about $15 billion with the mistaken belief (except in cases like Flint) that it's safer than tap water. Worldwide, some

200 billion liters (53 billion gallons) of bottled water are consumed annually. Public health experts say that municipal water is often safer than bottled water because most large cities test their water supplies every hour for up to 25 different chemicals and pathogens, while the requirements for bottled water are much less rigorous. About half of all bottled water in the United States is simply reprocessed municipal water, and much of the rest is drawn from groundwater aquifers, often with no legal monitoring requirements. A recent survey of bottled water in China found that two-thirds of the samples tested had dangerous levels of pathogens and toxins. Traditionally, Chinese drank boiled tea water, which is probably still a safer beverage to drink.

Some 80 percent of the bottles purchased in the United States end up in a landfill, even though the plastic is easily recyclable. Overall, the average energy cost to make the plastic, fill the bottle, transport it to market, and then deal with the waste would be "like filling up a quarter of every bottle with oil," says water expert Peter Gleick. It also takes three to five times as much water to make the bottles as they hold. Furthermore, if water is held in a plastic bottle for weeks or months, plasticizers and other toxic chemicals can leach from the bottle into the water (see chapter 8).

Whereas municipal water is monitored and treated, bottled water mainly benefits the corporations that sell it, and its safety is less sure than city water. Generally, drinking tap water will do a favor for your environment, your budget, and, possibly your health.

Groundwater is hard to monitor and clean

About half the people in the United States, including 95 percent of those in rural areas, depend on underground aquifers for their drinking water. This vital resource is threatened in many areas by overuse and pollution and by a wide variety of industrial, agricultural, and domestic contaminants. For decades it was widely assumed that groundwater was impervious to pollution because soil would bind chemicals and cleanse water as it percolated through. Springwater or artesian well water was considered to be the definitive standard of water purity. But that is no longer true in many areas.

The U.S. EPA estimates that every day some 4.5 trillion liters (1.2 trillion gal) of contaminated water seep into the ground in the United States from septic tanks, cesspools, municipal and industrial landfills and waste disposal sites, surface impoundments, agricultural fields, forests, and abandoned wells (fig. 18.18). The most toxic of these are probably waste disposal sites. Agricultural chemicals and wastes are responsible for the largest total volume of pollutants and area affected. Because deep underground aquifers often have residence times of thousands of years, many contaminants are extremely stable once underground. It is possible, but expensive, to pump water out of aquifers, clean it, and then pump it back.

In farm country, especially in the Midwest's Corn Belt, fertilizers and pesticides commonly contaminate aquifers and wells. Herbicides such as atrazine and alachlor are widely used on corn and soybeans and show up in about half of all wells in Iowa, for example. Nitrates from fertilizers often exceed safety standards in rural drinking water. These high nitrate levels are particularly dangerous to infants (nitrate combines with hemoglobin in the blood

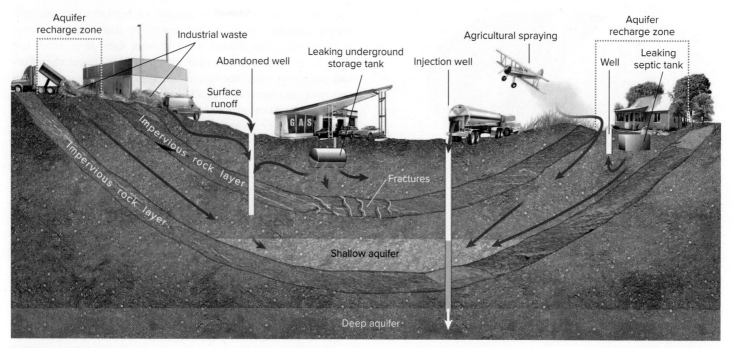

FIGURE 18.18 Sources of groundwater pollution. Septic systems, landfills, and industrial activities on aquifer recharge zones leach contaminants into aquifers. Wells provide a direct route for injection of pollutants into aquifers.

and results in "blue baby" syndrome). They also are transformed into cancer-causing nitrosamines in the human gut. In Florida, 1,000 drinking water wells were shut down by state authorities because of excessive levels of toxic chemicals, mostly ethylene dibromide (EDB), a pesticide used to kill nematodes (roundworms) that damage plant roots.

There are few controls on ocean pollution

Coastal zones, especially bays, estuaries, shoals, and reefs near large cities or the mouths of major rivers, often are overwhelmed by human-caused contamination. Suffocating and sometimes poisonous blooms of algae regularly deplete ocean waters of oxygen and kill enormous numbers of fish and other marine life. High levels of toxic chemicals, heavy metals, disease-causing organisms, oil, sediment, and plastic refuse are adversely affecting some of the most attractive and productive ocean regions. The potential losses caused by this pollution amount to billions of dollars each year.

Discarded bits of plastic debris are lightweight and non-biodegradable. They are carried thousands of miles on ocean currents and last for years (fig. 18.19). Even the most remote beaches of distant islands are likely to have bits of polystyrene foam containers or polyethylene packing material that were discarded half a world away. It has been estimated that some 6 million metric tons of plastic bottles, packaging material, and other litter are tossed from ships every year into the ocean, where they ensnare and choke seabirds, mammals (fig. 18.20), and even fish. In one day, volunteers in Texas gathered more than 300 tons of plastic refuse from Gulf Coast beaches. For further discussion of ocean pollution, see chapter 21.

FIGURE 18.19 Beach pollution, including garbage, sewage, and contaminated runoff, is a growing problem associated with ocean pollution.
John Lund/The Image Bank/Getty Images

FIGURE 18.20 A deadly necklace. Marine biologists estimate that castoff nets, plastic beverage yokes, and other packing residue kill hundreds of thousands of birds, mammals, and fish each year.
Source: Joe Lucas/Marine Entanglement Research Program/National Marine Fisheries Service NOAA

Section Review

1. How has the Clean Water Act reduced sewage discharge into public waterways?
2. Which countries have had the greatest success in providing safe water to their citizens?
3. What geographic region lacks safe water for the greatest remaining proportion of its population?

18.3 WATER POLLUTION CONTROL

- *Municipal sewage treatment involves primary, secondary, and tertiary treatment.*
- *Natural or artificial wetlands can treat wastes economically.*
- *Bioremediation can be effective in urban settings.*

The cheapest and most effective way to reduce pollution is usually to avoid producing it or releasing it in the first place. Elimination of lead from gasoline has dramatically cut the amount of lead in surface waters in the United States. Banning DDT and PCBs in the 1970s has resulted in significant reductions in levels in wildlife. Mercury in the U.S. waterways has declined along with emissions from coal power plants, the source of over half of U.S. mercury emissions. Better land management has succeeded in reducing nutrient loads and sediment in critical watersheds, as in the watersheds of New York's Catskills mountains, which provides drinking water to New York City. In the Catskills, economic incentives to help farmers control runoff have allowed New York City to avoid building filtration plants that would have cost billions of dollars.

Industry can reduce pollution by recycling or reclaiming materials that otherwise might be discarded in the waste stream. Both of these approaches usually have economic as well as environmental benefits. It turns out that a variety of valuable metals can be recovered from industrial wastes and reused or sold for other purposes. The company benefits by having a product to sell, and the municipal sewage treatment plant benefits by not having to deal with highly toxic materials mixed in with millions of gallons of other types of wastes.

Controlling nonpoint sources requires land management

Among the greatest remaining challenges in water pollution control are diffuse, nonpoint pollution sources. Unlike point sources, such as sewer outfalls or industrial discharge pipes, which represent both specific locations and relatively continuous emissions, nonpoint sources have many origins and numerous routes by which contaminants enter ground and surface waters. It is difficult to identify—let alone monitor and control—all these sources and routes. Some main causes of nonpoint pollution are:

- *Agriculture:* The EPA estimates that 60 percent of all impaired or threatened surface waters are affected by sediment from eroded fields and overgrazed pastures; fertilizers, pesticides, and nutrients from croplands; and animal wastes from feedlots.
- *Urban runoff:* Pollutants carried by runoff from streets, parking lots, and industrial sites contain salts, oily residues, rubber, metals, and many industrial toxins (fig. 18.21). Yards, golf courses, park lands, and urban gardens often are treated with

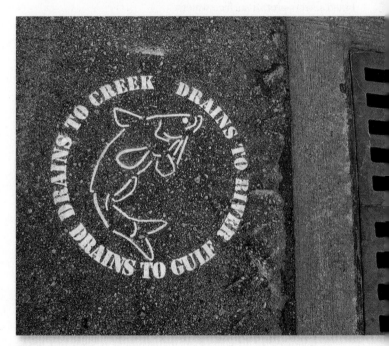

FIGURE 18.21 People often dump waste oil and other pollutants into street drains without thinking about where their wastes go. Painting reminders, such as this one, is a good project for students and youth groups.
William P. Cunningham

far more fertilizers and pesticides per unit area than farmlands. Excess chemicals are carried by storm runoff into waterways.

- *Construction sites:* New buildings and land development projects such as highway construction affect relatively small areas but produce vast amounts of sediment, typically 10 to 20 times as much per unit area as farming.

- *Land disposal:* When done carefully, land disposal of certain kinds of industrial waste, sewage sludge, and biodegradable garbage can be a good way to dispose of unwanted materials. Some poorly run land disposal sites, abandoned dumps, and leaking septic systems, however, contaminate local waters.

Generally, agricultural soil conservation methods (chapter 9) also help protect water quality. Applying precisely determined amounts of fertilizer, irrigation water, and pesticides saves money and reduces contaminants entering the water. Preserving wetlands, which act as natural processing facilities for removing sediment and contaminants, also helps protect surface water and groundwater.

Combined sewer overflows pollute surface waters

In many American cities, storm sewers (which carry runoff from streets and parking lots) are connected to sanitary sewers, which drain bathrooms and kitchens. Historically, this was a good idea because runoff from streets, yards, and parking lots contains a tremendous amount of contaminants, such as litter, fertilizers, pesticides, oils, rubber, tars, gasoline residue, and pet waste. Treating these contaminants, rather than routing them straight to the nearest river or lake, protects surface waters. In moderate rainfall, combined sewers protect surface waters.

As cities have grown, however, combined sewer systems are often overloaded, especially in heavy storms. Treatment plants overwhelmed by storm runoff then release everything untreated—raw sewage and toxic surface runoff—directly into lakes and rivers. These events are called **combined sewer overflows (CSOs)**, and in many cities, CSOs are the principal cause of surface water contamination. To prevent CSOs, cities are spending hundreds of millions of dollars to separate storm and sanitary sewers. These are huge, disruptive projects. When they are finished, surface runoff is diverted into surface waters, where other pollution problems may arise, but direct release of raw sewage is less frequent.

Reducing materials carried by storm runoff is critical for protecting water quality. All of us can try to reduce the debris, litter, and waste that enters storm drains—that includes everything from candy wrappers to piles of autumn leaves, which contribute nutrients to lakes and rivers. We should also minimize the use of fertilizers and pesticides in yards, which become water contaminants, and never pour into street drains any waste oil, paint, or other substances that you wouldn't want to see fish swimming in. Regular street sweeping greatly reduces water contamination.

A major effort in watershed management is seen in Chesapeake Bay, the United States' largest estuary. Once fabled for its abundant oysters, crabs, shad, striped bass, and other valuable fisheries, the Bay had deteriorated seriously by the early 1970s. Citizens' groups, local communities, state legislatures, and the federal government together established an innovative pollution-control program that made the bay the first estuary in America targeted for protection and restoration.

Controlling runoff from agricultural and urban lands is central to the Chesapeake Bay restoration plan. Pollution prevention measures such as banning phosphate detergents also are important, as are upgrading wastewater treatment plants and making sure plants are managed well. Since the 1980s, annual phosphorous discharges into the bay dropped 40 percent. Nitrogen levels, however, have remained constant or have even risen in some tributaries. The bay is still decades away from meeting its goals, which include reducing nitrogen and phosphate levels and restoring fisheries, but as former EPA administrator Carol Browner has said, the project demonstrates the "power of cooperation" in environmental protection.

Human waste disposal occurs naturally when concentrations are low

As we have already seen, human and animal wastes usually create the most serious health-related water pollution problems. More than 500 types of disease-causing (pathogenic) bacteria, viruses, and parasites can travel from human or animal excrement through water.

In the poorer countries of the world, rural people have traditionally gone out into the fields and forests to relieve themselves. Where population densities are low, natural processes eliminate wastes quickly, making this a feasible method of sanitation. The high population densities of cities make this practice unworkable, however. Even major cities of many less-developed countries are often littered with human waste that has been left for rains to wash away or for pigs, dogs, flies, beetles, or other scavengers to consume. This is a major cause of disease, as well as being extremely unpleasant. Studies have shown that a significant portion of the airborne dust in Mexico City is actually dried, pulverized human feces.

Where intensive agriculture is practiced—especially in wet rice paddy farming in Asia—it has long been customary to collect "night soil" (human and animal waste) to be spread on the fields as fertilizer. This waste is a valuable source of nutrients, but it is also a source of disease-causing pathogens in the food supply. It is the main reason that travelers in less-developed countries must be careful to surface sterilize or cook any fruits and vegetables they eat. Collecting night soil for use on farm fields was common in Europe and America until about 100 years ago, when the association between pathogens and disease was recognized.

Septic systems work in low densities

Until about 70 years ago, most rural American families and many residents of towns and small cities depended on a pit toilet or "outhouse" for waste disposal. Untreated wastes tended to seep into the ground, however, and pathogens sometimes contaminated drinking water supplies. The development of septic tanks and properly constructed drain fields considerably improved public health (fig. 18.22). In a typical septic system, wastewater is first drained into a septic tank. Grease and oils rise to the top and solids settle to the bottom, where they are subject to bacterial decomposition. The

FIGURE 18.22 A domestic septic tank and drain field system for sewage and wastewater disposal. To work properly, a septic tank must have healthy microorganisms, which digest toilet paper and feces. For this reason, antimicrobial cleaners and chlorine bleach should never be allowed down the drain.

clarified effluent from the septic tank is channeled out through a drainfield of small perforated pipes embedded in gravel just below the surface of the soil. The rate of aeration is high in this drainfield so that pathogens (most of which are anaerobic) will be killed, and soil microorganisms can metabolize any nutrients carried by the water. Excess water percolates up through the gravel and evaporates. Periodically, the solids in the septic tank are pumped out into a tank truck and taken to a treatment plant for disposal.

Where land is available and population densities are not too high, this can be an effective method of waste disposal. It is widely used in rural areas, but aging, leaky septic systems can be a huge cumulative problem. The Chesapeake Bay watershed has 420,000 individual septic systems, which constitute a major source of nutrients that create a dead zone over much of the Bay. Maryland alone plans to spend $7.5 million annually to upgrade failing septic systems.

Municipal treatment plants remove pathogens

Over the past 100 years, sanitary engineers have developed ingenious and effective municipal wastewater treatment systems to protect human health, ecosystem stability, and water quality. This topic is an important part of pollution control, and is a central focus of every municipal government; therefore, let's look more closely at how a typical municipal sewage treatment facility works.

Primary treatment is the first step in municipal waste treatment. It physically separates large solids from the waste stream.

As raw sewage enters the treatment plant, it passes through a metal grating that removes large debris (fig. 18.23). A moving screen then filters out smaller items. Brief residence in a grit tank allows sand and gravel to settle. The waste stream then moves to the primary sedimentation tank, where about half the suspended organic

1. **Screening** removes large solids
2. **Settlement tanks** remove most of the remaining solids
3. **Bacteria** in beds or tanks purify the solids

Solids and sludge are treated and sent to a landfill or incinerator, and sometimes sold as fertilizer

The water may be disinfected with ultraviolet light → 4. **Water** is returned to the environment

FIGURE 18.23 The process of conventional sewage treatment.

solids settle to the bottom as sludge. Many pathogens remain in the effluent, which is not yet safe to discharge into waterways or onto the ground.

Secondary treatment consists of biological degradation of the dissolved organic compounds. The effluent from primary treatment flows into a trickling filter bed, an aeration tank, or a sewage lagoon. The trickling filter is simply a bed of stones or corrugated plastic sheets through which water drips from a system of perforated pipes or a sweeping overhead sprayer. Bacteria and other microorganisms in the bed catch organic material as it trickles past and aerobically decompose it.

Aeration tank digestion is also called the activated sludge process. Effluent from primary treatment is pumped into the tank and mixed with a bacteria-rich slurry. Air pumped through the mixture encourages bacterial growth and decomposition of the organic material (fig. 18.24a and b). Water flows from the top of the tank, and sludge is removed from the bottom. Some of the sludge is added to incoming primary effluent, to inoculate it with useful bacteria. The remainder would be valuable fertilizer if it were not contaminated by metals, toxic chemicals, and pathogenic organisms. The toxic content of most sewer sludge necessitates disposal by burial in a landfill or incineration. Sludge disposal is a major cost in most municipal sewer budgets. In some communities, this is accomplished by land farming, composting, or anaerobic digestion, but these methods don't inactivate metals and some other toxic materials.

Where space is available for sewage lagoons, the exposure to sunlight, algae, aquatic organisms, and air does the same job more slowly but with less energy cost. Effluent from secondary treatment processes is usually disinfected with chlorine, UV light, or ozone to kill harmful bacteria before it is released to a nearby waterway.

Tertiary treatment removes plant nutrients, especially nitrates and phosphates, from the secondary effluent. Although wastewater is usually free of pathogens and organic material after secondary treatment, it still contains high levels of inorganic nutrients, such as nitrates and phosphates. When discharged into surface waters, these nutrients stimulate algal blooms and eutrophication. To preserve water quality, these nutrients also must be removed. Passage through a wetland or lagoon can accomplish this. Alternatively, chemicals often are used to bind and precipitate nutrients.

Low-cost systems use natural processes

The municipal sewage systems used in developed countries are often too expensive to build and operate in the developing world where low-cost, low-tech alternatives for treating wastes are needed (Exploring Science, p. 423). One option is **effluent sewerage,** a hybrid between a traditional septic tank and a full sewer system. A tank near each dwelling collects and digests solid waste just like a septic system. Rather than using a drainfield, however, to dispose of liquids—an impossibility in crowded urban areas—effluents are pumped to a central treatment plant. The tank must be emptied once a year or so, but because only liquids are

(a)

(b)

FIGURE 18.24 In conventional sewage treatment, aerobic bacteria digest organic materials in trickling filter beds in high-pressure aeration tanks. This is described as secondary treatment.
(a) *Thinkstock Images/Getty Images*, (b) *Steve Allen/Brand X Pictures/Alamy*

treated by the central facility, pipes, pumps, and treatment beds can be downsized and the whole system is much cheaper to build and run than a conventional operation.

Another alternative is to use natural or artificial wetlands to dispose of wastes. **Constructed wetlands** can cut secondary treatment costs to one-third of mechanical treatment costs, or less. Variations on this design are now operating in many places

EXPLORING SCIENCE

Inexpensive Water Purification

When Ashok Gadgil was a child in Mumbai, India, five of his cousins died in infancy from diarrhea spread by contaminated water. Although he didn't understand the implications of those deaths at the time, as an adult he realized how heartbreaking and preventable those deaths were. After earning a degree in physics from the University of Bombay, Gadgil moved to the University of California at Berkeley, where he was awarded a PhD in 1979 and is now a professor of civil and environmental engineering. His research focuses on solar energy and pollution issues.

But Dr. Gadgil wanted to do something about the problem of waterborne diseases and pollutants in India and other developing countries. Although progress has been made in bringing clean water to poor people in many countries, at least 780 million people worldwide still lack access to safe drinking water. After studying ways to purify water, he decided that UV light treatment had the greatest potential for removing infectious agents. It requires far less energy than boiling, and it takes less sophisticated chemical monitoring than chlorination.

There are many existing UV water treatment systems, but they generally involve water flowing around an unshielded fluorescent lamp. Minerals in the water collect on the glass lamp and must be removed regularly to maintain effectiveness. Regular disassembly, cleaning, and reassembly of the apparatus

A woman carries a jug with clean water from the village Water-Health kiosk. More than 6 million people's lives have been improved by this innovative system of water purification.
Subhash Sharma/Polaris/Newscom

are difficult in primitive conditions. The solution, Gadgil realized, was to mount the UV source above the water where it couldn't develop mineral deposits. He designed a system in which water flows through a shallow, stainless steel trough. The apparatus can be gravity fed and requires only a car battery as an energy source.

Another crisis in South Asia is arsenic contamination in well water. About 100 million people in Bangladesh and West Bengal are exposed to dangerous levels of arsenic in their drinking water (see section 18.1). Arsenic can be removed from water by being absorbed by iron or aluminum oxides. Professor Gadgil

and his team have devised an inexpensive, effective way to do this. They pass a weak electric current (less than 3 volts) through iron plates as the water passes over them. This releases iron oxides (rust) that bind to arsenic. The heavy iron particles can then be filtered out easily. A combined UV treatment and electrochemical arsenic precipitation system can purify 15 liters (4 gallons) of water per minute, killing more than 99.9 percent of all bacteria and viruses and removing an equal percentage of arsenic. This produces enough clean water for a village of 1,000 people at a cost of only about 5 cents per ton (950 liters).

WaterHealth International, the company founded to bring this technology to market, now makes several versions of Gadgil's treatment system for different applications. A popular version provides a complete water purification system, including a small kiosk, jugs for water distribution, and training on how to operate everything. A village-size system costs about $5,000. Grants and loans are available for construction, but villagers own and run the facility to ensure there is local responsibility. Each family in the cooperative pays about $1 per month for pure water. These systems have been installed in thousands of villages in India, Bangladesh, Africa, and the Philippines. Currently, about 6.6 million people are getting clean, healthy water at an easily affordable price from the simple systems Dr. Gadgil invented.

(header photo): David L. Hansen

(fig. 18.25). Effluent from these operations can be used to irrigate crops or even raise fish for human consumption if care is taken to first destroy pathogens. Usually 20 to 30 days of exposure to sun, air, and aquatic plants is enough to make the water safe. These systems also can make an important contribution to human food supplies. A 2,500-ha (6,000-acre) waste-fed aquaculture facility in Kolkata, India, for example, supplies about 7,000 metric tons of fish annually to local markets. The World Bank estimates that some 3 billion people will be without sanitation services by the middle of the next century under a business-as-usual scenario (fig. 18.26). With investments in innovative programs, however, sanitation could be provided to about half those people and thus a great deal of misery and suffering could be avoided.

Water remediation may involve containment, extraction, or phytoremediation

Remediation means finding remedies for problems. Just as there are many sources for water contamination, there are many ways to clean it up. New developments in environmental engineering are providing promising solutions to many water pollution problems.

Containment methods confine or restrain dirty water or liquid wastes *in situ* (in place) or cap the surface with an impermeable layer to divert surface water or groundwater away from the site and to prevent further pollution. Where pollutants are buried too deeply to be contained mechanically, materials sometimes can be injected to precipitate, immobilize, chelate, or solidify them. Bentonite clay, for example, can be injected to stabilize liquids in

FIGURE 18.25 This constructed wetland purifies water and provides an attractive landscape at the El Monte Sagrado Resort in Taos, New Mexico.

William P. Cunningham

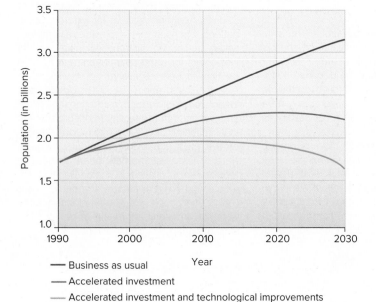

- Business as usual
- Accelerated investment
- Accelerated investment and technological improvements

FIGURE 18.26 World population without adequate sanitation—three scenarios in the year 2030. If business as usual continues, more than 3 billion people will lack safe sanitation. Accelerated investment in sanitation services could lower this number. Higher investment, coupled with technological development, could keep the number of people without adequate sanitation from growing even though the total population increases.

Source: World Bank estimates based on research paper by Dennis Anderson and William Cavendish, "Efficiency and Substitution in Pollution Abatement: Simulation Studies in Three Sectors."

porous substrates. Similarly, straw or other absorbent material is spread on surface spills to soak up contaminants.

Extraction techniques pump out polluted water so it can be treated. Many pollutants can be destroyed or detoxified by chemical reactions that oxidize, reduce, neutralize, hydrolyze, precipitate, or otherwise change their chemical composition. Where chemical techniques are ineffective, physical methods may work. Solvents and other volatile organic compounds, for instance, can be stripped from solution by aeration and then burned in an incinerator. Some contaminants can be removed by semipermeable membranes or resin filter beds that bind selectively to specific materials.

Often, living organisms can be used effectively and inexpensively to clean contaminated water. We call this bioremediation (chapter 21). For instance, restored wetlands along streambanks or lake margins can be very effective in filtering out sediment and removing pollutants. They generally cost far less than mechanical water treatment facilities, and they provide wildlife habitat as well.

Lowly duckweed (*Lemna* sp.), the green scum you often see covering the surface of eutrophic ponds, grows fast and can remove large amounts of organic nutrients from water. Under optimal conditions, a few square centimeters of these tiny plants can grow to cover nearly a hectare (about 2.5 acres) in four months. Large duckweed lagoons are being used as inexpensive, low-tech sewage treatment plants in developing countries. Where conventional wastewater purification typically costs $300 to $600 per person served, a duckweed system can cost one-tenth as much. The duckweed can be harvested and used as feed, fuel, or fertilizer. Up to 35 percent of its dry mass is protein—about twice as much as alfalfa, a popular animal feed.

"Living machines" use plants to capture contaminants

Where space for open lagoons is unavailable, bioremediation can be carried out in tanks or troughs. This has the advantage of controlling conditions more precisely and doesn't release organisms into the environment. Some of the most complex, holistic systems for water purification are designed by Ocean Arks International (OAI) in Falmouth, Massachusetts. Their "living machines" combine living organisms—chosen to perform specific functions—in contained environments. In a typical living machine, water flows through a series of containers, each with a distinct ecological community designed for a particular function. Wastes generated by the inhabitants of one vessel become the food for inhabitants of another. Sunlight provides the primary source of energy.

OAI has created water treatment plants in a dozen states and foreign countries. Designs range from remediating toxic wastes at Superfund sites to simply treating domestic wastes. Starting with microorganisms in aerobic and anaerobic environments where different kinds of wastes are metabolized or broken down, water moves through a series of containers containing hundreds of different kinds of plants and animals, including algae, rooted aquatic plants, clams, snails, and fish, each chosen to provide a particular service. Technically, the finished water is drinkable, although few people feel comfortable drinking it. More often, the final effluent

What Can You Do?

Steps You Can Take to Improve Water Quality

Individual actions have important effects on water quality. Here are some steps you can take to make a difference.

- Compost your yard waste and pet waste. Nutrients from decayed leaves, grass, and waste are a major urban water pollutant. Many communities have public compost sites available.
- Don't fertilize your lawn or apply lawn chemicals. Untreated grass can be just as healthy, and it won't poison your pets or children.
- Make sure your car doesn't leak fluids, oil, or solvents on streets and parking lots, from which contaminants wash straight into rivers and lakes. Recycle motor oil at a gas station or oil-change shop.
- Create a "rain garden" to capture and filter surface runoff. This helps recharge groundwater aquifers and keeps nutrients and toxins out of rivers and lakes (fig. 18.28).
- Don't buy lawn mowers, personal watercraft, or other vehicles with two-cycle engines, which release abundant fuel and oil into air and water. Instead, buy more efficient four-stroke engines.
- Visit your local sewage treatment plant. Often, public tours are available or group tours can be arranged, and these sites can be fascinating.
- Keep informed about water policy debates at local and federal levels. Policies change often, and public input is important.

FIGURE 18.28 A rain garden is a shallow depression situated to collect runoff from streets or parking lots. It's planted with species that can survive in saturated soils. This vegetation helps evaporate and cleanse runoff, while temporary storage in the basin allows groundwater recharge. You might build a rain garden in your yard, or on your campus, or elsewhere in your city.
William P. Cunningham

is used to flush toilets or for irrigation. This approach can save resources and money, as well as help clean up our environment (fig. 18.27).

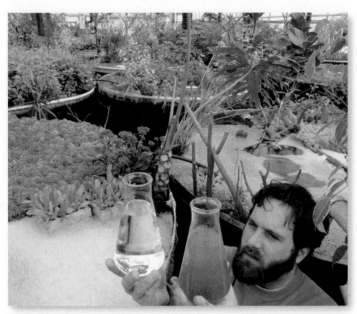

FIGURE 18.27 Bioreactors, such as these "living machines" from Ocean Arks International use living communities that mimic natural ecosystems to treat water. Polluted water (right flask) can be purified to drinking water quality (left flask).
Peter Essick/Aurora/Getty Images

Section Review

1. Describe the process of municipal sewage treatment.
2. What is a constructed wetland?
3. How does bioremediation clean water?

18.4 WATER LEGISLATION

- *The Clean Water Act has been successful in many ways.*
- *The Clean Water Act is disputed and needs updating.*
- *Other laws also help protect water quality.*

As the opening case study for this chapter shows, water pollution control has been among the most broadly popular and effective of all environmental legislation in the United States. It has not been without controversy, however. In this section, we will look at some of the major issues concerning water quality laws and their provisions (table 18.2).

The Clean Water Act was ambitious, bipartisan, and largely successful

Passage of the U.S. Clean Water Act of 1972 was a bold, bipartisan step determined to "restore and maintain the chemical, physical, and biological integrity of the Nation's waters" that made clean water a

Table 18.2 Some Important U.S. and International Water Quality Legislation

1. *Federal Water Pollution Control Act* (1972). Established uniform nationwide controls for each category of major polluting industries.

2. *Marine Protection Research and Sanctuaries Act* (1972). Regulates ocean dumping and established sanctuaries for the protection of endangered marine species.

3. *Ports and Waterways Safety Act* (1972). Regulates oil transport and the operation of oil handling facilities.

4. *Safe Drinking Water Act* (1974). Requires minimum safety standards for every community water supply. Among the contaminants regulated are bacteria, nitrates, arsenic, barium, cadmium, chromium, fluoride, lead, mercury, silver, pesticides; radioactivity and turbidity also are regulated. This act also contains provisions to protect groundwater aquifers.

5. *Resource Conservation and Recovery Act (RCRA)* (1976). Regulates the storage, shipping, processing, and disposal of hazardous wastes and sets limits on the sewering of toxic chemicals.

6. *Toxic Substances Control Act (TOSCA)* (1976). Categorizes toxic and hazardous substances, establishes a research program, and regulates the use and disposal of poisonous chemicals.

7. *Comprehensive Environmental Response, Compensation, and Liability Act (CERCLA)* (1980) and Superfund Amendments and Reauthorization Act (SARA) (1984). Provide for sealing, excavation, or remediation of toxic and hazardous waste dumps.

8. *Clean Water Act* (1985) (amending the 1972 Water Pollution Control Act). Sets as a national goal the attainment of "fishable and swimmable" quality for all surface waters in the United States.

9. *London Dumping Convention* (1990). Calls for an end to all ocean dumping of industrial wastes, tank washing effluents, and plastic trash. The United States is a signatory to this international convention.

---

national priority. Along with the Endangered Species Act and the Clean Air Act, this is one of the most significant and effective pieces of environmental legislation ever passed by the U.S. Congress. It also is an immense and complex law, with more than 500 sections, regulating everything from urban runoff, industrial discharges, and municipal sewage treatment to land-use practices and wetland drainage.

The ambitious goal of the Clean Water Act was to return all U.S. surface waters to "fishable and swimmable" conditions. For specific "point" sources of pollution, such as industrial discharge pipes or sewage outfalls, the act requires discharge permits and **best practicable control technology (BPT).** It sets national goals of **best available, economically achievable technology (BAT)** for toxic substances and zero discharge for 126 priority toxic pollutants. As we discussed earlier in this chapter, these regulations have had a positive effect on water quality. Although surface waters are not yet swimmable or fishable everywhere, surface-water quality in the United States has significantly improved, on average, over the past quarter century. Perhaps the most important result of the act has been the investment of $54 billion in federal funds and more than $128 billion in state and local funds for municipal sewage treatment facilities.

Not everyone is happy with the Clean Water Act. Industries, state and local governments, farmers, land developers, and others who have been forced to change their operations or spend money on water protection often object to the costs of controlling pollution. One of the most controversial provisions of the act has been Section 404, which regulates the draining or filling of wetlands. Although the original bill mentions wetlands only briefly, this section has evolved through judicial interpretation and regulatory policy to become one of the principal federal tools for wetland protection. Many people applaud the protection granted to these ecologically important areas that were being filled in or drained at a rate of about half a million hectares per year before the passage of the Clean Water Act. Farmers, land developers, and others who are prevented from converting wetlands to other uses often are outraged by what they consider the "taking" of private lands.

Another sore point for opponents of the Clean Water Act is what are called "unfunded mandates," or requirements for state or local governments to spend money that is not repaid by Congress. The $128 billion already spent by cities to install sewage treatment and to improve sewer systems to meet federal standards far exceeds the $54 billion in congressional assistance for these projects. Estimates are that local units of government could be required to spend another $130 billion to finish the job without any further federal funding. Small cities that couldn't afford or chose not to participate in earlier water quality programs, in which the federal government paid up to 90 percent of the costs, are especially hard hit by requirements that they upgrade municipal sewer and water systems. They now are faced with carrying out those same projects entirely with their own funds.

Clean water reauthorization remains contentious

In the decades since the Clean Water Act was passed in 1972, new challenges have emerged, and funding provisions have expired. Between 1989 and 1994, the CWA provided $7.2 billion to help communities build and maintain wastewater treatment facilities. Since 1994, that funding has not been reauthorized, leaving local governments to cover most costs. Available federal funding has depended on annual extensions, which are variable and unreliable, depending on the budget, or mood, of Congress. The loss of this funding is a problem, as many of the nation's water treatment plants, now over 40 years old, need updating, repairs, or expansion to serve the cities growing around them.

At the same time, new contaminants have emerged, such as organic compounds discussed earlier in this chapter. We still have no standard policies for controlling nonpoint runoff from streets and farm fields, which have become our most important water pollutants in many areas. There is need for better use of new stormwater management systems, such as rain gardens or permeable paving to inexpensively reduce runoff. In addition, climate change is altering water resources. In some areas, storms are becoming more intense, increasing stress on water systems. In other areas, water supplies are dwindling, making pollutants more concentrated than in the past.

The Clean Water Act was updated in 1977, 1981, and 1987, and there have been calls since then to update and reauthorize the law. Persistent, deep disputes over polluters' responsibility, and the powers of regulators, have prevented new updates. Groups such as the Natural Resources Defense Council have argued that enforcement of current laws is weak, that required reporting is often incomplete. In recent years, political impasses have made effective compromises difficult to achieve.

A variety of rules protect water quality

In addition to the Clean Water Act, a variety of U.S. laws establish water quality standards or seek to control pollution entering water. Among these is the Safe Drinking Water Act, which regulates water quality in commercial and municipal systems. This policy allows

for public oversight and protection of local water quality, although critics complain that standards and enforcement policies are too lax, especially for rural water districts and small towns. Some researchers report pesticides, herbicides, and lead in drinking water at levels they say should be of concern. A number of policies govern the handling of toxic and hazardous waste, such as the Resource Conservation and Recovery Act (RCRA), which seeks to prevent improper disposal of hazardous compounds.

The Superfund program for remediation of toxic waste sites was created in 1980 by the Comprehensive Environmental Response, Compensation, and Liability Act (CERCLA). This law directs the EPA to oversee cleanup of some of our worst sources of water (and air and soil) contamination. Sites listed for Superfund cleanup are generally orphan sites, or owned by compaines that have gone bankrupt. This important aspect of water protection is discussed in chapter 21.

Section Review

1. Identify five important examples of water quality legislation.
2. What is best practicable control technology?
3. Why does the Clean Water Act need reauthorization to be effective?

Connecting the Dots

Water pollution affects people everywhere, especially where populations are dense and where legal controls are poorly enforced or nonexistent. The Ganges River, lifeblood of over a billion people, shows all the classic issues of protecting surface waters. Because the river is so important culturally, religiously, personally, and economically, people are motivated to see it clean. But challenges to protecting it are steep and persistent.

The impact of pollutants on human health and on ecosystems depends a great deal on the nature of pollutants in question. Effective treatment also varies greatly among different types of pollutants. Nutrient enrichment causes algae blooms; sediments cause warming and reduce clarity; pathogens cause disease; metals cause nerve damage; organic compounds interfere with nerve

or hormone function. Regulations to control point sources of these pollutants is a strategy that has dramatically lowered these contaminants in many places. Controlling non-point sources, or surface runoff, is a much more difficult challenge even in places with relatively strong pollution-control rules, as in the United States.

Water issues in the Ganges River also remind us of the importance of water pollution in developing regions. We know how to reduce pollutants in water, and we know how to capture and remove pollutants. Finding ways to implement policies and pay for treatment is often difficult, though, even in wealthy countries. As developing regions struggle to raise health conditions and quality of life for their residents, finding ways to protect water quality will be a key need.

Reviewing Key Terms

Can you define the following terms in environmental science?

atmospheric deposition 18.1
best available, economically achievable technology (BAT) 18.4
best practicable control technology (BPT) 18.4
biochemical oxygen demand (BOD) 18.1

coliform bacteria 18.1
combined sewer overflow (CSO) 18.3
constructed wetlands 18.3
cultural eutrophication 18.1
dissolved oxygen (DO) content 18.1
effluent sewerage 18.3

eutrophic 18.1
nonpoint sources 18.1
oligotrophic 18.1
oxygen sag 18.1
point sources 18.1
primary treatment 18.3
red tide 18.1
secondary treatment 18.3

tertiary treatment 18.3
thermal plume 18.1
total maximum daily load (TMDL) 18.2
water pollution 18.1

Critical Thinking and Discussion Questions

1. Cost is the greatest obstacle to improving water quality. How would you decide how much of the cost of pollution control should go to private companies, government, or individuals?

2. How would you define *adequate sanitation*? Think of some situations in which people might have different definitions for this term.

3. What sorts of information would you need to make a judgment about whether water quality in your area is getting better or worse? How would you weigh the different sources, types, and effects of water pollution?

4. Imagine yourself in a developing country with a severe shortage of clean water. What would you miss most if your water supply were suddenly cut by 90 percent?

5. Proponents of deep well injection of hazardous wastes argue that it will probably never be economically feasible to pump water out of aquifers more than 1 kilometer below the surface. Therefore, they say, we might as well use those aquifers for hazardous waste storage. Do you agree? Why or why not?

6. Arsenic contamination in Bangladesh results from geological conditions, World Bank and U.S. aid, poverty, government failures, and other causes. Who do you think is responsible for finding a solution? Why? Would you answer differently if you were a poor villager in Bangladesh?

Data Analysis

Examining Pollution Sources

Water pollution has many sources and there are numerous approaches to reduce or remediate various contaminants. We have reviewed lots of these sources and approaches in this chapter. To understand these issues in greater depth, it's helpful to study and compare the graphs in this chapter that list the causes and consequences of water pollution. Visit Connect to find a set of questions about specific figures that will test your comprehension of this topic.

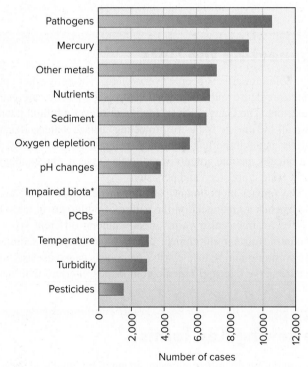

Leading causes of surface-water impairment in the United States. *Note: The * represents undetermined causes.*
Source: Data EPA, 2015

▲Fossil fuels have provided a large majority of our energy needs for two centuries, but now worries about air pollution, toxic emissions, environmental damage, and climate change are forcing us to look for alternative energy sources.
Clynt Garnham Energy/Alamy Stock Photo

Learning Outcomes

After studying this chapter, you should be able to:

19.1 Define *watt*, *joule*, and other units used to measure energy.

19.2 Explain how a coal plant produces electricity and explain why coal use is declining.

19.3 Describe several ways we use oil.

19.4 Identify the advantages and disadvantages of natural gas.

19.5 Summarize the potential and risk of nuclear power.

"The pessimist complains about the wind; the optimist expects it to change; the realist adjusts the sails."

– William Arthur Ward

The End of Coal?

In 2010, with great fanfare, Arch Coal, the second-largest U.S. coal company, announced a land lease in Montana's Powder River Basin, for what they promised would be the nation's largest open-pit coal mine. Arch calculated that the Otter Creek property held 1.4 billion tons of coal, worth $250 million annually to the state of Montana, or $5 billion over the 20-year life of the project. The economic value of the lease hinged on plans for new rail lines linking the mine to the giant deepwater shipping terminals planned in Oregon and Washington. From there, Montana coal would be sold to the fast-growing markets of China and India.

But the overseas export project met surprising resistance. Tribal leaders from the adjacent Northern Cheyenne Reservation, Montana ranchers, farmers, and conservationists fought the plans. Communities along the proposed rail line and at the West Coast terminal sites objected to the traffic congestion, noise, and dust from shipping. Climate activists objected to the disastrous implications of burning 1.4 billion tons of coal from the mine. After half a decade of lawsuits and protests, all six of the proposed West Coast coal terminals either had been rejected by courts or were mired in legal appeals.

The export terminal was essential for Arch Coal and other major coal companies. Domestic demand has been falling, in the face of competition from natural gas and renewables. Aging coal-fired power plants are expensive to operate, and power companies are retiring them around the country. Without new overseas markets, coal producers couldn't make a profit. Just six years after Arch Coal's celebrated expansion in Montana, the company declared bankruptcy.

When Arch Coal filed for bankruptcy in 2016, its opponents celebrated. But while bankruptcy for a coal company results from financial stress, it is also a strategy to shed debt and financial responsibility. In the past decade or so, 26 major coal companies, including the largest, Peabody Coal, have filed for bankruptcy. In doing so, a company can eliminate contracts with union labor, abandon pension commitments for retired miners, duck environmental responsibilities, and unload liabilities to creditors. While coal mining jobs have plummeted with mechanization, restructuring of coal industries further undermines workers and their communities. Even as coal companies are weakening, they are passing many of the costs on to coal workers and their communities.

However, the wave of bankruptcies does signal seismic shifts in the energy industry. Even while policies aimed at rescuing coal have dominated much environmental and energy politics in recent years, many analysts say this is the death struggle of an industry that no longer makes sense. Over 264 U.S. mines have closed in recent years. Coal plants, many built 50 years ago, cost more to operate every year and are closing steadily.

Cheap natural gas is displacing coal, meeting environmental protections in old coal plants is increasingly expensive, and opposition has grown. Few investors see much future in U.S. coal when cleaner, safer renewables also become cheaper and quicker to build.

Are we looking at the end of the coal era? Coal powered the Industrial Revolution, starting in the eighteenth century. It built export economies and employed millions of people worldwide. Coal also killed untold thousands of workers in mining disasters and with black lung disease—a consequence of breathing coal dust. Coal is responsible for ecological damage from acid rain and for health impacts in urban centers around the world. Coal also bears the largest responsibility for climate-changing greenhouse gases: In addition to being a dirty fuel, with impurities like mercury and sulfur emitted during combustion, coal produces more CO_2 per unit of heat than other fossil fuels. Coal supplied more than half of U.S. electricity a few years ago, but the number has fallen to less than 21 percent, slightly less than renewables (including hydropower).

While U.S. coal declines, though, coal is still growing globally. China and India are continuing to ramp up construction of coal plants, many of them overseas in South Africa, Vietnam, or Indonesia. This is a matter of intense global debate.

The world has seen energy systems shift repeatedly. We have transitioned from wood to coal to oil, gas, hydropower, and now wind and solar. The transition can be tense, as communities, investors, and owners fight to save their disappearing livelihoods. This makes it important for all of us to understand the benefits and costs of various energy sources. In this chapter, we examine fossil fuels and other conventional sources of energy. In chapter 20, we'll discuss renewable energy options.

FIGURE 19.1 Over half the coal produced in the United States comes from the northern Great Plains, where it's excavated almost entirely from open-pit surface mines such as this.
Moritz Werthschulte/Getty Images

Benchmark Data

Among the ideas and values in this chapter, the following are a few worth remembering.

80%	Global energy from fossil fuels (approx)
20%	Global energy from wind, water, solar, nuclear (approx)
2/3	Fraction of energy lost as heat in producing electricity from fossil fuel combustion
45%	Share of U.S. energy for transportation

19.1 ENERGY RESOURCES AND USES

- *Fossil fuels provide most of the world's energy.*
- *Transportation uses the greatest amount of energy in the United States.*
- *Energy consumption is rising rapidly in the developing world.*

Many of our most important questions in environmental science, from air pollution, climate change, and mining impacts, involve energy resources. Energy resources are also a rapidly changing story. The systems that drive the twenty-first century are sure to be different from the systems we know today.

We have undergone many energy transitions, from wood heat and draft animals to the coal-fired steam engines that powered the industrial revolution in the 1700s. Coal-powered steam became the leading source of electricity by the mid-twentieth century, while oil (and the gasoline derived from it) became powered transportation. In this chapter we explore these conventional energy sources, which still dominate our energy systems. In chapter 20 we discuss renewable energy and the ways it is evolving.

The future of energy is not the past

A central energy question in environmental science has long been, when will we run out of oil? Now the question is, how rapidly can we transition away from it? Oil is still our single most important energy source, in terms of both volume consumed and diversity of uses. It runs our cars, trains, and airplanes, powers factories, and heats our homes. Electricity production is shifting from dirtier coal fuel to natural gas in much of the world. Renewables make up about 20 percent of global electricity (10 percent of global energy), but renewables are the fastest-growing energy sector. The share of coal and natural gas in power production (electricity) is expected to decline steadily in coming decades (fig. 19.2).

Transitioning away from fossil fuels will be difficult, but there are good reasons to do it. The first is climate change, driven mainly by burning of fossil fuels. Costs of climate change will be about 10 percent of our economic output, a substantial loss. Another is the high environmental cost of oil spills, coal mining, and air pollution associated with fossil fuels (discussed in section 19.3). Still another reason is geopolitical instability. Many of the most important oil-producing regions are politically unstable, often with ongoing violence and persistent political corruption. Often

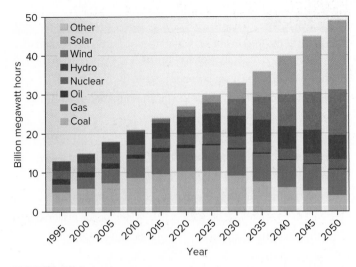

FIGURE 19.2 World electricity demand will grow rapidly as we electrify transportation and heating. Fossil fuels now supply about two-thirds of global electricity, but that share is expected to decline to one quarter by 2050.
Source: McKinsey Energy Insights 2019

the struggle to control oil resources is the root cause of conflict and corruption, and the world spends billions of dollars every year on wars and on military control in oil-producing regions.

Economic arguments support an energy transition. Annual subsidies for oil, gas, and coal far outweigh supports for renewables. The size of subsidies depends on what factors we consider: These might include tax supports, direct price supports, social and environmental costs, or even the cost of wars fought to defend overseas oil fields. Estimates range from around $550 billion per year to $1.9 trillion for fossil fuel subsidies, compared to around $100 billion per year for wind, solar, and other renewables, according to the International Energy Agency.

How do we describe energy?

Studying energy systems requires getting used to some terminology. When we discuss energy, we sometimes refer to total energy (for transportation, heating, electricity production, industry, and other uses). Power, on the other hand, refers to electricity. Renewables, such as wind and solar, mainly produce electricity, so we usually discuss these mainly in the context of power (electricity) sources.

We also have multiple terms to describe energy and what it does. **Work** is the application of force over distance, and we measure work in **joules** (table 19.1). **Energy** is the capacity to do work. **Power** is the rate of energy flow or the rate of work done: For example, one **watt** (W) is one joule per second. If you use a 100-watt lightbulb for 10 hours, you have used 1,000 watt-hours, or one kilowatt-hour (kWh; table 19.2). Most Americans use about 11,000 kWh worth of energy per year for household activities.

A typical power plant might supply about 1,000 megawatts (MW), or 1 gigawatt (GW) of electricity. This is enough for about 640,000 U.S. households, or about 1.3 million households in Europe, where households use about half as much energy. Some larger power

Table 19.1   Some Energy Units
1 joule (J) = the force exerted by a current of 1 amp per second flowing through a resistance of 1 ohm
1 watt (W) = 1 joule (J) per second
1 kilowatt-hour (kWh) = 1 thousand (103) watts exerted for 1 hour
1 megawatt (MW) = 1 million (106) watts
1 gigawatt (GW) = 1 billion (109) watts
1 petajoule (PJ) = 1 quadrillion (1015) joules
1 PJ = 947 billion Btu, or 0.278 billion kWh
1 British thermal unit (Btu) = energy to heat 1 lb of water 1°F
1 standard barrel (bbl) of oil = 42 gal (160 liter) or 5.8 million Btu
1 metric ton of standard coal = 27.8 million Btu or 4.8 bbl oil

Table 19.2 Some Energy Uses	
Uses	**kWh/year***
Computer	100
Television	125
100 W light bulb	250
15 W fluorescent bulb	40
Dehumidifier	400
Dishwasher	600
Electric stove/oven	650
Clothes dryer	900
Refrigerator	1100

* Averages shown; actual rates vary greatly.
Source: U.S. Department of Energy.

plants are designed to supply 2,000 MW or more. Wind turbines produce around 1–8 MW, so it takes a lot of wind turbines to add up to one conventional power plant. But it takes a lot of coal mines, gas fields, train lines, and pipelines to supply fuel to a conventional power plant. Commercial-scale solar plants typically produce around 150–500 MW of electricity.

The energy industry in the United States also uses traditional measures such as the British thermal unit (Btu), the amount of energy it takes to heat one pound of water one degree F. Oil is measured in barrels (bbl). One barrel equals 42 gallons (160 l). Natural gas, which is a gas like air, not liquid like gasoline, is measured in cubic meters or cubic feet (m^3 or ft^3). Coal is measured in tons (2,000 lb) or metric tons (1,000 kg, or 2,200 lb).

Fossil fuels still supply most of the world's energy

Fossil fuels (petroleum, natural gas, and coal) still supply about 80 percent of world commercial energy but that percentage is expected to decline in the next few decades. Oil supplies about 35 percent of our current supply, followed by natural gas (25 percent) and coal (20 percent). Renewable sources—solar, wind, geothermal, biomass, and hydro—now make up about 10 percent of our total energy supply, while nuclear reactors provide a little over 8 percent. Renewables are the most rapidly growing energy sector, but they have a long way to go to replace fossil fuels entirely.

Until recently, the richer countries, with about 20 percent of the world population, consumed roughly 80 percent of all commercial energy, while the other 80 percent of the world had only 20 percent of the total supply. That situation is changing now. By 2035, energy experts predict that emerging economies such as China and India will be consuming about 60 percent of all commercial energy.

Transportation consumes the largest, and fastest growing, share of primary energy in the United States (fig. 19.3). Almost three-quarters of transportation energy is used by motor vehicles, almost all of it from petroleum.

Industry is the second-largest major category of energy use. Mining, milling, smelting, and forging of primary metals consume about one-quarter of that industrial energy share. The chemical industry uses a large amount of fossil fuels, but only half of that is for energy generation. The remainder is raw material for plastics, fertilizers, solvents, lubricants, and hundreds of thousands of organic chemicals in commercial use. The manufacture of cement, glass, bricks, tile, paper, and processed foods also consumes large amounts of energy.

Residential and commercial customers use roughly 18 percent of the primary energy consumed in the United States, mostly for space heating, air conditioning, lighting, and water heating. Transportation has risen to 45 percent of U.S. energy use (fig. 19.4). About 98 percent of that energy comes from petroleum products refined into gasoline and diesel fuel, and the remaining 2 percent is provided by natural gas and electricity.

Producing and transporting energy also consumes energy. More than half the energy in primary fuels is lost during conversion to more useful forms, while being shipped to the site of end use, or during use. Electricity is generally promoted as a clean, efficient source of energy because, when it is used to run a resistance heater or an electrical appliance, almost 100 percent of its energy is converted to useful work and no pollution is given off. But as we have noted, about two-thirds of energy in fossil fuels is lost as heat in combustion. And about 10 percent more is lost during transmission and stepping down to household voltages.

How much energy do we use?

On average, Americans use about 4,500 kWh of electricity per person per year for household activities, twice what we used in 1970. In 1960 we used just 1,000 kWh per person per year. Europeans use about 2,000 kWh per year, and Chinese about 500 kWh. In low-income countries, people use less than 100 kWh per year—about what Americans use in a week.

But electricity is only about half of the residential energy we use. Including household heating, air conditioning, and other uses,

FIGURE 19.3 Sources and uses of energy in the United States in 2017. Units are quads, or quadrillion (10^{15}) British thermal units (BTUs). Line widths are proportional to the amount of energy. "Rejected energy" is lost, mainly as heat in fossil fuel combustion, rather than providing useful energy services. Rejected energy accounts for two-thirds of energy use.

Source: Lawrence Livermore National Laboratory and U.S. Department of Energy, 2018

Americans use about 11,000 kWh worth of energy per year. And personal transportation consumes slightly more, on average, than household uses.

Abundant energy resources are clearly associated with comfort and convenience in our lives. Having electric lights and automobiles makes life easier in many ways. But above some basic level of energy use, quality of life has little to do with energy consumption. Electricity consumption in developed countries with similar incomes, health care, and education ranges from around 1,500 to 10,000 kWh/year. Many European countries have higher standards of health care, education, employment, and leisure time, compared to most Americans, with half the energy use. In 2016, Sweden announced it intends to become entirely oil free, with an interim goal of carbon neutrality by 2045. Combined strategies such as wind, solar, and hydroelectricity, biogas (methane) for transportation, biomass, and efficiency make this goal possible.

Section Review

1. What are the major sectors of energy use shown in figure 19.3?
2. What are the major sources of commercial energy worldwide and in the United States?
3. How does energy use in the United States compare with that in other countries?

FIGURE 19.4 U.S. energy consumption by sector.
Source: Energy Information Agency 2018

19.2 COAL

- *We have a huge amount of coal.*
- *Burning coal releases many pollutants.*
- *Clean coal technology remains uneconomical.*

Coal is fossilized plant material compacted over millions of years into a rock-like, carbon-rich fuel (see fig. 14.12). Because it derives from fibrous plant residue, coal contains larger and more complex carbon compounds than oil. Coal is found in every geologic system since the Silurian Age 400 million years ago, but most deposits date to the Carboniferous period (286 million to 360 million years ago) when the earth's climate was warmer and wetter than it is now. Because coal takes so long to form, it is essentially a nonrenewable resource. Because they originate in anoxic swamp beds, coal deposits contain sulfur (common in low-oxygen biological systems), as well as metals such as mercury and cadmium, which plants accumulate from soils (see section 14.2). These become important pollutants when coal is burned.

Coal resources are greater than we can use

World coal deposits are enormous, ten times greater than conventional oil and gas resources combined. Global coal reserves are widely distributed across the world's regions, with some of the largest countries having the largest reserves. The United States, Russia, and China account for two-thirds of global reserves. The other third is distributed across Asia, South America, Europe, and Africa. Coal seams can be 100 m thick and can extend across tens of thousands of square kilometers. The total resource is estimated to be 10 trillion metric tons. At current rates of use, this would amount to several thousand years' supply. But it's clear that we couldn't survive if all that carbon (and associated pollutants) were emitted into the atmosphere.

Coal mining is a dirty, dangerous activity. Underground mines are notorious for cave-ins, explosions, and lung diseases, such as black lung suffered by miners. Surface mines (called strip mines, where large machines scrape off overlying sediment to expose coal seams) are cheaper and generally safer for workers than tunneling, but leave huge holes where coal has been removed and vast piles of discarded rock and soil.

An especially damaging technique employed in Appalachia is called mountaintop removal. Typically, the whole top of a mountain ridge is scraped off to access buried coal (fig. 19.5). In 2010 the EPA announced it would ban "valley fill," in which waste rock is pushed into nearby valleys, but existing operations were "grandfathered in" and allowed to continue (see Chapter 14 for further discussion).

How do we produce electricty from coal? Like all thermal power plants (that is, those using heat, such as coal, gas, or nuclear plants), coal plants generate steam, and the steam pressure spins a generator. The generator creates an electric field by spinning bundles of copper wires past magnets, which produce electric current that passes into power lines for distribution (fig. 19.6).

Burning coal releases huge amounts of air pollution. Every year the roughly 1 billion tons of coal burned in the United States (83 percent for electric power generation) releases close to a trillion metric tons of CO_2. This is about half of the industrial CO_2 released by the United States each year.

Coal also contains toxic impurities, such as mercury, arsenic, chromium, lead, and uranium. These contaminants were deposited in the ancient swamps that produced the coal, and they are released into the air during combustion. The coal burned every year in the United States releases 18 million metric tons of sulfur dioxide (SO_2), 5 million metric tons of nitrogen oxide (NO_x), 4 million metric tons of airborne particulates, 600,000 metric tons of hydrocarbons and carbon monoxide, and 40 tons of mercury.

FIGURE 19.5 One of the most environmentally destructive methods of coal mining is mountaintop removal. Up to 100 m of the mountain is scraped off and pushed into the valley below, burying forests, streams, farms, cemeteries, and sometimes houses.
Melissa Farlow/Getty Images

This is about three-quarters of the SO_2 and one-third of the NO_x released by the United States each year.

Another concern with coal combustion is that it produces vast amounts of toxic ash, which is expensive to manage safely. In 2009 an earthen dam, which held a pond for storing coal ash, broke in eastern Tennessee and released a billion gallons (3.8 billion liters) of coal ash sludge into a tributary of the Tennessee River. The ash contained dangerous levels of arsenic, mercury, and toxic hydrocarbons. This impoundment was only one of hundreds of equally risky coal ash dumps across the country.

Coal use is declining in the United States and Europe

Recent improvements in technology have made other choices, from gas and oil to wind and solar, cheaper than coal. At the same time, most of our coal-fired power plants are aging, and they are increasingly expensive to maintain. The owners of coal plants and coal mines have pushed for laws and subsidies that keep them operating, especially with help from the Trump administration, which has included a number of powerful coal lobbyists. However, despite supports and continued delays on clean air rules, coal use continues to decline. Gas-powered plants are cleaner, with less acidic emissions; equipment lasts longer, and operation and maintenance are less expensive.

Over 40 percent of global coal plants are operating at a loss, according to the group CarbonTracker, which monitors coal consumption. Between high maintenance and fuel costs and competition from gas and renewables, they are no longer the safe investment they once were. By 2040, 70 percent of global coal plants could be uneconomical.

European countries have depended on coal for centuries, but this dependence is also ending. The United Kingdom, the original

FIGURE 19.6 A coal-fired power plant works by generating steam, which turns an electrical generator. Nearly all electric power, apart from solar photovoltaic energy, is produced by spinning a turbine that turns a generator. Until recently, coal plants produced most of our electricity.

1 A conveyor belt carries crushed coal to a combustion chamber; where water circulating in pipes turns to steam.

2 High-pressure steam spins a turbine, which turns a generator to produce electricity.

Turbine

Generator

Cooling tower

Condenser cooling water

3 Water circulated from a lake, river, or cooling tower cools and condenses steam.

home of coal power, has closed its last coal mine and is decommissioning its coal plants. Germany, Europe's biggest coal user, has promised to eliminate coal power by 2040, as a necessary step to meet its climate goals.

With coal consumption falling in North America and Europe, coal companies are looking desperately for new markets overseas. As discussed in the opening case study for this chapter, Arch Coal and other western mining companies hoped to ship 20 million tons of coal per year to ports on the West Coast (fig. 19.7). But port communities, and those along the rail corridors, resisted vigorously. Shipping this much coal would require as much as 50 or 60 heavy, mile-long trains moving through a community every day—averaging one every 20 to 30 minutes, night and day. Communities objected to the noise, disruption, and dust the trains would produce.

Asian policies on coal have been shifting. After decades of rapid building, especially in China, development and permitting slowed in 2016. Plans have accelerated again since then, and China has permitted or begun building some 266 GW of new coal capacity, slightly more than the entire U.S. capacity. India is developing another 93 GW of new coal capacity (fig. 19.8). In addition, Chinese firms are building or financing 240 GW of coal overseas, mainly in southern Africa, Vietnam, Indonesia, Bangladesh, and Pakistan.

Why are these countries deepening investments in coal? One reason is that developing countries are hungry for electricity on a massive scale, and coal is readily available. It is also familiar, and while it's expensive, can power large industries. Momentum is another reason. Once a construction and banking sector has developed, with expertise in building, financing, and operating coal plants, that industry wants to grow. To do so, it must cultivate new

FIGURE 19.7 Many communities in the United States are opposed to the traffic disruptions, noise, dust, and other irritations associated with coal shipments. Exports from coastal coal-loading terminals would only exacerbate these conflicts.
Kevin Burke/Digital Stock/Corbis

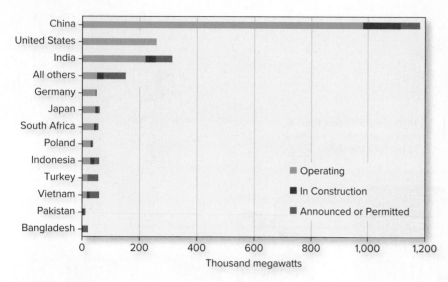

FIGURE 19.8 Coal plant capacity in development or in operation. Capacity (the amount plants are able to produce) in the United States is 260 GW, in China 980 GW, in India 220 GW. Actual production is often much less.

markets. A generation ago, U.S. corporations built and financed coal and hydropower power plants overseas. Now Chinese companies are doing the same.

Chinese companies are also building, and often overbuilding, at home. As regions compete for growth and industry, and as coal developers seek to expand, China's coal plants are now operating at 49 percent of capacity. This is a good thing for the climate, as they are burning less coal than promised. The economics of this overproduction appear risky, though.

Water shortages are another serious limitation for China's coal use. Water is needed for every aspect of coal use, from mining and processing to steam production and cooling in a power plant. More than half the country's coal mines and thermal power plants are in the arid north, where water has always been scarce. A decades-long drought has made the situation more dire. There isn't enough water in the north for coal mines, processing facilities, power plants, and coal-to-chemicals industry as well as people. China is building diversion projects costing billions of dollars to pump water thousands of kilometers from the well-watered south to the desert north, but these efforts aren't enough to support expansion of coal use at the rate planned.

Is clean coal technology an option?

There have been many attempts to develop clean coal technology. U.S. agencies have spent billions of dollars on this effort, especially for **carbon capture and storage (CCS),** in which CO_2 is captured and pumped into deep geologic formations. Climate analysts increasingly say that carbon capture is a necessary strategy for controlling climate change. However, despite ongoing innovation, CCS coal plants remain uneconomical. This technology has been used for many years to enhance oil recovery from depleted oil wells, but it remains too expensive to capture and transport the huge volumes of CO_2 produced by power plants to oil wells more than a few kilometers away.

In 2003, President George W. Bush described his FutureGen project as the world's first zero-emission coal plant, but abandoned it when it became too expensive. In 2010, President Obama pinned his hopes on a $2.4 billion project in project in Kemper County, Mississippi. By 2017, the cost had risen to $7.5 billion, and the operators had decided to switch to natural gas. Rather than bring economic development and jobs to this impoverished county, rising electric rates have driven away business and burdened residents with one of the most expensive power plants ever built. China claims to have about a dozen carbon-capture coal gasification plants, but observers report that although these facilities are capturing CO_2, none are actually storing it because they aren't required to by the central government. Instead, it appears they simply vent it to the air. Some people describe this program as "catch and release."

Section Review

1. Where are the largest coal deposits located?
2. Why does burning coal produce pollutants such as sulfur dioxide?
3. How does a coal-powered plant produce electricity?

19.3 OIL

- *Tar sands, deep ocean deposits, and tight shale formations that once were thought impossible to access have expanded our oil supplies.*
- *The Arctic Ocean could be the second-largest oil source in the world.*
- *Oil refineries supply many useful products but produce hazardous air pollutants.*

Like coal, petroleum is derived from organic molecules created by living organisms millions of years ago and buried in sediments, where high pressures and temperatures concentrated and transformed them into energy-rich compounds. While coal originates as vegetation in peat swamps, though, oil derives mainly from marine algae and plankton in coastal seas (see fig. 14.13). Oil therefore tends to be associated with sedimentary sand or shale (mud) deposits. Like coal, oil often contains sulfur because of the anoxic environment in which it was deposited. Depending on its age and history, a petroleum deposit will have varying mixtures of oil, gas, and solid tarlike materials. Extracting and processing these deposits can be expensive, complex, and often dangerous (fig. 19.9). But the different carbon molecules in oil are readily used and transformed, once extracted. This makes oil useful for countless purposes, from gasoline and other fuels to the raw material for a host of petrochemicals, including plastics and pesticides.

In an oil deposit, liquid and gaseous hydrocarbons can migrate up through cracks and pores out of the sediments

FIGURE 19.9 Oil still provides more than half our energy, but extraction is dangerous work. Here, workers thread pipe on Brazil's Enchova platform, on which 84 workers have died in explosions and fires. Safety concerns remain widespread in the industry.
Stephanie Maze/Getty Images

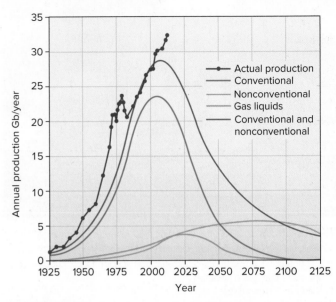

FIGURE 19.10 Hubbert's curve describes petroleum production. Dots show actual production. The area under the curve represents the amount of economically recoverable oil (Gb = billion barrels).
Source: Jean Laherrer, www.hubbertpeak.org

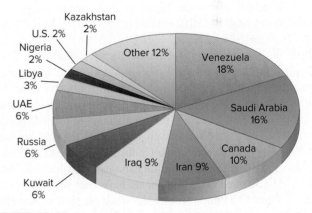

FIGURE 19.11 Proven oil reserves. The Middle East now accounts for a much smaller percent of the known, economically recoverable oil than in previous years. The numbers here add to more than 100 percent due to rounding.
Source: CIA Factbook, 2016

in which they formed and into surrounding rock layers. Oil and gas move upward (over millions of years) because they are less dense than the surrounding rock. Oil and gas deposits generally accumulate under layers of shale or other impermeable sediments, especially where folding and deformation of systems create pockets that trap upward-moving, low-density hydrocarbons. In these folds, oil fills spaces in a porous sandstone or limestone, much like water saturating a sponge. Many major oil fields include thousands of wells that dominate the landscape over very large areas.

In the 1940s, Dr. M. King Hubbert, a Shell Oil geophysicist, predicted that oil production in the United States would peak in the 1970s, based on estimates of U.S. reserves at the time. Until recently, Hubbert's prediction of **peak oil** seemed correct. Discovery of new oil declined, and the United States went from being the world's greatest oil exporter to being the largest importer.

Recently, though, global production and estimates of our recoverable oil supplies have expanded dramatically as we've developed techniques for obtaining oil from ever more remote and extreme places (fig. 19.10). Even the United States, which has been an oil importer for decades, is once again an oil and gas exporter. Not long ago, countries in the Middle East accounted for nearly 90 percent of all known oil deposits. Huge new discoveries such as Canada's Athabascan **tar sands,** Brazil's Santos Basin, Venezuela's Orinoco Belt, Angola's Kwanza Basin, the United States' Bakken formation, and the Arctic sea bed, for example, dramatically increase supplies. Between 1993 to 2003, proven (commercially extractable) reserves grew by nearly two-thirds, from 1.04 trillion barrels to 1.69 trillion barrels. Places that we never thought would have oil have become major suppliers (fig. 19.11). Venezuela and Canada may have nearly as much petroleum in their tar sands and heavy oils as the previously estimated world supply. Supplies must have some limit, but in recent years, instead of facing the end of oil, producing countries have

been struggling with overproduction and low prices. At present rates of consumption, a global oil peak remains an important question. The more important question now, though, is how to wean ourselves off of climate-disrupting oil long before it is all burned.

Extreme oil has extended our supplies

Drilling technology has made tremendous strides in recent years. It's now possible to drill into the extreme temperatures and pressures encountered at a depth of 12,000 m (40,000 ft). Drilling can extend horizontally up to 32 km (20 mi) from the well head, and as many as 24 individual wells can be supported from a single drilling pad. **Hydraulic fracturing ("fracking")** can release oil

FIGURE 19.12 In 2010, the oil drilling rig *Deepwater Horizon* exploded and sank, spilling at least 5 million barrels (800 million liters) of crude oil into the Gulf of Mexico. It was drilling in water 1 mile (1.6 km) deep, but other wells are now more than twice as deep.
U.S. Coast Guard/Getty Images

FIGURE 19.13 Alberta tar sands are now the largest single source of oil for the United States, but there are severe environmental and social costs of extracting this oil.
dan_prat/Getty Images

from "tight" formations through which passage of liquid would otherwise be obstructed. Fracking is also now used extensively in the quest for natural gas. We'll discuss some of the impacts of this technology in section 19.4.

Most people had never heard of deep ocean oil wells in remote places until the 2010 explosion and sinking of the *Deepwater Horizon* in the Gulf of Mexico (fig. 19.12). The rig was drilling the Macondo well for BP in 1-mi (1.6-km) deep water. At least 5 billion barrels (800 million liters) of oil were spilled during the 4 months it took to plug the leak. A massive effort was made to scoop up the oil, burn it, or disperse it with chemicals. At least 6.8 million liters (1.8 million gallons) of toxic dispersant was sprayed on the ocean. Eleven workers on the drill rig were killed, and the fishing and tourism industry lost billions of dollars. The damage to marine and wildlife habitats is incalculable. Altogether, criminal and civil settlements have cost BP and its contractors about $70 billion. It was the worst oil spill in American history. But much deeper wells in much more remote areas are being drilled today.

Melting of Arctic sea ice has triggered a scramble for territorial claims in the Arctic ocean. All the surrounding countries have asserted ownership of the seafloor out to (or beyond) their continental shelf. It's thought that the Arctic may have 15 percent of the world's undiscovered oil and 30 percent of all natural gas. Exploratory drilling has already started. In 2012, Shell Oil sent two floating drill rigs into the Arctic ocean. They were so badly damaged by storms and ice, however, that one had to be towed to Korea and the other all the way to Singapore for repairs.

This demonstrates the difficulties of working in the Arctic. Although most of the pack ice may be melted in the summer, it's still a very remote and dangerous place. A leak similar to the BP spill in the Gulf of Mexico would be very difficult to contain in the frigid and often stormy Arctic thousands of kilometers from the nearest supply base.

Canada's Athabasca oil sands show additional dangers and costs of our search for oil. These deposits are estimated to contain 1.7 trillion barrels of oil lying beneath some 141,000 km^2 (54,000 mi^2, or about the size of Greece) of boreal forest in northern Alberta. Extracting this oil produces severe environmental impacts. A typical facility producing 125,000 bbl of oil per day will consume several square miles of forest for mine pits and tailings ponds over its lifetime (fig. 19.13). It also creates about 15 million m^3 of toxic sludge, releases 5,000 tons of greenhouse gases, and consumes or contaminates at least 75 million liters (20 million gal) of water each year. Native Cree, Chipewyan, and Metis people are protesting the cancer-causing water pollution and destruction of forests and wildlife, which are eliminating their traditional ways of life.

Large deposits of solid organic material called **kerogen** (a solid, waxy form of oil), occur in sedimentary rocks in many places in the United States. When heated to about 480°C (900°F), the kerogen liquefies and can be extracted from the stone. Kerogen-containing **oil shale** beds up to 600 m (1,800 ft) thick occur in the Green River Formation in Colorado, Utah, and Wyoming, and lower-grade deposits are found over large areas of the eastern states. These deposits might yield the equivalent of several trillion barrels of oil. But mining and extracting kerogen is even more difficult than tar sands. It is expensive, extraction requires tremendous amounts of heat and vast quantities of water (a scarce resource in the arid West), and it produces about twice as much CO$_2$ per unit of energy as conventional oil. Thus far, it hasn't been economically feasible to mine these oil shales. Nevertheless, the Interior Department has discussed opening leases for oil shale research and development in Colorado and Utah.

Oil is a boom and bust industry

Oil prices are often extremely unstable. In July 2008, for example, on the eve of the "great recession," the global oil price reached U.S. $147 per barrel, and Americans saw gasoline prices over $4 per gallon. But then the housing market imploded, the world banking system teetered on the brink of collapse, and within months, the price of oil plummeted to $30 per barrel. The price surged

Water Protectors at Standing Rock

Energy policy and the environmental effects of energy use affect all of us in one way or another, but most of us don't know what we can do about it. We rarely know how to take action even if we feel strongly about energy and our environment. But sometimes a particular challenge galvanizes people, and new strategies for citizen action are learned.

The 2016 protest at the Standing Rock Sioux Reservation was one of these watershed moments. Thousands of protesters camped on the route of the Dakota Access Pipeline, being built from the Bakken oil fields in western North Dakota to southern Illinois. At issue: The pipeline threatened ancient burial grounds and sacred sites, and it crossed under the Missouri River just upstream from the reservation's water supply. A pipeline rupture would endanger drinking water and could poison vast reaches of the river.

The pipeline's original route crossed the Missouri just upstream of Bismark, North Dakota, but residents there objected to its proximity to residential areas and the risk of spills. The Army Corps of Engineers ordered the Texas-based pipeline company, Energy Transfer Partners, to move the route downstream, to just above the Standing Rock water intake instead. The tribe sued, arguing that they hadn't been consulted, as required by federal law. But construction continued apace.

In April 2016, a group of young people from multiple tribes, the International Indigenous Youth Council, decided to take action. They set up a prayer camp, which they called the Seven Council Fires camp in honor of seven major Dakota and Lakota groups, in the path of the pipeline construction. Their rallying cry was "Mni Wiconi!" (water is life!).

Through the summer and fall, members of more than 300 Native American tribes and indigenous people from as far away as Peru, Norway, and New Zealand joined the Water Protectors' camp. Thousands of supporters, drawn through social media networks, came from all over North America. This was the largest, most diverse tribal gathering since the battle of the Little Bighorn, which took place 141 years earlier and 1,000 km (600 mi) west of the site. The camps were remarkably harmonious. Many of the tribes were traditional enemies, but participants emphasized peace, spirituality, respect, and nonviolent civil disobedience (fig. 1).

Reactions from Energy Transfer Partners was neither peaceful nor respectful. Tensions exploded on September 3, when protestors blocked a road to protect sacred cultural sites. Private security guards attacked the unarmed protesters with police dogs, pepper spray, and clubs. Videos, seen around the world on YouTube, showed a grim similarity to earlier civil rights, labor, and anti-war protests. Police departments from across the Dakotas and neighboring states sent more armed soldiers and more

FIGURE 1 Although the Standing Rock Camp contained thousands of people from hundreds of tribes and indigenous groups from across North America and around the world, the encampment was remarkably peaceful and orderly. Protestors were committed to nonviolent resistance and cooperative decision making.
Helen H. Richardson/Denver Post/Getty Images

police with riot gear and military equipment. In early November, in freezing weather, police attacked protestors with water cannons, tear gas, rubber bullets, and concussion grenades, injuring hundreds.

The power of peaceful protest was demonstrated in December, when President Obama finally declared a stop to construction. But a month later, two days after his inauguration in January 2017, President Donald Trump ordered immediate completion of the pipeline. Within six months of starting operation, the pipeline had eight spills. There are no plans for what to do if, or when, crude oil leaks into the Missouri River.

Oil interests won this battle, but the protests focused global attention on the need for indigenous rights and environmental protection. Most important, a movement like this teaches a new generation that they can— and must—influence the political process. It teaches organization, social networking, media, and policy. Engaging in energy policy is the public's business, and movements like this often produce new candidates, policymakers, writers, and leaders, all key elements to the democratic process. What do you think? How would you intervene if you faced issues like these? What difference does it make that young people began this movement?

---

above $100 per barrel in mid-2014, only to plunge again to just over $30 per barrel in early 2016. Since then the price has bounced between about $40 and $80 per barrel.

Some of this volatility is driven by market forces, but much is geopolitical. Saudi Arabia, for example, flooded the market with cheap oil to win market share and to drive out competitors producing oil from expensive deposits. Sharply rising production in North Dakota have also disrupted oil markets. The Bakken formation underlying parts of North Dakota, Montana, and Saskatchewan is composed of "tight shale" deposits that are unsuited to conventional wells. Horizontal drilling and fracking, however, unleashed a major oil boom in this formation. North Dakota,

which ranked 38th in the country in per capita income in 2002, saw its economy double in a decade. By 2012, North Dakota had overtaken Alaska to become the number two oil-producing state in the country (after Texas). At its peak, the state was shipping more than a billion barrels of oil per day and had a budget surplus of $1 billion. Then prices plummeted to $30, thousands of workers lost their jobs, and towns struggled to pay for roads and housing started during the boom.

Because of price volatility and uncertain futures, many oil-producing nations are seeking to diversify. Even Saudi Arabia is trying to develop alternatives. Unpredictable prices motivate this move: When prices plummet, the kingdom's lavish expenses are difficult to maintain. In 2016, the Saudis offered to sell part of its US$10 trillion national oil company (the most valuable corporation in the world). They realize it's wise to diversify and not depend on a single commodity.

Low oil prices can be bad for the environment because they encourage people to buy inefficient vehicles and drive more. But they can also be an opportunity to avoid dangerous and long-lasting projects, such as pipelines and power plants. It's easier to leave fossil fuels in the ground when they aren't worth the cost of extraction.

Indigenous groups have challenged pipelines

As with coal, getting oil to market is a key challenge. Most oil is delivered by a network of thousands of kilometers of pipelines, many aging and under high pressure, which cross rivers, wetlands, farmlands, and communities. North American Indigenous and First Nations groups, whose land and water have often been most at risk, have often been responsible for raising visibility of pipelines and their risks.

The first nationally prominent standoff was over the Keystone XL pipeline, a network of large-volume pipelines designed to carry Alberta's tar sands oil to Houston for refining and export. Canadian First Nations groups had been protesting tar sand mining for years, but the pipeline to deliver it from Alberta to Texas for overseas shipping brought much wider attention to the project. Portions of the line have been in operation since 2010, but other sections continue to be challenged.

Of particular concern has been a proposed route across the Ogallala Aquifer, a vast underground water resource that irrigates farmland across the central Great Plains. Blowouts and leaks on the 1,400-km (875-mi) Keystone pipeline route are a serious concern. Pipelines carrying tar sands oil have a much higher rupture rate than those for conventional oil. The residual sand is abrasive, the oil is acidic and corrosive, and because it is dense, it requires high temperature and pressure to keep moving in the line. All these factors make tar sands pipelines more accident prone than other lines.

In 2016, after years of protests, President Obama announced that he would not approve the Keystone pipeline. In 2017, within days of his inauguration President Trump approved both the Keystone XL and the Dakota Access pipeline, built to deliver North Dakota's Bakken oil to refineries in Illinois and Texas. The Indigenous Environmental Network reports that lawsuits by native communities continue to challenge the plan (What Do You Think?, p. 439).

TransCanada, the company behind Keystone, also has been pursuing a Northern Gateway route that would cross the Canadian Rockies on its way to a terminal in the fjords of British Columbia's Great Bear Rainforest. In 2016, Canadian courts rejected the Gateway pipeline because it ignored First Nation's concerns. Legal debates between oil-dependent Alberta and environmental and First Nations groups continue.

Refineries produce useful products and hazardous pollutants

Oil produces a host of useful products. Crude oil is "cracked" or broken into smaller molecules by heating. Distilling operates on the principle that progressive cooling separates larger molecules from smaller ones (fig. 19.14).

FIGURE 19.14 Distillation operates on the principle that longer-chain hydrocarbons condense at a higher temperature than shorter-chain hydrocarbons. Fuel oils, for example, have very long carbon chains (20–70 C atoms in a molecule, shown as C_{20} to C_{70}), while natural gas has only 4 (C_4).

Some people argue that rather than burn petroleum for energy, we should save it for useful products that can be created from it, such as plastics, lubricating oils, paints, and other chemicals. It is also important to recognize that refining—at least as it's currently done in the United States and most other regions—releases a great deal of volatile organic compounds into the air. Some of the worst air quality in America is found near the heavy concentrations of petrochemical industries in Texas, Louisiana, and New Jersey. High levels of cancer, asthma, and other diseases often occur in these areas. BP, which is blamed for most of the bad decisions that led to the 2010 Gulf oil spill, has a long record of refinery failures. In the past three years, according to the Center for Public Integrity, BP accounted for 97 percent of all "flagrant violations" found in the refining industry. In 2005, for example, an explosion at a BP Texas City refinery killed 15 workers and injured 170 others. This is just one instance where BP's disregard for safety regulations proved disastrous, and deadly.

Section Review

1. How have our views about the size of world oil resources changed in the past few decades?

2. In what ways does oil production require water resources?

3. What are tar sands, and what problems are associated with their mining, processing, and shipping?

19.4 NATURAL GAS

- *We're finding large new supplies of natural gas in many places.*

- *Falling gas prices have challenged other energy sources.*

- *Fracking (hydraulic fracturing) has greatly expanded our gas supplies but has serious drawbacks.*

Natural gas is the world's third-largest commercial fuel, currently making up about one-quarter of global energy consumption. Natural gas consists of small, volatile organic compounds, mainly methane (CH_4), that accumulate at the top of oil or coal deposits. Because it consists of simpler, purer compounds than coal or oil, gas burns more cleanly than either coal or oil, and it produces only half as much CO_2 as an equivalent amount of coal during combustion. Natural gas's rapidly falling prices are also a large part of the recent decline in coal use. Many people hope this shift from coal to gas will help slow climate warming. Others are not so sure it will have that effect.

Most of the world's currently known natural gas is in a few countries

More than half of all the world's proven natural gas reserves are in the Middle East and Russia (fig. 19.15). Both eastern and western Europe are highly dependent on imported gas. The total ultimately recoverable natural gas resources are thought to be 10,000 trillion ft³, corresponding to about 80 percent as much energy as the estimated recoverable reserves of crude oil. The proven world

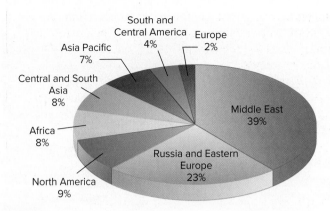

FIGURE 19.15 Proven natural gas reserves by region.
Source: U.S. Energy Information Agency, 2018

reserves of natural gas are 6,200 trillion ft³ (176 million metric tons). Because gas consumption rates are only about half of those for oil, current proven gas reserves represent roughly a 60-year supply at present usage rates.

Methane is often found with coal, which slowly breaks down into simpler CH_4 molecules. Accumulation of this explosive gas is one of the things that makes coal mining so dangerous. In many places where mining coal seams isn't economically feasible, "coal-bed" methane deposits still provide relatively cheap and accessible gas. In Wyoming's Powder River Basin, for example, 140,000 wells have been proposed for methane extraction. Together with the vast network of roads, pipelines, pumping stations, and service facilities, this industry is having a serious impact on ranching, wildlife, and recreation in formerly remote areas. In Wyoming's Upper Green River Basin, for example, 50,000 pronghorn antelope and 10,000 elk migrate through a narrow corridor every year on their way between summer and winter ranges. The Jonah Gas Field (fig. 19.16) lies across this migration route, and biologists worry that the noise, traffic, polluted wastewater pits, and activity around the wells may interrupt the migration and doom the herd.

Water consumption and pollution are also huge problems in the arid West. It takes large amounts of water to drill the wells, and once in production, each well can produce up to 75,000 liters of salty water per day. If this waste enters streams, it poisons wildlife and domestic livestock. In several western states, ranchers, hunters, anglers, conservationists, water users, and renewable energy activists have banded together in an unlikely coalition to fight against coal-bed gas extraction, calling on Congress to protect private property rights, preserve water quality, and conserve sensitive public lands. "It may be a clean fuel," says one rancher, "but it's a dirty business."

Deep gas-bearing shale formations underlie vast areas of Appalachia, Texas, and other regions (fig. 19.17). The Marcellus and Devonian Shales, underlying Pennsylvania and New York, have been the focus of considerable debate. The U.S. Geological Survey estimates that the Marcellus/Devonian formation may contain a 100-year supply for the United States at current consumption rates. With such large amounts of gas now available, prices

FIGURE 19.16 Gas wells in the Jonah Field in Wyoming's Upper Green River Basin.

Peter Aengst/The Wilderness Society/Lighthawk

have fallen sharply in recent years, making the resource much more attractive.

These shale deposits are generally "tight" formations through which gas doesn't flow easily. But pumping a mixture of water, sand, and various chemicals into rock formations at extremely high pressure has produced a flood of new, cheap gas. This has been a boon to consumers, but the multitude of wells, water pollution, and threats to water supplies on which millions of people depend, raise thorny problems (What Do You Think?, p. 444). Many utilities are converting old, dirty coal-burning power plants to cleaner natural gas. Cheap gas lowers enthusiasm for the development of renewable sources, such as solar, wind, and geothermal energy.

An extreme example of the risks of gas leaks occurred in California in October 2015. A well pipe ruptured in a depleted oil well being used for gas storage in Porter Ranch, a Los Angeles suburb. In the four months it took to find and plug the leak, more than 97,000 tons (5 billion ft^3) of methane and ethane were emitted from the well. In terms of greenhouse effects, this was equal to the annual emissions from about 600,000 cars.

Other countries are now developing their fracking industries. Poland has started fracking extensive shale deposits, and Russia, China, and Australia are developing similar projects. On the other hand, several European countries, including England, Germany, and Spain have banned fracking. Spread of this technology has major impacts on both energy prices and global climate.

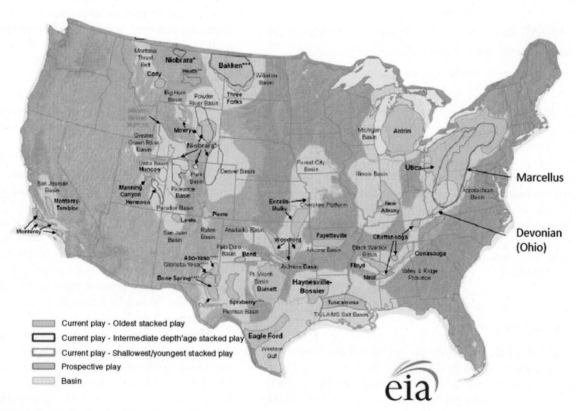

Current play - Oldest stacked play
Current play - Intermediate depth'age stacked play
Current play - Shallowest/youngest stacked play
Prospective play
Basin

FIGURE 19.17 Shale gas plays in the United States. The cost of extracting these resources varies with depth.

Source: Energy Information Association

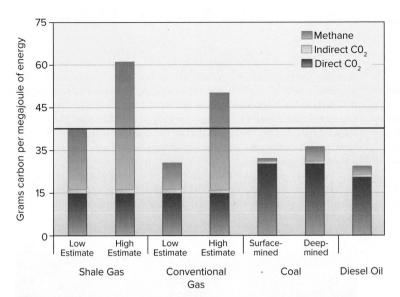

FIGURE 19.18 Comparison of greenhouse gas emissions from different fuel sources over a 20-year period. High and low estimates are provided for shale and conventional gas due to variations in field measurements. Under either scenario, shale gas has far higher climate-changing potential than any other fuel source.

Source: Data from Howarth, et al. 2011, https://www.fe.doe.gov/programs/gasregulation/authorizations/Orders_Issued_2012/75._Howarth-EtAl-2011.pdf

Fracking and other gas-drilling operations leak methane to the atmosphere, from fractured rock formations, cracks in well casings, and leaking valves and joints in pipelines. Nobody knows how much CH_4 and CO_2 are released by gas well drilling and fracking, but observers have estimated fugitive emissions from gas sales and sample measurements. Estimates are that 3 to 8 percent of gas might escape this way. Even a fraction of this would have a large impact, because methane is more than 20 times more effective, per molecule, in retaining heat in the atmosphere (fig. 19.18). Even under the most optimistic estimates, shale gas is thought to have a larger climate impact than either coal or oil.

Getting gas to market is a challenge

In many places, gas and oil are found together in sediments, and both can be recovered at the same time. In remote areas, however, where there are no gas-shipping facilities, the gas often is simply flared (burned) off—a wasteful practice, but an economical one where capturing and storing gas is expensive. The World Bank estimates that 100 billion m^3 of gas, or 1.5 times the amount used annually in Africa, are flared every year. Increasingly, however,

---

Quantitative Reasoning

As you can estimate from figure 19.17, shale gas mining releases only half as much CO_2 as surface-mined coal, but about 20 times as much methane. If methane has an infrared absorbing capacity 20 times as high as CO_2, what is the relative climate impact of these two fuels?

---

these "stranded" gas deposits are being captured and shipped to market.

World consumption of natural gas is growing by about 2.2 percent per year, considerably faster than either coal or oil. Much of this increase is in the developing world, where concerns about urban air pollution encourage the switch to cleaner fuel. Gas can be shipped easily and economically through buried pipelines. The United States has been fortunate to have abundant gas resources accessible by an extensive pipeline system. Until 2001, Canada was the primary source of natural gas for the United States, providing about 105 billion m^3 per year. **Liquefied natural gas (LNG)** imports were once expected to supply a significant portion of the U.S. supply. Since fracking has created a U.S. surplus, these ports are now being used as export, rather than import facilities.

In other places, gas lines have been subject to political or economic pressures. Russia, for example, has in the past cut off gas supplies to Ukraine and Belarus in a dispute over prices and has threatened shipments to northern Europe over policy differences. Recent political unrest in North Africa made southern Europe nervous about its access to gas.

Intercontinental gas shipping can be difficult and dangerous because gas is volatile and easily ignited, and because it is transported under pressure, in liquid form, in order to make shipping economical. At $-160°C$ ($-260°F$), the liquid's volume is about 1/600 that of gas. Special refrigerated ships transport LNG (fig. 19.19), but finding sites for terminals to load and unload these ships is difficult. Many cities are unwilling to accept the risk of an explosion of the volatile cargo. A fully loaded LNG ship

FIGURE 19.19 Global trade in natural gas involves shipments of pressurized, liquefied gas in specialized ships, such as this one at an Australian terminal. The potential hazards of handling volatile gas under pressure makes it difficult to find sites for these ports.
Regis Martin/The New York Times/Redux Pictures

The Fracking Debate

Vast amounts of methane may lie in relatively shallow sediments under large areas of North America. This gas burns more cleanly than coal or oil, is easier to ship, and produces less CO_2 during combustion. But much of the resource is held in "tight" formations through which gas doesn't flow easily. To boost well output, mining companies rely on hydraulic fracturing ("fracking"). A mixture of water, sand, and potentially toxic chemicals is pumped into the ground and rock formations at extremely high pressure. The pressurized fluid cracks sediments and releases the gas. Fracturing rock formations often disrupts aquifers, however, and contaminates water wells.

Drilling companies have generally refused to reveal the chemical composition of the fluids they use in fracking. They claim it's a proprietary secret, but it's well known that a number of petroleum distillates, such as diesel fuel, benzene, toluene, xylene, polycyclic aromatic hydrocarbons, glycol ethers, as well as hydrochloric acid or sodium hydroxide, may be used. Many of these chemicals are known to be toxic to humans and wildlife. The U.S. EPA recently forced mining companies to reveal the contents, but not the specific fractional composition, of their fracking fluids. Because hydraulic fracturing has a special exemption from the federal Safe Drinking Water Act (in an amendment known as the Halliburton clause, after the oil services company that is thought to have written it), it's up to states and local units of government to protect public health. Several states and many towns and cities have demanded details on the chemicals being pumped into the ground.

A study released in 2011 by the National Academy of Sciences reported that drinking water samples from shallow wells near methane drilling sites in Pennsylvania and New York had 17 times as much methane as those from sites far from drilling. And a study by researchers at Dartmouth concluded that 3 to 8 percent of the methane from shale gas wells escapes to the atmosphere in leaks and venting over the life of the well. These methane emissions are up to twice those from conventional gas wells.

Homeowners near fracking operations sometimes report that so much methane has seeped into their wells that they can ignite water from their faucets. Furthermore, a study published in 2013 reported that pumping fracking wastewater into a deep well near Youngstown, Ohio, triggered nearly 12 earthquakes per month. After water injection ended, the earthquakes stopped. Fracking has caused similar seismic activity in Oklahoma (see figure 14.21). And fracking in the Marcellus Shale often releases radioactivity from uranium in the deposits. A number of cities have passed ordinances prohibiting drilling within city limits or near schools or hospitals.

What do you think? Does having access to natural gas justify the social and environmental costs of its extraction? If you were voting on this issue, what restrictions would you impose on the companies drilling wells in your hometown?

There are huge deposits of natural gas in North America, but the costs of extracting this gas could be unacceptable in many instances.
Source: Courtesy Mike Williams, Ohio Department of Natural Resources, Division of Mineral Resources Management

---

contains about as much energy as a medium-size atomic bomb. Furthermore, huge amounts of seawater are used to warm and regasify the LNG. Water withdrawals and temperature changes can damage coastal ecosystems. In the United States, to override local objections, the federal government has assumed jurisdiction over LNG terminal siting.

Methane hydrates occur in deep ocean sediment

Among the many "unconventional" sources of gas that have been explored but not yet developed are methane hydrate deposits in arctic permafrost and beneath deep ocean sediments. **Methane hydrates** are composed of small bubbles or individual molecules of natural gas trapped in a crystalline matrix of frozen water. At least 50 oceanic deposits and hundreds of land deposits are known. Altogether, they are thought to hold some 10,000 gigatons (10^{13} tons) of carbon, or twice as much as the combined amount of all coal, oil, and conventional natural gas. There are already reports of methane bubbling up over wide areas of the melting Arctic Ocean and adjacent permafrost regions. This could trigger a catastrophic feedback spiral of global warming, because methane has much greater heat-absorbing capacity than CO_2. And as the climate warms, even more methane will be emitted.

In 2013, Japan reported successful gas production from methane hydrate deposits located about 300 m (nearly 1,000 ft) below the seafloor in the Eastern Nankai Trough off its Pacific Coast. This could represent a 100-year supply for the energy-poor island nation. There are serious concerns about whether this volatile gas can be safely extracted and shipped to markets. It is clear, though, that, mining the vast supplies of frozen methane worldwide would have disastrous consequences for efforts to control climate change.

Section Review

1. What is fracking, and why is it important?
2. What are "fugitive emissions" of methane?
3. What are methane hydrates, and why are they of concern?

19.5 NUCLEAR POWER

- *Nuclear power produces abundant, low-carbon, but expensive energy.*
- *Pressurized water reactors are by far the most common design.*
- *We haven't yet agreed on the best way to safely, and permanently, store nuclear waste.*

Nuclear power has been our most actively debated energy source. It is technologically complex and its waste products can be used for nuclear weapons. Specters of power plant meltdowns are frightening, but for decades, most plants have operated safely. Only a few catastrophic events have actually occurred, most notably at Chernobyl, in Ukraine, where a plant exploded and burned in 1986, and at Japan's Fukushima Daiichi plant, which melted down after a tsunami destroyed its cooling system in 2011. While these events were major disasters, which have made large regions uninhabitable for generations, they have been rare. On the other hand, managing spent fuel and other waste has been an unsolved problem for decades, and the costs of nuclear power far outweigh most other options.

In the past decade, a growing number of climate activists have argued that nuclear power is necessary as an alternative to coal and gas-powered energy. Nuclear power plants don't burn carbon-based fuel, so they produce no greenhouse gases at the point of power generation. Of course, the mining, processing, and transportation of uranium fuel does have severe environmental impacts, including carbon emissions. Nuclear opponents argue that energy conservation and ramping up renewables are cheaper and more sustainable than maintaining aging nuclear plants. Proponents argue that we don't know how to ramp up renewables and conservation as fast as we need to do. While nuclear accidents are a severe regional but low-probability risk, they point out, climate change is a certain and global risk, and all strategies are needed.

At its inception, nuclear power was intended to be cheap and abundant, as well as safe. In 1953, when President Dwight Eisenhower presented his "Atoms for Peace" speech to the United Nations, he promised that the United States would build nuclear-powered electrical generators to provide clean, abundant energy. Perceived shortages of oil and gas were invoked to promote abundant power "too cheap to meter" for continued industrial expansion of both the developed and developing world. It would be a supreme example of "beating swords into plowshares." Enthusiasm for nuclear power rose rapidly, and in the 1960s and 1970s, U.S. utilities ordered 140 new reactors for power plants.

Enthusiasm waned as costs rose and safety concerns increased. Many plants came in late and far over-budget. An antinuclear

FIGURE 19.20 New York's Indian Point nuclear power plant is ranked the riskiest in the country by the U.S. Nuclear Regulatory Commission due to its age and location on the Hudson River just 24 miles (38 km) north of Manhattan Island.
William P. Cunningham

movement formed over concerns about nuclear weapons proliferation, over corruption and lack of oversight in the nuclear industry, and over potential hazards of radioactive leaks. These fears were fed by irresponsible military use and disposal of radioactive materials. Many of our worst hazardous waste sites today involve radioactive materials improperly disposed of in the 1950s and 1960s. Waste management is a perpetual problem (discussed below). Of the 140 reactors on order in 1975, 100 were never built. Others are now being decommissioned because of safety and cost concerns at the aging facilities (fig. 19.20).

How do nuclear reactors work?

As with coal- and gas-powered thermal electric plants, nuclear plants produce electricity by heating water to produce steam, which turns a generator. The difference is that nuclear plants create heat by breaking up uranium atoms, rather than by burning carbon-based molecules. The radioactive subatomic particles released from uranium atoms are the reason nuclear energy is dangerous and needs to be closely contained.

The most commonly used fuel in nuclear power plants is uranium^{235}, a naturally occurring radioactive isotope of uranium. Ordinarily, U^{235} makes up only about 0.7 percent of uranium ore, too little to sustain a chain reaction in most reactors. It must be purified and concentrated to be useful. Mining and processing uranium to create nuclear fuel is even more dirty and dangerous than coal mining. This was especially true in past decades, when worker safety was often poor. In some uranium mines, 70 percent of the workers—most of whom were Native Americans—died from lung cancer caused by high radon and dust levels. In addition, mountains of radioactive tailings and debris have been left unprotected at fuel preparation plants.

When the U^{235} concentration reaches about 3 percent, the uranium is formed into cylindrical pellets slightly thicker than a pencil and about 1.5 cm long. These small pellets pack an amazing amount of energy. Each 8.5-gram pellet is equivalent to a ton of coal or four barrels of crude oil.

The pellets are stacked in hollow metal rods approximately 4 m long. About 100 of these rods are bundled together to make a **fuel assembly.** Thousands of fuel assemblies containing 100 tons of uranium are bundled together in a heavy steel vessel called the reactor core. Radioactive uranium atoms are unstable—that is, when struck by a high-energy subatomic particle called a neutron, they undergo **nuclear fission** (splitting). This splitting releases energy and more neutrons. When uranium is packed tightly in the reactor core, the neutrons released by one atom will trigger the fission of another uranium atom and the release of still more neutrons (fig. 19.21). Thus, a self-sustaining **chain reaction** is set in motion, and vast amounts of energy are released.

The chain reaction is moderated (slowed) in a power plant by a neutron-absorbing cooling solution that circulates between the fuel rods. In addition, operators can slow or stop the chain reaction by inserting **control rods** of neutron-absorbing material, such as cadmium or boron, into spaces between fuel assemblies. A cooling fluid, usually water, circulates around fuel rods, absorbing heat. In a pressurized water reactor (fig. 19.22), circulating water is kept under extremely high pressure and reaches a temperature of 317°C (600°F). This super-heated, high pressure water is then pumped to a steam generator, where it heats a secondary water-cooling loop. Steam from the secondary loop drives a turbine generator to produce electricity. Both the reactor vessel and the steam generator are contained in a thick-walled concrete and steel containment building that prevents radiation from escaping and is designed to withstand high pressures and temperatures in case of accidents.

Reactor designs vary in safety

The greatest danger in one of these complex machines is a cooling system failure. If the pumps fail or pipes break during operation, the nuclear fuel quickly overheats. If extreme heat breaches containment structures, deadly radioactive material can escape. A nuclear power plant cannot explode like a nuclear bomb, but the radioactive releases from a worst-case disaster like the 2011 meltdown at Japan's Fukushima Daiichi nuclear complex can be just as devastating. In that case, a failure of cooling pumps (after a tsunami wiped out the emergency generators) led to a meltdown in four of the plant's six reactors. An area tens of square kilometers around the site may not be safely inhabitable for centuries.

Overlapping layers of safety mechanisms are designed to prevent accidents, but these fail-safe controls make reactors very

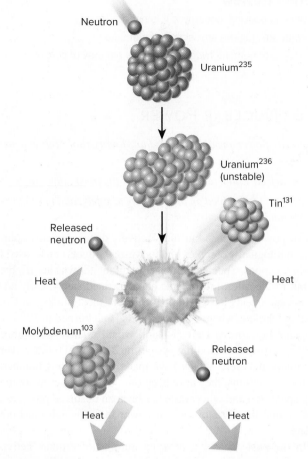

FIGURE 19.21 The process of nuclear fission is carried out in the core of a nuclear reactor. In the sequence shown here, the unstable isotope, uranium[235], absorbs a neutron and splits to form tin[131] and molybdenum[103]. Two or three neutrons are released per fission event and continue the chain reaction. The total mass of the reaction product is slightly less than that of the starting material. The residual mass is converted to energy (mostly heat).

expensive and very complex. A typical nuclear power plant has 40,000 valves, compared to one-tenth as many in a fossil-fuel-fired plant of similar size. In some cases, the controls are so complex that they confuse operators and cause accidents rather than prevent them. Under normal operating conditions, a PWR releases very little radioactivity and is probably less dangerous for nearby residents than a coal-fired power plant.

A simpler but dirtier and more dangerous reactor design is the boiling water reactor (BWR). In this model, water from the reactor core boils to make steam, which directly drives the turbine generators. This means that highly radioactive water and steam leave the containment structure. Controlling leaks is difficult, and the chances of releasing radiation in an accident are very high, as was the case in the BWR in Fukushima Daiichi, Japan.

In Britain, France, and Russia, a common reactor design uses graphite, both as a moderator and as the structural material

1 Fuel rods containing Uranium 235 are placed in water in reactor core, interspersed with neutron-absorbing control rods, containing boron or cadmium.

2 When control rods are raised, a chain reaction of nuclear fission occurs in fuel rods, creating heat.

3 High-pressure, superheated water flows to a heat exchanger, which generates steam in a separate, secondary loop.

4 Steam turns a turbine, which turns an electricity generator.

5 Cooling water condenses steam to water, which returns to the heat exchanger.

Reactor core
Primary loop
Turbine
Secondary loop
Generator
Fuel rods
Condenser cooling water
Cooling tower

FIGURE 19.22 Pressurized water nuclear reactor. Water is superheated and pressurized as it flows through the reactor core. Heat is transferred to nonpressurized water in the steam generator. The steam drives the turbogenerator to produce electricity.

for the reactor core. In the British MAGNOX design (named after the magnesium alloy used for its fuel rods), gaseous carbon dioxide is blown through the core to cool the fuel assemblies and carry heat to the steam generators. Russian designs use a graphite-moderated, water-cooled reactor), in which low-pressure cooling water circulates through the core in thousands of small metal tubes.

These designs were originally thought to be very safe, because graphite has high capacity for both capturing neutrons and dissipating heat. However, the small cooling tubes are quickly blocked by steam if the cooling system fails, and the graphite core burns when exposed to air. Two of the most disastrous reactor accidents in the world involved fires in graphite cores that allowed the nuclear fuel to melt and escape into the environment. A 1956 fire at the Windscale Plutonium Reactor in England contaminated hundreds of square kilometers of countryside. Similarly, burning graphite in the Chernobyl nuclear plant in Ukraine made the fire much more difficult to control than it might have been in another reactor design.

The most serious accident at a North American commercial reactor occurred in 1979 when the Three Mile Island nuclear plant near Harrisburg, Pennsylvania, suffered a partial meltdown of the reactor core. The containment vessel held in most of the radioactive material. No deaths or serious injuries were verified, but the accident was a serious blow to nuclear development.

Breeder reactors could extend the life of our nuclear fuel

For more than 30 years, nuclear engineers have been proposing high-density, high-pressure **breeder reactors** that produce fuel rather than consume it. These reactors create fissionable plutonium and thorium isotopes from the abundant, but stable, forms of uranium (fig. 19.23). The starting material for this reaction is plutonium reclaimed from spent fuel from conventional fission reactors. After about 10 years of operation, a breeder reactor would produce enough plutonium to start another reactor. Sufficient uranium currently is stockpiled in the United States to produce electricity for 100 years at present rates of consumption, if breeder reactors can be made to work safely and dependably.

Several problems have held back the breeder reactor program in the United States. Safety is the main concern. The reactor core of the breeder must be at a very high density for the breeding reaction to occur. Water cannot physically absorb the high amount of

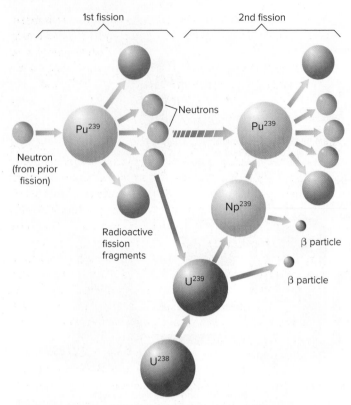

FIGURE 19.23 Reactions in a "breeder" fission process. Neutrons from a plutonium fission change U^{238} to U^{239} and then to Pu^{239} so that the reactor creates more fuel than it uses.

heat produced in the core, so liquid sodium generally is used as a coolant. Liquid sodium is very corrosive and difficult to handle. It burns with an intense flame if exposed to oxygen, and it explodes if it comes into contact with water. Because of its intense heat, a breeder reactor will melt down and self-destruct within a few seconds if the primary coolant is lost, as opposed to a few minutes for a normal fission reactor.

Another very serious concern about breeder reactors is that they produce excess plutonium that can be used for bombs. It is essential to have a spent-fuel reprocessing industry if breeders are used, but the existence of large amounts of weapons-grade plutonium in the world would surely be a dangerous and destabilizing development. The chances of some of that material falling into the hands of terrorists or other troublemakers are very high.

A proposed $1.7 billion breeder-demonstration project in Clinch River, Tennessee, was on and off for 15 years. At last estimate, it would cost up to five times the original price if it is ever completed. In 1986, France put into operation a full-sized commercial breeder reactor, the SuperPhénix, near Lyons. It cost three times the original estimate to build and produces electricity at twice the cost per kilowatt of conventional nuclear power. After only a year of operation, a large crack was discovered in the inner containment vessel of the SuperPhénix, and in 1997 it was shut down permanently.

We lack safe storage for radioactive wastes

The first disposal problem in the nuclear fuel cycle is the enormous piles of radioactive mine wastes and abandoned mill tailings at uranium mines. Producing 1 ton of uranium fuel typically generates 100 tons of tailings and 3.5 million liters of liquid waste. There now are approximately 200 million tons of radioactive waste in piles around mines and processing plants in the United States. This material is carried by the wind or washed into streams, contaminating areas far from its original source.

In addition to the leftovers from fuel production, the United States has about 100,000 tons of low-level waste (contaminated tools, clothing, building materials, and so on) and about 77,000 tons of high-level (very radioactive) wastes. The high-level wastes consist mainly of spent fuel rods from commercial nuclear power plants and assorted wastes from nuclear weapons production. While they're still intensely radioactive, spent fuel assemblies are stored in deep, water-filled pools at the power plants. These pools were originally intended only as temporary storage until the wastes were shipped to reprocessing centers or permanent disposal sites.

In 1987, after a years-long search the U.S. Department of Energy announced plans to build the first high-level waste repository under a barren desert ridge called Yucca Mountain in Nevada. Waste would be buried deep in the ground, where it was hoped it would remain unexposed to groundwater and earthquakes for the thousands of years required for the radioactive materials to decay to a safe level. But objections from Nevadans, who objected to receiving waste from far-away states, and continuing worries about the stability of the site led the Obama administration to cut off funding for the project in 2009 after 20 years of research and $100 billion in exploratory drilling and development. For the foreseeable future, the high-level wastes are being held in storage pools and dry casks located at 131 sites in 39 states (fig. 19.24), a long-term temporary, and not very secure, strategy.

FIGURE 19.24 Spent fuel is being stored temporarily in large, aboveground "dry casks" at many nuclear power plants.
Courtesy of Department of Energy/Office of Civilian Radioactive Waste Management

Russia has offered to store nuclear waste from other countries. Plans are to transport wastes to the Mayak in the Ural Mountains. The storage site is near Chelyabinsk, where an explosion at a waste facility in 1957 contaminated about 24,000 km² (9,200 mi²) of countryside. The region is now considered the most radioactive place on earth, so the Russians think it can't get much worse. They expect that storing 20,000 tons of nuclear waste should pay about US$20 billion.

Some nuclear experts believe that monitored, retrievable storage would be a much better way to handle wastes. This method involves holding wastes in underground mines or secure surface facilities where they can be watched. If canisters begin to leak, they could be removed for repacking. But safeguarding the wastes would be expensive, and the sites might be susceptible to wars or terrorist attacks. If the owners of nuclear facilities had to pay the full cost for fuel, waste storage, and insurance against catastrophic accidents, no one would be interested in this energy source. Rather than being too cheap to meter, nuclear power would be too expensive to matter.

Decommissioning nuclear plants is costly

Old power plants themselves eventually become waste when they have outlived their useful lives. Most plants are designed for a life of only 30 years. After that, pipes become brittle and untrustworthy because of the corrosive materials and high radioactivity to which they are subjected. Plants built in the 1950s and early 1960s already are reaching the ends of their lives. You don't just lock the door and walk away from a nuclear power plant; it is much too dangerous. It must be taken apart, and the most radioactive pieces have to be stored just like other wastes. This includes not only the reactor and pipes but also the meter-thick, steel-reinforced concrete containment building. The pieces are cut apart by remote-control robots because it's too dangerous to work on them directly.

Altogether, the U.S. reactors now in operation might cost somewhere between $200 billion and $1 trillion to decommission. No one knows how much it will cost to store the debris for thousands of years or how it will be done. These aren't the only reactors needing decommissioning: Plutonium weapons production plants and nuclear submarines also have to be decommissioned. Originally, the Navy proposed to just tow old submarines out to sea and sink them. The risk that the nuclear reactors would corrode away, however, and leak radioactive isotopes into the ocean makes this method of disposal unacceptable.

Decommissioning nuclear plants is expensive, but in some cases it has become cheaper than keeping them running. In 2016 Pacific Gas and Electric announced that it would close California's last nuclear power plant, Diablo Canyon, when its license expires in 2025. The plant has been controversial because of its vulnerability to earthquakes and tsunamis, but the reason for decommissioning it is economic. The cost of decommissioning the plant is expected to be at least $3.8 billion, but PG&E concludes that it is cheaper to build and operate renewables than it is to upgrade and maintain Diablo Canyon.

The cost of producing one MWh of electricity at the Diablo Canyon plant is $70, far more than the $50/MWh now needed to build and operate commercial-scale solar in California. To make a profit, PG&E needs to keep its production costs low, so it makes sense to close the nuclear plant. By planning ahead, the company can work with unions and employees to build new solar or wind capacity and efficiency measures while it prepares to shut down and deconstruct the nuclear plant.

Opinions about nuclear futures vary

High costs and management challenges have plagued U.S. nuclear power generation, but in Europe, commitments to nuclear power remain strong. France, in particular, gets about 70 percent of its energy from nuclear power. Support for nuclear power partly involves lower investments in oil, gas, and coal in France and other countries with strong nuclear programs. Support has also grown recently with the need to meet climate commitments.

China is also continuing to invest in nuclear power. Just as with coal, China is building nuclear plants faster than any other country, in order to power growing industries and cities. Fifteen new plant are under construction, on top of the 45 already in operation. However, China has also seen project delays and cost overruns, and no new projects have been planned since 2016.

Because China's energy policy largely determines our trajectory in climate change, many analysts worry about waning support for nuclear power. If China invests in coal or gas instead, that could spell disaster. Alternatively, China has already overbuilt its coal plants, and the growth in energy demand in coming years seems likely to slow, compared to the rapid industrial expansion of the past 20 years. If the country focuses on conservation and increasingly cheap renewables instead of nuclear or coal, climate impacts could still be moderated.

Another tantalizing proposal is nuclear fusion, as an alternative to nuclear fission. **Nuclear fusion** energy is released when two smaller atomic nuclei fuse into one larger nucleus (in contrast, nuclear *fission* involves breaking the nucleus apart). Nuclear fusion reactions are the energy source for the sun and for hydrogen bombs. The fuels for these reactions are deuterium and tritium, two heavy isotopes of hydrogen. At temperatures of 100 million °C and pressures of several billion atmospheres, fusion of deuterium and tritium will occur. Under these conditions, the electrons are stripped away from atoms and the forces that normally keep nuclei apart are overcome. As nuclei fuse, some of their mass transforms into energy, some of it in the form of heat. In theory, the advantages of fusion reactions would include production of fewer radioactive wastes, the elimination of products that could be made into bombs, and a fuel supply (deuterium and tritium) that is much more abundant and safer than uranium. Because this energy source would be so promising, more than 50 years of research and $25 billion have been invested in it. Thus far, however, fusion reactors have not yet been able to produce more energy than they consume.

Section Review

1. How does nuclear energy produce heat, steam, and electricity?
2. Why is cooling a nuclear reactor important? What are some mechanisms for cooling or slowing reactions?
3. What are some major obstacles to expansion of nuclear power?

Connecting the Dots

Energy has driven the prosperity we know today. Our use of it is also transforming our world, in terms of water, biodiversity, and pollution, and especially climate change. How we use energy, and which types we subsidize and invest in, will determine whether we prosper in the future. We are now making critical decisions about energy policies, both in wealthy countries and in developing regions, where energy demand is growing rapidly. Despite growing demand, our main energy challenge is no longer whether we have enough resources, but whether we will destroy our prosperity by using too much of them.

The uncertain prospects of coal power point to important changes in energy systems. The case of Arch Coal and its battle to export coal to Asia show that coal is no longer the obvious answer.

Communities want other power sources with smaller impacts. Yet we remain largely invested in oil, gas, and coal, both at home and abroad. What will be the costs to society as coal companies, as well as oil and gas companies, fight to survive?

Extraction and use of conventional energy has important environmental effects beyond climate impacts. Mining and drilling consume and contaminate water, land, and air. Shipping fuel to markets also puts water and land at risk, an issue many communities have resisted. It's also important to remember that thermal (steam-based) energy sources require abundant water for fuel extraction and electricity production. This is one more reason for seeking alternatives to current energy consumption patterns. In the next chapter, we discuss some of these options.

Reviewing Key Terms

Can you define the following terms in environmental science?

breeder reactor 19.5	fossil fuels 19.1	liquefied natural gas	peak oil 19.3
carbon capture and storage	fuel assembly 19.5	(LNG) 19.4	power 19.1
(CCS) 19.2	hydraulic fracturing	methane hydrates 19.4	tar sands 19.3
chain reaction 19.5	("fracking") 19.3	nuclear fission 19.5	watt 19.1
control rods 19.5	joule 19.1	nuclear fusion 19.5	work 19.1
energy 19.1	kerogen 19.3	oil shale 19.3	

Critical Thinking and Discussion Questions

1. We have discussed a number of different energy sources and energy technologies in this chapter. Each has advantages and disadvantages. If you were an energy policy analyst, how would you compare such different problems as the risk of a nuclear accident versus air pollution effects from burning coal?

2. If your local utility company were going to build a new power plant in your community, what kind would you prefer? Why?

3. The nuclear industry is placing ads in popular magazines and newspapers claiming that nuclear power is environmentally friendly because it doesn't contribute to the greenhouse effect. How do you respond to that claim?

4. Our energy policy effectively treats some strip-mine and well-drilling areas as national sacrifice areas, knowing they will never be restored to their original state when extraction is finished. How do we decide who wins and who loses in this transaction?

5. Storing nuclear wastes in dry casks outside nuclear power plants is highly controversial. Opponents claim the casks will inevitably leak. Proponents claim they can be designed to be safe. What evidence would you consider adequate or necessary to choose between these two positions?

6. The policy of the United States has always been to make energy as cheap and freely available as possible. Most European countries charge three to four times as much for gasoline as we do. Who benefits and who or what loses in these different approaches? How have our policies shaped our lives? What does existing policy tell you about how governments work?

Data Analysis

Comparing Energy Use and Standards of Living

In general, income and standard of living increase with energy availability. This makes sense because cheap energy makes it possible to obtain goods and services that enrich our lives, and to use machines to extend our productivity. However, energy use per capita isn't strictly tied to quality of life. Some countries use energy extravagantly without corresponding increases in income or standard of living. Go to Connect to examine a graph comparing per capita income and energy consumption and to answer questions about the relationship between these factors.

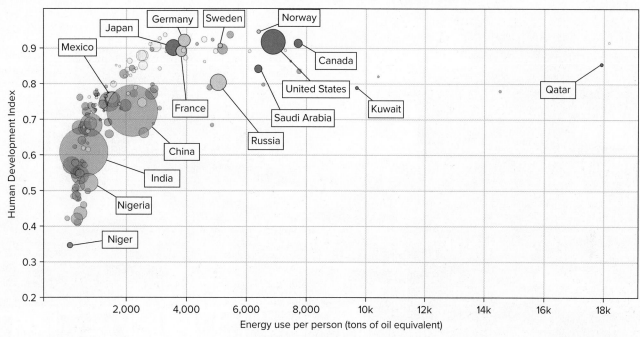

Source: gapminder.org

20

Sustainable Energy

▲Germany has the most solar power per person in the world. Renewable energy provides nearly 60 percent of peak energy demand for the country.
peart/E+/Getty Images

Learning Outcomes

After studying this chapter, you should be able to:

20.1 Describe strategies for energy efficiency.

20.2 Explain different ways we can use solar energy.

20.3 Describe how wind power production has changed.

20.4 Discuss the potential of biomass and fuel cells.

20.5 Summarize the costs and benefits of hydropower.

"We know the country that harnesses the power of clean, renewable energy will lead the twenty-first century."

– Barack Obama

A Renewable Energy Transition

In an effort to combat climate change, Germany has made ambitious commitments to a renewable energy transition, or *Energiewende*. The country aims to cut both fossil fuel use and carbon emissions 80 percent by 2050, and to replace most current energy use with conservation and sustainable energy. How will they accomplish this formidable task? And can others follow their lead?

Despite its cloudy, cool climate, Germany has enough solar and wind potential to fuel its economy. Germany also has an abundant supply of coal, which long dominated electricity generation, but most of that coal has to stay in the ground if climate goals are to be met. Germany is also eager to minimize reliance on natural gas, which it imports mainly from Russia. Gas costs the country tens of billions of euros every year, a cost they would be glad to shed.

Germany has been an early adopter and leader in both solar and wind. In 2014, Germany had nearly as much solar power as all the rest of Europe, and it had more solar than all the rest of the world outside of Europe. Along with Denmark and the United Kingdom, Germany has led a revolution in offshore wind power, a key part of global energy strategies. The transition to renewables has become cheaper with increasing production, and the price per watt for new renewables has declined well below that of conventional power. Critics predicted that big spending on renewables would bankrupt the country, but this hasn't happened. The economy has doubled since 1990, while renewables have gone from below 4 percent to over 40 percent of electricity (fig. 20.1).

By 2018, renewables produced more electric power than coal. Wind produced 20 percent of all German electricity in 2018, followed by solar, hydropower, and biomass (20 percent). By 2050, Germany expects to get 100 percent of its electricity from renewable sources, and coal will fall to 8 percent of the total energy supply.

Renewables also provide jobs. Where coal mining jobs have declined steadily with mechanization, jobs in solar, wind, and biomass are growing. Over 330,000 Germans work in renewable energy, twice as many as in the fossil fuel industry.

Germany started its *Energiewende* transition by getting local communities and voters on board. As an incentive, they instituted feed-in tariffs, a premium price for energy produced and fed into the grid from community producers. These tariffs were a subsidy that helped farmers build wind turbines and helped initiate community solar projects in towns and villages across Germany. Tariffs have declined steadily as renewables have become more abundant and more cost-effective. A focus on community participation also sidestepped large utilities, which, committed to coal and gas, initially resisted the shift to renewables.

What obstacles has Germany seen? The biggest is energy distribution. In summer, excess solar power must travel north from the sunny southern states; in winter and at night, wind power is most abundant in the North. Plans to expanding the electricity grid by just 5 percent (in kilometers of power lines) cause considerable controversy. On the other hand, the grid already is extensive. European states trade power constantly as they balance surpluses of Norwegian hydropower, Danish wind, French nuclear, and German or Spanish solar power during a week, a day, or even an hour.

U.S. power companies, like those in Germany, are invested in conventional energy and have resisted both community participation and renewables. A difference in Germany has been that the parliamentary government allows a Green Party to thrive, and voters have been able to band together to support its climate-protection policies. But U.S. power companies are watching the profits of their European counterparts, and their interest is growing.

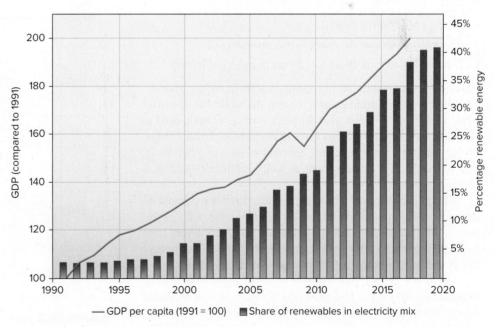

FIGURE 20.1 Germany has continued its economic growth while increasing renewable electricity sources from 5 to 40 percent in 20 years.
Source: Fraunhofer Institute and Heinrich Boll Foundation, 2019

CASE STUDY *continued*

Perhaps the greatest impact of Germany's *Energiewende* is that shows the world that renewables are realistic. They provide energy, and they support growth. Following Germany's example, China has charged ahead as the world's leader in renewables. As China has scaled up production, costs have fallen in the United States and worldwide.

Germany has also shown that people are eager to adopt sustainable energy, especially if policies lower obstacles. Communities like clean power and the local, skilled jobs it brings. Learning from early adopters might help us all find more sustainable energy paths.

Benchmark Data	
Among the ideas and values in this chapter, the following are a few worth remembering:	
>40%	Portion of Germany's electricity that was renewable in 2018
80%	Targeted reduction in greenhouse gases in New York by 2050
2,500	Solar potential in Arizona, kilowatt-hours/m^2/year
1,500	Solar potential in Germany, kilowatt-hours/m^2/year
100%	Percentage of global energy that could be provided by renewables

20.1 ENERGY EFFICIENCY

- *Saving energy saves money.*
- *Green buildings can cut energy costs dramatically.*
- *Transportation is an important factor in energy efficiency.*

Ensuring a sustainable future requires a shift in energy systems. Although we know from chapter 19 that the world has abundant fossil fuels, we also know that 80 percent of those fuels need to remain in the ground if we are to avoid catastrophic climate change. Fortunately, we have abundant alternative sources (fig. 20.2), and the technology to use them is well established—and it is improving steadily.

Energy systems are changing rapidly. Efficiency improvements, electrifying transportation and heating, wind, solar, water power, and biomass are leading strategies for transitioning from fossil fuels to renewables. We examine each of these in this chapter, as well as exploring likely aspects of our energy future.

Often the easiest and cheapest ways to reduce our energy impacts is simply to use less. Germany, for example, has increased energy efficiency by more than 50 percent over the past 25 years, while decreasing its primary energy consumption (fig. 20.3). These improvements reduce Germany's reliance on imported energy from distant and politically volatile places.

Energy conservation is the first step

Conservation is partly about behavior—turning off lights and adjusting thermostats—but it's also a matter of what technology is available. Sometimes inefficiency is a matter of convention: Nearly all U.S. households, for example, use air conditioning, but many could cut energy use over 99 percent and stay cool and save money by using electric fans. Sometimes efficiency comes with new technology. **LEDs (light-emitting diodes),** for example, have transformed lighting systems. Compared to compact fluorescent bulbs, LEDs use about half as much electricity, and last about five times as long. Compared to a standard incandescent bulb, LEDs use one-tenth as much energy and last up to 50 times longer. Many cities have replaced traffic lights and street lights with

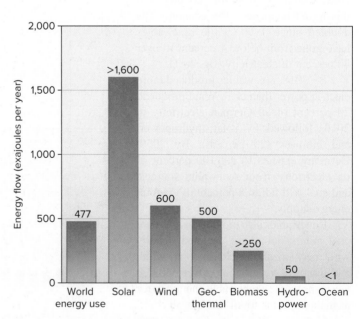

FIGURE 20.2 Potential energy available from renewable resources using currently available technology in presently accessible sites. Together, these sources could supply more than six times the current world energy use.
Source: Adapted from UNDP and International Energy Agency

454   Environmental Science

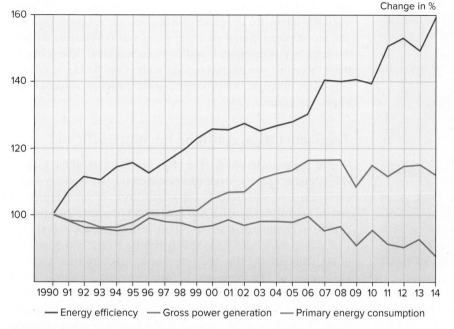

Change in %

— Energy efficiency   — Gross power generation   — Primary energy consumption

FIGURE 20.3 Changes in Germany's efficiency, power generation, and consumption, relative to 1990. Efficiency improvements are giving more value from less energy.
Source: Heinrich Böll Foundation 2015

LEDs. These retrofits typically pay for themselves (in dramatically lowered energy costs) in about two years.

We can also reduce energy use by paying attention to hidden energy uses such as appliances in standby mode. You may think you've turned off your TV, DVD player, cable box, or printer, but they're really continuing to draw power in an "instant-on" mode. By some estimates, these "vampire currents" can amount to a quarter of the monthly electric bill. Putting your computer to sleep saves about 90 percent of the energy it uses when fully on, but turning it completely off is even better. Plugging appliances into an inexpensive power strip allows you to switch them off when not in use.

Green buildings cut energy costs

Innovations in green building have transformed both commercial and household construction. Much of the innovation has occurred in large commercial structures, which have bigger budgets—and more to save through efficiency—than most homeowners have. Insulation protects a building from heat gain in summer as well as heat loss in winter (fig. 20.4). Insulating attics and sealing foundations usually pays for itself in a few years. Inexpensive measures such as sealing leaks with caulk or weather stripping is one of the quickest and most cost-effective things you can do to save energy, and to make living spaces more comfortable.

High-efficiency buildings are becoming the norm in Europe. Many regions in Germany require that new buildings conform to **passive house standards.** These standards are targets for extremely low energy use, achieved mainly

through thoughtful design and construction (fig. 20.5). Passive buildings use just 10 percent of the energy normal buildings use. Many cities and states across Europe have begun to follow Germany in either requiring or encouraging passive standards. In the United States, incentives, such as faster permitting or tax breaks to building projects that meet passive standards, are increasingly common.

Relatively rare a few years ago, passive standard houses and office buildings are becoming more common in Europe, North America, and Asia. Because they require less heating and cooling infrastructure—fewer copper gas lines, for example—the cost to build a passive house can be just 2-3 percent more than a standard house, and that extra cost pays for itself in a few years of reduced energy bills. Renovation to high energy standards is also increasingly common, with insulation, sealing, and better windows. Residents say passive houses are more comfortable, with even heating, better natural light, and less noise than conventional housing (fig. 20.6).

What does passive housing involve? These are the main principles:

1. No thermal bridges (wood or metal components conduct heat through walls, roof, or foundation)
2. Roof and wall insulation 24–32 cm (10–14 in) thick
3. Well-sealed windows and doors to prevent heat loss or gain
4. Heat exchangers, to warm or cool fresh air as it is exchanged with stale air leaving the building
5. Triple-pane windows that reduce radiative heat loss or gain
6. Windows well positioned to let in daylight

FIGURE 20.4 Energy loss from buildings is visble in an infrared image. Improving building efficiency is one of the most important ways to reduce energy consumption.
Mary Ann Cunningham

FIGURE 20.5 The Cornell Tech campus in New York City has some of the world's largest and tallest structures built to passive standards. These buildings use less than 10 percent as much energy as comparable buildings.
Mary Ann Cunningham

When a passive building produces electricity—for example, with solar panels on the roof—it can become a "net-zero" building that uses no more energy than it produces. Some of these are even "energy-plus" buildings, producing more energy than they consume on an annual basis. Additional measures are becoming common in many areas: Appliances often have timers so they can run overnight, when energy use is low. Thermostats can automatically adjust the temperature when you are not home or when you are sleeping. Smart metering gives you information on how much energy is being used and how much it costs. Using one of these systems, you might program your water heater to operate only after midnight, when surplus wind power is available (see Exploring Science, p. 457).

Transportation could be far more efficient

One of the areas in which most of us can accomplish the greatest energy conservation is in our transportation choices. You can decide every day how you travel to school, to work, or for shopping or entertainment. Automobiles and light trucks account for 40 percent of U.S. oil consumption, so our collective decisions are important.

Many outings don't necessarily require a car: Twenty percent of our daily driving trips are a mile or less, according to the U.S. Department of Transportation. Yet every year we drive more, often for shorter trips. Some of that is due to the design of our cities (chapter 22). Suburban subdivisions make travel without a car difficult. Shopping areas are surrounded by busy streets and vast parking lots that are unfriendly to pedestrians. But sometimes we use fuel inefficiently simply out of habit. The Census Bureau reports that three-quarters of all workers commute alone in private vehicles. About 20 percent carpool, while less than 5 percent use public transportation, and a mere 0.38 percent walk or travel by bicycle.

What can you do if you want to reduce energy use? The cheapest and healthiest alternative for short trips is walking or bicycling. You need to get some exercise every day, so can you make walking part of it? For trips less than 2 km, it's sometimes quicker to go by bicycle than to find a parking space for your car.

Public transportation saves energy, both because it's more efficient than personal vehicles and because it can run on renewable energy. City bus systems, for example, are increasingly electric, with renewable power sources, or at least hybrid-electric. Some even run on biogas (fig. 20.7). Public transit also reduces a city's need for parking and makes space for non-vehicle land uses.

Electric bikes are increasingly popular in some places (fig. 20.8). With a small, easily rechargeble battery, these bikes give an assist to pedaling, make it easy for more riders to stay out of cars and to get around cheaply. Especially in hilly terrain or for long distances, this is an efficient, low-carbon option.

Vehicle efficiency dramatically influences how much fuel you buy and

FIGURE 20.6 Energy-efficient buildings can lower energy costs dramatically. Many features can be added to older structures. New buildings that start with energy-saving features (such as SIPs or advanced framing) can save even more money.

*Labels in figure:*

Shade trees
Cut cooling cost

High-efficiency furnace and air conditioner

Photovoltaic roofing
Solar water heating

Roof
12 in. (30 cm) insulation = R36-48

Overhang blocks summer sun

Winter sun warms house

Weather stripping
Doors, windows, vents, fixtures
Prevents air leaks

Windows
Double or triple panes;
Well-placed for light

SIPs
Structural insulated panels for extra insulation

Heat exchanger
Saves on heating/cooling;
prevents moisture buildup

Advanced framing
Prevents conductive heat loss

Geothermal
Heating and cooling

Greening Gotham: Can New York Reach Its 80 by 50 Goal?

New York City has a gritty reputation, and it's not known as a world leader in energy conservation. As the biggest city in the United States, New York uses more energy than almost any other, almost all of it derived from oil, natural gas, and nuclear power. As a northern city, New York needs about six months of heating and four months of air conditioning each year for its aging, often leaky building stock, and it has limited solar potential.

At the same time, New York City planners know that climate change threatens the city with rising sea levels, increased hurricane frequency, intense heat waves, and economic instability as a warming climate undermines agriculture, transportation, health, and other underpinnings of the city's economic foundations. A serious heat wave can be especially deadly for low-income or elderly urban residents.

If these threats are to be reduced or averted, New York and other cities have to do their share to cut greenhouse gas emissions, and they need to do it soon. The Intergovernmental Panel on Climate Change (IPCC) reports that to avoid the worst impacts of climate change, the world needs to reduce its greenhouse gas emissions 80 percent by the year 2050. This is a daunting goal for any city. New York's energy infrastructure, like most of its building stock, was established half a century ago, or more. Can the city replace or upgrade its aging infrastructure on a tight budget and a tight timeline?

In 2014, the city's Office of Sustainability published a new policy stating that New York would aim for the IPCC's goal of 80 percent reductions by 2050, often called an "80 by 50" goal. The goal is ambitious, and the path to get there is not entirely clear, but Mayor Bill de Blasio and his administration agreed that it was necessary to protect the city's long-term interests. They also knew that if New York could lead the way, other cities would take the idea seriously and follow.

(header photo): David L. Hansen

Knickerbocker Commons, in Brooklyn, is the first midsized apartment building designed to passive house standards in the United States.

© Architect Chris Benedict, R.A. 803 Knickerboker Avenue

New York already uses less energy per person than any other major American city. Densely built apartments and offices share walls and conserve heat. Public transportation uses a minute fraction of the energy private cars do, per rider. Bicycle infrastructure has transformed the city in recent years, providing cheap and efficient transportation options. Many people walk to work, school, and stores because distances are short. People with small kitchens also tend to buy and waste less food. In New York, the biggest challenges are leaky, thin-walled buildings and a lack of easy places to plan wind, solar, or other alternative energy sources.

New York is starting with its building stock. Financial incentives are helping landlords upgrade heating and cooling systems. The public housing authority has begun installing geothermal systems, better windows and building sealing, and more efficient heating systems. Upgrades save New Yorkers money as well as energy, and they make buildings more comfortable. The city has even begun building apartment buildings that meet passive house standards, which prescribe buildings so tightly made that they use almost no energy for heating, with more natural lighting and more comfortable design than the standard housing of a generation ago.

The city also works closely with large institutions in finding ways to reduce electricity consumption at peak hours; help them finance improved lighting, windows, and thermostats; and employ measures to systematically turn off lights and turn down thermostats. Landlords are also investing in cogeneration—generating electricity with waste heat from heating and cooling systems—which reduces energy demand on public utilities.

New policies are as important as new technology and tighter buildings. Reducing paperwork and bureaucratic obstacles to upgrades saves time and money. New rules at the state level can give utilities new incentives to integrate solar, wind, geothermal, and other renewable energy sources into their energy production system. The city also collaborates with public advocacy organizations to help define new incentives for the public to invest in solar and wind energy, as well as energy conservation. The city is also negotiating with the state of New York to develop new rules that promote shared solar energy investments, and the easier permitting of offshore wind. Because, without some investment in alternatives, all the insulation in the world won't do the job that needs to be done.

In many ways, the story of energy in the future is going to be different from the story in the past. Just as oil and automobiles transformed life a century ago, alternative energy sources, consumption practices, and ownership policies are likely to transform how we live and work. Mayor de Blasio and his team are betting that the future scenario will have exciting co-benefits, including safer, more reliable, and more affordable energy, if we can just commit to doing it right.

FIGURE 20.7 Public transportation is one of the best ways to improve efficiency and reduce pollution, especially with electric or hybrid buses. This Swedish bus runs on biogas (methane) from agricultural waste and landfills.
Mary Ann Cunningham

use—which is why energy companies vigorously fight efficiency rules. In response to the 1970s oil price shocks, automobile gas-mileage averages in the United States more than doubled from 13.3 mpg in 1973 to 25.9 mpg in 1988. Then falling fuel prices in the 1990s discouraged further conservation, and mileage hardly improved for nearly 20 years. In 2012, however, the Department of Transportation and the Environmental Protection Agency issued regulations requiring cars and light trucks to achieve an average of 54.5 mpg (23 km/liter) by 2025. The agencies estimate

FIGURE 20.8 Many cities treat bicycles as a standard form of trans-portation that saves energy, pollution, and space--parking this many cars would use a lot of land. Electric-assist bikes, like these shared bikes in Copenhagen, make it even easier to leave the car behind.
Mary Ann Cunningham

this efficiency standard should save consumers $1.7 trillion in fuel costs, reduce oil consumption by 2 million barrels per day, and eliminate 6 billion metric tons of CO_2 emissions by 2025, if we follow through on these rules.

Transportation is electrifying

For vehicle efficiency, the leading choice has long been diesel-fueled engines. Diesel fuel has longer hydrocarbon chains, whose bonds contain more energy than gasoline. However, diesel also contains sulfur and emits more particulates than gasoline when burned. European carmakers are moving away from diesel and toward electric vehicles (EVs) as air quality rules have become more rigorous.

Electrifying transportation is central to most greenhouse gas reduction plans. An internal combustion engine converts about 83 percent of energy in fuel to heat of combustion. Only about 17 percent is converted to forward motion of your car. A battery-powered vehicle, in contrast, converts about 90 percent of power to motion. This makes a battery vehicle a far more efficient system. Electric motors also have more torque, for a faster pickup, which makes them sporty, as well as clean and quiet.

Buying an EV is, thus far, more expensive than buying a conventional vehicle (unless you're buying a luxury model). But operating a battery car is far cheaper. The cost per mile is about one quarter the cost of a standard gas vehicle. A Chevy Bolt, for example, runs at about 4 miles/kWh per mile (or equivalent to 100–120 miles per gallon, according to the EPA). They are also easier to maintain. A Bolt EV has just 24 moving parts, whereas a Volkwagen Golf has 129, according to an MIT analysis, and the Bolt never needs an oil change.

EVs take longer to charge than a gas car takes to refuel, but for nearly all daily uses, this is unlikely to matter. Most Americans drive less than 30 miles per day, and less than one percent of household trips are more than 100 miles; both are far less than the 200-mile range increasingly common in EVs.

Hybrid gas-electric vehicles also can have good efficiency. A Toyota Prius hybrid is rated at around 50 miles per gallon (over 10 km/liter). Plug-in hybrids give still more flexibility, freeing most people from buying gas at all but still having the gas engine when needed. If you source your electricity from wind or solar, you could drive almost entirely pollution-free.

An increasing number of countries, as well as U.S. states, are providing networks of charging stations and free parking areas for EVs, in order to encourage their adoption. China has particularly strong incentives. The government there is motivated to reduce air pollution, a major benefit of EVs, as well as to reduce energy

---

Quantitative Reasoning

In 2016, the combined corporate average fuel economy for cars and light trucks in the United States was 34.1 mpg. If the 190 million drivers traveled an average of 10,000 miles that year, how much gasoline did they use? If burning a gallon of gasoline creates 14 pounds of CO_2, how much CO_2 did they produce in their automobile travel?

What Can You Do?

Steps You Can Take to Save Energy

1. Drive less: make fewer trips; use telecommunications and mail instead of going places in person.

2. Walk, ride a bicycle, or use public transportation.

3. Use stairs instead of elevators.

4. Join a car pool or drive a smaller, more efficient car; reduce speeds.

5. Insulate your house or add more insulation to the existing amount.

6. Turn thermostats down in the winter and up in the summer, and dress appropriately for the season.

7. Weather-strip and caulk around windows and doors; add storm windows or plastic sheets over windows.

8. During the winter, close windows and drapes at night; during summer days, close windows and drapes if using air conditioning.

9. Turn off lights, television sets, and computers when not in use.

10. Take shorter, cooler showers; install water-saving faucets and showerheads.

11. Recycle glass, metals, and paper; compost organic wastes.

12. Eat locally grown food in season.

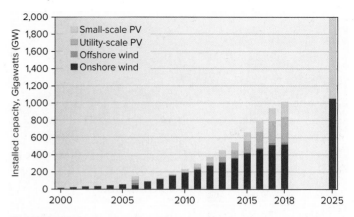

FIGURE 20.9 Cumulative installed capacity for wind and solar power, with estimates for 2025.

Source: Bloomberg New Energy Finance, 2018; Power Technology, 2018

consumption. In Shanghai and Beijing, getting a license for a standard vehicle is either expensive or nearly impossible; licenses for EVs, though, are free and easy to get. Nearly half of all EVs sold in 2018 were in China.

Transportation choices are ways many of us can adjust energy use (What Can You Do?, above). Which approaches are easiest for you to consider doing?

Section Review

1. List five ways we could conserve energy individually or collectively.

2. What are some ways building design can reduce energy loss?

3. Why is transportation efficiency important in energy conservation?

20.2 SOLAR ENERGY

- *Solar energy can be collected passively or actively.*
- *Solar thermal systems create steam to generate electricity.*
- *Photovoltaics generate electricity directly.*

Solar power was a cornerstone of Germany's energy transition, which helped kick-start solar globally (fig. 20.9). Germany's installed solar capacity grew from 6 GW in 2008 to 46 GW in 2018. U.S. capacity also grew from less than 1 GW to nearly 70 GW in that same 20 years. China

grew at a phenomenal rate over the same years, from 0.1 GW to 170 GW. Analysts project that global capacity will approach 970 GW by 2025.

Solar photovoltaic (PV) electricity is popular in part because it can be installed almost anywhere—on homes, warehouses, roadsides, field margins, and parking lots. The cost effectiveness—the value of electricity produced per panel—depends on available sunshine. Sunny Arizona, for example, receives around 2,500 kilowatt-hours per square meter per year (about half of what a German household uses in a year). Most of Germany, in contrast, receives around 1,500 kwh/m²/year (fig. 20.10). PV cells cannot capture all of this energy, however. Sunlight is a diffuse, low-intensity energy source (compared to gasoline, for example), and solar cells can only convert about 15–20 percent of received sunlight to usable electricity. Still, this small percentage can produce a great deal of power.

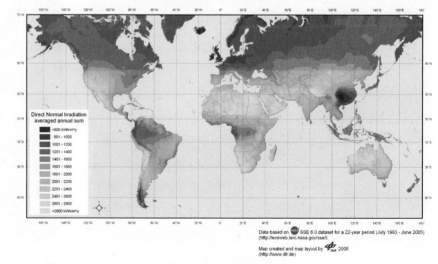

Direct Normal Irradiation (DNI)

FIGURE 20.10 Cumulative average annual solar radiation. Within six hours, deserts receive more energy from the sun than humans consume in a year.

Source: German Aerospace Center, 2012

FIGURE 20.11 Taos Pueblo in northern New Mexico uses adobe construction to keep warm at night and cool during the day.
William P. Cunningham

Solar heat collectors can be passive or active

Most discussions of solar energy today involve photovoltaic, or solar electric, energy. This is a flexible, easily distributed energy source when we have a grid and electrical appliances. But we use solar energy in many other forms as well. Our simplest and oldest use of solar energy is **passive heat absorption,** using natural materials or absorptive structures with no moving parts to simply gather and hold heat. For thousands of years people have built thick-walled stone and adobe dwellings that slowly collect heat during the day and gradually release that heat at night (fig. 20.11). After cooling at night, these massive building materials maintain a comfortable daytime temperature inside, even as they absorb external warmth.

A modern adaptation of this principle is a glass-walled "sun-space" or greenhouse on the south side of a building. Massive energy-storing materials, such as brick walls, stone floors, or barrels of heat-absorbing water, collect heat to be released slowly at night.

Active solar systems pump a heat-absorbing fluid (air, water, or an antifreeze solution) through a collector, such as a flat, glass-covered black surface, or glass vacuum tubes (fig. 20.12). Water (or another fluid) pumped through the collector picks up heat for space heating or to provide hot water. Also called solar thermal systems, they can be placed adjacent to or on top of buildings, and they can be readily retrofitted onto existing buildings. A collector with about 5 m² of surface can reach 95°C (200°F) and can provide enough hot water for an average family. China, which has historically had limited energy resources, produces about 80 percent of the world's solar water heaters, which cost about $160 each and pay for themselves in only two or three years. At least 30 million Chinese homes get hot water and/or space heat from solar energy.

These systems are relatively uncommon in the United States both because plumbing for fluids costs labor and money, and because other sources of heating are cheap and conventional.

High-temperature solar thermal plants can produce enough heat to spin a turbine and produce electricity. One of the most common designs for **concentrating solar power (CSP) systems** uses long trough-shaped parabolic mirrors to reflect and concentrate sunlight on a central tube containing a heat-absorbing fluid (fig. 20.12*b*). The fluid reaches extremely high temperatures, then passes through a heat exchanger, where it heats water and generates steam to turn a turbine to produce electricity. Because these systems are large and expensive to build, they are sited in desert areas with abundant sunshine.

CSP plants have important advantages. Heat from the transfer fluid can be stored in a medium, such as molten salt, for later use.

(a)

(b)

FIGURE 20.12 Active solar thermal systems heat a fluid as it circulates through heat collectors. These systems can be used for community heating, as in Marstal, Denmark (a) or for producing steam to run a generator (b).
(a) Tom Finkle (b) Source: Warren Gretz/NREL/U.S. Dept. of Energy

This allows the system to continue to generate electricity on cloudy days or at night. Some CSP plants can produce power nearly around the clock.

Some of the oil-rich Persian Gulf states are building CSP plants as they prepare for a fossil fuel–free future. The United Arab Emirates, for example, built a 100 MW CSP plant in Abu Dhabi in 2014. The 258,000 mirrors in the Shams (Arabic for Sun) plant can provide enough electricity for about 20,000 homes. Shams I is expected to be followed by units II and III for an eventual total of 300 MW. And Saudi Arabia has pledged US$109 billion for solar power. If carried through, this investment could produce 41 GW of solar energy and an additional 21 GW of geothermal and wind power. The Saudi decision was triggered by a 2012 estimate that the kingdom could become a net oil importer by 2038 if it didn't diversify its energy portfolio.

A number of U.S. CSP facilities have been planned in desert sites that are biologically rich and vulnerable, and that recover very slowly from disturbance. In 2010, state regulators approved 13 large solar thermal facilities and wind farms for California's Mojave Desert. Most were canceled or delayed over concerns that they threatened rare or endangered species such as the desert tortoise. As with all energy development, thoughtful siting is critical to minimize environmental impacts.

Another high-temperature system uses thousands of mirrors arranged around a tall central **power tower** (fig. 20.13). The mirrors track the sun and focus light on a heat absorber at the top of the "power tower," where a transfer medium is heated to temperatures as high as 500°C (1,000°F), which then drives a steam-turbine electric generator. Under optimal

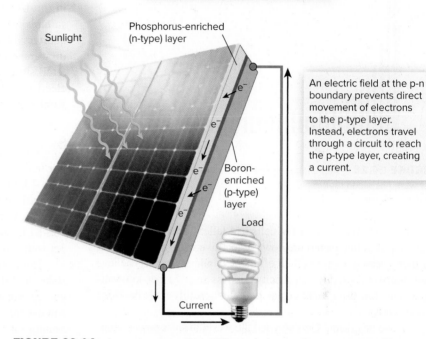

Photons striking the panel surface excite electrons (e⁻), which want to move toward the positive, p-type layer.

Sunlight

Phosphorus-enriched (n-type) layer

An electric field at the p-n boundary prevents direct movement of electrons to the p-type layer. Instead, electrons travel through a circuit to reach the p-type layer, creating a current.

Boron-enriched (p-type) layer

Load

Current

FIGURE 20.14 In a solar photovoltaic (PV) panel, the n-type silicon layer contains traces of an element such as phosphorus, with more outer-shell electrons than silicon, while the p-type silicon layer contains an element such as boron, with fewer outer-shell electrons. When sunlight strikes the panel, photons dislodge electrons, which travel through a circuit, creating an electrical current.

conditions, a 50 ha (130 acre) mirror array should be able to generate 100 MW of clean, renewable power. The Ivanpah Solar Station, built by Bright Source in California's Mojave Desert 60 km (40 mi) southwest of Las Vegas, with a capacity of 392 MW, is currently the world's largest CSP installation. Covering 4,000 acres (1,600 ha), the plant has 173,500 heliostats focusing energy on three towers, and produces enough power for 140,000 homes.

Photovoltaic cells generate electricity directly

Photovoltaic (PV) cells capture solar energy and convert it directly to electrical current by separating electrons from their parent atoms and accelerating them across a one-way electrostatic barrier formed by the junction between two different types of semiconductor material (fig. 20.14). The first PV cells were made by slicing thin wafers from giant crystals of extremely pure silicon.

Over the past 25 years the efficiency of energy captured by PV cells has increased from less than 1 percent of incident light to more than 15 percent under field conditions and over 75 percent in the laboratory. Increasing production volume has made solar PV far cheaper and more available in recent years (fig. 20.15). Some energy analysts suggest that solar collectors could provide 25 percent of global electricity in a decade.

FIGURE 20.13 A power tower is a form of concentrated solar thermal electrical generation. Thousands of movable mirrors focus sunlight on the central tower, where a liquid is heated to drive a steam turbine.
Kevin Burke/Getty Images

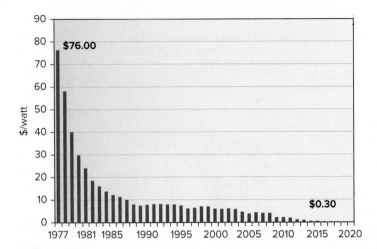

FIGURE 20.15 The plummeting cost of PV cells, now below 30 cents per watt, has aided their rapid expansion.

Source: Bloomberg New Energy Finance

By 2019, the price of PV cells had fallen below 30 cents per watt, and installed panels are now competitive with fossil fuels. As further research improves their efficiency, industry experts believe they will soon produce electricity for less than 5¢ per kilowatt-hour. This low price makes solar suitable for utility-scale baseload power arrays.

As we have seen, Germany led the world in solar power adoption. In 2015, however, China vaulted ahead of Germany to a total of 43 GW of PV power (compared to 40.1 in Germany and 26.8 GW in the United States). Japan and India also have moved aggressively into solar energy. Global PV installations are also growing rapidly. Over 950 GW of installed capacity is expected by 2025, compared to 225 GW in 2015 and about 5 GW in 2005.

Solar energy is evolving rapidly, as researchers and companies work to improve efficiency and innovative uses of PV systems. For example, flexible roof tiles with amorphous silicon collectors layered on their surface can be installed on residential rooftops (fig. 20.16a). In developing countries, solar power could make electricity available to some of the nearly 2 billion people who aren't served by a power grid.

Rooftop solar energy has huge potential. It's estimated that in the United States more than 1,000 mi^2 (2,590 km^2) of roofs suitable for photovoltaic systems could generate about three-quarters of our present electrical consumption. Just the roofs of warehouses, shopping malls, office buildings, and big-box retail stores (fig. 20.16b) could produce an estimated 62 GW of electricity. Already, Walmart, Target, IKEA, Apple, Costco, and Kohl's have installed solar arrays on their buildings.

Solar works at household or community scales

For individual households, a photovoltaic array of about 30–40 m^2 (3 to 4 kw) will generate enough electricity for an efficient house. The initial cost of this technology has been a barrier for most home-owners. When solar was $5 per watt installed, a home-sized system cost $15,000 to $20,000. But these prices are falling rapidly. At $1 per watt, a roof-top array might pay for itself in five to seven years.

Homeowners, who own their own roofs and have a personal stake in cost-savings, largely established the U.S. market for solar energy. But a majority of Americans either rent or don't have a roof suitable for solar. An emerging alternative for these situations is community solar, also called solar farms. Widely implemented in early stages of Germany's *Energiewende*, these are community-shared solar arrays connected to the utility grid (fig. 20.16c). Members buy power from the utility as usual, but they get a rebate from the garden based on their share of ownership. As with power on a private roof, members can either pay up front for a share, or pay on a monthly basis. There are many different business models for solar gardens.

Solar gardens can also give access to solar power, and solar cost savings, to people who don't own their home, or who can't afford to get involved in solar energy otherwise. As they have

(a) Flexible, thin-film solar tiles

(b) Roof-top solar array

(c) Community solar

FIGURE 20.16 Solar-photovoltaic energy is highly versatile and is evolving rapidly. (a) Thin-film PV collectors can be printed on flexible backing and used like ordinary roof tiles. (b) Millions of square meters of roof tops on schools and commercial buildings could hold solar panels. (c) Community solar expands access to more participation.

(a) Source: Stellar Sun Shop/NREL/U.S. Dept. of Energy; (b) Source: Craig Miller Productions/NREL/U.S. Department of Energy; (c) Monty Rakusen/Getty Images

Quantitative Reasoning

According to the International Energy Agency, total world subsidies for oil, coal, and natural gas amounted to US$493 billion in 2014. This was at least four times as much as for renewable energy. How much was this per person, if there were 320 million Americans?

become available in the United States since about 2008, community solar systems have become very popular. By the end of 2018, over 1.3 GW of community solar was installed, enough for over 200,000 homes. Because agreements with utilities have restricted the allowable size of community solar, this represents hundreds of separate systems. Even so, community solar is expected to grow rapidly.

Energy policies in some states include **feed-in tariffs,** which require utilities to buy surplus power from small producers at a slightly raised price. These tariffs (payments) expand access to solar power by making small and medium-sized installations more economical. Much of the costs of installing solar today involves planning, permitting, labor, and grid connections—the cells themselves are now a small part of the total cost. Small systems therefore cost more per watt than large systems. Feed-in tariffs can be a big incentive for small producers.

Section Review

1. What is the difference between active and passive solar energy?
2. How do photovoltaic cells generate electricity?
3. What are some policy strategies to support solar energy?

20.3 WIND

- *Wind power could meet all our energy needs.*
- *China is now the world's largest producer of wind turbines.*
- *Turbines pose risks to wildlife, like most energy sources, but careful placement and operation can reduce risks.*

Wind energy provides the largest share of renewable energy, roughly twice that of solar energy (fig. 20.17). The total installed capacity of wind and solar systems is similar (see fig. 20.9), but wind still leads solar in total generation, because wind turbines can spin day and night, while solar produces during the day. Over the past decade, China has become the world's leading producer of wind turbines, and this technology provides more than 2.5 million Chinese jobs. China now has over 200 GW of installed wind capacity, or more than all of Europe put together (fig. 20.18). The United States is second in the world with 100 GW of wind power, while Germany is third with 60 GW. India (35 GW) and the United Kingdom (20 GW) fill out the top five wind power leaders.

Modern **wind turbines** are far different from those employed a generation ago. The largest wind turbines are 40 times more powerful today than 20 years ago. Some now have towers up to

150 m tall, with 82-m-long blades that reach as high as a 50-story building. Large turbines can generate 8 MW of electricity, or enough for 3,000 typical American homes. These giant machines are mostly used offshore. Out of commission for

FIGURE 20.17 Proportion of global renewable power generation for 2020, as proportions of the nearly 7,500 terrawatt-hours (TWh) produced. Concentrated solar, biomass, and geothermal sum to about 0.2 percent and are not shown.
Data source: U.S. IEA 2019

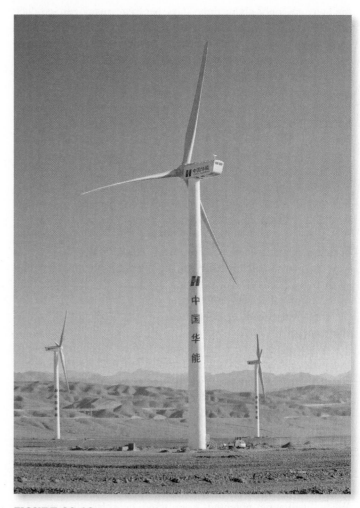

FIGURE 20.18 China is the world's leading producer of wind turbines, and now has about one-third of all installed wind power. This clean technology provides more than 2.5 million jobs in China.
Paul Springett 10/Alamy Stock Photo

FIGURE 20.19 Offshore wind is one of the fastest-growing power sources. High winds at sea allow these turbines to produce most of the year at high efficiency.
wangsong/123RF

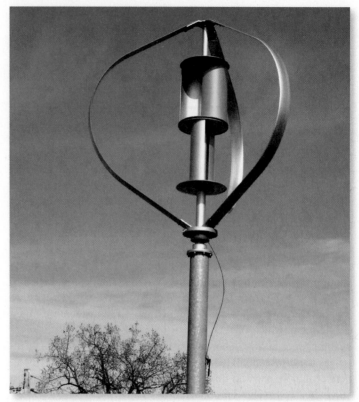

FIGURE 20.20 Vertical axis wind turbines are lighter and cheaper than similarly sized solar arrays. They can operate in slower wind speeds, and can be sized to be more appropriate for a single home or small business.
William P. Cunningham

maintenance only about three days per year, many can produce power 90 percent of the time. Theoretically up to 60 percent efficient, modern wind turbines typically produce about 35 percent of peak capacity under field conditions (fig. 20.19).

Germany currently gets over 20 percent of its electricity from wind tubines, but that number is growing steadily. Onshore wind farms are the cheapest source of *new* power generation, costing as little as 3¢/kWh, compared to 4–5¢/kWh for coal, and five times that much for nuclear fuel. Offshore wind is achieving similar costs now, as turbine design improves and the scale of turbines increases. Often one of the largest costs of installing a wind farm is the time and expense involved in licensing, planning, and logistics. As more systems are built, and processes become more familiar, these costs are falling steadily.

American states are now starting to get into the offshore wind business. Rhode Island was first, with a small 5-MW system, but New York has plans for 9 GW offshore wind by 2040, and a joint study by northeastern states anticipates that 8 GW could be built in the coming decade, providing around 36,000 jobs. Nationally, the U.S. Department of Energy has set a goal of 54 GW of offshore energy by 2030.

While giant modern turbines are most efficient, they don't fit everywhere. Vertical axis wind turbines (VAWTs) are lighter and cheaper than standard turbines, although they capture energy less efficiently. They can operate in slower wind speeds, and can be sized to be more appropriate for a single home or small business (fig. 20.20).

Capacity and efficiency are important questions in power production

When evaluating potential from wind or solar, or any other energy source, a central question is how much energy does, say, 1 megawatt of installed *capacity* produce? Energy sources vary in **conversion efficiency,** or how much of input energy is converted to useful work. They also vary in **capacity factor**—how much of the theoretical maximum is actually produced over time.

Efficiency is never 100 percent. In a coal plant or natural gas power plant, about 66 percent of the energy in the coal or gas fuel is converted to heat, which is dissipated to the environment through cooling towers or by circulating river or lake water through the plant. Conversion of fuel to electricity, then, is about 34 percent (see fig. 19.2). A hydropower plant can be around 90 percent efficient. A wind turbine converts about 34 percent of the force of wind into usable electricity, and it is relatively cheap and quick to build and suitable sites are plentiful. This explains why wind has become one of our cheapest power sources. A solar PV cell converts about 15–20 percent of incoming solar energy into electricity. Of course, the "fuels" for wind and solar have no cost, since wind and sunshine are free; in contrast, fuel for coal or gas power have high monetary, environmental, and social costs, so inefficiency has substantial impacts.

The capacity factor is the ratio between actual production and maximum possible production. A coal plant, for example, might have a "nameplate capacity," a calculated maximum, of 1 megawatt. If the plant ran at full potential for every hour of a year (8,760 hours), it would produce 8,760 megawatt-hours of power. But normally a plant is down sometimes for maintenance, and it might power down during off-peak, low demand periods. On average over a year, coal plants have a capacity factor of 56 percent, according to the U.S. Energy Information Agency. So the plant is likely actually to produce about 4,900 MWh of power.

For natural gas (combined cycle) power plants, the capacity factor averages 53 percent. Nuclear plants, which are expensive and difficult to start up and shut down, are designed to provide constant baseload power. These have an average capacity factor of 92 percent. Some gas plants, on the other hand are built just to provide peak-hour energy. These "peaking plants" have a capacity factor of about 7 percent.

Capacity factors for wind and solar power are limited by availability of wind and sunshine. Arizona's Agua Caliente PV facility reports a capacity factor of 29 percent, higher than the U.S. average of 25 percent for solar PV. In northern, cloudy places such as Germany, the capacity factor may be just 15 percent. The capacity factor for wind generators varies from about 30 to 60 percent, depending on the location of the turbines, and whether they are required to shut down at off-peak times. Hydropower, designed to shut down at off-peak times, averages 39 percent.

Estimating the number of wind turbines needed, then, and the cost of providing them or another power source, depends on efficiency, capacity factors, and the cost of fuel, as well as the cost of construction.

Wind could meet all our energy needs

Wind power is the world's fastest-growing energy source and could replace all the commercial energy we now use. The Global Wind Energy Council calculates that wind could supply 40 times the current world electricity supply and five times all global energy consumption if all the potential were tapped. Wind has a number of advantages over most other power sources. Wind farms have much shorter planning and construction times than fossil fuel or nuclear power plants. Wind farms are modular (more turbines can be added if loads grow) and they have no fuel costs or air emissions.

Wind does have limitations, however. Like solar energy, it is an intermittent source. Furthermore, not every place has strong enough or steady enough wind to make this an economical resource. Although modern wind turbines are more efficient than those of a few years ago, it takes a wind velocity between 7 and 27 m per second (16 to 60 mph) to generate useful amounts of electricity.

Large areas of the Great Plains and mountain states have persistent winds suitable for commercial development (fig. 20.21). Thirty-seven states now have utility-scale wind farms. Texas leads the nation, followed by California, Iowa, Illinois, and Oregon, but a huge potential remains to be tapped. Wind power is now so plentiful in Texas that on some occasions, the price has dropped into negative territory. That is, wind farms have to pay consumers to use their excess power.

Wind is a source of rural income

There is growing demand for wind projects from farmers, ranchers, and rural communities because of the economic benefits that wind energy brings. One thousand megawatts of wind power

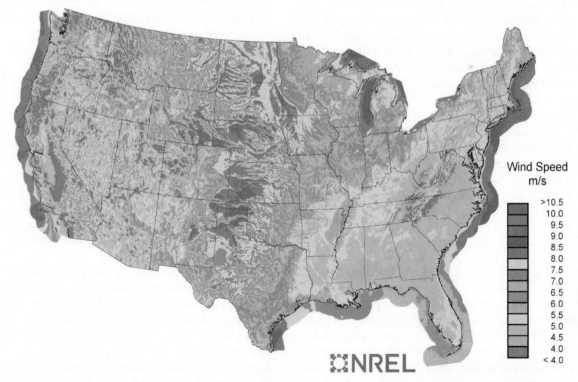

Wind Speed
m/s

>10.5
10.0
9.5
9.0
8.5
8.0
7.5
7.0
6.5
6.0
5.5
5.0
4.5
4.0
< 4.0

FIGURE 20.21 U.S. wind resource map. Mountain ranges and areas of the High Plains have the highest wind potential, but much of the country has fair to good wind supply.
Source: U.S. Department of Energy

FIGURE 20.22 Wind power is a good source of income in many rural areas, and it can coexist with agricultural production.

Dave G Kelly/Getty Images

Concern about wind turbines is important, but oil, gas, coal, and nuclear power cause more wildlife mortality than renewables do. Oil development is the leading cause of habitat destruction in much of western South America and West Africa. In North America, the vulnerable Sage Grouse and dwindling pronghorn antelope populations are threatened by development of vast oil and gas fields. Oil spills such as that in the Gulf of Mexico in 2010 (see chapter 19) cause uncounted deaths of birds and fish, impacting marine ecosystems worldwide every year. Ethanol and biofuels, meanwhile, are responsible for destruction of millions of hectares of wildlife habitat across North America, Malaysia, Ecuador, Congo, and other regions (see section 20.4).

Section Review

1. How much energy could we obtain from wind?
2. What regions of the United States have the strongest and most reliable wind?
3. What state currently has the greatest installed wind capacity?

(equivalent to one large nuclear or fossil fuel plant) can create more than 3,000 permanent jobs, while paying about $4 million in rent to landowners and $3.6 million in tax payments to local governments.

Wind power can integrate with other land uses. Wind takes about one-third as much area and creates about five times as many jobs to create the same amount of electrical energy as coal, when the land consumed by mining is taken into account. Furthermore, with each tower taking only about 0.1 ha (0.25 acre) of cropland, farmers can continue to cultivate 90 percent of their land while getting $2,000 or more in annual rent for each wind machine (fig. 20.22). An even better return results if the landowner builds and operates the wind generator, selling the electricity to the local utility. Annual profits can be as much as $100,000 per turbine, a wonderful bonus for the use of 10 percent of your land.

Cooperatives are springing up to help landowners finance, build, and operate their own wind generators. About 20 Native American tribes, for example, have formed a coalition to study wind power. Together their reservations (which were sited in the windiest, least productive parts of the Great Plains) could generate at least 350,000 MW of electrical power, equivalent to about half of the current total U.S. installed capacity.

Energy production has environmental impacts

In some places, high bird and bat mortality has been reported around wind farms. This was especially serious at the first wind farms in California, where rows of generators were placed at the summits of mountain passes, areas where wind velocities are high but where migrating hawks could easily fly into rotating blades. New generator designs and more careful tower placement seems to have reduced this problem in most areas. Shutting down turbines during migrations can help, too. Wildlife biologists and wind producers are working to understand rates of mortality and strategies for reducing them.

20.4 HYDROPOWER, BIOMASS, AND GEOTHERMAL ENERGY

- *A few countries dominate hydropower development.*
- *Massive hydropower dams produce abundant electricity but also environmental concerns.*
- *We use biomass in many different ways.*

Discussion of energy futures focus largely on emerging solar and wind power, but water was our first large-scale power source. Water power drove the early industrial revolution, and the invention of hydroelectric turbines in the nineteenth century, which produced electricity by spinning generators, greatly increased the efficiency of hydropower. By 1925, falling water generated 40 percent of the world's electric power. Since then, hydroelectric production capacity has multiplied, but fossil fuel use has risen faster, so that water power is now only 20 percent of total electrical generation. Still, many countries produce most of their electricity from falling water. Norway, for instance, depends on hydropower for 99 percent of its electricity.

Total world hydroelectric production is about 4,500 terawatt-hours. Six countries—Canada, Brazil, the United States, China, Russia, and Norway—account for more than half that total. In fact, of the approximately 50,000 dams in the world taller than 15 m (45 ft), roughly half are in China. Both Chinese and American corporations and banks are now reaching out to develop hydropower in Latin America, Central Africa, and India. Most of the world's major rivers have been dammed in the past century, and hydropower is an established and important part of our energy mix.

The great advantage of large hydro is that a single project produces a vast amount of electricity. And once built, the falling water is a carbon-free source of electricity. The disadvantages of large dams are their ecological and cultural impacts, and their staggering costs. On the other hand, this cost makes them appealing and profitable to financiers and builders. The Grand Ethiopian Renaissance Dam, under construction on the Blue Nile River since 2011 and the largest in Africa, is expected to cost nearly $5 billion when it is finished. Critics question whether there is enough water to fill the reservoir and whether the dam size exceeds Ethiopia's needs. Supporters point out that Ethiopia is an impoverished and energy-poor country that needs power to develop. Similarly, India's Sardar Sarovar Dam on the Narmada River was a long and controversial project. Initiated in 1979, the dam was inaugurated in 2017, at a cost of $8 billion. Like other large dams, these have been controversial because they flood river valleys and farmlands that are culturally important, economically valuable, and biologically critical.

Dams on transboundary rivers also cause tensions. Ethiopia's dam on the Nile has angered Sudan and Egypt, both of which worry about water resources downstream of the dam. Similarly, Turkey's Ataturk Dam on the Euphrates River worries downstream neighbors in arid Iraq and Syria.

Most hydroelectricity comes from large dams

Much of the hydropower development in recent decades has been in enormous dams. These tremendous undertakings are highly prestigious for their builders, and they facilitate industrial and urban expansion in developing countries. In fact, dams are often purpose-built for particular industries or factories. Iceland's controversial Kárahnjúkar Dam, for example was built explicitly to encourage an Alcoa aluminum refinery to move to Iceland.

The largest hydroelectric dam in the world is the Three Gorges Dam on China's Yangtze River, which spans 2 km (1.2 mi) and is 185 m (600 ft) tall. Designed to generate 25,000 MW of power, this dam produces as much energy as 25 large nuclear power plants (fig. 20.23). It now accounts for about 2 percent of China's total electricity demand.

The Three Gorges Dam was controversial because its reservoir displaced at least 1.5 million people and submerged 5,000 archaeological sites. China is now in the process of constructing a sequence of dams on the Mekong River, which runs from Tibet to Cambodia and Vietnam (see fig. 17.21). Chinese dams now block the seasonal floods that traditionally sustained the fisheries and farms of the river's lower reaches in Laos and Cambodia. Critics argue that these dams are not needed, and that the country has other sources to meet its energy needs. The main reason for building the dams, they say, is to funnel public money into company coffers.

This critique has been leveled against U.S. dams—historian Marc Reisner has called many western dams "cash register dams," whose main purpose was to draw federal dollars to fund more dams, and whose power was not really needed. Similar objections have been made for dams in Brazil and elsewhere.

FIGURE 20.23 China's Three Gorges Dam, completed in 2014, is the world's largest, and perhaps most debated dam.
menabrea/iStockphoto/Getty Images

In tropical regions, where vegetation is abundant, rotting trees and other organic material in reservoirs can have disastrous effects on water quality. When Lake Brokopondo in Suriname flooded a large region of uncut rainforest, underwater decomposition of the submerged vegetation produced hydrogen sulfide that killed fish and drove out villagers over a wide area. Acidified water from this reservoir ruined the turbine blades, making the dam useless for power generation. Decaying vegetation also creates methane, an important greenhouse gas (fig. 20.24). A recent study of one reservoir in Brazil suggested that decaying vegetation produced more greenhouse gases (carbon dioxide and methane) than would have come from generating an equal amount of energy by burning fossil fuels.

FIGURE 20.24 Tropical dams often flood heavily vegetated river valleys, where submerged vegetation produces abundant methane as it decays.
Muzhik/Shutterstock

Dams in arid regions, meanwhile, can evaporate much of a river's flow, reducing resources for farms and cities. This has been an especially serious problem on the Nile in Egypt and on the Colorado River in the southwestern United States (see chapter 17).

Smaller dams avoid many of the environmental and social problems of larger dams, although they also produce less power. **Low-head hydropower** generators, which can extract energy from lower dams, cause much less damage than larger projects. Modern, high-efficiency turbines can even operate on run-of-the-river flow. Submerged directly in the stream and small enough not to impede navigation in most cases, these turbines don't require a dam or diversion structure and can generate useful power with a current of only a few kilometers per hour. They also cause minimal environmental damage and don't interfere with fish movements, such as spawning migration.

Tides and waves contain significant energy

Ocean tides and waves contain enormous amounts of energy, and many projects have sought to harness this energy for useful work. An approach that has been effective in many tidal river areas is a tidal power station, basically a hydropower dam, with its turbines spinning as the tide flows through them. A high-tide/low-tide differential of several meters is required to spin the turbines. These systems have operated for several decades in some areas.

More ubiquitous but harder to harness is ocean wave energy. Dutch researchers estimate that 20,000 km of ocean coastline are suitable for harnessing wave power. Among the best places in the world for doing this are the west coasts of Scotland, Canada, the United States (including Hawaii), South Africa, and Australia. Wave energy specialists rate these areas at 40 to 70 kW per meter of shoreline. Altogether, it's calculated, if the technologies being studied today become widely used, wave power could amount to as much as 16 percent of the world's current electrical output. This huge potential has kept experimentation alive, although implemented projects have not lasted.

Biomass is an ancient and modern energy source

Biomass (meaning material derived from plants) provides a modest portion of our global energy supply, but in some regions it is extremely important. It is also versatile, with different forms supporting diverse needs.

In many countries, firewood gathering and charcoal production is a dominant cause of forest destruction and habitat degradation. Firewood gathering is also a socially unequal burden. Poor people often spend a high proportion of their income on cooking and heating fuel. In rural families, women and children may spend hours every day searching for fuel. Development agencies are working to design and distribute highly efficient stoves, both as a way to improve the lives of poor people and to reduce forest losses.

These crops also are carbon neutral—that is, they absorb as much CO_2 in growing as they emit when burned.

Some utilities are installing **flex-fuel boilers** in their power plants that can burn wood chips, agricultural waste, or other biomass fuels (fig. 20.25). A district heating plant in St. Paul, Minnesota, for example, uses 275,000 tons of wood per year (mostly from urban trees killed by storms and disease) to provide heating, air conditioning, and electricity to 25 million square feet of offices and living space in 75 percent of all downtown buildings. Although the efficiency of electrical generation in this plant is less than 40 percent (as it is in most power plants), the net yield is about 80 percent because waste heat is used rather than discarded.

Plants that produce steam heating can also produce energy with excess steam, a process known as **cogeneration**. Cogeneration is most often used to increase efficiency of fossil fuel-burning heating plants, but it can be used with any facility that produces heat by burning fuel.

Methane from biomass can be clean and efficient

People in many developing areas burn dung for household fuel. Dry dung burns cleanly, and where wood and other fuels are in short supply, livestock often supply a ready source of high-quality fuel (fig. 20.26a). In India, for example, cows produce more than 800 million tons of dung per year, more than half of which is dried and burned in cooking fires. Some development experts worry that this manure is needed to fertilize fields. By some estimates, applying that dung to fields as fertilizer could boost crop production enough to feed about 40 million people. However, people need fuel, too, and without dung they might seek more firewood or kerosene.

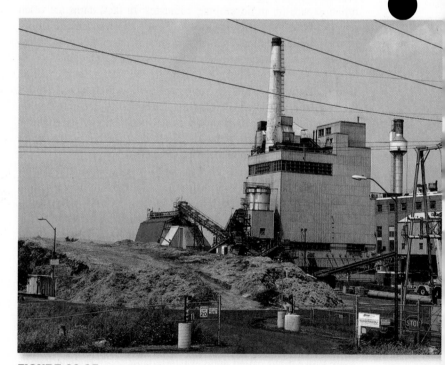

FIGURE 20.25 This Michigan power plant uses wood chips to fuel its boilers. Where wood supplies are nearby, this is a good choice both economically and environmentally.
William P. Cunningham

(a) Dung patties ready for burning

(b) A methane digestion plant at a dairy farm

FIGURE 20.26 Animal dung and other organic waste can be used directly as fuel (a) or digested to produce methane (b). Both are clean burning and locally important fuel sources.
(a) ephotocorp/Alamy Images; (b) Rudmer Zwerver/Shutterstock

In other regions, dung and other wet organic matter is collected and digested to produce methane, a clean-burning gas. In China, animal waste is widely used to produce methane gas in simple, inexpensive methane digesters designed for villages and homes. **Methane** (CH_4) is the main component of natural gas. It is produced by anaerobic decomposition (digestion by anaerobic bacteria) of any moist organic material. Many people are familiar with the fact that swamp gas is explosive. Swamps are simply large methane digesters, basins of wet plant and animal wastes sealed from the air by a layer of water. Under these conditions, organic materials are decomposed by anaerobic (oxygen-free) rather than aerobic (oxygen-using) bacteria, producing methane instead of carbon dioxide. This same process may be reproduced artificially by placing organic wastes in a container and providing warmth and water. After gas is removed, the remaining solid material can then be used as clean fertilizer.

Methane digesters are increasingly widespread in developed countries as well, especially in Switzerland, Germany, and other parts of Western Europe, where organic household waste is collected separately from recycling and other garbage, and where agricultural waste is abundant (see fig. 21.14). Any kind of organic waste material can be used to generate biogas—livestock manure, kitchen and garden scraps, and even municipal garbage and sewage. In the United States, municipal biogas production is rare, but dairies and hog farms increasingly produce methane from their abundant supplies of animal manure. The Haubenschild farm in central Minnesota, for instance, uses manure from 850 Holsteins to generate all the power needed for their dairy operation and still have enough excess electricity for an additional 80 homes. In a typical January, the farm saves about 35 tons of coal and 1,200 gallons of propane, while making more than $4,000 from electric sales (fig. 20.26b). Cattle feedlots and chicken farms in the United States also offer tremendous potential fuel sources. Feedlot wastes and collectible crop residues contain more energy than U.S. farmers use each year.

Municipal landfills, which contain vast amounts of buried (anaerobic) organic waste, are also active sites of methane production, contributing as much as 20 percent of the annual output of methane to the atmosphere. This is a waste of a valuable resource and a threat to the environment because methane is an important greenhouse gas (chapter 15). About 300 landfills in the United States currently burn methane and generate enough electricity together for a million homes. Another 600 landfills have been identified as potential sources for methane development.

A number of colleges around the United States are using biomass and biogas to wean themselves off fossil fuels. In Vermont, Middlebury College feeds locally harvested wood chips into a gasification plant (which heats chips to produce burnable gas) that provides both heat and electricity to the campus, and the college is buying methane from a local dairy as well. These steps have led the way in helping Middlebury become carbon neutral. And the University of New Hampshire in Durham plans to provide 80 percent of its heating and electrical needs by buying and burning methane gas given off by a landfill a few miles away. At the University of Minnesota's Morris campus, a gasification plant uses about 1,700 tons of corn stover (stalks and cobs) and other local agricultural material every year to provide as much as 80 percent of the school's heating and cooling needs (fig. 20.27), while a wind turbine provides most of its electricity. Together these alternative energy systems replace at least $1 million per year in fossil fuels (mostly natural gas).

U.S. policy supports ethanol and biodiesel

Biofuels, such as ethanol and biodiesel, are a means of using biomass for transportation fuel (fig. 20.28). Globally, production of these two fuels is booming. Brazil gets about 40 percent of its transportation energy from ethanol generated from sugarcane. In Southeast Asia, millions of hectares of tropical forest have been cleared for palm oil plantations. In the United States, about one-fifth of the corn (maize) crop is used to make

FIGURE 20.27 Biomass crops can provide heat and electricity as well as liquid fuels. Cellulosic ethanol avoids using food crops for fuel, but production is expensive and complex.
William P. Cunningham

ethanol. Crops with a high oil content, such as soybeans, sunflower seed, rapeseed (also called canola), and palm oil fruits are relatively easy to make into biodiesel. In some cases, the oil needs only minimal cleaning to be used in a standard diesel

FIGURE 20.28 Liquid biofuels, mixed here with gasoline, make use of nearly half the U.S. corn crop and much of the soybean crop.
Source: U.S. Department of Energy

engine. Yields per hectare for many of these crops are low, however. It would take a very large land area to meet our transportation needs with soy or sunflowers, for example. Furthermore, diversion of these oils for vehicles deprives humans of an important source of edible oils.

Oil palms are eight to ten times more productive per unit area than soy or sunflower but this biofuel has important environmental costs. Millions of hectares of species-rich forests in Southeast Asia are being destroyed to create palm oil plantations. Indonesia already has 6 million ha of palm oil plantations and Malaysia has nearly as much. Together these two countries produce nearly 90 percent of the world's palm oil. The burning of Indonesian forests and the peat lands on which they stand currently releases some 1.8 billion tons of carbon dioxide every year. Indonesia is currently third in the world—behind the United States and China—in human-made greenhouse gas emissions. At least 100 species, including orangutans, Sumatran tigers, and the Asian rhinoceros, are threatened by habitat loss linked to palm oil expansion.

Crops such as sugarcane and sugar beets have a high sugar content that can be fermented into ethanol, but sugar is expensive and the yields from these crops are generally low, especially in temperate climates. Starches in grains, such as corn, have higher yields and can be converted into sugars that can be turned by yeast into ethanol (this is the same process used to make drinking alcohol), butanol (which burns in engines much like gasoline), or methanol. The idea of burning ethanol in vehicles isn't new. Henry Ford designed his 1908 Model T to run on ethanol, which at the time was easier to acquire and much safer than gasoline.

U.S. policymakers have strongly subsidized ethanol, arguing that it will reduce our dependence on foreign oil. Since 1980, more than 200 new ethanol refineries have been built, and U.S. annual production capacity has grown from about 500 million liters to 60 billion liters (15 billion gallons; fig. 20.29). The United States and Brazil now produce about 95 percent of all the ethanol in the world.

There has been a great deal of debate about the net energy yield and environmental costs of ethanol from corn, which requires a great deal of irrigation, pesticides, and fertilizer, and which causes much of the soil erosion in the United States. There is greater agreement that **cellulosic ethanol,** made from woody plant tissue, would be more environmentally sustainable, if we can find ways to produce it economically. Most plants put the bulk of the energy they capture from the sun into cellulose. Chemists have sought for decades to find efficient ways to release those simple sugars so they can be fermented into ethanol. But it's difficult: If it were easy for microbes to dismantle woody material, we probably wouldn't have standing forests or prairies—the landscape would simply be covered with green slime. Extracting sugars from cellulosic materials involves chopping or shredding biomass, followed by treatment with bacteria or fungi (or enzymes derived from them).

In Denmark, where fuel costs are high and government subsidies are abundant, cellulosic ethanol is produced for commercial use. So far, there are no commercial-scale cellulosic ethanol factories operating in North America, but the Department of Energy has provided $385 million in grants for six cellulosic biorefinery

FIGURE 20.29 More than 200 ethanol plants distill biofuels from corn and other crops in the United States.
William P. Cunningham

plants. These pilot projects will test a wide variety of feedstocks, including rice and wheat straw, milo stubble, switchgrass hay, almond hulls, corn stover, and wood chips. A favorite potential fuel source is switchgrass (*Panicum virgatum*), a tall grass native to the Great Plains that is easy to plant mechanically. Switchgrass is a perennial, with deep roots that store carbon (and thus capture atmospheric greenhouse gases).

Could algae be an efficient energy source?

Algae are aquatic, photosynthetic plants that produce lipids (oils), the basic substances that led to ancient oil deposits we use today. In theory, all algae need to grow are nutrients (such as waste from a sewage plant), carbon dioxide (such as effluent from a power plant), sunlight, and water. In principle, algae could be an extremely productive biofuel source. This is partly because single-celled algae can grow 30 times as fast as higher plants. Furthermore, some algae species store up to half their total mass as oil. The challenges have been in developing algae varieties that both produce oil efficiently grow well, and in developing growth systems that scale up to produce reliably in quantity.

Ideally, algae could be grown in facilities next to conventional power plants, where CO_2 from burning either fossil fuels or biomass could be captured and used for algal growth. Thus, they could be carbon negative, providing a net reduction in atmospheric carbon while also creating useful fuel. Unfortunately, despite decades of experimentation around the world, commercially viable strains of algae remain elusive.

High-temperature geothermal produces electricity

Where volcanic activity brings the earth's internal heat near the surface, high-temperature **geothermal energy** has transformed energy systems. Most notably, Iceland and Costa Rica rely mainly on geothermal for heating and for electricity production, with their abundant geothermal heat. These resources have allowed both of these countries to approach carbon neutrality. High-temperature systems contrast with more widely available geothermal, or ground-source heat pumps, which concentrate heat from the moderate-temperature of the ground surface (see section 20.6).

In the American West, areas like Yellowstone National Park show that geothermal resources are present. Drilling 2,000 m (6,000 ft) reaches into rocks above 100°C. Fracturing deep rock to expose more surface area, and pumping water in can produce enough steam to run an electrical generator at a cost significantly lower than conventional fossil fuel or nuclear power. The well is no more expensive than most oil wells, and the resource is essentially perpetual. About 60 geothermal energy projects have been or are being developed in the United States. However, there are concerns that fracturing (or fracking) can contaminate groundwater aquifers. And some large geothermal projects have been canceled over concerns that they seemed to be triggering earthquakes.

Section Review

1. What are some advantages and disadvantages of large hydroelectric dams?
2. How is methane produced from waste material?
3. What is the primary use of ethanol and biodiesel?

20.5 WHAT DOES AN ENERGY TRANSITION LOOK LIKE?

- *A transition from fossil to renewable energy involves electrifying more energy uses.*
- *Energy storage, a key challenge, is growing and diversifying.*
- *Wind, water, and solar could meet all our needs with existing technology.*

Transitioning to new energy systems involves rethinking policy, energy use practices, and technology. It also has co-benefits in cost savings and cleaner air—and, of course, in working to avoid extreme climate change. This transition is expected to create millions of jobs and spur economic development. Reduced reliance on oil and gas, much of it sourced from politically unstable regions such as the Middle East, means that repowering with carbon-free energy would also reduce security concerns.

Back in 2012, the U.S. National Renewable Energy Laboratory concluded that with available, affordable technology, renewable energy could supply 80 percent of total U.S. electricity generation by 2050 while meeting hourly electricity demand in every region of the country. By the end of the twenty-first century, it said, renewable sources could provide all our energy needs if we took the necessary steps soon.

The grid will need improvement

Electricity is delivered through vast networks of power lines—in the U.S. grid, more than 1 million km (600,000 mi) of transmission lines. The grid will need to expand as more energy uses, especially heating and transportation, become electrified. Germany, for example, anticipates expanding its long-distance, high-voltage grids by about 6 percent. The amount of expansion in different countries and regions depends on whether efficiency improvements reduce energy demand.

The grid also provides opportunities for balancing demand and production: The sun shines and the wind blows at different times in different regions, so exchanging between high- and low-producing regions can distribute supply and demand. U.S. policy divides the grid in two sections (fig. 20.30). Many renewable energy proponents argue that improving connectivity would greatly increase flexibility for renewables.

Upgrading to smart grid technologies with digital information flow will make it easier to monitor and maintain the grid; it will also make it possible for users to monitor their usage more easily, to be charged less for off-peak times, or to be paid for contributing electricity production or storage. Grid upgrades will be an expensive but long needed part of grid maintenance and modernization.

Storage options are changing rapidly

Increasing reliance on electricity will require improved systems for storing it. For decades, a principal means of storing electricity has been pumped hydropower. This approach uses excess electricity—for example, from a solar plant at midday—to pump water up to an elevated reservoir. At night, that water can be released to run a hydropower generator. Alternatively, thermal storage, such as heated water or salt, can provide heat to run steam

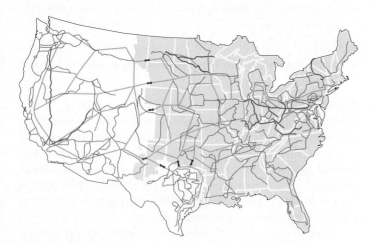

FIGURE 20.30 New high-voltage power lines (orange) will be needed if the United States is to make effective use of its renewable energy potential. The pink area served by the eastern electrical grid needs to be connected to the West by interlinks (black dots) for maximum efficiency.

turbines. Storage is an area of rapid change in renewable energy. The energy research group Irena reported in 2017 that electricity storage capacity appeared set to triple by 2030, to 12–15 terrawatt hours.

Battery storage is growing and diversifying rapidly. Commercial-scale battery banks, unheard of a few years ago, are becoming increasingly common: In 2018 the Hawaiian island of Kauai installed 52 MWh of batteries at a 13-MW solar park, at a cost of 11 cents per kWh. This was considerably less than the 15 cents/kWh that diesel cost on the island. A 65-MW solar farm in Arizona, coupled with a 50-MW battery storage system, is now able to provide electricity cheaper than conventional gas production. The U.S. Energy Information Agency reports that by 2018 U.S. utility-scale battery systems added up to over 900 MWh. These systems are far safer than storing and transporting gas, as well as being better for the climate. Battery systems are quick to install—just a few months for utility-scale systems—and now they are becoming cheaper than fossil fuels as well.

Lithium ion batteries have been standard for years because energy input and output are rapid. Battery research is active, however, including hydrogen, nanowires, solid-state lithium ion, graphene, and other technologies. The story is likely to keep changing. Batteries are also transforming the electric vehicle market, with new varieties promising to increase range and shorten recharge times dramatically.

Fuel cells release electricity from chemical bonds

Fuel cells, like batteries, have long been used for storing electricity. Fuel cells are similar to batteries except that they are recharged by adding a fuel or a source of hydrogen atoms, while a battery is recharged by adding electrical current. Fuel cells are not new; the basic concept was recognized in 1839 by William Grove, who was studying the electrolysis of water. He suggested that rather than use electricity to break apart water and produce hydrogen and oxygen gases, it should be possible to reverse the process by joining oxygen and hydrogen to produce water and electricity.

Fuel cells consist of a positive electrode (the cathode) and a negative electrode (the anode) separated by an electrolyte, a material that allows the passage of charged atoms, or ions, but is impermeable to electrons (fig. 20.31). In the most common systems, hydrogen or a hydrogen-containing fuel is passed over the anode while oxygen is passed over the cathode. At the anode, a reactive catalyst, such as platinum, strips an electron from each hydrogen atom, creating a positively charged hydrogen ion (a proton). The hydrogen ion can migrate through the electrolyte to the cathode, but the electron is excluded. Electrons flow through an external circuit, creating an electrical current. At the cathode, the electrons and protons are reunited and combined with oxygen to make water.

Most fuel cells currently use hydrogen from hydrocarbon fuels such as natural gas (mainly CH_4) or ammonia (NH_3). Methane effluents from landfills and wastewater treatment plants can be used as a fuel source. But the source can also be water (H_2O), whose molecules can be split (electrolyzed) with an electrical

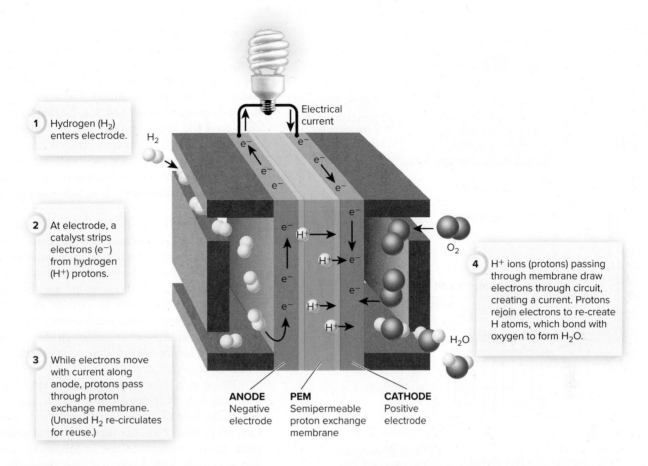

1. Hydrogen (H₂) enters electrode.

2. At electrode, a catalyst strips electrons (e⁻) from hydrogen (H⁺) protons.

3. While electrons move with current along anode, protons pass through proton exchange membrane. (Unused H₂ re-circulates for reuse.)

4. H⁺ ions (protons) passing through membrane draw electrons through circuit, creating a current. Protons rejoin electrons to re-create H atoms, which bond with oxygen to form H₂O.

Electrical current

H₂

e⁻   e⁻

e⁻   e⁻

e⁻   e⁻

O₂

H⁺

H⁺ → e⁻

e⁻

H⁺ → e⁻

H⁺

H₂O

ANODE
Negative electrode

PEM
Semipermeable proton exchange membrane

CATHODE
Positive electrode

FIGURE 20.31 Fuel cell operation. Electrons are removed from hydrogen atoms at the anode to produce hydrogen ions (protons) that migrate through a semipermeable electrolyte medium to the cathode, where they reunite with electrons from an external circuit and oxygen atoms to make water. Electrons flowing through the circuit connecting the electrodes create useful electrical current.

charge. If the source of electricity is wind or solar power, then a fuel cell can provide completely renewable power.

A fuel cell that runs on pure oxygen and hydrogen produces no waste products except water and heat. Other fuels create some pollutants (most commonly carbon dioxide), but the levels are typically far less than conventional fossil fuel combustion in a power plant or an automobile engine. Although the theoretical efficiency of electrical generation of a fuel cell can be as high as 70 percent, the actual yield is closer to 40 or 45 percent.

A number of cells can be stacked together until the desired power level is reached. A fuel cell stack that provides almost all of the electricity needed by a typical home (along with hot water and space heating) would be about the size of a refrigerator. A 200-kilowatt unit fills a medium-size room and provides enough energy for 20 houses or a small factory (fig. 20.32).

Heat pumps provide efficient, electric-powered cooling and heating

A transition to renewables involves electrifying many activities, including heating. Heat pumps are one of the main strategies for doing this. A study of 100 options for resolving the climate crisis,

FIGURE 20.32 Fuel cells can provide commercial or industrial scale backup power.
Peter Casolino/Alamy Stock Photo

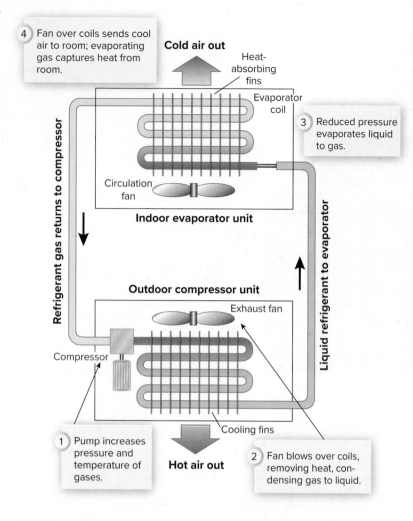

4 Fan over coils sends cool air to room; evaporating gas captures heat from room.

Cold air out

Heat-absorbing fins

Evaporator coil

3 Reduced pressure evaporates liquid to gas.

Refrigerant gas returns to compressor

Circulation fan

Indoor evaporator unit

Liquid refrigerant to evaporator

Outdoor compressor unit

Exhaust fan

Compressor

1 Pump increases pressure and temperature of gases.

Cooling fins

Hot air out

2 Fan blows over coils, removing heat, condensing gas to liquid.

FIGURE 20.33 A heat pump compresses and expands gases to cool a building. Run in reverse, the heat pump provides heating. Running on electric pumps, this system can heat and cool buildings without fossil fuels.

Household heating and cooling can be done on the same principles. A pump compresses a refrigerant between two heat exchanger coils (that is, coiled tubes containing the refrigerant). In one coil, the electric pump compresses the gas, which becomes warmer under pressure (see fig. 20.33). The warm gas travels to the condenser coils, where a fan blowing on the coils cools the heated gas, enough to condense it to a liquid state. The liquid is then pumped into expansion coils inside the refrigerator (or inside the house). As it expands (evaporates) to gas, the refrigerant absorbs heat from its surroundings. It is then pumped back to the compressor.

This system is reversible: In summer, the refrigerant evaporates inside your home and is condensed (releasing heat) outside. In winter, it is condensed inside the home, releasing heat, and it is evaporated (capturing heat) outside. Ground-source heat pumps use the ground's near-constant temperatures to provide or store heat (fig. 20.34). New air-source heat exchangers can capture heat even in cold weather—although the compressor pumps work harder and consume more electricity when winter air is cold.

This system is increasingly used for large buildings and campuses. Ball State University, in Indiana, was among the first to transition to geothermal. Skidmore College, in New York, has been 40 percent geothermal since 2014. Carleton College, in Minnesota, will heat almost entirely with geothermal heat pumps by 2020. This is just one of many ways colleges and universities can take action (table 20.1). Using climate-safe refrigerants is critical, as more of the world demands additional

called *Project Drawdown* (see section 15.6) listed heat pumps as one of the top solutions for heating and cooling. Heat pumps concentrate heat from the surrounding air (air-source heat pumps) or ground (ground-source heat pumps) and deliver it to indoor space. Or they capture heat from a room and deliver it outside. Depending on the circumstances, heat pumps cut heating and cooling costs, as well as decarbonizing your heating and cooling system if they use renewable electricity.

How do heat pumps work? Like a refrigerator, basically. They work on two physical principles. The first is that increasing pressure raises the temperature of a gas (and decreasing pressure cools it). The second principle is that when a gas changes phase and becomes a liquid, it releases heat to the surrounding environment (and when liquid evaporates to gas, it absorbs heat). In your refrigerator, cooling occurs as a refrigerant gas (such as CO_2, ammonia, or fluorine gas) is compressed to liquid form, and then decompressed inside the refrigerator, capturing heat as it returns to gas phase. Outside your refrigerator, the refrigerant fluid releases heat as it is compressed to liquid again.

Basement heat pump

Horizontal earth loop

FIGURE 20.34 Geothermal energy can cut heating and cooling costs by half in many areas. In summer (shown here), warm water is pumped through buried tubing (earth loops), where it is cooled by constant underground temperatures. In winter, the system reverses and the relatively warm soil helps warm the house.

Table 20.1	How Can Colleges and Universities Transition to Renewables?

Sightlines and the University of New Hampshire Sustainability Institute have published assessments of the progress toward carbon reductions across U.S. colleges and universities. These are some recommendations in their 2017 report.

1. Identify efficiency as a policy priority and a funding priority.

2. Plan a transition to electricity-based heating systems (such as air-source or ground-source heat pumps).

3. Think of energy efficiency as an investment, which can yield three to five times the returns of traditional endowments.

4. Collect and assess data: Efficiency only improves if energy use is monitored.

5. Buildings are growing faster than student populations. Focus on ways to right-size buildings by regularly evaluating usage and space needs.

6. Use power purchase agreements to buy renewable energy.

7. Work to develop on-site renewable energy, or buy renewables.

8. Where high-tech building equipment is required, focus on design and sustainability at the start.

Source: UNH Sustainability Institute and Sightlines, 2017 State of Sustainability in Higher Education 2017.

air conditioning, and as heat pumps become more common for heating and cooling (fig. 20.35).

Wind, water, and solar are good answers

It is increasingly acknowledged that reducing climate change means ending reliance on fossil fuels. This transition may be hard to imagine when more than 80 percent of energy use is oil, gas, and coal. But the Paris Accord and the IPCC have both cited safe levels of atmospheric CO_2 and temperature change that are only possible if we shift entirely to renewables.

How is this possible? Multiple strategies have been offered. Some of the most comprehensive and detailed have been done by Mark Jacobson, of Stanford University, and Mark Delucchi, of Santa Cruz, and their colleagues. Over a decade ago, they pointed out that renewables are complementary, and that wind, water, and solar meet our needs in combination (fig. 20.36). Since then, Jacobson and colleagues have explored further how we could use just wind, water, and solar (WWS) to meet all our heating, cooling, transportation, and electrical energy demands. For the United States, they have found that currently available wind, water, and solar technologies could supply 80 percent of energy needs by 2030 and 100 percent by 2050. They calculated that it would take 400,000 5-MW wind turbines, 75 million rooftop photovoltaic systems, 2.8 million commercial-scale PV facilities, and 2,000 concentrated solar power plants and industrial-sized photovoltaic arrays (fig. 20.37).

Building and installing all that technology would be a huge effort, but it's not without precedent. Jacobson points out that society has achieved massive transformations before. In 1956 the United States began building the Interstate Highway System, which now extends 47,000 mi (75,600 km) and has changed commerce, landscapes, and society. And every year roughly 60 million new cars and trucks are added to the world's highways. So there are precedents of dramatic shifts in technology and production.

The WWS models assume that most energy uses become electrified, including transportation, building heating and cooling, and manufacturing. Battery-electric vehicles, such as the Chevy Bolt or Tesla's battery-powered semitrailer truck, and electric

FIGURE 20.35 Carleton College, in Minnesota, is transitioning to geothermal heating, using ground-source heat pumps. With wind and solar electricity, the college aims to be 100 percent fossil-free by 2050.
Tom Finkle

FIGURE 20.36 The wind doesn't blow all the time, nor is sunshine always available, but a mix of renewable resources could supply all the energy we need, especially if distant facilities are linked together. This graph shows hypothetical energy supplies for California on a typical July day.

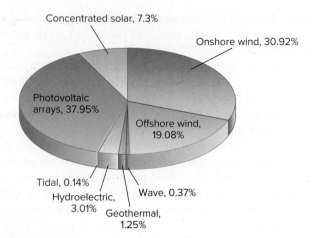

FIGURE 20.37 A 100 percent wind, water, and solar scenario for 2050.

Source: Jacobson, M., et al. 2015. "100% Clean and Renewable Wind, Water, and Sunlight (WWS) All-Sector Energy Roadmaps for the 50 United States." Energy Environ. Sci. 2015, no. 8: 20193–2117

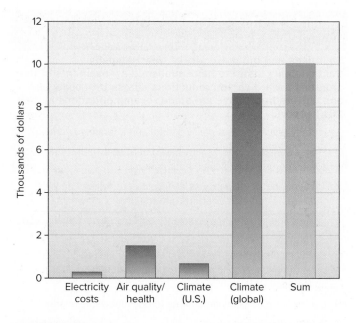

FIGURE 20.38 Calculated cost savings for the United States, per person per year, resulting from a transition from fossil fuels to wind, water, and solar energy. Climate costs include economic damage at both U.S. and global scales.

Source: Jacobson, M., et al. 2015. "100% Clean and Renewable Wind, Water, and Sunlight (WWS) All-Sector Energy Roadmaps for the 50 United States." Energy Environ. Sci. 2015, no. 8: 20193–2117

trains can provide a large share of transportation. Hydrogen fuel cells, powered by renewable electricity, can support transportation, energy storage, or other electric uses. Building heating and cooling can be done with air-source or ground-source heat pumps.

Fortunately, electricity is efficient for most uses, so it would take a great deal less total energy to meet our needs with sun, wind, and water than to continue using fossil fuels—about 39 percent less. This is mainly because conventional energy production through fuel burning loses a great deal of energy as heat. Operating a car on gasoline uses only about 17 percent of the fuel energy for moving the vehicle. The rest is lost as heat. Jacobson and colleagues point out that a battery-electric car, running on hydrogen fuel cells powered by solar energy, is more efficient than a car burning gasoline. Similarly, producing electricity by burning coal, oil, or gas uses only about one-third of the fuel to turn a turbine; the rest is lost as "rejected energy"—that is, heat, which then requires a great deal of water and infrastructure for cooling (see fig. 19.2).

Electricity is similarly more efficient for running pumps, heating water, or other uses. Furthermore, much of the energy from renewable sources could often be produced closer to where it's used, reducing losses in transmission and distribution of electricity in the grid.

Will it be expensive to install so much new technology? Yes, it will be, but the costs of continuing our current dependence on fossil fuels would be much higher. It's estimated that investing $700 billion per year now in clean energy will avoid 20 times that much in a few decades from the damages of climate change (fig. 20.38). To learn more about these energy scenarios, look online for the "Solutions Project."

Section Review

1. Why is an improved grid important?
2. How can geothermal energy be used for home heating?
3. What is a heat pump, and how does it work?
4. What are likely to be dominant energy sources of the future?

Connecting the Dots

Renewable energy is the primary means of reducing our dependence on fossil fuels. The expected transition is both daunting in scale and exciting in its opportunities. The approaches taken in Germany, and in other countries that are rapidly adopting renewables, shows that these power sources can run industry, transportation, even heating systems. Transitioning to renewables involves a great number of changes, though, in energy distribution, financing, and policy.

Of course, energy conservation and improving energy efficiency are the first step in reducing our dependence on conventional energy. Conservation has been a central aspect of Germany's transition to renewables. There are many steps you can take, that your school can adopt, and that societies can follow to make energy conservation easier.

Solar and wind production capacity has been growing exponentially, worldwide. Falling prices make it easier for more people and businesses to get involved. Production will need to continue accelerating, though, to meet climate protection goals. Hydropower is still the largest component of our carbon-free energy mix. Large-scale dams are often controversial, but they do provide electricity on a very large scale.

Analysts find that renewable and sustainable energy systems do have the capacity to meet our needs. Whether we have the political will to invest in them is the main question. Future developments are likely to include enhanced storage options, a more interconnected transmission grid, and relatively novel technologies such as heat pumps for heating and cooling buildings. The good news is that many of these technologies can save money as well as pollution. Their potential for transforming our energy use is tremendous.

Reviewing Key Terms

Can you define the following terms in environmental science?

active solar system 20.2

biofuel 20.4

biomass 20.4

capacity factor 20.3

cellulosic ethanol 20.4

cogeneration 20.4

concentrating solar power (CSP) system 20.2

conversion efficiency 20.3

feed-in tariffs 20.2

flex-fuel boilers 20.4

fuel cells 20.5

geothermal energy 20.4

light-emitting diodes (LEDs) 20.1

low-head hydropower 20.4

methane 20.4

passive heat absorption 20.2

passive house standards 20.1

photovoltaic (PV) cell 20.2

power tower 20.2

wind turbines 20.3

Critical Thinking and Discussion Questions

1. What alternative energy sources are most useful in your region and climate? Why?

2. What can you do to conserve energy where you live? In personal habits? In your home, dormitory, or workplace?

3. Do you think building wind farms in remote places, parks, or scenic wilderness areas would be damaging or unsightly? What is the most fair or reasonable way to decide where this or other energy sources should be located?

4. If you were the energy czar of your state, where would you invest your budget?

5. What could (or should) we do to help developing countries move toward energy conservation and renewable energy sources? How can we ask them to conserve when we live so wastefully?

Data Analysis

Energy Calculations

Suppose that you were debating between a high-mileage car, such as a Nissan leaf (rated at the equivalent of 114 mpg), or a Ford F150 4WD E85 (rated at 12 mpg). How do the energy requirements of these two purchases measure up? To put it another way, how long could you run a computer on the energy you would save by buying a Leaf rather than an F150?

Go to Connect to find information to help you calculate energy savings from various automobile models, and then compare that to the savings from running your computer for a year.

21

Solid, Toxic, and Hazardous Waste

▲ An endangered Hawaiian monk seal is disentangled from abandoned fishing nets in the Papahānaumokuākea Marine National Monument.
Source: Courtesy of Doug Helton, NOAA/NOS/ORR/ERD

Learning Outcomes

After studying this chapter, you should be able to:

21.1 Identify the components of solid waste, and describe our different methods of waste management.

21.2 Explain how we might shrink the waste stream.

21.3 Investigate hazardous and toxic wastes.

"We have no knowledge, so we have stuff; but stuff without knowledge is never enough."

– Greg Brown

Plastic Seas

The Papahānaumokuākea Marine National Monument is the largest marine reserve in the world, larger than all U.S. national parks combined. Established by President George W. Bush in 2006, the sanctuary was expanded by President Obama in 2016 to encompass a chain of islands, atolls, and reefs extending across 583,000 mi^2 northwest of Hawaii. The monument protects some of the most pristine and diverse deep coral reefs and over 7,000 marine species, including rare and endangered species such as the Laysan albatross and the Hawaiian monk seal. The string of isolated islets and coral atolls makes up the world's largest tropical seabird rookery, supporting 14 million nesting seabirds. The preserve is also home to a wealth of cultural and historic heritage sites, including shipwrecks and World Heritage cultural sites for native Hawaiians.

Despite its remote location, Papahānaumokuākea* remains vulnerable to the flotsam and jetsam of modern life. The islands and reefs lie within the vast circulating currents known as the Pacific gyre. These swirling currents, driven by winds and the Coriolis effect (chapter 15), concentrate nutrients, organic debris, and, in recent decades, an ocean of plastic trash. Often called the Great Pacific Garbage Patch or the Pacific Garbage Gyre, this region of floating plastic debris is really a drifting cloud of plastic particles, soda bottles and caps, disposable shopping bags, packaging, discarded fishing nets, and other debris. Much of it consists of tiny fragments floating just below the surface, but some pieces are large and recognizable, and some float 20–30 m deep. The greatest concentrations of plastic debris occur in the eastern Pacific, between California and Hawaii, and in the western Pacific near Japan. But the trash field extends across the ocean, with lesser aggregations near the Papahānaumokuākea preserve. Similar garbage patches have been identified in the Atlantic and elsewhere in the world's oceans, but the Pacific cases are the best studied.

Most of this plastic litter originates on land—plastic shopping bags blowing into streams and washing to the ocean, beverage bottles, take-out food containers, straws, and so on. Poor disposal and management of garbage—and rising use of single-use plastic—is the root cause. The problem is especially severe in developing regions, where garbage production is growing faster than waste management facilities. One study found that 90 percent of ocean plastic originated in Asia (40 percent in China).

The Pacific Garbage Gyre is thought to contain more than 100 million tons of plastic. In some areas, this debris outweighs the living biomass. Fish have been found with stomachs full of plastic fragments. Seabirds gulp down plastic fragments, then regurgitate them for their chicks. With stomachs blocked by indigestible bottle caps, disposable lighters, and other items, chicks starve to death. In one study of Laysan albatrosses, 90 percent of the carcasses of dead albatross chicks contained plastic fragments (fig. 21.1). Seals, turtles, porpoises, and seabirds become ensnared in ghost fishing nets and drown, or they die from ingesting indigestible materials. Oceanographers worry that this debris is slowly starving ocean ecosystems.

Surveys at sea and on beaches indicate that 50–80 percent of the floating material originates onshore. The rest is discarded or lost at sea. Stray shopping bags, drink containers, fast-food boxes, and other refuse fall from dumpsters, wash away from landfills, or are discarded on the street, then wash into storm sewers and streams. Eventually these items travel to the sea, where they gradually break into smaller pieces as they join the great global masses of ocean plastic.

The problem has been extraordinarily difficult to address because it is widespread, diffuse, abundant, and constantly replenished by careless or incomplete disposal of waste onshore and at sea. But growing awareness is starting to make a difference. Cleanup cruises in Papahānaumokuākea have collected more than 700 metric tons of discarded fishing gear that had clogged reefs.

FIGURE 21.1 A Laysan albatross chick, which died after being fed plastic debris rather than fish. Starvation after plastic ingestion is a leading cause of death for these albatross chicks.
Source: National Marine Sanctuary, photographer Claire Fackler/NOAA

In Papahānaumokuākea and elsewhere, marine debris has also caught the public's attention, and widespread beach cleanups are having a visible effect. And people are becoming active worldwide. According to the Ocean Conservancy, an organization that conducts regular coastal cleanups around the world and then measures and weighs the trash, nearly 9 million volunteers had collected almost 66,000 tons of debris between 1990 and 2015.

You can help protect the oceans, no matter how far from them you are. Next time you see plastic debris that might wash into a storm sewer, remember that everything ends up eventually in the

---

*Pronounced Pa-pa-ha-nao-Mo-kua-kea. To hear the pronunciation, visit the monument's website, www.papahanaumokuakea.gov.

ocean. Pick it up if you can, and try to prevent your own plastic bottles, caps, and packaging from escaping into waterways. (And never release helium balloons into the sky.) You can also try to reduce the amount of disposable containers, bottles, and packaging you buy.

The remote atolls of northwestern Hawaii show us that no place is too far away to be affected by our waste production and disposal. The materials we buy and the ways we manage our garbage can have dramatic impacts on living systems everywhere.

At the same time, responses to the problem have shown that people everywhere have an interest in taking care of the land and oceans and in keeping them beautiful. Often, the obstacles and volumes of waste seem insurmountable, but cleanup efforts in Hawaii have shown that progress can be real if we keep at it. In this chapter, we'll examine the waste we produce, our methods to dispose of it, and strategies to reduce, reuse, and recycle it.

21.1 WHAT DO WE DO WITH WASTE?

- *Municipal solid waste is the portion of the waste stream for which we are most directly responsible.*
- *Much of the world's waste continues to be disposed of inappropriately, especially e-waste and hazardous waste.*
- *Sanitary landfills dominate U.S. waste management, but a number of other options exist.*

Waste is everyone's business. We all produce wastes in nearly everything we do. Are there ways we can produce less and handle waste better? As you read this chapter, keep in mind the "three Rs:" reduce, reuse, and recycle. Where in our production and management of waste can we improve our performance on the three Rs?

Municipal waste—a combination of household and commercial refuse—amounts to roughly 250 million metric tons per year in the United States (fig. 21.2). That's approximately two-thirds of a ton for each of us every year, twice as much per capita as Europe or Japan and five to ten times as much as most developing countries.

Industrial waste, other than mining and mineral production, amounts to some 400 million metric tons per year in the United States. That makes around 1.2 tons per person per year. Most of this material is recycled, converted to other forms, destroyed, or disposed of in private landfills or deep injection wells. About 60 million metric tons of industrial waste falls into a special category of hazardous and toxic waste, which we will discuss later in this chapter.

Mineral processing waste (such as mine tailings and smelter slag) and agricultural waste (crop residue, manure) amounts to about 11 billion tons per year. Most of this is stored or recycled near its source, on mines or farms. When mismanaged, though, this waste can cause serious and widespread pollution. Runoff from uncontrolled agricultural waste, for example, is our single largest source of nonpoint air and water pollution.

The waste stream is everything we throw away

Think for a moment about how much we discard every year. There are organic materials, such as yard and garden wastes, food wastes, and sewage sludge from treatment plants; junked cars; worn-out furniture;

(a) Waste generation rates

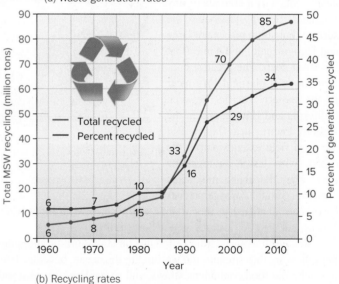

(b) Recycling rates

FIGURE 21.2 Bad news and good news in solid waste production in the United States. Per capita waste has risen to more than 2 kg/person/day, but it shows signs of declining (a). Recycling rates are rising (b), although they remain lower than in other developed economies.

(a) *Konstantin Faraktinov/123RF*

and consumer products of all types. Newspapers, magazines, catalogs, and office refuse make paper one of our major wastes (fig. 21.3). In spite of recent progress in recycling, many of the 200 *billion* metal, glass, and plastic food and beverage containers used every year in the United States end up in the trash. Wood, concrete, bricks, and glass come from construction and demolition sites, dust and rubble from landscaping and road building. All of this varied and voluminous waste has to arrive at a final resting place somewhere.

The **waste stream** is the steady flow of varied wastes that we all produce, from domestic garbage and yard wastes to industrial, commercial, and construction refuse. Many of the materials in our waste stream would be valuable resources if they were not mixed with other garbage. Unfortunately, our collecting and dumping processes mix and crush everything together, making separation an expensive and sometimes impossible task.

When we put hazardous materials in the waste stream, they become dispersed through thousands of tons of miscellaneous garbage. This is especially a problem when mixed waste is incinerated. When pesticides, batteries (zinc, lead, or mercury), cleaning solvents, smoke detectors containing radioactive material, plastics, and wet food scraps are burned with paper and other nontoxic materials, incineration can produce dioxins and PCBs, airborne heavy metals, and acidic particulates and aerosols.

Open dumps pollute air and water

Open, unregulated dumps were historically the way nearly everyone dealt with waste, and these are still the main disposal method in most developing countries (fig. 21.4). There, fast-growing

Quantitative Reasoning

Figure 21.2 shows an increase in waste production per capita. What is the percentage increase per capita from 1960 to 2015? (*Hint:* Calculate (4.4 − 2.7) ÷ 2.7.) What might account for this increase? Is there a relationship between waste production and our quality of life?

megacities have enormous garbage problems. Mexico City, one of the largest cities in the world, generates some 13,000 tons of trash *each day*. Most of this waste goes into giant landfills that aren't much different from open dumps. Thousands of people live and work on these mountains of refuse, scavenging for recyclable items or edible food scraps. The government would like to close these dumps, but finding other disposal methods has been a challenge.

Most developed countries now forbid open dumping, at least in metropolitan areas, but illegal dumping is still a problem. You may have seen trash accumulating along roadsides and in vacant, weedy lots in the poorer sections of cities. This is not just a question of aesthetics. Plastic and other waste products are long-lasting and sometimes dangerous. Discarded oil and solvents from cars, paints, and household chemicals are also toxic. An estimated 200 million liters of waste motor oil are poured into the sewers or allowed to soak into the ground every year in the United States. This is about five times as much as was spilled by the *Exxon Valdez* in Alaska in 1989. No one knows the volume of solvents and other chemicals disposed of by similar methods.

Ocean dumping is mostly uncontrolled

The oceans are vast, but they're not large enough to absorb our waste without harm. Calculating the amount of plastic we put in the oceans is difficult, but a 2015 study by Jenna Jambeck and

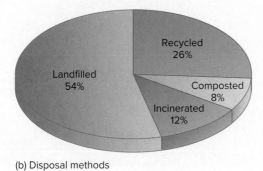

FIGURE 21.3 Composition of municipal solid waste in the United States by weight, before recycling, and disposal methods.
Source: Data from U.S. Environmental Protection Agency, Office of Solid Waste Management, 2015

FIGURE 21.4 Trash disposal has become a crisis in the developing world, where people have adopted cheap plastic goods and packaging but lack good recycling or disposal options.
William P. Cunningham

FIGURE 21.5 Plastic trash dumped on land and at sea kills wildlife as well as contaminating remote beaches.
sOulsurfing - Jason Swain/Getty Images

colleagues estimated that between 5 and 13 million tons of plastic waste enters the oceans every year. This trash washes up beaches even in remote regions (fig. 21.5). Another 150,000 tons of fishing gear—including more than 1,000 km (660 mi) of nets—are lost or discarded at sea each year. An estimated 50,000 northern fur seals are entangled in this refuse and drown or starve to death every year in the North Pacific alone (see opening case study).

Much of the problem is mismanaged onshore disposal, as well as carelessly discarded litter and uncontained garbage, but there is also deliberate disposal at sea. Until recently, many U.S. cities dumped municipal refuse, industrial waste, sewage, and sewage sludge into the ocean. Federal legislation now prohibits this dumping. New York City, the last to stop offshore sewage sludge disposal, finally ended this practice in 1992. Oceangoing cruise ships and container ships also deliberately, and illegally, dispose of waste at sea when no one is watching.

Marine ecologists are now discovering that microplastics and nanoparticles pose especially insidious threats to marine ecosystems. Microplastics, fragments less than 5 mm long, often result from the breakdown of larger plastic fragments. But they are also used as components of facial scrubs, toothpaste, ceramic glazes, and other consumer products. These tiny particles wash down the drain, and too small to be captured by wastewater treatment plants, they enter rivers and lakes, and eventually the ocean. When marine organisms consume these fragments and die, the marine food web can become depleted.

Nanoparticles, with a diameter of less than 100 nanometers (nm), are small enough to be taken up by cells in your body or by organisms in the environment. They can carry hazardous substances, adhered to their surface, which then enter cells. We're not yet certain what the ecological or health effects of these materials are, but there are serious concerns about their growing presence in freshwater and marine ecosystems.

Oceanographers are trying to find ways of controlling this debris, but most is too fine-grained to capture easily, and it is distributed widely around the world's oceans. Growing awareness and prevention are the best strategies for resolving the problem.

Landfills receive most U.S. waste

Of the municipal solid waste produced in U.S. cities, we send about 54 percent to landfills, recycle about 33 percent, and incinerate 13 percent. We have a long way to go in controlling waste, but this is a dramatic change from 1960, when 94 percent was landfilled and only 6 percent was recycled.

A modern **sanitary landfill** is designed to keep waste contained. Operators are required to compact the refuse and cover it every day with a layer of dirt, to decrease smells and windblown litter and to discourage insects and rats. This cover helps control pollution, but the dirt fill also takes up as much as 20 percent of landfill space. Since 1994, all U.S. landfills have been required to control hazardous substances such as oil, chemical compounds, and toxic metals, which seep through piles of waste along with rain water. To prevent leakage to groundwater and streams, landfills require an impermeable clay and/or plastic lining (fig. 21.6). Drainage systems are installed in and around the liner to catch drainage and to help monitor chemicals that leak out. Modern municipal solid waste landfills now have many of the safeguards of hazardous waste repositories described later in this chapter.

Sanitary landfills also must manage methane, a combustible gas produced when organic material decomposes in the anaerobic conditions deep inside a landfill. Methane is also a key greenhouse gas, and landfills are one of the largest anthropogenic sources of methane in the United States. Globally, landfills produce more than 700 million metric tons of methane annually. Because methane is over 30 times as potent at absorbing heat as CO_2, this represents about 12 percent of all greenhouse gas emissions. Until recently almost all this landfill methane was simply vented into the air. Now about half of all landfill gas in the United States is either flared (burned) on site or collected and used as

FIGURE 21.6 A plastic liner being installed in a sanitary landfill. This liner and a bentonite clay layer below it prevent leakage to groundwater. Trash is also compacted and covered with earth fill every day.
Doug Sherman/Geofile

Quantitative Reasoning

Examine figure 21.2 closely. How has waste production changed from 1960 to today, in pounds per person? What are some of the reasons for this change? Think of someone you know who lived in the United States in 1960. Ask them if they can explain the difference in production per person. Ask these same questions about recycling, in terms of percentage recycled.

fuel to generate electricity. Methane recovery in the United States produces 440 trillion Btu per year and is equivalent to removing 25 million vehicles from the highway. Some operators now pump water into landfills in order to speed up production of this valuable fuel.

Historically, landfills provided a convenient and cheap means of disposal. This was because we had much less waste to deal with—we produced only a third as much in 1960 as today—and because there were few regulations about disposal sites and methods. Since 1984, when stricter financial and environmental protection requirements for landfills took effect, roughly 90 percent of landfills in the United States have closed. Nearly all of those landfills lacked environmental controls to keep toxic materials from leaking into groundwater and surface waters. Many areas now suffer groundwater and stream contamination from decades-old unregulated dump sites.

With new rules for public health protection, landfills are becoming fewer, larger, and more expensive. Cities often truck their garbage hundreds of miles for disposal, a growing portion of the $10 billion per year that we spend to dispose of trash. A decade from now, it may cost us $100 billion per year to dispose of our trash and garbage. On the other hand, rising landfill costs make it more economical to pursue alternatives strategies, including waste reduction and recycling.

We often export waste to countries ill-equipped to handle it

Most industrialized nations agreed to stop shipping hazardous and toxic waste to less-developed countries in 1989, but the practice still continues. In 2006, for example, 400 tons of toxic waste was illegally dumped at 14 open dumps in Abidjan, the capital of the Ivory Coast. The black sludge—petroleum wastes containing hydrogen sulfide and volatile hydrocarbons—killed ten people and injured many others. At least 100,000 city residents sought medical treatment for vomiting, stomach pains, nausea, breathing difficulties, nosebleeds, and headaches. The sludge—which had been refused entry at European ports—was transported by an Amsterdam-based multinational company on a Panamanian-registered ship and handed over to an Ivoirian firm (thought to be connected to corrupt government officials) to be dumped in the Ivory Coast. The Dutch company agreed to clean up the waste and pay the equivalent of (U.S.) $198 million to settle claims.

Most of the world's obsolete ships are now dismantled and recycled in poor countries. The work is dangerous, and old ships often are full of toxic and hazardous materials, such as oil, diesel fuel, asbestos, and heavy metals. On India's Alang Beach, for example, more than 40,000 workers tear apart outdated vessels using crowbars, cutting torches, and even their bare hands. Metal is hauled away for recycling. Plastic and oil waste is often simply burned on the beach, where ashes and oily residue wash back into the water.

Discarded electronics, or **e-waste,** is one of the greatest sources of toxic material currently going to developing countries. At least 2 billion television sets and personal computers are in use globally. Televisions often are discarded after only about 5 years, while computers, cell phones, and other electronics become obsolete even faster. An estimated 50 million tons of e-waste is discarded every year worldwide. Only about 20 percent of the components are recycled, and the rest generally goes to open dumps or landfills. This waste stream contains at least 2.5 billion kg of lead, as well as mercury, gallium, germanium, nickel, palladium, beryllium, selenium, arsenic, and valuable metals, such as gold, silver, copper, and steel.

Until recently, most of this e-waste went to China, where villagers, including young children, broke it apart to retrieve valuable metals. Often this scrap recovery was done under primitive conditions where workers had little or no protective gear. Health risks in this work are severe, especially for growing children. Soil, groundwater, and surface-water contamination at these sites is extremely high. Food grown in contaminated soils often contains toxic levels of lead and other metals.

Shipping e-waste to China is now officially banned, although illegal smuggling continues. In 2018, China extended its ban on trash imports to many more categories, including many materials commonly collected for recycling in the United States (What Do You Think?, p. 484).

With tighter regulation in China, informal e-waste recycling has shifted to India, Ghana, and other countries where weak environmental regulation make irresponsible waste handling profitable (fig. 21.7a). Adding to the difficulty of this problem, these developing areas will soon be producing more e-waste than wealthier countries with better regulation (fig. 21.7b). Will those developing countries be able to defend public health with this increase in waste production?

One group monitoring global e-waste shipments and working conditions is the Basel Action Network. Their studies, including GPS tracking of waste shipments, continues to find increasing amounts of illicit e-waste exports around the world. This includes increasing shipments from China to Southeast Asia and elsewhere.

Incineration produces energy from trash

Faced with growing piles of garbage and a lack of available landfills at any price, many cities have built waste incinerators to burn municipal waste. Incineration reduces the volume of waste by 80 to 90 percent. Residual ash and unburnable residues, representing 10 to 20 percent of the original volume, usually go to a landfill for disposal, but landfilling costs are greatly reduced.

Most incinerators do **energy recovery,** using the heat derived from incinerated refuse to heat nearby buildings or to produce

Who Will Take Our Waste?

Have you ever thrown something into recycling that you shouldn't have—perhaps a greasy pizza box or a partially filled soda can? If you have, you aren't alone. According to industry professionals, wishful recyclers are their biggest problem. A high percentage of our recycling stream is contaminated with items that shouldn't be there.

Most of our recycling is shipped overseas, rather than processed at home, so contaminated recycling is an international issue. Since the 1990s, China has been both the world's manufacturer and its leading waste importer. Trash recycling provided valuable resources for China's manufacturing sector along with thousands of low-skill (and low-paying) jobs. Recyclable waste also fills container ships that would otherwise return empty to China, after delivering consumer goods to North America or Europe. But recycling without safety regulations often releases toxic or dangerous materials. Air and water quality along with workers' health have suffered.

In 2018, China announced that it would no longer accept 24 kinds of low-quality, or dangerous, solid waste, including mining slag, unsorted paper, plastic scrap, discarded textiles, and a variety of commonly accepted recycling products. China is trying to clean up its environment and it wants to show its citizens that progress is being made. Furthermore, China has become a world power, and it no longer wants to be the world's waste dump.

To enforce these new restrictions, China is requiring much stricter examination of waste coming into the country. A batch of recycling with more than 1 percent contamination will be rejected, a much lower threshold than the 15 percent contamination usually accepted in the United States.

Until this ban, China accepted more than half of all the world's used plastic and paper. The United States, which is the world's leading waste source, shipped about 19 million tons of waste paper and 2 million tons of plastic to China in 2017. China's restrictions on foreign garbage imports have thrown American waste collectors and shippers into panic. If we can't send our waste to China, what will we do with it? In many places, landfills are filling up rapidly, and new sites are in short supply.

Could the American public be persuaded to use fewer disposable or single-use products? Could we start sorting recycling more carefully? Could you be more vigilant about what you throw in the recycling bin? Or will we simply start shipping our waste to some other poor country in Asia or Africa where standards are lower and people are more desperate? What do you think, what is the most ethical solution to this form of waste imperialism?

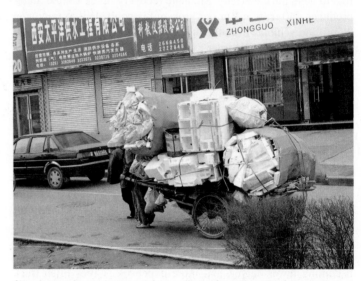

China's recycling industry employs millions, formally and informally, but still struggles to meet demand.
William P. Cunningham

(a)

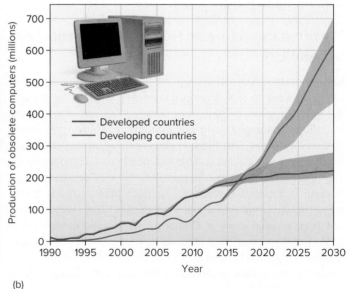

(b)

FIGURE 21.7 Demanufacturing of e-waste often occurs in places with weak health and environmental protections, as here in Accra, Ghana (a). Demanufacturing exposes workers and communities to hazardous materials, but production of e-waste is growing, with an increasing share from developing countries (b).
(a) Maniglia Romano/Pacific Press/LightRocket via Getty Images; (b) Source: Yu et al. Environmental Science and Technology *(2010).*

steam and generate electricity. Internationally, well over 1,000 waste-to-energy plants in Brazil, Japan, and western Europe generate energy. In some areas, this is a major source of municipal heat and electricity production.

Municipal incinerators are specially designed plants capable of burning thousands of tons of waste per day. To burn cleanly, refuse needs to be sorted. Unburnable or recyclable materials must be removed, especially metals, batteries, wallboard that contains sulfur, and wet material. This sorted waste is called **refuse-derived fuel,** because the enriched burnable fraction has a higher energy content than mixed trash. Another approach, called a **mass burn,** is to dump everything into a giant furnace, unsorted, and burn as much as possible (fig. 21.8). This technique avoids the expensive and unpleasant job of sorting, but it produces more unburned ash, and often it produces more air pollution as plastics, batteries, and other mixed substances are burned.

The cost-effectiveness of garbage incinerators is the subject of persistent debates. Initial construction costs are high—usually between $100 million and $300 million for a typical municipal facility. Ironically, one worry about incinerators is whether enough garbage will be available to feed them. Incinerators also compete with recyclers for paper and plastics, which are high-energy fuel. Cities usually have contracts guaranteeing certain amounts of waste daily. Some communities with successful recycling programs have had to buy garbage from neighbors to meet contractual obligations with waste-to-energy facilities.

Incinerators work best when regulatory agencies, such as the EPA, can ensure that they burn well and that emissions are closely monitored. Clean emissions require that municipalities sort their waste. Wet organic material (such as food) burns poorly and should be composted; batteries from toys and electronics produce airborne metals; incompletely burned plastics produce carcinogenic dioxins and furans; gypsum board, which contains sulfur, produces sulfur dioxide.

European countries rely heavily on incineration (fig. 21.9), in part because they lack landfill space. Studies of hazardous air pollutants in Europe have shown that although municipal incinerators were major sources of hazardous emissions in 1990, modern incinerators are among the smallest sources now. They now produce one-eightieth as much emissions as metal processors do, and one-twentieth as much as household fireplaces emit, because of tighter permitting processes, better inspections, and high burn temperatures.

To prevent pollution emissions, municipal waste systems in many European cities sort waste to remove both recyclables and toxic materials. Increasingly, European cities are banning plastics from incinerator waste and requiring households to separate plastics from other garbage. Waste separation is expected to eliminate

FIGURE 21.8 A municipal garbage incinerator burns waste material, ideally sorted to remove unburnables. Steam produced in the boiler can be used to generate electricity or heat nearby buildings.

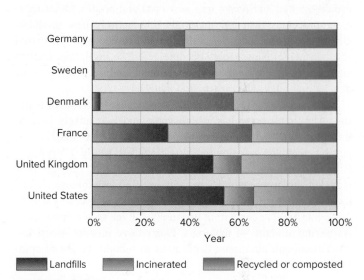

Landfills Incinerated Recycled or composted

FIGURE 21.9 Reliance on landfills, recycling/composting, and incineration varies considerably among countries.
Source: Data Source Eurostat; U.S. EPA 2014

nearly all dioxins and other combustion by-products. Separation also helps avoid the expense of installing costly pollution-control equipment.

In places with weaker oversight, it may be less certain that pollution-control equipment is correctly used or maintained. Opponents of incineration argue that we should be investing in more recycling rather than building expensive incinerators to deal with waste.

Section Review

1. Why is e-waste difficult to dispose of properly?
2. How does a sanitary landfill differ from an open dump?
3. Why is it good to divert yard waste and organic material from the general waste stream?
4. Why do many northern European countries incinerate much of their waste?

21.2 SHRINKING THE WASTE STREAM

- *Reducing waste is the most important of the three Rs.*
- *Recycling and reuse save materials and energy.*
- *Composting is an important component of recycling.*

Reducing waste in the first place can save money right through the waste disposal chain. For the waste we produce, the best way to handle it may be to re-imagine it as a resource. In nature, everything is reused, and the waste produced by one organism is a resource for another. Can we do the same with our own wastes?

In terms of solid waste management, **recycling** is the reprocessing of discarded materials into new, useful products (fig. 21.10). This contrasts with *reusing* materials, such as refillable beverage containers, which are returned to a factory to be refilled. Some recycling processes reuse materials for the same purposes; for instance, old aluminum cans and glass bottles are usually melted and recast into new cans and bottles. Other recycling processes turn old materials into entirely new products. Old tires, for instance, are shredded and turned into rubberized road surfacing. Newspapers become cellulose insulation, kitchen wastes become a valuable soil amendment, and steel cans become new automobiles and construction materials.

How much we recycle usually depends on the value of material. Aluminum scrap has a high value and is relatively light to transport. It's also expensive to produce from raw materials. A large percentage of aluminum is recycled nearly everywhere (fig. 21.11). About half of all aluminum beverage cans are now recycled, up from only 15 percent in 1970. Aluminum can be recycled rapidly: Half of the cans now on grocery shelves will be made into another can within two months. Even so, we throw away staggering amounts of materials. Every three months Americans landfill enough aluminum drink cans to rebuild the entire commercial airline fleet.

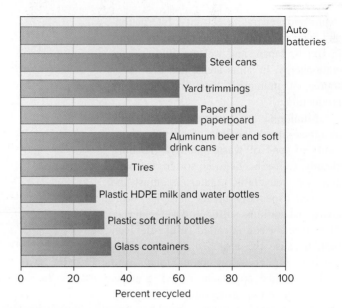

FIGURE 21.11 Recycling rates for selected materials in the United States. Car battery recycling is required by law, to keep lead out of the waste stream.
Source: U.S. Environmental Protection Agency 2015

Recycling has multiple benefits

Recycling saves money, energy, raw materials, and landfill space, while also reducing pollution. Recycling also encourages individual awareness and responsibility for the refuse produced. Household sorting is the bedrock of many recycling systems, but many cities now have single-stream recycling, in which mixed, (but clean!) recyclable materials are delivered to sorting facilities. Sorting at a centralized site eliminates the problem of poor sorting at home. It can also increase participation because it is convenient for users.

Curbside pickup of recyclables costs around $35 per ton, as opposed to the $80 paid to dispose of them at an average metropolitan landfill. Many recycling programs cover their own expenses with materials sales and may even bring revenue to the community. Although landfills continue to dominate American waste disposal, recycling (including composting) has quadrupled since 1980 (fig. 21.12).

Recycling cuts our waste volumes drastically. Philadelphia has invested in neighborhood collection centers that recycle enough to eliminate the need for a previously planned, high-priced incinerator. New York, which exports its daily 11,000 tons of waste by truck, train, and barge, to New Jersey, Pennsylvania, Virginia, South Carolina, and Ohio, has set ambitious recycling goals of 50 percent waste reduction, although the city still recycles less than 30 percent of its household and office waste. In contrast, Minneapolis and Seattle recycle nearly 60 percent of domestic waste, Los Angeles and Chicago over 40 percent. In 2002, New York Mayor Michael Bloomberg raised a national outcry by canceling most of the city's recycling program. He argued that the program didn't pay for itself and the money should be spent to balance the city's budget. A year later, Bloomberg relented after

FIGURE 21.10 Creating a stable, economically viable market for recycled products is essential for recycling success. Consumers can help by buying recycled products.
David Trevor/Alamy Images

realizing that it cost more to ship garbage to Ohio than to recycle. Recycling was reinstated for nearly all recyclable materials.

Japan is probably the world's leader in recycling. Short of land for landfills, Japan recycles about half its municipal waste and incinerates about 30 percent. The country has begun a push to increase recycling, because incineration costs almost as much. Some communities have raised recycling rates to 80 percent, and others aim to reduce waste altogether by 2020. This level of recycling is most successful when waste is well sorted. In Yokohama, a city of 3.5 million, there are now 10 categories of recyclables, including used clothing and sorted plastics. Some communities have 30 or 40 categories for sorting recyclables.

Perhaps most important, recycling lowers our demands for raw resources. In the United States, we cut down 2 million trees every day to produce newsprint and paper products, a heavy drain on our forests. Recycling the print run of a single Sunday issue of *The New York Times* would spare 75,000 trees. Recycling 1 ton of aluminum saves 4 tons of bauxite (aluminum ore) and 700 kg (1,540 lbs) of petroleum coke and pitch, as well as keeping 35 kg (77 lbs) of aluminum fluoride out of the air.

Recycling also reduces energy consumption and air pollution. Plastic bottle recycling can save 50 to 60 percent of the energy needed to make new ones. Making new steel from old scrap offers up to 75 percent energy savings. Producing aluminum from scrap instead of bauxite ore cuts energy use by 95 percent, yet we still throw away more than a million tons of aluminum every year. If aluminum recovery were doubled worldwide, more than a million tons of air pollutants would be eliminated every year.

Recycling has countless benefits, but it's critical to avoid "wishful recycling." That is, don't throw stuff in the recycling bin just because you hope it's recyclable, or because you don't have a trash bin nearby. As you know by now, a single piece of unrecyclable material can make a whole dumpsterful worthless. If in doubt, throw it out! Put questionable items in the garbage, and you'll help recycling in the long run.

Plastic recycling rates are low

We often fail to recycle plastic because we are hasty, or recycling isn't convenient. But another part of the problem is that plastic is tricky to reuse and recycle. Plastic recycling is easily contaminated if materials are mixed. Most of the 24 billion plastic soft drink bottles sold every year in the United States are made of PET (polyethylene terephthalate), which can be melted and remanufactured into carpet, fleece clothing, plastic strapping, and nonfood packaging. However, even a trace of vinyl—a single PVC (polyvinyl chloride) bottle in a truckload, for example—can make PET useless.

Although most bottles are now marked with a recycling number, it's hard for consumers to remember which is which. Another

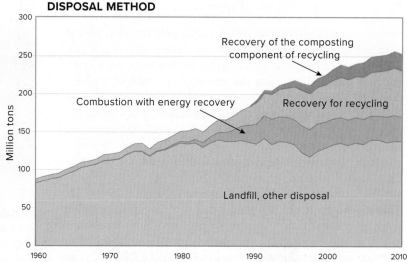

FIGURE 21.12 Disposal of municipal solid waste, 1960–2010. Landfills remain the dominant destination, but recycling and composting are increasing.
Source: U.S. Environmental Protection Agency 2012

obstacle is that many soft drink bottles are sold and consumed on the go, and never make it into recycling bins. As a consequence, Americans have an extremely low recovery rate for plastics (fig. 21.13).

Reducing litter is an important benefit of recycling. Ever since disposable paper, glass, metal, foam, and plastic packaging began to accompany nearly everything we buy, these discarded wrappings have collected on our roadsides and in our lakes, rivers, and oceans. Without incentives to properly dispose of beverage cans, bottles, and papers, it often seems easier to just toss them aside when we have finished using them. Litter is a costly as well as unsightly problem. The United States pays an estimated 32 cents for each piece of litter picked up by crews along state highways, which adds up to $500 million every year. "Bottle bills" requiring deposits on bottles and cans have reduced littering in many states.

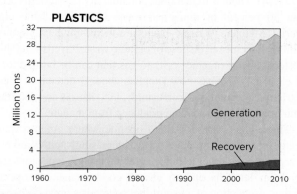

FIGURE 21.13 Recycling of plastics in the United States is improving but remains extremely low. This helps explain why so much plastic ends up in the ocean.
Source: U.S. Environmental Protection Agency 2012

Recycling has financial obstacles

Despite the efficiency of recycling, public policies often tend to favor extraction of new raw materials: Energy, water, and raw materials are often sold to industries below their real cost, in order to create jobs and stimulate the economy. A pound of recycled clear PET, the material in most soft drink bottles, is worth about 40¢, while a pound of virgin PET costs about 25¢. Setting the prices of natural resources at their real cost would tend to encourage efficiency and recycling.

Price fluctuations are a constant challenge for businesses trying to make an income on recycling. As with any primary materials, prices can vary dramatically. Some years, copper is so valuable that copper pipes, wires, and flashing are stolen from houses; other years, prices and demand are low. States and cities have often helped to stabilize markets by requiring government agencies to purchase a minimum amount of recycled materials. Each of us can play a role in creating markets, as well. If we buy things made from recycled materials—or ask for them if they aren't available—we will help make it possible for recycling programs to succeed. Growing world demand, with expanding consumer economies around the world, is also likely to increase the value of recycled materials.

Compost and biogas are useful products

Composting is a type of recycling that converts organic waste to soil-enriching organic fertilizer. Many U.S. cities have municipal compost facilities, which keep yard clippings and leaves out of landfills and provide gardeners with a valuable soil amendment. Composting uses natural aerobic (oxygen-rich) decomposition to reduce organic debris to a nutrient-rich soil amendment. Home gardeners have always composted yard and garden waste in their backyards, but cities and towns increasingly see this as a way to reduce waste disposal costs. Organic debris such as yard waste makes up 13 percent of the waste we generate (see fig. 21.3). Almost two-thirds of our yard waste is composted.

While compost is a useful material, its market value is modest. Another option is biogas, or methane, which is produced by fermenting organic waste in air-tight, liquid-filled containers (fig. 21.14). There is tremendous potential in these systems because about 30 percent of municipal waste is biodegradable organic material. Methane "biogas" plants are increasingly common ways for European cities to dispose of organic waste. Over 14,500 biogas digesters convert organic waste to gas that can be burned for heat and electricity. Biogas now supplies about 3 percent of German electricity consumption. Increasingly, U.S. dairies are also starting to run operations on methane produced from manure.

These plants are easiest to operate where organic waste is already separated from other waste. Increasingly, cities are requiring separation in order to reduce greenhouse gas emissions (methane) from landfills and to ensure good incinerator operation. Efficient production of a useful fuel source is a great co-benefit.

Demolition and construction debris is another major source of waste. Every year, thousands of tons of debris from building sites heads to landfills, but recycling facilities are beginning to collect,

FIGURE 21.14 Organic waste from agriculture, households, and food-processing factories is converted to methane (biogas) in these enclosed digesting tanks. This system diverts a vast amount of waste from landfills while producing gas that is used to power city buses and to heat city buildings. *Global Warming Images/Shutterstock*

sort, and resell increasing portions of this debris. Recyclers routinely sort and resell 97 percent of demolition debris, diverting it from landfills. This work is economical partly because of the large scale of materials, and partly because there are relatively unmixed resources. Trees are ground up and converted to mulch for landscaping. Dirt from stumps is screened and sold as clean garden soil. Mixed materials are sorted into recyclable glass, metals, and plastics. Construction debris is sorted and ground: Broken drywall is ground to fresh gypsum, which is sold to drywall producers; wood is composted or burned; and bricks are crushed for fill and construction material. Organic waste that can't be separated, such as food-soaked paper, can be sent to a gasifier, an enclosed, oxygen-free pressure cooker, which converts biomass to natural gas. Gas can be used to power or heat the recycling facility. The 3 percent of incoming waste that doesn't get recycled is mainly mixed plastics, which currently are landfilled.

Appliances and e-waste must be demanufactured

Demanufacturing is the disassembly and recycling of obsolete products, such as TV sets, computers, refrigerators, and air conditioners. As we mentioned earlier, electronics and appliances are among the fastest-growing components of the global waste stream. Americans throw away about 54 million household appliances, such as stoves and refrigerators, 12 million computers, and uncounted cell phones each year. Most office computers are used only 3–5 years; televisions last 5 years or so; refrigerators last longer, an average of 12 years. In the United States, an estimated 300 million computers await disposal in storage rooms and garages.

Demanufacturing is key to reducing the environmental costs of e-waste and appliances. A single personal computer can contain 700 different chemical compounds, including toxic materials (mercury, lead, and gallium), and valuable metals (gold, silver, copper), as well as brominated fire retardants and plastics. A typical personal computer has about $6 worth of gold, $5 worth of copper, and $1 of silver. Approximately 40 percent of lead entering

U.S. landfills, and 70 percent of heavy metals, comes from e-waste. Batteries and switches in toys and electronics make up another 10 to 20 percent of heavy metals in our waste stream. These contaminants can enter groundwater if computers are landfilled, or the air if they are incinerated. When collected, these materials can become a valuable resource—and an alternative to newly mined materials.

To reduce these environmental hazards, the European Union now requires cradle-to-grave responsibility for electronic products. Manufacturers now have to accept used products or fund independent collectors. An extra $20 (less than 1 percent of the price of most computers) is added to the purchase price to pay for collection and demanufacturing. Manufacturers selling computers, televisions, refrigerators, and other appliances in Europe must also phase out many of the toxic compounds used in production. Japan is rapidly adopting European environmental standards, and some U.S. companies are following suit, in order to maintain their international markets. In the United States, at least 29 states have passed, or are considering, legislation to control disposal of appliances and computers, in order to protect groundwater and air quality.

Reuse is more efficient than recycling

Even better than recycling or composting is cleaning and reusing materials in their present form, thus saving the cost and energy of remaking them into something else. We do this already with some specialized items. Auto parts are regularly sold from scrap yards, especially for older car models. In some areas, stained glass windows, brass fittings, fine woodwork, and bricks salvaged from old houses bring high prices. Some communities sort and reuse a variety of materials received in their dumps (fig. 21.15).

In some places, glass and plastic bottles are routinely returned to beverage producers for washing and refilling. The reusable, refillable bottle is the most efficient beverage container we have. This is better for the environment than remelting and more

profitable for local communities. A reusable glass container makes an average of 15 round-trips between factory and customer before it becomes so scratched and chipped that it has to be recycled. Reusable containers also favor local bottling companies and help preserve regional differences.

Since the advent of cheap, lightweight, disposable food and beverage containers, many small, local breweries, canneries, and bottling companies have been forced out of business by huge national conglomerates. These big companies can afford to ship food and beverages great distances as long as it is a one-way trip. If they had to collect their containers and reuse them, canning and bottling factories serving large regions would be uneconomical. Consequently, the national companies favor recycling rather than refilling, because they prefer fewer, larger plants and don't want to be responsible for collecting and reusing containers. In some circumstances, life-cycle assessment shows that washing and decontaminating containers takes as much energy and produces as much air and water pollution as manufacturing new ones.

Reducing waste is the best option

Most of our attention in waste management focuses on recycling. But slowing the consumption of throw-away products is by far the most effective way to save energy, materials, and money. In the 3R waste hierarchy—reduce, reuse, recycle—the most important strategy is reuse. Industries are increasingly finding that reducing saves money. Soft drink makers use less aluminum per can than they did 20 years ago, and plastic bottles use less plastic. Individual action is essential too (What Can You Do?, below).

Excess packaging of food and consumer products is one of our greatest sources of unnecessary waste. Paper, plastic, glass, and metal packaging material make up 50 percent of our domestic trash by volume. Much of that packaging is primarily for marketing and has little to do with product protection (fig. 21.16).

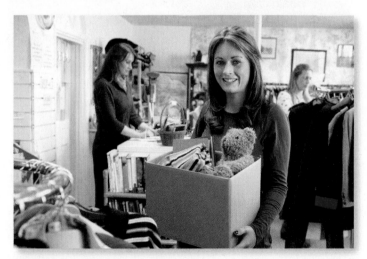

FIGURE 21.15 Reusing discarded products is a creative and efficient way to reduce wastes. Donating things you no longer need helps others and reduces clutter in your life. Shopping at garage sales and thrift stores can save you money and help our environment.
SpeedKingz/Shutterstock

What Can You Do?

Reducing Waste

1. Help your school establish zero-waste (or low-waste) policies.
2. Buy foods that come with less packaging; shop at farmers' markets or co-ops, using your own containers.
3. Carry and use your own refillable beverage container, instead of buying new ones.
4. Avoid single-use plastic shopping bags. Try to buy products in recyclable containers.
5. Minimize use of hard-to-recycle products such as aluminum foil.
6. Separate your cans, bottles, papers, and plastics for recycling.
7. Compost yard and garden wastes, leaves, and grass clippings.
8. Write to your elected representatives and urge them to vote for container deposits, recycling, and safe incinerators or landfills.

Source: Minnesota Pollution Control Agency

FIGURE 21.16 How much more do we need? Where will we put what we already have?

JIM BORGMAN/Cincinnati Enquirer. Reprinted with permission of UNIVERSAL UCLICK. All rights reserved.

Manufacturers and retailers might be persuaded to reduce these wasteful practices if consumers ask for products without excess packaging. Canada's National Packaging Protocol (NPP) recommends that packaging minimize depletion of virgin resources and production of toxins in manufacturing. The preferred hierarchy is (1) no packaging, (2) minimal packaging, (3) reusable packaging, and (4) recyclable packaging.

We can reduce the volume of waste in landfills somewhat by using materials that are compostable or degradable. **Photodegradable plastics** break down when exposed to ultraviolet radiation in sunlight. **Biodegradable plastics** incorporate materials such as cornstarch that can be decomposed by microorganisms. These degradable plastics are an imperfect solution, though. Many don't decompose completely; they only break down to small particles that remain in the environment. In doing so, they can release toxic chemicals into the environment. In modern, lined landfills, they cannot decompose at all. Furthermore, they make recycling less feasible and may lead people to believe that littering is okay.

Section Review

1. Describe some general trends in landfilling, recycling, and incineration of waste.
2. Why are plastics difficult to recycle?
3. What is demanufacturing, and why is it used for some types of waste?

21.3 HAZARDOUS AND TOXIC WASTES

- *Hazardous waste is dangerous in small doses, easily combustible, or corrosive.*
- *Handling and cleaning up hazardous waste is very expensive, so federal legislation enforces cleanup and safe handling.*
- *Hazardous waste can be used, recycled, decontaminated, or stored permanently.*

The most dangerous aspect of the waste stream we have described is that it often contains highly toxic and hazardous materials that are injurious to both human health and environmental quality. We now produce and use a vast array of flammable, explosive, caustic, acidic, and highly toxic chemical substances for industrial, agricultural, and domestic purposes (fig. 21.17). According to the EPA, industries in the United States generate about 265 million metric tons of *officially* classified hazardous wastes each year, slightly more than 1 ton for each person in the country. In addition, considerably more toxic and hazardous waste material is generated by industries or processes not regulated by the EPA. At least 40 million metric tons (22 billion lbs) of toxic and hazardous wastes are released into the air, water, and land in the United States each year. The biggest source of these toxins are the chemical and petroleum industries (fig. 21.18).

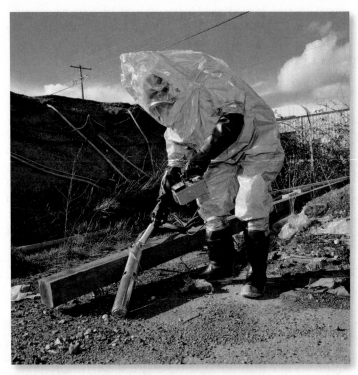

FIGURE 21.17 Hazardous waste is dangerous even with small exposure. Here a worker tests for radioactive soil.

Arthur S. Aubry/Getty Images

FIGURE 21.18 Producers of hazardous wastes in the United States.

Source: U.S. EPA

Legally, a **hazardous waste** is any discarded material, liquid or solid, that contains substances known to be (1) fatal to humans or laboratory animals in low doses, (2) toxic, carcinogenic, mutagenic, or teratogenic to humans or other life-forms, (3) ignitable with a flash point less than 60°C, (4) corrosive, or (5) explosive or highly reactive (undergoes violent chemical reactions either by itself or when mixed with other materials). Notice that this definition includes both toxic and hazardous materials as defined in chapter 8. Certain compounds are exempt from regulation as hazardous waste if they are accumulated in less than 1 kg (2.2 lbs) of commercial chemicals or 100 kg of contaminated soil, water, or debris. Even larger amounts (up to 1,000 kg) are exempt when stored at an approved waste treatment facility for the purpose of being beneficially used, recycled, reclaimed, detoxified, or destroyed.

Hazardous waste must be recycled, contained, or detoxified

To protect the public from hazardous wastes, most is recycled, converted to nonhazardous forms, stored, or disposed of on-site by the generators. Most producers of these wastes are chemical companies, petroleum refiners, and other large industrial facilities. Still, the hazardous waste that does enter the waste stream or the environment represents a serious environmental problem.

Orphan wastes, abandoned by industries that have gone out of business, are an especially tricky threat to both environmental quality and human health. For years, little attention was paid to this material. Forgotten wastes stored on private property, buried, or allowed to soak into the ground were considered of little concern to the public. An estimated 5 billion metric tons of highly poisonous chemicals were improperly disposed of in the United States between 1950 and 1975, before regulatory controls became more stringent.

Federal legislation requires waste management

Two important federal laws regulate hazardous waste management and disposal in the United States. The Resource Conservation and Recovery Act (RCRA, pronounced "rickra") of 1976 is a comprehensive program that requires rigorous testing and management of toxic and hazardous substances. A complex set of rules requires generators, shippers, users, and disposers of these materials to keep meticulous account of everything they handle and what happens to it from generation (cradle) to ultimate disposal (grave) (fig. 21.19).

The Comprehensive Environmental Response, Compensation and Liability Act (CERCLA or Superfund Act), passed in 1980 and modified in 1984 by the Superfund Amendments and Reauthorization Act (SARA), is aimed at rapid containment, cleanup, or remediation of abandoned toxic waste sites. This statute authorizes the Environmental Protection Agency to undertake emergency actions when a threat exists that toxic material will leak into

EPA or state agency office

Secure landfill

Transporter

Hazardous waste generator

Transporter

Treatment facility

FIGURE 21.19 Toxic and hazardous wastes must be tracked from "cradle to grave" by detailed shipping manifests.

Quantitative Reasoning

What proportion of hazardous waste producers are chemical and petroleum industries (fig. 21.18)? What proportion are metal processing and mining? What kinds of contaminants are most likely to occur at these different types of sites? Can you think of two or three locations where chemical and petroleum industries might be concentrated?

the environment. The agency is empowered to bring suit for the recovery of its costs from potentially responsible parties such as site owners, operators, waste generators, or transporters.

SARA also established that communities have a right to know about toxic substances that are produced or used nearby, which could be released into air or water. To give the public access to this information, SARA established a reporting system, the **Toxic Release Inventory.** More than 20,000 manufacturing facilities are required to report annually on the use, release, or transfer of toxic substances. This inventory is not always maintained completely, but it is the best available source of public information on exposure risk. You can find it on the EPA website and see what potential sources are in your neighborhood.

The government does not have to prove that anyone violated a law or what role they played in a Superfund site. Rather, liability under CERCLA is "strict, joint, and several," meaning that anyone associated with a site can be held responsible for the entire cost of cleaning it up no matter how much of the mess they made. In some cases, property owners have been assessed millions of dollars for removal of wastes left there years earlier by previous owners. This strict liability has been a headache for the real estate and insurance business, but it also allows for protection of public health.

Superfund sites are listed for federal cleanup

The EPA estimates that there are at least 36,000 seriously contaminated sites in the United States. The General Accounting Office (GAO) places the number much higher, perhaps more than 400,000 when all are identified. By 2016, some 1,170 sites were on the National Priority List (NPL) for cleanup with financing from the federal Superfund program. This number was down from 1,680 sites eight years earlier. The decline represents removal of sites from the list, mostly due to the completion of cleanup projects. Fifty-two new sites had been proposed for NPL listing.

The **Superfund** is a revolving pool designed to (1) provide an immediate response to emergency situations that pose imminent hazards, and (2) to clean up or remediate abandoned or inactive sites. Without this fund, sites would languish for years or decades while the courts decided who was responsible to pay for the cleanup. Originally, a $1.6 billion pool, the fund peaked at $3.6 billion. From its inception, the fund was financed by taxes on producers of toxic and hazardous wastes. Industries opposed this "polluter pays" tax, because current manufacturers are often not the ones responsible for the original contamination. In 1995, Congress agreed to let the tax expire. Since then, the Superfund has dwindled, and the public has picked up an increasing share of the bill. In the 1980s, the public covered less than 20 percent of the Superfund; now public tax dollars from the general fund cover most of the cost of toxic waste cleanup, in combination with fines collected from polluters (fig. 21.20).

Reliance on general funds makes cleanup progress vulnerable to political winds in Congress. In some years, Congress is able and willing to cover the costs needed to protect the public from exposure to hazardous substances; in other years, it can't or won't provide sufficient funding.

Total costs for hazardous waste cleanup in the United States are estimated between $370 billion and $1.7 trillion, depending on how clean sites must be and what methods are used. For years, Superfund money was spent mostly on lawyers and consultants, and cleanup efforts were often bogged down in disputes over liability and the best cleanup methods. Progress depends largely on the priorities of the president and his aids: During the Reagan and the first Bush administrations, there was little investment in cleanup; the Clinton administration significantly accelerated work: From 1993 to 2000, the number of completed NPL cleanups jumped from 155 to 757, almost half the 1,680 sites on the list at that time. From 2000 to 2008, progress slowed again, due to underfunding and a lack of interest from the federal administration. After 2008, the Obama administration accelerated cleanup activity again.

What qualifies a site for the NPL? These sites are considered to be especially hazardous to human health and environmental quality because they are known to be leaking or have a potential for leaking supertoxic, carcinogenic, teratogenic (causing birth defects), or mutagenic (cancer causing) substances (see chapter 8, Environmental Health and Toxicology for discussion of these terms). The ten substances of greatest concern or most commonly detected at Superfund sites are lead, trichloroethylene, toluene,

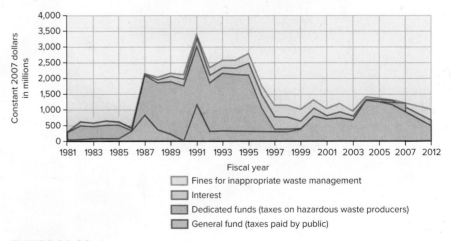

FIGURE 21.20 Congress has gradually shifted the funding sources for the Superfund, used to clean up toxic and hazardous waste sites, from industrial producers of hazardous waste to taxpayers.

benzene, PCBs, chloroform, phenol, arsenic, cadmium, and chromium. These and other hazardous or toxic materials are known to have contaminated groundwater at 75 percent of the sites now on the NPL. In addition, 56 percent of these sites have contaminated surface waters, and airborne materials are found at 20 percent of the sites.

Where are these thousands of hazardous waste sites, and how did they get contaminated? Old industrial facilities such as smelters, mills, petroleum refineries, and chemical manufacturing plants are highly likely to have been sources of toxic wastes. Regions of the country with high concentrations of aging factories such as the "rust belt" around the Great Lakes or the Gulf Coast petrochemical centers have large numbers of Superfund sites (fig. 21.21). Mining districts also are prime sources of toxic and hazardous waste. Within cities, factories and places such as railroad yards, bus repair barns, and filling stations where solvents, gasoline, oil, and other petrochemicals were spilled or dumped on the ground often are highly contaminated.

Some of the most infamous toxic waste sites were old dumps where many different materials were mixed together indiscriminately. For instance, Love Canal in Niagara Falls, New York, was an open dump used by both the city and nearby chemical factories as a disposal site. More than 20,000 tons of toxic chemical waste was buried under what later became a housing development. Another infamous example occurred in Hardeman County, Tennessee, where about a quarter of a million barrels of chemical waste were buried in shallow pits that leaked toxins into the groundwater. In other sites, liquid wastes were pumped into open lagoons or abandoned in warehouses.

Studies of populations living closest to Superfund and toxic release inventory sites reveal that minorities often are overrepresented in neighborhoods near waste sites. Charges of environmental racism have been made, but they are difficult to prove conclusively.

- Hazardous waste site
- Aquifers

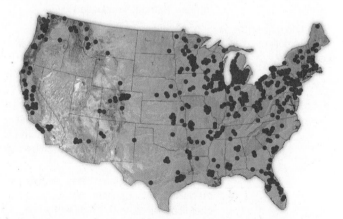

FIGURE 21.21 Hazardous waste sites are often located on aquifer recharge zones, making groundwater contamination a common risk.
Source: U.S. EPA

Brownfields present both liability and opportunity

Many cities have unused or abandoned industrial sites, with soil or water contaminated with toxic and hazardous substances. Because these contaminants are health risks, it is unsafe—and legally risky—to rebuild on these **brownfields** without expensive cleanup projects. As a result, brownfields remain unused eyesores, and a drain on a city's productive tax base, for years or decades. Brownfield redevelopment has become a priority for many cities, and in recent decades federal funding has become available to help pay for the millions of dollars of remediation before new building can even begin.

One of the thorny problems with brownfield redevelopment is the question of how clean it needs to be. Must a site be cleaned to the point that young children could play in the dirt safely, and food could be grown in the soil? Is it possible to clean a site somewhat without planning for every unexpected use far in the future? Might rules become more strict in the future?

These questions were tested in a brownfield site in Columbia, Mississippi. For many years, the 35-ha (81-acre) site in Columbia was used for turpentine and pine tar manufacturing. Soil tests showed concentrations of phenols and other toxic organic compounds exceeding federal safety standards. The site was added to the Superfund National Priority List, and remediation was ordered. Some experts argued that the best solution was to simply cover the surface with clean soil and enclose the property with a fence to keep people out. The total costs would have been about $1 million.

Instead, the EPA ordered Reichhold Chemical, the last known property owner, to excavate more than 12,500 tons of soil and haul it to a commercial hazardous waste dump in Louisiana at a cost of some $4 million. The intention was to make the site safe enough to be used for any purpose, and to remove risk from exposure to anybody, even children, who might be exposed to soil on the site.

Similarly, in places where contaminants have seeped into groundwater, the EPA generally demands that cleanup be carried to drinking water standards, because we can't be sure that people won't be drinking the water sometime in the future. Many critics believe that these pristine standards are unreasonable. Former congressman Jim Florio, a principal author of the original Superfund Act, says, "It doesn't make any sense to clean up a rail yard in downtown Newark so it can be used as a drinking water reservoir."

Over time, brownfield cleanup policies have found compromise positions. Some brownfields are designated specifically as business sites, not for residential use. For organic chemicals, soil might be dug up and put into a mobile incinerator on-site, to break down contaminants, rather than trucked away for storage. Sometimes a visible, impermeable membrane can be laid over a contaminated site, then covered by clean soil (fig. 21.22). The membrane prevents access to contaminated soil, alerts future excavators to the unsafe deeper layers, and allows rapid rebuilding on the site. These steps don't really clean a site, but they do contain the hazards and prevent further contamination. They also allow cities to redevelop near the urban core, and to slow sprawl, with its many environmental costs, at the urban margin.

What Can You Do?

Alternatives to Hazardous Household Chemicals

Most household cleaning solutions can be replaced with inexpensive, non-toxic, safe alternatives. Here are some examples.

Chrome cleaner: Use vinegar and nonmetallic scouring pad.

Floor cleaner: Mop floors with 1 cup vinegar mixed with 2 gallons of water.

Silver polish: Rub with toothpaste on a soft cloth.

Furniture polish: Rub with olive, almond, or lemon oil.

Drain opener: Use a plumber's snake; pour boiling water in drain; keep fats and hair out of drains.

Upholstery cleaner: Clean stains with club soda.

Window cleaner: Mix 1/3 cup ammonia, 1/4 cup white vinegar in 1 quart warm water. Spray on window. Wipe with soft cloth.

Toilet cleaner: Pour 1/2 cup vinegar or liquid chlorine bleach into toilet bowl. Let stand for 30 minutes, scrub with brush, flush.

Garden pest control: Spray plants with soap-and-water solution for aphids, mealybugs, and other insects. Plant pest-repellent plants such as marigolds, coriander, thyme, yarrow, rue, and tansy. Introduce natural predators such as ladybugs or lacewings.

Hazardous waste can be recycled or contained

We have four general strategies for managing hazardous waste: reduce, recycle, convert to less hazardous forms, and contain safely in perpetuity. Some of these options apply to our personal uses, as well as to public and industrial uses of substances. At home, you can reduce the amount of toxic materials you buy, and you can buy just as much as you need, not extra. You can replace some toxic substances with safer and cheaper alternatives (What Can You Do?). You can't recycle household hazardous materials, but you can take them to a collection center for proper, safe disposal (table 21.1).

In public and industrial contexts, too, the safest and cheapest way to minimize hazardous waste problems is to avoid creating the wastes in the first place. Manufacturing processes can be modified to reduce or eliminate waste production. In Minnesota, the 3M Company reformulated products and redesigned manufacturing processes to eliminate more than 140,000 metric tons of solid and hazardous wastes, 4 billion liters (1 billion gal) of wastewater, and 80,000 metric tons of air pollution each year. Often, these new processes also saved money by using less energy and fewer raw materials.

Recycling and reusing materials can also eliminate hazardous wastes and pollution. Many waste products of one process or industry are valuable commodities in another. Already, about 10 percent of the wastes that would otherwise enter the waste stream in the United States are sent to surplus material exchanges where they are sold as raw materials for use by other industries. This figure could probably be raised substantially with better waste management. In Europe, at least one-third of all industrial wastes are exchanged through clearinghouses where beneficial uses are found. This represents a double savings: The generator doesn't have to pay for disposal, and the recipient pays less for raw materials.

Substances can be converted to safer forms

Several processes are available to convert hazardous materials to less dangerous forms. *Physical treatments* capture or isolate substances. Charcoal or resin filters absorb toxins. Distillation separates hazardous components from aqueous solutions. Precipitation

FIGURE 21.22 This brownfield site on the Hudson River has soil contaminated with arsenic from lumber treatment. To make the site safe for reuse, contaminated soil is being covered with an orange, impermeable (waterproof) sheet, which is then being covered by several feet of clean soil.
Mary Ann Cunningham

Table 21.1 How Should You Dispose of Household Hazardous Waste?	
Flush to sewer system (drain or toilet)	Cleaning agents with ammonia or bleach, disinfectants, glass cleaner, toilet cleaner
Put dried solids in household trash	Cosmetics, putty, grout, caulking, empty solvent containers, water-based glue, fertilizer (without weed killer)
Save and deliver to a waste collection center	*Solvents:* cleaning agents (drain cleaner, floor wax-stripper, furniture polish, metal cleaner, oven cleaner), paint thinner and other solvents, glue with solvents, varnish, nail polish remover
	Metals: mercury thermometers, button batteries, NiCad batteries, auto batteries, paints with lead or mercury, fluorescent light bulbs/tubes/ballasts, electronics and appliances
	Poisons: bug spray, pesticides, weed killers, rat poison, insect poison, mothballs. Other chemicals: antifreeze, gasoline, fuel oil, brake fluid, transmission fluid, paint, rust remover, hair spray, photo chemicals

Source: U.S. EPA

and immobilization in ceramics, glass, or cement isolate toxins from the environment so they become essentially nonhazardous. One of the few ways to dispose of metals and radioactive substances is to fuse them in silica at high temperatures to make a stable, impermeable glass that is suitable for long-term storage.

Incineration is a quick way to dispose of many kinds of hazardous waste. Incineration is not necessarily cheap—nor always clean—unless it is done correctly. Wastes must be heated to over 1,000°C (2,000°F) for a sufficient length of time to complete destruction. The ash resulting from thorough incineration is reduced in volume up to 90 percent and often is safer to store in a landfill or other disposal site than the original wastes. Incineration remains highly controversial, though, if only because it has a long history of releasing toxic contaminants to the air (fig. 21.23).

Several sophisticated features of modern incinerators improve their effectiveness. Liquid injection nozzles atomize liquids and mix air into the wastes so they burn thoroughly. Fluidized bed burners force air up from the bottom of the furnace, separating and aerating waste, to allow for rapid and complete combustion. Before heading to the smoke stack, gases are passed by electrically charged plates, which capture particulates in the flue gases.

Chemical processing can transform materials so they become non-toxic. Included in this category are neutralization, removal of metals or halogens (chlorine, bromine, etc.), and oxidation. The Sunohio Corporation of Canton, Ohio, for instance, has developed a process called PCBx in which chlorine in such molecules as PCBs is replaced with other ions that render the compounds less toxic. A portable unit can be moved to the location of the hazardous waste, eliminating the need for shipping them.

Biological waste treatment or **bioremediation** taps the great capacity of microorganisms to absorb, accumulate, and detoxify a variety of toxic compounds (Exploring Science, p. 496). Bacteria in activated sludge basins, aquatic plants (such as water hyacinths or cattails), soil microorganisms, and other species remove toxic materials and purify effluents. Recent experiments have produced bacteria that can decontaminate organic waste metals by converting them to harmless substances. After the *Deepwater Horizon* oil spill in the Gulf of Mexico in 2010, most of the waste remediation was performed by naturally occurring bacteria, which were adapted to metabolizing oil compounds. Bioremediation holds exciting possibilities for addressing many organic pollutants, including oils, PCBs, and other toxic compounds.

Permanent storage is often needed

Inevitably, there will be some materials we can't destroy, make into something else, or otherwise cause to vanish. We will have to store them out of harm's way. There are differing opinions about how best to do this.

Retrievable storage allows for access later, in case we need to recover wastes or use new treatment methods. Often this means **permanent retrievable storage.** Waste storage containers are placed in a secure building, salt mine, or bedrock cavern where they can be inspected periodically and retrieved, if necessary, for repacking or for transfer if a better means of disposal is developed. This technique is more expensive than burial in a landfill because the storage area must be guarded and monitored continuously to prevent leakage, vandalism, or other dispersal of toxic materials.

Remedial measures are much cheaper with this technique, however, and it may be the best system in the long run.

Secure landfills are one of the most popular solutions for hazardous waste disposal. Although, as we saw earlier in this chapter, many such landfills have been environmental disasters; newer techniques make it possible to create safe, modern **secure landfills** that are acceptable for disposing of many hazardous wastes. The first line of defense in a secure landfill is a thick, bottom cushion of compacted clay that surrounds the pit like a bathtub (fig. 21.24).

FIGURE 21.23 Actor Martin Sheen joins activists protesting a hazardous waste incinerator in East Liverpool, Ohio. Concerns over hazardous emissions worry residents near incinerators.
Piet van Lier

FIGURE 21.24 A secure landfill for toxic waste. A thick plastic liner and two or more layers of impervious compacted clay enclose the landfill. A gravel bed between the clay layers collects any leachate, which can then be pumped out and treated. Well samples are tested for escaping contaminants, and methane is collected for combustion.

EXPLORING SCIENCE

Phytoremediation: Cleaning Up Toxic Waste with Plants

Getting contaminants out of soil and groundwater is one of the most widespread and persistent problems in waste cleanup. Once leaked into the ground, solvents, metals, radioactive elements, and other contaminants are dispersed and difficult to collect and treat. The main method of cleaning up contaminated soil is to dig it up, then decontaminate it or haul it away and store it in a landfill in perpetuity. At a single site, thousands of tons of tainted dirt and rock may require incineration or other treatment. Cleaning up contaminated groundwater usually entails pumping vast amounts of water out of the ground—hopefully extracting the contaminated water faster than it can spread through the water table or aquifer. In the United States alone, there are tens of thousands of contaminated sites on factories, farms, gas stations, military facilities, sewage treatment plants, landfills, chemical warehouses, and other types of facilities. Cleaning up these sites is expected to cost at least $700 billion.

Recently, a number of promising alternatives have been developed using plants, fungi, and bacteria to clean up our messes. *Phytoremediation* (remediation, or cleanup, using plants) can include a variety of strategies for absorbing, extracting, or neutralizing toxic compounds. Certain types of mustards and sunflowers can extract lead, arsenic, zinc, and other metals (*phytoextraction*). Poplar trees can absorb and break down toxic organic chemicals (*phytodegradation*). Reeds and other water-loving plants can filter water tainted with sewage, metals, or other contaminants. Natural bacteria in groundwater, when provided with plenty of oxygen, can neutralize contaminants in aquifers, minimizing or even eliminating

the need to extract and treat water deep in the ground. Radioactive strontium and cesium have been extracted from soil near the Chernobyl nuclear power plant using common sunflowers.

How do the plants, bacteria, and fungi do all this? Many of the biophysical details are poorly understood, but in general, plant roots are designed to efficiently extract nutrients, water, and minerals from soil and groundwater. The mechanisms involved may aid extraction of metallic and organic contaminants. Some plants also use toxic elements as a defense against herbivores—locoweed, for example, selectively absorbs elements such as selenium, concentrating toxic levels in its leaves. Absorption can be extremely effective. Braken fern growing in Florida was found to contain arsenic at concentrations more than 200 times higher than the soil in which it was growing.

Plants can absorb, concentrate, and even decompose toxic contaminants in soil and groundwater.

Genetically modified plants are also being developed to process toxins. Poplars have been grown with a gene borrowed from bacteria that transform a toxic compound of mercury into a safer form. In another experiment, a gene for producing mammalian liver enzymes, which specialize in breaking down toxic organic compounds, was inserted into tobacco plants. The plants succeeded in producing the liver enzymes and breaking down toxins absorbed through their roots.

These remediation methods are not without risks. As plants take up toxins, insects could consume leaves, allowing contaminants to enter the food web. Some absorbed contaminants are volatilized, or emitted in gaseous form, through pores in plant leaves. Once toxic contaminants are absorbed into plants, the plants themselves are usually toxic and must be landfilled. But the cost of phytoremediation can be less than half the cost of landfilling or treating toxic soil, and the volume of plant material requiring secure storage ends up being a fraction of a percent of the volume of the contaminated dirt.

Cleaning up hazardous and toxic waste sites will be a big business for the foreseeable future, both in the United States and around the world. Innovations such as phytoremediation offer promising prospects for business growth as well as for environmental health and saving taxpayers' money.

(header photo): David L. Hansen

Moist clay is flexible and resists cracking if the ground shifts. It is impermeable to groundwater and will safely contain wastes. A layer of gravel is spread over the clay liner, and perforated drain pipes are laid in a grid, so as to collect any seepage that escapes from the stored material. A thick polyethylene liner, protected from punctures by soft padding materials, covers the gravel bed. A layer of soil or absorbent sand cushions the inner liner and the wastes are packed in drums, which then are placed into the pit, separated into small units by thick berms of soil or packing material.

When the landfill has reached its maximum capacity, a cover much like the bottom sandwich of clay, plastic, and soil—in that order—caps the site. Vegetation stabilizes the surface and improves its appearance. Sump pumps collect any liquids that filter through the landfill, either from rainwater or leaking drums. This leachate is treated and purified before being released. Monitoring wells check groundwater around the site to ensure no toxins have escaped.

Transportation of hazardous wastes to disposal sites is of concern because of the risk of accidents. Emergency preparedness officials conclude that the greatest risk in most urban areas is not nuclear war or natural disaster but crashes involving trucks or trains carrying hazardous chemicals through densely occupied urban corridors. Another worry is who will bear financial responsibility for abandoned waste sites and for landfills over the long term. As is the case with nuclear wastes (see chapter 19, Conventional Energy), we may need new institutions to ensure perpetual care of these wastes.

Section Review

1. What is a legal definition of hazardous waste?
2. What are brownfields? Why are they both important and difficult to clean up?
3. How might hazardous wastes be converted to safer forms?
4. What is bioremediation?

Connecting the Dots

Waste is a global problem. Each year, we consume more materials and produce more waste. Finding ways to dispose of all our garbage and hazardous substances is a constant challenge. Open dumping releases hazardous materials into the air and water. Everything we dump into the sewer eventually ends up in the ocean. Millions of tons of discarded plastic accumulate in giant gyres killing wildlife and contaminating beaches. In the United States, recycling rates are improving, but we still landfill more than half our municipal waste. Many other countries, especially those short on landfill space, recycle over half their waste. Modern landfills seek to keep trash from contaminating air and groundwater. These sites are a great improvement over the past, but they are often remote from major cities, which must transport garbage long distances for disposal. Incineration is a costly but widely used alternative to landfilling.

Waste disposal is expensive, but our policies are often better set up for landfilling or incinerating waste than for recycling. Government policies and economies of scale make it cheaper and more convenient to extract virgin raw materials than to reuse or recycle. But the increasing toxicity of modern products, including e-waste, makes it more urgent that we reduce, reuse, and recycle

materials worldwide. For decades, China recycled more than half the world's used plastic and paper, but in 2018 the Chinese government announced that it would no longer accept low-quality, or contaminated solid waste. This leaves waste collectors in a panic. What will we do with our waste? Could we do a better job of separating our wastes? Strategies and opportunities for recycling are expanding, including reuse markets, bioenergy generation, deconstructing demolition debris, and composting yard waste. Creating gold from garbage—making money from biogas, compost, and efficient waste disposal—is a growing industry, especially in Europe.

Hazardous and toxic waste remain a serious health threat and environmental risk. Many abandoned and derelict sites must be cleaned up by the Superfund, although that fund has dwindled since it lost its main source of support, industry contributions, in 1995. These sites threaten public health, especially for minority groups, and are a serious problem. At the same time, we are producing new hazardous substances that require safe disposal. We can all help by thinking carefully about what we buy, use, and dispose of in our own communities.

Reviewing Key Terms

Can you define the following terms in environmental science?

biodegradable plastics 21.2

bioremediation 21.3

brownfields 21.3

composting 21.2

demanufacturing 21.2

energy recovery 21.1

e-waste 21.1

hazardous waste 21.3

mass burn 21.1

permanent retrievable storage 21.3

photodegradable plastics 21.2

recycling 21.2

refuse-derived fuel 21.1

sanitary landfills 21.1

secure landfills 21.3

Superfund 21.3

Toxic Release Inventory 21.3

waste stream 21.1

Critical Thinking and Discussion Questions

1. A toxic waste disposal site has been proposed for the Pine Ridge Indian Reservation in South Dakota. Many tribal members oppose this plan, but some favor it because of the jobs and income it will bring to an area with 70 percent

unemployment. If local people choose immediate survival over long-term health, should we object or intervene?

2. There is often a tension between getting your personal life in order and working for larger structural changes in society.

Evaluate the trade-offs between spending time and energy sorting recyclables at home compared to working in the public arena on a bill to ban excess packaging.

3. Should industry officials be held responsible for dumping chemicals that were legal to dump when they did it but are now known to be extremely dangerous? When can we argue that they should have known about the hazards involved?

4. Look at the discussion of recycling or incineration presented in this chapter. List the premises (implicit or explicit) that underlie the presentation, as well as the conclusions (stated or not) that seem to be drawn from them. Do the conclusions necessarily follow from these premises?

5. The Netherlands incinerates much of its toxic waste at sea by a shipborne incinerator. Would you support this as a way to dispose of our wastes as well? What are the critical considerations for or against this approach?

Data Analysis

How Much Do You Know about Recycling?

As people become aware of waste disposal problems in their communities, more people are recycling more materials. Some things are easy to recycle, such as newsprint, office paper, or aluminum drink cans. Other things are harder to classify. Most of us give up pretty quickly and throw things in the trash if we have to think too hard about how to recycle them.

1. Take a poll to find out how many people in your class know how to recycle the items in the table shown at the end of these questions. Once you have taken your poll, convert the numbers to percentages: Divide the number who know how to recycle each item by the number of students in your class, and then multiply by 100.

2. Now find someone on your campus who works on waste management. This might be someone in your university or college administration, or it might be someone who actually empties trash containers. (You might get more interesting and straightforward answers from the latter.) Ask the following questions: (1) Can this person fill in the items your class didn't know about? (2) Is there a college/university policy about recycling? What are some of the points on that policy? (3) How much does the college spend each year on waste disposal?

How many tuition payments does that total? (4) What are the biggest parts of the campus waste stream? (5) Does the school have a plan for reducing that largest component?

Item	Percentage Who Know How to Recycle
Newspapers	
Paperboard (cereal boxes)	
Cardboard boxes	
Cardboard boxes with tape	
Plastic drink bottles	
Other plastic bottles	
Styrofoam food containers	
Food waste	
Plastic shopping bags	
Plastic packaging materials	
Furniture	
Last year's course books	
Leftover paint	

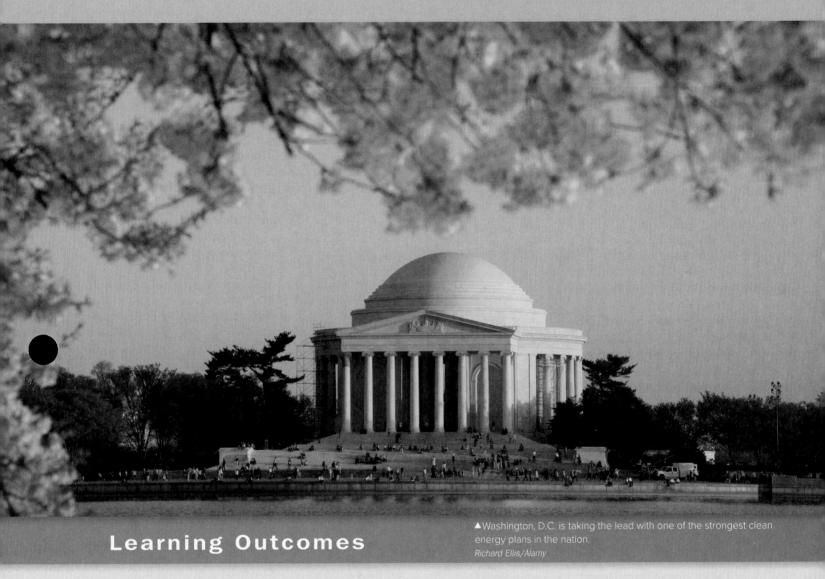

22

Urbanization and Sustainable Cities

▲ Washington, D.C. is taking the lead with one of the strongest clean energy plans in the nation.
Richard Ellis/Alamy

Learning Outcomes

After studying this chapter, you should be able to:

22.1 Describe trends in urban growth, and reasons for it.
22.2 Understand urban challenges in the developing world.
22.3 Identify urban challenges in the developed world.
22.4 Explain smart growth and urban sustainability.

*"What kind of world do you want to live in?
Demand that your teachers teach
you what you need to know
to build it."*

– Peter Kropotkin

Cities Show the Way in Climate Policy

Cities are taking the lead in combating climate change. Even when national-level leadership has fallen short, cities and towns everywhere are stepping up to the challenge. Although the United States was a world leader when the Kyoto Protocol on climate change was negotiated in the 1990s, our support for climate action has been mixed and often backsliding. In 2017, President Trump declared his intention to withdraw from the Paris Accord, in which 195 nations agreed to reduce greenhouse gas emissions. When agencies in his government issued a major report detailing the dangers of climate change, Trump dismissed the science in the report, saying, "I don't believe it."

In response to the climate denial at the federal level, over 1,000 American mayors have pledged to do their part in honoring the promise to cut greenhouse emissions 26 percent by 2025. States, businesses, and institutions have signed similar pledges. The Mayor's Climate Protection Agreement sponsors a variety on climate-related reports and projects.

In 2018, the Washington, D.C., city council unanimously approved the most ambitious urban energy policy in the country. The plan requires utility providers to generate 100 percent of their energy supply from renewable sources by 2032. Hawaii and California have both pledged state-wide goals of 100 percent renewable energy for electricity by 2045, and several smaller cities have already reached 100-percent goals, but Washington is by far the largest individual city to make such a commitment.

The city cites several motivations for this plan, including improvements in jobs and the economy, health and wellness of residents, and equity for diverse populations, as well as climate and the environment. Washington, D.C., is one of the most diverse cities in the country. It's home to more than 170 different nationalities and ethnic groups. Extreme wealth and poverty exist side by side among its 650,000 residents. Highly paid government officials and lobbyists share the city with families caught in chronic unemployment. The poverty rate is 18.6 percent, significantly higher than the U.S. average, but the median property value is nearly $600,000, about three times the national average. The city is home to some of the most prestigious educational institutions in the country, but its public schools have only a 56 percent graduation rate. These disparities make it difficult to accomplish grand policy initiatives.

FIGURE 22.1 Washington, D.C.'s Mayor Muriel Browser.
Source: Government of the District of Columbia. Muriel Browser, Mayor

Much of the credit for recent climate progress in the city belongs to the mayor, Muriel Browser (fig. 22.1). Serving since 2015, the mayor has worked to create pathways to the middle class for all citizens in her city. She has taken bold steps to end homelessness and to create a more transparent and open government. She has supported green buildings, promoted energy and water efficiency, and fostered renewable energy projects including financing for solar and wind power projects for both commercial buildings and private homes.

Much of the new plan focuses on improving buildings, which account for 74 percent of D.C.'s carbon emissions. The new policy assists with building performance and sets efficiency standards for large buildings. The city is also assisting residents with buying renewable electricity, including the option of aggregated purchasing that saves money as well as cutting carbon. The plan also addresses public transportation. By 2045, all public transportation and privately owned vehicle fleets in the city will be electric-powered and will emit zero carbon.

How will this plan be financed? Utility companies serving the District already collect fees from customers who use natural gas and electricity. Some of these fees will be used to fund energy efficiency. Utilities that fail to meet standards for renewable energy will be assessed penalties that will go into the city's Renewable Energy Development Fund (REDF). Money from this fund will help low-income families pay energy bills and upgrade energy efficiency—lowering household energy costs. Job training in renewable energy fields will also be subsidized by REDF as will efficiency upgrades in city-owned affordable housing.

All these steps show what can be done when leaders have a positive vision for their communities. And when policies are informed by environmental science. Even when federal leadership is captured by oil money and other special interests, cities can make real change. New policies like climate action are challenging, but they are exciting when they focus on common interests of the diverse communities in the city.

For these and many other reasons, cities are critical players as we try to answer our persistent environmental challenges. In this chapter, we examine ways cities can be designed to make them environmentally sound, socially equitable, and economically sustainable.

Benchmark Data

Among the ideas and values in this chapter, the following are a few worth remembering.

60	Percentage of world population in cities by 2030
10	Number of world's largest cities in Europe and North America, 1900
0	Number in 2100
7 million	Population of Lagos, Nigeria, in 2000
88 million	Projected population of Lagos in 2100
25 cm/year	Rate at which Jakarta, Indonesia, is sinking
#1	Jakarta's ranking among fastest sinking cities in the world
600%	Growth in area of Las Vegas, while population tripled

FIGURE 22.2 Shanghai grew from 13 million to 24 million in just 20 years. The Pudong district above was mostly farmland in 1990.
charleschen/Shutterstock

22.1 URBANIZATION

- *The world is becoming more urban, and cities are growing larger.*
- *The largest cities are now in developing countries.*
- *Economic policies can push people toward cities.*

Cities are places of crisis and opportunity. All the challenges of development and uneven wealth occur there. But in times of social and political change, cities are places where ideas flow and evolve easily. This is part of the reason climate action and climate planning happen mainly in urban areas like Washington, D.C. Innovative thinkers can share knowledge and resources to resolve common problems.

Cities also are far more efficient than suburbs and rural areas in terms of resource use and environmental impact per person, at least in wealthy countries. Concentrating populations in urban areas leaves open space available for farming and biodiversity. Living together in apartment buildings saves energy. Cities can offer public transportation, and they provide amenities within walking or biking distance. This greater efficiency also makes climate action goals easier to envision and to achieve.

But cities also have concentrations of poverty, pollution, and displaced members of society. Environmental quality in cities can be poor, especially in low-income areas. And in developing areas, cities can have a relatively large ecological footprint, as urban residents often have more money than rural people, and so can buy more consumer goods.

More than half the world's population now lives in urban areas. Many cities are burdened by pollution and poverty, but cities are also places of opportunity and creativity. Most young people today plan to live in cities, because of their educational, employment, and cultural resources. You may or may not plan to live in a city, but understanding and building better cities, and making them more just and socially equitable, is essential if we are to live sustainably, and if we are to protect the environmental quality, biodiversity, and ecological services on which we depend.

For most of human history, the majority of people lived in rural areas or small villages. Since the beginning of the Industrial Revolution about 300 years ago, cities have grown rapidly in both size and power (fig. 22.2). By 2050, over two-thirds of us will live in urban areas (table 22.1)—in fact, many of the world's regions already are 70 to 80 percent urban.

Demographers predict that 90 percent of the human population growth in this century will occur in developing countries, and that almost all of that growth will occur in cities (fig. 22.3). Already, huge **urban agglomerations** (multiple expanding and converging municipalities) are forming throughout the world. Some 370 million people live in urban areas with populations of 10–30 million people, which are often called **megacities.** This explosive growth is a challenge in developing areas, where cities often fail to keep up with rapidly growing demand for housing, fresh water, sanitation, and other services.

Cities have specialized functions

Since their earliest origins, cities have been centers of education, religion, commerce, recordkeeping, communication, and political power. As cradles of civilization, cities have influenced

Table 22.1	Urban Share of Total Population (Percentage)		
	1950	**2000**	**2030***
Africa	18.4	35.0	48.4
Asia	19.3	37.5	56.7
Europe	56.0	71.1	77.5
Latin America	40.0	75.5	83.6
North America	63.9	79.1	84.7
Oceania	32.0	68.3	68.9
World	38.3	46.7	60.4

*Projected

Source: Data from United Nations Population Division 2018.

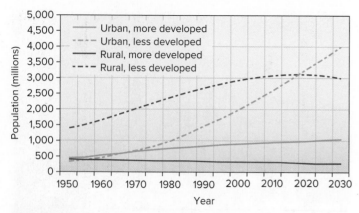

FIGURE 22.3 Growth of urban and rural populations in more-developed regions and in less-developed regions.
Source: United Nations Population Division

culture and society far beyond their proportion of the total population (fig. 22.4). Until about 1900, only a small percentage of the world's people lived permanently in urban areas, and even the greatest cities of antiquity were small by modern standards.

Just what makes up an urban area or a city? Definitions differ. The U.S. Census Bureau considers any incorporated community to be a city, regardless of size, and defines any city with more than 2,500 residents as urban. More meaningful definitions are based on *functions*. In a **rural area,** most residents depend on agriculture or other ways of harvesting natural resources for their livelihood. In an **urban area,** a majority of people work in commerce, manufacturing, governance, or information-based occupations.

A **village** is a collection of rural households linked by culture, custom, family ties, and association with the land (fig. 22.5). A **city,** by contrast, is larger and has more diversified populations, cultures, economies, and functions. While the rural village often gives a sense of security and connection, small communities may

FIGURE 22.4 Since their earliest origins, cities have been centers of education, religion, commerce, politics, and culture. They have also been centers of resource consumption, pollution, and inequality.
Lefteris/Getty Images

FIGURE 22.5 This village in Chiapas, Mexico, is closely tied to the land through culture, economics, and family relationships. While the timeless pattern of life here gives a great sense of identity, it can also be stifling and repressive.
William P. Cunningham

also resist changes that might lead to economic or social progress. A city offers more freedom to experiment, to be upwardly mobile, and to break from restrictive traditions, for better and for worse.

Large cities are expanding rapidly

Very large cities are becoming more numerous, and they are shifting to developing areas. In 1900, only 13 cities in the world had populations over 1 million (table 22.2). All of those cities except Tokyo and Beijing were in more developed countries in Europe and North America. London was the only city in the world with more than 5 million residents. By 2019, there were over 500 cities with more than 1 million residents. Fifteen metropolitan areas had more than 20 million residents, none of them in Europe. By 2035, most of the world's largest cities will be in Africa or Asia (fig. 22.6). In just the next 25 years, Mumbai and Delhi (India), Karachi (Pakistan), Manila (Philippines), and Jakarta (Indonesia) are all expected to grow by at least 50 percent.

Large, regional urban agglomerations, including tens of millions of people, are increasingly common. In the United States, a region stretching from Boston to Washington, D.C., has merged into a more or less continuous urbanized area. This "megalopolis," sometimes called Bos-Wash, contains about 35 million people. Because these agglomerations have expanded beyond what we normally think of as a city, they are considered

Quantitative Reasoning

How many of the large cities in table 22.2 were in developing countries in 1900? How many cities on the list in 1900 were still on the list in 2016? How much did Tokyo grow between 1900 and 2016? New York? Beijing? How might these changes affect quality of life, for better or for worse?

Table 22.2 The World's Largest Urban Areas (Populations in Millions)

1900		2000		2100 (est.)	
London, England	6.6	Mexico City, Mexico	26.3	Lagos, Nigeria	88.3
New York, USA	4.2	Sao Paulo, Brazil	24.0	Kinshasa, Congo	83.5
Paris, France	3.3	Tokyo, Japan	17.1	Dar Es Salaam, Tanzania	73.7
Berlin, Germany	2.4	Calcutta, India	16.6	Mumbai, India	67.2
Chicago, USA	1.7	Bombay, India	16.0	Delhi, India	57.3
Vienna, Austria	1.6	New York, USA	15.5	Khartoum, Sudan	56.6
Tokyo, Japan	1.5	Seoul, Korea	13.5	Niamey, Niger	56.1
St. Petersburg, Russia	1.4	Shanghai, China	13.5	Dhaka, Bangladesh	54.2
Phihdelphia, USA	1.4	Rio de Janeiro, Brazil	13.3	Kolkata, India	50.3
Manchester, England	1.3	Delhi, India	13.3	Kabul, Afghanistan	50.2

Source: Data from T. Chandier, Three Thousand Years of Urban Growth, 1974; UN Population Division. The World's Cities in 2006.

urbanized **core regions** that dominate the social, political, and economic life of most countries (fig. 22.7). Similar urban agglomerations can be found around the world. The Tokyo-Yokohama-Osaka-Kobe megalopolis, for example, has about 50 million residents, while the Beijing-Tianjin-Hebei agglomeration ("Jing-Jin-Ji") in northeastern China will have 150 million inhabitants by 2050.

Developing areas have urbanized rapidly

In its transition from a rural, agrarian society to an urban, industrial country, China has seen the largest demographic shift in human history (chapter 7). Since the end of collectivized farms and factories in 1986, at least 500 million people have moved from rural areas into China's 662 main cities. The country is now about 60 percent urban, and demographers expect that by 2025, more than 1 billion people will inhabit China's cities. More than 220 of these cities will have 1 million or more residents.

Such rapid growth is a struggle for developing regions. Fast-growing countries often urbanize before they have the infrastructure or economies to support large cities. In Europe and the Americas, industrialization and urbanization took centuries to occur. (That urbanization was not a pretty process, fraught with urban poverty and overcrowded housing, as described by writers such as Charles Dickens.) Many developing countries today are urbanizing in a matter of decades. Cities in the developing world struggle to provide housing, roads, schools, clean water, and sanitation for growing populations. Mazes of informal settlements often develop, with extreme levels of poverty, while wealthy districts invest in walls and private security. Analysts project that many of

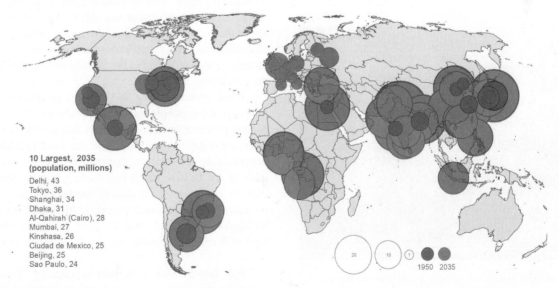

10 Largest, 2035
(population, millions)

Delhi, 43
Tokyo, 36
Shanghai, 34
Dhaka, 31
Al-Qahirah (Cairo), 28
Mumbai, 27
Kinshasa, 26
Ciudad de Mexico, 25
Beijing, 25
Sao Paulo, 24

FIGURE 22.6 By 2035, most of the largest cities will be in Africa and Asia. Rapid growth makes it difficult for cities to provide essential resources such as water, housing, and transportation.

Data source: United Nations Population Division. 2018. World Urbanization Prospects.

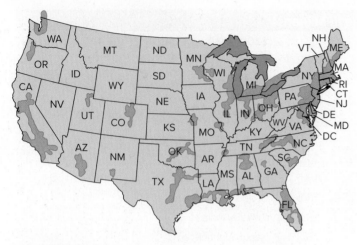

FIGURE 22.7 Urban core agglomerations (*lavender areas*) are forming megalopolises in many areas. While open space remains in these areas, the flow of information, capital, labor, goods, and services links each into an interacting system.
Source: U.S. Census Bureau

these cities will reach astonishing sizes by the end of this century (see table 22.2), some with more than 50 million residents.

Consider the case of Niamey, Niger. It may be the largest city of which you've never heard. For centuries, Niamey was a sleepy fishing village on the Niger River about 500 mi (800 km) upstream from Timbuktu. When Niger became independent in 1926 Niamey became its capital, even though its population was only a few thousand people. The population grew slowly to about 30,000 by 1960. Severe droughts in the 1970s and 1980s, together with civil wars and famine, triggered a mass migration of rural people to the city. By 2000, Niamey had grown to a vast tent city of nearly 1 million migrants (fig. 22.8). Most dwellings lack clean water and sanitation. The city's dusty streets are choked with traffic, and food shortages are perennial. Crowded refugee camps and

FIGURE 22.8 Niamey, once a sleepy fishing village on the Niger River, now has more than 1 million refugees. By 2100, the city could have 50 million residents if current trends continue.
ISSOUF SANOGO/Staff/Getty Images

transient settlements surround the city, and residents face severe poverty and often violence.

By 2100, Niamey is expected to have a population of more than 50 million. The city has practically no rain from mid-October to April and it is one of the hottest major cities on the planet. Temperatures reach 38°C (100°F) at least some days in every month in the year. One of the poorest countries in the world, Niger is second to last on the UN Human Development Index ranking. How will a city of this size provide for a growing population with so few resources?

Push and pull factors motivate people to move to cities

Urban populations grow in two ways: by natural increase (more births than deaths) and by immigration. Natural increase is fueled by improved food supplies, better sanitation, and advances in medical care that reduce death rates and cause populations to grow both within cities and in the rural areas around them (chapter 7). In Latin America and East Asia, natural increase is responsible for two-thirds of urban population growth. In Africa and West Asia, immigration is the largest source of urban growth.

Immigration to cities can be caused both by **push factors** that force people out of the country and by **pull factors** that draw them into the city. Common push factors are population growth and unemployment in rural areas, which drive people to cities in search of jobs, food, and housing. Economic forces can also be push factors: Trade policies might make farming unprofitable, or production of export crops can displace food production and deplete rural food supplies. Political, racial, or religious conflicts also drive people from their homes and villages. These events can actually depopulate the countryside. The United Nations High Commission on Refugees (UNHCR) reported in 2018 that more than 65 million people were displaced from their homes by economics, politics, persecution, or war. About 40 million of those people were internal refugees within their own country, while another 25 million moved (or attempted to move) to another country. Many of these refugees end up in the already overcrowded megacities of the developing world.

Land tenure patterns and changes in agriculture also play a role in pushing people into cities. Often wealthy or foreign landowners purchase or lease farmlands, displacing subsistence farmers who cannot pay as much to buy or rent land. In addition, the same pattern of agricultural mechanization that eliminated most farm labor in the United States is spreading now to developing countries. In many areas, the cost of mechanization is leading to the increasingly concentrated ownership of land by fewer and wealthier landowners. These changes force subsistence farmers off the land so it can be converted to grazing lands or monoculture cash crops.

Pull factors also draw people to cities. Young people are drawn by the excitement, vitality, and opportunity to meet others like themselves. Cities offer jobs, housing, entertainment, and freedom from the constraints of village traditions. The city provides opportunities for upward social mobility, prestige, and power not ordinarily available in the country. Cities support specialization in arts, crafts, and professions for which markets don't exist elsewhere. Attractive images of urban life on television also draw people to the city. And even for the desperately poor and homeless, cities often provide social services that are sometimes preferable to what rural areas can provide.

Section Review

1. What proportion of the world's population is now urban?
2. List 5 of the largest cities in 2018.
3. What factors can promote migration to cities?

22.2 URBAN CHALLENGES IN THE DEVELOPING WORLD

- *Low-income cities cannot build infrastructure to keep up with rapid urban growth.*
- *Poor water and air quality are especially important concerns.*
- *Housing and waste management are also widespread problems.*

Large cities in both developed and developing countries face similar challenges in accommodating the needs and by-products of dense populations. The problems are most intense, however, in rapidly growing cities of developing nations.

Cities in developing nations are also where most population growth will occur in coming decades (see fig. 22.6). These cities already struggle to supply food, water, housing, jobs, and basic services for their residents. The unplanned and uncontrollable growth of those cities poses urgent social and environmental challenges. What responsibilities might people in richer countries have to help people in these developing areas?

Pollution and water shortages affect developing cities

First-time visitors to a megacity—particularly in a developing country—often are overwhelmed by the immense crush of pedestrians and vehicles of all sorts jostling for space in the streets. The noise, congestion, and confusion of traffic make it seem suicidal to venture onto the street. Delhi, India, for instance, is one of the most densely populated cities in the world (fig. 22.9). Traffic is chaotic and often gridlocked. People often spend 3 or 4 hours each way commuting to work from outlying areas.

In 2018, the World Health Organization (WHO) declared that Delhi had the worst air quality of any of the 1,600 major cities it surveyed around the world. Beijing, China, is infamous for bad air quality, but, on average, Delhi's air is twice as bad as Beijing's. India has the world's highest death rate from chronic respiratory diseases and asthma, according to the WHO. An estimated 2.5 million people die every year in India from diseases related to air pollution (see chapter 16).

FIGURE 22.9 Motorized rickshaws, cycle rickshaws, scooters, bicycles, and buses compete for space on this crowded street in Delhi. Providing infrastructure to meet growing demand is a challenge.
travelwild/Shutterstock

Wastewater treatment is also limited in fast-growing, developing countries. The World Bank estimates that only one-third of urban residents in developing countries have satisfactory sanitation services. By some estimates, nearly half of India's 1.2 billion people have no toilet at home, and many people defecate in open fields or railway corridors. Less than 1 percent of India's 500,000 towns and villages have even partial sewage systems or water treatment facilities. In Egypt, Cairo's sewer system was built about 50 years ago to serve a population of 2 million people. It's now being overwhelmed by more than 10 million users.

Water shortages can be a dire threat, especially in a warming climate. Cape Town, South Africa, for example, came close to running out of water in 2018. Falling reservoir levels, a consequence of reduced rainfall, rapid growth, and inefficient practices, brought water resources to just 20 percent of normal. Emergency water conservation measures, including rationing water to 50 liters per day, averted the crisis, but many people around the world watched this event and began to wonder what it would be like for a major city to run dry.

Cities often pump groundwater to provide for growing populations, but this can cause the ground to subside, or sink, as groundwater is depleted. Jakarta, Indonesia, built on a low, soggy coastal marsh as a commercial trading port, is one of these. Today Jakarta has 30 million residents. As these residents extract groundwater, the city is gradually sinking. The many drainage canals crossing the city become more stagnant and contaminated, raising risks of waterborne illnesses (fig. 22.10), and major rainstorms flood increasingly large areas of the city. Rising sea levels compound these issues. Jakarta is just one of many major cities facing this combination of subsiding ground and rising sea levels (Exploring Science, p. 506).

Sinking Cities Amid Rising Seas

Around the world, major coastal cities are sinking into the sea. Hundreds of millions of lives and trillions of dollars in infrastructure are at risk. Look at a world map, and you'll see that many of the major world cities are located on or near a coastline. That makes sense for trading seaports, but low-lying coastal cities are increasingly vulnerable to subsidence, intense storms, and rising sea levels.

Jakarta, Indonesia, shows some of the causes and consequences of this existential problem (see fig. 22.10). This is the fastest-sinking city in the world, subsiding about 25 cm (10 in.) per year. Part of the problem is that residents have sunk illegal wells and withdrawn so much groundwater that the surface is collapsing. At the same time, the growing skyscrapers compress the boggy ground. About one-third of the city now floods regularly, sometimes for months at a time.

Dhaka, Bangladesh has similar issues. Built 1,200 years ago on soft sediments of the Ganges River delta, the city is threaded by tidal rivers and lagoons. With 20 million people, Dhaka is one of the fastest-growing urban areas in the world. But sea level in the Bay of Bengal is rising about ten times faster than the world average, while the land surface is sinking steadily. And climate change is causing stronger and more frequent storms. In spring 2017, nearly 10 m (30 ft) of rain fell on the city. The flat landscape, poorly drained under the best of circumstances, can't accommodate all that water. If that much rain coincided with a storm surge, much of the city could be underwater for months.

These problems don't just affect developing countries. Tokyo, Japan is one of the richest (and most expensive) cities in the world and is home to 38 million people. The original name for the city, Edo, meant "estuary," and like Dhaka and Jakarta, Tokyo sits on a broad delta where dozens of rivers empty into Tokyo Bay. And as elsewhere, Japan's storms are becoming stronger. In 2018 Typhoon Jebi struck Japan with 216 kmh (135 mph) winds, the most intense in 25 years. Tokyo avoided a direct hit, but more than half a meter (20 in.) of rain fell in some areas. About one third of Tokyo lies below sea level, largely because of ground water overpumping (as in Jakarta). Tokyo has built levees along its river-banks to prevent flooding, but major rains can fill the city like a bathtub.

Tokyo has spent $2 billion on a network of giant, underground caverns to store storm-water, as well as giant pumps and moveable barriers to prevent storm surges from running up river from Tokyo Bay. But in a combined heavy rain and a typhoon storm surge, disposing of water would be a major challenge.

Tokyo's situation will be familiar to anyone from New Orleans, Louisiana, another subsiding city in a bowl ringed by levees. New Orleans also has storm surge barriers and giant pumps to expel floods. But when the levees failed in Hurricane Katrina, in 2006, the city's bathtub-walls held in the flood for weeks.

In Houston, Texas, uncontrolled building has filled in natural drainage channels and covered the land with impermeable surfaces. When Hurricane Harvey dumped up to 1.5 m (60 in.) of rain on the city in just a few days in 2017, massive flooding resulted (fig. 1). Miami, Florida, now floods regularly in high tides. By 2100, Miami may be a series of hilly islands surrounded by ocean. New York City is also struggling to plan for storm surges and for high tides with rising sea levels. If Manhattan floods, trillions of dollars and millions of people will be at risk.

These predicaments help explain why many cities are enthusiastic about climate action. As in the case of Washington, D.C. (opening case study), cities like New York and Tokyo take sea level rise as a personal threat. They can see which neighborhoods will suffer, and they can calculate the dollar-cost of inaction. New York City has even sued the federal government for climate-related costs of Superstorm Sandy, whose storm floods cost the city billions of dollars.

Understanding the causes of sinking cities involves understanding geology, water resources, and urban development. Understanding storm surges involves an awareness of climate systems and climate change. Increasingly, cities are putting these pieces together as they face costly and life-threatening flood events.

FIGURE 1 Houston, Texas, experienced disastrous flooding after Hurricane Harvey dumped up to 1.5 m of rain on the city in 2017.

Joe Raedle/Getty Images

(header photo): David L. Hansen

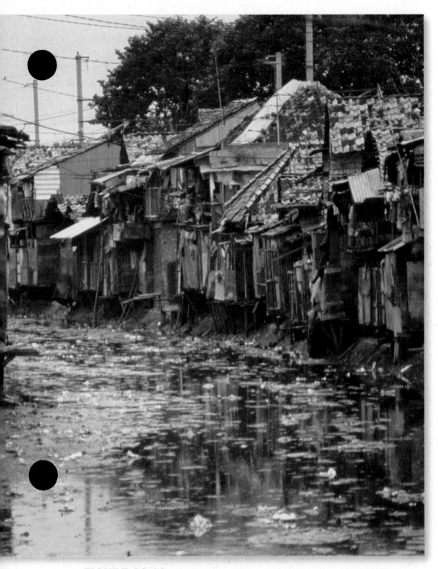

FIGURE 22.10 This tidal canal in Jakarta serves as an open sewer. Although most of the city has modernized in recent years, millions of people still lack access to modern sanitation systems.
William P. Cunningham

Many cities lack adequate housing

The United Nations estimates that at least 1 billion people—nearly 15 percent of the world's population—live in crowded settlements that ring the outskirts of most developing world cities. Around 100 million people have no home at all. In Mumbai, India, for example, it is thought that a million people sleep on the streets, sidewalks, and traffic circles because they have no other place to live. In Brazil, perhaps a million children live on the streets, surviving however and wherever they can.

Informal settlements is a general term for unplanned residential districts that develop organically as people migrate steadily to the city from rural areas. The term slum is also used, although it can be seen as more judgemental. Lacking organized roads, drainage, or water treatment, these settlements

FIGURE 22.11 Homeless people have built shacks along this busy railroad track in Jakarta, Indonesia. It's a dangerous place to live, with many trains per day using the tracks, but for the urban poor, there are few other choices.
Ruth Burke

can be unhealthy for residents. Lacking police and security, they can be violent and unsafe. (On the other hand, police forces don't always protect the poorest populations.) At the same time, these districts provide access to employment, and poor populations often have no place else to live.

Sometimes these settlements originated as government-built housing for workers, which then became overcrowded. In the tenements of Mumbai, for example, most families occupy a single room, often with shared kitchens. This level of crowding can be unsafe, especially for children. Shantytown is another term for informal settlements that develop on the margins of urban neighborhoods, parks, railway margins, and other sparsely used parts of a city (fig. 22.11).

These populous but unauthorized settlements usually lack sewers, clean water supplies, electricity, and roads. Often, the land on which they are built was not previously used because it is unsafe or unsuitable for habitation. In Rio de Janeiro, La Paz (Bolivia), Guatemala City, and Caracas (Venezuela), shantytowns are perched on landslide-prone hills. São Paulo, Brazil, is the country's wealthiest city, and the hub of finance and business, but still has at least 2 million people living in favelas or slums (fig. 22.12).

As desperate as conditions are in informal settlements, many people do more than merely survive there. They keep themselves clean, raise families, educate their children, find jobs, and save a little money to send home to their parents. People are adaptable and resilient, and life in rapidly growing cities is, in many ways, comparable to life in the early industrial cities of Europe and America a century ago. Conditions then were often squalid,

FIGURE 22.12 São Paulo, Brazil, is South America's largest urban area. Traffic congestion is legendary, and the wealthy residents increasingly depend on helicopters to avoid crime and traffic jams. With more than 193 heliports, and more than 700 flights daily, the city has the world's largest helicopter fleet.
William P. Cunningham

but they improved over time. Continuing development may bring better conditions to cities of the developing world—as it has for many in the developed world.

Section Review

1. Why are traffic congestion and water pollution problems in most fast-growing cities in developing countries?
2. What are informal settlements, and why are they common?
3. List some of the world's major sinking cities.

22.3 URBAN CHALLENGES IN THE DEVELOPED WORLD

- *Cities in developed countries suffer from pollution and poverty.*
- *Urban sprawl undercuts urban sustainability.*

For the most part, the rapid growth of central cities that accompanied industrialization in nineteenth-and early twentieth-century Europe and North America has now slowed or even reversed. Wealthier cities in developed countries, with more funding for planning, infrastructure, and maintenance, have fewer of the worst environmental conditions that they had in the past. Environmental quality varies across a city, though. Even in wealthy cities people of color, low-income populations, or immigrants often live in unhealthy conditions.

In most developed countries, improved sanitation and medical care have reduced or eliminated many of the communicable diseases that once afflicted urban residents. Air and water quality have improved dramatically as heavy industry such as steel smelting and chemical manufacturing have moved to developing countries. In consumer and information economies, workers no longer need to be concentrated in central cities. They can live and work in dispersed sites. Automobiles now make it possible for much of the working class to enjoy amenities such as single-family homes, yards, and access to recreation that once were available only to the elite.

In the United States, many post-industrial cities (older, densely built manufacturing centers), have declined in population as industry has moved overseas and residents have moved to the suburbs. Philadelphia, Baltimore, Detroit, and other cities have seen these demographic shifts, as both businesses and workers have moved to other regions or to larger cities (fig. 22.13). Most of the rapidly growing cities such as Seattle, San Francisco, or Atlanta are high-tech centers. Technology companies often provide high-paying jobs; they also bring traffic congestion, exploding property values, increasing disparity between the rich and poor, and a loss of local culture and identity. For many high-tech companies, being located near industrial centers and shipping is less important than a good climate, ready access to air travel, and a substantial creative class.

Urban sprawl consumes land and resources

While the move to suburbs and rural areas has brought many benefits to the average citizen, it also has caused numerous urban problems. Cities that once were compact now spread over the landscape, consuming open space and wasting resources. This pattern of urban growth is known as **sprawl.** While there is no universally accepted definition of the term, sprawl generally

FIGURE 22.13 Two fire damaged houses in Detroit, Michigan. Since the deindustrialization of the city, the population has declined more than 60 percent and at least 100,000 abandoned homes and buildings have been demolished.
Suzanne Tucker/Shutterstock

Quantitative Reasoning

Compare the photos in figures 22.13 and 22.22*a*. How many families would fit in a city block in these two developments? How would your daily time budget differ in getting to school, work, shopping, and socializing in these two contexts?

includes the characteristics outlined in table 22.3. As former Maryland governor Parris N. Glendening said, "In its path, sprawl consumes thousands of acres of forests and farmland, woodlands and wetlands. It requires government to spend millions extra to build new schools, streets, and water and sewer lines." And Christine Todd Whitman, former New Jersey governor and head of the Environmental Protection Agency, added, "Sprawl eats up our open space. It creates traffic jams that boggle the mind and pollute the air. Sprawl can make one feel downright claustrophobic about our future."

In most American metropolitan areas, new residential development has primarily been in vast suburban, or ex-urban, areas that leapfrog out beyond the edge of the city in a search for inexpensive rural land with few restrictions on land use or building practices (fig. 22.14). The U.S. Department of Housing and Urban Development estimates that urban sprawl consumes some 200,000 ha (roughly 500,000 acres) of farmland each year. Because cities often are located in fertile river valleys or shorelines, much of that land would be especially valuable for producing crops for local consumption. But with planning authority divided among many small, local jurisdictions, metropolitan areas have no way to regulate growth or provide for rational, efficient resource use. Small towns and township or county officials generally welcome this growth because it profits local landowners and businesspeople. Although the initial price of suburban homes often is less than

FIGURE 22.14 Huge houses on sprawling lots consume land, alienate us from our neighbors, and make us ever more dependent on automobiles. They also require a lot of lawn mowing!

Regents of the University of Minnesota. All Rights Reserved. Used with permission of the Metropolitan Design Center

comparable urban property, there are external costs in the form of new roads, sewers, water mains, power lines, schools, shopping centers, and other extra infrastructure required by this low-density development.

Landowners, builders, real estate agents, and others who profit from this crazy-quilt development pattern generally claim that growth benefits the suburbs in which it occurs. They promise that adding additional residents will lower the average taxes for everyone, but in fact the opposite often is true. In a study titled *Better Not Bigger*, author Eben Fodor analyzed the costs of medium-density and low-density housing. In suburban Washington, D.C., for instance, each new house on a quarter acre (0.1 ha) lot cost $700 more in services than it paid in taxes. A typical new house on a 5-acre (2-ha) lot, however, cost $2,200 more than it paid in taxes, because of higher expenses for infrastructure and services. Expanding suburbs also raise health care costs: car-dependent areas have been called "obesigenic" suburbs, because they produce an unhealthy lifestyle that involves sitting in cars much of the day.

In a study of 58 large American urban areas, David Rusk, an author and a former mayor of Albuquerque, found that between 1950 and 1990, populations grew 80 percent, while land area grew 305 percent. In Atlanta, Georgia, the population grew 32 percent between 1990 and 2000, while the total metropolitan area increased by 300 percent. The city is now more than 175 km across. Atlanta loses an estimated $6 million to traffic delays every day. By far the fastest-growing urban area in the United States is Las Vegas, Nevada, which tripled its population while increasing its area by 600 percent (fig. 22.15*a* and *b*).

Table 22.3 Characteristics of Urban Sprawl
1. Unlimited outward extension.
2. Low-density residential and commercial development.
3. Leapfrog development that consumes farmland and natural areas.
4. Fragmentation of power among many small units of government.
5. Dominance of freeways and private automobiles.
6. No centralized planning or control of land uses.
7. Widespread strip malls and "big-box" shopping centers.
8. Great fiscal disparities among localities.
9. Reliance on deteriorating older neighborhoods for low-income housing.
10. Decaying city centers as new development occurs in previously rural areas.

Source: Excerpt from a speech by Anthony Downs at the CTS Transportation Research Conference, as appeared on the website by Planners Web, Burlington, Vermont, 2001

(a) 1972

(b) 2018

 FIGURE 22.15 Satellite images of Las Vegas, Nevada, in 1972 (a) and 2018 (b). The metropolitan area grew about 600 percent in 46 years.
Source: (a) National Aeronautics and Space Administration (NASA); (b) National Aeronautics and Space Administration (NASA)

Transportation is crucial in city development

Getting people around within a large urban area has become one of the most difficult problems that many city officials face. A century ago, most American cities were organized around transportation corridors. First horse-drawn carriages, then electric streetcars provided a way for people to get to work, school, and shops. Everyone, rich or poor, wanted to live as close to the city center as possible.

The introduction of automobiles allowed people to move to suburbs, and cities began to spread over the landscape. The U.S. Interstate Highway System was the largest construction project in human history. Originally justified as necessary for national defense, it was really a huge subsidy for the oil, rubber, automobile, and construction industries. Its 72,000 km (45,000 mi) of freeways probably did more than anything to encourage sprawl and change America into an auto-centered society.

Because many Americans now live far from where they work, shop, and recreate, most consider it essential to own a private automobile. The average U.S. driver spends about 443 hours per year behind a steering wheel. This means that for most people, the equivalent of one full 8-hour day per week is spent sitting in an automobile. Of the 5.8 billion barrels of oil consumed each year in the United States (60 percent of which is imported), about two-thirds is burned in cars and trucks. As discussed in chapter 16, about two-thirds of all carbon monoxide, one-third of all nitrogen oxides, and one-quarter of all volatile organic compounds emitted each year from human-caused sources in the United States are released by automobiles, trucks, and buses.

Building the roads, parking lots, filling stations, and other facilities needed for an automobile-centered society takes a vast amount of space and resources (fig. 22.16). In some metropolitan areas, it is estimated that one-third of all land is devoted to the automobile. To make it easier for suburban residents to get from their homes to jobs and shopping, we provide an amazing network of freeways and highways. At a cost of several trillion dollars to build, the interstate highway system was designed to improve transportation speed and efficiency. As more and more drivers clog

FIGURE 22.16 Freeways give us the illusion of speed and privacy, but they consume land, encourage sprawl, and create congestion as people move farther from the city to get away from traffic and then have to drive to get anywhere.
Comstock Images/Alamy

the highways, however, the reality is far different. In Los Angeles, for example, which has the worst congestion in the United States, the average speed in 1982 was 58 mph (93 km/hr), and the average driver spent less than 4 hours per year in traffic jams. In 2018, the average highway speed in Los Angeles was only 35.6 mph (57.3 km/hr), and the average driver spent 82 hours per year waiting for traffic. Although new automobiles are much more efficient and cleaner operating than those of a few decades ago, the fact that we drive so much farther today and spend so much more time idling in stalled traffic means that we burn more fuel and produce more pollution than ever before.

Altogether, it is estimated that traffic congestion costs the United States $78 billion per year in wasted time and fuel. A common answer to traffic jams is to build more freeways, or widen them with more traffic lanes. While this consumes a great deal of land and money, studies show that expanding highway lanes rarely resolves congestion. Instead, it increases traffic. Easier access to suburbs encourages more drivers to travel farther on a daily basis. Rather than ease congestion and save fuel, more freeways can exacerbate the problem.

Sprawl impoverishes central cities from which residents and businesses have fled. With a reduced tax base and fewer civic leaders living or working in downtown areas, the city is unable to maintain its infrastructure. Streets, parks, schools, and civic buildings fall into disrepair at the same time that these facilities are being built at great expense in new suburbs. The poor who are left behind when the upper and middle classes abandon the city center often can't find jobs where they live and have no way to commute to the suburbs where jobs are now located. About one-third of Americans are too young, too old, or too poor to drive. For these people, car-oriented development causes isolation and makes daily tasks like grocery shopping very difficult. Parents, especially mothers, spend long hours transporting young children. Teenagers and aging grandparents are forced to drive, often presenting a hazard on public roads.

Finally, sprawl fosters uniformity and alienation from local history and natural environment. Housing developments often are based on only a few standard housing styles, while shopping centers and strip malls everywhere feature the same national chains. You could drive off the freeway in the outskirts of almost any big city in America and see exactly the same brands of fast-food restaurants, motels, stores, filling stations, and big-box shopping centers.

Public transit can make cities more livable

Many American cities are now rebuilding the public transportation systems that were abandoned in the 1950s (fig. 22.17). There are many strategies for this. An innovative and inexpensive approach to public transit is found in Curitiba, Brazil. High-speed, bi-articulated buses, each of which can carry 270 passengers, travel on dedicated roadways closed to all other vehicles (fig. 22.18). These bus-trains are linked to 340 feeder routes extending throughout the city. The system is designed to provide bus service within walking distance to everyone in the city, with frequent,

FIGURE 22.17 Many American cities are now rebuilding light rail systems that were abandoned in the 1950s when freeways were built. Light rail is energy-efficient and popular, but it can cost up to $100 million per mile ($60 million per kilometer) to install.
William P. Cunningham

convenient, affordable service. Curitiba's buses carry some 1.9 million passengers per day, or about three-quarters of all personal trips within the city. Working with existing roadways for the most part, the city was able to construct this system for one-tenth the cost of a light rail system or freeway system, and one-hundredth the cost of a subway.

Curitiba's level of coordination is relatively uncommon, though, and many developing countries are adopting the automobile-centered model of the United States rather than Curitiba's model. Traffic accidents have become the third-largest cause of years of lost life worldwide. The number of vehicles increased eight-fold in Nigeria and sixfold in Pakistan between 1980 and 2000,

FIGURE 22.18 The bus-rapid-transit system of Curitiba, Brazil, features high-speed, bi-articulated buses, each of which can carry 270 passengers, and travels on dedicated roadways closed to all other vehicles. This system provides three-quarters of all passenger trips in the city.
William P. Cunningham

while the road networks in those countries expanded by only 10 to 20 percent in the same time.

The recent introduction of the Tata Nano, an inexpensive, abundantly produced personal car, in India raises nightmares for both urban planners and energy experts. Costing less than $2,000 brand new, these tiny vehicles put car ownership within reach for millions who could never afford it before. But they will probably increase gasoline consumption greatly and result in huge traffic jams as inexperienced drivers take to the road for the first time.

Section Review

1. Describe urban sprawl, and explain some reasons for it.
2. Identify several costs associated with expansive urban growth.
3. How did Curitiba, Brazil, make public transit work?

22.4 SUSTAINABLE URBANISM AND SMART GROWTH

- *Better environmental planning involves mixed uses, alternative transportation, and urban growth boundaries.*
- *Garden cities and new urbanism aimed to make cities appealing.*
- *Smart growth emphasizes alternative energy and transportation and compact growth.*

Smart growth involves strategies for well-planned developments that make efficient and effective use of land resources and existing infrastructure. An alternative to haphazard, poorly planned sprawling developments, smart growth involves thinking ahead to develop pleasant neighborhoods while minimizing the wasteful use of space and tax dollars for new roads and extended sewer and water lines.

Smart growth aims to make land-use planning democratic. Public discussions allow communities to guide planners. Mixing land uses, rather than zoning exclusive residential areas well away from commercial areas, makes living in neighborhoods more enjoyable. By planning a range of housing styles and costs, smart growth allows people of all income levels, including young families and aging grandparents, to find housing they can afford. Open communication between planners and the community helps make urban expansion fair, predictable, and cost-effective (table 22.4).

Table 22.4   Goals for Smart Growth
1. Create a positive self-image for the community.
2. Make the downtown vital and livable.
3. Alleviate substandard housing.
4. Solve problems with air, water, toxic waste, and noise pollution.
5. Improve communication between groups.
6. Improve community member access to the arts.

Source: Vision 2000, Chattanooga, Tennessee

Garden cities and new towns were early examples of smart growth

The twentieth century saw numerous experiments in building "new towns" for society at large that try to combine the best features of the rural village and the modern city. One of the most influential of all urban planners was Ebenezer Howard (1850–1929), who not only wrote about ideal urban environments but also built real cities to test his theories. In *Garden Cities of Tomorrow,* written in 1898, Howard proposed that the congestion of London could be relieved by moving whole neighborhoods to **garden cities** separated from the central city by a greenbelt of forests and fields.

In the early 1900s, Howard worked with architect Raymond Unwin to build two garden cities outside of London: Letchworth and Welwyn Garden. Interurban rail transportation provided access to these cities. Houses were clustered in "superblocks" surrounded by parks, gardens, and sports grounds. Streets were curved. Safe and convenient walking paths and overpasses protected pedestrians from traffic. Businesses and industries were screened from housing areas by vegetation. Each city was limited to about 30,000 people to facilitate social interaction. Housing and jobs were designed to create a mix of different kinds of people and to integrate work, social activities, and civic life. Trees and natural amenities were carefully preserved, and the towns were laid out to maximize social interactions and healthful living. Care was taken to meet residents' psychological needs for security, identity, and stimulation.

Today, Letchworth and Welwyn Garden each have 70 to 100 people per acre. This is a true urban density, about the same as New York City in the early 1800s and five times as many people as most suburbs today. By planning the ultimate size in advance and choosing the optimum locations for housing, shopping centers, industry, transportation, and recreation, Howard believed he could create a hospitable and satisfying urban setting while protecting open space and the natural environment. He intended to create parklike surroundings that would preserve small-town values and encourage community spirit in neighborhoods.

Planned communities also have been built in the United States following the theories of Ebenezer Howard, but most plans have been based on personal automobiles rather than public transit. Radburn, New Jersey, was designed in the 1920s, and two highly regarded new towns of the 1960s are Reston, Virginia, and Columbia, Maryland.

Mixed uses make cities more livable

Some urban designers focus on a neotraditionalist approach. These designers attempt to recapture some of the best features of small towns and livable cities of the past. They prioritize modest-sized urban communities and neighborhoods that integrate houses, offices, shops, and civic buildings. In theory, everyone should live within a 5-minute walk from a neighborhood center with a convenience store, a coffee shop, a bus stop, and other amenities. A mix of apartments, townhouses, and detached houses in a variety of price ranges ensures that neighborhoods will include a diversity of ages and income levels.

Other "sustainable urbanists" are concerned less with city size and more with ensuring sustainable transportation, energy efficiency, and urban green space. These are some of the principles of many of these planners:

- Maintain greenbelts in and around cities. These provide recreational space, promote efficient (high-density) land use, improve air quality, moderate temperature, and reduce water pollution.

- Plan where development will take place. Avoiding disorganized development protects property values, historical sites, agricultural lands, ecosystem services, clean water, and groundwater replenishment.

- Locate everyday shopping and services so people can meet daily needs with greater convenience. This helps avoid stress, automobile dependency, and waste of time and energy (fig. 22.19). Small-scale commercial development near residential areas can increase community interactions.

- Encourage walking and bicycling. Compact cities, safe bike lanes, allowing bikes on buses, reducing parking spaces, and convincing people that biking is a safe and realistic means of transportation.

- Promote diverse housing styles, rather than just the detached, single-family houses prioritized in many cities. In-fill building reduces travel distances, supports local businesses, and supports diverse households and income ranges. Cooperatives and co-housing are especially efficient.

- Promote sustainability by growing food locally, recycling wastes and water, developing renewable energy, and encouraging community gardening in inner-city spaces or greenbelts (fig. 22.20).

- Equip buildings with "green roofs" or rooftop gardens. These improve air quality and reduce heating and cooling costs, stormwater management costs, noise, and urban heat island effects (fig. 22.21).

- Preserve urban habitat. This supports biodiversity as well as mental health, recreation, and access to nature.

These planning principles aren't just a matter of aesthetics. Dr. Richard Jackson, former director of the National Center for Environmental Health in Atlanta, points out a strong association between urban design and our mental and physical health. As our cities have become ever more spread out and impersonal, we have fewer opportunities for healthful exercise and socializing. Chronic diseases, such as cardiovascular diseases, asthma, diabetes, obesity, and depression, are becoming the predominant health concerns in the United States.

"Despite common knowledge that exercise is healthful," Dr. Jackson says, "fewer than 40% of adults are regularly active, and 25% do no physical activity at all. The way we design our communities makes us increasingly dependent on automobiles for the shortest trip, and recreation has become not physical but observational." Long commutes and a lack of reliable mass transit and walkable neighborhoods mean that we spend more and more time in stressful road congestion. "Road rage" isn't imaginary. Every commuter can describe unpleasant encounters with rude drivers.

FIGURE 22.19 This walking street in Queenstown, New Zealand, provides opportunities for shopping, dining, and socializing in a pleasant outdoor setting.
William P. Cunningham

FIGURE 22.20 Many cities have large amounts of unused open space that could be used to grow food. Residents often need help decontaminating soil and gaining access to the land.
William P. Cunningham

FIGURE 22.21 This award-winning green roof on the Chicago City Hall is functional as well as beautiful. It reduces rain runoff by about 50 percent, and keeps the surface as much as 30°F cooler than a conventional roof on hot summer days.
Tim Boyle/Staff/Getty Images

Urban design that offers the benefits of more walking, more social contact, and surroundings that include water and vegetation can provide healthful physical exercise and psychic respite.

Smart growth and sustainable urbanism place a strong focus on efficiency and reinvesting in cities:

- Focus on in-fill development—filling in the inner city so as to help preserve green space in and around cities. Where possible, focus on **brownfield developments,** building on abandoned, reclaimed industrial sites. Brownfields have been eyesores and environmental liabilities in cities for decades, but as urban growth proceeds, they are becoming an increasingly valuable land resource.

- Build high-density, attractive, low-rise, mixed-income housing near the center of cities or near public transportation routes (fig. 22.22*a*). Densely packed housing saves energy as well as reduces the infrastructure costs per person.

- Provide incentives for alternative transportation, such as reserved parking for shared cars (fig. 22.22*b*) or bicycle routes and bicycle parking spaces. Figure 22.22*c* shows an 8,000-bicycle parking garage at the train station in Leiden, the Netherlands. An 8,000-car parking garage at the station would cut out the heart of the city. Discourage car use by minimizing the amount of space devoted to driving and parking cars, or by charging for parking space, once realistic alternatives are available.

- Encourage green building techniques, including energy efficiency, water conservation systems, solar water heating, and efficient appliances.

- Encourage high-density development and co-housing—groups of households clustered around a common green space that share child care, gardening, maintenance, and other activities. Co-housing can reduce consumption of space, resources, and time while supporting a sense of community.

- Provide facilities for recycling organic waste, building materials, appliances, and plastics, as well as metals, glass, and paper.

- Invite public participation in decision making. Emphasize local history, culture, and environment to create a sense of community and identity. Coordinate regional planning through metropolitan boards that cooperate with but do not supplant local governments.

Open-space design preserves landscapes

Traditional suburban development typically divides land into a checkerboard layout of nearly identical 1- to 5-ha parcels with no designated open space (fig. 22.23, *top*). The result is a sterile landscape consisting entirely of house lots and streets. This style of development, which is permitted—or even required—by local zoning and ordinances, consumes agricultural land and fragments wildlife habitat. Many of the characteristics that people move to

(a)

(b)

(c)

FIGURE 22.22 Green urbanism includes (a) concentrated, low-rise housing, (b) car-sharing clubs that receive special parking allowances, and (c) alternative transportation methods. These examples are from Amsterdam and Leiden, the Netherlands.
(a-c): © Mary Ann Cunningham

Vauban: A Car-Free Neighborhood

We have noted that cities can be centers of climate action. Can this extend to life without a car? In Vauban, a district in Freiburg, Germany, residents have shown that a low-carbon, car-free life can be not just easy, but economical. In Vauban, getting around by tram, bicycle, and on foot is easy, so there is little need for cars. The community provides stores, banks, schools, and restaurants in easy walking distance of homes. Jobs and office space are available nearby, and trams to the city center run every few minutes through the center and around the edges of Vauban district. Residential streets are narrow and vehicle free, making a great place for bicycles and playing children (fig. 1).

Residents who own cars can park them in a municipal ramp, but cars aren't encouraged: Buying a space in the ramp costs $40,000. (Most cities, in contrast, cover the cost of on-street parking, at a rate of around $10,000 per parking spot, plus maintenance.) Nearly three-quarters of Vauban's families don't own a car, and more than half sold their car when they moved there.

Fewer vehicles means less energy consumption and less pollution. But most families don't move to Vauban for environmental reasons. They move there because they find a car-free lifestyle is healthier for children and for themselves. The neighborhoods are quiet, and the air is clean. Schools, childcare services, playgrounds, and sports facilities are within easy biking or walking distance. Children can play outside and can walk or bike to school without having to cross busy streets.

In most American cities, one-third or more of the land area is dedicated to cars, mainly for parking and roads. In Vauban, reducing car dependence has saved so much space that neighborhoods have abundant green space, gardens, and play areas while still being small enough for easy walking.

A successful and growing car-sharing program makes it still easier not to own a car. The city's car-sharing program began in about 1992 and now has more than 2,500 members, who save money and parking space by using shared cars. The cars are available all around town, and they can be reserved online or by mobile phone. In addition, a single monthly bus ticket covers all regional trains and buses, making it especially easy to get around by public transportation.

A car-free lifestyle makes economic sense. Even in the United States, car ownership costs about $9,000 per year in purchase costs, maintenance, insurance, and fuel. In Germany, costs are still higher, and living without a car frees up that money for other uses.

Vauban's comfortable row houses, with balconies and private gardens, are built with **passive house standards** that conserve energy but maximize quality of life. Well-designed space, large, triple-pane windows, warm woodwork, and large balconies make the small-footprint homes feel spacious. Many houses are so efficient that they don't need a heating system at all. In addition, a highly efficient woodburning, district

FIGURE 1 Car-free roads provide a cleaner, safer, healthier environment for residents of the Vauban district in Freiburg, Germany.
Martin Specht/The New York Times/Redux Pictures

heat and power plant provides much of the space heating and electricity for the area, and rooftop solar collectors and photovoltaic panels provide hot water and power for individual homes. Many of the houses produce more energy than they consume.

Vauban's example provides an attractive alternative to suburban sprawl. In most places government policies, such as easy and abundant public funding for road building, underfunding of public transportation, and low tax rates on the edge of town, have encouraged people to suburbanize. As cities expand, though, they become more costly to maintain and occupy. We tend to disregard costs of suburban expansion: high transportation costs, lack of easy access to schools, shopping, or jobs, time spent in travel. Suburbs are especially difficult for people too young, too old, or too poor to have a car of their own. Suburban expansion also destroys farmland and woodlands on the edge of town, and everyone bears the costs of extending fire protection, water service, road maintenance, sewers, school buses, and other service to outlying areas.

City planners are increasingly recognizing some of these costs, though. Across, Europe, car-free urban development is encouraged, for both financial and environmental reasons. A growing number of young Americans are also moving to cities with public transport systems and choosing to live car-free. Many cities are also reinvesting in older central districts, which were built from the start on the principle of easy access to jobs, shopping, and entertainment.

What might it take to make this lifestyle acceptable for citizens in your hometown? What policy changes would be necessary to bring such a transition about?

---

the country to find—space, opportunities for outdoor recreation, access to wild nature, a rural ambiance—are destroyed by dividing every acre into lots that are "too large to mow but too small to plow."

An alternative known as **conservation development,** cluster development, or open-space zoning preserves at least half of a subdivision as natural areas, farmland, or other forms of open space.

Among the leaders in this design movement are landscape architects Ian McHarg, Frederick Steiner, and Randall Arendt. They have shown that people who move to the country don't necessarily want to own a vast acreage or live miles from their nearest neighbor. What they most desire is long views across an interesting landscape, an opportunity to see wildlife, and access to walking paths through woods or across wildflower meadows.

FIGURE 22.23 Conventional subdivision (*top*) and an open space plan (*bottom*). Although both plans provide 36 home sites, the conventional development allows for no public space. Cluster housing on smaller lots in the open space design preserves at least half the area as woods, prairie, wetlands, farms, or other conservation lands, while providing residents with more attractive vistas and recreational opportunities.

By clustering houses on smaller lots, a conservation subdivision can provide the same number of buildable lots as a conventional subdivision and still preserve 50 to 70 percent of the land as open space (fig. 22.23, *bottom*). This not only reduces development costs (less distance to build roads, lay telephone

FIGURE 22.24 Jackson Meadows, an award-winning cluster development near Stillwater, Minnesota, groups houses at sociable distances and preserves surrounding open space for walking, gardening, and scenic views from all houses.
William P. Cunningham

lines, sewers, power cables, and so on) but also helps foster a greater sense of community among new residents. Walking paths and recreation areas get people out of their houses to meet their neighbors. Home owners have smaller lots to care for, and yet everyone has an attractive vista and a feeling of spaciousness (fig. 22.24).

Urban habitat can make a significant contribution toward saving biodiversity. In a groundbreaking series of habitat conservation plans designed to protect the endangered California gnatcatcher, some 85,000 ha (210,000 acres) of coastal scrub near San Diego was protected as open space within the rapidly expanding urban area. This is an area larger than Yosemite Valley, and will benefit many other species, as well as humans.

Section Review

1. List several reasons for smart growth.
2. Identify some priorities for making cities more liveable.
3. What are some benefits of clustering houses in a development?

Connecting the Dots

The case of Washington, D.C., illustrates the critical social and environmental challenges cities face, as well as the extraordinary opportunities cities offer for innovation and progress. Climate action plans in Washington and many other cities are leading the way, inventing new paths toward sustainability and toward more equitable access to resources and opportunities.

More than half of us now live in cities, so understanding their growth and their environmental impacts is important to understanding environmental questions from climate action to land conservation to population growth and resource use. Increasingly, the world's largest cities are emerging in developing countries. With limited resources, these cities cannot keep up with growing demands for housing, water, sanitation, and other infrastructure. Millions of people live in growing informal settlements. These cities will become still larger urban agglomerations in the next generation or two, some with over 50 million residents. Impacts on environment and quality of life are hard to predict.

Urban planning can reduce environmental costs of growth. Although urban sprawl and automobile infrastructure consumes farmland, forests, energy, and money, planners offer many ways to reduce sprawl. Smart growth, sustainable urbanism, public transportation, and high-density development can all help build efficient, livable cities. Green jobs, renewable energy, and efficient housing also help make cities more equitable and sustainable.

Urban environments are an important theme, then, in environmental science. Every year, more of us live in cities, and cities create both environmental damage and inspiring innovations, in climate action and many other issues. Envisioning what we want cities to be, and for whom, can help us produce the kinds of future cities we want to see.

Reviewing Key Terms

Can you define the following terms in environmental science?

brownfield development 22.4	garden city 22.4	pull factors 22.1	sprawl 22.3
city 22.1	informal settlement 22.2	push factors 22.1	urban agglomerations 22.1
conservation development 22.4	megacity 22.1	rural area 22.1	urban area 22.1
core region 22.1	passive house standards 22.4	smart growth 22.4	village 22.1

Critical Thinking and Discussion Questions

1. Picture yourself living in a rural village or a developing world city. What aspects of life there would you enjoy? What would be the most difficult for you to accept?

2. Why would people move to one of the megacities of the developing world, if conditions are so difficult there?

3. A city could be considered an ecosystem. Using what you learned in chapters 3 and 4, describe the structure and function of a city in ecological terms.

4. Look at the major urban area(s) in your state. Why were they built where they are? Are those features now a benefit or drawback?

5. Weigh the costs and benefits of automobiles in modern American life. Is there a way to have the freedom and convenience of a private automobile without its negative aspects?

6. Boulder, Colorado, has been a leader in controlling urban growth. One consequence is that the city has stayed small and charming, but housing prices have skyrocketed and working people have been driven out. If you lived in Boulder, what solutions might you suggest? What do you think is an optimum city size?

Data Analysis

Plotting Urban and Economic Indicators

Urbanization and economic growth are two closely related changes going on in societies today. How are these two processes related? How do they compare in different parts of the world? Is the United States more urbanized or less so than other regions? How do economic growth rates compare in different regions?

Gapminder.org is a rich source of data on global population, health, and development that you may have already examined in the Data Analysis exercise for chapter 7. The site includes animated graphs showing change over time. Go to Connect to find and analyze this animated Gapminder graph, and to answer questions about what they tell you.

23

Ecological Economics

▲British Columbians are justly proud of their beautiful environment, and worried about the effects of global climate change. The province is using tax policy to shift the economy to a carbon-constrained future.
Pierre Leclerc Photography/Moment/Getty Images

Learning Outcomes

After studying this chapter, you should be able to:

23.1 Identify some assumptions of classical and neoclassical economics.

23.2 Explain key ideas of ecological economics.

23.3 Describe relationships among population, technology, and scarcity.

23.4 Understand ways we measure growth.

23.5 Summarize how market mechanisms can reduce pollution.

23.6 Discuss the importance of green development and business.

"Underground nuclear testing, destruction of rain forests, toxic waste . . . Let's put it this way: if the world were a big apartment, we wouldn't get our deposit back."

– John Ross

Using Economics to Fight Climate Change

British Columbians are justifiably proud of their spectacular environment, and they want to protect it. One of the greatest challenges they—and all of us—face is global climate change. How can we move our economy to a low-carbon future?

The province has taken a bold and important step in that direction by adopting a tax on carbon fuels. In 2008, British Columbia (BC) passed the first broad-based carbon tax in North America. Conservatives claimed the tax would "destroy jobs and growth," but the opposite has happened. The latest numbers from Statistics Canada show that the carbon tax has been a tremendous environmental and economic success. Since the tax came into effect, fossil fuel use has dropped by 16 percent in BC compared to a 3.5 percent increase in the rest of Canada. At the same time, the BC economy has grown faster and unemployment has been lower than any other province.

The BC carbon tax is designed to be revenue neutral. That is, every penny taken in by the tax is used to reduce other taxes. Roughly 65 percent of the income is used to decrease corporate taxes. About 35 percent goes to lower personal taxes. Half that amount is targeted to low-income families and individuals for whom the carbon tax creates a particular disadvantage. And about half the corporate tax reductions are targeted for industries, such as clean energy, digital media and film, international businesses, and investment capital that the government wants to encourage. The result is that BC has the lowest personal income tax rate in Canada, and one of the lowest corporate rates in North America. Tax reductions have actually been greater than the amount taken in by the carbon tax. The difference is made up by increased taxes in sectors stimulated by tax shifts (fig. 23.1).

Passing this carbon tax was easier in BC than in other places because approximately 95 percent of the province's electricity comes from hydroelectric dams, which don't need carbon fuels. Still, most people heat their homes with natural gas, and fuel their vehicles with petroleum-based fuels. The carbon tax started out low (C$10 per metric ton of CO_2) to ease people into the new system, and then rose by C$5 to the current C$30 per ton. In 2012, the tax was frozen so that BC wouldn't be at a disadvantage compared to other provinces.

Nearly 60 percent of BC's residents approved of the carbon tax when it was first proposed. That support has slipped somewhat since then, but most people are still keenly aware of how urgent greenhouse gas reductions are. They know that insect infestations caused by warmer winters have devastated their forests, and that droughts, heat waves, and violent storms impact their lives. Fewer are probably aware of less direct costs, as air pollution caused by burning fossil fuels costs thousands of lives and adds more than $8 billion per year to Canada's health expenses.

The idea of using tax policy to discourage harmful actions (pollution, climate change) and to reward beneficial sectors (such as renewable energy) is a classic in economic theory. This fiscal reform is especially appealing when it may bring a double dividend: reducing pollution and stimulating economic growth in renewable energy and conservation. So far, most people find their total tax bill has gone down with this measure. Low-income and rural communities have benefited from programs supported by the tax.

In 2016, British Columbia released the results of a Climate Leadership Plan for reducing greenhouse gas emissions while supporting a strong economy. Thousands of organizations and individuals contributed to the plan. As a result of this effort, more than 150 corporate leaders called for a $10 per ton increase in the carbon tax. They argued that the current price of $30 per ton is far below the real social, economic, and environmental costs of global climate change.

Interest in carbon taxes is spreading. Seven U.S. states have had proposals for carbon taxes, and Washington State was the first to vote on a carbon fee in 2018. Enthusiasm for the bill was high, but after oil companies spent $30 million fighting it, Washington voters rejected the proposal.

There are many ways to put a price on carbon, or to use economic policy to achieve social goals. In this chapter, we'll look at ecological economics and how we value both natural resources and human well-being. Finally, we'll examine other ways that market mechanisms can help us solve environmental problems, and how businesses can contribute to sustainability.

FIGURE 23.1 Revenues from British Columbia's carbon tax and personal and corporate tax reductions made possible by this revenue.
Data from Duke/Nicholas Institute

Benchmark Data

Among the ideas and values in this chapter, the following are a few worth remembering.

$173 trillion	2012 estimate of total biosphere ecological services
$78 trillion	2012 global GDP
$500,000	Average cost to restore 1 ha of coral reef
<$1 million	Average annual benefit from 1 ha of coral reef per year.
50,000	Number of coal miners in the United States in 2019
473,000	Number of direct solar and wind jobs in the United States in 2019
17 million	Chinese workers in green economy in 2019

23.1 PERSPECTIVES ON THE ECONOMY

- *Economics involves understanding how we evaluate and use resources.*
- *Resources can be nonrenewable, renewable, or intangible.*
- *Classical and neoclassical economics differ in how they define and analyze resources, growth, and price setting.*

Economy is the management of resources, ideally to meet our needs as efficiently as possible. The terms *ecology* and *economy* share a common root, *oikos* (ecos), the Greek word for "household." Economics is the *nomos,* or counting, of the household resources. Ecology is the *logos,* or logic, of how the household works.

Much of our economy involves using natural resources—such as minerals, soil, water, or solar energy—to produce goods (fig. 23.2). Some resources are renewable, others are not. Knowledge, experience, and creativity, for example, are valuable

FIGURE 23.2 Our economy depends on using resources to produce marketable goods. How we understand and value "resources" is a central question in environmental economics.
William P. Cunningham

resources that aren't exhausted by use. Economics has traditionally been involved with choices and trade-offs, between goods and services that are nonrenewable. Understanding the balance of costs and benefits of these choices is a major part of classical economics. Environmental economics, like environmental science, tends to ask questions about long-term resource use: Are we using resources efficiently? Are the costs of our resource use reflected in the prices we pay for goods? Are there alternative strategies that could help us produce goods and services with fewer resources?

Can development be sustainable?

One of the most important questions in environmental science is how we can continue to improve human welfare within the limits of the earth's natural resources and biological systems. Does our use of resources limit the opportunities of others—either future generations or people in other regions—to lead healthy and productive lives? This question often comes up in discussions of development around the world. As we saw in Chapter 1, *development* means improving people's lives, usually through increased access to goods (such as food) or services (such as education). *Sustainability* means living on the earth's renewable resources without damaging the ecological processes that support us all (table 23.1). **Sustainable development** is an effort to marry these two ideas. A definition developed by the World Commission on Environment and Development in 1987 is that "sustainable development is development that meets the needs of the present without compromising the ability of future generations to meet their own needs."

Is this possible? Not at our present population and rates of consumption. Some observers insist that there is no way more people can live at a high standard of living without irreversibly degrading our environment. Others say that as natural resources become scarce, we will simply find alternatives. Still others argue that there's enough for everyone if we can just share equitably and consume less. Much of this debate depends on how we define resources and economic growth.

Resources can be renewable or nonrenewable

A **resource** is anything with potential use in creating wealth or giving satisfaction. Natural resources can be either renewable or nonrenewable. In general, **nonrenewable resources** are materials present in fixed amounts in the environment, especially earth

Table 23.1	Goals for Sustainable Natural Resource Use

- Harvest rates for renewable resources (those like organisms that regrow, or those like fresh water that are replenished by natural processes) should not exceed regeneration rates.
- Waste emissions should not exceed the ability of nature to assimilate or recycle those wastes.
- Nonrenewable resources (such as minerals) may be exploited by humans, but only at rates equal to the creation of renewable substitutes.

FIGURE 23.3 Nonrenewable resources, such as the oil from this forest of derricks in Huntington Beach, California, are irreplaceable. Once they're exhausted (as this oil field was half a century ago), they will never be restored on a human time scale.
William P. Cunningham

FIGURE 23.4 Renewable resources, such as clean water, fresh air, plants, and animals, replenish themselves—unless we overuse and exhaust them. Often, we think of renewable resources in terms of ecosystem services. For instance, the freshwater marsh shown here provides purified water and air, freshwater storage, and biodiversity.
William P. Cunningham

resources such as minerals, metals, and fossil fuels (fig. 23.3). Many of these resources, such as oil and coal, are renewed or recycled over geological time, but on a human time scale they are not renewable. Predictions abound that we are in imminent danger of running out of one or another of these exhaustible resources. Supplies of metals and other commodities, however, have frequently been extended by more efficient use, recycling, substitution of one material for another, or new technologies that can extract resources from dilute or remote sources.

Renewable resources are things that can be replenished or replaced. These include living organisms, fresh water from rain and snow, and sunlight—our ultimate energy source. These systems also provide essential ecological services on which we depend, although most of us don't think of these resources very often (fig. 23.4). We discuss these ideas further in section 23.2.

Because biological organisms and ecological processes are self-renewing, we often can harvest surplus organisms or take advantage of ecological services without diminishing future availability, if we do so carefully. But persistence of these resources depends on how rapidly we exploit or degrade them. Even once-vast populations such as passenger pigeons, American bison, and Atlantic cod were exhausted by overharvesting in only a few years. Similarly, we are now reducing renewable water resources (from rainfall and snowpack) in many regions by modifying the climate system. This modification of a renewable resource is leading to drought and reduced crop production in dry regions (chapter 15). Mismanagement of renewable resources, then, often makes them more ephemeral and limited than nonrenewable resources.

We also depend on **intangible resources,** such as open space, beauty, serenity, wisdom, and diversity (fig. 23.5). Paradoxically, these resources can be both infinite *and* exhaustible. There is no upper limit to the amount of beauty, knowledge, or compassion that can exist in the world, yet they can be easily destroyed. A single piece of trash can ruin a beautiful vista, or a single cruel remark can spoil an otherwise perfect day. On the other hand, unlike tangible resources that usually are reduced by use or sharing, intangible resources often are increased by use and

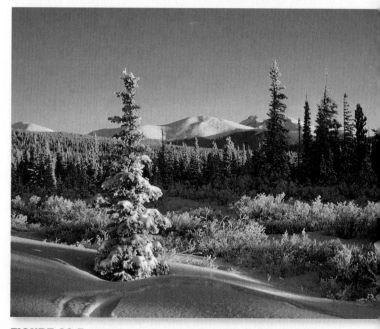

FIGURE 23.5 Intangible resources, such as the scenic beauty or solitude of this Colorado wilderness area, are widely treasured but are hard to evaluate in economic terms.
William P. Cunningham

multiplied by being shared. Nonmaterial assets can be important economically. Information management and tourism—both based on intangible resources—have become two of the largest and most powerful industries in the world.

Another term used to describe resources is **capital,** often defined as wealth that can be used to produce more wealth. Usually, capital refers to something that has been built up or accumulated over time. There can be many forms of capital. Banks provide *financial capital* (money) that small businesses need to start or grow. Economists also consider *manufactured* or *built capital* (tools, infrastructure, and technology), *natural capital* (goods and services provided by nature), *human* or *cultural capital* (knowledge, experience, and ideas about how to make or do things), and even *social capital* (shared values, trust, cooperative spirit, and community organization). All these kinds of capital may be needed to produce marketable goods and services.

Classical economics examines supply and demand

Originally, **classical economics** was a branch of moral philosophy concerned with how individual interest and values intersect with larger social goals. We trace many of our ideas about economy to Adam Smith (1723–1790), a moral philosopher concerned with individual freedom of choice. Smith's landmark book *Inquiry into the Nature and Causes of the Wealth of Nations,* published in 1776, argued:

> Every individual endeavors to employ his capital so that its produce may be of the greatest value. He generally neither intends to promote the public interest, nor knows how much he is promoting it. He intends only his own security, only his own gain. And he is in this led by an *invisible hand* to promote an end which was no part of his intention. By pursuing his own interests he frequently promotes that of society more effectually than when he really intends to.

This statement often is taken as a foundation for the capitalist system, in which willing sellers and fully informed buyers agree on a fair price for goods in the market. Smith proposed that this agreement would bring about the greatest efficiency of resource use, because efficiency is necessary to produce goods at an acceptably low price. Assuming that all buyers and sellers are free to make any choice, this system also ensures individual liberty (fig. 23.6). The British economist John Maynard Keynes summarized faith in free markets and pricing this way: "Capitalism is the astounding belief that the most wickedest of men will do the most wickedest of things for the greatest good of everyone."

In a real market, producing goods at a low cost often requires that some costs are **externalized,** or passed off to someone else. Environmental costs and social costs are often supported by communities. For example, producing electricity at a power plant requires a stable and educated work force to run the plant, and the cost of educating workers normally is borne by society in general. Transportation networks, mining and oil drilling, and insurance costs are often supported by the public. Tax breaks also represent a subsidy because they transfer costs from one taxpayer to other sectors of society. Producing electricity also usually involves

FIGURE 23.6 Informal markets such as this one in Bali, Indonesia, may be the purest example of willing sellers and buyers setting prices based on supply and demand.
William P. Cunningham

some pollution, and a power company allows the public to absorb the cost of that pollution (such as health care cost or reduced crop production). All these costs are involved in producing and selling electricity, but they are external to the company's cost calculations. Many economists, both conservative and liberal, argue that subsidies are a problem because they mask the real costs of production and lead to inefficiency in our economy.

David Ricardo (1772–1823), a contemporary of Adam Smith, introduced a description of the relationship between supply and demand in economics. **Demand** is the amount of a product or service that consumers are willing and able to buy at various possible prices, assuming they are free to express their preferences. **Supply** is the quantity of that product being offered for sale at various prices, other things being equal. Classical economics proposes that there is a direct, inverse relationship between supply and demand (fig. 23.7).

With increasing quantity of production, supply increases and prices fall. With decreasing quantity, prices rise and demand falls. The difference between the cost of production and the price buyers are willing to pay, Ricardo called "rent." Today we call it profit.

In an idealized free market of independent and rational buyers and sellers, an optimal price is achieved at the intersection of the supply and demand curves (fig. 23.7). This intersection is known as the **market equilibrium.** In real life, prices are not determined strictly by total supply and demand as much as by what economists call **marginal costs and benefits.** Sellers ask themselves, "What would it cost to produce one more unit of this product or service? Suppose I add one more worker or buy an extra supply of raw materials, how much more profit could I make?" Buyers ask themselves similar questions, "How much would I benefit and

FIGURE 23.7 Classic supply/demand curves. When price is low, supply is low and demand is high. As prices rise, supply increases but demand falls. Market equilibrium is the price at which supply and demand are equal.

FIGURE 23.8 The neoclassical model of the economy focuses on the flow of goods, services, and factors of production (land, labor, capital) between business and individual workers and consumers. The social and environmental consequences of these relationships are irrelevant in this view.

what would it cost if I bought one more widget?" If both buyer and seller find the marginal costs and benefits attractive, a sale is made.

There are exceptions to this theory of supply and demand. Consumers will buy some things regardless of cost. Raising the price of cigarettes, for instance, doesn't necessarily reduce demand. When prices don't change with supply and demand, we call it price *inelasticity*. Other items have **price elasticity:** Price always changes as supply and demand change.

Neoclassical economics emphasizes growth

Toward the end of the nineteenth century, the field of economics divided into two broad camps. **Political economy** continued the tradition of moral philosophy and concerned itself with social structures and relationships among the classes. This group included reformers such as Karl Marx and later E. F. Schumacher, who argued that unregulated capital accumulation inevitably leads to inequity, which leads to instability in society. The other camp, **neoclassical economics,** strove to adapt methods of modern science, and to be mathematically rigorous, noncontextual, abstract, and predictive. Neoclassical economists claim to be objective and value-free, leaving social concerns to other disciplines. Like their classical predecessors, they retain an emphasis on scarcity and the interaction of supply and demand in determining prices and resource allocation (fig. 23.8).

Constant economic growth is considered necessary and desirable, in the neoclassical view. Growth keeps people happy by always offering more income or goods than they had last year. In a growing population, economic growth is seen as the only way to maintain full employment and avoid class conflict arising from inequitable distribution of wealth. Growth is also essential because businesses borrow resources to operate and grow. Few lenders are willing to share their money without a promise of greater returns later. Thus, businesses must continue to expand in order to increase profits and maintain the confidence of shareholders, whose money they are using to run their operations.

John Stuart Mill (1806–1873), a classical economist and philosopher, argued that perpetual growth in material well-being is neither possible nor desirable. Economies naturally mature to a steady state, he proposed, leaving people free to pursue nonmaterialistic goals. He didn't regard this equilibrium state to be necessarily one of stagnation or poverty. Instead, he wrote that in a stable economy, "there would be as much scope as ever for all kinds of mental culture, and moral and social progress; as much room for improving the art of living, and much more likelihood of its being improved when minds cease to be engrossed by the art of getting on."

Some neoclassical economists point out that not all growth involves increased resource consumption and pollution. Growth based on education, entertainment, and nonconsumptive activities, as suggested by Mill, can still contribute to economic expansion.

Neoclassical economics tends to view natural resources as interchangeable. As one resource becomes scarce, neoclassical economists believe, substitutes will be found. Oil, for example, has substitutes in natural gas, in wind and solar power, and other energy sources. The need to reduce oil consumption has helped drive innovation in these alternatives, making them easier and cheaper to produce. Labor is also substitutable, by mechanization or technological change. Because materials and labor are substitutable, they are not considered indispensable. Debates about the nature of growth, consumption, and resource scarcity have become increasingly active lately, in response to recent developments in economics, including environmental economics, the concept of a steady-state economy, and sustainable development.

Section Review

1. Define nonrenewable, renewable, and intangible resources.
2. What does it mean to externalize costs?
3. Why does neoclassical economics emphasize growth?

23.2 ECOLOGICAL ECONOMICS

- *Ecological economics incorporates ecological principles.*
- *Accounting for externalized costs is an important challenge.*
- *Ecosystem services regulate, supply, provide, or produce resources we need. Often these ideas overlap somewhat.*

Classical and neoclassical economics shape most of our economic activities, but the externalized costs and social inequity of these approaches have led to an interest in new ways to evaluate economic progress. **Ecological economics** has emerged as a way to understand the relationship between our economy and the ecological systems that support it. These economists point out that even Adam Smith wrote that the profit motive is good at guiding decision making in the short term, but it is ill-suited to long-term decision making. Ecological economists argue that we need to improve our long-term decision making if we intend to be here for the long term.

Ecological economists have followed two main approaches to resolving the short-term view of conventional economics: One approach is to question the necessity of constant growth. Another is to identify externalities and calculate their costs. If we know these costs, we can make the price reflect the total costs of production, or we can reduce those costs by changing the ways we make things.

Ecological economics draws on ecological concepts such as systems (chapter 2), thermodynamics, and material cycles (chapter 3). In a system, all components are interdependent. Disrupting one component (such as climate conditions) risks destabilizing other components (such as agricultural production) in unpredictable and possibly catastrophic ways. Thermodynamics and material cycles teach that energy and materials are continually reused. One organism's waste is another's nutrient or energy source. This perspective allows us to consider relationships among the many parts of our environment, our economy, and our own well-being (fig. 23.9).

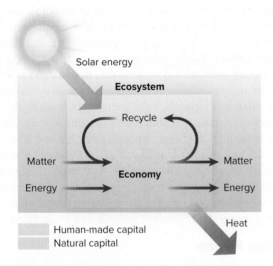

FIGURE 23.9 Ecological economics includes services such as recycling and resource provision and economic accounting. An effort is made to internalize, rather than externalize, natural resources and services.

Systems analysis also proposes that the earth has a limited carrying capacity for human populations. A limited carrying capacity means that unlimited economic growth is not possible. Many ecological economists, such as Herman Daly, argue for a **steady-state economy,** characterized by low human birth rates and death rates, the use of renewable energy sources, material recycling, and an emphasis on durability, efficiency, and stability. **Throughput,** the volume of materials and energy consumed and of waste produced, should be minimized. Most neoclassical growth models have assumed constantly increasing consumption and waste production, with resources being substituted as they run out. The steady-state idea, in contrast, reflects the notion of an ecosystem in equilibrium, or a population below its carrying capacity, where overall conditions remain generally stable over time.

Ecological economics accounts for the value of ecosystems

Ecological economics seeks to draw natural systems and ecological processes into economic accounting. Classical and neoclassical economics usually have focused on the value of human capital and built capital and resources, such as buildings, roads, or labor. Natural systems are considered free of cost—in part because in fact they have been free, abundant, or unregulated. Rivers absorb wastewater, bacteria decompose waste, and winds carry away smoke, but until recently there were few mechanisms for monitoring these processes or assigning them a value.

Consider energy production again, in which producers weigh their profits against the costs of coal, labor, buildings, and furnaces. These are **internal costs,** part of the normal expenses of doing business. Costs such as the climate's absorption of carbon dioxide, sulfur dioxide, or mercury are external costs, outside the accounting system. Similarly, if populations downwind suffer from air pollution, incurring lost workdays or illness, traditional economics has argued that society should bear the responsibility.

Can we internalize these costs? This is basically the idea of a carbon tax (opening case study)—assigning a price that helps account for costs to society and the environment for each unit of fuel used. More exact efforts to assign prices are not easy, but they are increasingly common. A groundbreaking study by Paul Epstein and his colleagues in 2011 produced an unusually complete estimate of the externalized costs of coal power, for example. Epstein and his colleagues examined the costs of land

---

Quantitative Reasoning

What do you suppose are the internal costs (those calculated into the price) when you buy a gallon of gas? What might be some external costs? How many of those costs are geographically expressed where you live? Are there any costs that might be paid for through your taxes? Are there any that are not paid for by your taxes? (*Hint:* Think of the conditions necessary to get oil safely delivered to your gas station. How are those conditions created or maintained?)

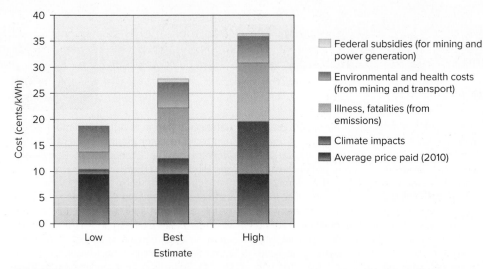

FIGURE 23.10 Estimates of externalized costs of coal power, in ¢/kilowatt-hour (kWh). Externalized costs (red colors) are shown added to the average price paid in 2010 (blue).

Legend for figure:
- Federal subsidies (for mining and power generation)
- Environmental and health costs (from mining and transport)
- Illness, fatalities (from emissions)
- Climate impacts
- Average price paid (2010)

Although many ecological processes have no direct market value, we can estimate replacement costs, contingent values, shadow prices, and other methods of indirect assessment to determine a rough value. For instance, we now dispose of much of our wastes by letting nature detoxify them. How much would it cost if we had to do this ourselves?

Estimates of the annual value of all ecological goods and services provided by nature vary over a wide range. A widely quoted 1997 estimate placed the value at $33 trillion per year. A more recent estimate calculated the value of nature to be $173 trillion per year, or more than twice the current global GDP (Exploring Science, p. 526).

disturbance, methane emissions from mines, fatalities during coal transportation, climate impacts, downwind costs to health, child development, cancers, direct federal subsidies to coal miners and energy producers, the costs of restoring abandoned mines, and other costs. Their best estimate of externalized costs was approximately 18¢/kilowatt-hour (kWh). The average cost of electric power in 2010 was about 9.5¢/kWh, or just half of the externalized costs. In other words, the real cost was about triple the price paid for electricity (fig. 23.10).

High and low estimates were also calculated in this study, to account for uncertainties in the data. These suggested that the public absorbs about 9¢ to 28¢ for every kWh of coal-based electricity. This study focused on coal because it has long been the world's dominant source of electric power, but similar accounting could be done for any power source or economic activity.

Accounting for all costs should make production more efficient, because an accurate price can help the public make more informed decisions. In general, the cost of cleaning up a power plant usually is lower than the cost of health care and lost productivity. In economic terms, the extra costs of illness and environmental damage are "market inefficiencies": They represent inefficient overall use of resources (money, time, energy, materials) because of incomplete accounting of costs and benefits.

Ecosystem services include provisioning, regulating, and aesthetic values

Ecosystem services is a general term for the resources provided and waste absorbed by our environment. These services are often grouped into four general classes (table 23.2): regulation (of climate, water supplies, and other factors), provision (of foods and other resources), supporting or preserving (of crop pollinators, nutrient cycling), and aesthetic or cultural benefits (fig. 23.11).

Table 23.2	Important Ecological Services
1.	*Regulate* global energy balance; chemical composition of the atmosphere and oceans; local and global climate; water catchment and groundwater recharge; production, storage, and recycling of organic and inorganic materials; maintenance of biological diversity.
2.	*Provide* space and suitable substrates for human habitation, crop cultivation, energy conversion, recreation, and nature protection.
3.	*Produce* oxygen, fresh water, food, medicine, fuel, fodder, fertilizer, building materials, and industrial inputs.
4.	*Supply* aesthetic, spiritual, historic, cultural, artistic, scientific, and educational opportunities and information.

Source: From Jansson, AnnMari, Monica Hammer, Carl Folke, and Robert Costanza. Investing in Natural Capital. *Washington, DC: Island Press, 1994.*

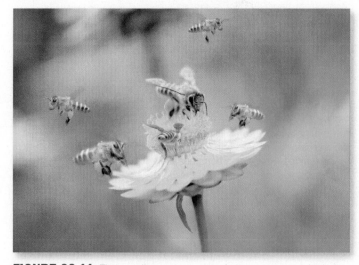

FIGURE 23.11 These pollinators, as part of an ecosystem or biological community, provide services essential for maintaining both our crops and the natural world on which we depend.
zefei/123RF

What's the Value of Nature?

The opening case study of this chapter discussed a carbon tax, which aimed to help put a price on carbon pollution because without a price, people disregard the importance of something like pollution, or clean air and climate stability. Another approach to assigning prices to otherwise intangible values is the idea of ecosystem services. We have always known that we depend on a healthy and resilient environment, but we have rarely had a means of knowing how much it was worth when it comes to decisions about preventing pollution, protecting forests, or defending biodiversity. Sometimes calculating value is easy: Loss of clean water or air has health impacts that we can quantify. Sometimes calculating value is hard: How can we quantify the value of biodiversity? Of butterflies? Of the cooling effect of standing forests?

We rarely appreciate or understand many of the ecosystem-regulating services from which we benefit. For example, natural predators that control insect pests may be mostly invisible to us. Or flood protection provided by wetlands or coastal barrier islands may be underappreciated or misunderstood. Climatic forces that regulate temperature, rainfall, winds, and sunshine may be unknown or misinterpreted by the public that benefits from their operation.

How do we estimate the value of ecosystem services? For some natural resources we can simply gather data from current consumption. How much do we pay presently for seafood or wood products? For intangible benefits or services that are now free, however, the calculation is more difficult. What's the value of a walk in the woods or the ability to gaze at a beautiful sunset? One approach is to question people about their "willingness-to-pay" for these opportunities. How much compensation would you accept if those benefits were denied to you? However, individuals may not have an accurate assessment of

how valuable a resource or an opportunity is until it's actually gone.

In some cases, we can measure the economic losses from storms or floods when coastal wetlands are destroyed or damaged, or we can calculate what it would cost to build substitute protections. But some natural resources, such as clean air or water, have no realistic substitute. We simply can't live without them. They're worth everything we have. Some people argue that it's offensive to try to put a price on existential resources. They are literally priceless. Nevertheless, it can help us make better decisions if we acknowledge how important these ecosystem services are.

A 2017 summary written by a large group of distinguished ecological economists estimated the value of all global ecosystem services to be equivalent to U.S. $173 trillion in current dollars. This study estimated values per unit area of ten major biomes based on a

database drawn from 320 publications from a number of research groups, including the UN's program on The Economics of Ecosystems and Biodiversity (TEEB).

Drawing on more than 600 individual estimates, the authors suggest that the value of the bundle of services that could be provided by these biomes can vary from $490/ha/yr for open oceans to almost $350,000/ha/yr for an "average" hectare of coral reefs (fig. 1). There is a large uncertainty in these estimates because most of the benefits are intangible and have no market price. Still, the "average" unit value for each of these biomes gives us some measure of the importance of protecting nature.

For more information, see Constanza, R., et al. 2017. "Twenty years of ecosystem services: How far have we come and how far do we still need to go?" *Ecosystem Services* 28 (2017) 1–16.

(header photo): David L. Hansen

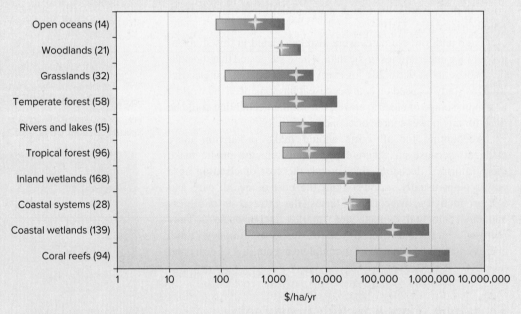

FIGURE 1 Estimated value of ecosystem services for major biomes in 2007 dollars/hectare. The number of estimates is shown in parentheses. Bars show the range of values, and stars show average estimates.
de Groot, et al. 2012. Ecosystem Services 1:50–61.

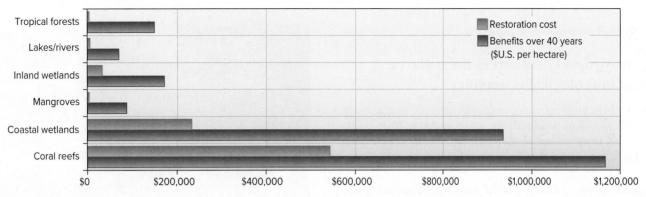

FIGURE 23.12 Can we afford to restore biodiversity? It's harder to find money to restore ecosystems than to destroy them, but the benefits over time greatly exceed the average cost of restoration.

Accounting for ecosystem services is a focus of several global initiatives on sustainable development. A UN program called The Economics of Ecosystems and Biodiversity (TEEB) has been working to improve estimates of the value of ecosystem services. TEEB studies have shown that preserving ecosystems is far more cost-effective than using up their resources. Even restoring already-damaged ecosystems has enormous paybacks (fig. 23.12). Calculating a price for carbon storage in natural ecosystems has been the aim of REDD (Reducing Emissions from Deforestation and Forest Degradation) programs. These efforts are discussed in chapter 12. A 2003 economic study from Cambridge University (UK) estimated that protecting a series of nature reserves representing samples of all major biomes would cost $45 billion per year, but would preserve ecological services worth 100 times that cost—$4.4 trillion to $5.2 trillion annually.

Section Review

1. What is ecological about ecological economics?
2. What are ecosystem services? Give several examples.
3. List several externalized costs of power production with coal.
4. What is TEEB, and what are its aims?

23.3 POPULATION, SCARCITY, AND TECHNOLOGY

- *The question of how to avoid depleting finite resources, such as oil, has long worried ecologists and economists.*
- *The tragedy of the commons is one way to understand scarce resources; communal resource management is another.*
- *There is disagreement on whether resources are fixed.*

Despite changing perspectives on resources, many analysts continue to ask the question that worried Adam Smith and Thomas Malthus (chapter 7): Are we about to run out of essential natural resources? It stands to reason that if we consume a fixed supply of nonrenewable resources at a constant rate, eventually we'll use up all the economically recoverable reserves. The dismal Malthusian prospects of diminishing resources, starvation, and social decay inspire many people to call for immediate changes in our consumption rates. Others respond by saying that we simply need to redefine how we think about resources. In this section, we consider some of the ways we might understand resource limits and growth.

Are we about to run out of fossil fuels?

In the last half of the twentieth century, many economists argued that the world was going to run out of oil. This prediction was generally associated with M. King Hubbert, a Shell Oil geologist who predicted that oil production would start to decline at the beginning of the twenty-first century (see fig. 19.10).

Predictions of "peak oil" and declining supplies have dominated oil supply projections for decades, but they account poorly for resource substitution and for technological change. Hubbert couldn't have anticipated the effects of new drilling techniques and the discoveries of unconventional oil, including fracking, that have dramatically increased known supplies. He also didn't account for new evidence about the high cost of extracting and using all that oil, which have driven innovation in alternative energy technologies.

Even without concerns about climate change, transitions to new energy technologies, such as wind or solar, might have reduced our reliance on fossil fuels before supplies are entirely depleted. It's often said that the stone age didn't end because we ran out of stones. And the fossil fuel age may not end because we're out of fossil fuels, but because we choose to not use them all.

Climate change, then, does provide further support for resource substitution. It also gives a new perspective of the Hubbert curve. That is, we simply cannot use most of our remaining fossil fuels. Most of the trillions of dollars on the books of major oil and gas companies represents resources still in the ground. It is increasingly clear that these resource cannot be used without catastrophic consequences. It matters little how many years' supplies we have if we cannot use it anyway. Many economists now argue that if oil and gas resources are unburnable (without endangering life as we know it), then its value—and the value of the companies and the shares it supports—will disappear.

Common property resources are a classic problem in ecological economics

One of the difficulties of economics and resource management is that there are many resources we all share but nobody owns. Clean air, fish in the ocean, clean water, wildlife, and open space are all natural amenities that we exploit but that nobody clearly controls.

In 1968, biologist Garret Hardin wrote an influential article entitled "The Tragedy of the Commons." He argued that any commonly held resource inevitably is degraded or destroyed because the narrow self-interest of individuals tends to outweigh public interest. Hardin offered as a metaphor the common woodlands and pastures held by most colonial New England villages. In deciding how many cattle to put on the commons, Hardin explained, each villager would attempt to maximize his or her own personal gain. Adding one more cow to the commons could mean a substantially increased income for an individual farmer. The damage done by overgrazing, however, would be shared among all the farmers (fig. 23.13). This is known as the "free rider" problem, where individuals take more than their fair share of commonly held resources. Hardin concluded that the only solutions would be either to give coercive power to the government or to privatize the resource.

Hardin intended this parable to warn about human overpopulation and overexploitation of resources. Other authors have used his metaphor to explain such diverse problems as famines, air pollution, or collapsing fisheries.

What Hardin was really describing, however, was an **open-access system** in which there are no rules to manage resource use. In fact, the work of Nobel laureate Elinor Ostrom and others shows that many communal resources have been successfully managed for centuries by cooperative arrangements among users. Some examples include Native American management of wild rice beds and hunting grounds; Swiss village-owned mountain forests and pastures; Maine lobster fisheries; communal irrigation systems in Spain, Bali, and Laos; and nearshore fisheries almost everywhere in the world (fig. 23.14).

A large body of literature has developed to describe how **common property management** works. Features of these systems, also known as communal resource management systems, include the following: (1) community members have lived on the land or used the resource for a long time and anticipate that their children and grandchildren will as well, thus giving them a strong interest in sustaining the resource and maintaining bonds with their neighbors; (2) the resource has clearly defined boundaries; (3) the community group size is known and enforced; (4) the resource is relatively scarce and highly variable so that the community is forced to be interdependent; (5) management strategies appropriate for local conditions have evolved over time and are collectively enforced; that is, those affected by the rules have a say in them; (6) the resource and its use are actively monitored, discouraging

FIGURE 23.13 Adding more cattle to the Brazilian Cerrado (savanna) increases profits for individual ranchers, but is bad for biodiversity and environmental quality.
William P. Cunningham

anyone from cheating or taking too much; (7) conflict resolution mechanisms reduce discord; and (8) incentives encourage compliance with rules, while sanctions for noncompliance keep community members in line.

Rather than being the only workable solution to problems in common-pool resources, privatization and increasing external controls often prove to be disastrous. Where small villages have

FIGURE 23.14 These rice terraces on the island of Bali have been cultivated and managed communally for centuries without depleting productivity.
William P. Cunningham

owned and managed jointly held forests or fishing grounds for generations, privatization has led to shortsighted decision making, leading to rapid destruction of both society and ecosystems. A tragic example of this was the forced privatization of Indian reservations in the United States. Where communal systems once enforced restraint over harvesting, privatization encouraged narrow self-interest. With individuals making decisions for personal, near-term benefit, many people chose to sell their resources to outsiders, who could easily take advantage of the weakest members of the community. Failing to recognize or value local knowledge and forcing local people to participate in a market economy allowed outsiders to disenfranchise native people and resulted in disastrous resource exploitation. Distinguishing between open-access systems and communal property regimes is important in understanding how best to manage natural resources.

Scarcity can lead to innovation

In a pioneer or frontier economy, methods for harvesting resources and turning them into useful goods and services tend to be inefficient and wasteful. The history of logging in the United States, for example, is a classic case of inefficient resource exploitation. Between about 1860 and 1930, the supply of American forests was vast and unregulated. Logging companies cleared eastern states, then swarmed across the Great Lakes forests. As prime timber in the northern forests was depleted, the companies simply shifted to the Rocky Mountains and the Pacific Northwest. At each stage, logging wasted a vast amount of wood, but this inefficiency seemed unimportant because the supply of trees was so great. Labor, financial capital, and transportation to market were the scarce resources.

Today our use of forest products is considerably more efficient. We have smaller supplies and greater demand, and we have developed better technology and methods that allow us to create the same amount of goods and services using fewer resources. Instead of using giant old-growth timbers for building, we use laminated beams, chipboard, and other products that can be produced from what once was scrap wood. We also require that logging companies replant new forests after logging, because they can no longer simply move on to a new region.

Scarcity often is a catalyst for innovation and change (fig. 23.15). As materials become more expensive and difficult to obtain, it becomes cost-effective to discover new supplies or to use available ones more efficiently. Several important factors play a role in this cycle of technological development:

- Technical inventions can increase the efficiency of extraction, processing, use, and recovery of materials.

- Substitution of new materials or commodities for scarce ones can extend existing supplies or create new ones.

- Trade makes remote supplies of resources available and may also bring unintended benefits in information exchange and cultural awakening.

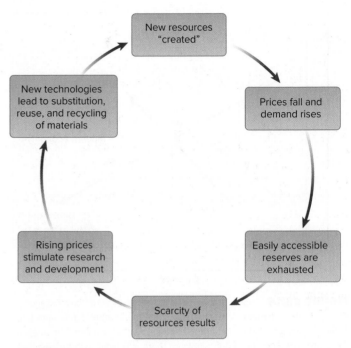

FIGURE 23.15 Scarcity/development cycle. Paradoxically, resource use and depletion of reserves can stimulate research and development, the substitution of new materials, and the effective creation of new resources.

- Discovery of new reserves through better exploration techniques, more investment, and looking in new areas becomes rewarding as supplies become limited and prices rise.

- Recycling becomes feasible and accepted as resources become more valuable. Recycling now provides about 37 percent of the iron and lead, 20 percent of the copper, 10 percent of the aluminum, and 60 percent of the antimony consumed each year in the United States.

Increasing technological efficiency can dramatically shift supply-and-demand relationships. As technology makes goods and services cheaper to produce, the quantity available at a given price can increase greatly. The market equilibrium, or the point at which supply and demand equilibrate, will shift to lower prices and higher quantities as a market matures (fig. 23.16).

Carrying capacity is not necessarily fixed

Despite repeated warnings that rapidly growing populations and increasing affluence are bound to exhaust natural resources and result in rapid price increases, technological developments have resulted in price decreases for most raw materials over the last hundred years. Consider copper, for example. Twenty years ago, worries about impending shortages led the United States to buy copper and store it in strategic stockpiles. Estimated demand for this important metal, essential for electric motors, telephone lines, transoceanic cables, and other uses, far exceeded known reserves. It looked as if severe shortages and astronomical price increases

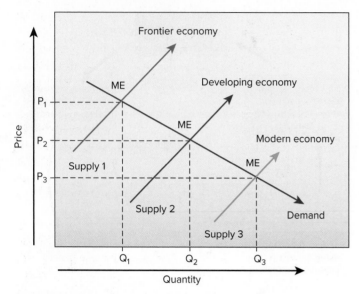

FIGURE 23.16 Supply and demand curves at three different stages of economic development. At each stage, there is a market equilibrium point at which supply and demand are in balance. As the economy becomes more efficient, the equilibrium shifts so there is a larger quantity available at a lower price than before. (P = price, Q = quantity, ME = market equilibrium)

were inevitable. But then aluminum power lines, satellites, fiber optic cables, integrated circuits, microwave transmission, and other inventions greatly diminished the need for copper. Although prices are highly variable, the general trend for most materials was downward in the twentieth century.

Recent reports have warned that increasing demand for consumer goods and infrastructure in China will raise demand, and prices, to previously unimagined levels. Will this increase lead to shortages and crisis? Or will it lead to innovation and resource substitution? Economists generally believe that substitutability and technological development will help us avoid catastrophe. Ecologists generally argue that there are bound to be limits to how much we can consume.

A noted example of this debate occurred in 1980. Ecologist Paul Ehrlich bet economist Julian Simon that increasing human populations and growing levels of material consumption would inevitably lead to price increases for natural resources. They chose a package of five metals—chrome, copper, nickel, tin, and tungsten—priced at the time at $1,000. If, in ten years, the combined price (corrected for inflation) was higher than $1,000, Simon would pay the difference. If the combined price had fallen, Ehrlich would pay. In 1990, Ehrlich sent Simon a check for $576.07; the price for these five metals had fallen 47.6 percent.

Does this prove that resource abundance will continue indefinitely? Hardly. Ehrlich argued that the timing and set of commodities chosen simply were the wrong ones. The fact that we haven't yet run out of raw materials doesn't mean that it will never happen. Many ecological economists now believe that some nonmarket resources such as ecological processes may be more irreplaceable than tangible commodities like metals. What do you think? Are we approaching limits to consumption? Which resources, if any, do you think are most likely to be limiting in the future?

Economic models compare growth scenarios

In the early 1970s, an influential study of resource limitations was funded by the Club of Rome, an organization of wealthy business owners and influential politicians. The study was carried out by a team of scientists from the Massachusetts Institute of Technology headed by the late Donnela Meadows. The results of this study were published in the 1972 book *Limits to Growth*. A complex computer model of the world economy was used to examine various scenarios of different resource depletion rates, growing population, pollution, and industrial output.

Given the Malthusian assumptions built into this model, catastrophic social and environmental collapse seemed inescapable (fig. 23.17a). Food supplies and industrial output rise as population grows and resources are consumed. Once past the carrying capacity of the environment, however, a crash occurs as population, food production, and industrial output all decline precipitously. Pollution continues to grow as society decays and people die, but, eventually, it also falls. Notice the similarity between this

(a) Business as usual

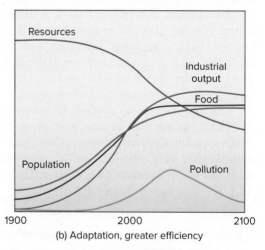

(b) Adaptation, greater efficiency

FIGURE 23.17 Models of resource consumption and scarcity. Running the model with assumptions of Malthusian limits and high consumption causes food, productivity, and populations to crash, while pollution increases (left). Running the same model with assumptions of slowing population growth and consumption, with better technologies, produces stable output and population (right). Which of these models will we follow?

set of curves and the "boom and bust" population cycles described in chapter 6.

Many economists criticized these results because the models discount technological development and factors that might mitigate the effects of scarcity. In 1992, the Meadows group published updated computer models in *Beyond the Limits* that include technological progress, pollution abatement, population stabilization, and new public policies that work for a sustainable future. If we adopt these changes sooner rather than later, all factors in the model stabilize sometime in this century at an improved standard of living for everyone (fig. 23.17*b*). Of course, neither of these computer models shows what will happen, only what some possible outcomes *might* be, depending on the choices we make.

Section Review

1. Explain the tragedy of the commons.
2. What is common property management?
3. How do the two "limits to growth" graphs (fig. 23.17) differ and why?

23.4 MEASURING GROWTH

- *GNP and GDP are common, though imperfect, measures of productivity.*
- *GPI, EPI, HDI, and WHI are some alternative measures of progress.*
- *Cost–benefit analysis is used to assess policies, although assigning costs accurately is difficult.*

How do we monitor our economic progress? In order to know if conditions in general are getting better or worse, economists have developed indices that countries or regions can use to monitor change over time. These indices track a variety of activities and values, to produce an overall picture of the economy. Which factors we choose to monitor, though, reflect judgments about what is important in society, and those judgments can vary substantially.

GNP is our dominant growth measure

The most common way to measure a nation's output is **gross national product (GNP).** GNP can be calculated in two ways. One is the money flow from households to businesses in the form of goods and services purchased. The other is to add up all the costs of production in the form of wages, rent, interest, taxes, and profit. In either case, a subtraction is made for capital depreciation, the wear and tear on machines, vehicles, and buildings used in production. Some economists prefer **gross domestic product (GDP),** which includes only the economic activity within national boundaries. Thus, the vehicles made and sold by Ford in Europe don't count in American GDP.

Both GNP and GDP have been criticized as measures of real progress or well-being because they don't attempt to distinguish between beneficial activities and harmful activities. A huge oil spill that pollutes beaches and kills wildlife, for example, shows up as a positive addition to GNP because it generates economic activity in the costs of cleanup. Or if many people die from cancer or AIDS, there is a great deal of economic activity for doctors and undertakers.

Ecological economists also argue that GNP doesn't account for natural resources used up or ecosystems damaged by economic activities. Robert Repeto of the World Resources Institute estimates that soil erosion in Indonesia, for instance, reduces the value of crop production about 40 percent per year. If natural capital were taken into account, Indonesian GNP would be reduced by at least 20 percent annually.

Similarly, Costa Rica experienced impressive increases in timber, beef, and banana production between 1970 and 1990. But decreased natural capital during this period represented by soil erosion, forest destruction, biodiversity losses, and accelerated water runoff add up to at least $4 billion or about 25 percent of annual GNP.

Alternate measures account for well-being

A number of systems have been proposed as alternatives to GNP that reflect genuine progress and social welfare. In their 1989 book, Herman Daly and John Cobb proposed a genuine progress index (GPI) that takes into account real per capita income, quality of life, distributional equity, natural resource depletion, environmental damage, and the value of unpaid labor. They point out that while per capita GDP in the United States nearly doubled between 1970 and 2000, per capita GPI increased only 4 percent (fig. 23.18). Some social service organizations would add to this index the costs of social breakdown and crime, which would decrease real progress even further over this time span.

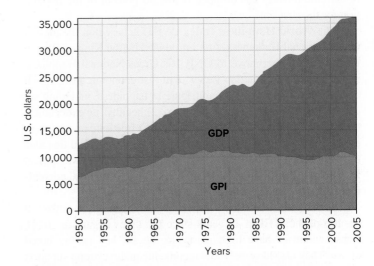

FIGURE 23.18 Although per capita GDP in the United States nearly doubled between 1970 and 2000 in inflation-adjusted dollars, a genuine progress index that takes into account natural resource depletion, environmental damage, and options for future generations hardly increased at all.

A newer measure is the **Environmental Performance Index (EPI)** created by researchers at Yale and Columbia Universities to evaluate national sustainability and progress toward achievement of the United Nations Millennium Development Goals. The EPI is based on 16 indicators tracked in six categories: environmental health, air quality, water resources, productive natural resources, biodiversity and habitat, and sustainable energy. The top-ranked countries—New Zealand, Sweden, Finland, the Czech Republic, and the United Kingdom—all commit significant resources and effort to environmental protection. In 2006, the United States ranked 28th in the EPI, or lower than Malaysia, Costa Rica, Columbia, and Chile, all of which have a GDP that is between 6 and 15 times lower than the United States.

The United Nations Development Programme (UNDP) uses a benchmark called the **Human Development Index (HDI)** to track social progress. HDI incorporates life expectancy, educational attainment, and standard of living as critical measures of development. Gender issues are accounted for in the gender development index (GDI), which is simply HDI adjusted or discounted for inequality or achievement between men and women.

HDI statistics note encouraging progress in some areas. Child death rates in developing countries as a whole have been more than halved. Average life expectancy has increased by 30 percent while malnutrition rates have declined by almost a third. The proportion of children who lack primary school has fallen from more than half to less than a quarter. And the share of rural families without access to safe water has fallen from nine-tenths to about one-quarter.

Poverty has also declined in most countries. In China and other Asian countries, the proportion of people living below the poverty rate has fallen by half. Globally, according to the World Bank, the number of people living in extreme poverty (less than U.S. $1.25/day) fell from 1.8 billion in 1990 to about 1.3 billion in 2018. Even in sub-Saharan Africa, the number of people living in extreme poverty has dropped below 50 percent of the population for the first time ever.

The **World Happiness Index (WHI)** is based on the idea of Gross National Happiness, first introduced by the King of Bhutan in 1972 as an alternative to GNP. His Gross National Happiness index looks at satisfaction rather than just economic progress. It considers factors known to contribute to happiness, including six closely tied to well-being: (1) income, (2) healthy life expectancy, (3) low corruption, (4) social networks, (5) freedom, and (6) generosity (fig. 23.9). In 2011, the United Nations borrowed this idea to evaluate progress. Every year, the Sustainable Development Division of the UN publishes the WHI, a survey of global contentment and well-being, as a measure of achievement.

Scandinavian countries, especially Denmark, have ranked at the top of the WHI in all years since the index began (in 2012). These countries have high marks in the six key factors noted above. They also have very low income inequality, strong supports for new parents, free or low-cost childcare and early childhood education, universal health care, gender equity, a healthy lifestyle, a generally positive outlook, and dignity with meaningful activities for all ages.

FIGURE 23.19 Strong social networks, including cultural heritage and well-supported families, contribute to gross national happiness.
William P. Cunningham

Many countries in recent years have achieved economic growth at the cost of sharply rising inequality, entrenched social exclusion, and grave damage to the natural environment. Paying attention to economic, social, and environmental objectives advances happiness and general satisfaction for both present and future generations.

Cost–benefit analysis aims to optimize benefits

One way to evaluate public projects is to analyze the costs and benefits they generate in a **cost–benefit analysis (CBA).** This process attempts to assign values to resources, as well as to the social and environmental effects of carrying out or not carrying out a given undertaking. It tries to find the optimal efficiency point at which the marginal cost of pollution control equals the marginal benefits (fig. 23.20).

CBA is one of the main conceptual frameworks of resource economics and is used by decision makers around the world as a way of justifying the building of dams, roads, and airports, as well as in considering what to do about biodiversity loss, air pollution, and global climate change. This approach became entrenched in U.S. policy making in the 1980s, when President Reagan ordered cost-benefit analysis for all regulatory decisions and legislative proposals. Opponents of regulation have seen CBA as a way to eliminate environmental protections. Often, though, CBA has provided strong evidence for the value and efficacy of rules that protect public health, such as the Clean Air Act (chapter 16).

The first step in CBA is to identify who or what might be affected by a particular plan. What are the potential outcomes and results? What alternative actions might be considered? After identifying and quantifying all effects, an attempt is made to assign monetary costs and benefits to each one. Usually, the direct expenses of a project are easy to ascertain. How much will you

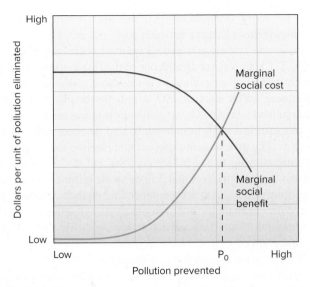

FIGURE 23.20 To achieve maximum economic efficiency, regulations should require pollution prevention up to the optimum point (P_0) at which the costs of eliminating pollution just equal the social benefits of doing so.

have to pay for land, materials, and labor? The monetary value of lost opportunities—to swim or fish in a river, or to see birds in a forest—is much harder to appraise (fig. 23.21). How would you put a price on good health or a long life? It's also important to ask who will bear the costs and who will reap the benefits of any proposal. Are there external costs that should be accounted for? Eventually, the decision maker compares all the costs and benefits to see whether the project is justified or whether some alternative action might bring more benefits at less cost.

FIGURE 23.21 What is the value of solitude or beauty? How would you assign costs and benefits to a scene such as this?
William P. Cunningham

Because of the difficulty of assigning monetary prices to intangible or public resources we value, many people object to CBA. In analyzing the costs and benefits of a hydroelectric dam, for example, economists often cannot assign suitable values to land, forests, streams, fisheries and livelihoods, and community. Ordinary people often cannot answer questions about how much money they would pay to save a wilderness or how much they would accept to allow it to be destroyed.

Cost–benefit analysis is also criticized for its absence of standards and inadequate attention to alternatives. Who judges how costs and benefits will be estimated? How can we compare things as different as the economic gain from cheap power with loss of biodiversity or the beauty of a free-flowing river? Critics claim that placing monetary values on everything could lead to a belief that only money and profits count and that any behavior is acceptable as long as you can pay for it. Sometimes speculative or even hypothetical results are given specific numerical values in CBA and then treated as if they are hard facts. Risk-assessment techniques (see chapter 8) may be more appropriate for comparing uncertainties.

Section Review

1. What are GNP and GPI? How do the two differ?
2. List several factors that contribute to the WHI.
3. What is the purpose of cost–benefit analysis?

23.5 CAN MARKETS REDUCE POLLUTION?

- *Setting a price for emissions can create incentives for reducing them.*
- *Some pollutants are easier to price effectively than others, and setting the right price can be difficult.*

We are becoming increasingly aware that our environment and economy are mutually interconnected. Natural resources and ecological services are essential for a healthy economy, and a vigorous economy can provide the means to solve environmental problems. In this section, we'll explore some of these links.

Economists are gaining confidence and precision in calculating financial costs of environmental damage. This increased confidence was one of the key points in the Fourth National Climate Assessment (NCA4), issued late in 2018 by 13 U.S. federal agencies. The NCA4 reported that in just three years, climate change had cost the United States nearly $400 billion, owing to extreme temperatures, intense droughts and storms, and flooding. Economic losses will reach hundreds of billions of dollars by 2100 if warming continues at the current pace. Loss of crops, real estate, infrastructure, and other sectors will lower U.S. GDP by 10 percent.

The report noted that climate change is already transforming where we live, our quality of life, and our economy. It is no longer a future threat. Coastal states and major cities are especially vulnerable, with rising sea level, greater storm surges, more intense storms, and falling property values. The NCA4 emphasized that more than 90 percent of observed warming is caused by human activities.

The Trump administration rejected the conclusions of its own agencies and tried to bury the report by releasing the contents late on the Friday after Thanksgiving. Ironically, though, this step gave the public a leisurely holiday weekend to read, discuss, and report on its findings.

The NCA4 built on previous studies on economic impacts of climate change. Most notably, the British economist Sir Nicholas Stern has warned that global losses are likely to reach 20 percent of global GDP, whereas preventing major losses would cost about 1 percent of global GDP. Bold climate action, according to a 2018 Stern report, could have $26 trillion in economic benefits by 2030.

Reducing greenhouse gas emissions and adapting to climate change will create significant business opportunities as new markets are created in low-carbon energy technologies and services (fig. 23.22). These markets could create millions of jobs and be worth hundreds of billions of dollars every year (Exploring Science, p. 535). Already, Europe has more than 5 million jobs in renewable energy, and the annual savings from solar, wind, and hydro power are saving the European Union about $10 billion per year in avoided oil and natural gas imports. Being leaders in the fields of renewable energy and carbon reduction gives pioneering countries a tremendous business advantage in the global marketplace. Markets for low-carbon energy could be worth $500 billion per year by 2050, according to the Stern report.

Sulfur trading offers a good model

In addressing climate change, pollution, or other issues, economists generally hold market forces are the best way to determine where, when, and by whom pollution reductions should be made. In theory, competition in a free market will always find the most efficient balance between costs and benefits.

Emissions trading is a leading example of this approach. The 1990 U.S. Clean Air Act created one of the first market-based systems for reducing air pollution. It mandated a decrease in acid-rain-causing sulfur dioxide (SO_2) from power plants and other industrial facilities. An SO_2 targeted reduction was set at 10 million tons per year, leaving it to industry to find the most efficient way to do this. The government expected that meeting this goal would cost companies up to $15 billion per year, but the actual cost has been less than one-tenth of that. Prices on the sulfur exchange have varied from $60 to $800 per ton depending on the availability and price of new technology, but most observers agree that the market has found much more cost-effective ways to achieve the desired goal than rigid rules would have required.

This program is regarded as a shining example of the benefits of market-based approaches. There are complaints, however, that while nationwide emissions have come down, "hot spots" remain where local utilities have paid for credits rather than install pollution abatement equipment. If you're living in one of these hot spots and continuing to breathe polluted air, it's not much comfort to know that nationwide average air quality has improved. Currently, credits and allowances of more than 30 different air pollutants are traded in international markets.

Emissions trading rewards efficiency

The Kyoto Protocol, which was negotiated in 1997, and has been ratified by every industrialized nation in the world except the United States and Australia (oil-rich Canada also withdrew in 2012), sets up a mechanism called **emissions trading** to control greenhouse gases. This is also called a **cap-and-trade** approach. The first step is to mandate upper limits (the cap) on how much each country, sector, or specific industry is allowed to emit. Companies that can cut pollution by more than they're required to can sell the **carbon credits** to other companies that have more difficulty meeting their mandated levels.

Suppose you've just built a state-of-the-art power plant that allows you to capture and store CO_2 for about $20 per ton, and that allows you to cut your CO_2 emissions far below the amount you are permitted to produce. Suppose, further, that your neighboring utility has a dirty, old coal-fired power plant for which it would cost $60 per ton to reduce CO_2 emissions. You might strike a deal with your neighbor. You reduce your CO_2 emissions, and he pays you $40 for each ton you reduce, so he doesn't have to reduce. You make $20 per ton, and your neighbor saves $20 per ton. Both of you benefit. On the other hand, if your neighbor can find an even cheaper way to offset his carbon emissions (that is, to pay for someone else to reduce emissions), he's free to do so. This creates an incentive to continually search for ever more cost-effective ways to reduce emissions.

Opportunities are increasing for all of us to buy personal carbon offsets. When you buy an airplane ticket, for example, some airlines offer you the chance to pay a few extra dollars, which will be used to pay for projects to reduce greenhouse gas emissions.

FIGURE 23.22 Markets for low-carbon energy could be worth $500 billion per year by 2050, and could create millions of high-paying jobs.

jorgenjacobsen/E+/Getty Images

Green Jobs Versus Fossil Fuels

President Trump has said that he "loves fossil fuels" and promises to bring back "so many beautiful jobs in coal mining," but the evidence suggests that he's betting on the wrong horse in this race. In 2018, solar energy employed more than twice as many Americans as coal. Wind and bioenergy engaged half again the total number in coal. Mechanization and competition from cheap natural gas have eliminated thousands of jobs in coal mining. In 1923, the United States employed nearly 863,000 coal miners. But in 2016, although total coal production had nearly doubled, the number of miners had fallen 95 percent to only about 50,000.

More than 373,000 Americans work part or full time in solar energy. These jobs are distributed in communities across the country, making them available to a wide variety of people. The solar sector has more jobs across the country, but coal is still a more important employer in a few places. Five states—Wyoming, West Virginia, Kentucky, Illinois, and Pennsylvania—produce almost three-fourths of all coal mined in the United States. Not surprisingly, those states represent some of the strongest political support for President Trump. According to a Columbia University study, Kentucky, a state that voted overwhelmingly for Trump, lost 64 percent of its coal jobs between 2011 and 2016. Most of Trump's attempts to restore those jobs have centered on reducing regulations (such as

mine safety, pollution prevention, and medical care for workers), but critics claim that overturning rules helps mine owners while reducing security for employees.

Wind power is the fastest-growing energy sector in the United States, increasing by about 20 percent to more than 100,000 jobs in 2018. The U.S. Department of Energy projects that by 2050 more than 600,000 wind-related jobs, including manufacturing, installation, and

elenathewise/123RF

maintenance, will be available. Although a majority of these jobs will be in the Midwest and Great Plains where winds are strongest, some workers will be needed in every state to support this rapidly growing sector. Wind turbine technician isn't just one of the fastest-growing jobs in clean energy—it's the single fastest-growing

occupation in America, according to the Bureau of Labor Statistics. And it pays well, with a median income of $52,000 per year, which is set to double by 2024, the agency estimates.

Solar installation is another rapidly growing U.S. job class. More than 73,000 jobs in this industry were added in 2016, a 25 percent increase over the previous year. A benefit of solar power is that jobs exist wherever the sun shines. You aren't limited to a few coal-producing states for where you live. Growth in solar has sparked an array of opportunities for other supporting industries that make or distribute solar equipment or batteries. Solar power is expanding 17 times faster than the total U.S. economy, according to the International Renewable Power Agency. Although men have most of the jobs currently in solar installation, that is starting to change, especially in the sales business. Women now hold 28 percent of solar jobs, up from 19 percent in 2013.

Workers who have lost jobs in the fossil fuel industry deserve compassion, but it makes little sense to prop up industries that contribute to irreversible climate change. Old industries always have suffered when new technologies replace them. Alternatives can retool the economy and create jobs in sustainable occupations that allow workers and their communities to transition to a new system.

(header photo): David L. Hansen

You can also buy carbon offsets if you have an old, inefficient car: For about $20 per ton (or about $100 per year for the average American car), you can pay for someone elsewhere to plant trees, build a windmill, or provide solar lights to a village in a developing country to compensate for your emissions. You can take pride in being carbon-neutral at a far lower price than buying a new automobile.

Emissions trading is widely implemented, most notably in the European Union, in China, and in several U.S. states and Canadian

provinces. The approach involves minimal regulation, so economists and conservatives approve, but it addresses emissions, so liberals approve. The tricky issue is that markets often fail to produce prices high enough to inspire reductions. Economists consider about US$30 per ton of CO_2 equivalent to be the threshold to trigger a shift to renewable energy. The real social cost of carbon emissions (taking all the damage from climate change into account) is probably well over US$100 per ton. But the traded price of carbon has often been $10 per ton or less. This weakness in carbon markets occurs when regulators issue too many pollution permits, so that supply exceeds demand. Economic downturns, like that after 2008, and the rapid decrease in renewable energy prices, also weaken carbon markets.

In the early stages of emissions marketing, more than 80 percent of the international emissions payments went to just four countries, and nearly two-thirds of those payments were for reductions of the refrigerant HFC-23 (fig. 23.23). Critics of our current emissions markets point out that (so far) carbon trading has produced a great deal of wealth for banks and stock brokers, but has done little to reduce actual carbon emissions or help reduce inequity. Critics charge that rather than finance renewable energy in developing countries, marketing emission credits is primarily benefiting bankers, consultants, and factory owners so far and is leading to short-term fixes rather than fundamental, long-term solutions.

Are carbon taxes a better answer?

A criticism of cap-and-trade is that it has no mechanism to evaluate or price external costs of complex pollutants, such as greenhouse gases. Corporations calculate how much it costs to capture pollutants, not how much it may cost future generations to live in a hotter, more unstable world.

These are among the reasons British Columbia (opening case study) decided that a carbon tax could be an effective way to achieve the goals of reducing GHG emissions, as well as financing other social programs. The BC carbon tax covers a wide base. It started low to ease the transition, and rose to a more substantive level, roughly in line with estimates of the damage per ton or the "social cost of carbon." Rather than have the profits from pollution reduction go into corporate coffers, as they do in emissions trading, the tax revenue can be used to promote social goals or to finance clean industries. The revenues in British Columbia, for example, are used to lower income taxes on businesses and households, as well as to benefit disadvantaged low-income households and residents of remote, rural areas. Reporting of the sources and uses of carbon tax funds is subject to a highly transparent process, in which

politicians and the public can track how the revenues are used. Do these taxes cost jobs? No. The four Canadian provinces with a price on pollution lead the country in both economic and job growth.

Sweden also has used a carbon tax to reduce greenhouse gas emissions since 1991. Although Sweden has employed a suite of other policies to reduce emissions, its environment ministry estimates that the carbon tax has led to cuts in emissions of 20 to 25 percent above cuts due to regulations. Sweden's carbon emissions have decreased 60 percent since 1991 while its economy has grown 25 percent. Its carbon tax has been credited with spurring innovation and development in the field of low-carbon energy technologies such as green-heating technologies that have led to significant reductions in the use of oil for home heating. Sweden's carbon tax has also been credited in part for putting the country on target to achieve and honor its commitment under the Kyoto Protocol. Unfortunately, carbon taxes have met resistance in the United States. Thus far, few senators and Congress members have had the courage to back the idea. Fossil fuel companies and their lobbyists remain too powerful. Ironically, in the case of climate action, anti-regulatory policymakers frequently use regulations to prevent market forces from working.

Section Review

1. How did the cap-and-trade approach work to control sulfur?
2. List some arguments for and against carbon trading.
3. Why do some economists believe a carbon tax is a better policy?

23.6 GREEN DEVELOPMENT AND BUSINESS

- *Trade increases the value of local products, but producers often get only a small share of that value.*
- *The WTO, IMF, World Bank, and NAFTA are some of the many organizations promoting large-scale development and trade.*
- *Green business models emulate ecological ideas of efficiency, diversity, and life-cycle thinking.*

Trade can be a powerful tool in making resources available and raising standards of living. Think of the things you now enjoy that might not be available if you had to live exclusively on the resources available in your immediate neighborhood. Too often, the poorest, least powerful people suffer in this global marketplace. To balance out these inequities, nations can deliberately invest in economic development projects. In this section, we'll look at some aspects of trade, development, business, and jobs that have impacts on our environment and welfare.

International trade brings benefits but also intensifies inequities

The banking and trading systems that regulate credit, currency exchange, shipping rates, and commodity prices were set up by the richer and more powerful nations in their own self-interest. The General Agreement on Tariffs and Trade (GATT) and World Trade Organization (WTO) agreements, for example, negotiated

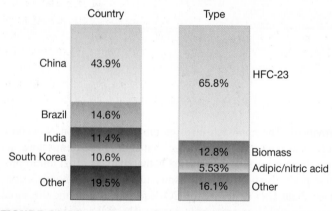

FIGURE 23.23 Distribution of payments for emissions reduction, deriving from emissions trading markets. Four countries collect 80 percent of payments, and two-thirds of payments reward the incineration of relatively cheap HFC-23, making this a bonanza for some corporations.

primarily between the largest industrial nations, regulate 90 percent of all international trade.

These systems tend to keep the less-developed countries in a perpetual role of resource suppliers to the more-developed countries. The producers of raw materials, such as mineral ores or agricultural products, get very little of the income generated from international trade (fig. 23.24).

Policies of the WTO and the IMF have provoked criticism and resistance in many countries. As a prerequisite for international development loans, the IMF frequently requires debtor nations to adopt harsh "structural adjustment" plans that slash welfare programs and impose cruel hardships on poor people. The WTO has issued numerous rulings that favor international trade over pollution prevention or protection of endangered species. Trade conventions such as the North American Free Trade Agreement (NAFTA) have been accused of encouraging a "race to the bottom" in which companies can play one country against another and move across borders to find the most lax labor and environmental protection standards.

No single institution has more influence on financing and policies of developing countries than the World Bank. Of some $25 billion loaned each year for development projects by international agencies, about two-thirds comes from the World Bank. Founded in 1945 to fund the reconstruction of Europe and Japan, the World Bank shifted its emphasis to aid developing countries in the 1950s. Many of its projects have had adverse environmental and social effects, however. Its loans often go to corrupt governments and fund ventures such as nuclear power plants, huge dams, and giant water diversion schemes. Former U.S. treasury secretary Paul O'Neill said that these loans have driven poor countries "into a ditch" by loading them with unpayable debt. He said that funds should not be loans, but rather grants to fight poverty.

Microlending helps the poorest of the poor

Global aid from the WTO usually aids banks and industries more than it helps the impoverished populations who most need assistance. Often, structural adjustment leads the poorest to pay back loans negotiated by their governments and industries. These concerns led Dr. Muhammad Yunus of Bangladesh to initiate the micro-loan plan of the Grameen Bank.

One of the most important innovations of the Grameen Bank is that borrowers take out loans in small groups. Everyone in the group is responsible for each other's performance. The group not only guarantees loan repayment, it helps businesses succeed by offering support, encouragement, and advice. Where banks depend on the threat of foreclosure and a low credit rating to ensure debt repayment, the Grameen Bank has something at least as powerful for poor villagers—the threat of letting down your neighbors and relatives. Even in the United States, organizations assist micro-enterprises with loans, grants, and training. The Women's Self-Employment Project in Chicago, for instance, teaches job skills to single mothers in housing projects. Similarly, "tribal circle" banks on Native American reservations successfully finance microscale economic development ventures.

The most recent venture for the Grameen Bank is providing mobile phone service to rural villages. Supplying mobile phones to poor women not only allows them to communicate, it provides another business opportunity. They rent out their phone to neighbors, giving the owner additional income, and linking the whole village to the outside world. Suddenly, people who had no access to communication can talk with their relatives, order supplies from the city, check on prices at the regional market, and decide when and where to sell their goods and services. This is a great example of "bottom-up development." Founded in 1996, Grameen Phone now has 2.5 million subscribers and is Bangladesh's largest mobile phone company.

Green business involves efficiency and creative solutions

At home, as well as abroad, business leaders are increasingly discovering that they can save money and protect our environment by greening up their business practices. They can save money through fuel efficiency and reducing electricity consumption. These steps also cut greenhouse gases. Recycling waste and minimizing the use of hazardous materials saves on disposal costs. In addition, these companies win public praise and new customers by

20¢ to retailers

25¢ to processors and wholesalers

28¢ to traders, brokers, and shippers

19¢ to producing countries

8¢ to growers

FIGURE 23.24 What do we really pay for when we purchase a dollar's worth of coffee?

Quantitative Reasoning

Suppose the World Bank has $25 billion to lend for development aid and that there are about 5 billion residents of developing countries. How would you distribute that aid? In what ways is that $25 billion more or less effective than the $71 million distributed by Kiva to 171,000 smaller entrepreneurs?

---

demonstrating an interest in our shared environment. By conserving resources, they also help ensure the long-term survival at their own corporations.

The green business movement has had some remarkable successes and presents an encouraging pathway for how we might achieve both environmental protection and social welfare. Some of the leaders in this new approach to business include Paul Hawken, William McDonough, Ray Anderson, Amory Lovins, David Crockett, and John and Nancy Todd.

These individuals and others have shown that operating in a socially responsible manner that is consistent with the principles of sustainable development and environmental protection can be good for employee morale, public relations, and the bottom line simultaneously. Environmentally conscious or "green" companies such as the Body Shop, Patagonia, Aveda, Malden Mills, Johnson and Johnson, and Interface, Inc., consistently earn high marks from community and environmental groups. Conserving resources, reducing pollution, and treating employees and customers fairly may cost a little more initially but can save money and build a loyal following in the long run.

New business models adopt concepts of ecology

Paul Hawken's 1993 book, *The Ecology of Commerce,* was a seminal influence in convincing many people to reexamine the role of business and economics in environmental and social welfare. Basing his model for a new industrial revolution on the principles of ecology, Hawken points out that almost nothing is discarded or unused in nature. The wastes from one organism become the food of another. Industrial processes, he argues, should be designed on a similar principle (table 23.4). Rather than a linear pattern in which we try to maximize the throughput of material and minimize labor, products and processes should be designed to:

- be energy efficient;
- use renewable materials;
- be durable and reusable or easily dismantled for repair and remanufacture, nonpolluting throughout their entire life cycle;

Table 23.4 Goals for an Eco-Efficient Economy

- Introduce no hazardous materials into the air, water, or soil.
- Measure prosperity by how much natural capital we can accrue in productive ways.
- Measure productivity by how many people are gainfully and meaningfully employed.
- Measure progress by how many buildings have no smokestacks or dangerous effluents.
- Make the thousands of complex governmental rules unnecessary that now regulate toxic or hazardous materials.
- Produce nothing that will require constant vigilance from future generations.
- Celebrate the abundance of biological and cultural diversity.
- Live on renewable solar income rather than fossil fuels.

- provide meaningful and sustainable livelihoods for as many people as possible;
- protect biological and social diversity;
- use minimum and appropriate packaging made of reusable or recyclable materials.

We can do all this and at the same time increase profits, reduce taxes, shrink government, increase social spending, and restore our environment, Hawken claims. Recently, Hawken has served as chairperson for The Natural Step in America, a movement started in Sweden by Dr. K. H. Robert, a physician concerned about the increase in environmentally related cancers. Through a consensus process, a group of 50 leading scientists endorsed a description of the living systems on which our economy and lives depend. More than 60 major European corporations and 55 municipalities have incorporated sustainability principles (table 23.5) into their operations.

Another approach to corporate responsibility is called the **triple bottom line.** Rather than reporting only net profits as a measure of success, ethically sensitive corporations include environmental effects and social justice programs as indications of genuine progress.

Corporations committed to eco-efficiency and clean production include such big names as Monsanto, 3M, DuPont, Duracell, and Johnson & Johnson. Following the famous three Rs—reduce, reuse, recycle—these firms have saved money and gotten welcome publicity. Savings can be substantial. Slashing energy use and redesigning production to use less raw material and to produce less waste is reported to have saved DuPont $3 billion over the past decade, while also reducing its greenhouse emissions 72 percent.

Efficiency starts with product design

Our current manufacturing system often is incredibly wasteful. On average, for every truckload of products delivered in the United States, 32 truckloads of waste are produced along the way. The automobile is a typical example. Industrial ecologist, Amory Lovins, calculates that for every 100 gallons (380 liters) of gasoline burned in your car engine, only one percent (1 gal or 3.8 liters) actually moves passengers. All the rest is used to move the vehicle itself. The wastes produced—carbon dioxide, nitrogen oxides, unburned hydrocarbons, rubber dust, heat—are spread through the environment where they pollute air, water, and soil.

Table 23.5 The Natural Step: System Conditions for Sustainability

1. Minerals and metals from the earth's crust must not systematically increase in nature.
2. Materials produced by human society must not systematically increase in nature.
3. The physical basis for biological productivity must not be systematically diminished.
4. The use of resources must be efficient and just with respect to meeting human needs.

Table 23.6 McDonough Design Principles

Inspired by the way living systems actually work, Bill McDonough offers three simple principles for redesigning processes and products:

1. *Waste equals food.* This principle encourages elimination of the concept of waste in industrial design. Every process should be designed so that the products themselves, as well as leftover chemicals, materials, and effluents, can become "food" for other processes.

2. *Rely on current solar income.* This principle has two benefits: First, it diminishes, and may eventually eliminate, our reliance on hydrocarbon fuels. Second, it means designing systems that sip energy rather than gulping it down.

3. *Respect diversity.* Evaluate every design for its impact on plant, animal, and human life. What effects do products and processes have on identity, independence, and integrity of humans and natural systems? Every project should respect the regional, cultural, and material uniqueness of its particular place.

FIGURE 23.25 The award-winning Gap, Inc. corporate offices in San Bruno, California, demonstrate some of the best features of environmental design. A roof covered with native grasses provides insulation and reduces runoff. Natural lighting, an open design, and careful relation to its surroundings all make this a pleasant place to work.
Steve Proehl/Getty Images

Architect William McDonough urges us to rethink design approaches (table 23.6). In the first place, he says, we should question whether the product is really needed. Could we provide the same service in a more eco-efficient manner? According to McDonough, products should be divided into three categories:

1. *Consumables* are products like food, natural fabrics, or paper that can harmlessly go back to the soil as compost.

2. *Service products* are durables such as cars, TVs, and refrigerators. These products should be leased to the customer to provide their intended service, but would always belong to the manufacturer. Eventually, they would be returned to the maker, who would be responsible for recycling or remanufacturing the product. Knowing that they will have to dismantle the product at the end of its life will encourage manufacturers to design for easy disassembly and repair.

3. *Unmarketables* are compounds like radioactive isotopes, persistent toxins, and bioaccumulative chemicals. Ideally, no one would make or use these products. But because eliminating their use will take time, McDonough suggests that in the meantime these materials should belong to the manufacturer and be molecularly tagged with the maker's mark. If they are discovered to be discarded illegally, the manufacturer would be held liable.

Following these principles, McDonough Bungart Design Chemistry has created non-toxic, easily recyclable materials to use in buildings and for consumer goods. Among some important and innovative "green office" projects designed by the McDonough and Partners architectural firm are the Environmental Defense Fund headquarters in New York City, the Environmental Studies Center at Oberlin College in Ohio (see fig. 20.11), the European Headquarters for Nike in Hilversum, the Netherlands, and the Gap Corporate Offices in San Bruno, California (fig. 23.25). Intended to promote employee well-being and productivity, as well as eco-efficiency, the Gap building has

high ceilings, abundant skylights, windows that open, a full-service fitness center (including a pool), and a landscaped atrium for each office bay that brings the outside in. The roof is covered with native grasses. Warm interior tones and natural wood surfaces (all wood used in the building was harvested by certified sustainable methods) give a friendly feeling. Paints, adhesives, and floor coverings are low toxicity and the building is one-third more energy efficient than strict California laws require. A pleasant place to work, the offices help recruit top employees and improve both effectiveness and retention. As for the bottom line, Gap, Inc. estimates that the increased energy and operational efficiency will have a four- to eight-year payback.

Green consumerism gives the public a voice

Consumer choice can play an important role in persuading businesses to produce eco-friendly goods and services (What Can You Do?, p. 540). Increasing interest in environmental and social sustainability has caused an explosive growth of green products. You can find eco-travel agencies, telephone companies that donate profits to environmental groups, entrepreneurs selling organic foods, shade-grown coffee, efficient houses, paint thinner made from orange peels, sandals made from recycled auto tires, earthworms for composting, sustainable clothing, shoes, rugs, balm, shampoo, and insect repellent. Although these eco-entrepreneurs represent a tiny sliver of the $19-trillion-per-year U.S. economy, they often serve as pioneers in developing new technologies and offering innovative services.

Personally Responsible Economy

There are many things that each of us can do to lower our ecological impacts and support green businesses through responsible consumerism and ecological economics.

- Practice living simply. Ask yourself if you really need more material goods to make your life happy and fulfilled.
- Minimize consumption of resources, to save personal and global costs of electricity, gas, metals, plastics, and other resources. Recycle or reuse products, and avoid excessive packaging.
- Look at the amount of your garbage on trash day. Is that the amount of throughput you would like to produce?
- Support environmentally friendly businesses. Consider spending a little more for high-quality, fairly produced goods, at least some of the time. Ask companies what they are doing about environmental protection and human rights.
- Buy green products. Look for efficient, high-quality materials that will last and that are produced in the most environmentally friendly manner possible. Subscribe to clean-energy programs if they are available in your area.
- Think about the total life-cycle costs of the things you buy, including environmental impacts, service charges, energy use, and disposal costs, as well as initial purchase price.
- Invest in socially and environmentally responsible mutual funds or green businesses when you have money for investment.
- Try making a Kiva or similar micro-loan. You may find that it's fun and educational, and it can feel good to help others.
- Vote thoughtfully. Think carefully about the long-term versus short-term social and environmental impacts of economic policies, and work with others in your community to push elected representatives to act in ways that safeguard resources for the generations to come.

In some industries, eco-entrepreneurs have found profitable niches within a larger market. In other cases, once a consumer demand has built up, major companies add green products or services to their inventory. Natural foods, for instance, have grown from the domain of a few funky, local co-ops to a $50 billion market segment. Most supermarket chains now carry some organic food choices. Similarly, natural-care health and beauty products are now more than $13 billion per year in the United States. By supporting these products, you can ensure that they will continue to be available and, perhaps, even help expand their penetration into the market.

Walmart, the large-volume discount price chain, has established a name as the world's largest seller of organic products and energy-efficient lightbulbs, among other green products. The incentive for Walmart has been that green products and green production processes are often less wasteful, and thus cheaper (when produced in volume), than other products and processes.

Does this mean that Walmart has internalized all its costs and produced sustainable relationships with suppliers? Some critics say no, and that Walmart has diluted the idea of sustainable and organic production. But supporters point out that the chain has also helped legitimize these ideas for the public. Evidently, Walmart shoppers also are enthusiastic about contributing to environmental solutions while they shop, because they eagerly buy the organic and energy-efficient products offered.

Environmental protection creates jobs

For years, business leaders and politicians have portrayed environmental protection and jobs as mutually exclusive. Pollution control, protection of natural areas and endangered species, and limits on use of nonrenewable resources, they claim, will strangle the economy and throw people out of work. Ecological economists dispute this claim, however. Their studies show that only 0.1 percent of all large-scale layoffs in the United States in recent years were due to government regulations (fig. 23.26). Environmental protection, they argue, not only is necessary for a healthy economic system, it actually creates jobs and stimulates business.

Recycling, for instance, makes more new jobs than extracting virgin raw materials. This doesn't necessarily mean that recycled goods are more expensive than those from virgin resources. We're simply substituting labor in the recycling center for energy and huge machines used to extract new materials in remote places.

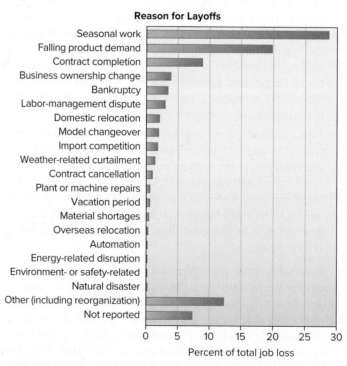

FIGURE 23.26 Although opponents of environmental regulation often claim that protecting the environment costs jobs, studies by economist E. S. Goodstein show that only 0.1 percent of all large-scale layoffs in the United States were the result of environmental laws.
Source: E. S. Goodstein, Economic Policy Institute, Washington, D.C.

Could We Have a Green New Deal?

The Green New Deal (GND) is a proposed economic stimulus plan for the United States intended to address both economic inequality and climate change. The name comes from President Franklin Roosevelt's 1930s New Deal legislation that helped jump-start the economy, provide employment for vast numbers of people, and pull the United States out of the worst depression in its history. Under this plan, millions of WPA workers built highways, bridges, and public buildings, while youth in the Civilian Conservation Corps carried out reforestation projects and built trails and picnic shelters in national parks that we still use and enjoy.

The proposed GND would shift our economy away from polluting industries, embrace energy efficiency and resource conservation, and produce 100 percent of our energy from renewable resources. This would produce millions of new, clean jobs, and provide training in renewable technologies for many people who are now stuck in low-paying, dead-end jobs. It's meant to transform not just the energy sector but the entire economy. It will both decarbonize the economy and make it fairer and more just.

The idea of green jobs isn't new. The Blue-Green Alliance has been advocating a workforce redevelopment for decades. Journalist Tom Friedman was the first to use this terminology in a 2007 essay. But the arrival in Congress in 2019 of the most racially diverse, youthful, progressive and highly female cohort in the country's history promises to change things radically. In particular, newly elected Representatives including Alexandria Ocasio-Cortez, Ayanna Pressley, Rashida Tlaib, Ilhan Omar, and Antonio Delgado have championed a massive GND. They have been supported in their campaign by a youth-led organization called the Sunrise Movement. After a sit-in at the House Speaker's office as committee assignments were being established, more than 40 members of Congress—including several potential presidential candidates—signed on to support a GND Select Committee in the 116th Congress.

Among the proposals advanced for a GND include the following:

- Low-carbon infrastructure redevelopment in order to create jobs;
- A directed tax on the profits of oil and gas companies with proceeds being invested in renewable energy and energy efficiency;
- End fossil fuel subsidies and correct other market distortions that prop up "brown economies"
- Financial incentives for green investment and reduced energy usage, including low interest rates for green investment
- A Global Marshall Plan Initiative to create money to fund the "great transition" to a society free of fossil fuels and other measures that aim to preserve the biosphere.
- A massive jobs-training program to teach useful skills to unemployed or underemployed people so they can participate in the new, sustainable economy.

To produce an economy based entirely on renewable energy, we'd need to build millions of wind turbines and install billions of solar panels. Is this possible? Doubters claim it isn't. But remember that during the four years of WWII, the United States built 600,000 airplanes, 7,000 fighting ships, 120,000 tanks, 30,000 tons of bombs, and a vast amount of other military paraphernalia. And most of that stuff was blown up or discarded after the war. With a GND, we can produce stuff we'll actually use for positive purposes for generations.

What do you think? Is a GND feasible? Would you like to be a part of it?

As chapter 20 shows, China has emerged as the world leader in sustainable energy. Recognizing the multibillion-dollar economic potential of "green" business, China is investing at least $8 billion per year on research and development, and now it is selling about $12 billion worth of new technology equipment and services per year worldwide. Altogether, green businesses in China now employ about 17 million workers. Japan, also, is marketing advanced waste incinerators, pollution-control equipment, alternative energy sources, and water treatment systems. Superefficient "hybrid" gas-electric cars are helping Japanese automakers flourish, while U.S. corporations that once dominated the world slide into bankruptcy. Letting go of fossil fuel-based industries, economies, and ideas from past generations is always a challenge, and resistance to change has strongly influenced U.S. politics in recent years. But it is increasingly clear that moving into a green economy is where the future is. The question is how we will negotiate this (What Do You Think?, above).

During the eight years of the Obama presidency, we made encouraging progress on green initiatives including renewable energy, conservation, climate change abatement, mass transit, passive buildings, and green-jobs training. Unfortunately almost all that progress has been abandoned or reversed under the Trump administration.

The International Monetary Fund (IMF) estimated the global distortion for fossil-fuel subsidies alone at $5.3 trillion per year, or 6.5 percent of global GDP. This could be corrected by a carbon tax, and the money could be redirected toward a more equitable and sustainable economy. Morgan Stanley, a global financial services firm, estimates that global sales from clean energy alone could grow to as much as $1 trillion per year by 2030. Already authors are rushing books to publication giving advice on how to make a fortune investing in green corporations and renewable energy technologies. For students contemplating career choices, clean energy and conservation could be good areas to explore.

Section Review

1. Describe some of the features of a Green New Deal.
2. Why might microlending produce more widespread results than World Bank lending?
3. What does it mean that green business adopts ideas from ecology?
4. List several benefits and weaknesses of green consumerism.

Connecting the Dots

The carbon tax in British Columbia demonstrates one of the ways economics can address environmental issues. Economists argue that this approach maximizes efficiency. But policymakers have often had difficulty implementing it.

Economic policies are often at the root of the success or failure of many human endeavors. These policies build from some basic sets of assumptions about the nature of resources. Classical economics assumes that resources are finite, so that we compete to control them. Neoclassical economics assumes that resources are based on capital, which can include knowledge and social capital as well as resources, and that constant growth is both possible and essential. Natural resource economics extends neoclassical ideas to internalize the value of ecological services in economic accounting.

Is stewardship of our climate a "tragedy of the commons," or can we make it a successful case of common property management? The tragedy metaphor points to failure; BC's carbon tax is an effort toward management for the long term. Success in common property management depends on a number of factors, including an ability to monitor the resource and clear, agreed-upon rules of use. BC is one of many places developing clear rules and incentives for responsible management of our climate and environment.

According to the United Nations Environment Program, by the end of 2018, 46 countries and 25 sub-national jurisdictions were pricing carbon, accounting for around 20 percent of global greenhouse-gas emissions and raising $82 billion in revenue. The IMF estimated that removing fossil-fuel pricing distortions would cut global carbon emissions by 21 percent, reduce deaths from fuel-related air pollution by 55 percent and raise extra revenue equivalent to 4 percent of global GDP.

Carbon taxes, carbon trading, and cutting environmentally harmful subsidies are some of the ways to finance a Green New Deal. These funds could support better infrastructure for renewable energy, more sustainable urban development, and research into clean energy. They could also be used to raise the minimum wage, provide payments or retraining for displaced workers, and reduce burdens for vulnerable households affected by the green transition. Redirecting money toward desirable outcomes, and away from undesirable ones, is how economics can help create a sustainable and equitable future.

Reviewing Key Terms

Can you define the following terms in environmental science?

cap-and-trade 23.5	ecological economics 23.2	gross national product	neoclassical	steady-state
capital 23.1	ecosystem services 23.2	(GNP) 23.4	economics 23.1	economy 23.2
carbon credits 23.5	emissions trading 23.5	Human Development	nonrenewable	supply 23.1
classical economics 23.1	Environmental	Index (HDI) 23.4	resources 23.1	sustainable
common property	Performance Index	intangible resources 23.1	open-access system 23.3	development 23.1
management 23.3	(EPI) 23.4	internal costs 23.2	political economy 23.1	throughput 23.2
cost–benefit analysis	externalized costs 23.1	marginal costs and	price elasticity 23.1	triple bottom line 23.6
(CBA) 23.4	gross domestic product	benefits 23.1	renewable resources 23.1	World Happiness Index
demand 23.1	(GDP) 23.4	market equilibrium 23.1	resource 23.1	(WHI) 23.4

Critical Thinking and Discussion Questions

1. When ecologists warn that we are using up irreplaceable natural resources, and economists respond that ingenuity and enterprise will find substitutes for most resources, what underlying premises and definitions shape their arguments?

2. How can intangible resources be infinite and exhaustible at the same time? Isn't this a contradiction in terms? Can you find other similar paradoxes in this chapter?

3. What would be the effect on the developing countries of the world if we were to change to a steady-state economic system? How could we achieve a just distribution of resource benefits while still protecting environmental quality and future resource use?

4. Resource-use policies bring up questions of intergenerational justice. Suppose you were asked: "What has posterity ever done for me?" How would you answer?

5. If you were doing a cost–benefit study, how would you assign a value to the opportunity for good health or the existence of rare and endangered species in faraway places? Is there a danger or cost in simply saying some things are immeasurable and priceless and therefore off limits to discussion?

6. If natural capitalism or eco-efficiency has been so good for some entrepreneurs, why haven't all businesses moved in this direction?

Data Analysis

Evaluating the Limits to Growth

The graphs shown in figure 23.17 contrast two views of resource supply and scarcity. Models like these are very influential in both describing and informing our view of resources, scarcity, consumption patterns, and the kind of society we may live in, in the future. They also represent specific economic ideas and assumptions. Review the text for this figure, then go to Connect to demonstrate your understanding of these ideas.

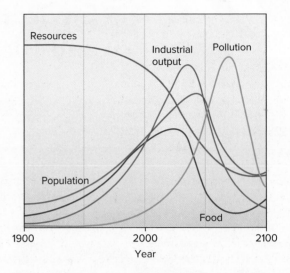

Models of resource consumption and scarcity are used both to describe our economic system and to predict future resource scarcity.

▲Green sea turtles are one of the great success stories of the Endangered Species Act. In 1978 when they were declared endangered, turtle populations were down an estimated 90 percent from historic highs. Now, the greatest density of this species in the world nests every summer on Florida's Atlantic coast.
Douglas Klug/Getty Images

Learning Outcomes

After studying this chapter, you should be able to:

24.1 List several basic concepts in policy.

24.2 Describe several major environmental laws.

24.3 Identify ways that executive, judicial, and legislative bodies shape policy.

24.4 Explain the purposes of international treaties and conventions.

24.5 Describe community-based and environmental approaches to policy.

"The power to command frequently causes failure to think."

– Barbara Tuchman

Turtles Return to Archie Carr

A female green sea turtle (*Chelonia mydas*) glides gracefully through the shallow waters off Florida's east coast. It's early July, and she's headed back to the beach in the Archie Carr National Wildlife Refuge, just south of Cape Canaveral, where she hatched many years ago. When night falls, she'll lumber ashore to dig a deep hole in the sand above the high tide line, where she'll lay more than 100 round eggs the size of ping-pong balls. After several hours of hard work, the mother turtle will cover the nest and shuffle back to the water and swim away, never to see her offspring. About 50 days later, the baby turtles will claw their way out of the sand and scramble down to the surf (fig. 24.1).

The fact that there are turtles once again in the Wildlife Refuge is a testament to the Endangered Species Act (ESA), one of the most effective policies in U.S. environmental protection. When green sea turtles were added to the list for protection in 1978, fewer than 300 of them nested on the Florida shore, and populations worldwide were thought to be down 90 percent from a couple of centuries earlier. By 2017, more than 39,000 nests were recorded in Florida, most of them in the wildlife refuge named after Archie Carr, the nation's foremost sea turtle expert and a tireless champion for their preservation.

In 2016, the green sea turtle's status was downlisted from endangered to threatened, in recognition of the dramatic success in protecting and restoring this charismatic species. But they're still at risk from many factors. Beach erosion, rising sea levels, and coastal development destroy habitat. Artificial light disorients baby turtles, drawing them away from the safety of the ocean. Beach traffic, meat and egg hunters, natural predators, and domestic cats and dogs kill turtles on land. Longline and trawler fishing snare turtles at sea, while pollution, plastic trash, oil spills, and climate change all continue to endanger the species. Despite these threats, many U.S. sea turtle populations are either stable or slowly rebuilding their numbers.

The ESA is the strongest law for protecting biodiversity passed by any nation. It has broad public support. More than two-thirds of all Americans want the act strengthened or left alone, not weakened. The ESA has also set a global example, inspiring action by other nations around the world. In addition to heartening progress along Florida's coastline, protections in other countries are showing promising results. In Costa Rica's Tortuguero National Park, researchers reported a few years ago that every single egg laid was immediately stolen by egg hunters. By 2017, the park had an estimated 22,500 green turtle nests, most guarded by turtle protectors. Similarly, Raine Island, on Australia's Great Barrier Reef, now has the healthiest sea turtle population in the South Pacific.

Sea turtles are just one of many species that have benefited from ESA protections. The whooping crane, peregrine falcon, black-footed ferret, bald eagle, southern sea otter, humpback whale, American alligator, nēnē goose, and brown pelican all have come back from the brink of extinction after being listed as endangered. More than 1,459 imperiled species of plants, fish, and wildlife have been protected by the ESA, and only 10 have gone extinct after being listed. Furthermore, many of the species singled out by this act are keystone or umbrella organisms in their ecosystem. Protecting habitat for these species helps protect vast numbers of other organisms. As we discussed in chapter 23, some ecological economists calculate the value of ecosystem services provided by this wonderful panoply of life to be at least $173 trillion per year, or more than twice the global GDP.

In this chapter, we'll examine how environmental policies such as the ESA are created, and we'll discuss the structures involved in shaping, refining, and enforcing laws, at national, local, and international scales.

FIGURE 24.1 Newly hatched baby sea turtles crawl toward the surf. They'll swim hundreds or thousands of kilometers to feeding grounds before returning in 5 to 7 years to this same beach. Only one in 1,000 will survive to adulthood.
photostock77/Shutterstock

24.1 Basic Concepts in Policy

- *Environmental policy includes rules and regulations as well as public opinion.*
- *Money influences policy.*
- *Public awareness and action shape policy.*

The basic idea of **policy** is a statement of intentions and rules that outline acceptable behaviors or accomplish some end. In the case of the Endangered Species Act, these rules were formally established by Congress, but policy can also be less formally defined. You might have an informal policy to always do your homework; a church might have an open-door policy for visitors; most countries have policies to ensure the welfare of citizens. Laws, such as the ESA or the U.S. Clean Water Act (chapter 18), are formal statements of national policy. Like the rules that define how you play a board game, policies define agreed-upon limits of behavior. Because it sets the rules we live by, policy making is a contentious and extremely important process.

We will take **environmental policy** to mean official rules and regulations concerning the environment, as well as public opinion that helps shape environmental policy. Ideally, environmental policy serves the needs of human health, economic stability, and ecosystem health. Often, these interests have been seen as contradictory, leading to bitter disputes in policy making. Increasingly, however, the public and policymakers acknowledge that these interests overlap. Protecting species, for example, supports the economy and is often a source of pride for communities. Controlling pollution protects human health, saves money in health care, and preserves essential resources.

The passage of the ESA in 1973 was part of a wave of environmental protections that started in many countries in the 1960s and 1970s. Increasing evidence of the damage caused by unconstrained resource use and pollution, following the industrial expansions of World War II, helped foster widespread interest in pollution control, species protection, health and human safety, and other safeguards. Since then, global interest in environmental quality has remained high and has even strengthened in recent years, as more people become aware of our dependence on a healthy environment. A 2015 survey by the Pew Research Center, for example, found that climate action was a priority for 77 percent of respondents in 40 countries surveyed. Even though less than half thought climate change would impact them personally, most still favored action on climate protection (fig. 24.2). Most people are enthusiastic about environmental protection, then. The question is how policies ensure those protections in the face of economic interests.

Basic principles guide environmental policy

Protection of fundamental rights is usually one aim of policy making. For many groups, these rights include a right to a safe, sustainable environment (fig. 24.3). The 1987 World Commission on Environment and Development, for example, stated, "All human beings have the fundamental right to an environment adequate for their health and well-being." Of the 195 independent nations in the world, 109 now have constitutional provisions for protection of

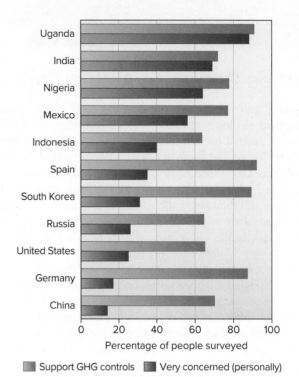

FIGURE 24.2 Support for international agreements to curb greenhouse gas (GHG) emissions is strong, even in countries where few people are very concerned about personal impacts.
Source: Pew Research Center, 2015

the environment and natural resources. Often such policies are not legally enforceable, but they provide a statement of intentions that can guide subsequent laws.

In theory, democratic societies try to establish policies that are fair to everyone, at least in principle, and that defend everyone's basic needs. Ideally, the majority interests are served, but

FIGURE 24.3 How do we ensure a safe and healthy environment for everyone? Policies of individuals, communities, and governing institutions are all responsible.
Digital Vision/Getty Images

interests of the minority are also defended. Fairness in policy making requires that rules are transparent and decided by public debate and input from many groups in society.

Cost–benefit analysis is often involved in policy making. The aim is to assign standard values, such as dollar value, to competing concerns, then compare the costs and benefits of a plan. This way, we can decide if benefits outweigh the costs of a policy. In principle, this is a utilitarian approach to ensuring an objective, methodical decision process. In practice, there are many difficulties in implementing rational cost–benefit analysis:

- Many conflicting values and needs cannot be compared: They are fundamentally different or we lack complete information.

- We often lack agreed-upon broad societal goals; benefits to different groups and individuals often conflict.

- Policymakers undervalue societal goals and over-value factors that aid their own power, status, money, or reelection.

- Past investments in existing programs and policies create "path dependence" (a tendency to follow established ways) and hidden costs that prevent policymakers from considering good alternatives foreclosed by previous decisions.

- Policymakers often lack the information or models to calculate costs and benefits accurately. Some values, such as health, freedom, or scenery, are very hard to price in dollars.

Another consideration is the **precautionary principle:** When an activity threatens to harm health or the environment, we should fully understand risks before initiating that activity. According to this principle, for example, we shouldn't mass-market new chemicals, new cars, or new children's toys until we're sure they are safe. There are four widely accepted tenets of this principle:

- People have a duty to take steps to prevent harm. If you anticipate harm, you have an obligation to try to stop it.

- The burden of proof for safety of a new technology, process, activity, or chemical lies with the proponents, not with the general public.

- Before using a new technology, process, or chemical, or starting a new activity, people have an obligation to examine a full range of alternatives, including the alternative of not using it.

- Decisions using the precautionary principle must be open and democratic and must include the affected parties.

The European Union has adopted this precautionary principle as the basis of its environmental policy. In the United States, opponents of this approach claim that it slows down productivity and innovation. However, many American firms that do business in Europe—including virtually all of the largest corporations—are having to change their manufacturing processes to adapt to more careful E.U. standards. For example, lead, mercury, and other hazardous materials must be eliminated from electronics, toys, cosmetics, clothing, and a variety of other consumer products. A proposal being debated by the E.U. would require testing of thousands of chemicals, would cost industry billions of dollars, and lead to many more products and compounds being banned as they are shown to be unsafe to the public. What would you do about this? Is this proposal just common sense, or is it an invitation to decision paralysis?

Money influences policy

Politics is often seen as a struggle for power among competing interest groups, which strive to shape public policy to suit their own agendas. Usually, money is a key to political power, so the issue of money in politics is a matter of perennial and contentious debate. Even in a democratic society, money is an important factor in what rules are made, or not made. Often those with the most money can advertise to promote their point of view, draw voters to their side, and win the most legislative seats. Wealthy individuals and corporations can buy influence and friends in government agencies. Large donations are now essential to success in a political campaign, so wealthy donors can influence policymakers with generous funding, or by threatening to withhold funds.

Debates about power and money in politics in the United States have increased since 2010, when the U.S. Supreme Court decided in the *Citizen's United* case that corporations can spend unlimited amounts of money in political advertising (section 24.3). The *Citizens United* case opened the door to a flood of money, often from anonymous sources, in political campaigns and elections. If you were a politician, how might this influence your policymaking? Would it matter if the company were a foreign-owned corporation working to shape U.S. laws?

Public awareness and action shape policy

Although power and money are important forces, they can't account for our many policies that serve the public interest. Public participation by scientists and citizens is also an essential force in policy formation. Consider the ESA, or the international policy built on it, the Convention on Trade in Endangered Species (CITES). Both policies have very broad support. Both protect species with little economic value but often with cultural or biodiversity value. Both were established on the work of concerned scientists, communities, and policymakers who objected to poaching and wanton destruction of wildlife. Both the ESA and CITES have had significant success in raising awareness about, and often protecting, threatened and endangered species.

Within the United States, civic action has led to many of our strongest environmental and social protections, such as the Clean Water Act, the Clean Air Act, and the Voting Rights Act, which defend the interests of all citizens. Public citizenship is the most important force in protecting and enacting democratic government.

Policy changes often start with protests over environmental contamination, pollution, or resource waste. Media attention helps publicize these protests and the problems they are challenging. One of the most lauded U.S. environmental rules, the Clean Water Act, followed shocking and widely broadcast images in 1969 of burning industrial waste in the Cuyahoga River in Cleveland. The same year, TV news broadcast a disastrous oil spill on the coast of Santa Barbara, California, following an oil well blowout just off the coast. An estimated 100,000 barrels (16,000 m^3) of black,

FIGURE 24.4 Beach cleanup efforts after the Santa Barbara oil spill in 1969 made excellent media material and had an important role in U.S. environmental policy.
Esther Henderson/Science Source

gooey crude oil was spilled, and much of it washed up onto beautiful Southern California beaches. Dead birds, dolphins, and sea lions caught public attention, and cleanup crews rallied to try to protect the beaches (fig. 24.4). For weeks, TV news broadcast daily updates on the spill nationwide, using live satellite feeds, for which the technology had just become available.

These images helped to build national concern for environmental protections. Images of young volunteers, smudged with oil, trying vainly to sweep gooey oil off a beautiful beach, were ideal for TV. Like the Cuyahoga fires, the Santa Barbara oil spill played an important role in mobilizing public opinion and was a major factor in passage of the 1972 U.S. Clean Water Act.

Opportunities for public awareness and participation have increased in recent years. There is a growing diversity of news outlets. YouTube, Twitter, and other social media let anyone post images and news. In addition, there are many groups committed to encouraging public involvement in environmental policy. One of these is the Environmental Working Group, which monitors policy and budget issues in Congress. Another is the League of Conservation Voters, which publishes a scorecard for all members of Congress, recording how they voted on environmental laws. You can see how your own representatives are doing by looking at their website: scorecard.lcv.org/scorecard.

Section Review

1. Why were green sea turtles declared endangered in 1978?
2. What is the precautionary principle?
3. How did the 1969 Santa Barbara oil spill help lead to environmental legislation?

24.2 MAJOR ENVIRONMENTAL LAWS

- *NEPA (1969) establishes public oversight.*
- *The Clean Air Act (1970) regulates air emissions.*
- *The Clean Water Act (1972) protects surface water.*
- *The Endangered Species Act (1973) protects wildlife.*
- *The Superfund Act (1980) lists hazardous sites.*

We depend on many different laws to protect resources such as clean water, clean air, safe food, and biodiversity (table 24.1). Here we'll review a few of these laws, and in the sections that follow we'll examine how laws are created. We will focus mainly on U.S. laws in this section, but other countries have followed the early lead of the United States in environmental policy formation. In recent years, other countries and the international community have increasingly led the way, as global concern for environmental policy has expanded.

Most of the laws we take for granted now are relatively recent. For most of its history, U.S. policy has had a laissez-faire or hands-off attitude toward business and private property. Pollution and environmental degradation were regarded as the unfortunate but necessary costs of doing business. There were some early laws forbidding gross interference with another person's property or rights—the Rivers and Harbors Act of 1899, for example, made it illegal to dump so much refuse in waterways that navigation was blocked. But in general there were few rules regarding actions on private property, even when those actions impaired the health or resources of neighbors.

Many of these attitudes and rules changed in the 1960s and 1970s, which marked a dramatic turning point in our understanding of the dangerous consequences of pollution. Rachel Carson's *Silent Spring* (1962) and Barry Commoner's *Closing Circle* (1971) alerted the public to the ecological and health risks of pesticides, hazardous wastes, and toxic industrial effluents. Public activism in the civil rights movement and protests against the war in Vietnam carried over to environmental protests and demands for environmental protection. These also led to the celebration of the first Earth Day in 1970 (see chapter 25).

Rising public concern and activism about environmental issues, such as DDT-poisoned birds, water pollution, and rising smog levels, led to more than 27 major federal laws for environmental protection and hundreds of administrative regulations established in the environmental decade of the 1970s. Among the most important were the establishment of the Environmental Protection Agency (EPA) and the National Environmental Policy Act (NEPA), which requires environmental impact statements for all major federal projects. Because of their power, the EPA and NEPA have both been the targets of repeated attacks by groups that dislike regulations imposed on pollution emissions or resource uses.

In the initial phase of this environmental revolution, the main focus was on direct regulation and lawsuits to force violators to obey the law. In recent years, attention has shifted to pollution prevention and collaborative methods that can provide win–win

Table 24.1 Major U.S. Environmental Laws

Legislation	Provisions
Wilderness Act of 1964	Established the national wilderness preservation system.
National Environmental Policy Act of 1969	Declared national environmental policy, required Environmental Impact Statements, created Council on Environmental Quality.
Clean Air Act of 1970	Established national primary and secondary air quality standards. Required states to develop implementation plans. Major amendments in 1977 and 1990.
Clean Water Act of 1972	Set national water quality goals and created pollutant discharge permits. Major amendments in 1977 and 1996.
Federal Pesticides Control Act of 1972	Required registration of all pesticides in U.S. commerce. Major modifications in 1996.
Marine Protection Act of 1972	Regulated dumping of waste into oceans and coastal waters.
Coastal Zone Management Act of 1972	Provided funds for state planning and management of coastal areas.
Endangered Species Act of 1973	Protected threatened and endangered species, directed FWS to prepare recovery plans.
Safe Drinking Water Act of 1974	Set standards for safety of public drinking-water supplies and to safeguard groundwater. Major changes made in 1986 and 1996.
Toxic Substances Control Act of 1976	Authorized EPA to ban or regulate chemicals deemed a risk to health or the environment.
Federal Land Policy and Management Act of 1976	Charged the BLM with long-term management of public lands. Ended homesteading and most sales of public lands.
Resource Conservation and Recovery Act of 1976	Regulated hazardous waste storage, treatment, transportation, and disposal. Major amendments in 1984.
National Forest Management Act of 1976	Gave statutory permanence to national forests. Directed USFS to manage forests for "multiple use."
Surface Mining Control and Reclamation Act of 1977	Limited strip mining on farmland and steep slopes. Required restoration of land to original contours.
Alaska National Interest Lands Act of 1980	Protected 40 million ha (100 million acres) of parks, wilderness, and wildlife refuges.
Comprehensive Environmental Response, Compensation and Liability Act of 1980	Created $1.6 billion "Superfund" for emergency response, spill prevention, and site remediation for toxic wastes. Established liability for cleanup costs.
Superfund Amendments and Reauthorization Act of 1994	Increased Superfund to $8.5 billion. Shares responsibility for cleanup among potentially responsible parties. Emphasizes remediation and public "right to know."

Source: Vig, N., and M. Kraft. "Environmental Policy in the 1990s." 3rd Congressional Quarterly Press

solutions for all stakeholders. At the same time, many corporations now recognize that unregulated pollution is unacceptable, and that cleaner, more efficient practices are widely understood to be good for profits as well as for the environment.

Environmental, health, and public safety laws, like other rules, impose burdens for some people and provide protections for others. The Clean Water Act, for example, requires that industries take responsibility for treating waste, rather than discharging it into public waters for free. Cities have had to build and maintain expensive sewage treatment plants, instead of pouring untreated sewage into lakes and rivers. These steps internalize costs that previously had been left to the public to deal with. These national laws are intended to protect public health and shared resources for all areas and all citizens. In fact, enforcement varies, but the existence of national, legally enforceable laws allows some recourse for victims when environmental laws are broken.

You can see the text of these laws, together with some explanation, on the EPA's website: www.epa.gov/lawsregs/laws/index .html#env. If you have never examined the text of a law, you should take a look at these.

NEPA (1969) establishes public oversight

Signed into law by President Nixon in 1970, the **National Environmental Policy Act (NEPA)** is the cornerstone of U.S. environmental policy.

NEPA does three important things: (1) it authorizes the Council on Environmental Quality (CEQ), the oversight board for general environmental conditions; (2) it directs federal agencies to take environmental consequences into account in decision making; and (3) it requires an **environmental impact statement (EIS)** be published for every major federal project likely to have an important impact on environmental quality (fig. 24.5). NEPA doesn't forbid environmentally destructive activities if they comply otherwise with relevant laws, but it demands that agencies admit publicly what they plan to do. Once embarrassing information is revealed, however, few agencies will bulldoze ahead, ignoring public opinion. And an EIS can provide valuable information about government actions to public interest groups, which wouldn't otherwise have access to this information.

FIGURE 24.5 Every major federal project in the United States must be preceded by an Environmental Impact Statement that examines alternative approaches.
Stephen Reynolds

What kinds of projects require an EIS? The activity must be federal and it must be major, with a significant environmental impact. Evaluations are always subjective as to whether specific activities meet these characteristics. Each case is unique and depends on context, geography, the balance of beneficial versus harmful effects, and whether any areas of special cultural, scientific, or historical importance might be affected. A complete EIS for a project is usually time-consuming and costly. The final document is often hundreds of pages long and generally takes six to nine months to prepare. The EIS process provides a rare opportunity for the public to see evidence for or against a project. It also gives the public time to rally opposition when resources or communities are at risk. If agencies don't agree to prepare an EIS voluntarily, citizens can petition the courts to force them to do so.

Every EIS must contain the following elements: (1) purpose and need for the project, (2) alternatives to the proposed action (including taking no action), (3) a statement of positive and negative environmental impacts of the proposed activities. In addition, an EIS should make clear the relationship between short-term resources and long-term productivity, as well as any irreversible commitment of resources resulting from project implementation.

Many lawmakers in recent years have tried to ignore or limit NEPA in forest policy, energy exploration, and marine wildlife protection (chapter 12). The "Healthy Forest Initiative," for example, eliminated public oversight of many logging projects by bypassing EIS reviews and prohibiting citizen appeals of forest management plans. Similarly, when the Bureau of Land Management proposed 140,000 coal-bed methane wells in Wyoming and Montana, promoters claimed that water pollution and aquifer depletion associated with this technology didn't require environmental review (chapter 19). And in the 2005 Energy Bill, Congress inserted a clause that exempts energy companies from NEPA requirements in a number of situations, with the aim of speeding energy development on federal land.

The Clean Air Act (1970) regulates air emissions

The first major environmental legislation to follow NEPA was the Clean Air Act (CAA) of 1970. Air quality has been a public concern at least since the beginning of the industrial revolution, when coal smoke, airborne sulfuric acid, and airborne metals such as mercury became common in urban and industrial areas around the world (fig. 24.6). Sometimes these conditions produced public health crises: One infamous event was the 1952 Great Smog of London—several days of cold, still weather that trapped coal smoke in the city and killed some 4,000 people from infections and asphyxiation. Another 8,000 died from respiratory illnesses in the months that followed (see chapter 16).

Although crises of this magnitude have been rare, chronic exposure to bad air has long been a leading cause of illness in many areas. The Clean Air Act provided the first nationally standardized rules in the United States to identify, monitor, and reduce air contaminants. The core of the act is an identification and regulation of major "criteria pollutants," also known as "conventional pollutants." These include sulfur oxides, lead, carbon monoxide, nitrogen oxides (NO_x), particulates (dust), volatile organic compounds, and metals and halogens (such as mercury and bromine compounds). Recent revisions to the Clean Air Act require the EPA to monitor carbon dioxide, our dominant greenhouse gas, which endangers the public through heat stress, drought, and other mechanisms (chapter 15).

FIGURE 24.6 The Clean Air Act has greatly reduced the health and economic losses associated with air pollution.
William P. Cunningham

Two significant regulations under the Clean Air Act are the Mercury and Air Toxics Standards and the Clean Power Plan. In 2016, after 20 years of study and discussion, the EPA issued final regulations for mercury and air toxics from power plants. These standards aimed to cut releases of these toxins from about 600 power plants by 75 percent each year. They will force utilities to install and operate equipment that removes mercury and fine particulate material from their emissions. The EPA estimated the rules would cost utilities about $9.6 billion per year, but would save the public between $37 billion and $90 billion annually, mostly by avoiding premature deaths and reducing health problems.

President Obama's Clean Power Plan, proposed in 2014, was designed to improve air quality and also help the U.S. meet its greenhouse gas reduction targets promised at the Paris climate meetings in 2015. The plan aimed to cut carbon emissions from power plants by 30 percent compared to 2005 levels, while creating thousands of new jobs in the clean power sector. It also would reduce soot and smog levels by at least 25 percent by 2030. The Environmental Protection Agency estimated that the plan would result in climate and health benefits worth between $55 billion and $93 billion by 2030. It should also avoid around 5,000 premature deaths and 150,000 asthma attacks.

The Clean Power Plan set state-by-state emissions reduction targets, but it let states decide how to achieve those reductions. However, these accommodations to state's rights weren't enough for critics. In 2018, Andrew Wheeler, a former coal lobbyist and President Trump's head of the EPA, overturned the plan, saying that it would cost coal jobs, and that any health benefits were only incidental and inapplicable as justification for the plan. Ironically, most utilities pleaded with the government to retain the rule, as they had already spent the money on pollution controls. Eliminating the requirements would put them at a disadvantage compared to bad actors that had refused to comply with the law.

The Clean Water Act (1972) protects surface water

Water protection has been a goal with wide public support, in part because clean water is both healthy and an aesthetic amenity. The act aimed to make the nation's waters "fishable and swimmable"; that is, healthy enough to support the propagation of fish and shellfish that could be consumed by humans, and low enough in contaminants that they were safe for children to swim and play in them.

The first goal of the Clean Water Act (CWA) was to identify and control point source pollutants, end-of-the-pipe discharges from factories, municipal sewage treatment plants, and other sources. Discharges are not eliminated, but water at pipe outfalls must be tested, and permits are issued that allow moderate discharges of low-risk contaminants such as nutrients or salts. Metals, solvents, oil, high counts of fecal bacteria, and other more serious contaminants must be captured before water is discharged from a plant.

By the late 1980s, point sources were increasingly under control, and the CWA was used to address nonpoint sources, such as runoff from urban storm sewers. The act has also been used to promote watershed-based planning, in which communities and agencies collaborate to reduce contaminants in their surface waters. As with the CAA, the CWA provides funding to aid pollution-control projects. Those funds have declined in recent years, however, leaving many municipalities struggling to pay for aging and deteriorating sewage treatment facilities. For more details on the CWA and water pollution control, see chapter 18.

The Endangered Species Act (1973) protects both plants and animals

Preventing the loss of charismatic species such as sea turtles (opening case study) has tremendous public support, and we depend on the Endangered Species Act (ESA) for this protection. This law provides a structure for identifying and listing species that are vulnerable, threatened, or endangered. Once a species is listed as endangered, the ESA provides rules for protecting it and its habitat, ideally to make recovery possible (fig. 24.7). Listing of a species can be a controversial process because habitat conservation can get in the way of land development. Many ESA controversies arise when developers want to put new housing or other developments in scenic areas where the last remnants of a species reside. Often, however, disputes have been resolved by negotiation and by more creative planning, which can sometimes allow for both development and species protection.

FIGURE 24.7 The Endangered Species Act is charged with protecting species and their habitat. The black-footed ferret was declared extinct in 1979, but a remnant population was discovered in 1981, and captive breeding programs have restored the species to more than 1,000 mature, wild-born animals in eight states.
Jeff Vanuga/Digital Stock/Corbis

There is considerable collaboration between the ESA and the Convention on International Trade in Endangered Species (CITES) established by the UN, in listing and monitoring endangered species. The ESA maintains a worldwide list of endangered species, as well as a U.S. list. The ESA also provides for grants and programs to help landowners protect species. The responsibility for studying and attempting to restore threatened and endangered species lies mainly with the Fish and Wildlife Service and the National Oceanic and Atmospheric Administration.

Industry and developers have long criticized the ESA because species protections sometimes restricted logging, ranching, drilling, mining, and land development. The sympathetic Trump administration, in particular, accelerated their efforts to "modernize" the ESA, with new barriers to listing species for protection, and delisting species that impeded policy objectives such as oil and gas development. Ecologists, policymakers, and politically engaged citizen groups, meanwhile, sought to protect the act. This tug of war is constant in environmental politics. It shows the critical importance of civic action and voter engagement.

The Superfund Act (1980) lists hazardous sites

Most people know this law as the Superfund Act because it created a giant fund to help remediate abandoned toxic sites. The proper name of this law is informative, though: the Comprehensive Environmental Response, Compensation, and Liability Act (CERCLA). The act aims to be comprehensive, addressing abandoned sites, emergency spills, or uncontrolled contamination, and it allows the EPA to try to establish liability, so that polluters help to pay for cleanup. It's much cheaper to make toxic waste than to clean it up, so we have thousands of chemical plants, gas stations, and other sites that have been abandoned because they were too expensive to clean properly. The EPA is responsible for finding a contractor to do cleanup, and the Superfund was established to cover the costs, which can be in the billions of dollars. Until recently, the fund was financed mainly by contributions from industrial producers of hazardous wastes. In the 1990s, however, Congress voted to end that source, and the Superfund was allowed to dwindle to negligible levels. Site cleanup is now funded by taxpayer dollars.

According to the EPA, one in four Americans lives within 3 miles of a hazardous waste site. The Superfund program has identified more than 47,000 sites that may require cleanup. The most serious of these (or the most serious for which proponents have been sufficiently vigorous) have been put on a National Priorities List. Some 1,680 sites have been put on the list, and about half have been cleaned up. The total cost of remediation is thought to be somewhere between $370 billion and $1.7 trillion. To read more, see chapter 21.

Section Review

1. Describe the major provisions of NEPA, the Clean Air Act, the Clean Water Act, the Endangered Species Act, and the Superfund Act.

2. What are some reasons for opposing or supporting these policies?

24.3 How Are Policies Made?

- *Congress and legislatures vote on statutory laws.*
- *Judges decide case law.*
- *Executive agencies make rules and enforce laws.*

The general process by which policies are developed is often called the *policy cycle*, because rules are developed, enacted, and revised repeatedly (fig. 24.8). The cycle starts when a problem is identified as a priority. In the case of endangered species protections (both CITES and the ESA), for example, the public became concerned about accelerating damage to species and ecosystems. Citizen groups and wildlife advocates initiated public debates on the issue. Communities, students, and environmental groups worked to organize stakeholders, choose tactics, and aggregate related issues into a case for species preservation that was of interest to a broad range of groups and communities.

Citizens often draw attention to their cause by organizing protests, marches, street theater, or other kinds of public events (fig. 24.9). Getting involved in local election campaigns, writing letters, or making telephone calls to legislators also influences the decision process. You'd be surprised at how few letters or calls legislators receive from voters, even on important national issues. Your voice can have an important impact (see chapter 25). Getting a proposal enacted as a law or a rule takes persistence, negotiation, and usually years of effort.

Every country has different legislative and legal processes. The discussion that follows focuses on the U.S. system, which shares many general similarities with other democratic systems.

Congress and legislatures vote on statutory laws

Elected legislative bodies, such as Congress, state legislatures, or town councils, debate and vote on policies that become legally enforceable laws. Federal laws (statutes) are enacted by Congress

FIGURE 24.8 The policy cycle is a path through which rules are developed, enacted, tested, and revised.

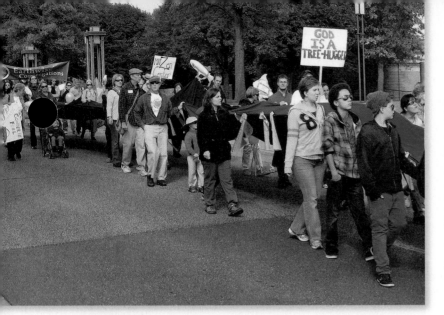

FIGURE 24.9 Marches, sit-ins, and other peaceful demonstrations can help call attention to your cause and encourage your neighbors to join the effort.
William P. Cunningham

FIGURE 24.10 Nevada college students testify against higher education cuts at a Congressional field hearing. Positive participation in an event such as this can influence both legislators and the general public.
Cathleen Allison/AP Photo

and signed by the president. They originate as legislative proposals called "bills," which are usually drafted by the congressional staff, often in consultation with representatives of various interest groups.

Thousands of bills are introduced every year in Congress. Some are very narrow, providing funds to build a specific road or bridge or to help a particular person, for instance. Others are extremely broad, perhaps overhauling the social security system or changing the entire tax code.

After a bill is introduced, it goes to a committee, where it is discussed and debated. Most hearings take place in Washington, but if the bill is controversial, field hearings may be held closer to the site of the controversy. The public often has an opportunity to give testimony at field hearings (fig. 24.10). Elected officials may be swayed by public opinion, and they need public support in policy making.

The language of a bill is debated, revised, and negotiated in a committee until it is considered widely acceptable enough to send to the full House of Representatives or Senate. Compromises are necessary to make the bill acceptable to different parties. By the time an issue has passed through both the House and Senate, the versions approved by the two bodies are likely to be different. They then go to conference committee to iron out any differences between them. After going back to the House and Senate for confirmation, the final bill goes to the president, who may either sign it into law or veto it. If the president vetoes the bill, it may still become law if two-thirds of the House and Senate vote to override the veto.

Each step of this convoluted process is recorded, and you can track it any time through the official congressional website, www. thomas.gov. It can take practice to find the legislation you want, but it's a rich repository of public records.

Legislative riders sidestep public debate

There are two types of legislation: authorizing bills become laws, while appropriation bills provide the money for federal agencies and programs. Eliminating funding for an agency in an

appropriation bill is often an effective way to prevent laws from being implemented. Appropriation bills are not supposed to make policy, but merely fund existing plans and projects. Legislators who can't muster enough votes to pass pet projects through regular channels often will try to add authorizing amendments called **riders** into completely unrelated funding bills. Even if they oppose the riders, other members of Congress have a difficult time voting against an appropriation package for disaster relief or to fund programs that benefit their districts. This often happens in conference committee because when the conference report goes back to the House and Senate, the vote is either to accept or reject with no opportunity to debate or amend further.

Starting with the 104th Congress (in 1995) industry groups began using this tactic to roll back environmental protections and gain access to natural resources. For instance, the 2016 appropriation bills had more than 113 anti-environmental riders that attempted among other things to block climate change research; require the government to burn more fossil fuels; prohibit pesticide regulations; discourage the use of efficient light bulbs; defund clean water protection; restrict endangered salmon and sea otter protection; reduce public participation in land-use decisions; forbid dam removal on U.S. rivers; open millions of acres of wilderness to mining, drilling, and off-road motor use; and exempt grazing on public lands from environmental laws. Often, these amendments are added after all other debate on the particular bill is finished so that other legislators don't even know what they're voting for or against.

Lobbying influences government

Citizens and organizations seek to influence policy in a variety of ways, most visibly through public demonstrations (fig. 24.11). But visiting lawmakers to speak to them directly about a specific piece of legislation, a practice known as lobbying, can be even more important. The term *lobbying* derives from the practice of waiting in hallways and lobbies of Congress to catch the elbow of a passing legislator.

Citizens often make trips to Washington—or to state capitals, county seats, or city halls—to try to personally persuade elected officials on upcoming votes. This direct contact is a basic part of the democratic process, but it can sometimes work unevenly because most people can't abandon work or school and fly to Washington to lobby.

Because it's often hard to travel to meet legislators, many people join organizations that can collectively send representatives, or hire professional lobbyists, to make sure their message is heard. Most major organizations now have lobbyists in Washington. The biggest single citizen lobbying group is probably the American Association of Retired Persons (AARP), which actively lobbies for issues considered of interest to senior citizens. The National Rifle Association (NRA), Union of Concerned Scientists, and many other groups participate in lobbying. Environmental organizations such as the Natural Resources Defense Council, Audubon, and the Sierra Club lobby on many environmental bills. Lobbyists and volunteer activists attend hearings, draft proposed legislation, and meet with officials. The range of interests involved in lobbying is astounding. Business organizations, workers, property owners, religious and ethnic groups, are all there. Walking the halls of Congress, you see an amazing mixture of people attempting to be heard.

Lobbying firms often hire former legislators, military officers, and agency staff. This "revolving door" ushers retiring legislators into lucrative lobbying jobs, rewards them for supporting key legislation, and aids a lobbying company in catching the ear of voting members of Congress.

Industry groups have especially large rewards to gain through lobbying. *The Washington Post* reported on a case in which a group of corporations invested $1.6 million lobbying for a special low overseas tax rate, and the effort saved them over $100 billion in tax payments. In another case, the Carmen Group, a lobbying firm, charged $500,000 to lobby for insurance claims following

the September 11, 2001, attacks on the World Trade Center, and as a result the government agreed to cover $1 billion in insurance premiums for its clients.

Often, lobbyists write the bills that a legislator introduces. For example, in 2013 Congress passed a continuing resolution to fund the federal government for a few additional months. Tucked into the 587-page bill was a brief provision worth millions of dollars for genetic engineering firms. It exempted firms like Monsanto from judicial review, allowing them to sell and plant genetically engineered crops even if a court of law orders them to stop. It turned out that Monsanto lobbyists wrote this section of the bill.

The number of lobbyists registered in Washington more than doubled between 2000 and 2005, from 16,000 to almost 35,000. The biggest industry lobbying firms, such as The Federalist Group, can charge $20,000 to $40,000 per month for their services. Lobbying is, by its nature, about tipping the tables in the favor of an interest group. But lobbying is also something that many people see as necessary, as part of getting voices heard in a democratic process. What do you think? Is corporate lobbying important? Is it necessary? How would you distinguish the actions of an oil industry's lobbyists from those of an environmental or community group's lobbyists?

Judges decide case law

Often, environmental policies are established when groups bring complaints to the courts, involving damage to property or health, failure to enforce existing laws, or infringement of rights. Judges or juries decide these cases by determining whether written law (statutes) or customary law (common law) has been violated. The body of legal opinions built up by many court cases is called case law. Cases might involve criminal prosecutions, claims against the federal government, or complaints in which parties come from multiple states.

Case law is important because legislation is often written in vague and general terms, which help make a bill widely enough accepted to gain passage. Congress often leaves it to the courts to "fill in the gaps," especially in environmental laws. As one senator said when Congress was about to pass the Superfund legislation, "All we know is that the American people want these hazardous waste sites cleaned up . . . Let the courts worry about the details."

Lawsuits with very far-reaching implications are decided by the United States Supreme Court, a group of nine justices whose job is to judge whether a law is consistent with the U.S. Constitution, or whether a policy is consistent with a law as written by Congress. States also have Supreme Courts for deciding cases at the state level. The Supreme Court can only study and judge a few cases a year, and for many people the Court and its actions are little known and poorly understood. But this is a body that makes pivotal and far-reaching decisions, many of them affecting our national environmental policies (fig. 24.12). In theory, the court follows the Constitution and established law, but in recent years, it has become more partisan, rewriting many established rules and enhancing the power of corporations.

FIGURE 24.11 Public protests bring attention to policy issues.
©Robert van Waarden/Survival Media Agency

FIGURE 24.12 The Supreme Court decides pivotal cases, many of them bearing on natural resources or environmental policies.
IM LO SCALZO/EPA-EFE/Shutterstock

Landmark cases have vast impacts

The most sweeping rule change the Supreme Court has made in recent years was the 2010 decision in *Citizens United v. Federal Election Commission*. For nearly a century, anticorruption laws had limited corporate and union spending on political campaigns, generally on the grounds that corporations have more money than individuals, which gives them unfair influence in elections. In a hotly debated, 5–4 split vote, the Court decided that these laws limited free speech (in the form of political advertising), which is protected under the Constitution. The five-person majority argued that because U.S. law gives corporations the same rights and protections as individual people, corporate advertising could not be limited. Within months of the decision, a coalition of oil, gas, and coal corporations announced that they would substantially increase their campaign contributions in the next congressional races.

Many commentators, including dissenting members of the Supreme Court, argue that the *Citizens United* decision protects corporations from a wide range of public oversight and legal restrictions. Among other concerns, this decision opens the door to new levels of corporate influence in justice, as well as politics. For example, a coal company could easily fund a judge's election campaign. If a community filed a suit against the coal company for polluting streams, or for unsafe working conditions, could they be sure to receive a fair judgment?

A dissenting opinion from the four justices who voted against the *Citizens United* decision stated that "corporations have no consciences, no beliefs, no feelings, no thoughts, no desires . . . and their 'personhood' often serves as a useful legal fiction. But they are not themselves members of 'We the People' by whom and for whom our Constitution was established." The dissent states further, "The fact that corporations are different from human beings might seem to need no elaboration, except that the majority opinion almost completely elides it."

One clear effect of the *Citizens United* case was to increase the amount of spending in campaigns. Every election cycle now produces new records in campaign spending, as backers compete to sway elections.

Law suits require legal standing

Before a trial can start, the litigants must establish that they have **standing,** or a right to stand before the bar and be heard. The main criterion for standing is a valid interest in the case. Plaintiffs must show that they are materially affected by the situation they petition the court to redress. This is an important point in environmental cases. Groups or individuals often want to sue a person or corporation for degrading the environment. But unless they can show that they personally suffer from the degradation, courts are likely to deny standing.

In a landmark 1969 case, *Sierra Club v. Morten,* the Sierra Club challenged a decision of the Forest Service and the Department of the Interior to lease public land in California to Walt Disney Enterprises for a ski resort. The land in question was a beautiful valley that cut into the southern boundary of Sequoia National Park. Building a road into the valley would have required cutting down a grove of giant redwood trees within the park (fig. 24.13). The Sierra Club argued that it should have standing in the case to represent the trees, animals, rocks, and mountains that couldn't defend their own interests in court. After all, the club pointed out, corporations—such as Disney—are treated as persons and represented by attorneys in the courts. Why not grant trees the same representation? An influential legal brief by Christopher D. Stone titled *Should Trees Have Standing?* proposed that organisms as well as ecological systems and processes should have legally recognized rights.

The case went all the way to the Supreme Court, which ruled that the Sierra Club failed to show that its members would be materially affected by the development. However, the Court established a key precedent in stating that "aesthetic and environmental well-being, like economic well-being, are important ingredients of the quality of life" and are "deserving of legal protection."

FIGURE 24.13 In 1969, the Sierra Club sued the U.S. Department of the Interior on behalf of the giant redwood trees in Sequoia National Park. The U.S. Supreme Court ruled that the Sierra Club didn't have legal standing in this case, but agreed that "aesthetic and environmental well-being" deserve protection.
Royalty-Free/Corbis

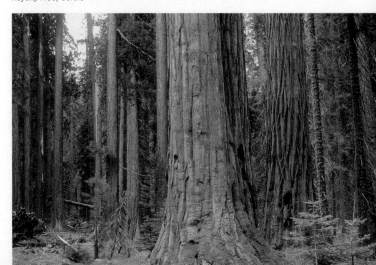

Criminal law prosecutes lawbreakers

Violation of many environmental statutes constitutes criminal offenses. In 1975, the U.S. Supreme Court ruled that corporate officers can be held criminally liable for violations of environmental laws if they were grossly negligent, or the illegal actions can be considered willful and knowing violations. Theoretically (though rarely) criminal violations of environmental law can lead to jail sentences. In 1982, the EPA created an Office for Criminal Investigation. Under the Clinton administration, prosecutions for environmental crimes rose to nearly 600 per year. They fell below 150 per year under George W. Bush, but the Obama administration again increased new prosecutions to about 400 per year. During the Trump administration, prosecutions for environmental crimes dropped to less than 100 cases per year.

Civil lawsuits, involving civil rights and property, can lead to fines for polluters. In 2000, the Koch oil company, one of the largest pipeline and refinery operators in the United States, agreed to pay $35 million in fines and penalties to state and federal authorities for negligence in more than 300 oil spills in Texas, Oklahoma, Kansas, Alabama, Louisiana, and Missouri between 1990 and 1997. Koch also agreed to spend more than $1 billion on cleanup and improved operations. In the aftermath of the 2010 Gulf oil spill, BP (formerly British Petroleum) agreed to pay a total of $20.8 billion to settle lawsuits, criminal penalties, and fines for violations of the U.S. Clean Water Act. The cleanup costs added another $40 billion to this amount.

Lawsuits can also be used to stop public interest groups from challenges to industry. Because defending a lawsuit is so expensive, the mere threat of litigation can be a chilling deterrent. Increasingly environmental advocates are being harassed with **strategic lawsuits against public participation (SLAPP).** Citizens who criticize businesses that pollute or government agencies that neglect their responsibility to protect the public are often sued in retaliation.

Most of these preemptive strikes are groundless and ultimately dismissed, but defending yourself against them can be cripplingly expensive and they can halt progress on the issue in question. Public interest groups and individual activists—many of whom have little money to defend themselves—often are intimidated from taking on polluters. For example, when a West Virginia farmer wrote an article about a coal company's pollution of the Buckhannon River, the company sued him for $200,000 for defamation. Similarly, citizen groups fighting a proposed incinerator in upstate New York were sued for $1.5 million by their own county governments. A Texas woman called a nearby landfill a dump—and her husband was named in a $5 million suit for failing to "control his wife."

Executive agencies make rules and enforce laws

More than 100 federal agencies and thousands of state and local boards and commissions oversee environmental policies. They set rules, adjudicate disputes, and investigate misconduct and operate parks and refuges (fig. 24.14). In the federal government, the president's cabinet (a group of department heads in charge of various tasks) includes the heads of major departments such as Agriculture, Interior, or Justice. Rules that these agencies make and enforce decide many of our most important environmental, resource, and health issues.

The public can influence agency rule-making by giving comments on a proposed rule. Comments can be made at regional hearings, which agencies are required to offer, and by mail or e-mail during a "public comment period," usually a few months, that precedes the acceptance of a new rule or action. Often, thousands of comments are collected, and these can substantially influence whether or not a proposed rule or plan is enacted.

Executive orders also can be agents for change. President Barack Obama, for example, used the Antiquities Act more than any other president, establishing 23 new national monuments and expanding dozens of existing national parks and wildlife refuges. Altogether, Obama ordered protection for 107 million ha (265 million acres) of land and water.

Rules and policies made by executive decree in one administration, however, can be quickly undone in the next one. As noted in chapter 12, shortly after taking office, President Trump announced draconian reductions in the most recently established National Monuments. Bear's Ears was to be reduced by 85 percent, and more than half of Grand Staircase-Escalante was to be opened to mining, ranching, and other destructive uses. Grand Staircase has coal deposits that industry covets, and Bear's Ears may have uranium. Presidential rules, then can be easy for a president to implement but just as easy for the next president to undo.

An important administrative rule that has been reversed repeatedly is a ban on building new logging roads into nearly 24 million hectares of roadless public land. The Clinton administration wrote a rule establishing that these "de facto" wilderness

FIGURE 24.14 Federal agencies administer public lands and enforce environmental regulations in places such as Glacier National Park in Montana.
William P. Cunningham

areas should be protected. The pro-industry Bush administration reversed Clinton's rules; then Obama overturned Bush's reversals. The Trump administration then eliminated these rules for some of the areas. Roadless area rules may be in contention for some time to come.

Regulatory agencies oversee policies

The EPA is the primary agency with responsibility for protecting environmental quality. It was created in 1970, at the same time as NEPA, and its head is appointed by the president as part of his cabinet, the small body of top administrators who work directly with the president. Like other cabinet positions, the EPA is strongly influenced by political winds. EPA funding, enforcement activity, and rule-making strongly reflect the views of the president's political advisers. Under the Nixon and Carter administrations, the EPA grew rapidly and enforced air and water quality standards vigorously. EPA activity and funding declined sharply during the Reagan administration, recovered under Bill Clinton, declined again during the Bush administrations, rose again under Obama, and fell under Trump.

The Departments of the Interior and Agriculture manage natural resources. The National Park Service, which is responsible for more than 376 national parks, monuments, historic sites, and recreational areas, is part of the Interior Department. Other Interior agencies are the Bureau of Land Management (BLM), which administers some 140 million ha of land, mostly in the western United States, and the Fish and Wildlife Service, which operates more than 500 national wildlife refuges and administers endangered species protection.

The Department of Agriculture is home to the U.S. Forest Service, which manages about 175 national forests and grasslands, totaling some 78 million ha (fig. 24.15). With 39,000 employees, the Forest Service is nearly twice as large as the EPA. The Department of Labor houses the Occupational Safety and Health Agency (OSHA), which oversees workplace safety, including exposure to toxic substances. In addition, several independent agencies that are not tied to any specific department also play a role in environmental protection and public health. The Consumer Products Safety Commission passes and enforces regulations to protect consumers, and the Food and Drug Administration is responsible for the purity and wholesomeness of food and drugs.

All of these agencies have a tendency to be "captured" by the industries they are supposed to be regulating.

Regulatory capture undermines agency work

When the executive branch objects to agency functions, such as water quality monitoring or wildlife protections, the executive can designate agency heads who are antagonistic to the mission of the agency. Through this process, known as **regulatory capture,** opponents can prevent an agency from doing its work, choose not to pursue policies, or actively change rules. This is a long tradition in politics, but the Trump administration provided particularly notable cases.

FIGURE 24.15 Smokey Bear is the public face of an executive agency, the U.S. Forest Service.
William P. Cunningham

For example, Scott Pruitt, who became head of the EPA in 2017, had been attorney general of Oklahoma, where he sued the EPA more than a dozen times to overturn clean air and water rules. In his first 4 months in office, Pruitt reversed or delayed dozens of air, water, and climate regulations. He dismissed members of his scientific review board and replaced them with industry representatives whose pollution the agency is supposed to control. In 2018, Andrew Wheeler, a chief lobbyist for the coal industry, replaced Pruitt as EPA head. Wheeler almost immediately overturned an Obama-era mercury limit for burning coal. The rule, which had reduced mercury emissions by 90 percent, was credited for preventing 11,000 premature deaths and over 100,000 asthma and heart attacks each year in addition to protecting millions of children from neurological damage. The estimated health benefits were $37 billion to $90 billion annually while costs to industry were estimated to be $7.4 billion to $9.6 billion per year. Nevertheless, Wheeler (and President Trump) regarded the cost to the coal industry to be unacceptable, and the rule was overturned.

Similarly, Rick Perry, head of the Department of Energy, had said as a candidate that he wanted to eliminate the whole agency, and Ryan Zinke, head of the Department of the Interior, promised to open coastal waters to oil drilling, shrink national monuments, lift fossil fuel regulations, and reduce wildlife protection, all steps he implemented rapidly upon taking office. In his first budget proposal, Trump cut renewable energy research by 72 percent and reduced the EPA budget by one-third. And his $200 billion infrastructure plan released in 2018 was much less about rebuilding roads and bridges than about reducing environmental reviews for fossil fuel developments and privatizing public lands.

At the end of 2018, the *New York Times* collected a list of 90 environmental rules and regulations that President Trump reversed or attempted to reverse during his first two years in office. Most of these rollbacks would allow more air pollution and toxic emissions. Many others overturned efforts to reduce global climate change. Since President Trump and many Republicans rejected the science of climate change, they regarded these measures as needless obstacles.

Trump also rescinded water pollution rules aimed at protecting rivers, lakes, and groundwater. He ordered opening of millions of acres of public land, national parks, wildlife refuges, and offshore areas for drilling and mining. In particular, he rejected any rules that regulated or limited the fossil fuel industry, a major funder of conservative election campaigns. Federal courts quickly overturned some of the rule changes, but others remain in effect.

How much government do we want?

In his 1981 inaugural address, President Ronald Reagan famously said, "Government is not the solution to our problem; government is the problem." In this, he invoked a perennial debate in American politics: Is government a power that undermines personal liberties? Or is government a representative of the people and a defender of personal liberties against bullies?

The answer sometimes depends on when you ask. During political campaigns, many candidates decry the size and cost of government agencies. But in a crisis most of us assume the government will be there to help us out with rescue teams, emergency aid, and federal funding for reconstruction (fig. 24.16).

Debates about the proper size and role of government are common. We value self-reliance and rugged individualism. Yet we also want someone to protect us from contaminated food and drugs, to educate our children, and to provide roads, bridges, and safe drinking water. President Reagan was among those who favor "free market" capitalism, with businesses unfettered by rules such as the Clean Air Act or the Clean Water Act. Political strategist Grover Norquist, president of the Americans for Tax Reform, famously said he'd like to "shrink the government down to the size where we can drown it in the bathtub." Other observers note that while Reagan's language focused on freedom for individuals and small business owners, the elimination of public health and safety regulations tends to undermine the interests of individuals and small businesses. Advantages are often given to the biggest players in a game without rules.

Part of the reason these disputes persist may be that both views are partially correct. Regulations, such as those imposing expensive pollution abatement technologies for polluters, require private businesses to protect, rather than damage, public resources. Businesses are squeezed between shareholders' demands for ever-higher profits and agency demands for safer operating standards. Viewed another way, regulations require businesses to clean up their own messes. Opponents of agency regulations point out that corporations produce the economic vitality on which our prosperity depends. Proponents of regulations point out that corporations couldn't prosper without subsidies, tax breaks, transportation infrastructure, and a healthy and educated workforce. These costs are necessary to doing business but are normally external to the accounting of business costs and profits.

These debates probably will always be with us. Which view is most correct depends on many factors: your interest group, life experience, philosophical perspective, economic position, the time frame you analyze, and other priorities. What factors influence your view on these issues? Do you think there is room for compromise? If not, why not? If yes, where?

Section Review

1. Describe the path of a bill through Congress. When are riders and amendments attached?
2. What is legal standing, and why is it important?
3. What is regulatory capture? Give an example.

24.4 INTERNATIONAL CONVENTIONS

- *Many international agreements protect our environment.*
- *International enforcement often depends on national honor.*

Growing interconnections in our global environment and economy have made nations increasingly interested in signing on to international agreements (or "conventions") for environmental protection. A principal motivation for participating in these treaties is the recognition that countries can no longer act alone to protect their resources and interests. Water resources, the atmosphere, trade in endangered species, and many other concerns cross international boundaries. Over time, the number of parties taking part in negotiations has grown sharply (fig. 24.17). The speed at which agreements take force has also increased. The Convention on International Trade in Endangered Species (CITES), for example, was not enforced until 14 years after its ratification in 1973, but the Convention on Biological Diversity (1991) was enforceable after just one year, with 160 nations signing the agreement just four years after its introduction.

Hundreds of other conventions and agreements address concerns ranging from intercontinental shipping of hazardous waste, to deforestation, overfishing, trade in endangered species, and wetlands protection.

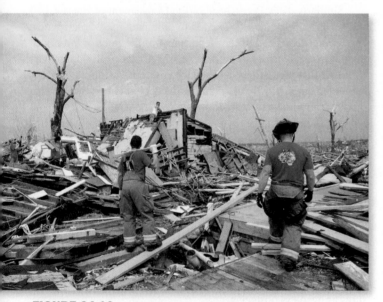

FIGURE 24.16 When tornadoes, floods, hurricanes, or other natural disasters afflict us, we expect government agencies to help us.
Julie Denesha/Getty Images

- ▬ Ramsar Convention, wetland protection
- ▬ World Heritage Program, protecting cultural sites
- ▬ Stockholm Declaration
- ▬ CITES
- ▬ Convention on Migratory Species
- ▬ UN Convention on Law of the Sea
- ▬ Montreal Protocol, ozone protection
- ▬ Basel Convention on hazardous waste
- ▬ Convention on Biological Diversity
- ▬ UN Convention on Combat Desertification
- ▬ UN Framework Convention on Climate Change
- ▬ Kyoto Protocol on climate change
- ▬ Rotterdam Convention, restricting industrial chemicals
- ▬ Cartagena Protocol on biotechnology

FIGURE 24.17 Major international environmental agreements, listed in order of ratification dates.

Major international agreements

International accords and conventions have emerged slowly but fairly steadily from meetings such as those in Stockholm and Rio (table 24.2). A few of the important benchmark agreements are discussed here.

The **Convention on International Trade in Endangered Species** (CITES, 1973) declared that wild flora and fauna are valuable, irreplaceable, and threatened by human activities. To protect disappearing species, CITES maintains a list of threatened and endangered species that may be affected by trade. As with most international agreements, this one takes no position on movement or loss of species within national boundaries, but it establishes rules to restrict unauthorized or illegal trade across boundaries. In particular, an export permit is required specifying that a state expert declares an export is legal, that it is not cruel, and that it will not threaten a wild population.

The **Montreal Protocol** (1987) protects stratospheric ozone. This treaty committed signatories to phase out the production and use of several chemicals that break down ozone in the atmosphere. The ozone "hole," a declining concentration of ozone (O_3) molecules over the South Pole, threatened living things: Ozone high in the atmosphere blocks cancer-causing ultraviolet radiation, keeping it from reaching the earth's surface. The stable chlorine- and fluorine-based chemicals at fault for reducing ozone are used mainly as refrigerants. Alternative refrigerants have since been developed, and the use of chlorofluorocarbons (CFCs) and related

molecules has plummeted. Although the ozone "hole" has not disappeared, it has declined as predicted by atmospheric scientists since the phase-out of CFCs. The Montreal Protocol is often held up as an example of a highly successful and effective international environmental agreement.

The Montreal Protocol was effective because it bound signatory nations not to purchase CFCs or products made using them from countries that refused to ratify the treaty. This trade restriction put substantial pressure on producing countries. Initially, the protocol called for only a 50 percent reduction in CFC production, but subsequent research showed that ozone was being depleted faster than previously thought (chapter 16). The protocol was strengthened to an outright ban on CFC production, in spite of the objection of a few countries.

The **Basel Convention** (1992) restricts shipment of hazardous waste across boundaries. The aim of this convention, which has 172 signatories, is to protect health and the environment, especially in developing areas, by stating that hazardous substances should be disposed of in the states that generated them. Signatories are required to prohibit the export of hazardous wastes unless the receiving state gives prior informed consent, in writing, that a shipment is allowable. Parties are also required to minimize production of hazardous materials and to ensure that there are safe disposal facilities within their own boundaries. This convention establishes that it is the responsibility of states to make sure that their own corporations comply with international laws. The Basel Convention was enhanced by the Rotterdam Convention (1997),

Table 24.2 Some Important International Environmental Treaties
CBD: Convention on Biological Diversity 1992 (1993)
CITES: Convention on International Trade on Endangered Species of Wild Fauna and Flora 1973 (1987)
CMS: Convention on the Conservation of Migratory Species of Wild Animals 1979 (1983)
Basel: Basel Convention on the Transboundary Movements of Hazardous Wastes and Their Disposal 1989 (1992)
Ozone: Vienna Convention for the Protection of the Ozone Layer and Montreal Protocol on Substances that Deplete the Ozone Layer 1985 (1988)
UNFCCC: United Nations Framework Convention on Climate Change 1992 (1994)
CCD: United Nations Convention to Combat Desertification in Those Countries Experiencing Serious Drought and/or Desertification, Particularly in Africa 1994 (1996)
Ramsar: Convention on Wetlands of International Importance Especially as Waterfowl Habitat 1971 (1975)
Heritage: Convention Concerning the Protection of the World Cultural and Natural Heritage 1972 (1975)
UNCLOS: United Nations Convention on the Law of the Sea 1982 (1994)

which places similar restrictions on unauthorized transboundary shipment of industrial chemicals and pesticides.

The 1994 **UN Framework Convention on Climate Change (UNFCCC)** directs governments to share data on climate change, to develop national plans for reducing greenhouse gas (GHG) emissions, and to cooperate in planning for adaptation to climate change. Under the UNFCCC, the Kyoto Protocol (1997) set binding targets for signatories to reduce greenhouse gas emissions to less than 1990 levels by 2012. The idea of binding targets was strong, but few countries achieved their goals, and some, such as the United States, declined to be subjected to international policies, particularly because Kyoto set tighter restrictions on industrialized countries, which are responsible for roughly 90 percent of GHG emissions up to the present, than on developing countries. In 2012, Canada also withdrew from the agreement, in part because energy-intensive development of oil shale makes it impossible for Canada to meet its targets.

In 2015, the UNFCCC led to the Paris Agreement, which set a climate change target of "well below" 2°C or less, far preferably 1.5°C, and in which the 195 signatory countries agreed to establish their own national determined contributions to this goal (see chapter 15). Within two years, over 190 countries signed on to the Paris Agreement. A few leaders have since declared an intent to withdraw; however, withdrawal requires a three-year waiting period and cannot be done immediately or by executive decree (fig. 24.18).

Enforcement often depends on national pride

In the Paris Agreement and many other international accords, there is no clear means of forcing parties to make good on promises, so enforcement often depends on the fact that countries care about their international reputation. In general, states are wary of interfering with the internal sovereignty of other states, except in extreme cases such as genocide (for which the global community will sometimes send an external police). However, most countries are reluctant to appear irresponsible or immoral in the eyes of the international community, so moral persuasion and public embarrassment can be effective enforcement strategies. Shining a spotlight on transgressions will often push a country to comply with international agreements.

Trade sanctions can be an effective tool to compel compliance with international treaties. The Montreal Protocol used the threat of trade sanctions very effectively to cut CFC production dramatically. On the other hand, trade agreements also can work against environmental protection. The World Trade Organization (WTO) was established in 1995 to promote free international trade and to encourage economic development. The WTO's emphasis on unfettered trade, however, has led to weakening of local environmental rules. For example, in 1990 the United States banned the import of tuna caught using methods that kill thousands of dolphins each year. Shrimp caught with nets that kill endangered sea turtles were also banned. Mexico filed a complaint with the WTO, contending that dolphin-safe tuna laws represented an illegal barrier to trade. Thailand, Malaysia, India, and Pakistan filed a similar suit against turtle-friendly shrimp laws. The WTO ordered the United States to allow the import of both tuna and shrimp from countries that allow fisheries to kill dolphins and turtles. Environmental advocates point out that the WTO has never ruled against a corporation because it is composed of industry leaders. As such, the WTO mainly defends the interests of the business community, not the broader public interest.

Section Review

1. Describe five important international environmental treaties.
2. Why has most international environmental protection been based on an honor system?

FIGURE 24.18 German Chancellor Angela Merkel and other world leaders confront a defiant President Trump at the 2018 G7 meeting in Canada over his refusal to recognize climate change.
Jesco Denzel/German Federal Government/AP

24.5 NEW APPROACHES TO POLICY

- *Community-based planning uses local knowledge.*
- *Green plans outline goals for sustainability.*
- *Bolivia set a world precedent with its Law of Mother Earth.*

As we gain experience with environmental governance, new policy strategies are being developed. These approaches are growing in part because environmental protection has remained a high priority for the public (fig. 24.19). One of the key changes has been to seek win–win compromises in environmental debates. Dispute resolution and mediation are strategies for reaching agreements without the mutual suspicions and hostility inherent in a lawsuit. Dispute resolution can avoid the time, expense, and winner-take-all confrontation inherent in lawsuits. These techniques encourage compromise and workable solutions with which everyone can live.

Arbitration is a formal process of dispute resolution somewhat like a trial. There are stringent rules of evidence, cross-examination of witnesses, and the process results in a legally binding decision. The arbitrator can actively work to find creative

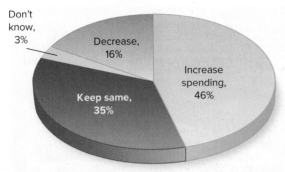

FIGURE 24.19 Americans persistently favor maintaining or increasing spending on environmental protections.

FIGURE 24.20 Adaptive management recognizes that we need to treat management plans for ecosystems as scientific experiments in which we monitor, evaluate, and adjust our policies to fit changing conditions and knowledge.

resolutions to the dispute. **Mediation** is generally less formal. Disputants are encouraged to sit down and talk to see if they can come up with a solution. In face-to-face meetings, people are often more willing to see their opponent's viewpoint and seek solutions.

Less rigid strategies for rule making are also being developed. One of these is **adaptive management,** or "learning by doing." This approach proposes that management should be experimental. Environmental policies should be designed from the outset to test clearly formulated hypotheses about the ecological, social, and economic impacts of the actions being undertaken (fig. 24.20). What initially seemed to be the best policy may not always be best, so we need to carefully monitor how conditions are changing. And we need to be able and willing to revise plans if our initial assumptions don't hold up over time.

Ecological principles also suggest that policymakers should plan for resilience—that is, for changes and recovery in a system (table 24.3). This means, for example, that protected forests must be large enough to allow for disturbances (such as fires or pest outbreaks) and recovery, or that policymakers should anticipate the possibility that the climate or species abundance may change over time.

Community-based planning uses local knowledge

Over the past several decades, natural resource managers have come to recognize the value of holistic planning that acknowledges multiple users and perspectives. Involving all stakeholders and interest groups early in the planning process can help find agreeable plans before adversaries become entrenched in non-negotiable positions. Working with local communities can tap into traditional knowledge and gain acceptance for management plans that finally emerge from policy planning. There are many reasons to use collaborative approaches:

- Incorporating a variety of perspectives early in the process is more likely to lead to the development of acceptable solutions in the end. Public buy-in to an idea is likely to be better if many people have a voice from the start.

- Two heads are better than one. Involving multiple stakeholders and multiple sources of information enriches the process.

- Community-based planning provides access to information from and experience of local residents.

- Participation is an important management tool. Stakeholders are more likely to cooperate over the long term when they can participate throughout the planning process.

- Experience builds capacity to carry out projects. Local participation in later phases of a project often works only when the community is actively involved, and has a chance to gain experience, early on.

Community-based planning can be seen in the Atlantic Coastal Action Programme (ACAP) in eastern Canada. The purpose of this project is to develop blueprints for the restoration and maintenance of environmentally degraded harbors and estuaries in ways that are both biologically and socially sustainable. Officially established under Canada's Green Plan and supported by Environment Canada, this program created 13 community groups, some rural and some urban, with membership in each dominated by local residents. Federal and provincial government agencies are

Table 24.3 Planning for Resilience
1. Interdisciplinary, integrated modes of inquiry are needed for adaptive management of wicked problems.
2. We must recognize that these problems are fundamentally nonlinear and that we need nonlinear approaches to them.
3. We must attend to interactions between long-term processes, such as climate change or soil erosion in the American Corn Belt, and rapid events, such as the collapse of Antarctic ice sheets or the appearance of a dead zone in the Gulf of Mexico.
4. The spatial and temporal scales of our concerns are widening. We must consider global interconnections in our planning.
5. We need adaptive management policies that focus on building resilience and the capacity of renewal both in ecosystems and in human institutions.

represented primarily as nonvoting observers and resource people. Each community group is provided with core funding for full-time staff who operate an office in the community and facilitate meetings.

Four of the 13 ACAP sites are in the Bay of Fundy, an important and unique estuary lying between New Brunswick and Nova Scotia. Approximately 270 km long, and with an area of more than 12,000 km^2, the bay, together with the nearby Georges Bank and the Gulf of Maine once formed one of the richest fisheries in the world (fig. 24.20a). With the world's highest recorded tidal range (up to 16 m at maximum spring tide), the bay still sustains a great variety of fishery and wildlife resources, and provides habitat for a number of rare or endangered species. Now home to more than 1 million people, the coastal region is an important agricultural, lumbering, and paper-producing region. Pollution and sediment damage harbors and biological communities. Overfishing and introduction of exotic species have resulted in endemic species' decline. The collapse of cod, halibut, and haddock fishing has had devastating economic effects on the regional economy and the livelihoods of local residents. To cope with these complex, intertwined social and biological problems, ACAP is bringing together different stakeholders from around the bay to create comprehensive plans for ecological, economic, and social sustainability, including citizen monitoring and adaptive management.

Green plans outline goals for sustainability

Several national governments have undertaken integrated environmental planning that incorporates community roundtables for vision development. Canada, New Zealand, Sweden, and Denmark all have so-called **green plans** or comprehensive, long-range national environmental strategies. The best of these plans weave together complex systems, such as water, air, soil, and energy, and mesh them with human factors such as economics, health, and carrying capacity. Perhaps the most thorough and well-thought-out green plan in the world is that of the Netherlands (fig. 24.20b).

Developed in the 1980s through a complex process involving the public, industry, and government, the 400-page Dutch plan contains 223 policy changes aimed at reducing pollution and establishing economic stability. Three important mechanisms have been adopted for achieving these goals: integrated life-cycle management, energy conservation, and improved product quality. These measures should make consumer goods last longer and be more easily recycled or safely disposed of when no longer needed. For example, auto manufacturers are now required to design cars so they can be repaired or recycled rather than being discarded.

Among the guiding principles of the Dutch green plan are: (1) the "stand-still" principle that says environmental quality will not deteriorate, (2) abatement at the source rather than cleaning up afterward, (3) the "polluter pays" principle that says users of a resource pay for the negative effects of that use, (4) the prevention of unnecessary pollution, (5) application of the best practicable means for pollution control, (6) carefully controlled waste disposal, and (7) motivating people to behave responsibly.

(a)

(b)

FIGURE 24.21 Innovations in environmental policy and planning can be found worldwide. Examples include community planning in Canada's Bay of Fundy (a) and the Dutch green plan that restores ecosystems (b). *(a): Rolf Hicker/All Canada Photos/Getty Images (b): symbiot/Shutterstock*

The Netherlands have invested billions of euros in implementing this comprehensive plan. Some striking successes already have been accomplished. Between 1980 and 1995, emissions of sulfur dioxide, nitrogen oxides, ammonia, and volatile organic compounds were reduced more than 30 percent; pesticide use had been reduced 25 percent; chlorofluorocarbon use had been virtually eliminated; and industrial wastewater discharge into the Rhine River was down 70 percent. Some 250,000 ha (more than 600,000 acres) of former wetlands that had been drained for agriculture are being restored as nature preserves and 40,000 ha (99,000 acres) of forest are being replanted. This is remarkably generous and foresighted for Europe's most densely populated country.

Section Review

1. What is community-based planning?
2. What are green plans?
3. Describe the Dutch green plan.

Connecting the Dots

Policy is a complex business, but the heartening recovery of green sea turtles described in the opening case study for this chapter shows an example of effective public policy. Once nearly extinct, turtles now thrive again on Florida's Atlantic coast because they are protected by a regulations that preserve their habitat, protect eggs and hatchlings, and reduce unintended deaths from fishing nets and longlines. This success has encouraged other countries protect endangered turtle populations elsewhere.

Many other important environmental policies have been passed over the past 50 years in both the United States and other countries. Often these policies have strong public support although businesses complain about the cost and inconvenience of complying. Laws such as the Clean Water Act have been among the most effective tools that conservationists have had to protect biodiversity and habitat. But there is a constant struggle between those who want to strengthen environmental laws and those who want to reduce or eliminate them.

Policy is daunting, but you can contribute in many ways. Probably the best way to participate in environmental policy formation or passage of environmental laws is to join a group that shares your concerns. Being part of a group amplifies your influence. But even as an individual, you can write to or call your legislator. They do pay attention to constituents.

On a smaller scale, community planning and community knowledge have also become key parts of policy formation. Understanding all these aspects of policy is a first step toward learning about environmental policy, and toward empowering yourself to influence the health of the environment in which you live.

Reviewing Key Terms

Can you define the following terms in environmental science?

adaptive management 24.5

arbitration 24.5

Basel Convention 24.4

Convention on International Trade in Endangered Species (CITES) 24.4

cost–benefit analysis (CBA) 24.1

environmental impact statement (EIS) 24.2

environmental policy 24.1

green plans 24.5

mediation 24.5

Montreal Protocol 24.4

National Environmental Policy Act (NEPA) 24.2

policy 24.1

precautionary principle 24.1

regulatory capture 24.3

riders 24.3

standing 24.3

strategic lawsuits against public participation (SLAPP) 24.3

UN Framework Convention on Climate Change (UNFCC) 24.4

Critical Thinking and Discussion Questions

1. In your opinion, how much environmental protection is too much? Think of a practical example in which some stakeholders may feel oppressed by government regulations. How would you justify or criticize these regulations?

2. Among the steps in the policy cycle, where would you put your efforts if you wanted influence in establishing policy?

3. Do you believe that trees, wild animals, rocks, or mountains should have legal rights and standing in the courts? Why or why not? Are there other forms of protection you would favor for nature?

4. It's sometimes difficult to determine whether a lawsuit is retaliatory or based on valid reason. How would you define a SLAPP suit and differentiate it from a legitimate case?

5. Create a list of arguments for and against an international body with power to enforce global environmental laws. Can you see a way to create a body that could satisfy both reasons for and against this power?

6. Identify a current environmental problem, and outline some policy approaches that could be used to address it. What strengths and weaknesses would different approaches have?

Data Analysis

Examine Your Environmental Laws

The federal government publicizes the text of laws in multiple locations on the Internet. Reading about these laws is a good way to get a sense of the structures of environmental regulation, and to understand some of the compromises and the complexity of making rules that apply to thousands of different cases across the country. The primary way to access government rules and laws is through www.thomas.gov. A more direct source for environmental legislation is to go to the EPA website: www.epa.gov/lawsregs/laws/index.html#env.

Go to the EPA website, and then log in to Connect to find suggestions about how to examine a specific law.

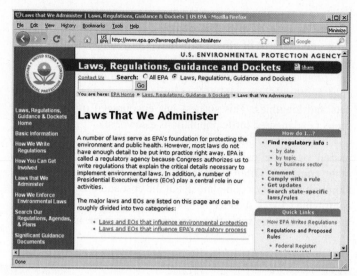

The web address listed above gives you direct access to federal laws that define how the U.S. environment and resources are managed.
United States Environmental Protection Agency (U.S. EPA)

What Then Shall We Do?

▲ Concerned citizens gather to petition their government on Earth Day 1970.
William P. Cunningham

Learning Outcomes

After studying this chapter, you should be able to:

25.1 Explain why environmental education is important.
25.2 Evaluate what individuals can do.
25.3 Review how we can work together.
25.4 Investigate campus greening.
25.5 Define the challenge of sustainability.

"When spiders unite, their web can tie down an elephant."

– African proverb

The Dawn of a New Era

Earth Day, a global celebration of environmental action and awareness, occurred for the first time on April 22, 1970. Many regard this event as the beginning of a new era. On the first Earth Day in 1970, events took place in more than 2,000 colleges and universities and in at least 10,000 primary and secondary schools. The day brought out 20 million Americans in peaceful demonstrations for environmental protection. It awakened a concern about environmental issues along with a sense of optimism that citizens working together could bring about meaningful change. And the collective energy extended far beyond the day itself: Many of our most important environmental protection laws were passed in the decade following the first Earth Day (chapter 24).

A number of factors contributed to the first Earth Day's success. The country had gone through years of civil rights and anti-war demonstrations that helped establish techniques for public organization and effective political action. Those movements showed that concerned citizens could change public opinion and influence history.

Environmental crises at the time were obvious and very local. You could easily see streams choked with sludge that ran black, or red, or green. Or urban air so filled with smog that you could only see for a few hundred meters. The publication of *Silent Spring* by Rachel Carson in 1962 awakened the public to the threats of pollution and toxic chemicals to humans as well as other species. Students today generally have no idea how bad environmental conditions were in America 50 years ago. Undoubtedly, you have heard about how dangerous the air and water quality are in Beijing and New Delhi today. But in 1970, air and water quality in some U.S. industrial cities was nearly as bad.

Several highly publicized events in the years leading up to 1970 made people aware that pollution was widespread and dangerous. In January and February 1969, an oil well blew out just offshore from Santa Barbara, California. Oil seeped ashore and sullied pristine beaches. Photos of young people working to clean the beaches filled the evening news.

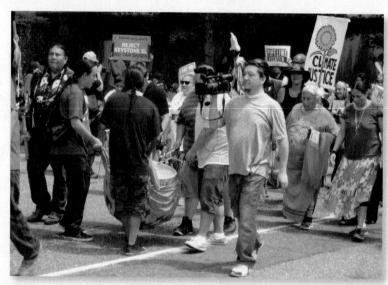

FIGURE 25.1 Many environmental struggles have become more diverse in 50 years. Here Canadian First Nations people join U.S. environmental protests against the Keystone XL pipeline.
Madeleine Desrochers

In June 1969, the Cuyahoga River caught fire. Decades of industrial dumping had created an oil slick that caught fire and incinerated several bridges. When fire trucks arrived they pumped water out of the river to put out the blaze, which just made the conflagration worse. And this wasn't even the first time the river had burned. People realized that something was seriously wrong.

U.S. Senator Gaylord Nelson, of Wisconsin, is often regarded as the founder of the first Earth Day. While the original Earth Day was focused on the United States, Denis Hayes, who was the national coordinator in 1970, took the movement global 20 years later with events in 141 nations. On Earth Day 2016, the landmark Paris Agreement was signed by 120 nations including the United States and China. And in April 2017, the People's Climate March mobilized several hundred thousand people in more than 300 locations including Washington D.C.'s National Mall to protest the lack of action by the United States to combat climate change.

Some things have changed in 50 years. Many environmental threats, such as smog or flaming rivers, were more immediate and obvious then. Current issues like climate change or energy extraction in distant regions are complex and remote. On the other hand, we have gained experience in motivating groups and individuals. Awareness of issues is strong, and environmental concern is widely shared. Even more important, more voices are being heard, compared to 50 years ago. Indigenous groups and people of color, who often suffer most immediately from environmental degradation, are leading in efforts to protect the land and water we all depend on (fig. 25.1). And many people now see that concerned citizens can bring about change. We may not agree how far we need to go to protect nature, but it's rare for anyone to say publically that they don't care what happens to our environment.

Clearly, bringing communities together to learn, celebrate, and act is key to protecting our shared environment. In this chapter, we'll look at some of the ways individuals and groups are working to reduce our environmental impacts and build a sustainable future.

25.1 MAKING A DIFFERENCE

- *Environmental stewardship is everyone's responsibility.*
- *Environmental literacy is a national goal and a lifelong process.*

We are surrounded by environmental challenges, from climate change to biodiversity to energy policy debates. Biodiversity is disappearing at the fastest rate in human history; major ocean fisheries have collapsed; within 50 years, it is expected that two-thirds of countries will experience water shortages, and 3 billion people may live in slums. As we have come to understand these problems, many exciting innovations have been developed to deal with them. New irrigation methods reduce agricultural water use; bioremediation provides inexpensive methods to treat hazardous waste; new energy sources, including wind, solar, geothermal, and even pressure-cooked garbage, offer strategies for weening our society from its dependence on oil and gas. Green consumerism has developed markets for recycled materials, low-energy appliances, and organic foods. Population growth continues, but its rate has plummeted from a couple of generations ago.

Stewardship for our shared resources is increasingly understood to be everybody's business. The environmental justice movement has shown that people of color and the poor frequently suffer more from pollution than wealthy white people. African Americans, Latinos, Native Americans, and other groups have a clear interest in pursuing environmental solutions. Religious groups are voicing new concerns about preserving our environment (chapter 1). Farmers are seeking ways to save soil and water resources (chapter 10). Loggers are learning about sustainable harvest methods (chapter 12). Business leaders are discovering new ways to do well by doing good work for society and the environment (chapter 23). These changes are exciting, although many challenges remain.

Whatever your skills and interests, you can contribute to understanding and protecting our common environment. If you enjoy science, many disciplines contribute to this cause. As you know by now, biology, chemistry, geology, ecology, climatology,

FIGURE 25.2 Helping children develop a sense of wonder is a first step in protecting nature. As the Senegalese poet Baba Dioum said, "In the end, we conserve only what we love. We will love only what we understand. We will understand only what we are taught."
William P. Cunningham

geography, demography, and other sciences all provide ideas and data that are essential to understanding our environment. Environmental scientists usually focus on one of these disciplines, but their work also serves the others. An environmental chemist, for example, might study contaminants in a stream system, and this work might help an aquatic ecologist understand changes in a stream's food web.

You can also help seek environmental solutions if you prefer writing, art, working with children (fig. 25.2), history, politics, economics, or other areas of study. As you have read, environmental science depends on communication, education, good policies, and economics in addition to science.

In this chapter, we discuss some of the steps you can take to help find solutions to environmental dilemmas. You have already taken the most important step, educating yourself. When you understand how environmental systems function—from nutrient cycles and energy flows to ecosystems, climate systems, population dynamics, agriculture, and economies—you can develop well-informed opinions and help find useful answers.

Environmental literacy has lasting importance

In 1990, Congress recognized the importance of environmental education by passing the National Environmental Education Act. The act established two broad goals: (1) to improve understanding among the general public of the natural and built environment and the relationships between humans and their environment, including global aspects of environmental problems, and (2) to encourage postsecondary students to pursue careers related to the environment. Objectives include developing an awareness and appreciation of our natural and social/cultural environment, knowledge of basic ecological concepts, acquaintance with a broad range of current environmental issues, and experience in using investigative,

Benchmark Data	
Among the ideas and values in this chapter, the following are a few worth remembering.	
2/3	Proportion of countries expected to experience water shortages in 50 years
20 million	Americans who participated in Earth Day 1970
$6 trillion	Total global divestment (so far) from fossil fuels
80%	Proportion of fossil fuel reserves that must stay in the ground to avoid catastrophic climate change
77%	Percentage of residents of 40 countries who support climate action
1 million	U.S. workers in green energy
2x	Goods and services consumed by the average American compared to 50 years ago

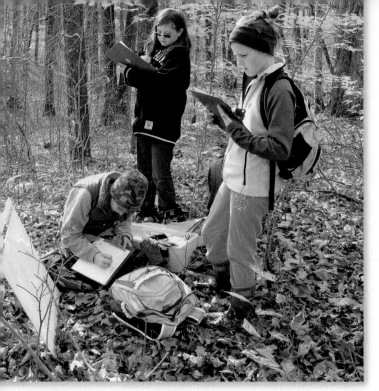

FIGURE 25.3 Environmental education helps develop awareness and appreciation of ecological systems and how they work.
William P. Cunningham

critical-thinking, and problem-solving skills in solving environmental problems (fig. 25.3). Several states, including Arizona, Florida, Maryland, Minnesota, Pennsylvania, and Wisconsin, have incorporated these goals into their curricula (table 25.1).

The importance of environmental education was established in the National Environmental Education Act. Former EPA administrator William K. Reilly, speaking in favor of this act, called for broad **environmental literacy** in which every citizen is fluent in the principles of ecology and has a "working knowledge of the basic grammar and underlying syntax of environmental wisdom." Environmental literacy, according to Reilly, can help establish a

stewardship ethic—a sense of responsibility to care for and manage wisely our natural endowment and our productive resources for the long haul.

Pursuing your own environmental literacy is a lifelong process. Some of the most widely influential environmental books suggest solutions (table 25.2). To this list we'd add some personal favorites: *The Singing Wilderness* by Sigurd F. Olson, *My First Summer in the Sierra* by John Muir, *Encounters with the Archdruid* by John McPhee, and *The Life and Death of Great American Cities* by Jane Jacobs.

An important part of environmental education that you shouldn't forget is experience in enjoying your environment. Get outside, have fun, explore, and observe for yourself. You can do this anywhere—in an urban park or waterway, in suburban roadsides and yards, or in remote wilderness areas. The more places you explore, the better. There's always something to see and learn. As author Edward Abbey wrote,

> "It is not enough to fight for the land; it is even more important to enjoy it. While it is still there. So get out there and mess around with your friends, ramble out yonder and explore the forests, encounter the grizz, climb the mountains. Run the rivers, breathe deep of that yet sweet and lucid air, sit quietly for a while and contemplate the precious stillness, that lovely mysterious and awesome space."

Table 25.1	Outcomes from Environmental Education
The natural context: An environmentally educated person understands the scientific concepts and facts that underlie environmental issues and the interrelationships that shape nature.	
The social context: An environmentally educated person understands how human society is influencing the environment, as well as the economic, legal, and political mechanisms that provide avenues for addressing issues and situations.	
The valuing context: An environmentally educated person explores his or her values in relation to environmental issues; from an understanding of the natural and social contexts, the person decides whether to keep or change those values.	
The action context: An environmentally educated person becomes involved in activities to improve, maintain, or restore natural resources and environmental quality for all.	

Source: A Greenprint for Minnesota, Minnesota Office of Environmental Education, 1993.

Table 25.2	The Environmental Scientist's Bookshelf
What are some of the most influential and popular environmental books? In a survey of environmental experts and leaders around the world, votes for the top 12 books on nature and the environment were as follows:	
A Sand County Almanac by Aldo Leopold (100)[1]	
Silent Spring by Rachel Carson (81)	
State of the World by Lester Brown and the Worldwatch Institute (31)	
The Population Bomb by Paul Ehrlich (28)	
Walden by Henry David Thoreau (28)	
Wilderness and the American Mind by Roderick Nash (21)	
Small Is Beautiful: Economics as if People Mattered by E. F. Schumacher (21)	
Desert Solitaire: A Season in the Wilderness by Edward Abbey (20)	
The Closing Circle: Nature, Man, and Technology by Barry Commoner (18)	
The Limits to Growth: A Report for the Club of Rome's Project on the Predicament of Mankind by Donella H. Meadows et al. (17)	
The Unsettling of America: Culture and Agriculture by Wendell Berry (16)	
Man and Nature by George Perkins Marsh (16)	

[1]Indicates number of votes for each book. Because this was a survey of American readers, the preponderance of respondents were from the United States (82 percent), and American books are overrepresented.

Source: Robert Merideth, The Environmentalist's Bookshelf: A Guide to the Best Books, *1993, by G. K. Hall, an imprint of Macmillan, Inc.*

Doing Citizen Science with eBird

eBird is a collaborative venture between the Cornell Lab of Ornithology and the National Audubon Society. It provides wonderful tools for amateur birders, but it also provides a rich database about bird abundance and distribution at a variety of spatial and temporal scales for researchers. Every observation, whether made by a recreational birder or highly trained professionals, adds to the resource. Collectively, the thousands of participants in this project are amassing one of the largest and fastest-growing biodiversity data resources in existence. On May 5, 2018, for example, more than 28,000 people in 170 countries found 6,899 species, or about two-thirds of the world's known species in a single day.

You can participate even if you don't yet know how to identify birds. For starters, you can find a group to go out with. Your school may have a bird-watching club, or your city recreation department may sponsor bird outings. Nearly every state has an Audubon Club chapter that offers free local trips. Most birders are happy to share their passion with others. And an extra set of eyes is always welcome. When a group

Most birders are happy to share their interest and expertise with others. It's a great way to get outdoors and learn about your environment.
William P. Cunningham

works together, it always sees more than a single person would. You can also explore online learning resources, such as Cornell's allaboutbirds.org.

To get your sightings included in the database, they have to be entered in eBird as one or more checklists. Go to the eBird web page (ebird.org) to learn more. Don't forget, every bird counts. Whether you only

saw some pigeons and starlings eating spilled grain along the railroad tracks or spied a critically endangered species on a remote island, eBird and the world want to know about your observations. The more data in the collection, the richer and more meaningful it becomes.

What if you don't know where to go to see birds? You can find the best places in your neighborhood, city, state, or country to see birds on the "Trip Planning" tab at http://ebird.org/ebird/places. An interactive map of local "hot spots" can help you decide where to go. Species maps tell you both what birds you can expect to see in a given location as well as when during the year they're most likely to be there.

Perhaps best of all, you can feel good about contributing to real science. Researchers compile and analyze results from the database to follow bird movements and population dynamics. In this way, contributions made to eBird increase our understanding of the distribution, richness, and uniqueness of the biodiversity of our planet.

(header photo): David L. Hansen

Citizen science lets everyone participate

While university classes often are theoretical and abstract, many students are discovering they can make authentic contributions to scientific knowledge through active learning and undergraduate research programs. Internships in agencies or environmental organizations are one way of doing this. Another is to get involved in organized **citizen science** projects in which ordinary people join with established scientists to answer real scientific questions.

Community-based research was pioneered in the Netherlands, where several dozen research centers now study environmental issues ranging from water quality in the Rhine River,

to cancer rates by geographic area, to substitutes for harmful organic solvents. In each project, students and neighborhood groups team with scientists and university personnel to collect data. Their results have been incorporated into official government policies.

You can participate in a collaboration between Audubon and the Cornell Ornithology Lab, for example (Exploring Science, above). Groups such as Earthwatch offer more organized and focused opportunities to take part in research. Every year, hundreds of Earthwatch projects each field a team of a dozen or so volunteers who spend a week or two working on issues ranging from loon nesting behavior to archaeological digs.

Environmental careers range from engineering to education

The need for both environmental educators and environmental professionals opens up many job opportunities in environmental fields. Scientists are needed to understand the natural world and the effects of human activity on the environment. Lawyers and other specialists are needed to develop government and industry policy, laws, and regulations to protect the environment. Engineers are needed to develop technologies and products to clean up pollution and to prevent its production in the first place. Economists, geographers, and social scientists are needed to evaluate the costs of pollution and resource depletion and to develop solutions that are socially, culturally, politically, and economically appropriate for different parts of the world. Businesses increasingly seek environmentally literate and responsible leaders who appreciate how products sold, and services rendered, affect our environment.

Trained people are essential in these professions at every level, from technical and clerical support staff to top managers. Perhaps the biggest national demand over the next few years will be for environmental educators to help train an environmentally literate populace.

We urgently need teachers at every level who are trained in environmental education and environmental science. In the classroom and outdoors, experience in natural sciences and environmental issues are employable. These topics can be incorporated into reading, writing, arithmetic, and every other part of education.

Green business and technology are growing fast

Businesses increasingly consider environmental protection and sustainability to be a strategic advantage. Most large corporations now have an environmental department. Those that don't have many opportunities to reduce waste and improve efficiency by planning for sustainability. There is growing interest in designing products and manufacturing processes to minimize environmental impacts right from the start. In the long run, this can save money and make their businesses more competitive in future markets.

The rush to "green up" business is good news for those looking for jobs in environmental fields, which are predicted to be among the fastest-growing areas of employment during the next few years. More than 1 million workers in the United States are now employed in energy efficiency, solar, wind, and alternative vehicle jobs. And more than 3 million people work in energy-efficiency construction or manufacturing of sustainable products. Altogether, these green jobs support more than 15 times all the workers in fossil-fuel industries (fig. 25.4).

How can you prepare yourself to enter this market? The best bet is to get some technical training: Environmental engineering, analytical chemistry, microbiology, ecology, limnology, groundwater hydrology, or computer science all have great potential. Environmental law, planning, and business administration are also rapidly expanding fields.

FIGURE 25.4 Many interesting, well-paying jobs are opening up in environmental fields. Here an environmental technician takes a sample from a monitoring well for chemical analysis.
William P. Cunningham

Good communicators, artists, and especially journalists are also needed. A liberal arts education will help you develop skills such as communication, critical thinking, balance, vision, flexibility, and caring that should serve you well. Employers need a wide variety of people; small companies need a few people who can do many things well. Society as a whole, too, needs planners, health professionals, writers, teachers, and policymakers who understand environmental science and sustainability.

Section Review

1. What is environmental literacy?
2. Why are corporations interested in hiring people with environmental science training?
3. How can artists and writers, as well as chemists and biologists, work for environmental quality and environmental health?

25.2 What Can Individuals Do?

- *Green consumerism is one way to practice sustainability.*
- *Active citizenship includes participation in your community.*

The most important step you can take is what you are doing right now: learning to understand environmental science and the ways it informs environmental practices. And as an informed consumer you can work to reduce overconsumption and promote sustainability in your community.

The average American consumes twice as many goods and services as 50 years ago. The average house is now more than twice as big as it was 50 years ago, even though the typical family has half as many people. We need more space to hold all the stuff we buy. One study found that three-fourths of American garages couldn't hold a car because they were too full of stuff. Shopping has become the way many people define themselves. A century ago, economist and social critic Thorstein Veblen, in his book *The Theory of the Leisure Class,* coined the term **conspicuous consumption** to describe buying things we don't want or need just to impress others. Does that idea remain relevant today?

Sometimes it becomes unclear whether we control our things, or if our things control us. Some critics argue that consumerism, and earning money to buy all these consumer goods, has eclipsed family, ethnicity, even religion in our lives. With so much attention on earning and spending money, we don't have time to have real friends, to cook real food, to have creative hobbies, or to do work that makes us feel we have accomplished something with our lives. Social critics have called this drive to possess stuff **"affluenza."** We find ourselves stuck in a vicious circle: We work frantically at a job we don't like to buy things we don't need, so we can save time to work even longer hours (fig. 25.5).

All choices are environmental choices

Everything we do has consequences, so it's worth practicing awareness of the impacts of our choices, both positive and negative. Choosing to live near public transit or far from it, for example, determines a great deal of your environmental impact and carbon footprint. Buying efficient appliances or lightbulbs makes you more environmentally responsible without thinking about it. Buying inefficient appliances, or more than you need, makes it extremely hard to reduce your environmental impacts, no matter how good your intentions. Turning off lights, appliances, and computers, and turning down furnaces and air conditioners all reduce your environmental costs.

FIGURE 25.5 Is this our highest purpose? Used with permission from www.CartoonStock.com

Bradford Veley/www.CartoonStock.com

A study at the University of Chicago found that switching from a diet based on red meat to a vegetarian one can reduce your carbon footprint as much as trading in a normal-size sedan for a hybrid Prius. But it's not necessary to give up meat altogether: "Meatless Mondays" are an easy way to reduce your impact, and it could lead you to discover healthy food choices, too. You can also reduce your impact by buying chickens or farm-raised fish (preferably vegetarian ones, such as tilapia or catfish) more often than beef. This is because beef takes about 10 times as much energy to produce as a chicken or fish does. Eating locally grown, grass-fed beef has far less environmental impact than eating beef from confined feeding operations. This choice also has a positive impact on your local economy (What Can You Do?, below).

What Can You Do?

Reducing Your Impact

Purchase Less

Ask yourself whether you really need more stuff, and avoid buying things you don't need or won't use.

Use items as long as possible (and don't replace them just because a new product becomes available).

Make gifts, or give nonmaterial gifts.

Reduce Excess Packaging

Carry reusable bags when shopping and refuse bags for small purchases.

Buy items in bulk or with minimal packaging; avoid single-serving foods. Choose packaging that can be recycled or reused.

Avoid Disposable Items

Use cloth napkins, handkerchiefs, and towels.

Bring a washable cup to meetings; use washable plates and utensils rather than single-use items.

Choose items built to last and have them repaired; you will save materials and energy while providing jobs in your community.

Conserve Energy and Water

Walk, bicycle, or use public transportation. Carpool and combine trips to reduce car mileage.

Turn off or turn down lights, water, heat, and air conditioning.

Save money with clotheslines or racks instead of a clothes dryer.

Use water-saving devices and fewer flushes with toilets.

Turn off running water when washing hands, food, dishes, and teeth.

Vote

Participation in local policy making gives you a voice and influence that can have widespread impacts.

Join civic organizations for fun and to influence policy.

Green consumerism encourages corporations to have an environmental conscience

Much of our global impact involves things we choose to buy. As long as corporations know that consumers want sustainably or fairly produced goods, they will take steps to use "green" production methods—or at least they will try to appear to do so. To help consumers make informed choices, several national programs have been set up to independently and scientifically analyze the environmental impacts of major products. Germany's Blue Angel, begun in 1978, is the oldest of these programs. Other familiar certification systems include fair trade and organic certification programs that trace production chains for foods, coffee, and other products to ensure that they meet established standards of environmentally or socially acceptable production. The U.S. EPA certifies appliances, such as refrigerators, as "Energy Star" to indicate accepted standards of efficiency. Manufacturers want to participate in certification programs because environmentally conscious shoppers have shown that they are willing to pay more for products they know have minimum environmental impacts. The best of these organizations attempt "cradle-to-grave" **life-cycle analysis** (fig.25.6) that evaluates material and energy inputs and outputs at each stage of manufacture, use, and disposal of the product.

Certification can be a powerful force, but its strength depends on how well standards are defined and enforced. USDA organic certification, for example, allows for a wide variety of chemicals on crops. Fair trade programs are generally superior to conventional trade, but they often fail to provide living wages to the workers they are meant to protect. The EPA Energy Star program can have modest efficiency standards.

Advertisers also use "greenwashing," false or meaningless claims, to exaggerate environmental benefits. The following are some types of greenwashing you might encounter:

- "Non-toxic" suggests that a product has no harmful effects on humans, but the term has many meanings and no legal definition. Substances safe for humans also can be harmful to other organisms.

- "Biodegradable," "recyclable," or "compostable" claims depend on how you dispose of a product. In a landfill, nothing is recycled or biodegraded.

- "Natural" and "organic" can connote different things in different places. Loopholes in standards allow many synthetic chemicals to be included in "organics," especially in shampoos and skin-care products. Some cigarette brands advertise that they're organic, but they're still toxic.

- "Environmentally friendly," "environmentally safe," and "won't harm the ozone layer" are often empty claims. Because there are no standards to define these terms, anyone can use them. How much energy and nonrenewable material are used in manufacturing, shipping, use, or excess packaging?

Sometimes it's hard to be sure we're making good choices as consumers. Is it better to buy organic foods that are produced on an industrial scale and shipped thousands of miles across the

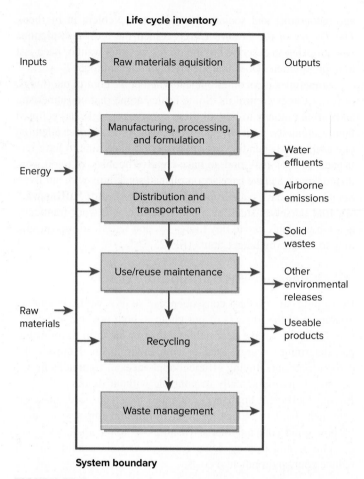

Life cycle inventory

FIGURE 25.6 At each stage in its life cycle, a product receives inputs of materials and energy, produces outputs of materials or energy that move to subsequent phases, and releases wastes into the environment.

country, or is it better to eat only locally grown products? Maybe some of both strategies are needed. Should you buy shampoo and skin-care products that contain environmentally harmful components, such as antibacterial triclosan or environmental estrogens (see chapter 18)? Can you find alternatives that are safer?

You are a citizen, as well as a consumer

We often frame our options as individuals in terms of consumption, and in public discourse we often describe ourselves as consumers. But you also participate, one way or another, in a community. People around you, possibly including yourself, take action to shape policies about energy consumption, urban development, educational policy, and many other matters of global importance.

Participating in political campaigns, or running for office yourself, is an obvious way you can influence environmental policy and sustainability. You can also participate in petition drives, lobbying, and public meetings and hearings. Policy makers watch these events carefully as they gauge public opinion on policy matters about energy, water treatment, development policy, and countless other issues that bear on environmental questions in your area.

Politics aren't the only way to participate, though. Joining a club, a civic organization, a library board, a church, a school organization, or a neighborhood group also gives you a chance to learn about your area and the issues going on. One of the aims of these activities can be to get to know your neighbors and, ideally, have some fun together. This social networking is one of the keys to strengthening the fabric of civil society. That fabric is necessary to influence public policy. Nobel Prize–winning sociologist and economist Elinor Ostrom found that healthy social networks were an essential element in good management of common resources. Everyone can participate in that networking, and it can be enriching, too.

Voting is something every member of a democracy should do. Freedom and democracy take constant tending to remain strong. It may not always be easy, but getting out to vote, even for small and local elections, can give you pride in your community and yourself.

You can learn leadership

All of the environmental progress you have read about has occurred because individuals decided to organize and lead groups in order to promote change. Changes to environmental policy, such as the Clean Water Act or the Clean Air Act, happened because citizen groups pushed their legislators to back new rules. These changes also happened because people with an interest in public health and environmental quality decided to run for office, and others stood with them and helped their campaigns. These activities take energy and commitment. They are not easy. But they are energizing and often exciting for those who choose to take action.

Leadership is something most of us learn only through practice. How do you organize a group or set goals and priorities? How do you learn to stand up and speak to a group, to get others motivated and directed toward useful action? These are lessons you can practice in student groups, environmental organizations, and civic organizations wherever you are (table 25.3).

Student organizations are an especially important place to learn leadership. You have peers who share your viewpoint, and who can help you plan. You have other peers who disagree but who can challenge you to debate and to sharpen your arguments. You can publish your arguments in student newspapers and explore them in classes. And all these experiences can help you work toward goals you care about—and even jobs you hope to get after you graduate.

You can own this class

It may seem self-evident, but one of the most important things you can do is to think carefully about ways your classes can help you become better at bringing about the world you hope to see. Many students take classes because they are required as part of their major, or because they seem unusual and amusing. But just about any class has something that contributes to your larger goals, and identifying those contributions can make you a more purposeful and successful student and citizen.

Some classes push you to write better. Some encourage you to read more closely and to examine arguments and form arguments of your own. Some teach artistic communication, which can capture

Table 25.3 Organizing an Environmental Campaign
1. *What do you want to change?* Can you specify why your goals are focused and reasonable?
2. *What and who will be needed to get the job done?* What resources do you have; who can help you?
3. *Who are the stakeholders in this issue?* Who are your allies and constituents? How can you work with them?
4. *How will your group make decisions and set priorities?* Will you operate by consensus, majority vote, or informal agreement?
5. *Have others already worked on this issue?* What can you learn from their successes or failures?
6. *Who has the power to resolve your issue?* Which individuals, organizations, corporations, or elected officials should be targeted by your campaign, and what arguments will they listen to?
7. *What tactics will be effective?* Finding the right tactics will avoid alienating decision makers and convince people your ideas are sound.
8. *How should you make yourself presentable to your audience?* Gaining trust involves dressing or speaking in a way your audience understands and respects. Personal presentation can be important in delivering a message.

the imagination of people around you. Some explain fundamental ideas of chemistry, biology, and other sciences, without which it is impossible for us to fully understand problems of water pollution, pesticides, food production, climate change, and other issues in which environmental science and policy intersect.

Section Review

1. What is "affluenza"?
2. Describe the benefits and challenges of green consumerism.
3. Why is it important to vote?
4. What are some ways you can learn leadership, and why is this important?

25.3 How Can We Work Together?

- *National groups can be effective in promoting resource conservation.*
- *New groups emerge as new actors and issues become important.*
- *International nongovernmental organizations pursue diverse causes.*

Working together gives you encouragement, energy, and useful information from meeting regularly with others who share your interests. It's easy to get discouraged by the slow pace of change, so having a support group is important. And once you've made a commitment to do some small local project, it becomes easier to follow up with a bigger task (fig. 25.7). In this section, we will look at some environmental organizations that have been influential in environmental policy change.

FIGURE 25.7 Working with others can give you energy, inspiration, and a sense of accomplishment.
William P. Cunningham

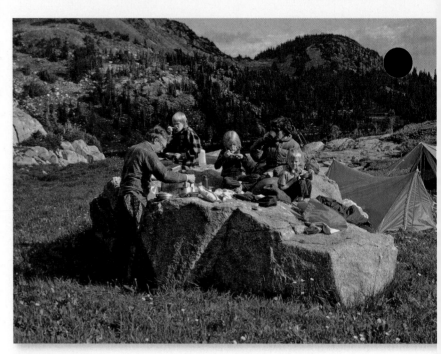

FIGURE 25.8 We can learn about environmental systems almost anywhere. The first step is getting outside to have fun.
William P. Cunningham

National organizations influence policy

Public organizations often arise in response to environmental and social challenges. Many of the groups we consider establishment institutions today started as associations of young radicals fighting powerful political and economic forces. The Audubon Society organized to protect egrets and other birds, which were being slaughtered for their plumes to decorate ladies' hats. The Sierra Club organized to protect Yosemite Valley and the giant redwood trees of the surrounding Sierra Nevada mountains. It helped establish many of our most cherished national parks. It also educated the public about nature through outings (fig. 25.8). Ducks Unlimited formed during the Dust Bowl years to protect ducks and their habitat, in a time when farmers were draining wetlands as fast as they could and converting them to croplands. The Wilderness Society organized at about the same time to protect open space for all citizens, not just the wealthy elite who could afford private retreats in the mountains.

Large organizations have sometimes become less radical, as their interests have expanded, but they also have important influence in policy circles. In the United States, a "group of 10" large organizations include the National Wildlife Federation, the World Wildlife Fund, the Audubon Society, the Sierra Club, the Izaak Walton League, Friends of the Earth, Greenpeace, Ducks Unlimited, the Natural Resources Defense Council, and The Wilderness Society. They have professional staff that work with lawmakers in state and federal governments. Most members are passive and only occasionally involved in organizational activities. Some have argued that these organizations are increasingly professional and less volunteer-based. However, volunteers rarely have the time or expertise to give effective leadership in policy making, which is often the field of highly paid corporate lawyers. As lawyers representing corporations write and lobby for more and more legislation, citizens' groups need larger and more professional representation. Effective work often can't be done without dedicated, paid, and experienced staff.

Despite the activity of major environmental groups in Washington, most of these groups rely on a foundation of local chapters, which gather for social outdoor activities (as with Audubon and the Sierra Club), volunteer work (such as restoration at Nature Conservancy preserves), or political projects. Anybody can join these groups to learn more and meet people.

Established groups are thus influential in policy because their mass membership, professional staffs, and long history give them a degree of respectability and influence not found in newer, smaller groups. The Sierra Club, for instance, with more than 2.5 million members in 64 chapters, has a national staff of about 200, an annual budget of about $100 million, and up to 20 full-time professional lobbyists in Washington, D.C. These national groups have become a potent force in Congress, especially when they band together to push important legislation, such as the Endangered Species Act or the Clean Air Act.

In a survey that asked congressional staff and officials of government agencies to rate the effectiveness of groups that attempt to influence federal policy on pollution control, the top five were national environmental organizations. In spite of their large budgets and important connections, the American Petroleum Institute, the Chemical Manufacturers Association, and the Edison Electric Institute ranked far behind these environmental groups in terms of influence.

Some large environmental groups employ a large professional staff to carry out research and conservation. The Nature

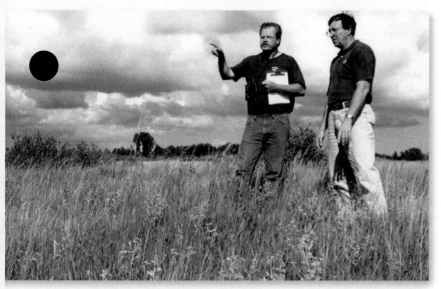

FIGURE 25.9 The Nature Conservancy buys land with high biodiversity or unique natural values to protect it from misuse and development.
David L. Hansen

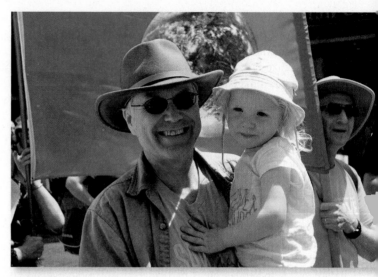

FIGURE 25.10 350.org has inspired thousands of people around the world to work to stop climate change. This isn't just a campaign for our own interests, but for our children and grandchildren.
Julia Frost Nerbonne

Conservancy, for example, buys and preserves land of high ecological value. With more than 3,200 employees and assets around $3 billion, this group manages 7 million acres in what it describes as the world's largest private sanctuary system (fig. 25.9). The Conservancy is controversial for some of its management decisions, such as allowing gas and oil drilling in some reserves, and for including executives from some questionable companies on its governing board and advisory council. The Conservancy replies that it is trying to work with these companies to bring about change rather than just criticize them.

New players bring energy to policy making

There's a new energy today in environmental groups. Science reporting in popular media has raised general awareness of issues such as climate change, ocean pollution, and disappearing pollinators. Social media also helps spread awareness of issues that historically might have been relegated to the environmental science classroom. An increasingly diverse, savvy, and passionate movement is taking shape.

The 350.org network, for example, has inspired tens of thousands of citizens to work to stop climate change. This global, grassroots movement uses social media to organize hundreds of student groups and allies around the world. Their campaign is for social justice for the poorest people in the world, who will suffer the most. It's also a matter of intergenerational equity (fig. 25.10). They're working to leave carbon in the ground, pressure governments to limit greenhouse gas emissions, encourage investors to divest from fossil fuels, stop offshore drilling, and find people-centered solutions to climate change. The divestment campaign has resulted in a larger and faster shifting of money than any similar movement in history. Although Washington insiders considered approval of the Keystone Pipeline to be a foregone conclusion, this

scrappy, student-led movement denied some of the largest, richest corporations a project they wanted badly.

350.org's web page is one of many that offer resources and suggestions for getting involved. You can explore an interactive map to find groups or projects close to you. You can also find sample posters, signs, videos, campaign plans, and DIY templates, stencils, and sample press-releases. Get inspired and "skilled up" at workshops, conferences, and seminars. Attend an action to meet people and get ideas. There's a world of ways to take part.

Many new advocacy and research organizations have formed over time, in response to new needs and opportunities. Some groups are large, national, and professional. Others are local and citizen-based. There is always room for improvement in these groups, and there is always room for more energy and participation, which can be rewarding and educational for anyone who can spare some time.

International NGOs mobilize many people

International **nongovernmental organizations (NGOs)** can be vital in the struggle to protect areas of outstanding biological value. Without this help, most local groups could never mobilize the public interest or financial support for major projects.

The rise in international NGOs in recent years has been phenomenal. At the Stockholm Conference in 1972, only a handful of environmental groups attended, almost all from fully developed countries. Twenty years later, at the Rio Earth Summit, more than 30,000 individuals representing several thousand environmental groups, many from developing countries, held a global Ecoforum to debate issues and form alliances for a better world.

Some NGOs are located primarily in the more highly developed countries of the north and work mainly on local issues. Others are headquartered in the north but focus their attention on the problems of developing countries in the south. Still others are truly global, with active groups in many different countries. A few are highly professional, combining private individuals with representatives of government agencies. Many work for political change, more specialize in gathering and disseminating information, and some undertake direct action to protect a specific resource.

Public education and raising consciousness—through protest marches, demonstrations, civil disobedience, and other participatory public actions and media events—are generally important tactics for these groups. Greenpeace, for instance, carries out well-publicized confrontations with whalers, seal hunters, toxic waste dumpers, and others who threaten very specific and visible resources. Greenpeace may well be the largest environmental organization in the world, claiming some 2.8 million contributing members.

In contrast to these highly visible groups, others choose to work behind the scenes, but their impact may be equally important. Conservation International has been a leader in debt-for-nature swaps to protect areas particularly rich in biodiversity. It also has some interesting initiatives in economic development: seeking products made by local people that will provide livelihoods along with environmental protection (fig. 25.11).

Section Review

1. What are some organizing goals of national environmental organizations?
2. What are some strategies of environmental organizations?
3. Identify an international environmental NGO and its goals.

25.4 CAMPUS GREENING

- *Many student organizations work to organize, promote change, and teach leadership.*
- *Students have considerable power to push their institutions toward more sustainable practices.*

Colleges and universities can be powerful catalysts for change. Across North America, and around the world, students and faculty study sustainability and exchange ideas about improving both technology and policy for environmental and social progress. Organizations for secondary and college students often are among our most active and effective groups for environmental policy.

The Sunrise Movement, for example, has mobilized students to act locally on national and global affairs having to do with green jobs, renewable energy, and climate protection.

This is one of many campus-based organizations. Another that has existed for decades and remains influential is the network of Public Interest Research Groups (PIRGs), which are active across the United States. Addressing a range of environmental and social policy issues, the PIRGs have been important forces for public education and policy influence. Joining a PIRG or similar organization is an important way to learn how many others share your concerns and want to work with you to bring about change. Joining an environmental campaign is also an important way to learn the power of working together to publicize policy issues (fig. 25.12). Participating in an organization also helps you learn

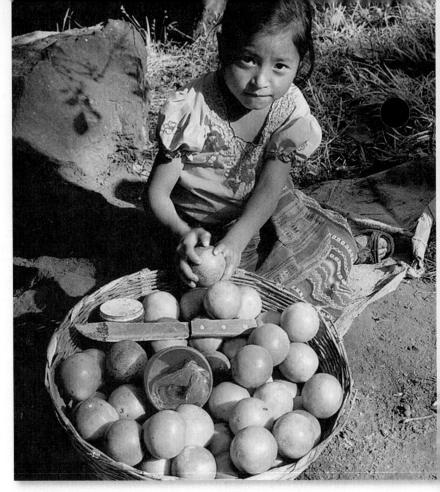

FIGURE 25.11 Developing areas produce marketable goods that provide income without natural resource destruction. Often international development projects try to target these alternative income sources, to aid both local economies and environments.
William P. Cunningham

FIGURE 25.12 Protests, marches, and public demonstrations can be an effective way to message to policymakers and administrators, and to influence legislators.
Tom Finkle

essential techniques, such as using media to get your message out (table 25.4).

Schools provide environmental leadership

Colleges and universities are important sources of ideas, information, and experimentation. Faculty have knowledge and experience to share, and students have the energy, enthusiasm, curiosity, and creativity to research new ideas. Colleges and universities also have a clear interest in the future of our shared resources, because students are actively preparing to influence the world they expect to inhabit in the future.

Students have led innovation on countless aspects of sustainability. At many schools, students have persuaded administration to buy locally produced food and to provide organic, vegetarian, and fair trade options in campus cafeterias. Increasing vegetarian options is a way to reduce energy consumption, since meat production involves a great deal of energy. Developing local and fair trade food sources also supports communities while teaching about alternative practices. Student-led

Table 25.4	Using the Media to Influence Public Opinion

Shaping opinion, reaching consensus, electing public officials, and mobilizing action are accomplished primarily through the use of the communications media. Here are some ways to use these resources:

1. *Write letters to the editor and news releases.* Submit these to local newspapers, blogs, and radio shows.

2. *Collect data on campus resource use or environmental attitudes.* Include this evidence in letters to papers and administrators.

3. *Write feature stories for local news sources.* These outlets may be happy to publish public interest items they don't have to pay for.

4. *Produce a public service announcement for local radio and TV stations.* Publically supported stations are required to air public service announcements, so why not yours?

5. *Find celebrities to support your position.* Ask them to give a concert or performance to attract attention to the issue. They might like to be associated with your cause.

6. *Hold a media event that is photogenic and newsworthy.* Clean up your local river and invite photographers to accompany you. Hold peaceful or amusing demonstrations. Avoid violence or conflict.

7. *If you hear negative remarks about your issue on TV or radio, ask for free time under the Fairness Doctrine to respond.* Stations must do a certain amount of public service and may be happy to accommodate you.

8. *Ask your local TV or newspaper to do a feature story about your issue or about your organization.* Articles can give valuable publicity and inspire others to participate.

fossil fuel divestment campaigns have transformed how we think about the impacts of investments (What Do You Think?, p. 578). Students also develop career paths that have an impact: At more than 100 universities and colleges across the United States, graduating students have taken a pledge for social and environmental responsibility. Could you introduce something similar at your school?

> I pledge to explore and take into account the social and environmental consequences of any job I consider and will try to improve these aspects of any organization for which I work.

The Association for the Advancement of Sustainability in Higher Education (AASHE) is one of the leading groups that offer resources to college and university students, faculty, and staff. The AASHE website offers information about greenhouse gas monitoring, ideas for efficiency, and reports on projects at other campuses. You can find examples of curriculum innovation, green building projects, purchasing standards for fuel-efficient vehicles, food service efforts, recycling, and many other projects.

Campus building projects are important for modeling sustainability research and development. Buildings are visible, they consume most energy on most campuses, and they offer teaching opportunities for students and the public. Many colleges and universities have environmental building standards or have committed to following standards of the U.S. Green Building Council, the organization that does Leadership in Energy and Environmental Design (LEED) certification. Often building standards are adopted because students ask administrators what they are doing to improve building performance. Examples of prize-winning sustainable designs can be found at Stanford University, Oberlin College in Ohio, and the University of California at Santa Barbara, and many other schools. Stanford students worked with the administration to develop *Guidelines for Sustainable Buildings,* a booklet that covers everything from energy-efficient lighting to native landscaping. Hampshire College, in western Massachusetts, has one of the first college buildings built to meet "living building" standards in its Kern Center: With solar panels, excellent insulation, and composting toilets, the center is designed to consume no energy or water not produced within its own footprint. The center is an important part of the college's strategy for achieving campus-wide carbon neutrality.

Sustainable buildings don't need to be especially expensive. UC Santa Barbara's Bren School of Environmental Science and Management looks like a normal institutional building, but it was the first academic laboratory building in the United States to get LEED Platinum certification, the highest level (fig. 25.13). It wasn't originally funded as an experimental green building, but planners found that some simple features like large windows for natural light and ventilation make the building both more functional and more appealing. Motion detectors control light levels, and sensors monitor and refresh the air when there is too much CO_2 putting students to sleep. More than 30 percent of interior materials are recycled. Solar panels supply 10 percent of the

Fossil Fuel Divestment

Students at many colleges and universities are calling for their institution to "decarbonize" their endowments by eliminating assets in fossil fuel-related companies and putting their money into clean energy. Oil, gas, and coal corporations have long been part of the backbone of many investment plans: They are huge corporations, worth a great deal of money. Their market is locked in, so their profits are more or less guaranteed, as long as manufacturing and transportation economies are strong, and as long as we limit alternative sources of energy.

Student environmental groups argue that educational institutions should be committed to the future well-being of their students. It seems hypocritical, they argue, to be simultaneously educating students and undermining the future they will inhabit, by investing in companies that bring on global climate change that threatens cities, farmland, water supplies, public health, biodiversity, and ecosystem services. Organizers have encouraged students to press their administrations to wean their endowments—the investments that colleges rely on to support operations—from this environmentally destructive sector.

Divestment supporters argue that this strategy helped topple South Africa's racist apartheid regime in the 1980s. Similarly, divestment from Sudan, with its predatory military government, helped to shift policy positions toward this regime in the 2000s. Divestment from tobacco companies was also a way that colleges made statements about reducing political tolerance of smoking-related illness.

Nevertheless, shifting assets has been a contentious issue. Opponents of divestment argue that it is essential to remain engaged, and to own shares in companies, in order to influence corporate practices. They also point out that most educational institutions have most of their money in large, mixed funds, in which it is impossible to tell precisely where investment dollars lie. Unlike personal investments, which can be placed in socially responsible funds or indexes, institutions often can't choose socially or environmentally responsible funds. Further, because the size of educational institutions' investments is minuscule compared to the size of the fossil fuel industry, selling these assets will have little practical

Burning fossil fuels causes climate change, damages our health, and pollutes our environment.

Rudmer Zwerver/Shutterstock

effect on those companies. In addition, many colleges and universities are already on delicate ground financially. Risking lower returns is something many administrators are unwilling to do.

Pro-divestment student organizations counter that investment in fossil fuels is no longer financially safe, and that divesting from direct holdings makes an important policy statement, and that if funds are mixed, then colleges and universities have little power to influence corporate practices. Acting as engaged shareholders is therefore not possible. Moreover, for companies whose business is to sell oil, gas, and coal, it is hard to imagine ways to push them to adjust corporate policy to ease their climate impacts. Divestment proponents also argue that even if selling shares has little impact on share value directly, the goal is to force public examination of our dependence on fossil fuels, and to question their influence on public policy, from public subsidies and tax breaks to international support for corrupt oil-producing regimes.

Divestment is spreading globally. Some 800 colleges, universities, cities, counties, states, and national governments representing $6 trillion in assets have pledged to eliminate their fossil fuel investments.

Many economists and asset managers now are persuaded by economic arguments that align with ethical reasons for dumping fossil fuel investments. Climate change associated with fossil fuels is not just another social issue; it is an existential threat to society and the global economy, with trillions of dollars at stake in agriculture, forest products, water resources, and other factors. Studies of fuel stocks suggest that 80 percent of all coal, oil, and gas reserves must stay in the ground if we are to avoid catastrophic levels of climate change. Already companies like Shell and BP are "writing down" the value of their in-the-ground assets—lowering estimates of their value in the company's net worth. As these concerns become more widely recognized, coal, oil, and gas companies are likely to become bad investments. And renewable energy is rapidly replacing fossil fuels and now offers an affordable alternative energy source.

The idea remains controversial, challenging, energizing, and uncertain. Divestment challenges colleges and universities to unite theory with practice—often a hard thing to do. What do you think? What arguments draw you to one side of the debate or the other? What information would you need to form a clearer opinion? Where could you get that information?

---

electricity, and the building exceeds federal efficiency standards by 30 percent. And all this cost only 2 percent more than a conventional design, an amount that was rapidly recovered in lower energy costs. "The overriding and very powerful message is it really doesn't cost any more to do these things," says Dennis Aigner, dean of Bren School.

A green campus is an educational opportunity

Green buildings offer important educational opportunities. At Carnegie Mellon University in Pittsburgh, students helped design a green roof for Hamershlag Hall. They now monitor how the living roof is reducing storm water drainage and improving water

FIGURE 25.13 The University of California at Santa Barbara's Bren School of Environmental Science and Management doesn't look radically different, but design features make it extremely sustainable and comfortable. *William P. Cunningham*

Section Review

1. Why are colleges and universities good places for exploring new strategies for sustainability?
2. What is fossil fuel divestment, and what are some arguments for and against it?
3. What are some ways students can promote sustainability on campus?

25.5 SUSTAINABILITY IS A GLOBAL CHALLENGE

- *Global goals of sustainability involve widespread commitment to development and to justice.*
- *The Millennium Development Goals are a set of quantifiable measures of sustainability and development.*

quality. A kiosk inside the dorm shows daily energy use and compares it to long-term averages. Classrooms within the dorm offer environmental science classes in which students can see sustainability in action. Green dorms are popular with students, who appreciate natural lighting, clean air, lack of allergens in building materials, and other features of LEED-certified buildings.

Concordia University in Austin, Texas, was the first college or university in the country to purchase all of its energy from renewable sources. The 5.5 million kilowatt-hours of "green power" it uses each year, mainly wind power, eliminate about 8 million pounds of CO_2 emissions annually, the equivalent of planting 1,000 acres of trees or taking 700 cars off the roads. Emory University in Atlanta, Georgia, is a leader in green building standards, with 11 buildings that are or could become LEED certified. Emory's Whitehead Biomedical Research Building was the first facility in the Southeast to be LEED certified. Like a number of other colleges, Carleton College in Northfield, Minnesota, has built its own wind turbines, which generate about two-thirds of the college's electricity. And an elaborate network of geothermal wells provides most of the heating and cooling for the campus (see chapter 20).

A recent study by the Sustainable Endowments Institute evaluated more than 100 of the leading colleges and universities in the United States on their green building policies, food and recycling programs, climate change impacts, and energy consumption. The report card ranked Dartmouth, Harvard, Stanford, and Williams as the top of the "A list" of 23 greenest campuses. Berea College in Kentucky got special commendation as a small school with a strong commitment to sustainability. Berea's "ecovillage" has a student-designed house that produces its own electricity and treats wastewater in a living system. The college has a full-time sustainability coordinator to provide support to campus programs, community outreach, and teaching.

As the developing countries of the world become more affluent, they are adopting the resource-intensive consumption patterns of richer countries. Automobile production in China, for example, is increasing at about 19 percent per year, or doubling every 3.7 years. Fuel consumption has risen in response, and Beijing, Shanghai, and other major cities have some of the worst peak levels of urban air pollution on record. Throughout the developing world, increasing ownership of refrigerators, air conditioners, televisions, and other appliances is driving up demand for electric power. Between 1995 and 2020, per capita emissions of CO_2 more than doubled in India and nearly tripled in China. At the same time, diseases associated with affluent lifestyles—such as obesity, diabetes, heart attacks, depression, and traffic accidents—are becoming the leading causes of morbidity and mortality worldwide.

On the other hand, there appears to be a dramatic worldwide shift in public attitudes toward environmental protection. In a BBC poll of 22,000 residents of 21 countries, 83 percent agreed that individuals would definitely or probably have to make lifestyle changes to reduce the amount of climate-changing gases they produce. Overall, 70 percent said they were personally ready to make sacrifices to protect the environment. In the United States, where some policymakers argue that the public opposes EPA pollution standards and greenhouse gas regulation, a 2016 bipartisan poll found that 77 percent of respondents supported stricter CO_2 standards, and 69 percent said that EPA scientists, not Congress, should set pollution standards.

We would all benefit by helping developing countries access more efficient, less-polluting technologies. Education, democracy, and access to information are essential for sustainability. In the 2015 Paris climate negotiations, in which 195 countries agreed on plans to reduce greenhouse gas emissions, one of the central priorities was building an investment fund through which wealthy countries would invest $100 billion in clean technology in developing countries. This agreement was about fairness, but it was also about improving our global environment and supporting regional stability and security. Overall, the costs to control population growth, develop renewable energy sources, stop soil erosion, protect ecosystems, and provide a decent standard of living for the world's poor are estimated at $350 billion per year. This is a great deal of money, but it is small compared to over $1 trillion per year spent on wars and military equipment.

Sustainable development means social, environmental, and economic goals

Sustainable development has been a theme running through much of this book. The idea emphasizes the use of renewable resources in harmony with ecological systems (fig. 25.14); the aim is to seek a combination of economic, well-being, and environmental priorities. A widely used definition of sustainable development, from the 1992 report *Our Common Future,* is "meeting the needs of the present without compromising the ability of future generations to meet their own needs" (see sections 1.2 and 1.3). These are some goals of sustainable development:

- A demographic transition to a stable world population of low birth and death rates.

- An energy transition to high efficiency in production and use, coupled with increasing reliance on renewable resources.

- A resource transition to reliance on nature's "income" without depleting its "capital."

- An economic transition to sustainable development and a broader sharing of its benefits.

- A political transition to global negotiation grounded in complementary interests between North and South, East and West.

- An ethical or spiritual transition to attitudes that do not separate us from nature or each other.

Sustainability is an even more important goal for those who live in wealthy countries than for those in developing countries (fig. 25.15). Because of our resource-intensive lifestyles, we have a much greater impact on our environment than the billions of people in poorer countries. Finding ways for all of us to live more lightly on the earth is a critically important goal.

In 2000, United Nations Secretary-General Kofi Annan called for a **millennium assessment** of the consequences of ecosystem change on human well-being. The assessment also explored the scientific basis for actions to enhance the conservation and sustainable use of environmental systems. More than 1,360 experts from around the world worked on technical reports about the conditions and trends of ecosystems, and scenarios for the future. As a result of this assessment, the United Nations developed a set of Millennium Development Goals for sustainable development, with targets for poverty reduction and environmental quality that countries were to achieve between 2000 and 2015.

Contrary to many expectations, considerable progress was made on these goals. Hunger, poverty, and child mortality dropped sharply. Access to education rose, as did access to clean water and sanitation. There were some improvements in democracy and women's rights, although this progress was mixed. Building on lessons learned from both successes and failures in the millennium goals, the UN launched a set of Sustainable Development Goals

FIGURE 25.14 A model for integrating ecosystem health, human needs, and sustainable economic growth.
McGraw-Hill Global Education Holdings, LLC

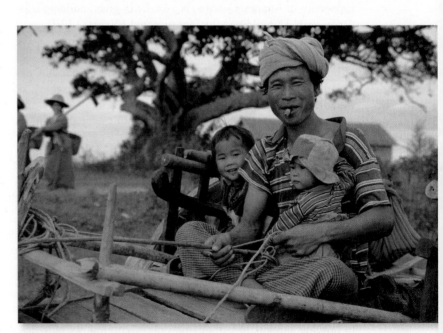

FIGURE 25.15 Sustainable development to ensure a healthy environment for all the world's people is the aim of the millennium development goals.
Jerry Alexander/Getty Images

for 2016–2030 (see chapter 1). These goals represent major challenges, but they are also opportunities to improve the world we share. Findings from the millennium assessment serve as a good summary for this book:

- All of us depend on nature and ecosystem services to provide the conditions for a decent, healthy, and secure life.

- We have made unprecedented changes to ecosystems in recent decades to meet growing demands for food, fresh water, fiber, and energy.

- These changes have helped improve the lives of billions, but at the same time they weakened nature's ability to deliver other key services, such as purification of air and water, protection from disasters, and the provision of medicine.

- Among the outstanding problems we face are the dire state of many of the world's fish stocks, the intense vulnerability of the 2 billion people living in dry regions, and the growing threat to ecosystems from climate change and pollution.

- Human actions have taken the planet to the edge of a massive wave of species extinctions, further threatening our own well-being.

- The loss of services derived from ecosystems is a significant barrier to reducing poverty, hunger, and disease.

- The pressures on ecosystems will increase globally unless human attitudes and actions change.

- Measures to conserve natural resources are more likely to succeed if local communities are given ownership of them, share the benefits, and are involved in decisions.

- Even today's technology and knowledge can reduce considerably the human impact on ecosystems. They are unlikely to be deployed fully, however, until ecosystem services cease to be perceived as free and limitless.

- Better protection of natural assets will require coordinated efforts across all sections of governments, businesses, and international institutions.

From what you've learned in this book, how do you think we could work—individually and collectively—to accomplish these goals? What are the ways in which you would like to contribute?

Section Review

1. List several goals or aspects of sustainability.
2. What is the source of the Millennium Development Goals?
3. Identify eight major areas in which quantifiable goals have been set for the Millennium Development Goals.

Connecting the Dots

All through this book you've seen evidence of environmental degradation and resource depletion, but there are also many cases in which individuals and organizations are finding ways to stop pollution, use renewable rather than irreplaceable resources, and even restore biodiversity and habitat. Sometimes all it takes is the catalyst of a pilot project to show people how things can be done differently to change attitudes and habits. The energy, enthusiasm, and organization generated by Earth Day and subsequent public demonstrations have helped shape public opinion and pass environmental regulations. In this chapter, you've seen practical approaches to living more lightly on the world individually, as well as working collectively to create a better world.

Public attention to issues in the United States seems to run in cycles. Concern builds about some set of problems, and people are willing to take action to find solutions, but then interest wanes and other topics come to the forefront.

Recently, however, approaches to the environment in much of the world seem to have reached a tipping point. Countries, cities, companies, and campuses all are divesting of their assets in fossil fuels and turning to renewable energy sources. This may be a very good time to work on social change and sustainable living. As the famous anthropologist Margaret Mead said, "Never doubt that a small group of thoughtful, committed people can change the world. Indeed, it is the only thing that ever has."

Reviewing Key Terms

Can you define the following terms in environmental science?

affluenza 25.2

citizen science 25.1

conspicuous consumption 25.2

environmental literacy 25.1

life-cycle analysis 25.2

millennium assessment 25.5

nongovernmental organizations (NGOs) 25.3

Critical Thinking and Discussion Questions

1. What green activities are ongoing at your school? How might you help promote sustainability on campus?

2. Reflect on how you learned about environmental issues. What have been the most important formative experiences or persuasive arguments in shaping your own attitudes? If you were designing an environmental education program for youth, what elements would you include?

3. How might it change your life if you were to minimize your consumption of materials and resources? Which aspects could you give up, and what is absolutely essential to your happiness and well-being? Does your list differ from your friends' and classmates' lists?

4. Have you ever been involved in charitable or environmental work? What were the best and worst aspects of that experience? If you haven't yet done anything of this sort, what activities seem appealing and worthwhile to you?

Data Analysis

Campus Environmental Audit

If you want to understand how sustainable your campus is, the first step is to gather data. A campus environmental audit can be a huge, professionally executed task, but you can make a reasonable approximation if you work with your class to gather some basic information.

Among the many groups that guide audits is the Association for the Advancement of Sustainability in Higher Education. You can find details of their full audit system online (https://stars.aashe.org/resources-support/technical-manual/). Below is a set of 30 questions that approximate the AASHE guidelines. Give one point for each Yes answer. Then, see what your score is out of 30 possible points. Which No answers would be easiest or most important to change? Which Yes answers still have room for improvement? What can student groups do to advocate for those improvements?

1. Education: Does your school offer courses focused on sustainability? Are there programs or facilities that encourage student research, monitoring, or data collection on sustainability?

2. Buildings and energy: Are any buildings LEED certified? Are LEDs or other efficient devices in use? Are windows double-paned? Are roofs insulated? Do dorms or classroom buildings have energy meters, so that energy can be monitored separately? Does your school purchase or use energy from renewable sources?

3. Climate: Has your campus inventoried greenhouse gas emissions?

4. Dining: Do dining services compost food waste? Is locally grown or vegan food available? Do disposable products contain recycled content, or are they recyclable? Are reusable take-out containers in use?

5. Grounds and purchasing: Does landscaping include native plantings or habitat? Is there a policy of minimizing pesticide and herbicide use? Are environmentally safe cleaning products used? Is office paper recycled?

6. Transportation: Does your school own fuel-efficient vehicles? Are there programs to encourage public transit or ride sharing? Are safe bicycle routes and sufficient bike racks available?

7. Waste: Are convenient recycling facilities available? Is there a mechanism for exchanging household goods when students move in and out of dorms? Is disposal of e-waste (computers, phones, etc.) tracked to recycling facilities?

8. Water: Does your school limit lawn watering, especially during daytime? Are waterless urinals and low-flush toilets widely available?

9. Administration: Does your administration actively promote and coordinate sustainability planning? Are energy conservation and sustainability integrated into campus management and building plans? Does your campus promote farmers' markets, public education, or other public engagement?

Glossary

Source: imageBROKER/Alamy Stock Photo

A

abiotic Nonliving.

abundance The number or amount of something.

acid deposition Acidic rain, snow, fog, or dry particulates deposited from the air due to an increased release of acids by anthropogenic or natural resources.

acids Substances that release hydrogen ions (protons) in water.

active solar system A mechanical system that uses moving parts or pumped fluids to collect, concentrate, distribute, or store solar energy.

acute effects Sudden, severe effects.

adaptation The acquisition of traits that allow a species to survive and thrive in its environment.

adaptive management A management plan designed from the outset to "learn by doing" and to actively test hypotheses and adjust treatments as new information becomes available.

administrative law Executive orders, administrative rules and regulations, and enforcement decisions by administrative agencies and special administrative courts.

aerosols Minute particles or liquid droplets suspended in the air.

aesthetic degradation Changes in environmental quality that offend our aesthetic senses.

affluenza An addiction to spending and consuming beyond one's needs.

albedo A description of a surface's reflective properties.

allergens Substances that activate the immune system.

allopatric speciation Speciation deriving from a common ancestor due to geographic barriers that cause reproductive isolation; *also called* geographic isolation.

ambient air The air immediately around us.

analytical thinking Thinking that breaks a problem down into its constituent parts.

anemia Low levels of hemoglobin due to iron deficiency or lack of red blood cells.

antigens Chemical compounds to which antibodies bind.

aquaculture Growing aquatic species in net pens or tanks.

aquifers Porous, water-bearing layers of sand, gravel, and rock below the earth's surface; reservoirs for groundwater.

arbitration A formal process of dispute resolution in which there are stringent rules of evidence, cross-examination of witnesses, and a legally binding decision made by the arbitrator that all parties must obey.

artesian well A pressurized aquifer from which water gushes without being pumped, due to the aquifer's intersecting the surface or being penetrated by a pipe or conduit; *also called* a spring.

atmospheric deposition Sedimentation of solids, liquids, or gaseous materials from the air.

atom The smallest unit of matter that has the characteristics of an element; consists of three main types of subatomic particles: protons, neutrons, and electrons.

atomic number The characteristic number of protons per atom of an element; used as an identifying attribute.

B

barrier islands Low, narrow, sandy islands that form offshore from a coastline. They protect the shore from storms.

Basel Convention Restricts shipment of hazardous waste across international boundaries.

bases Substances that bond readily with hydrogen ions.

Batesian mimicry Evolution by one species to resemble the coloration, body shape, or behavior of another species that is protected from predators by a venomous stinger, bad taste, or some other defensive adaptation.

benthic The bottom of a sea or lake.

best available, economically achievable technology (BAT) The best pollution control available.

best practicable control technology (BPT) The best technology for pollution control available at reasonable cost and operable under normal conditions.

binomials Two-part names (genus and species, usually in Latin) invented by Carl Linneaus to show taxonomic relationships.

bioaccumulation The selective absorption and concentration of molecules by cells.

biocentric preservation A philosophy that emphasizes the fundamental right of living organisms to exist and to pursue their own goods.

biochemical oxygen demand (BOD) A standard test of water pollution, measured by the amount of dissolved oxygen consumed by aquatic organisms over a given period.

biocide A broad-spectrum poison that kills a wide range of organisms.

biodegradable plastics Plastics that can be decomposed by microorganisms.

biodiversity The genetic, species, and ecological diversity of the organisms in a given area.

biodiversity hot spots Areas with exceptionally high numbers of endemic species.

biofuel Fuels such as ethanol, methanol, biodiesel, or vegetable oils from crops.

biological community The populations of plants, animals, and microorganisms living and interacting in a certain area at a given time.

biological controls Use of natural predators, pathogens, or competitors to regulate pest populations.

biomagnification An increase in the concentration of certain stable chemicals (for example, heavy metals or fat-soluble pesticides) in successively higher trophic levels of a food chain or web.

biomass The total mass or weight of all the living organisms in a given population or area; also a general term for material produced by living things, such as wood, leaves, and other material.

biome A broad, regional type of ecosystem characterized by distinctive climate and soil conditions and a distinctive kind of biological community adapted to those conditions.

bioremediation The use of biological organisms to remove or detoxify pollutants from a contaminated area.

biosphere reserves World heritage sites identified by the IUCN as worthy of national park or wildlife refuge status because of high biological diversity or unique ecological features.

biotic Pertaining to life; environmental factors created by living organisms.

bitumen A sticky, semi-solid hydrocarbon found in tar sands.

blind experiment Experiment in which those carrying out the experiment don't know, until after the gathering and analysis of data, which was the experimental treatment and which was the control.

body burden The sum total of all persistent toxins in our body that we accumulate from our air, water, diet, and surroundings.

bog An area of waterlogged soil that tends to be peaty; bogs are fed mainly by precipitation and have low productivity, and some are acidic.

boreal forest A broad band of mixed coniferous and deciduous trees that stretches across northern North America (and also Europe and Asia); its northernmost edge, the taiga, intergrades with the arctic tundra.

breeder reactor A nuclear reactor that produces fuel by bombarding isotopes of uranium and thorium with high-energy neutrons that convert inert atoms to fissionable ones.

bronchitis A persistent inflammation of bronchi and bronchioles (large and small airways in the lungs).

brownfield development Building on abandoned or reclaimed polluted industrial sites.

brownfields Abandoned or underused urban areas in which redevelopment is blocked by liability or financing issues related to toxic contamination.

C

cancer Invasive, out-of-control cell growth that results in malignant tumors.

cap-and-trade An approach to controlling pollution by mandating upper limits (the cap) on how much each country, sector, or specific industry is allowed to emit. Companies that can cut pollution by more than they're required to can sell the credit to other companies that have more difficulty meeting their mandated levels.

capital Any form of wealth, resources, or knowledge available for use in the production of more wealth.

carbon capture and storage (CCS) Carbon dioxide (generally from fuel combustion) is captured and stored in geological formations.

carbon credits Permits for pollution emissions that can be bought and sold. In theory, the trading of carbon credits produces pollution reductions for the lowest cost.

carbon cycle The circulation and reutilization of carbon atoms, especially via the processes of photosynthesis and respiration.

carbon monoxide (CO) A colorless, odorless, non-irritating, but highly toxic gas produced by incomplete combustion of fuel, incineration of biomass or solid waste, or partially anaerobic decomposition of organic material.

carbon neutral A system or process that doesn't release more carbon to the atmosphere than it consumes.

carbon sink Places of carbon accumulation, such as in large forests (organic compounds) or ocean sediments (calcium carbonate); carbon is thus removed from the carbon cycle for moderately long to very long periods of time.

carcinogens Substances that cause cancer.

carnivores Organisms that prey mainly upon animals.

carrying capacity The maximum number of individuals of any species that can be supported by a particular ecosystem on a long-term basis.

case law Precedents from both civil and criminal court cases.

cell Minute biological compartments within which the processes of life are carried out.

cellular respiration The process in which a cell breaks down sugar or other organic compounds to release energy used for cellular work; may be anaerobic or aerobic, depending on the availability of oxygen.

cellulosic ethanol Ethanol made from (plant) material.

chain reaction A self-sustaining reaction in which the fission of nuclei produces subatomic particles that cause the fission of other nuclei.

chaparral Thick, dense, thorny, evergreen scrub found in Mediterranean climates.

chemical energy Potential energy stored in chemical bonds of molecules.

chemosynthesis The process in which inorganic chemicals, such as hydrogen sulfide (HS) or hydrogen gas (H^2), serve as an energy source for synthesis of organic molecules.

chlorinated hydrocarbons Hydrocarbon molecules to which chlorine atoms are attached.

chlorofluorocarbons (CFCs) Chemical compounds with a carbon skeleton and one or more attached chlorine and fluorine atoms. Commonly used as refrigerants, solvents, fire retardants, and blowing agents.

chronic effects Long-lasting results of exposure to a toxin; can be caused by a single, acute exposure or a continuous, low-level exposure.

chronic obstructive pulmonary disease (COPD) Irreversible damage to the linings of the lungs caused by irritants.

chronically undernourished Those people whose diet doesn't provide the 2,200 kcal per day, on average, considered necessary for a healthy, productive life.

citizen science Projects in which trained volunteers work with scientific researchers to answer real-world questions.

city A differentiated community with a sufficient population and resource base to allow residents to specialize in arts, crafts, services, and professional occupations.

civil law A body of laws regulating relations between individuals, or between individuals and corporations, concerning property rights, personal dignity and freedom, and personal injury.

classical economics Modern, Western economic theories of the effects of resource scarcity, monetary policy, and competition on supply of and demand for goods and services in the marketplace. This is the basis for the capitalist market system.

climate A description of the long-term pattern of weather in a particular area.

climax community A relatively stable, long-lasting community reached in a successional series; usually determined by climate and soil type.

closed system An ecosystem that neither receives nor emits energy or matter to the surrounding environment.

closed-canopy forest A forest where tree crowns spread over 20 percent of the ground; has the potential for commercial timber harvests.

cloud forests High mountain forests where temperatures are uniformly cool and fog or mist keeps vegetation wet all the time.

coarse woody debris Logs, branches, and other woody debris in a stream.

coevolution The process in which species exert selective pressure on each other and gradually evolve new features or behaviors as a result of those pressures.

cogeneration The simultaneous production of electricity and steam or hot water in the same plant.

cold front A moving boundary of cooler air displacing warmer air.

coliform bacteria Bacteria that live in the intestines (including the colon) of humans and other animals; used as a measure of the presence of feces in water or soil.

command and control economies Economies in which government-set rules and standards guide production and consumption, rather than market forces.

commensalism A symbiotic relationship in which one member is benefited and the second is neither harmed nor benefited.

common law The body of court decisions that constitute a working definition of individual rights and responsibilities where no formal statutes define these issues.

common property management Resources managed cooperatively by a community for long-term sustainability. Also known as communal resource management systems.

complexity The number of species at each trophic level and the number of trophic levels in a community.

composting The biological degradation of organic material under aerobic (oxygen-rich) conditions to produce compost, a nutrient-rich soil amendment and conditioner.

compound A molecule made up of two or more kinds of atoms held together by chemical bonds.

concentrating solar power (CSP) system A facility that captures solar energy to generate steam, which turns a turbine to produce electricity.

conclusion A statement that follows logically from a set of premises.

confidence limits A statistical measure of the quality of data that tells you how close the sample's average probably is to the average for the entire population of that species.

confined animal feeding operation (CAFO) Facilities in which large numbers of animals spend most or all of their life in confinement.

conifers Needle-bearing trees that produce seeds in cones.

conservation development Consideration of landscape history, human culture, topography, and ecological values in subdivision design. Using cluster housing, zoning, covenants, and other design features, at least half of a subdivision can be preserved as open space, farmland, or natural areas.

conservation medicine A medical field that attempts to understand how environmental changes threaten our own health, as well as that of the natural communities on which we depend for ecological services.

conservation of matter In any chemical reaction, matter changes form; it is neither created nor destroyed.

conspicuous consumption A term coined by economist and social critic Thorstein Veblen to describe buying things we don't want or need in order to impress others.

constructed wetlands Artificially constructed wetlands used to purify water.

consumer An organism that obtains energy and nutrients by feeding on other organisms or their remains. *See also* heterotroph.

consumption The fraction of withdrawn water that is lost in transmission or that is evaporated, absorbed, chemically transformed, or otherwise made unavailable for other purposes as a result of human use.

contour plowing Plowing along hill contours; reduces erosion.

control rods Neutron-absorbing material inserted into spaces between fuel assemblies in nuclear reactors to regulate fission reactions.

controlled study Study in which comparisons are made between experimental and control populations that are (as far as possible) identical in every factor except the one variable being studied.

convection currents Rising or sinking air currents that stir the atmosphere and transport heat from one area to another. Convection currents also occur in water. *See* spring overturn.

Convention on International Trade in Endangered Species (CITES) The 1963 convention that regulates international trade in endangered species.

conventional or criteria pollutants The seven major pollutants (sulfur dioxide, carbon monoxide, particulates, hydrocarbons, nitrogen oxides, photochemical oxidants, and lead) identified and regulated by the U.S. Clean Air Act.

coral bleaching Whitening of corals caused by expulsion of symbiotic algae—often resulting from high water temperatures, pollution, or disease.

coral reefs Prominent oceanic features composed of hard, limy skeletons produced by coral animals; usually formed along edges of shallow, submerged ocean banks or along shelves in warm, shallow, tropical seas.

core (earth's) The dense, intensely hot mass of molten metal, mostly iron and nickel, thousands of kilometers in diameter at the earth's center.

core habitat Essential habitat for a species.

core region The primary industrial region of a country; usually located around the capital or largest port; has both the greatest population density and the greatest economic activity of the country.

Coriolis effect The influence of friction and drag on air layers near the earth; deflects air currents to the direction of the earth's rotation.

corridor A strip of natural habitat that connects two adjacent nature preserves to allow migration of organisms from one place to another.

cost–benefit analysis (CBA) An evaluation of large-scale public projects by comparing the costs and benefits that accrue from them.

cover crops Plants, such as rye, alfalfa, or clover, that can be planted immediately after harvest to hold and protect the soil.

creative thinking Asks, how could I do this differently.

criteria pollutants *See* conventional pollutants.

critical factor The environmental factor closest to a tolerance limit for a given species, which is the most important in determining distribution at a given time.

critical thinking An ability to evaluate information and opinions in a systematic, purposeful, efficient manner.

crude birth rate The number of births in a year divided by the midyear population.

crude death rate The number of deaths per thousand persons in a given year; *also called* crude mortality rate.

crust The cool, lightweight, outermost layer of the earth's surface that floats on the soft, pliable, underlying layers; similar to the "skin" on a bowl of warm pudding.

cultural eutrophication An increase in biological productivity and ecosystem succession caused by human activities.

cyclonic storms Storms with winds that spiral clockwise out of an area of high pressure in the Northern Hemisphere and counterclockwise into a low-pressure zone.

D

debt-for-nature swap Forgiveness of international debt in exchange for nature protection in developing countries.

deciduous Trees and shrubs that shed their leaves at the end of the growing season.

decomposer Fungi and bacteria that break down complex organic material into smaller molecules.

deductive reasoning Deriving testable predictions about specific cases from general principles.

demand The amount of a product that consumers are willing and able to buy at various possible prices, assuming they are free to express their preferences.

demanufacturing Disassembly of products so components can be reused or recycled.

demographic bottleneck A population founded when just a few members of a species survive a catastrophic event or colonize a new habitat that is geographically isolated from other members of the same species.

demographic transition A pattern of falling death rates and birthrates in response to improved living conditions; could be reversed in deteriorating conditions.

demography Vital statistics about people: births, marriages, deaths, etc.; the statistical study of human populations relating to growth rate, age structure, geographic distribution, etc., and their effects on social, economic, and environmental conditions.

density-dependent Factors affecting population growth that change as population size changes.

density-independent Factors that affect population growth, but that do not vary as population size changes.

dependency ratio The number of nonworking members compared to working members for a given population.

dependent variable A variable that is affected by the condition being altered in a manipulative experiment.

desert A biome characterized by low moisture levels and infrequent and unpredictable precipitation. Daily and seasonal temperatures fluctuate widely.

desertification Conversion of productive lands to desert.

detritivore Organisms that consume organic litter, debris, and dung.

disability-adjusted life years (DALYs) A measure of premature deaths and losses due to illnesses and disabilities in a population.

discharge The amount of water that passes a fixed point in a given amount of time; usually expressed as liters or cubic feet of water per second.

disease A deleterious change in the body's condition in response to destabilizing factors such as nutrition, chemicals, or biological agents.

dissolved oxygen (DO) content The amount of oxygen dissolved in a given volume of water at a given temperature and atmospheric pressure; usually expressed in parts per million (ppm).

disturbance Periodic destructive events such as fires or floods; changes in an ecosystem that affect (positively or negatively) the organisms living there.

disturbance-adapted species Species that depend on disturbances to succeed.

diversity The number of species present in a community (species richness), as well as the relative abundance of each species.

DNA (deoxyribonucleic acid) A giant molecule composed of millions or billions of nucleotides (sugars and bases called purines and pyramidines held together by phosphate bonds) that form a double helix and store genetic information in all living cells.

double-blind experiment A design in which neither the experimenter nor the subjects know, until after the gathering and analysis of data, which was the experimental treatment and which was the control.

E

e-waste Discarded electronic equipment, such as computers, cell phones, and television sets.

earthquake A sudden, violent movement of the earth's crust.

ecological diseases Emergent diseases (new or rarely seen diseases) that cause devastating epidemics among wildlife and domestic animals.

ecological economics A relatively new field that brings the insights of ecology to economic analysis.

ecological footprint A measure that computes the demands placed on nature by individuals and nations.

ecological niche The functional role and position of a species (population) within a community or an ecosystem, including what resources it uses, how and when it uses the resources, and how it interacts with other populations.

ecological restoration Bringing a landscape back to a former condition. Ecological restoration involves active manipulation of nature to re-create conditions that existed before human disturbance.

ecology The scientific study of relationships between organisms and their environment. It is concerned with the life histories, distribution, and behavior of individual species, as well as the structure and function of natural systems at the level of populations, communities, and ecosystems.

ecosystem management An integration of ecological, economic, and social goals in a unified systems approach to resource management.

ecosystem services Resources provided by nature, such as food, water, energy, fiber, waste disposal, and other conditions on which life depends.

ecotone A transition zone or boundary between different biomes, ecosystems, or ecological communities.

ecotourism A combination of adventure travel, cultural exploration, and nature appreciation in wild settings.

edge effects A change in species composition, physical conditions, or other ecological factors at the boundary between two ecosystems.

effluent sewerage A low-cost alternative sewage treatment for cities in poor countries that combines some features of septic systems and centralized municipal treatment systems.

El Niño A climatic change marked by the shifting of a large warm water pool from the western Pacific Ocean toward the east. Wind direction and precipitation patterns are changed over much of the Pacific and perhaps around the world.

electrostatic precipitators The most common particulate controls in power plants; fly ash particles pick up an electrostatic surface charge as they pass between large electrodes in the effluent stream, causing particles to migrate to the oppositely charged plate.

element A molecule composed of one kind of atom; cannot be broken into simpler units by chemical reactions.

emergent diseases A new disease or one that has been absent for at least 20 years.

emergent properties Characteristics of whole, functioning systems that are quantitatively or qualitatively greater than the sum of the system's parts.

emissions trading Programs in which companies that have cut pollution more than they're required to can sell "credits" to other companies that still exceed allowed levels.

endangered species A species considered to be in imminent danger of extinction.

endemic species Those found only in a particular place.

endocrine disrupters Chemicals that disrupt normal hormone functions.

energy The capacity to do work (that is, to change the physical state or motion of an object).

energy recovery Incineration of solid waste to produce useful energy.

entropy Disorder in a system.

environment The circumstances or conditions that surround an organism or group of organisms, as well as the complex of social or cultural conditions that affect an individual or community.

environmental health The science of external factors that cause disease, including elements of the natural, social, cultural, and technological worlds in which we live.

environmental impact statement (EIS) An analysis, required by provisions in the National Environmental Policy Act of 1970, of the effects of any major program a federal agency plans to undertake.

environmental justice A recognition that access to a clean, healthy environment is a fundamental right of all human beings.

environmental literacy Fluency in the principles of ecology that gives us a working knowledge of the basic grammar and underlying syntax of environmental wisdom.

Environmental Performance Index (EPI) A measure that evaluates national sustainability and progress toward achievement of the United Nations Millennium Development Goals.

environmental policy The official rules or regulations concerning the environment adopted, implemented, and enforced by some governmental agency.

environmental racism Decisions that restrict certain people or groups of people to polluted or degraded environments on the basis of race.

environmental science The systematic, scientific study of our environment, as well as our role in it.

enzymes Molecules, usually proteins or nucleic acids, that act as catalysts in biochemical reactions.

epigenetic effects Temporary changes in genetic material that affect gene expression.

epigenetic effects Temporary changes in genetic material that affect gene expression

estuary A bay or drowned valley where a river empties into the sea.

ethics A branch of philosophy concerned with right and wrong.

eutrophic Rivers and lakes rich in organisms and organic material (*eu* = truly; *trophic* = nutritious).

evolution A theory that explains how random changes in genetic material and competition for scarce resources cause species to change gradually.

existence value The importance we place on just knowing that a particular species or a specific organism exists.

exponential growth Growth at a constant rate of increase per unit of time; can be expressed as a constant fraction or exponent. *See* geometric growth.

externalized costs Expenses, monetary or otherwise, borne by someone other than the individuals or groups who use a resource.

extinction The irrevocable elimination of species; can be a normal process of the natural world as species outcompete or kill off others or as environmental conditions change.

F

family planning Controlling reproduction; planning the timing of birth and having the number of babies that are wanted and can be supported.

famines Acute food shortages characterized by large-scale loss of life, social disruption, and economic chaos.

feed-in tariffs Require utilities to buy surplus power from small producers at a fair price. These tariffs are generally essential to individual solar installations.

fen An area of waterlogged soil that tends to be peaty; fed mainly by ground water; low in productivity.

fetal alcohol syndrome A tragic set of permanent physical, mental, and behavioral birth defects that result from the mother's drinking alcohol during pregnancy.

first law of thermodynamics Energy can be transformed and transferred, but cannot be destroyed or created; i.e., energy is *conserved.*

flagship species Especially interesting or attractive organisms, such as giant pandas or killer whales, to which people react emotionally. These species can motivate the public to preserve biodiversity and contribute to conservation.

flex-fuel boilers Boilers that can use a variety of different fuels: coal, oil, biomass, etc.

food chain A linked feeding series; in an ecosystem, the sequence of organisms through which energy and materials are transferred, in the form of food, from one trophic level to another.

food security The ability of individuals to obtain sufficient food on a day-to-day basis.

food web A complex, interlocking series of individual food chains in an ecosystem.

forest Any area where trees cover more than 10 percent of the land.

fossil fuels Petroleum, natural gas, and coal created by geological forces from organic wastes and the dead bodies of formerly living biological organisms.

founder effect The effect on a population founded when just a few members of a species survive a catastrophic event or colonize a new habitat that is geographically isolated from other members of the same species.

fuel assembly A bundle of hollow metal rods containing uranium oxide pellets; used to fuel a nuclear reactor.

fuel cells Mechanical devices that use hydrogen or hydrogen-containing fuels such as methane to produce an electric current. Fuel cells are clean, quiet, and highly efficient sources of electricity.

fugitive emissions Substances that enter the air without going through a smokestack, such as dust from soil erosion, strip mining, rock crushing, construction, and building demolition.

fumigants Toxic gases such as methyl bromine that are used to kill pests.

fungicide A chemical that kills fungi.

G

gap analysis A biogeographical technique of mapping biological diversity and endemic species to find gaps between protected areas that leave endangered habitats vulnerable to disruption.

garden city A new town with special emphasis on landscaping and rural ambience.

generalist Species that can tolerate a wide range of conditions or exploit a wide range of resources.

genetic drift The gradual changes in gene frequencies in a population due to random events.

genetically modified organisms (GMOs) Organisms whose genetic code has been altered by artificial means such as interspecies gene transfer.

geographic isolation Speciation deriving from a common ancestor due to geographic barriers that cause reproductive isolation; *also called* geographic isolation.

geothermal energy Energy drawn from the internal heat of the earth, either through geysers, fumaroles, hot springs, or other natural geothermal features, or through deep wells that pump heated groundwater.

global environmentalism A concern for, and action to help solve, global environmental problems.

grasslands A biome dominated by grasses and associated herbaceous plants.

green plans Integrated national environmental plans for reducing pollution and resource consumption while achieving sustainable development and environmental restoration.

green pricing Setting prices to encourage conservation or renewable energy; plans that invite customers to pay a premium for energy from renewable sources.

green revolution Dramatically increased agricultural production brought about by "miracle" strains of grain; usually requires high inputs of water, plant nutrients, and pesticides.

greenfield developments Housing projects built on previously undeveloped farmlands or forests on the outskirts of large cities.

greenhouse effect Gases in the atmosphere are transparent to visible light but absorb infrared (heat) waves that are re-radiated from the earth's surface.

greenhouse gases Chemical compounds that trap heat in the atmosphere. The principal anthropogenic greenhouse gases are carbon dioxide, methane, chlorofluorocarbons, nitrous oxide, and sulfur hexafluoride.

gross domestic product (GDP) The total economic activity within national boundaries. Gross domestic product (GDP) is used to distinguish economic activity within a country from that of offshore corporations.

gross national product (GNP) The sum total of all goods and services produced in a national economy.

groundwater Water held in gravel deposits or porous rock below the earth's surface; does not include water or crystallization held by chemical bonds in rocks or moisture in upper soil layers.

gully erosion Removal of layers of soil, creating channels or ravines too large to be removed by normal tillage operations.

H

habitat The place or set of environmental conditions in which a particular organism lives.

habitat conservation plans (HCP) Agreements under which property owners are allowed to harvest resources or develop land as long as habitat is conserved or replaced in ways that benefit resident endangered or threatened species in the long run. Some incidental "taking" or loss of endangered species is generally allowed in such plans.

hazardous air pollutants (HAPs) Especially dangerous air pollutants, including carcinogens, neurotoxins, mutagens, teratogens, endocrine system disrupters, and other highly toxic compounds.

hazardous waste Any discarded material containing substances known to be toxic, mutagenic, carcinogenic, or teratogenic to humans or other life-forms; or ignitable, corrosive, explosive, or highly reactive alone or with other materials.

health A state of physical and emotional well-being; the absence of disease or ailment.

heat A form of energy transferred from one body to another because of a difference in temperatures.

herbicide A chemical that kills plants.

herbivore An organism that eats only plants.

HIPPO *H*abitat destruction, *I*nvasive species, *P*ollution, *P*opulation (human), and *O*verharvesting—the leading causes of extinction.

homeostasis The maintenance of a dynamic, steady state in a living system through opposing, compensating adjustments.

Hubbert curve A curve describing a peak and decline in production of natural resources, especially oil production, defined by M. King Hubbert in 1956.

Human Development Index (HDI) A measure of quality of life using life expectancy, child survival, adult literacy, childhood education, gender equity, and access to clean water and sanitation as well as income.

hurricanes Large cyclonic oceanic storms with heavy rain and winds exceeding 119 km/hr (74 mph).

hydraulic fracturing ("fracking") A mixture of water, sand, and toxic chemicals is pumped into rock formations at extremely high pressure to fracture sediments and release oil or gas.

hydrologic cycle The natural process by which water is purified and made fresh through evaporation and precipitation. This cycle provides all the freshwater available for biological life.

hypothesis A provisional explanation that can be tested scientifically.

I

I = PAT formula A formula that says our environmental impacts (I) are the product of our population size (P) times our affluence (A) and the technology (T) used to produce the goods and services we consume.

igneous rocks Crystalline minerals solidified from molten magma from deep in the earth's interior; basalt, rhyolite, andesite, lava, and granite are examples.

independent variable A factor not affected by the condition being altered in a manipulative experiment.

indicator species Species whose critical tolerance limits can be used to judge environmental conditions.

indicators Species whose critical tolerance limits can be used to judge environmental conditions.

inductive reasoning Inferring general principles from specific examples.

infiltration The process of water percolation into the soil and the pores and hollows of permeable rocks.

inherent value Ethical values or rights that exist as an intrinsic or essential characteristic of a particular thing or class of things simply by the fact of their existence.

inorganic pesticides Inorganic chemicals such as metals, acids, or bases used as pesticides.

insecticide A chemical that kills insects.

instrumental value The value or worth of objects that satisfy the needs and wants of moral agents. Objects that can be used as a means to some desirable end.

intangible resources Factors such as open space, beauty, serenity, wisdom, diversity, and satisfaction that cannot be grasped or contained. Ironically, these resources can be both infinite and exhaustible.

integrated pest management (IPM) An ecologically based pest-control strategy that relies on natural mortality factors, such as natural enemies, weather, cultural control methods, and carefully applied doses of pesticides.

Intergovernmental Panel on Climate Change (IPCC) An international organization formed to assess global climate change and its impacts. The IPCC is concerned with social, economic, and environmental impacts of climate change, and it was established by the United Nations Environment Programme and the World Meteorological Organization.

internal costs The expenses, monetary or otherwise, borne by those who use a resource.

interspecific competition In a community, competition for resources between members of *different* species.

intraspecific competition In a community, competition for resources among members of the *same* species.

invasive species Organisms that thrive in new territory where they are free of predators, diseases, or resource limitations that may have controlled their population in their native habitat.

ions Electrically charged atoms that have gained or lost electrons.

island biogeography The study of rates of colonization and extinction of species on islands or other isolated areas, based on size, shape, and distance from other inhabited regions.

isotopes Forms of a single element that differ in atomic mass due to having a different number of neutrons in the nucleus.

J

jet streams Powerful winds or currents of air that circulate in shifting flows; similar to oceanic currents in their extent and effect on climate.

joule A unit of energy. One joule is the energy expended in 1 second by a current of 1 amp flowing through a resistance of 1 ohm.

K

K-selected species Species that reproduce more slowly, occupy higher trophic levels, have fewer offspring, longer life spans, and greater intrinsic control of population growth than *r*-selected species.

kerogen A solid, energy-rich organic material similar to bitumen.

keystone species A species whose impacts on its community or ecosystem are much larger and more influential than would be expected from mere abundance.

kinetic energy Energy contained in moving objects, such as a rock rolling down a hill, the wind blowing through the trees, or water flowing over a dam.

kwashiorkor A widespread human protein-deficiency disease resulting from a starchy diet low in protein and essential amino acids.

Kyoto Protocol An international agreement to reduce greenhouse gas emissions.

L

La Niña The part of a large-scale oscillation in the Pacific (and, perhaps, other oceans) in which trade winds hold warm surface waters in the western part of the basin and cause upwelling of cold, nutrient-rich, deep water in the eastern part of the ocean.

landscape ecology The study of the reciprocal effects of spatial pattern on ecological processes. The study of how landscape history shapes the features of the land and the organisms that inhabit it, as well as our reaction to, and interpretation of, the land.

latent heat Stored energy in a form that is not sensible (cannot be detected by ordinary senses).

LD50 A chemical dose lethal to 50 percent of a test population.

life expectancy The average age that individuals born in a particular time and place can be expected to attain.

life-cycle analysis The evaluation of material and energy inputs and outputs at each stage of manufacture, use, and disposal of a product.

light-emitting diodes (LEDs) A semiconductor light source. When a hole recombines with an electron (from an electric current), light is emitted in an effect called electroluminescence. In a sense, this is the reverse of a photovoltaic panel. The color of the light is determined by the energy band gap of the semiconductor. These lights can be very small, but powerful. They take only a fraction of the energy of an incandescent bulb or a fluorescent light, and are said to last for years.

liquefied natural gas (LNG) When natural gas (mostly methane) is cooled to approximately −162°C (−260°F) it becomes a colorless, odorless liquid that takes up about 1/600th the volume of the gaseous state. This makes it much more convenient to ship and store.

locavore Someone who eats locally grown, seasonal food.

logical thinking Asks, can the rules of logic help understand this.

logistic growth Growth rates regulated by internal and external factors that establish an equilibrium with environmental resources.

low-head hydropower Small-scale hydro technology that can extract energy from small headwater dams; causes much less ecological damage.

LULUs *L*ocally *U*nwanted *L*and *U*ses, such as toxic waste dumps, incinerators, smelters, airports, freeways, and other sources of environmental, economic, or social degradation.

M

magma Molten rock from deep in the earth's interior; called lava when it spews from volcanic vents.

malnourishment A nutritional imbalance caused by a lack of specific dietary components or an inability to absorb or utilize essential nutrients.

Man and Biosphere (MAB) program A design for nature preserves that divides protected areas into zones with different purposes. A highly protected core is surrounded by a buffer zone and peripheral regions in which multiple-use resource harvesting is permitted.

managing the commons Systems for managing common resources.

mangroves Trees from a number of genera that live in shallow salt water.

manipulative experiment An experiment in which some conditions are deliberately altered while others are held constant to study cause-and-effect relationships.

mantle A hot, pliable layer of rock that surrounds the earth's core and underlies the cool, outer crust.

marasmus A widespread human protein-deficiency disease caused by a diet low in calories and protein or imbalanced in essential amino acids.

marginal costs and benefits The costs and benefits of producing one additional unit of a good or service.

market equilibrium The dynamic balance between supply and demand under a given set of conditions in a "free" market (one with no monopolies or government interventions).

market forces Dynamics by which prices or wages are set by negotiation between buyers and sellers.

marsh Wetland without trees; in North America, this type of land is characterized by cattails and rushes.

mass burn Incineration of unsorted solid waste.

mass wasting Mass movement of geological materials downhill caused by rockslides, avalanches, or simple slumping.

matter Anything that takes up space and has mass.

mean Average; a representation of the middle of a group; commonly calculated as the sum of values divided by the number of observations.

mediation An informal dispute resolution process in which parties are encouraged to discuss issues openly but in which all decisions are reached by consensus and any participant can withdraw at any time.

megacity *See* megalopolis.

metabolism All the energy and matter exchanges that occur within a living cell or organism; collectively, the life processes.

metamorphic rock Igneous and sedimentary rocks modified by heat, pressure, and chemical reactions.

metapopulation A collection of populations that have regular or intermittent gene flow between geographically separate units.

methane CH_4; the simplest hydrocarbon and the main component of natural gas.

methane hydrate Small bubbles or individual molecules of methane (natural gas) trapped in a crystalline matrix of frozen water.

micro-hydro generators Small power generators that can be used in low-level rivers to provide economical power for four to six homes, freeing them from dependence on large utilities and foreign energy supplies.

microbial agents Beneficial microbes (bacteria, fungi) that can be used to suppress or control pests; *also called* biological controls.

microbiome The complex community of microorganisms that live in and on your body. The microbiome is essential to digestion and other processes.

mid-ocean ridges Mountain ranges on the ocean floor created where molten magma is forced up through cracks in the planet's crust.

Milankovitch cycles Periodic variations in tilt, eccentricity, and wobble in the earth's orbit; Milutin Milankovitch suggested that these are responsible for cyclic weather changes.

millennium assessment A set of ambitious environmental and human development goals established by the United Nations in 2000.

Millennium Development Goals A set of goals established in 2000 by the United Nations that include ending poverty and hunger, universal education, gender equity, child health, maternal health, combating HIV/AIDS, environmental sustainability, and global cooperation in development efforts.

mineral A naturally occurring, inorganic, crystalline solid with definite chemical composition and characteristic physical properties.

minimum viable population size The number of individuals needed for long-term survival of rare and endangered species.

mitigation Repairing or rehabilitating a damaged ecosystem or compensating for damage by providing a substitute or replacement area.

models Simple representations of more complex systems.

molecule A combination of two or more atoms.

monoculture agroforestry Intensive planting of a single species; an efficient wood production approach, but one that encourages pests and disease infestations and conflicts with wildlife habitat or recreational uses.

monsoon A seasonal reversal of wind patterns caused by the different heating and cooling rates of the oceans and continents.

Montreal Protocol An international convention to eliminate chlorofluorocarbon production.

moral extensionism Expansion of our understanding of the inherent value or rights to persons, organisms, or things that might not be considered worthy of value or rights under some ethical philosophies.

moral value The value or worth of something based on moral principles.

morbidity Illness or disease.

mortality Death rate in a population; the probability of dying.

Müllerian mimicry Evolution of two species, both of which are unpalatable and have poisonous stingers, or some other defense mechanism, to resemble each other.

mutagens Agents, such as chemicals or radiation, that damage or alter genetic material (DNA) in cells.

mutation A change, either spontaneous or by external factors, in the genetic material of a cell. Mutations in the gametes (sex cells) can be inherited by future generations of organisms.

mutualism A symbiotic relationship between individuals of two different species in which both species benefit from the association.

mycorrhizal symbiosis An association between the roots of most plant species and certain fungi. The plant provides organic compounds to the fungus, while the fungus provides water and nutrients to the plant.

N

National Environmental Policy Act (NEPA) *National Environmental Policy Act*, the cornerstone of U.S. environmental policy. Authorizes the Council on Environmental Quality, directs federal agencies to take environmental consequences into account when making decisions, and requires an environmental impact statement for every major federal project likely to have adverse environmental effects.

natural experiment A study of events that have already happened.

natural organic pesticides "Botanicals" or organic compounds naturally occurring in plants, animals, or microbes that serve as pesticides.

natural selection The mechanism for evolutionary change in which environmental pressures cause certain genetic combinations in a population to become more abundant. Genetic combinations best adapted for present environmental conditions tend to become predominant.

negative feedback loop A situation in which a factor or condition causes changes that reduce that factor or condition.

negawatts Energy saved through conservation.

neoclassical economics A branch of economics that attempts to apply the principles of modern science to economic analysis in a mathematically rigorous, noncontextual, abstract, predictive manner.

neonicotinoid Pesticides with a chemical structure similar to nicotine. Thought to be a cause of massive bee deaths.

neurotoxins Toxic substances, such as lead or mercury, that specifically poison nerve cells.

nitric oxide NO.

nitrogen cycle The circulation and reutilization of nitrogen in both inorganic and organic phases.

nitrogen dioxide NO_2.

nitrogen oxides Highly reactive gases formed when nitrogen in fuel or combustion air is heated to over 650°C (1,200°F) in the presence of oxygen, or when bacteria in soil or water oxidize nitrogen-containing compounds.

nitrous oxide N_2O.

nongovernmental organizations (NGOs) A term referring collectively to pressure and research groups, advisory agencies, political parties, professional societies, and other groups concerned about environmental quality, resource use, and many other issues.

nonpoint sources Scattered, diffuse sources of pollutants, such as runoff from farm fields, golf courses, and construction sites.

nonrenewable resources Minerals, fossil fuels, and other materials present in essentially fixed amounts (within human time scales) in our environment.

novel ecosystem A specific biological community and its physical environment interacting in an exchange of matter and energy.

nuclear fission The radioactive decay process in which isotopes split apart to create two smaller atoms.

nuclear fusion A process in which two smaller atomic nuclei fuse into one larger nucleus and release energy; the source of power in a hydrogen bomb.

O

oak savanna Open grasslands with sparse tree cover.

obese Generally considered to be a body mass greater than 30 kg/m², or roughly 30 pounds above normal for an average person.

ocean wave energy Electricity generated by ocean wave power.

oil shale A fine-grained sedimentary rock rich in solid organic material called kerogen. When heated, the kerogen liquefies to produce a fluid petroleum fuel.

oligotrophic The condition of rivers and lakes that have clear water and low biological productivity (*oligo* = little; *trophic* = nutrition); usually clear, cold, infertile headwater lakes and streams.

omnivore An organism that eats both plants and animals.

open system A system that exchanges energy and matter with its environment.

open-access system A commonly held resource for which there are no management rules.

organic compounds Complex molecules organized around skeletons of carbon atoms arranged in rings or chains; includes biomolecules, molecules synthesized by living organisms.

organophosphates Organic molecules to which one or more phosphate groups are attached.

overgrazing Allowing livestock to eat so much forage that the ecological health of the habitat is damaged.

overharvesting Harvesting so much of a resource that its existence is threatened.

oxygen sag Oxygen decline downstream from a pollution source that introduces materials with high biological oxygen demands.

ozone A highly reactive molecule containing three oxygen atoms; a dangerous pollutant in ambient air. In the stratosphere, however, ozone forms an ultraviolet-absorbing shield that protects us from mutagenic radiation.

P

paradigm shift Dramatic changes in models that provide frameworks for interpreting observations.

parasitism An organism that lives in or on another organism, deriving nourishment at the expense of its host, usually without killing it.

particulate matter Atmospheric aerosols, such as dust, ash, soot, lint, smoke, pollen, spores, algal cells, and other suspended materials. The term originally was applied only to solid particles but now is extended to droplets of liquid.

passive heat absorption The use of natural materials or absorptive structures without moving parts to gather and hold heat; the simplest and oldest use of solar energy.

passive houses Highly insulated, virtually airtight buildings heated primarily by passive solar gain and by the energy released by people, electrical appliances, cooking, and other ordinary activities.

pastoralists People who make a living by herding domestic livestock.

peak oil A prediction, made about 1940 by Dr. M. King Hubbert, that oil production in the United States would peak in the 1970s and then decline.

pelagic Zones in the vertical water column of a water body.

permanent retrievable storage Placing waste storage containers in a secure building, salt mine, or bedrock cavern where they can be inspected periodically and retrieved if necessary.

persistent organic pollutants (POPs) Chemical compounds that persist in the environment and retain biological activity for a long time.

pest resurgence Rebound of pest populations due to acquired resistance to chemicals and nonspecific destruction of natural predators and competitors by broadscale pesticides.

pesticide Any chemical that kills, controls, drives away, or modifies the behavior of a pest.

pesticide treadmill A need for constantly increasing doses or new pesticides to prevent pest resurgence.

pH A value that indicates the acidity or alkalinity of a solution on a scale of 0 to 14, based on the proportion of H1 ions present.

phosphorus cycle The movement of phosphorus atoms from rocks through the biosphere and hydrosphere and back to rocks.

photochemical oxidants Products of secondary atmospheric reactions. *See* smog.

photodegradable plastics Plastics that break down to smaller particles when exposed to sunlight or to a specific wavelength of light.

photosynthesis The biochemical process by which green plants and some bacteria capture light energy and use it to produce chemical bonds. Carbon dioxide and water are consumed while oxygen and simple sugars are produced.

photovoltaic (PV) cell An energy-conversion device that captures solar energy and directly converts it to electrical current.

phytoplankton Microscopic, free-floating, autotrophic organisms that function as producers in aquatic ecosystems.

pioneer species In primary succession on a terrestrial site, the plants, lichens, and microbes that first colonize the site.

point sources Specific locations of highly concentrated pollution discharge, such as factories, power plants, sewage treatment plants, underground coal mines, and oil wells.

policy A societal plan or statement of intentions intended to accomplish some social good.

political economy The branch of economics concerned with modes of production, distribution of benefits, social institutions, and class relationships.

population A group of individuals of the same species occupying a given area.

population crash A sudden population decline caused by predation, waste accumulation, or resource depletion; *also called* a dieback.

population momentum A potential for increased population growth as young members reach reproductive age.

positive feedback loop A situation in which a factor or condition causes changes that further enhance that factor or condition.

potential energy Stored energy that is latent but available for use; for example, a rock poised at the top of a hill, or water stored behind a dam.

power The rate of energy delivery; measured in horsepower or watts.

power tower A high-temperature concentrating solar power system in which thousands of mirrors arranged in concentric rings around a tall central tower track the sun and focus light on a heat absorber to generate steam, which drives an electrical generator.

prairie potholes Small ponds on the prairie.

precautionary principle The decision to leave a margin of safety for unexpected developments.

predator An organism that feeds directly on other organisms in order to survive; live-feeders, such as herbivores and carnivores.

predator-mediated competition A situation in which predation reduces prey populations and gives an advantage to competitors that might not otherwise be successful.

premises Introductory statements that set up or define a problem. Those things taken as given.

prescribed burning Periodic fires that clean out brush and trees to maintain a prairie.

price elasticity A situation in which supply and demand of a commodity respond to price.

primary forest Forest composed primarily of native species, where there are no clearly visible indications of human activity and ecological processes are not significantly disturbed.

primary pollutants Chemicals released directly into the air in a harmful form.

primary productivity Synthesis of organic materials (biomass) by green plants using the energy captured in photosynthesis.

primary succession An ecological succession that begins in an area where no biotic community previously existed.

primary treatment A process that removes solids from sewage before it is discharged or treated further.

principle of competitive exclusion No two species can occupy the same ecological niche for long.

producer An organism that synthesizes food molecules from inorganic compounds by using an external energy source; most producers are photosynthetic.

productivity The synthesis of new organic material; synthesis done by green plants using solar energy is called primary productivity.

pronatalist pressures Influences that encourage people to have children.

property assessed clean energy (PACE) A program that uses city bonds to pay for renewable energy and conservation expenses, such as solar power and insulation.

pull factors (in urbanization) Conditions that draw people from the country into the city.

push factors (in urbanization) Conditions that force people out of the country and into the city.

Q

quality-of-life indicators Factors, such as infant mortality, life expectancy, income, sanitation, and education, that indicate quality of life in a country.

R

r-selected species Species that tend to have rapid reproduction and high offspring mortality. They frequently overshoot the carrying capacity of their environment and display boom-and-bust life cycles. They lack intrinsic population controls and tend to occupy lower trophic levels in food webs than do *k*-selected species.

rain shadow A dry area on the downwind side of a mountain.

re-creation (in ecology) Construction of an entirely new biological community to replace one that has been destroyed on that or another site.

reallocation To use a site (and its resources) to create a new and different kind of biological community rather than the existing one.

recharge zone An area where water infiltrates an aquifer.

reclamation Chemical, biological, or physical cleanup and reconstruction of severely contaminated or degraded sites to return them to something like their original topography and vegetation.

recycling Reprocessing of discarded materials into new, useful products; not the same as reuse of materials for their original purpose, but the terms are often used interchangeably.

red tide A population explosion or bloom of minute, single-celled marine organisms called dinoflagellates. Billions of these cells can accumulate in protected bays, where the toxins they contain can poison other marine life.

reflective thinking Asks, what does this all mean.

reformer A device that strips hydrogen from fuels such as natural gas, methanol, ammonia, gasoline, or vegetable oil so they can be used in a fuel cell.

refuse-derived fuel Processing of solid waste to remove metal, glass, and other unburnable materials; organic residue is shredded, formed into pellets, and dried to make fuel for power plants.

rehabilitate To rebuild elements of structure or function in an ecological system without necessarily achieving complete restoration to its original condition.

remediation Cleaning up chemical contaminants from a polluted area.

renewable portfolio standard A mandate that a utility company's generating capacity, or purchasing practices, include a specified amount of renewable energy. This policy regulation is designed to establish a market for alternative energy sources.

renewable resources Resources normally replaced or replenished by natural processes; resources not depleted by moderate use; examples include solar energy, biological resources such as forests and fisheries, biological organisms, and some biogeochemical cycles.

renewable water supplies Annual freshwater surface runoff plus annual infiltration into underground freshwater aquifers that are accessible for human use.

replication Repeating studies or tests to verify reliability.

reproducibility The capacity for a particular result to be observed or obtained more than once.

residence time The length of time a component, such as an individual water molecule, spends in a specific compartment or location before it moves on through a particular process or cycle.

resilience The ability of a community or ecosystem to recover from disturbances.

resource In economic terms, anything with potential use in creating wealth or giving satisfaction.

resource partitioning In a biological community, various populations sharing environmental resources through specialization, thereby reducing direct competition.

riders Amendments attached to bills in conference committee, often completely unrelated to the bill to which they are added.

rill erosion The removal of thin layers of soil by little rivulets of running water that gather and cut small channels in the soil.

risk The probability that something undesirable will happen as a consequence of exposure to a hazard.

risk assessment Evaluation of the short-term and long-term risks associated with a particular activity or hazard; usually compared to benefits in a cost–benefit analysis.

Roadless Rule A Clinton-era ban on logging, road building, and other development on the lands identified as deserving of wilderness protection in the Roadless Area Review and Evaluations (RARE).

rock A solid, cohesive aggregate of one or more crystalline minerals.

rock cycle The process whereby rocks are broken down by chemical and physical forces; sediments are moved by wind, water, and gravity, sedimented and reformed into rock, and then crushed, folded, melted, and recrystallized into new forms.

rotational grazing Confining animals to a small area for a short time (often only a day or two) before shifting them to a new location.

run-of-the-river flow Ordinary river flow not accelerated by dams, flumes, etc. Some small, modern, high-efficiency turbines can generate useful power with run-of-the-river flow or with a current of only a few kilometers per hour.

rural area An area in which most residents depend on agriculture or the harvesting of natural resources for their livelihood.

S

salinization A process in which mineral salts accumulate in the soil, killing plants; occurs when soils in dry climates are irrigated profusely.

salt marsh Shallow wetlands along coastlines that are flooded regularly or occasionally with seawater.

saltwater intrusion Movement of saltwater into freshwater aquifers in coastal areas where groundwater is withdrawn faster than it is replenished.

sample Noun: A small portion of a population, used to approximate characteristics of the population. Verb: to collect observations or a small portion of a population, to allow characterization of a population.

sanitary landfills A landfill in which garbage and municipal waste are buried every day under enough soil or fill to eliminate odors, vermin, and litter.

scavenger An organism that feeds on the dead bodies of other organisms.

science A process for producing knowledge methodically and logically.

scientific consensus A general agreement among informed scholars.

scientific theory An explanation supported by many tests and accepted by a general consensus of scientists.

second law of thermodynamics With each successive energy transfer or transformation, less energy is available to do work.

secondary pollutants Chemicals that acquire a hazardous form after entering the air or that are formed by chemical reactions as components of the air interact.

secondary succession Succession on a site where an existing community has been disrupted.

secondary treatment Bacterial decomposition of suspended particulates and dissolved organic compounds that remain after primary sewage treatment.

secure landfill A solid-waste disposal site lined and capped with an impermeable barrier to prevent leakage or leaching. Drain tiles, sampling wells, and vent systems provide monitoring and pollution control.

sedimentary rock Deposited material that remains in place long enough, or is covered with enough material, to compact into stone; examples include shale, sandstone, breccia, and conglomerates.

sedimentation The deposition of organic materials or minerals by chemical, physical, or biological processes.

selection pressures Factors in the environment that favor the successful reproduction of individuals possessing certain heritable traits and that reduce the viability and fertility of individuals that do not possess those traits.

shantytowns Settlements created when people move onto undeveloped lands and build their own shelters with cheap or discarded materials. Some are simply illegal subdivisions where a landowner rents land without city approval; others are land invasions.

sheet erosion The peeling off of thin layers of soil from the land surface; accomplished primarily by wind and water.

sick building syndrome Headaches, allergies, chronic fatigue, and other symptoms caused by poorly vented indoor air contaminated by pathogens or toxins.

significant figure Also known as significant digits, a meaningful degree of precision.

slums Legal but inadequate multifamily tenements or rooming houses; some are custom-built for rent to poor people, others are converted from some other use.

smart growth Efficient use of land resources and existing infrastructure.

smart meter A meter records energy use data and communicates with the utility company, so that you can be charged less for off-peak energy and more for peak energy.

social justice Equitable access to resources and the benefits derived from them; a system that recognizes inalienable rights and adheres to what is fair, honest, and moral.

solar garden A solar array shared by multiple owners, who help finance the facility and gain from the electricity produced. Also called a solar farm.

solar garden Energy derived from temperature differentials between warm ocean surface waters and cold deep waters. This differential can be used to drive turbines attached to electric generators.

species A population of morphologically similar organisms that can reproduce sexually among themselves but cannot produce fertile offspring when mated with other organisms.

sprawl Unlimited outward extension of city boundaries that lowers population density, consumes open space, generates freeway congestion, and causes decay in central cities.

squatter towns Shantytowns that occupy land without owner's permission. Some are highly organized movements in defiance of authorities; others grow gradually.

standing The right to take part in legal proceedings.

state shift A permanent or long-lasting change in a system to a new set of conditions and relations in response to a disturbance.

statistics Numbers that describe observations or groups of observations; also a field of study that provides methods of comparing descriptive numbers.

statute law Formal documents or decrees enacted by the legislative branch of government.

steady-state economy Characterized by low birth and death rates, use of renewable energy sources, recycling of materials, and emphasis on durability, efficiency, and stability.

stewardship A philosophy that holds that humans have a unique responsibility to manage, care for, and improve nature.

strategic lawsuits against public participation (SLAPP) Lawsuits that have no merit but are brought merely to intimidate and harass private citizens who act in the public interest.

stratosphere The zone in the atmosphere extending from the tropopause to about 50 km (30 mi) above the earth's surface; temperatures are stable or rise slightly with altitude; has very little water vapor but is rich in ozone.

stratospheric ozone The ozone (O_3) occurring in the stratosphere 10 to 50 km above the earth's surface.

stress-related diseases Diseases caused or accentuated by social stresses such as crowding.

subduction The process by which one tectonic plate is pushed down below another as plates crash into each other.

subsidence A settling of the ground surface caused by the collapse of porous formations that result from a withdrawal of large amounts of groundwater, oil, or other underground materials.

subsoil A layer of soil beneath the topsoil that has lower organic content and higher concentrations of fine mineral particles; often contains soluble compounds and clay particles carried down by percolating water.

sulfur dioxide A colorless, corrosive gas directly damaging to both plants and animals.

Superfund A fund established by Congress to pay for the containment, cleanup, or remediation of abandoned toxic waste sites. The fund is financed by fees paid by toxic waste generators and by cost recovery from cleanup projects.

supply The quantity of a product being offered for sale at various prices, other things being equal.

sustainable development Improving well-being and the standard of life over the long-term; that is, meeting the needs of the present without compromising the ability of future generations to meet their own needs.

Sustainable Development Goals (SDGs) A set of 17 goals adopted by the United Nations in 2015, which aim to end poverty, protect the planet, and ensure prosperity for all as part of a new sustainable development agenda. Each goal has specific targets to be achieved between 2015 and 2030.

swamp A wetland with trees, such as the extensive swamp forests of the southern United States.

symbiosis The intimate living together of members of two different species; includes mutualism, commensalism, and, in some classifications, parasitism.

sympatric speciation Species that arise from a common ancestor due to biological or behavioral barriers that cause reproductive isolation even though the organisms live in the same place.

synergistic effect An interaction in which one substance exacerbates the effects of another. The sum of the interaction is greater than the parts.

systems Networks of interactions among many interdependent factors.

T

taiga The northernmost edge of the boreal forest, including species-poor woodland and peat deposits; intergrading with the arctic tundra.

tar sands Sand deposits containing petroleum or tar.

technological optimists Those who believe that technology and human enterprise will find cures for all our problems; *also called* Promethean environmentalism.

tectonic plates Huge blocks of the earth's crust that slide around slowly, pulling apart to open new ocean basins or crashing ponderously into each other to create new, larger landmasses.

temperate rainforest The cool, dense, rainy forest of the northern Pacific coast; enshrouded in fog much of the time; dominated by large conifers.

temperature inversion A stable layer of warm air overlying cooler air, trapping pollutants near ground level.

teratogens Chemicals or other factors that cause abnormalities during embryonic growth and development.

terracing Shaping the land to create level shelves of earth to hold water and soil; requires extensive hand labor or expensive machinery, but enables farmers to farm very steep hillsides.

tertiary treatment The removal of inorganic minerals and plant nutrients after primary and secondary treatment of sewage.

thermal plume A plume of hot water discharged into a stream or lake by a heat source, such as a power plant.

thermocline In water, a distinctive temperature transition zone that separates an upper layer that is mixed by the wind (the epilimnion) and a colder, deep layer that is not mixed (the hypolimnion).

threatened species A species that is still abundant in parts of its territorial range but has declined significantly in total numbers and may be on the verge of extinction in certain regions or localities.

throughput The flow of energy and matter into, through, and out of a system.

tidal station A dam built across a narrow bay or estuary that traps tide water flowing both in and out of the bay. Water flowing through the dam spins turbines attached to electric generators.

tide pool Depressions in a rocky shoreline that are flooded at high tide but cut off from the ocean at low tide. They often have a rich collection of marine life.

tolerance limits Chemical or physical factors that limit the existence, growth, abundance, or distribution of an organism.

topsoil The uppermost layer of a soil, including the "O" and "A" layers, which are usually rich in organic material.

tornado A violent storm characterized by strong swirling winds and updrafts. Tornadoes form when a strong cold front pushes under a warm, moist air mass over the land.

total fertility rate The number of children born to an average woman in a population during her entire reproductive life.

total growth rate The net rate of population growth resulting from births, deaths, immigration, and emigration.

total maximum daily load (TMDL) The amount of particular pollutants that a water body can receive from both point and nonpoint sources and still meet water quality standards.

toxic colonialism Shipping toxic wastes to a weaker or poorer nation.

Toxic Release Inventory (TRI) A program created by the Superfund Amendments and Reauthorization Act of 1984 that requires manufacturing facilities and waste-handling and disposal sites to report annually on releases of more than 300 toxic materials.

toxic substances Poisonous chemicals that react with specific cellular components to kill cells or to alter growth or development in undesirable ways; often harmful, even in diluted concentrations.

Tragedy of the Commons An inexorable process of degradation of communal resources due to the selfishness of "free riders" who use or destroy more than their fair share of common property. *See* open access system.

triple bottom line Corporate accounting that reports social and environmental costs and benefits, as well as merely economic ones.

trophic level A step in the movement of energy through an ecosystem; an organism's feeding status in an ecosystem.

tropical rainforests Forests in which rainfall is abundant—more than 200 cm (80 in.) per year—and temperatures are warm to hot year-round.

tropical seasonal forest Semi-evergreen or partly deciduous forests tending toward open woodlands and grassy savannas dotted with scattered, drought-resistant tree species; distinct wet and dry seasons, hot year-round.

troposphere The layer of air nearest to the earth's surface; both temperature and pressure usually decrease with increasing altitude.

tsunami A giant seismic sea swell that moves rapidly from the center of a submarine earthquake; can be 10 to 20 meters high when it reaches shorelines hundreds or even thousands of kilometers from its source.

tundra Treeless arctic or alpine biome characterized by cold, harsh winters, a short growing season, and a potential for frost any month of the year; vegetation includes low-growing perennial plants, mosses, and lichens.

U

umbrella species Require large blocks of relatively undisturbed habitat to maintain viable populations. Saving this habitat also benefits other species.

UN Framework Convention on Climate Change (UNFCCC) Directs governments to share data on climate change, to develop national plans for controlling greenhouse gases, and to cooperate in planning for adaptation to climate change.

unburnable carbon Carbon fuels that we need to leave in the ground if we want to avoid disastrous climate change.

urban agglomerations An aggregation of many cities into a large metropolitan area.

urban area An area in which a majority of the people are not directly dependent on natural-resource-based occupations.

utilitarian conservation A philosophy that resources should be used for the greatest good for the greatest number for the longest time.

V

vertical zonation Terrestrial vegetation zones determined by altitude.

village A collection of rural households linked by culture, custom, and association with the land.

volatile organic compounds (VOCs) Organic chemicals that evaporate readily and exist as gases in the air.

volcano A vent in the earth's surface through which gases, ash, or molten lava are ejected. Also a mountain formed by this ejecta.

vulnerable species Naturally rare organisms or species whose numbers have been so reduced by human activities that they are susceptible to actions that could push them into threatened or endangered status.

W

warm front A long, wedge-shaped boundary caused when a warmer advancing air mass slides over neighboring cooler air parcels.

waste stream The steady flow of varied wastes, from domestic garbage and yard wastes to industrial, commercial, and construction refuse.

water pollution Anything that degrades water quality.

water scarcity When annual available freshwater supplies are less than 1,000 m^3 per person.

water stress A situation when residents of a country don't have enough accessible, high-quality water to meet their everyday needs.

water table The top layer of the zone of saturation; undulates according to the surface topography and subsurface structure.

waterlogging Water saturation of soil that fills all air spaces and causes plant roots to die from lack of oxygen; a result of overirrigation.

watt (W) The force exerted by 1 joule, or the equivalent of a current of 1 amp per second flowing through a resistance of 1 ohm.

weather Description of the physical conditions of the atmosphere (moisture, temperature, pressure, and wind).

weathering Changes in rocks brought about by exposure to air, water, changing temperatures, and reactive chemical agents.

wedge analysis Policy options proposed by R. Socolow and S. Pacala for reducing greenhouse gas emissions using existing technologies. Each wedge represents a cumulative reduction of the equivalent of 1 billion tons of carbon over the next 50 years.

wetland mitigation Replacing a wetland damaged by development (roads, buildings, etc.) with a new or refurbished wetland.

wetlands Ecosystems of several types in which rooted vegetation is surrounded by standing water during part of the year. *See also* swamp, marsh, bog, fen.

wind turbines Large windmills that produce electricity.

withdrawal A description of the total amount of water taken from a lake, river, or aquifer.

work The application of force through a distance; requires energy input.

world conservation strategy A proposal for maintaining essential ecological processes, preserving genetic diversity, and ensuring that the utilization of species and ecosystems is sustainable.

World Happiness Index (WHI) A measure of life-satisfaction rather than mere economic success.

Z

zero population growth (ZPG) The number of births at which people are just replacing themselves; *also called* the replacement level of fertility.

zone of aeration Upper soil layers that hold both air and water.

zone of saturation Lower soil layers where all spaces are filled with water.

Index

Source: imageBROKER/Alamy Stock Photo

antigens, 161
Antiquities Act (1906), 270, 556
Antofagasta, 316
appeal to authority, 2
appeal to ignorance, 2
aquaculture, 187, 188–189
aquifers, 397. *See also* groundwater
  characteristics, 385–386
  depletion, 391–393
arbitration, 560–561
Arch Coal, 430, 435
archæa, 57
Archie Carr National Wildlife Refuge
  (FL), 545 (case study)
arctic amplification, 342
Arctic ecosystems
  positive and negative feedback, 43
  sea ice loss, 237, 238, 343, 438
  tundra, 105–106
Arctic Ocean
  exploratory oil drilling in, 438
  methane hydrate release, 444
Arctic tundra, 105–106
Arendt, Randall, 515
aridisols, 202
Aristotle, 262
Arizona
  Grand Canyon, 267
  Hassayampa River Preserve, 279
  Lake Powell, water supply, 381–382,
    395, 396
  red squirrels, geographic isolation, 78
  solar energy production, 465
  solar farm battery storage, 472
  Sonoran Desert, critical factors and
    tolerance limits, 74–75
  water problems, 381–382
Arkansas, fracking-related earthquakes,
  318
Army Corps of Engineers, U.S.
  Dakota Access Pipeline route, 439
  Elwha River restoration, 293, 396
  Everglades water diversion, 294–295
  large dam projects, 396
  Mississippi River restoration, 293
  reclamation projects, 299
Arrhenius, Svante, 338
arsenic
  fossil fuels as source, 415
  as water pollutant, 411, 416, 423
artesian wells, 386, 417
Asia. *See also names of individual
    countries*
  air pollution, 13–14, 366–367
  arsenic contamination, 416
  biofuel production, 469–470
  deforestation, 255
  electronic waste shipments to, 483
  land degradation, 203–204, 206
  night soil collection, 420
  palm oil plantations, 470
  reforestation programs, 255
  urbanization, 503
  water conflicts, 397
Asia Pulp and Paper, 258
Asian Americans, environmental racism,
  29–30, 493
Asian Brown Cloud, 366–367
Association for the Advancement of
  Sustainability in Higher Education
  (AASHE), 577
Athabasca oil sands (Canada), 437, 438
Atlantic Coastal Action Programme
  (ACAP), 561–562
atmosphere, 326–329
  air pollution. *See* air pollution
  components, 326–327, 328, 339–341

energy balance, 328
energy redistribution through
  circulation, 329, 383
greenhouse effect, 329, 339–341
processes, 365–370
solar energy absorption by land surface,
  328
as water compartment, 384
weather patterns, 326, 330–335
zones, 327
atmospheric deposition, 407
atomic number, 50–51
atoms, 50–52
"Atoms for Peace" speech (1953), 445
atrazine, 213, 407, 412, 417
Audubon, John James, 238
Audubon Society, 569, 573–574
aurora australis, 327
aurora borealis, 327
Australia
  anthropogenic disturbances, 92
  chytridiomycosis in amphibians, 159
  coral reef "assisted evolution," 277
  coral reef destruction, 236, 269, 276
  drought conditions, 397
  indigenous title to land, 27
  invasive species, 241
  prickly pear cactus control, 220
  surface-water quality, 415
  wastewater recycling, 400
  water policy, 401
  wealth in, 20
Austria
  air pollution reduction, 377
  birth dearth, 145
automobiles
  air pollution controls, 374
  black carbon and, 340
  car-sharing, 515
  China, electric vehicles (EVs), 458–459
  diesel engines, 370–371, 470
  efficiency improvements, 456–459
  energy use calculations, 477
  ethanol use, 470–471
  fuel cell use, 475–476
  fuel efficiency, 458
  Germany, car-free suburb (Vauban),
    515
  hybrid/electric vehicles (EVs), 310,
    458–459, 475–476
  India, Tata Nano, 512
  Interstate Highway system, 289, 475,
    510
  "mining" scrapped cars, 312, 489
  urban development and, 510–512
Aveda, 538

B

"B" horizon (soil), 201
Babbitt, Bruce, 396
baby boom, 144–145
bacillus thuringiensis (Bt), 191–193,
  220–221
bacteria
  antibiotic-resistant, 78, 85, 159–160
  in bioremediation, 298
  in human microbiome, 85
  in soil, 200–201
  as water pollutant, 407–408
bacterial gene editing system (CRISPR),
  55, 193
Bakken formation, 437, 439, 440
bald eagles, 237, 243–244
Baloney Detection Kit, 45

Ban Ki-Moon, 24
bananas, 192, 209, 216, 221
Bangladesh
  Dhaka, sea-level rise, 506
  floods, displacement from, 320, 321
  Grameen Bank, 537
  "green bonds," 347
  monsoon rains, 333
  population, 140
  water purification systems, 423
barnacles, 84
Barnett, Tim, 381
barrier islands, 110
Bartholomew (patriarch), 29
Basel Action Network, 483
Basel Convention (1992), 559–560
bases (alkaline), 52–53, 411
Bates, H. W., 82
Batesian mimicry, 82
bathypelagic zone, 107
bats
  white-nose syndrome, 158–159
  wind farm mortality, 466
battery storage, 472–473, 475–476
Bayer, 194
beach erosion, 321–322
Beagle, H.M.S., 71, 73
bears
  black bears, 76
  grizzly bears, 128, 242
  polar bears, 216, 237, 238
Bears Ears National Monument (UT),
  270, 556
beaver, 298
bees
  colony collapse, 159, 214, 215
  pesticides and, 214, 215
beetles, pine bark, 236, 251
Begaye, Russell, 270
beluga whales, 216
Beni Biosphere Reserve (Andes), 259
benthic communities, 107
benthos, 111
best available, economically achievable
  technology (BAT), 426
best practicable control technology (BPT),
  426
Better Not Bigger (Fodor), 509
Beyond the Limits (Meadows et al.), 531
bias
  experimenter, 39–40
  in science, 171–172
Bible
  Noah's flood, 44
  stewardship and, 28–29
bicycles, 456, 457
BIDE factors, 123–125, 137
Bill and Melinda Gates Foundation, 160
binomials, 79–80
bio-organic compounds, 53–54
bioaccumulation, 166
biocentric preservation, 17–18
biochar, 209
biochemical oxygen demand (BOD),
  408–409
biocides, 211
biodegradable plastics, 490
biodiversity, 226–274
  benefits of, 231–233
  captive breeding and species survival
    plans, 246–248
  case studies, 227, 251–252
  defined, 228
  endangered species management, 227
    (case study), 241–246, 259

evolution and. *See* evolution
  food, 231
  of freshwater biomes, 110–112
  genetic diversity and survival, 127–128
  grasslands and, 262–265
  habitat destruction, 234
  hot spots, 230–231
  indigenous cultures and, 27
  invasive species, 235–236, 251–252
    (case study)
  kinds of, 228
  loss of, 12–13
  of marine biomes, 106–110
  nature of, 228
  overharvesting, 238–241
  parks and preserves, 187, 227,
    265–273, 288
  restoration ecology, 12–13
  species concept in, 87–88, 228–233
  of terrestrial biomes, 99–106
  threats to, 233–241
  tips for protecting, 240, 245
  world forests and, 251–261
biodiversity hot spots, 230–231
biofuels, 13, 184, 298, 456, 469–471
biogas, 13, 433, 456, 458, 469, 488
biological communities
  defined, 60
  disturbances, 91–92
  interactions, 83–86
  productivity, 72–73 (case study), 89–90
  resilience, 94–95
  structure, 88–90
  succession, 90–91
biological controls, 215
biological species concept, 228
biomagnification, 166
biomass, 468–471. *See also* methane
  (CH₄)
  accumulation, 89
  biofuels, 469–471
  burning, 340, 468–469, 482–483
  defined, 60, 468
  ecological pyramids, 62–64
  relative accumulation, 89
biomes, 97–115
  case study, 98
  defined, 98, 99
  freshwater, 110–112
  human disturbance, 112–113
  marine, 106–110
  overview (map), 100
  resilience variations, 99, 100–102, 103,
    108, 109, 112–113
  shifts in, 98 (case study)
  terrestrial, 99–106
biopiracy, 231
bioremediation
  brownfield, 299
  defined, 298, 424, 495
  phytoremediation, 298, 424–425, 496
  reclamation, 299
  restoration, 292–298
biosphere reserves, 271–272
biotic population factors, 124
biotic potential, 118
birds
  abundance and diversity of, 87
  American Ornithological Union, 281
  avian malaria, 236
  Bermuda, cahow reintroduction,
    282–283
  bioaccumulation of DDT, 166
  in deciduous forests, 104
  eBird collaboration, 569
  Florida, Pelican Island National Bird
    Reservation, 242, 281

deforestation. *See* deforestation
ecosystem management, 259–261
forest products, 254–255
habitat destruction, 234, 237–238
harvesting methods, 254–255
indigenous culture protection efforts, 257–259, 438
logging. *See* logging
management, 254–255
nutrient cycles, 100
old-growth forests, 12, 252–254, 259–260
prevalence, 252–254
protection, 257–259, 259–260
reducing forest impacts, 260–261
reforestation, 254–255, 279
road building in, 257, 260, 556–557
shelterwood harvesting, 254, 255
single tree selective harvesting, 254, 255
surface mining for oil, 438
tips for minimizing impact on, 261
types, 98, 99–102, 100–102, 104–105, 252–254. *See also* temperate forests; tropical forests
U.S. restoration projects, 284–285
water as product of, 255
formaldehyde, 161
fossil fuels, 13
alternative energy sources, 346, 347, 350
availability, 347, 437, 527–528
black carbon, 340
carbon tax, 519 (case study)
coal. *See* coal
in conventional farming, 207
in debate on climate change, 346
defined, 432
divestment campaigns, 575, 578
emissions trading, 349–350
formation, 309–311
fracking (hydraulic fracking), 318, 363, 390, 437–438
fugitive emissions, 357
green jobs *vs.,* 535
natural gas. *See* natural gas
oil. *See* oil
renewable sources *vs.,* 432, 450. *See also* sustainable energy
rollback of environmental protections, 245
subsidies, 346, 347, 349, 434, 541
transitioning from, 431
fossil water, 392
founder effect, 127
Fourth National Climate Assessment (NCA4), 533–534
foxes, 76, 78
"fracking." *See* hydraulic fracturing ("fracking")
France
nuclear power use, 448, 449
SuperPhénix breeder reactor, 448
Francis (pope), 29
Franklin, Jerry, 260
free-rider problem, 528
freshwater ecosystems
acid rain effects, 372, 411
biodiversity in, 110–112
disturbances, 113
restoration, 292–298
water compartments and, 384–387
Friedman, Tom, 541
Friends of the Earth, 574
frogs
carnivorous, 81
chytridiomycosis, 159
tree, behavioral isolation, 78

frugivores, 86
fruit flies, 79
fuel assembly, 446
fuel cells, 472–473, 475–476
fuelwood
global consumption, 254
usage data, 468
fugitive emissions, 357
Fukushima-Daiichi nuclear disaster (2011), 319, 445, 446, 447
fumigants, 214
fungal diseases, 158–159
fungi, as keystone species, 86
fungicides, 211
FutureGen project (2003), 436

G

Gadgil, Ashok, 423
Galápagos Islands
Darwin's studies of species, 73, 78, 127
finch speciation, 78, 127
natural selection research, 74
Gandhi, Mohandas, 135, 259
Ganges River (India)
arsenic in sediments, 411
pollution, 405–406 (case study), 416
sedimentation, 413
Gap, Inc., 539
gap analysis, 245–246
Gapminder.org, 151, 517
garden cities, 512
gasoline. *See also* automobiles; fossil fuels
banning leaded, 360
royalty payments, 316
Gates, Bill and Melinda, Foundation, 160
Gateway pipeline (Canada), 440
Gause, G. F., 77
gender
China, one-child policy, 132, 146
depression rates, 156
female education, 24, 144, 155
gender development index (GDI), 532
indoor air pollution exposure and, 365
life expectancy and, 142
pronatalist pressures, 143
women and pesticide exposure, 217
women in food production, 181
women in politics, 509, 541
women's rights, 23, 24, 28, 144, 155
Women's Self-Employment Project (IL), 537
gene editing system, CRISPR, 55, 193
General Accounting Office (GAO), 492
General Agreement on Tariffs and Trade (GATT), 536–537
General Mills, 258
General Mining Law (1872), 316
General Motors, 310
generalists/generalist species, 76, 91
genetic diversity, 127–128, 228
genetic drift, 127, 128
genetic engineering
gene editing, 55, 193
in pest control, 221
phytoremediation and, 496
techniques and safety, 191–193
genetically modified organisms (GMOs), 167, 191–193
Atlantic salmon, 193
Cerrado (Brazil) crops, 198 (case study)
defined, 191
gene editing, 55, 193
social costs, 194

genetics
conservation, 126–127
DNA sequencing for taxonomy, 229
Genghis Khan, 336
geographic isolation, 78
Geological Survey, U.S.
California, San Joaquin Valley subsidence, 392
historic water recommendations, 388
mercury levels in Pacific fish, 362
methane sources, 441–442
Nevada, mining data, 314
stream pollution data, 412
geology, 304–308. *See also* earth resources
composition of earth, 304–305
geologic time, 307
hazards, 317–322
ice cores, 335–336
recycling geologic resources, 311–312
rock cycle, 307–308
rock formation and composition, 306–307
subsidence due to groundwater use, 392–393, 505–506, 507
tectonic processes, 305–306, 317
weathering process, 308
Georgia
Atlanta, urban sprawl, 509
Emory University, green building standards, 579
geothermal energy, 471, 474–475
Germany
air pollution reduction, 377, 435
biogas production, 488
birth dearth, 145
Blue Angel program, 572
car-free suburb, Vauban, 515
coal use, 434–435
electricity grid upgrades, 472
Energiewende renewable energy transition, 453–454 (case study), 462
energy conservation and efficiency, 454, 455
energy transition, 453–454 (case study), 459
renewable electricity use, 350
solar power adoption, 459, 462
wind power use, 463, 464
Ghana, e-waste recycling, 483
GHGs. *See* greenhouse gases (GHGs)
giant panda, as dietary specialist, 76
glaciers
black carbon, 340
freshwater inputs from melting, 332
mechanical weathering by, 308
shrinking, 44, 332, 340, 341–342, 344, 384–385, 406
as water compartment, 384–385
glass
recycling, 486, 489
reusable, 489
Gleason, H. A., 91
Gleick, Peter, 417
Glendening, Parris N., 509
Global Change Research Program, U.S., 344–345, 352
global climate change
allopatric speciation, 78
Arctic tundra, 105–106
Australia, drought, 397
causes, 198, 338–342
climate action, 325–326 (case study), 346–350
climate change denial, 346, 500

combating, 345
components, 11
costs *vs.* prevention, 344–345
debate, 345–346
disease and parasite expansion, 159
as disturbance, 92, 98
effects, 12–13, 140, 189–190, 325–326, 332, 343–346, 381, 397
evidence of, 44–45, 343–346, 534
food and nutrition, 178, 181, 189–190
forests, 198, 251–252 (case study), 256, 258, 260–261
glacier melting, 385
greenhouse gases, 356–360, 363–364
habitat destruction, 234, 237–238, 276–277
human-caused, 112–113, 338–342
ice core data, 51, 335–336
Kigali Amendment (2016), 370
Kyoto Protocol (1997), 347, 370, 500, 534–536, 560
living organisms, 233–234
local solutions, 26
market-based forces in reducing pollution, 519 (case study), 533–536
mass extinctions, 233–234, 306, 317
mass migrations, 148
Montreal Protocol (1987), 369–370, 559
natural variability, 335–338, 346
Paris Climate Agreement. *See* Paris Climate Agreement (2016)
permafrost, 444
planetary boundaries, 26–27
resilience and, 94–95
sea-level rise, 140, 344, 345, 505, 506
simulation models, 40–41
support for fossil fuel substitutes, 527–528
Global Environmental Monitoring System (GEMS), 377
global environmentalism, 19
global hectares (gha), 15
Global Marshall Plan Initiative, 541
global warming. *See* global climate change
Global Wind Energy Council, 465
glucose, 54, 58–59
glufosinate (Liberty), 192
glyphosate (Roundup), 167–168, 192–193, 213, 216
GMOs. *See* genetically modified organisms (GMOs)
gnatcatchers, California, 229, 516
goiter, 183
gold mining, 313, 314, 316
Golden Rice, 192
Grameen Bank/Grameen Phone, 537
Grand Canyon (AZ), 267, 268
Grand Staircase-Escalante Monument (UT), 270, 556
Grandjean, Philippe, 162–163
granite, formation, 307
grasshopper effect, 216–217, 368
grasslands, 262–265
alternative approaches to grazing, 263–265
buffalo reintroduction, 288, 290–292
conversion to agriculture, 198 (case study)
desertification, 206
disturbances, 43
grazing and overgrazing, 41, 102, 206, 262–265
prairie restoration, 287–292
temperate, 103–104
tropical, 102

refrigerants, climate-safe, 474–475
refugees
   climate change, 14, 148, 345, 397
   famine, 181
   number (2018), 504
refuse-derived fuel, 485
regime shifts, 93–94
Regional Greenhouse Gas Initiative
   (RGGI), 349–350
regulatory capture, 557–558
rehabilitation, defined, 278
Reichhold Chemical, 493
Reilly, William K., 568
reintroduction, defined, 278
Reisner, Marc, 467
relative abundance, 87–88
release phase, 93
reliability, 38
religion
   Noah's flood, 44
   pronatalist pressures, 142–143
   stewardship and, 28–29
remediation
   defined, 278
   water, 423–425
remote sensing. *See* satellites
renewable/alternative energy sources, 432,
   450. *See also* sustainable energy
Renewable Power Agency, 535
renewable resources, 521
renewable water supplies, 387, 400
reorganization phase, 93
Repeto, Robert, 531
replacement fertility rate, 139–140, 149
replanting, 280
replication
   in evolution, 74, 79
   in science, 35
reproducibility, in science, 35–36
reproductive isolation, 228
reservoirs, sedimentation problems,
   396–397
residence time, 384
resilience, 43, 90, 94–95
   economic policy, in, 561
   variation by biome, 99, 100–102, 103,
      108, 109, 112–113
Resource Conservation and Recovery Act
   (RCRA, 1976), 427, 491, 549
resource management, 17–18
resource partitioning, 77–78
resource use, describing, 25–26
resources. *See also* earth resources
   defined, 520
   intangible, 521–522
   natural resources as
      interchangeable, 523
   nonrenewable, 520–521
   partitioning, 77–78
   renewable, 521
respiration, cellular, 59, 65–66. *See also*
   photosynthesis
restoration ecology. *See* ecological
   restoration
reverse osmosis desalination, 399
RGGI (Regional Greenhouse Gas
   Initiative), carbon trading, 349–350
rhinos, white and Javanese, 247
Rhode, Robert, 354–355
ribonucleic acid (RNA), 54
Ricardo, David, 522
rice
   crop supports, 194
   genetically modified, 192, 193
   as key food source, 184, 191
   methane production, 341
   "night soil" use, 420

sustainable farming of, 208
   water used for, 389
ricin, 170
riders, legislative, 553
rill erosion, 204
risk, 171–173
   acceptance of, 171, 172–173
   defined, 171
   health policy and, 173–174
   lifetime odds of death, 173
   "precautionary principle" for
      manufacturers, 174
   relative, 174
risk assessment, 171–172, 173, 174
rivers
   sedimentation, 297–298, 410
   storm sewer runoff, 420
   as water compartments, 386–387
   in wetlands restoration, 292–299
Rivers and Harbors Act (1899), 548
Roadless Area Review and Evaluation
   (RARE), 260, 556–557
Roadless Rule, wilderness guideline, 260,
   556–557
Robert, K. H., 538
Robinson, Frances, 4
rock cycle, 307–308
rocks
   composition, 306–307
   igneous, 307
   metamorphic, 307–308
   minerals in, 306–308
   sedimentary, 308
   weathering and sedimentation, 308
Rockström, Johan, 26–27
Rogers, Will, 177
Roosevelt, Franklin, 541
Roosevelt, Theodore, 16–17, 18, 270, 281
Ross, John, 518
rotational grazing, 264–265
Rotterdam Convention (1997), 559–560
Roundup-Ready crops, 168, 192–193, 213
rubber tires, recycling, 486
rule of 70, 119, 138
rural areas
   China
      aging population, 132
      one-child policy exemption, 146
   communications issues, 537
   deer population issues, 122
   defined, 502
   groundwater quality, 417–418
   pronatalist pressures, 142, 143
   push factors, 504–505
   urbanization issues, 132, 504–505,
      508–510
   wastewater treatment, 420–421
   water treatment, 416–417, 423
   wind energy as income source, 465–466
Rusk, David, 509
Russia. *See also* Soviet Union (former)
   birth dearth, 145
   Chelyabinsk, nuclear waste site
      explosion, 449
   coal deposits in, 434
   drug-resistant tuberculosis in, 157
   fishing quota enforcement, 242
   ice core drilling of Vostok ice sheet, 336
   life expectancies, 142
   natural gas reserves, 441, 443
   Norilsk, toxic air pollution, 377
   nuclear reactor design, 447
   old-growth forests, 254
   population growth, 140
   rare earth metals, 310
   wood and paper pulp production in, 254
Rwanda, reforestation program, 284

S

S-shaped growth curve, 120, 134, 135
saccharin, 170
Sachs, Jeffrey, 24, 160
Safe Drinking Water Act (1974), 549
Safe Drinking Water Act (1986)
   Halliburton clause (exemption for
      fracking), 444
   preventive measures, 417
Safe Harbor Policy, 244
Sagan, Carl, 45, 302
sage grouse, 244–245, 466
saguaro cacti, adaptation and
   specialization, 74–75
Sahara Desert (Africa)
   desertification near, 206
   locust population growth, 125
   overgrazing, 103
Sahel (Africa), 103, 104
   food security threat, 178 (case study)
   rotational grazing, 264
salinization, 206, 392, 396–397
salmon
   aquaculture issues, 188
   copper *vs.* in Alaska, 303–304 (case
      study)
   dams, effects of, 395–396
   Elwha River dam, 293, 396
   endangered, 242, 244
   genetically modified, 193
   pollution issues, 188, 303–304 (case
      study), 314
salt cedar, 236, 278
salt marshes, 109
saltwater intrusion, freshwater aquifer, 392
Salween dams, Burma and Thailand, 19
sample, statistical, 37–39
sampling, 126
Sand County Almanac, A (Leopold), 17,
   18, 280–281
sand dunes, barrier island, 110
sanitary landfills, 482–483
sanitation, 20
Sargasso Sea, 108
satellites
   air quality monitoring, 13, 360–361
   chlorophyll monitoring, 107
   coral reef surveys, 269
   drought monitoring, 198
   fishery monitoring, 187
   Gravity Recovery and Climate
      Experiment (GRACE), 394–395
   water measuring, 394–395
Saudi Arabia
   desalination in, 399
   oil prices and, 439–440
   preserves, 266
   solar energy, 461
savannas, 102, 103
   defined, 252
   oak, burning and restoration, 285–286
Savory, Allan, 264
scarcity
   fossil fuel availability, 527–528
   innovation and, 527–528, 529
   population and technology, 527–531
   water, 387–388
scavengers/scavenger organisms, 61–63, 81
Schaller, George, 247
schistosomiasis, 84, 86
Schneider, Steve, 350
Schumacher, E. F., 523
science, 35–41
   accuracy and precision, 35–36
   basic principles, 35

climate regions, 101
climographs, 101, 115
consensus and conflict in, 43–45
deductive and inductive reasoning
   in, 36
experimental design and, 37–40, 39–40
gene editing, 55, 193
hypotheses and theories, 36–37
models, 40–41
probability in, 37–39
pseudoscience detection, 44–45
public policy and, 37
statistics in, 38–39
systems in, 41–43
science literacy, 37
scientific consensus, 43–44
scientific method, 36–37
scientific theory, 37
Scolow, Robert, 348
Scott, J. Michael, 245, 246
sea ice
   loss of, 341–342, 343–344
   oil exploration, 438
   polar bears and, 237, 238
sea-level rise, 140, 344, 345, 505, 506
sea otters, 242
sea turtles, 545 (case study)
seafood. *See* fish/seafood
seagrass meadows, 72–73 (case study), 88,
   94, 96, 108
SeagrassSpotter project, 72–73 (case
   study), 96
seals
   Hawaiian monk, 479–480
   inbreeding, 128
   northern fur, 482
second law of thermodynamics, 57, 63
secondary pollutants, 357
secondary productivity, 60, 61, 62
secondary succession, 91, 92
secondary treatment, 422
secure landfills, 483, 495–497
sedimentary rocks, 308, 335
sedimentation
   in dam reservoirs, 396–397, 467
   defined, 308
   in streams and rivers, 297–298, 410
   as water pollutant, 413
selection pressures, 79, 81
selective breeding, 55
Sen, Amartya K., 181
Seneca, 152
septic tanks, 420–421
Sequoia National Park (CA), 28, 29,
   286–287, 555
Serengeti National Park ecosystem
   (Tanzania), 34 (case study), 39,
   44, 47
service products, 539
sewage treatment, 420–424
   biological, 422–423
   in developing countries, 415–416, 505
   effectiveness, 551
   infectious agents in human waste,
      407–408
   low-cost treatment, 423–424
   mechanical method, 421–422
   municipal, 422–423
   pet waste problem, 415
   requirements, 415
   septic tanks, 420–421
   wetland method, 422–423
   worldwide data, 415
sexual function
   bioaccumulation of DDT, 166
   environmental estrogens/
      androgens, 163